T0201506

DEEP CARBON

Carbon is one of the most important elements of our planet, and 90% of it resides inside Earth's interior. This book summarizes 10 years of research by scientists involved in the Deep Carbon Observatory, a global community of more than 1200 scientists. It is a comprehensive guide to carbon inside Earth, including its quantities, movements, forms, origins, changes over time, and impacts on planetary processes. Leading experts from a variety of fields, including geoscience, biology, chemistry, and physics, provide exciting new insights into the interconnected nature of the global carbon cycle and explain why it matters to the past, present, and future of our planet. With end-of-chapter problems, illustrative infographics, full-color images, and access to online models and data sets, it is a valuable reference for graduate students, researchers, and professional scientists interested in carbon cycling and Earth system science. This title is also available as Open Access on Cambridge Core at doi.org/10.1017/9781108677950.

BETH N. ORCUTT is Senior Research Scientist at Bigelow Laboratory for Ocean Sciences, USA. Her research focuses on understanding microscopic life at and below the seafloor. Having clocked over 600 days at sea on field missions, including dives to the seafloor in the *Alvin* submersible, she is an expert in ocean exploration. Orcutt has received a Kavli Frontiers in Science Fellowship and the 2018 Geobiology and Geomicrobiology Division Post-Tenure Award from the Geological Society of America.

ISABELLE DANIEL is Professor of Earth Sciences at the Université Claude Bernard Lyon 1, France. She is also affiliated with the Laboratoire de Géologie de Lyon and chairs the Observatoire de Lyon. Her research focuses on geobiology and minerals, rocks, and fluids under extreme conditions. She investigates serpentinization and serpentine minerals, fluid–rock interactions at high pressure, and microorganisms under extreme conditions. She is a fellow of the Mineralogical Society of America.

RAJDEEP DASGUPTA is Professor of Earth, Environmental and Planetary Sciences at Rice University, USA. His research focuses on the deep processes of Earth and planetary interiors, which he pursues using geochemical and petrological approaches. He is a recipient of the James B. Macelwane Medal and Hisashi Kuno Award from the American Geophysical Union, the F. W. Clarke Medal from the Geochemical Society, the Faculty Early Career Award from the US National Science Foundation, and the Packard Fellowship for Science and Engineering. He is also a fellow of the American Geophysical Union.

DEEP CARBON

Past to Present

Edited by

BETH N. ORCUTT
Bigelow Laboratory for Ocean Sciences

ISABELLE DANIEL
Université Claude Bernard Lyon 1

RAJDEEP DASGUPTA
Rice University

CAMBRIDGE
UNIVERSITY PRESS

University Printing House, Cambridge CB2 8BS, United Kingdom

One Liberty Plaza, 20th Floor, New York, NY 10006, USA

477 Williamstown Road, Port Melbourne, VIC 3207, Australia

314–321, 3rd Floor, Plot 3, Splendor Forum, Jasola District Centre, New Delhi – 110025, India

79 Anson Road, #06–04/06, Singapore 079906

Cambridge University Press is part of the University of Cambridge.

It furthers the University's mission by disseminating knowledge in the pursuit of education, learning, and research at the highest international levels of excellence.

www.cambridge.org
Information on this title: www.cambridge.org/9781108477499
DOI: 10.1017/9781108677950

First published 2020
Reprinted 2020

Printed in the United Kingdom by TJ International Ltd., Padstow Cornwall

A catalogue record for this publication is available from the British Library.

Library of Congress Cataloging-in-Publication Data
Names: Orcutt, Beth, 1980– editor. | Daniel, Isabelle, 1968– editor. | Dasgupta, Rajdeep, 1976– editor.
Title: Deep carbon : past to present / edited by Beth Orcutt (Bigelow Laboratory for Ocean Sciences, USA), Isabelle Daniel (Université Claude Bernard Lyon 1, France), Rajdeep Dasgupta (Rice University, Houston).
Description: Cambridge ; New York, NY : Cambridge University Press, 2020. |
Includes bibliographical references and index.
Identifiers: LCCN 2019019485 | ISBN 9781108477499 (hardback : alk. paper)
Subjects: LCSH: Carbon. | Carbon cycle (Biogeochemistry) | Earth (Planet)–Crust. | Earth (Planet)–Mantle.
Classification: LCC QD181.C1 D44 2020 | DDC 549/.27–dc23
LC record available at https://lccn.loc.gov/2019019485

ISBN 978-1-108-47749-9 Hardback

Additional resources for this publication at www.cambridge.org/deepcarbon

Contents

Contributors

Alessandro Aiuppa
Università di Palermo

Patrick Allard
Institut de Physique du Globe de Paris

Jan Amend
University of Southern California

Muriel Andreani
Université Claude Bernard Lyon 1

Sonja Aulbach
Goethe Universität

Jennifer F. Biddle
University of Delaware

Christiana Bockisch
Arizona State University

Frank E. Brenker
Goethe Universität

Yana Bromberg
Rutgers University

Hélène Bureau
Sorbonne Université

Antony D. Burnham
Australian National University

Mike Burton
University of Manchester

Carlo Cardellini
Università di Perugia

Simon Carn
Michigan Technological University

Pierre Cartigny
Institut de Physique du Globe de Paris

Valerio Cerantola
European Synchrotron Radiation Facility

Thomas Chacko
University of Alberta

Bin Chen
University of Hawai'i at Mānoa

Giovanni Chiodini
Istituto Nazionale di Geofisica e Vulcanologia

Charles S. Cockell
University of Edinburgh

David Cole
Ohio State University

Elizabeth Cottrell
Smithsonian Institution

Darlene Trew Crist
Crist Communications

Isabelle Daniel
Université Claude Bernard Lyon 1

Rajdeep Dasgupta
Rice University

Robert T. Downs
University of Arizona

Marie Edmonds
University of Cambridge

Ahmed Eleish
Rensselaer Polytechnic Institute

Charlene Estrada
Arizona State University

Paul G. Falkowski
Rutgers University

Kristopher Fecteau
Arizona State University

Tobias P. Fischer
University of New Mexico

Peter Fox
Rensselaer Polytechnic Institute

Daniel J. Frost
Universität Bayreuth

Fabrice Gaillard
Université d'Orleans

Matthieu Emmanuel Galvez
ETH Zurich

Bertrand García-Morena
Johns Hopkins University

Emmanuel Gardés
CIMAP Ganil

Sujoy Ghosh
Indian Institute of Technology

Donato Giovannelli
CNR-ISMAR

Joshua J. Golden
University of Arizona

Ian R. Gould
Arizona State University

Damanveer S. Grewal
Rice University

Bertrand Guillot
Université Pierre et Marie Curie

Tahar Hammouda
Université Clermont Auvergne

Hilairy Hartnett
Arizona State University

Erik H. Hauri
Carnegie Institution for Science

Robert M. Hazen
Carnegie Institution for Science

Daniel R. Hummer
Southern Illinois University

Grethe Hystad
Purdue University Northwest

Dorrit E. Jacob
Macquarie University

Steven D. Jacobsen
Northwestern University

Hehe Jiang
Rice University

Kristin Johnson
Earth-Life Science Institute

Sean P. Jungbluth
Department of Energy Joint Genome Institute

Katherine A. Kelley
University of Rhode Island

Ekaterina S. Kiseeva
University College Cork

Andrew H. Knoll
Harvard University

Simon C. Kohn
University of Bristol

Susan Q. Lang
University of South Carolina

Doug LaRowe
University of Southern California

Marion Le Voyer
Carnegie Institution for Science

Cin-Ty A. Lee
Rice University

Congrui Li
Rensselaer Polytechnic Institute

Jie Li
University of Michigan

Chao Liu
Carnegie Institution for Science

Karen G. Lloyd
University of Tennessee, Knoxville

Sergey S. Lobanov
GeoForschungsZentrum

Robert W. Luth
University of Alberta

Cara Magnabosco
Simons Foundation

Ananya Mallik
University of Rhode Island

Jared Marske
Carnegie Institution for Science

Malcolm Massuyeau
University of Johannesburg

Catherine A. McCammon
Universität Bayreuth

Bénédicte Ménez
Institut de Physique du Globe de Paris

Marco Merlini
University of Milan

Sami Mikhail
University of St. Andrews

Mainak Mookherjee
Florida State University

Eli K. Moore
Rutgers University

Guillaume Morard
Sorbonne Université

Shaunna M. Morrison
Carnegie Institution for Science

A.D. Muscente
Harvard University

Oded Navon
The Hebrew University

Fabrizio Nestola
University of Padova

Paolo Nimis
University of Padova

Beth N. Orcutt
Bigelow Laboratory for Ocean Sciences

Magdalena R. Osburn
Northwestern University

Mederic Palot
University of Alberta

D. Graham Pearson
University of Alberta

Anirudh Prabhu
Rensselaer Polytechnic Institute

Manuel Pubellier
École Normale Supérieure

Jolyon Ralph
Mindat.org

Guillaume Richard
Université d'Orleans

Kirtland Robinson
Arizona State University

Catherine A. Royer
Rensselaer Polytechnic Institute

Michelle Y. Rucker
University of Arizona

Simone E. Runyon
Carnegie Institution for Science

Alberto E. Saal
Brown University

Nicolas Sator
Sorbonne Université

Cody S. Sheik
University of Minnesota at Duluth

Kei Shimizu
Carnegie Institution for Science

Hiroshi Shinohara
Geological Survey of Japan

Jessie Shipp
Versum Materials

Steven B. Shirey
Carnegie Institution for Science

Everett Shock
Arizona State University

David Sifré
European Synchrotron Radiation Facility

Karen V. Smit
Gemological Institute of America

Evan M. Smith
Gemological Institute of America

Carl Spandler
James Cook University

Thomas Stachel
University of Alberta

Vincenzo Stagno
Sapienza University of Rome

Andrew Steele
Carnegie Institution for Science

Andrew D. Steen
University of Tennessee, Knoxville

Richard A. Stern
University of Alberta

Alberto Striolo
University College London

Emilie Thomassot
Université de Lorraine

Andrew R. Thomson
University College London

Mark Torres
Rice University

Jonathan M. Tucker
Carnegie Institution for Science

Katrina I. Twing
University of Utah

Michael J. Walter
Carnegie Institution for Science

Lisa A. Warden
Carnegie Institution for Science

Yaakov Weiss
Hebrew University of Jerusalem

Cynthia Werner
US Geological Survey

Lynda Williams
Arizona State University

Gregory M. Yaxley
Australian National University

Edward D. Young
University of California, Los Angeles

Hao Zhong
Rensselaer Polytechnic Institute

1

Introduction to *Deep Carbon: Past to Present*

BETH N. ORCUTT, ISABELLE DANIEL, RAJDEEP DASGUPTA, DARLENE TREW CRIST, AND MARIE EDMONDS

Carbon is one of the most important elements in Earth. Its behavior has important consequences for the global climate system, for the origin and evolution of life, for carbon-based energy resources, and for a vast array of carbon-based materials that are central to our daily lives. In short, carbon matters.

Carbon moves between Earth's surface and interior. The solid Earth, its fluids, and the subsurface biosphere are estimated to contain about 90% of the carbon on the planet. These deep reservoirs control the size and extent of surface carbon sinks. Plate tectonics and other hallmarks of an active planet drive change in the size of these surface carbon reservoirs over geologic time.

Deep carbon science emerged as a new field in the twenty-first century, aiming to better understand carbon's quantities, movements, forms, and origins in Earth's subsurface. Scientists set out to answer questions about where deep carbon comes from, what forms it takes, how much lies beneath the surface, and how it is transported into and out of Earth's interior reservoirs. Although deep carbon exerts fundamental control over the operation of Earth's well-studied surface carbon cycle, we have a limited understanding of carbon's sources, sinks, transfers, and quantities below the surface. Deep carbon scientists from a wide range of disciplines – including geoscience, biology, materials science, physics, and chemistry – came together to address this problem by researching carbon in a planetary context. This community of scientists was united under the umbrella of the Deep Carbon Observatory (DCO; http://deepcarbon.net), a decadal research effort seed funded by the Alfred P. Sloan Foundation to promote advances in what is known about deep carbon and its impact on planetary processes. The origins of the DCO and the questions that motivated this field were summarized in *Carbon in Earth*, published in 2013 (1), which is a companion primer for this volume.

This new book represents a synthesis of the transformational discoveries in interdisciplinary deep carbon science that have occurred over the past decade, highlighting some of the state-of-the-art instrumentation that has come online in the past decade to enable this work. Reflecting the four thematic scientific communities of the DCO, the chapters in this book highlight new understandings of:

(1) carbon-bearing phases and their origins under the *extreme physics and chemistry* of Earth's mantle and core (Chapters 2–7);

(2) the nature and origins of Earth's deep carbon-bearing fluids and the abiotic synthesis of carbon under the umbrella of *deep energy* (Chapters 12–15);
(3) Earth's carbon *reservoirs and fluxes*, with a particular focus on out-gassing and in-gassing of carbon in various magmatic and tectonic settings such as convergent and divergent plate margins (Chapters 8–10) and on how such deep interior carbon cycling affects the surface carbon cycle (Chapter 11);
(4) the surprising diversity, extent, and limits of Earth's subsurface *deep life* (Chapters 16–19).

The volume concludes with a look toward the future and at how data-driven discovery can reveal new knowledge about carbon's origins, forms, quantities, and movements in space and time (Chapter 20).

This synthesis effort exists because of the dedicated and voluntary efforts of many members of the deep carbon community, for which the editorial team is extremely grateful. Our deep thanks to all of the authors who contributed their thoughts and talents to this effort, especially to the chapter lead authors Muriel Andreani, David Cole, Fabrice Gaillard, Matthieu Emmanuel Galvez, Erik H. Hauri, Robert M. Hazen, Susan Q. Lang, Doug LaRowe, Cin-Ty A. Lee, Jie Li, Karen G. Lloyd, Cara Magnabosco, Steven B. Shirey, Everett Shock, Vincenzo Stagno, Cynthia Werner, Gregory Yaxley, and Edward D. Young. We also would like to extend our deepest thanks to Elizabeth Cottrell and Shaunna M. Morrison for their extra efforts toward the completion of their chapters. We thank the many scientists who contributed thoughtful reviews of chapters. Josh Wood of the DCO was instrumental in designing many of the graphics in this volume including the cover, and Craig Schiffries of the DCO Secretariat was a welcome champion of this project. Emma Kiddle and Sarah Lambert of Cambridge University Press were endlessly patient and supportive and helped guide this volume to completion. Finally, this entire effort would not have been possible without the support of the Alfred P. Sloan Foundation and DCO Science Advisor Jessie Ausubel.

We would like to dedicate this entire volume to the memories of Erik H. Hauri (Figure 1.1) of the Carnegie Institution of Science and Louise Kellogg of University of California at Davis, whose intellectual curiosity, scientific excellence, and good humor were inspirational, and who will always be remembered.

We acknowledge funding from the Sloan Foundation for making this book possible, and from the NSF (grant OCE-1338842), NASA (grant 80NSSC18K0828), and Rice University for supporting editorial efforts.

Figure 1.1 Erik H. Hauri (1966–2018) was a geochemist at the Carnegie Institution for Science in Washington, DC. He used isotope analysis, modeling, and seismic imaging to study planetary processes and volcanism, with a particular focus on the distribution of water on Earth, the Moon, and other celestial objects.

Photograph courtesy of Steven Jacobsen (Northwestern University).

Reference

1. Hazen R. M., Jones A. P., Baross J. A. Carbon in Earth. *Reviews in Mineralogy and Geochemistry*. 2013; 75: 1–698.

2

Origin and Early Differentiation of Carbon and Associated Life-Essential Volatile Elements on Earth

RAJDEEP DASGUPTA AND DAMANVEER S. GREWAL

2.1 Introduction

Earth's unique status as the only life-harboring planet in the inner Solar System is often linked to its orbit being in the habitable zone that resulted in the stabilization of liquid water on its surface. However, in addition to surface liquid water, the presence of carbon and other life-essential volatile elements (LEVEs) such as nitrogen and sulfur in the surface environment is also critical for the habitability of rocky planets. Earth's long-term equable climate and chemically habitable surface environment are therefore results of well-tuned fluxes of carbon and other LEVEs, involving the deep and the surficial Earth. Given that Earth developed into a volcano-tectonically active planet with both outgassing and ingassing mechanisms, the surface inventory of LEVEs over million- to billion-year timescales is maintained by interactions of the ocean–atmosphere system with the silicate fraction of the planet. Yet it remains uncertain how and when the initial inventory of LEVEs for the surface reservoir plus the silicate fraction (bulk silicate Earth (BSE) altogether), which set the initial boundary conditions for subsequent planet-scale volatile cycles, got established. The answers to these questions lie within Earth's formative years – in the building blocks, in the process of Earth's accretion, and in the early differentiation that made the major reservoirs that constitute the core, the mantle, and the crust–atmosphere.

Earth is a differentiated planet with a central metallic core and an outer shell of dominantly silicate rocks divided into the mantle and crust, which is overlain by the fluid envelope of the ocean–atmosphere system. Direct constraints on carbon and the other life-essential volatile abundance of Earth's core are lacking (Chapter 3, this volume), although significant concentrations of carbon, sulfur, hydrogen, and nitrogen can be in Earth's core given that the outer core is 10% lighter than pure Fe–Ni alloy liquid.[1–3] More is known about the carbon content of the present-day silicate fraction of the planet and its fluid envelope because Earth's mantle reservoir is sampled to some extent via mantle-derived melts (e.g. Refs. 4–9 and Chapter 9, this volume). Yet the estimates of bulk mantle carbon abundance from concentrations in basalts are uncertain partly because most basalts are partially degassed and basalt generation only directly samples typically the top 100–200 km of the mantle. Nonetheless, largely based on the carbon content of basalts (e.g. $C/^{3}He$, CO_2/Nb, CO_2/Ba, and $C/^{40}Ar$ ratios), the BSE abundance of carbon is estimated to be

~100–530 ppm.[7,9–13] Among this range of estimated C budget for the BSE, most estimates[7,9,10–12] have converged to a concentration of 90–130 ppm, and this will be used in this chapter. It can be surmised that this BSE budget of C is an end product of Earth's accretion and early differentiation,[8,14,16] but the uncertainty in the C budget of BSE makes testing such hypotheses quantitatively less rigorous. Further insight could be gained, however, from C to other LEVE ratios (i.e. C/X, where X = N, S, or H; e.g. Refs. 7, 10, 11, 16–18). Specifically, these volatile ratios for the terrestrial reservoirs in general and BSE in particular could be compared with the same ratios in various planetary building blocks and evaluated against the known processes of planet formation and early differentiation.

The key questions focus on what planetary building blocks were responsible for delivering C and other associated LEVEs to Earth and when such delivery processes took place. If the C abundance and the C/X ratios of the BSE as well as the isotopic compositions of all LEVEs are similar to any of the undifferentiated meteorites, it may be concluded that such building blocks were responsible for bringing these volatiles to Earth and no postdelivery process altered their ratios (Figure 2.1). On the other hand, if the C abundance and C/X ratios of the BSE are distinct from these known building blocks, one needs to investigate whether any postdelivery processes could have altered these quantities such that the BSE LEVE geochemistry could be established. For example, if core formation processes fractionate C and/or affect the C/X ratios, then could equilibrium core formation scenarios be responsible for establishing carbon, oxygen, hydrogen, nitrogen, and sulfur in BSE (Figure 2.1)? The timing of volatile delivery is also important. If the LEVEs were delivered during the main stage of accretion, then early differentiation processes like core–mantle segregation and magma ocean (MO) degassing would have played an important role in setting up the relative budgets of the LEVEs in the primary reservoirs (Figure 2.1).[8,11] The delivery mechanism and the timing of delivery may be intimately linked; however, less interrogation has taken place thus far regarding the delivery mechanism. For example, there is a growing recognition that following an initial period of accretion of undifferentiated planetesimals, differentiated planetary embryos with masses ranging from ~0.01 to $0.1M_E$ (where M_E is the mass of present-day Earth) were abundant in the inner Solar System;[19] therefore, accretion of these differentiated bodies significantly contributed to the growth of larger planets like Earth.[19,20] Did the impacts or merger of differentiated bodies deliver the LEVEs to the BSE and establish the C abundance and the C/X ratios[16,17] or did post-core formation addition of undifferentiated or differentiated bodies[21] establish the BSE LEVE budget? And what were the relative contributions of the impacts of large differentiated and smaller undifferentiated bodies to bringing in LEVEs versus causing a depletion of LEVEs via atmospheric mass loss?[22–24]

This chapter will discuss the state of the art on what is known about the origin of C relative to the other LEVEs in the BSE, highlighting open questions and gaps in knowledge. We will discuss constraints based on geochemical comparison of the potential terrestrial building blocks with Earth reservoirs and interrogate each key process of planet formation and early differentiation, such as core formation, MO crystallization, MO

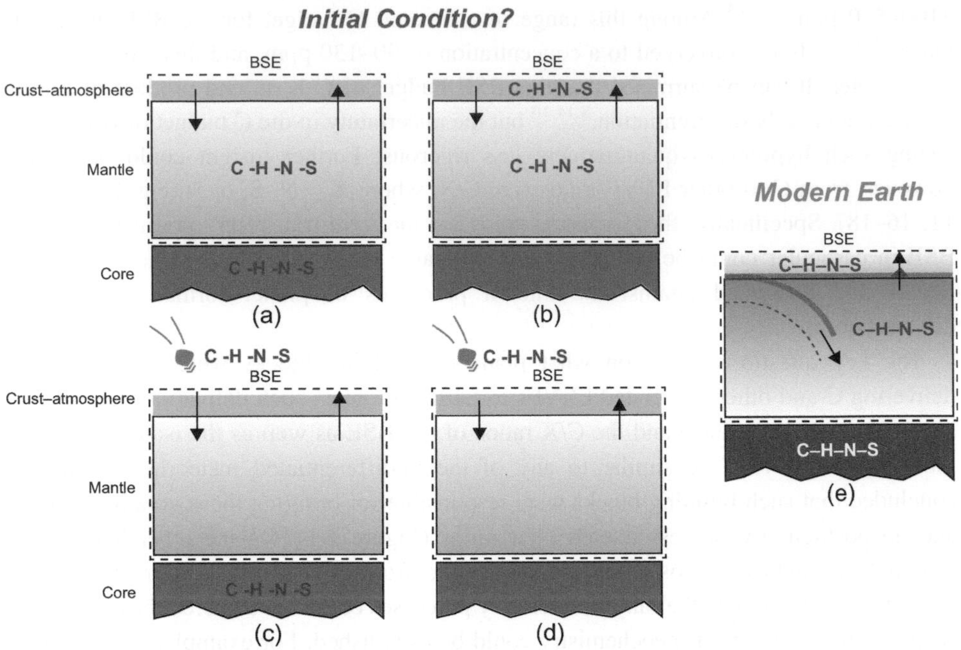

Figure 2.1 Plausible initial distributions of C, H, N, and S between Earth's major reservoirs at the end of accretion and core formation. (a) LEVEs are available during accretion and all LEVEs are initially sequestered in the core and the MO. (b) LEVE delivery during accretion is temporally separate from the main episodes of core formation and thus the core is effectively devoid of LEVEs. (c) Available LEVEs during accretion are effectively sequestered in the core or lost to space (not shown) and the BSE acquires LEVEs via later additions. (d) Core–mantle differentiation takes place in the absence of LEVEs and LEVEs are added via late additions. These distributions eventually evolved into the setup seen in the modern world (e), where these volatile elements are thought to exchange between Earth's mantle and the crust–ocean–atmosphere system (ingassing and outgassing) regulates the long-term habitability of the planet.

degassing, and atmospheric loss in shaping the absolute and relative abundances of C and the other LEVEs. The current understanding of the fractionation of C and other LEVEs during the core formation process will also be used to put constraints on the C budget of the core.

2.2 Constraints on the Compositions of Terrestrial Building Blocks

2.2.1 Constraints from Isotopes of Refractory Elements

Geochemical models based on elemental abundances and isotopic constraints predict the accretion of bulk Earth primarily from chondritic meteorites.[25,26] Because elemental and isotopic fractionation trends for bulk silicate samples and chondritic meteorites have well-defined relationships, it is widely assumed that the building blocks of Earth belong to a class of primitive objects that are sampled by chondritic meteorites.[27] Close similarity of the abundances of refractory lithophile elements between the BSE and CI led to an early development of a CI model for Earth.[25] Allègre et al.[27] further consolidated this model by arguing that volatile depletion trends in the BSE are matched most closely by CI chondrites. However, a distinct disparity of oxygen (second most abundant element in the BSE by mass) isotopes (i.e. ^{16}O, ^{17}O, and ^{18}O) between CI chondrites and the BSE, which could not be explained by any fractionation or differentiation models,[28,29] was one of the first observed isotopic variations that challenged the CI model for Earth. Over the past decades, isotopic dissimilarity between CI chondrites and the BSE has been extended from $\Delta^{17}O$ (deviation from a reference mass-dependent fractionation line in a triple isotope of oxygen plot[30]) to several radiogenic as well as non-radiogenic isotopic anomalies for a wide suite of refractory elements with contrasting geochemical behaviors: lithophile elements ($\varepsilon^{48}Ca$ (Ref. 31), $\varepsilon^{50}Ti$, $\mu^{142}Nd$), moderately siderophile elements ($\varepsilon^{54}Cr$, $\varepsilon^{64}Ni$, $\varepsilon^{92}Mo$), and highly siderophile elements (HSEs; $\varepsilon^{100}Ru$) (Refs. 32, 33 and references therein). Interestingly, the BSE isotopic compositions for all of these elements are almost indistinguishable from enstatite chondrites (E-chondrites). This has led to the development of an E-chondrite model for Earth.[32,34,35] Tracking the isotopic evolution of the growing BSE by taking into account the alloy–silicate partitioning behavior of several lithophile, moderately siderophile, and HSEs during core–mantle equilibration, Dauphas[32] proposed that half of the first 60% of accreted mass was composed of E-type chondrites, while the remaining 40% was all E-type chondritic materials. However, there are a few outstanding issues with an E-chondritic model for Earth, namely: (1) the Ca/Mg ratio of the BSE (0.118[36]) is distinctly higher than that of E-chondrites (0.080[37]), suggesting that the BSE has an enriched refractory elemental abundance relative to E-chondrites;[38] (2) E-chondrites have an Mg/Si ratio that is lower than that of the BSE sampled by upper-mantle rocks,[39] which can only be explained by either incorporation of Si into the core or via an unsampled lower mantle with a low Mg/Si ratio;[40] and (3) $\delta^{30}Si_{BSE}$ is 0.3–0.4‰ higher than $\delta^{30}Si$ of E-chondrites and aubrites.[41] Experimentally determined Si isotopic fractionation parameters[42,43] rule out the possibility of explaining the Si isotopic composition as well as the Mg/Si ratio of the BSE via incorporation of Si in the core if Earth was primarily accreted from E-chondrites.[41] Although there are studies that argue for a lower mantle or at least a portion of the lower mantle being richer in silica compared to Earth's upper mantle,[40,44] such findings are debated. To circumvent these issues, Dauphas et al.[45] proposed that the bulk Earth and the E-chondrites sample an isotopically similar reservoir with their subsequent chemical evolution offset by ensuing nebular fractionation and planetary differentiation processes. In summary, a large number of geochemical systematics point to the bulk Earth being made from E-chondrite-type materials, although questions on the major element and volatile element abundance mismatch remain.

2.2.2 Constraints from Isotopes of Highly Volatile Elements

In contrast to constraints from refractory elements, constraints from highly volatile elements (i.e. C, N and H (water)) paint a different picture on the potential sources of LEVEs.

Although ordinary chondrites (OCs), which sample S-type asteroids in the inner belt (2.1–2.8 AU), have similar amounts of bulk water as the BSE, the D/H ratio of the BSE is not similar to that of OCs,[46] but rather is strikingly similar to that of carbonaceous chondrites, especially CI chondrites,[47] which are representative of C-type asteroids (beyond 2.8 AU) (Figure 2.2a). Similarity of $^{15}N/^{14}N$ between the BSE and CI chondrites also points toward a carbonaceous chondritic origin of volatiles on Earth (Figure 2.3b).[13,48,49] $^{13}C/^{12}C$ of the BSE cannot distinguish between a chondritic or cometary origin of C (Figure 2.2b).[50,51] However, a clear distinction of the $^{15}N/^{14}N$ ratio of the BSE from the corresponding cometary values would necessitate C delivery by CI chondrites as well, assuming all the major volatiles (i.e. C, N, and water) were sourced from a similar parent body.

Although thus far direct matching of the isotopes of C, H, and N of the BSE with undifferentiated meteorites has been the approach[59] to determining the source of the LEVEs on Earth, there is a growing realization that there may be fractionation of stable isotopes during early terrestrial differentiation, such as devolatilization, MO degassing, and core–mantle fractionation. Indeed, such fractionation processes may be able to reconcile some of the observed differences between the LEVE isotopic compositions of the BSE and the chondrites.[60] For example, carbon isotopic compositions of many CI chondritic materials are distinctly lighter ($\delta^{13}C$, approximately –15 to –7‰[50]) than the average carbon isotope composition of Earth's mantle ($\delta^{13}C$, approximately –5‰[61,62]). Similarly, the sulfur isotope composition of Earth's mantle has also been argued to be non-chondritic.[63] Graphite–Fe–carbide melt carbon isotope fractionation experiments[64] showed that ^{12}C preferentially incorporates into the metallic phase, leaving a ^{13}C-enriched signature in the graphite or diamond. If similar fractionation behavior applies between silicate MO and core-forming alloy liquid, then core formation could explain how Earth's mantle evolved to have a heavier carbon isotope value compared to the chondritic building blocks.

2.2.3 Constraints from Theoretical Modeling

Because isotopic compositions of refractory elements suggest bulk Earth to be composed of materials similar to E-chondrites while volatile elements favor their delivery via CI chondritic material, the timing of admixing of C-rich material into the accreting zone of Earth becomes key. Such admixing could have taken place either during the primary stage of Earth's growth or after the main phase of accretion. Similarly, delivery of C-bearing materials may have occurred via accretion of undifferentiated bodies similar to chondrites or via amalgamation of differentiated bodies with C abundances in relevant reservoirs of those bodies set by core–mantle differentiation and atmospheric losses.[16–18] Theoretical models have been used to simulate the early evolution of the Solar System by accounting for dynamics of planetary accretion along with geochemical, cosmochemical, and chronological relationships between accreting and resulting bodies.[65,66] Constrains from Hf/W chronometry predict that Mars-sized planetary embryos were formed from mostly E-type

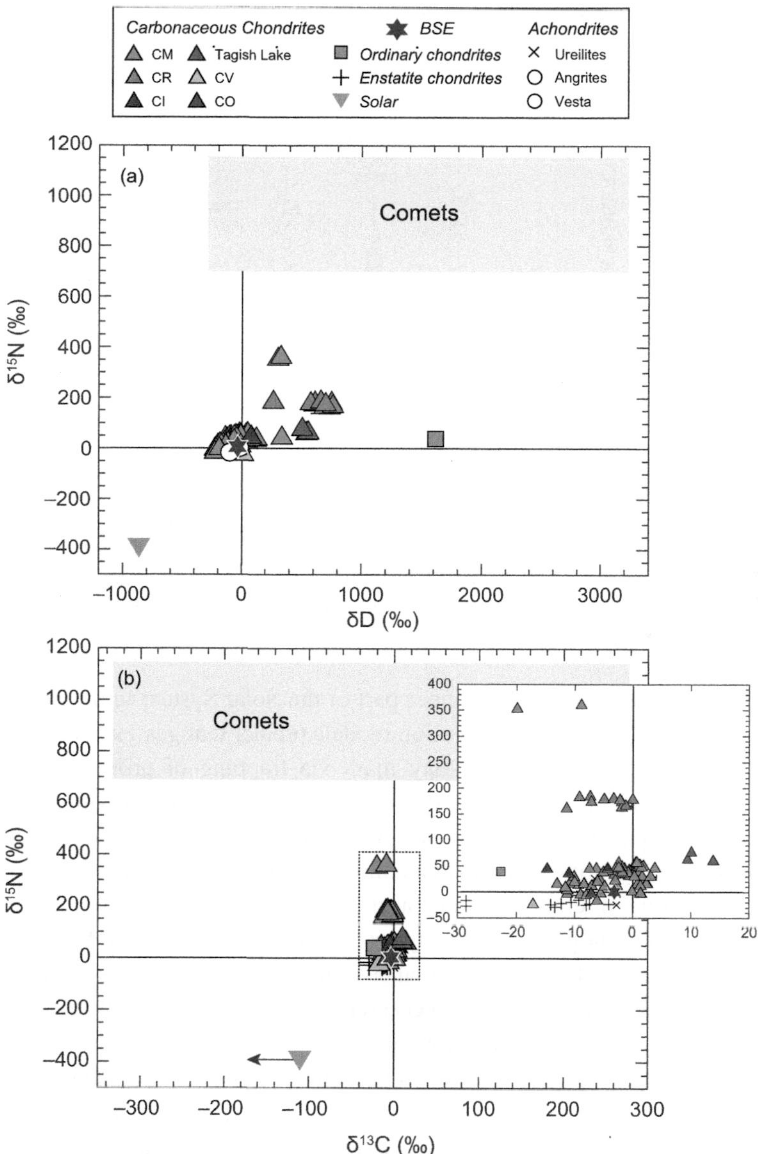

Figure 2.2 Comparison of isotopic compositions of (a) hydrogen and nitrogen and (b) carbon and nitrogen for several Solar System objects and reservoirs. The solar reservoir is depleted while the cometary reservoir is enriched in the heavier isotopes of nitrogen and hydrogen relative to all classes of meteorites as well as the BSE. Although $\delta^{13}C$ alone cannot distinguish between the meteoritic and cometary sources, the isotopic compositions of all major LEVEs suggest that the BSE had a similar parent reservoir as carbonaceous chondrites.

Data sources: Carbonaceous chondrites, Refs. 48, 49; E-chondrites, Refs. 52, 53; comets, Ref. 54; angrites and Vesta, Refs. 51, 55; and solar, Refs. 56–58.

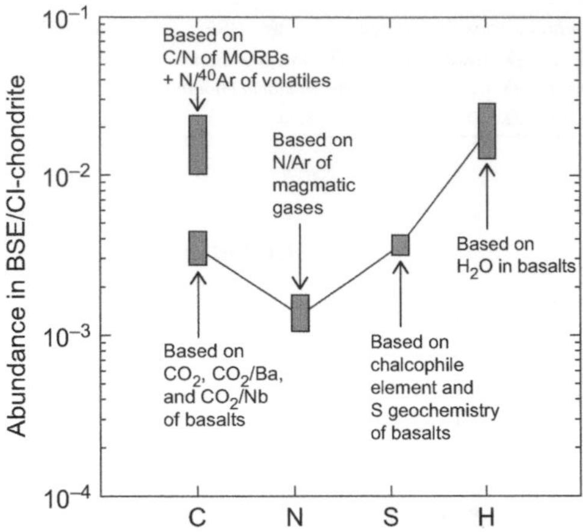

Figure 2.3 Comparison of the abundances of C, N, S, and H in the BSE normalized to their respective abundances in CI chondrites. In the BSE, C is enriched relative to N, has a similar abundance to S, and is depleted relative to H in comparison to its abundance in CI chondrites. Data from Refs. 10, 11, and 13. Also noted in the figure is the main geochemical information used to estimate the BSE concentrations of the respective elements in the original studies. MORB = mid-ocean ridge basalt.

chondritic material sourced from the inner part of the Solar System in the initial ~5 Ma.[67] Recent advancements like pebble accretion models predict that gas giants like Jupiter also grew synchronously to their present-day mass via trapping of proto-solar nebular gas within the lifetime of the protoplanetary disk.[68,69] However, the growth of gas giants is thought to rapidly deplete the asteroid belt, which would mean minimal interaction of C-rich carbonaceous chondritic material that condensed in the outer parts of the Solar System with the growth zone of terrestrial planet accretion within 2 AU.[19] With the consideration that the presence of giant planets in the system affects the mixing and delivery of LEVEs from the outer regions to the terrestrial planet-forming region in the disk, the migration of giant planets has been argued to be necessary to perturb the orbits of cometary and asteroid-like bodies, bringing a fraction of these to the inner regions, where they can collide with terrestrial planets, delivering additional volatile elements to the rocky planets.[19] A specific planetary migration model, the Grand Tack scenario,[19,20,65] postulated that the inward and outward movement of Jupiter likely repopulated the asteroid belt in the inner parts of the Solar System with C-type chondritic material. However, these models cannot constrain the net mass of volatile-rich material being delivered, as they allow for a total influx of volatile-rich material $\geq$0.5–2.0% of present-day Earth's mass, assuming volatile-rich CI chondrites have water in the range of 5–10 wt.%.[19,20] Delivery of a C budget that is larger than the present day in the BSE would require offsetting via later differentiation processes, while C delivery being restricted to the present-day budget of the BSE would have to be unaffected by later differentiation processes. A model such as Grand Tack does not directly constrain the relative timing of acquiring the C- and water-rich chondritic materials in the terrestrial planet-forming regions with respect to specific giant impact events. However, with the Moon formation being a specific event, similarity or a lack thereof between C and other LEVE budgets and volatile isotope signatures of Earth and Moon may shed light on when LEVE-rich materials were brought in with respect to the Moon-forming event. Little is known about the C budget of the Moon or the lunar mantle;[70,71] however, if water and carbon were co-delivered, the appreciable water content of lunar glasses and melt inclusions[72,73] and the E-chondritic nature of late veneer[32] can only be reconciled if volatiles were delivered to the Earth–Moon system before[74,75] or during the Moon-forming impact.[18] However, if C and other LEVEs were delivered during the main stage of Earth's accretion, then differentiation processes like core–mantle separation, as well as the delivery mechanism of volatiles – either by smaller, undifferentiated planetesimals composed of primarily CI-chondritic material or via relatively large, differentiated planetary embryos heterogeneously composed of material sourced from both inner and outer parts of the Solar System – can provide additional information on the origin of LEVEs in the BSE.

2.3 C and Other Volatiles: Abundances, Ratios, and Forms in Various Classes of Meteorites and Comparison with the BSE

The present-day BSE is estimated to be depleted in C and other highly volatile elements relative to all classes of undifferentiated meteorites by at least an order of magnitude and by as much as two orders of magnitude compared to the more primitive carbonaceous chondrites (Table 2.1).[13] For example, carbonaceous chondrites are C rich,[48,54,76] while enstatite and OCs are relatively C poor.[54,61] Among various carbonaceous chondrites, CI chondrites are the most C rich,[54,76–79] and they are also richer in other LEVEs such as N (0.19–0.32 wt.%[54,76,77]) and H (1.55–2.02 wt.%[54,77]) compared to other types of carbonaceous chondrites such as CV, CO, CM, and CR (Table 2.1). Severe depletion of C in the BSE relative to CI chondrites can be explained either by accretion of extremely small quantities of CI chondrites as late additions or via accretion of larger quantities of CI chondrites, but subject to subsequent loss during accretion or differentiation. Such loss of C from carbonaceous building blocks may be expected given that primary carbon is present as soluble as well as insoluble organic molecules in carbonaceous chondrites, which are expected to be unstable at the high temperatures occurring during inner-Solar System processes. Similarly, the carbonates that are products of secondary alterations in carbonaceous chondrites are also expected to volatilize during delivery and accretion in the inner-Solar systems bodies. Even though E-chondrites may have C abundances that are not much greater than that in the BSE, such concentrations are more likely to survive the delivery to the inner-Solar System bodies given their presence in the form of more refractory graphite, diamonds, and carbides.[80] It remains unclear, however, in what form and abundance C was in E-chondrite parent bodies before thermal metamorphism. It cannot be ruled out that unprocessed E-chondritic protoliths were more C rich and may have even contained

Table 2.1 *C content and C/S, C/N, and C/H weight ratios of major terrestrial reservoirs and chondritic building blocks*

Reservoirs	C (wt.%)	C/S	C/N	C/H
Terrestrial reservoirs				
BSE	0.011 ± 0.002	0.49 ± 0.14	40.00 ± 8.00	1.13 ± 0.20
Mantle	0.008 ± 0.002	0.36 ± 0.10	72.73 ± 37.72	2.00 ± 0.70
Crust–ocean–atmosphere	0.002 ± 0.000	8.89 ± 2.22	12.5 ± 3.22	0.51 ± 0.04
Core	0.50 (?)	0.32 (?)	85.00 (?)	8.40 (?)
Carbonaceous chondrites				
CO	0.63 ± 0.24	0.32 ± 0.12	21.74 ± 19.16	10.40 ± 0.00
CV	0.92 ± 0.43	0.42 ± 0.20	23.96 ± 24.16	10.20 ± 0.00
CM	1.89 ± 0.48	0.58 ± 0.15	21.01 ± 8.35	3.20 ± 0.00
CI	4.24 ± 0.77	0.72 ± 0.13	19.66 ± 11.69	4.80 ± 0.00
Enstatite chondrites				
EH	0.36 ± 0.13	0.06 ± 0.02	13.73 ± 7.26	11.00 ± 0.00
EL	0.48 ± 0.16	0.15 ± 0.05	24.43 ± 9.40	11.00 ± 0.00
Ordinary chondrites				
H	0.11	0.06	>104.00	2.50 ± 0.70
L	0.09	0.04	46.10 ± 1.00	3.00 ± 1.00
LL	0.09	0.05	51.22 ± 1.20	2.60 ± 0.50

The mantle, the crust-ocean-atmosphere, and the BSE averages and 1σ standard deviations are from Hirschmann and Dasgupta[7] for C/H, Bergin et al.[10] for C/N, and Hirschmann[11] for C/S. The carbonaceous chondrite data for C, H, and N are from Figure 2.3, E-chondrite C/N data are from Grady et al.[52] and Alexander et al.[53] OC C/N data are from Bergin et al.[10] and references therein. The C/S data for all chondrites are from Wasson and Kallemeyn,[37] and the C/H data for enstatite and OCs are from Hirschmann[11] and references therein. The core compositional estimates are using bulk Earth concentrations from McDonough,[83] and $D_x^{alloy/silicate}$ values are from Figure 2.5 and assuming complete core–mantle equilibration. The data for C contents are from Bergin et al.[10] for terrestrial reservoirs, from Alexander et al.[48] for carbonaceous chondrites, from Grady et al.[52] for E-chondrites, and from Wasson and Kallemeyn[37] for OCs.

organics. Despite the survival potential of refractory C-bearing phases in E-chondrites in the inner-Solar System processes, it is generally thought that the E-chondrites had much less water than the present-day BSE,[81] although the H budget of the BSE has significant uncertainty owing to poor constraints on the water budget of the mantle.[82]

In contrast to absolute abundances, the relative abundance of C with respect to other volatile elements in the BSE (i.e. C/H, C/N and C/S) is a powerful tracer to constrain the C abundance of Earth and provides additional constraints on the processes that must have fractionated the geochemical reservoirs relative to the cosmochemical reservoirs (Figure 2.3).

Due to the comparable 50% condensation temperatures of C and N as well as their abundances being positively correlated in various classes of chondrites, comparison of the

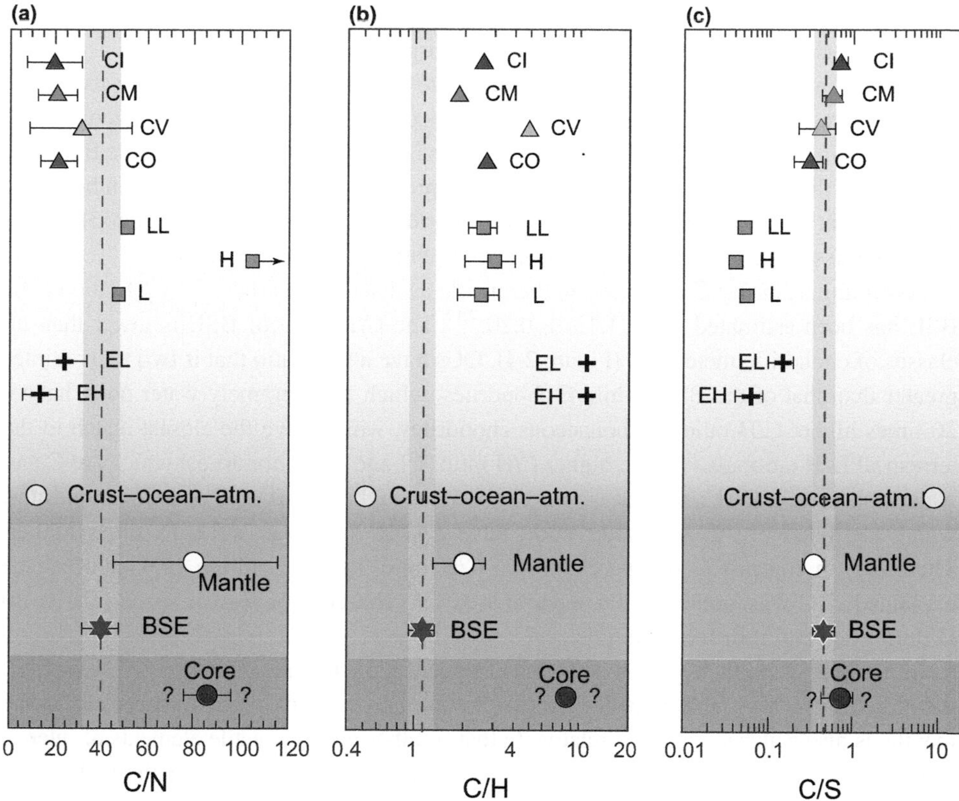

Figure 2.4 Comparison of bulk C/N (a), C/H (b), and C/S (c) weight ratios between different terrestrial and chondritic reservoirs. The data are from Table 2.1. CI = Ivuna-type Carbonaceous chondrite; CM = Mighei-type Carbonaceous chondrite; CV = Vigarano-type Carbonaceous chondrite; CO = Ornans-type Carbonaceous chondrite; LL = low total iron, low metal ordinary chondrite; H = highest total iron ordinary chondrite; L = low total iron ordinary chondrite; EL = (low enstatite) enstatite chondrite; EH = (high enstatite) enstatite chondrite.

C/N ratio of the BSE with that of the building blocks can help track the accretion and differentiation history of C in the early Earth. Recently, Bergin et al.[10] estimated the BSE C/N ratio to be 40 ± 8 by calculating the mantle C content using CO_2/Nb and CO_2/Ba ratios of possibly undegassed ocean island basalts (Figure 2.3).[84] This C/N ratio is similar to the estimate of Halliday,[12] but lower than the estimate of Marty.[13] We will adopt the C/N ratio from the latest studies.[11,18,85] The C/N ratios of volatile-rich CI and CM as well as slightly volatile-poor CO and CV chondrites are lower than the C/N ratio of the BSE by approximately a factor 2, while E-chondrites also have low C/N ratios (5–25; Figure 2.4a). Although OCs show large scatter in C/N ratios due to uncertainties in determining extremely low C and N contents, their average C/N ratios are higher than that of the BSE (Figure 2.4a). Because the average C/N ratio of all classes of chondrites lies in the range of 5–30, the superchondritic

C/N ratio of the BSE has been postulated to be a legacy of either preferential incorporation of N into the core or loss of an early N-rich atmosphere.[13] However, ureilites, a class of achondrites and assumed to be the mantle restite of an asteroidal body, are extremely C rich (average 3 wt.% C) with intergranular veins of graphite and traces of diamonds.[86,87] It is interesting to note that the C/N ratio of ureilites is much higher than that of the BSE;[11,88] therefore, if the C/N ratios of this achondrite are representative of an ureilite parent body and such a parent body significantly contributed to Earth's accretion,[89] then preferential loss of N as a necessary condition for bulk silicate reservoirs in Earth is precluded.

Assuming a similar C inventory in the mantle as discussed earlier, the C/H ratio of the BSE has been estimated to be 1.13 ± 0.20.[7,11] The C/H ratio of BSE is lower than all classes of chondritic meteorites (Figure 2.4). OCs have a C/H ratio that is two to five times greater than that of the BSE, while E-chondrites, which are extremely water poor, have a 20-times higher C/H ratio. Carbonaceous chondrites, which have the closest match to the terrestrial D/H ratio, also have a higher C/H ratio; CI and CM chondrites, which are C and water rich, have a C/H ratio that is four to six times higher; while CO and CV chondrites, which are relatively C and water poor, have a C/H ratio that is 12–15 times higher. Therefore, in contrast to a superchondritic C/N ratio, a subchondritic C/H ratio can be explained if C was preferentially incorporated into the core or lost to space relative to H during terrestrial differentiation.

Sulfur has a much higher solar nebula condensation temperature than C, N, and H; therefore, it is not a highly volatile element. Sulfur abundance in the BSE, especially in the mantle, is also better known[90–92] relative to that of other highly volatile elements of interest (Figure 2.3). Therefore, the C/S ratio can also act as an important tracer to track the evolution of C in the early Earth. The C/S ratio of the BSE is higher than that of C-depleted E-chondrites, but it is similar to or lower than that of C-rich carbonaceous chondrites (Figure 2.4c). Therefore, a near-chondritic C/S ratio in the BSE stipulates that either both C and S showed similar loss/gain behavior during early differentiation processes or they were primarily delivered in post-differentiation processes via undifferentiated, chondritic materials.

Because the C/H ratio of the BSE is subchondritic, the C/S ratio is near-chondritic, while the C/N ratio is superchondritic, H, N, and S have widely different geochemical behaviors relative to C during core formation or MO degassing and/or they track different stages of terrestrial accretion. The following sections will explore whether there are known processes that can explain the difference between the LEVE geochemistry of the BSE and those of carbonaceous chondrites.

2.4 Establishing LEVE Budgets of the BSE After Core Formation?

2.4.1 The Role of Late Accretion

The theory of the addition of undifferentiated meteorites after the completion of metal–silicate differentiation chiefly stems from the approximately chondritic relative abundance of HSE (e.g. Re, Os, Ir, Ru, Pt, Rh, Pd, and Au) concentrations in Earth's mantle and the

BSE, plus the fact that the BSE is not as depleted of HSEs as it would have been if the entire HSE inventory of the bulk Earth was available before equilibrium core formation took place.[93,94] Assuming that core formation originally left the silicate Earth virtually HSE free and later additions of chondritic materials elevated the HSE abundances to the present level, the chondritic mass added is estimated to be 0.5% of Earth's mass (0.005 M_E). Walker[93] noted that this estimate may be ~1.5 times higher or a factor of four lower[95] depending on the exact concentrations of HSEs in the chondritic materials being added. Willbold et al.[96] also calculated that in order to offset the pre-late accretion ^{182}W-excess composition of the BSE, 0.08 M_E chondritic mass needs to be added. Given that many carbonaceous chondrites are rich in LEVEs, many studies have surmised that late accretion of these materials brought all of the LEVEs to the BSE.[71,97–99] If this was the mechanism, then the core formation would have no influence on the relative abundance of various LEVEs. Indeed, the addition of 0.0030–0.0075 M_E chondritic materials with 3.48–3.65 wt.% C to the BSE would set the BSE C budget to ~100–250 ppm C (or ~360–900 ppm CO_2 in the mantle[8]). Although this C budget may be sufficient to match even the enriched basalt source regions, as briefly discussed by Dasgupta,[8] there are several challenges in invoking CI-type carbonaceous chondrite as the chief or sole source of the BSE C and other LEVEs. Section 2.2 already discussed which elemental and isotopic compositions of the BSE are not satisfied if the BSE is chiefly made up of carbonaceous chondrites. Section 2.2 also outlined how the elemental abundance of C and the C/S ratio of the BSE can be satisfied if CI-type carbonaceous chondrite accreted to Earth after core formation was complete, but the C/N and C/H ratios and possibly δ^{13}C and δ^{34}S of the BSE are difficult to explain. Here, we specifically discuss what challenges exist in invoking other known meteorites as the late-accreting materials to explain the entire inventory of carbon and other LEVEs in the BSE.

Given E-chondrites (e.g. (low enstatite) enstatite chondrite (EL) and (high enstatite) enstatite chondrite (EH)) are depleted in water and also poor in C and N,[52] bringing LEVEs in general and water in particular to Earth exclusively by E-chondrite is not thought to be a realistic mechanism. EH and EL chondrites also have distinctly lower C/N and C/S ratios compared to the BSE (Figure 2.4a and c). Hence, unprocessed E-chondrite, known from the studied samples, could not have brought LEVEs to Earth after core formation in the right proportions. This is at odds with recent propositions that, based on Ru isotopic composition, argued for the late veneers having E-chondritic character.[32,33] One issue in evaluating E-chondrites as plausible terrestrial building blocks based on volatile element geochemistry is that all known E-chondrites studied thus far are heavily metamorphosed; hence, before thermal metamorphism, these chondrites might have been more LEVE rich and C could have been originally organic.[100]

OCs are depleted in volatiles with respect to carbonaceous chondrites chiefly because the volatile-rich matrix volume percentage is less than that of carbonaceous chondrites.[54] In fact, bulk C content of most primitive OCs can be perfectly explained by its matrix material being entirely CI-type material.[48] OC matrices are drier, however, with bulk H of

OCs falling below the concentration of what would be expected if the entire OC matrix was made up of CI-type materials.[54] Therefore, the C/H ratio of OCs is higher than that of the BSE (Figure 2.4b). Similarly, the C/S ratios of OCs are lower and the C/N ratios of OCs are distinctly higher compared to the estimates for the BSE (Figure 2.4a and c). Hence after core formation, OC delivery cannot be considered as the chief process of origin of Earth's LEVEs.

Among all stony meteorites, ureilites are the only group of primitive achondrites that are rich in C and depleted in N, leading to higher C/N ratios.[101–103] Carbon in ureilites is also present in relatively non-labile phases (i.e. as graphites, nanodiamonds, and lonsdaleite) and hence is likely to survive high-temperature accretional processes. Primarily motivated by their high C/N ratios, recent studies[11,88] have suggested that C-rich achondrites such as ureilites may be responsible for establishing the abundances of LEVEs in the BSE through post-core formation addition. While the C/N ratio of the BSE could be heavily shaped by the addition of a ureilite-like achondrite, the C/S ratio of ureilite is significantly higher than that estimated for the BSE. Hence, any attempt to match the C/S ratios of the BSE through addition of a ureilite-like achondrite-rich late veneer requires a very unique consideration of the composition and extent of core–mantle equilibration of proto-Earth.[11] No model of ureilite-like late-veneer addition can explain the C/N and C/S ratios simultaneously. For example, scenarios of C-rich late-silicate addition in which the superchondritic C/N ratio can be established also result in a superchondritic C/S ratio.[11]

Overall, late accretion alone does not seem to account for the C budget and C to other LEVE ratios of the present-day BSE (Figure 2.5). Hence, other early Earth processes need to be considered.

2.4.2 The Role of Post-core Formation Sulfide Segregation

The segregation of sulfide melt has been shown to be a necessary process for explaining the HSE geochemistry of the BSE.[112] This would also deplete the BSE in terms of its S inventory, but owing to the low solubility of C in sulfide-rich melts (Figure 2.6a)[17,18,113–116] caused by strong non-ideality of mixing between C- and S-rich components in Fe-rich alloy melts, such sulfide segregation should also elevate the C/S ratio of the BSE. Hence, if late accretion establishes a chondritic C/S ratio, sulfide segregation should alter such a ratio. N also is not compatible in S-rich Fe-alloy melts,[18] hence the S/N ratio of the BSE would also drop after sulfide melt segregation. Similarly, it is also unclear what the effects of late-stage sulfide melt segregation might be on the H budget of the BSE. If H is stored in nominally anhydrous silicate minerals in solid, post-core formation BSE, partitioning of H between segregating sulfide melts and solid silicate minerals could influence the H inventory and hence the C/H ratio of the BSE. If H partitions preferentially into sulfide liquid over relevant nominally anhydrous silicate minerals, then the C/H ratio may also increase above the chondritic value.

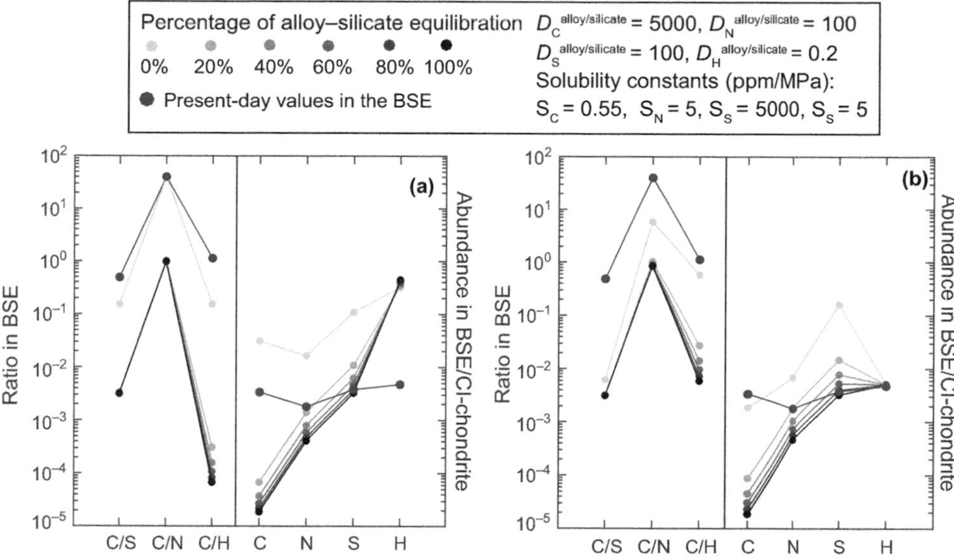

Figure 2.5 Application of alloy–silicate partition coefficients and solubilities of LEVEs in the silicate melts to examine the effect of core formation, with varying degrees of alloy–silicate equilibration, with or without loss of an early atmosphere formed via MO degassing on the remnant abundances of LEVEs in the bulk silicate reservoir. (a) LEVEs, when delivered as 0.015 M_E late-accreting materials (i.e. 0% alloy–silicate equilibration), cause the volatile abundance to be higher than the present-day BSE. Core formation with increasing degrees of alloy–silicate equilibration increasingly depletes the remnant MO in all LEVEs, with C being much more depleted than other LEVEs, leading to subchondritic C/N, C/H, and C/S ratios. (b) Combining early atmospheric loss with core formation cannot offset C loss to the core due to the lower solubility of C relative to the other LEVEs in the silicate MOs. Bulk Earth volatile abundance data are from McDonough,[83] while the alloy–silicate partition coefficients in a deep MO (P = 50 GPa, T = 3500 K; e.g. Siebert et al.[104]) for C, N, S, and H are from the parametrized relationships of Chi et al.,[71] Grewal et al.,[105] Boujibar et al.,[106] and Clesi et al.,[107] respectively. Solubility constant data for C, N, S, and H in the silicate melt are from Armstrong et al.,[108] Roskosz et al.,[109] O'Neill and Mavrogenes,[110] and Hirschmann et al.[111]

2.4.3 The Role of MO–Atmosphere Interactions and Atmospheric Loss

Because C and other LEVEs are atmophile elements, they were likely heavily concentrated in the proto-atmospheres at various stages of MO evolution and terrestrial accretion. Hence, it is worth considering under what circumstances silicate magma–atmosphere interactions and possible atmospheric loss through impact-driven blow-off[23] could influence the inventory of LEVEs in the BSE. The timing of LEVE delivery is of critical importance in determining the role of silicate–atmosphere interactions (and atmospheric blow-off). For example, if a large fraction of core formation takes place before the LEVEs were available to the growing Earth,[20,118] then such core formation would have had negligible influence on the BSE LEVE budget, with MO–atmosphere processes being more important. On the other hand, if LEVEs were delivered throughout the terrestrial

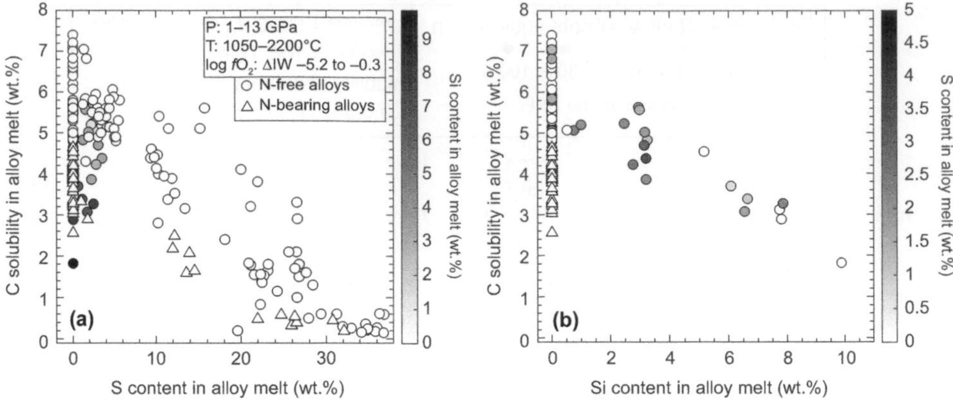

Figure 2.6 Experimental data showing the effects of S and Si contents in the Fe-rich alloy melt on C solubility. (a) Carbon solubility decreases monotonically with increases in sulfur content in the alloy melt. Si- and N-bearing alloys have lower C solubility for a given S content in the alloy melt. (b) Carbon solubility decreases linearly with increases in Si content in the alloy or decreases in oxygen fugacity (fO_2) of silicate–alloy systems. The presence of S in the alloy does not have a major effect on carbon solubility in silicon-bearing alloys.

Data for N-free alloys are from Dasgupta et al.,[99,114] Chi et al.,[71] Armstrong et al.,[108] and Li et al.,[16,117] while data for N-bearing alloys are from Dalou et al.[85] and Grewal et al.[18]

accretion history,[13,51,119] then both atmosphere–MO interactions and metal–silicate equilibration could have played roles in establishing the LEVE budget of the BSE.

If the BSE C–N–S–H budgets were established chiefly by loss of proto-atmosphere overlying MO, then N needed to be lost preferentially to C and C needed to be lost preferentially to H. In other words, the proto-atmosphere overlying MO has to be N rich relative to C, C rich relative to H, and with similar abundances of C and S. Indeed, the N-depleted nature of Earth has been explained previously by loss of an N-rich atmosphere.[120] But evaluation of whether an atmosphere with a concentration of LEVEs in the order N > C~S > H can be generated overlying a silicate magma reservoir depends on the mixed-volatile solubility in shallow MO as a function of oxygen fugacity (fO_2; effective partial pressure of oxygen). Hirschmann[11] considered a simple model evaluating the fractionation of C, N, H, and S between silicate MO, overlying atmosphere, and equilibrating core-forming liquid alloy with three distinct fO_2 conditions at the MO surface (i.e. IW – 3.5 (reduced), IW – 2 (intermediate fO_2), and IW + 1 (oxidized), where IW refers to the log fO_2 set by equilibrium of metallic iron and iron oxide (FeO)). Among these, the most oxidizing condition is not very realistic during or soon after core formation because, for a well-mixed MO with a near-constant Fe^{3+}/Fe_T ratio, the fO_2 gradient is such that the shallow MO is always more reduced.[121–123] Thus, if a well-mixed MO is close to equilibrium with an Fe-rich core at its base, its surface should be more reduced and therefore, for MO–atmosphere interactions, the relevant fO_2 would be less than IW. The only way this may be different is if at higher pressures the fO_2 gradient reverses (reduction with increasing depth) and iron disproportionation and subsequent Fe–metal segregation leads to oxidation of metal-free MO.[124] Solubility data for most LEVEs are lacking or sparse for peridotitic or ultramafic silicate melts due to difficulty in obtaining a glass and reliable chemical analyses for such compositions.[125] Based on the data available for silicic to mafic silicate liquids, for log fO_2 < IW – 0.5 and with decreasing fO_2, N becomes much more soluble (reaching the weight percentage level chiefly as N–H species[85,126,127]) than C (tens of ppm by weight mostly as CO_3^{2-} and to some extent as C–H and C–N species[18,71,99,108,117,125,128,129]), with the latter mostly decreasing with decreasing fO_2 and then leveling off with stabilization of C–H species. Similar to N, H also remains quite soluble in silicate melt in the form of OH^- and H_2O even in fairly reducing conditions such as log fO_2 of IW – 1 to IW – 2, with molecular hydrogen, H_2, contributing to some degree.[111] Therefore, at log fO_2 < IW – 1.5, atmospheric blow-off could lower the C/H ratio from a chondritic value, but this would result in an MO C/N ratio that is subchondritic. Similarly, S is much more soluble in mafic–ultramafic silicate melts (several thousands of ppm[110,130,131]) compared to C and hence loss of an atmosphere would result in a residual silicate MO with a C/S ratio distinctly lower than that of chondritic value. Therefore, no condition exists where a proto-atmosphere overlying an MO can attain all of the compositional characteristics necessary to leave an MO with C/N, C/S, and C/H ratios of the modern-day BSE (Figure 2.5). Hence, late accretion of chondritic materials to an alloy-free MO followed by an atmospheric loss cannot be the chief origin of BSE LEVEs.

An alternative suggestion[120] is for atmospheric loss to help define the BSE C/N and C/H ratios, but at conditions where C would be retained in the form of carbonate or bicarbonate ions either as dissolved aqueous species or in the crust, mediated by the presence of a liquid water ocean. According to this model, a superchondritic C/N ratio and a subchondritic C/H ratio could have been attained via loss of an N-rich atmosphere overlying an ocean. This mode of atmospheric loss is supported by the dynamical model of a giant impact that shows that such an impact would preferentially remove the atmosphere relative to the ocean.[132] However, it is unclear what the water chemistry would be in these oceans, which, if they existed, is expected to be highly anoxic and thus may not sequester significant dissolved carbonates.

2.5 Establishing the Volatile Budget of the BSE through Equilibrium Accretion and MO Differentiation

The discussion on the origin of LEVEs in the BSE thus far considered building blocks and processes after the core formation was mostly complete. This approach would be appropriate if all of the LEVEs were delivered through late accretions. The discovery of a solar component (e.g. neon) in deep plume-derived magma,[133,134] however, suggests that Earth might have trapped nebular gas early within the first few millions of years of the formation of the Solar System. Indeed, the possibility of acquiring volatile elements such as primordial noble gases directly from nebular gas has been proposed in a number of studies.[135,136] Although nebular gas can persist for ~10 Myr, its median lifetime (i.e. the time at which half of all systems lost their nebular gas disks) is ~2.5 Myr.[137] Given the mean time of terrestrial accretion (i.e. the time to have grown to 50% of the final mass) is ~11 Myr[138] while the mean accretion timescale of Mars is $1.9^{+1.7}_{-0.8}$ Myr,[67,139] it is likely that although the entire proto-Earth mass may not have equilibrated with the solar nebula, at least a few Mars-sized embryos that contributed to the initial proto-Earth mass likely inherited nebular volatiles. These constraints suggest that early accreting material may have been LEVE poor but likely not LEVE free. However, the LEVE compositions do not suggest direct Solar contribution to be important (Figure 2.2).

The $^{182}Hf/^{182}W$ ratio of iron meteorites suggests metal–silicate separation in protoplanetary bodies occurred as early as ~1 Myr after the formation of the Solar System. Accordingly, core formation would influence the initial distribution of LEVEs among the planetary reservoirs.[8,11,14,16,17,71,85,99,117] Over the past decade, a significant body of research has tried to understand the fate of C and other LEVEs (S, N, and H) during core formation processes and associated alloy–silicate equilibration.[14–18,71,85,99,106,109,117,140–142] These investigations chiefly focused on determining the partition coefficients of LEVEs between Fe-rich alloy melts and coexisting silicate melts (i.e. $D_x^{alloy/silicate}$ (= concentration of an element in the alloy melt divided by concentration of the same element in the equilibrium silicate melt), where x = C, N, S, etc.). If $D_x^{alloy/silicate}$ is >1 and if a given LEVE, x, is available during core formation, then it should be preferentially sequestered in the metallic core. On the other hand, if $D_x^{alloy/silicate} < 1$, the LEVE, x, present during core–mantle fractionation should preferentially concentrate in the silicate Earth. $D_x^{alloy/silicate}$ values provide the framework to understand how a given LEVE is expected to be distributed between the metallic and silicate portions of a differentiated planet if alloy and silicate melts equilibrated before separation.

The most robust constraints available on the alloy–silicate partitioning of LEVEs are obtained through high pressure–temperature experiments. Determinations of $D_C^{alloy/silicate}$ at high pressure–temperature exist chiefly for graphite/diamond-saturated conditions (i.e. experiments conducted in graphite capsules). Concentrations of C in Fe-rich alloy melts at graphite/diamond saturation and in the absence of any other light nonmetals show small dependence on pressure and temperature (not shown)[14,16] and are ~5–7 wt.%. However, C solubility in Fe-rich alloy melts decreases strongly with increasing S and Si content, and at S of content 22–36 wt.% or Si content >10–12 wt.%, C solubility is <1 wt.% (Figure 2.6a and b). Carbon solubility in Fe-rich alloy melts also diminishes with increasing Ni content in the S-free or S-poor alloy.[71,115,116,143] Unlike in alloy melts, C solubility in equilibrium silicate melts is tens to hundreds of ppm and mostly decreases with decreasing fO_2 (Figure 2.7). The MOs of Earth and other inner-Solar System bodies are expected to be ultramafic to mafic; that is, relatively poor in silica and rich in MgO.

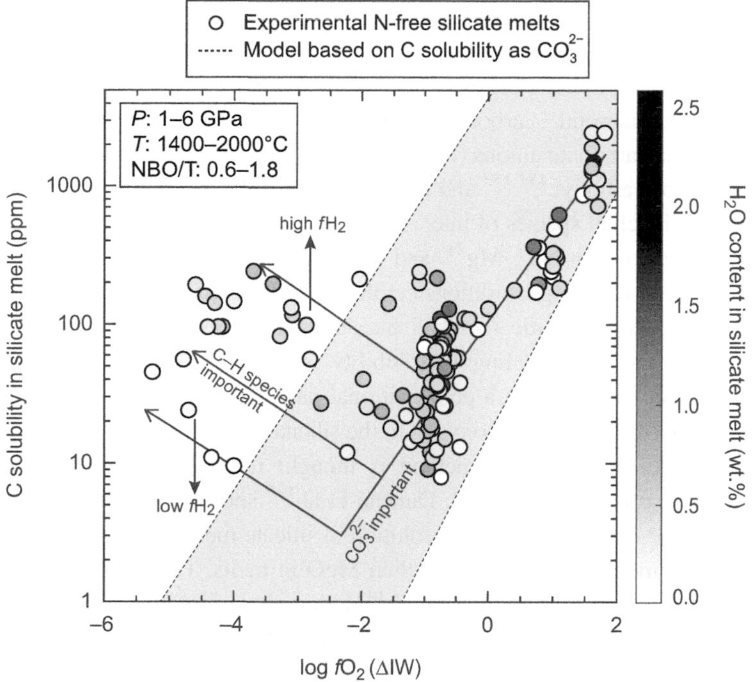

Figure 2.7 Experimental data showing the effects of fO_2 and water content in the silicate melt on C solubility. Carbon solubility decreases with decreasing log fO_2 until ~ΔIW of –2 to –1 followed by an increase below ~IW – 2 depending upon bulk water content in the silicate melt. This change is likely caused by the change in contributions of C as CO_3^{2-} at higher fO_2 and C as C–H species such as methyl groups at lower fO_2. The pressure–temperature–composition (expressed as NBO/T) condition of the experimental glasses are given in the legend. The shaded region represents predicted solubility of C as CO_3^{2-} at graphite saturation based on the model of Eguchi and Dasgupta[146] for the entire pressure, temperature, and X_{H2O} space of the experimental data, while the NBO/T parameter is extrapolated for peridotite-like silicate melts (NBO/T = 2.7).
Data for N-free silicate melts are from Refs. 16, 17, 71, 99, 108, 117, 125, 129, 152, and 157.

Because of the difficulty of rapidly cooling such depolymerized silicate liquids from well-constrained high pressures–temperatures to glasses at ambient conditions, which is necessary for reliable quantitative analyses of carbon and hydrogen, C solubility data in such ultramafic–mafic liquids at core-forming reduced conditions are sparse.[125] To address this, experimental studies have been conducted with a silicate melt of varying polymerization so that a meaningful extrapolation of C solubility and $D_C^{alloy/silicate}$ to desired silicate melt composition can be made.

2.5.1 Carbon Speciation in MO

$D_C^{alloy/silicate}$ and C solubility in silicate melts are affected by the incorporation mechanism of C in the silicate melt structure. Irrespective of the saturating phase such as CO_2-bearing vapor or graphite/diamond, carbon can dissolve in silicate melt as molecular CO_2 ($CO_2^{mol.}$ [144–148]), as carbonate anions (CO_3^{2-} [125,149,150]) bonded to cation modifiers, possibly as molecular CO or carbonyl,[151,152] and as possible C–H, C–N complexes.[18,117] For mafic–ultramafic melts, the chief species of interest is CO_3^{2-}, which complexes most strongly with Ca^{2+} cations and also with Na^+, Mg^{2+}, and maybe K^+.[125,146,153–156] Most experiments on carbon dissolution at reducing conditions employed mafic compositions of variable degrees of polymerization (i.e. tholeiitic to alkali basalts or even more silicic melts). The most common approach to extrapolating C solubility and $D_C^{alloy/silicate}$ to systems containing peridotitic compositions is to use a compositional parameter NBO/T, where "NBO" is the total concentration of non-bridging oxygen in the silicate melt and "T" is the total number of tetrahedral cations. Although this method is thought to give reasonable predictions of ultramafic melt C solubility as CO_3^{2-}, Duncan et al.[125] showed that the NBO/T approach may overpredict the extent of CO_3^{2-} dissolution in silicate melts if an increase of NBO/T is correlated with an increase in CaO rather than MgO in melts. This is because Ca^{2+} cations complex with CO_3^{2-} much more than Mg^{2+},[146] although peridotitic melts are still predicted to have higher CO_3^{2-} dissolution capacities than basalts. Figure 2.7 shows CO_3^{2-} dissolution as a function of fO_2 and shallow MO conditions. Also shown in Figure 2.7 are other possible species such as C–H that may become important at low fO_2.

2.5.2 $D_C^{alloy/silicate}$ and Its Impact on Carbon Distribution between BSE versus Core in Various Scenarios of Equilibrium Core Formation

Figure 2.8 provides the available data of $D_C^{alloy/silicate}$, showing their dependence on alloy composition and fO_2 at shallow MO conditions. Shallow MO $D_C^{alloy/silicate}$ at log fO_2 of IW – 2 to IW – 0.5 are in the order of ~10^3, with the values increasing with increasing pressure and decreasing fO_2 and decreasing with increasing temperature and melt depolymerization.[16,71,117] In more reducing conditions where the Si content becomes significant in the equilibrium Fe-rich alloy melts and where decreasing C dissolution as CO_3^{2-} is offset to some extent by the increasing fraction of C–H complexes (Figure 2.7), $D_C^{alloy/silicate}$ may be somewhat lower and may not monotonically increase with decreasing fO_2 (Figure 2.8b).[16,117] Application of empirical parameterizations of $D_C^{alloy/silicate}$ as a function of temperature, pressure, fO_2, silicate melt composition, and alloy composition (e.g. Ni content) to conditions that many studies considered relevant for the deep terrestrial MO (e.g. pressure = 45–65 GPa, temperature = 3500–4000 K, log fO_2 = IW – 2) yields values in the order of 10^3–10^4. The existing constraints on $D_C^{alloy/silicate}$ in graphite-saturated conditions therefore suggest that carbon is highly siderophile. If core formation was an equilibrium process (i.e. if the entire terrestrial core mass equilibrated with a global silicate MO), post-core formation silicate Earth would be severely depleted in carbon ($\ll$10 ppm),

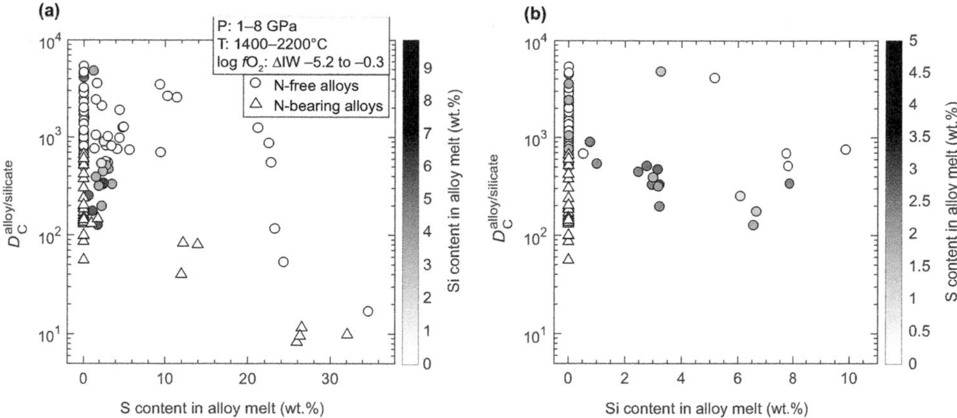

Figure 2.8 Experimental data showing the effects of S and Si in Fe-rich alloy melts on $D_C^{alloy/silicate}$. (a) $D_C^{alloy/silicate}$ decreases with increases in S content in the alloy melt. (b) $D_C^{alloy/silicate}$ decreases with increases in Si content in the alloy or indirect decreases in fO_2 of the alloy–silicate systems. The presence of S in Si-bearing alloy melts generally decreases $D_C^{alloy/silicate}$ in comparison to S-free systems. Data for N-free alloys are from Dasgupta et al.,[99,114] Chi et al.,[71] Armstrong et al.,[108] Li et al.,[16,117] and Tsuno et al.,[17] while data for N-bearing alloys are from Dalou et al.[85] and Grewal et al.[18]

with the extent of depletion being dependent upon the available bulk carbon that participates in such core–mantle fractionation.[8,17,71,99] This analysis does not imply that the core formation process was a single-stage alloy–silicate equilibration event, but rather what the C distribution between Earth's metallic and nonmetallic portions would be for an average condition of core–mantle separation.

Figure 2.9 shows more realistic scenarios of alloy–silicate equilibration and subsequent core segregation in terrestrial MO where the depth, temperature, and fO_2 evolve with the extent of accretion.[158] This style of core–mantle fractionation is described as multistage, equilibrium core formation. Competing arguments exist in the literature as to whether, with continuing accretion, proto-Earth's MO evolved from a relatively reduced to an oxidized condition[158,159] or a relatively oxidized to a reducing condition.[160] Figure 2.9a therefore shows various plausible fO_2 paths of MO evolution of a growing Earth and computes the change of $D_C^{alloy/silicate}$ along those paths, following Li et al.[16] It must be noted that the current data sets of $D_C^{alloy/silicate}$ do not show any effects of pressure–temperature in relatively reducing conditions (e.g. log fO_2 < IW − 1.5). In such reducing conditions, the controlling variables appear to be fO_2, the Si content of the metal, and the water content of the silicate melt. At ~IW − 2 < log fO_2 < IW, however, the key variables controlling $D_C^{alloy/silicate}$ are pressure, temperature, fO_2, and silicate melt composition. Figure 2.9b, however, shows that irrespective of MO condition, $D_C^{alloy/silicate}$ remains $\gg$1, and in fact deeper MO, toward the later stage of accretion, leads to even higher $D_C^{alloy/silicate}$ predictions than those experimentally measured at shallow MO conditions. Therefore, the multistage core formation model also leads to C being stripped off highly efficiently from the silicate MO, leaving behind the BSE that is <1–10 ppm for a bulk Earth with 1000 ppm C (Figure 2.9c).

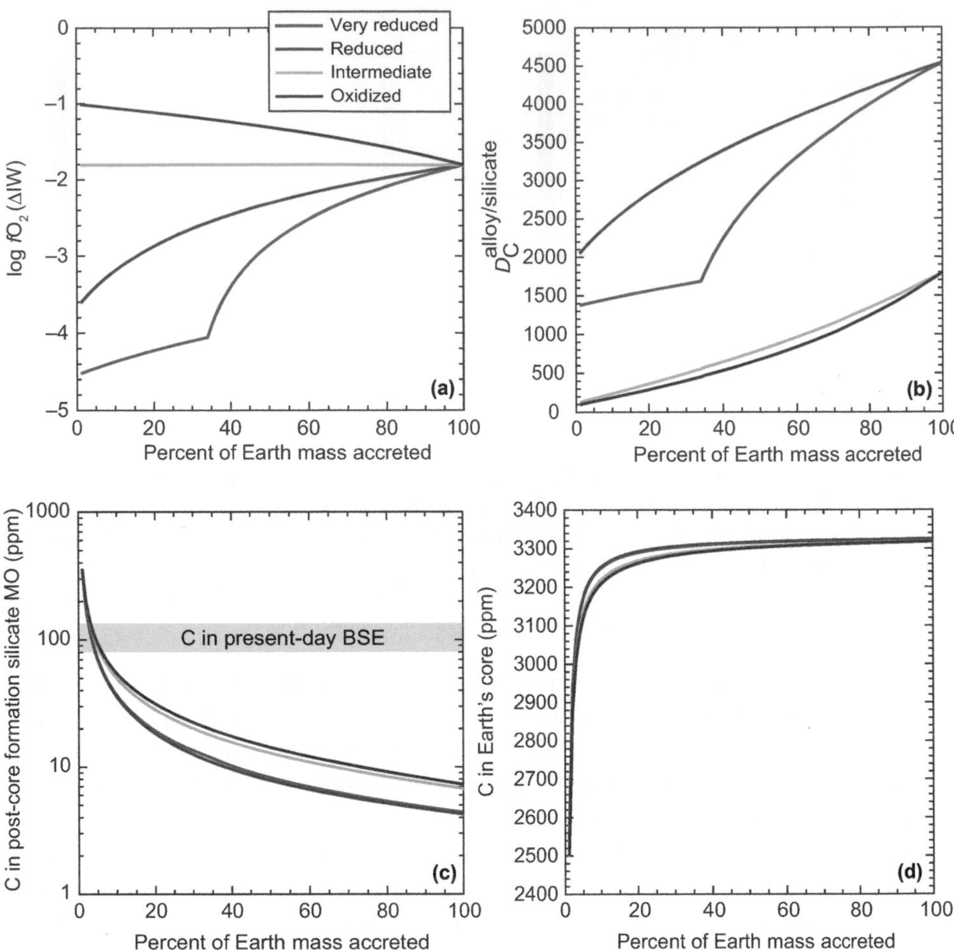

Figure 2.9 The effects of different accretion scenarios on the core–mantle partitioning behavior of carbon. (a) Four possible evolution paths for the FeO content of the mantle with its growth. The "very reduced" and "reduced" paths simulate the accretion of progressively oxidized material with time, and the "oxidized" path simulates accretion of progressively reduced material with time. All of the models converge at log fO_2 of IW – 1.8, or present-day FeO content (8 wt.%) of the primitive upper mantle. (b) $D_C^{alloy/silicate}$ at each step of accretion (i.e. accretion of 1% M_E), calculated using parametrized relationships from Li et al.[16] For the "very reduced" and "reduced" paths, where methyl (or alkyl) groups would be the predominant dissolved carbon species in the silicate melt, eq. (4) is chosen from Li et al.,[16] whereas for the "intermediate" and "oxidized" paths, where carbonate groups would be the predominant dissolved carbon species in the silicate melt, eq. (3) is chosen from Li et al.[16] (c) Post-core formation MO resulting from a bulk system with 1000 ppm C and assuming 100% alloy–silicate equilibration at each accretional step. All of the models lead to MO with distinctly less C that those estimated for the BSE. (d) Carbon content in Earth's core converges at ~3300 ppm at the end of all accretional models.

2.5.3 Comparison of $D_C^{alloy/silicate}$ with Alloy–Silicate Melt Partitioning of Other LEVEs

The preceding section argues that equilibrium core formation either via homogeneous or heterogeneous accretion results in a nonmetallic Earth that is too depleted in C compared to the present-day BSE. Explaining the origin of LEVEs in the BSE through equilibrium core formation is exacerbated when $D_C^{alloy/silicate}$ is compared with the alloy–silicate partition coefficients of other LEVEs such as S, N, and H.

The partition coefficient of sulfur between alloy and silicate melts, $D_S^{alloy/silicate}$, has been experimentally constrained in a number of studies,[106,140,142] but only a few recent studies measured $D_S^{alloy/silicate}$ and $D_C^{alloy/silicate}$ simultaneously.[16–18] These studies show that over the entire range of shallow MO conditions explored experimentally thus far, $D_C^{alloy/silicate} \gg D_S^{alloy/silicate}$, with $D_C^{alloy/silicate}/D_S^{alloy/silicate}$ varying between ~1000 and 20 as log fO_2 varies from IW – 4.5 to IW – 0.5. Therefore, under any conditions of alloy–silicate equilibration, C is more siderophilic than S and equilibrium core formation leaves a subchondritic C/S ratio for the silicate MO.

Experiments also exist that constrained $D_N^{alloy/silicate}$.[18,109,127] Among these, the recent studies that constrain both $D_C^{alloy/silicate}$ and $D_N^{alloy/silicate}$ from the same experiments[18,85] show that nitrogen is only mildly siderophile in relatively oxidizing conditions and possibly even lithophile in relatively reducing conditions (e.g. log fO_2 < IW – 3). Therefore, $D_C^{alloy/silicate}/D_N^{alloy/silicate}$ is ~6–120 at 1–7 GPa, 1400–1800°C, and fO_2 of IW – 3.5 to IW – 0.5, and hence any models of equilibrium core formation would also produce a subchondritic C/N ratio for the post-core formation BSE.

Alloy–silicate partitioning studies for H are limited, and two available studies[107,161] provide contrasting data. Okuchi[161] estimates the $D_H^{alloy/silicate}$ value at a fixed pressure of 7.5 GPa and as a function of temperature, but does not provide any pressure effect. Based on the experiments of Okuchi,[161] H is mildly siderophile. On the other hand, Clesi et al.[107] conducted experiments at 5, 10, and 20 GPa with C-rich alloy melts and measured $D_H^{alloy/silicate}$ of 0.04–0.77. Therefore, it remains unclear whether H is retained mostly in the core of differentiated planets or in the silicate fraction. Despite the discrepancy in the available experimental data on $D_H^{alloy/silicate}$, the BSE could attain a subchondritic C/H ratio via an equilibrium core formation process where much more C would be sequestered to the metallic core than H (Figure 2.4). However, if the $D_H^{alloy/silicate}$ data of Clesi et al.[107] are appropriate, then equilibrium core formation would lead to a BSE C/H ratio that is much lower than what is estimated for the present-day BSE.

2.6 C and Other LEVE Budgets of the BSE: A Memory of Multistage Accretion and Core Formation Process with Partial Equilibrium?

The discussion presented thus far suggests that no single process of volatile delivery such as via late accretion of any known class of meteorites or comets, through loss of an atmosphere overlying a MO,[23] being subjected to equilibrium core formation processes, or sulfide segregation to the core[112,162] can simultaneously explain carbon's absolute as well as relative abundance compared to other LEVEs. This does not mean that these processes did not impact the volatile inventory of the BSE, and it is plausible that all of these processes acted in tandem. However, the nature of the fractionation of C with respect to H, N, and S for each of these processes is such that even if they acted together, not all of the attributes of the BSE volatile budget can be satisfied. This brings us to consider other styles of terrestrial accretion that may have been essential in establishing the LEVE budget of modern Earth.

2.6.1 Disequilibrium Core Formation

Models of protoplanetary disk dynamics, dust and planetesimal accretion, and planetary migration show that although the first 50% of proto-Earth mass growth was continuous, rapid, and completed within the first few million years of Solar System formation, the protracted growth of Earth spanning ~100 Myr[163] involved few abrupt increases in the accreted mass via several giant impacts of Mars-sized or perhaps even larger bodies.[164] The final such impact is thought to be the Moon-forming giant impact.[165,166] A key aspect of accretion via these impacts is that proto-Earth collided with differentiated planetary embryos. Merger of proto-Earth with large differentiated bodies is thought to have taken place via near-disequilibrium merger of the core of the impactor with proto-Earth's core owing to limited emulsification of the impactor's core and the lack of complete mixing of the core with the entrained silicate melt.[167,168] Therefore, core formation involving giant impacts may leave a silicate fraction with a very different LEVE budget than that predicted by equilibrium core formation models. One critical question is whether any differentiated, large impactor could bring LEVE attributes that would help proto-Earth to evolve to the present-day BSE. A full-parameter space involving not only the LEVEs but also other nonvolatile, highly siderophile, and moderately siderophile elements, major elements, and stable isotope anomalies, as well as various accretion and core formation model assumptions, is not yet explored. However, recent studies[16–18] showed how having an S-rich or Si-rich Fe-alloy in the cores of planetary embryos allow attaining high C/N and C/S ratios in the mantles of such embryos. This would occur as the cores of these embryos reach the C solubility limit and expel graphite or diamond to the overlying silicate fraction (Figure 2.10).[16–18] Therefore, a merger of the silicate reservoirs of such embryos with the proto-silicate Earth reservoir could give rise to the BSE LEVE inventory. Using inverse Monte Carlo simulations, Grewal et al.[18] showed that if the merger of the impactor is in perfect disequilibrium, depending on the core size of the impactor and bulk S content, the impactor could be similar in size as Mars (~0.1 M_E; Figure 2.10). It is also showed that the bulk C content of such an impactor would not need to be enriched in carbon and could have a bulk C concentration (~1000–5000 ppm C)[17,18] similar to those of E-chondrites. These model results are intriguing because these potentially circumvent the long-standing challenge of reconciling the volatile element isotope constraints[51,59] and constraints from the isotopic anomalies of lithophile and siderophile elements[31,32] in the major building blocks

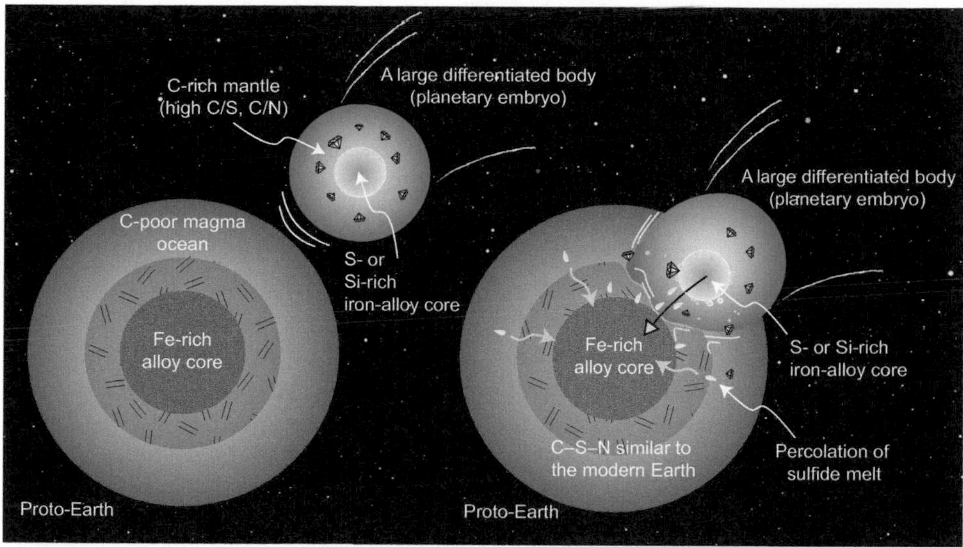

Figure 2.10 A schematic depiction of the late-stage accretion for a large rocky planet such as Earth where a large fraction of mass increase takes place via giant impacts of small planets and planetary embryos (0.1–0.2 M_E).[16–18] This mode of accretion creates a unique possibility of establishing the LEVE inventory in the degassable portion of the rocky planets.

of Earth. Some of these latest studies on the origins of terrestrial volatiles through late-stage accretion of a differentiated planetary embryo are also conceptually similar to models that call for the addition of a highly reduced rocky body to the growing Earth to potentially explain the superchondritic Sm/Nd and $^{142}Nd/^{144}Nd$ anomalies observed in Earth's present mantle.[169,170] The concept of bringing carbon and other LEVEs via accretion of the mantle of a differentiated embryo with little post-impact equilibration is also similar to the suggestion that Earth's Nb/Ta may be explained by accretion of a differentiated asteroid with a reduced S-rich core that efficiently sequestered Nb.[171] Grewal et al.[18] also posited that Earth's volatile-delivering impactor may have been the Moon-forming giant impactor, given the geochemical similarity, including volatiles, between the BSE and the Moon. This is again broadly consistent with Wade and Wood,[172] who demonstrated that near-disequilibrium accretion of a reduced planetary embryo of 0.1–0.2 M_E could explain the isotopic constraints, FeO contents, and Nb/Ta and Hf/W ratios of the silicate Earth and Moon.

2.7 Carbon as a Light Element in the Core

One by-product of our discussion thus far is constraint on the concentration of C in the core. One of the main uncertainties in constraining the C budget of Earth's core is the bulk C content in the growing Earth through time in the first ~100–120 Myr of the Solar System.

It is not realistic to conceive that Earth's building blocks and therefore the bulk C content did not change through the accretion history, but for simplicity here we assume a fixed C budget for the entire history of core formation or MO equilibration. If core formation was achieved by a single-stage equilibrium event at a deep MO, although not a physically realistic process, C content in the core as shown by Dasgupta[8] would be ~0.2 wt.% for a bulk Earth C content of ~730–1000 ppm. For a multistage core formation, with evolving depth–temperature–fO_2 conditions of alloy–silicate equilibration, the core would finally contain ~0.34 wt.% C (Figure 2.9d). Finally, in the model where the C–N–S budget of the BSE is attained via late-stage merger of a large differentiated planetary embryo, the impactor's S-rich core contributes ~100–700 ppm C to Earth's core.[17,18] Therefore, depending on the volatile budget of the growing Earth leading to the giant impact, from a completely C-free proto-Earth to a proto-Earth with 1000 ppm bulk C, the final core C content would be between 700 ppm and 0.4 wt.%. The upper bound derived here is similar to the previous constraint[14] placed using the mantle C abundance. Wood et al.[60] also attempted to place constraint on the C content of Earth's core based on the combined effects of C and S in the alloy on alloy–silicate partitioning of W and Mo, with the chief effect being W and Mo becoming more siderophile with increasing C in the alloy melt. By coupling S and C delivery in the final 13% of accretion, Wood et al.[60] showed that in order to obtain the known Mo and W abundances of the mantle, the bulk C content of the core needs to be <1 wt.%. Badro et al.[160] also reached a similar conclusion that, in order to simultaneously satisfy the geochemical and geophysical constraints, Earth's core composition cannot contain C as one of the major light elements. Therefore, all of the constraints thus far point to Earth's core being the largest C reservoir, yet C being a minor light element in the core.

2.8 Conclusion

The origins of Earth's LEVEs, including C, remain debated. Such debate stems in part from the uncertainty regarding the whole-Earth budget of C and other volatile elements and in part from the fate of these elements in putative planetary building blocks throughout their journey in the protoplanetary disk and planet formation process. While the abundance of each LEVE in the BSE is poorly constrained, the ratios of various LEVEs (along with their isotopic compositions) are more reliable geochemical indices to isolate key processes. Some of the key processes to consider are the accretion of undifferentiated chondritic meteorites, compound-specific condensation of the LEVEs, fractionation of the LEVEs during alloy–silicate differentiation, and MO–proto-atmosphere partitioning aided by redox-controlled solubility. Probing each of these processes in the light of the evolution of the LEVE ratios suggests that no single process or combination of processes can result in the currently estimated abundance of LEVEs in the BSE. However, a new paradigm is emerging where the late-stage growth of Earth via the impact and accretion of differentiated planetary embryos (0.1–0.2 M_E) may not only explain the abundances of C–N–S, but also satisfy the broad geochemical similarity of Earth and the Moon. Despite the similarity in the arguments put forward by many recent studies that call for the accretion of one or more large planetary embryos with little post-impact equilibration of the impactor's core with the terrestrial mantle/MO, differences remain in their details. For example, Grewal et al.[18] proposed accretion of a relatively oxidized planet with an S-rich core, whereas Wade and Wood[172] called for accretion of a reduced planet with S-rich core. Future studies, therefore, will have to investigate how to reconcile these differences while fulfilling all of the geochemical constraints, both for non-volatile lithophile and siderophile elements and LEVEs. Nonetheless, it appears inevitable that a large planet like Earth's LEVE budgets were shaped by its protracted growth history with episodes of planetesimals, planetary embryos, and planet accretion, whereas other smaller planets that grew quickly lacked some of these possible supply and modification mechanisms of key ingredients. Whatever the accretion mechanism may be, as long as mostly volatile-depleted materials accreted on Earth throughout its core formation history, C as well as N and H are going to be minor elements in the bulk core. As the search continues to find other rocky, habitable worlds beyond our Solar System, the origins of C–H–N–S on Earth provide a guiding framework.

2.9 Limits of Knowledge and Unknowns

The limits of knowledge on the origin of Earth's C and other LEVEs derive from current unknowns about many fundamental questions regarding the formation and evolution of Earth and other rocky planets in the Solar System, as well as about the compositions and processes in Earth's deep interior. The condensation temperature of C that dictates the formation of solid C-bearing compounds from gases is thought to be extremely low,[173] leading to the idea that most C brought to the inner Solar System at the early stage of planet formation is largely lost from the building blocks. However, it remains unknown whether some of the refractory C-bearing phases such as graphite and carbide could have contributed to the C budget of Earth during its early growth stage.

Experimental data on alloy–silicate elemental partitioning for mixed-volatile systems, especially in conditions approaching the deep MO of Earth, are still lacking and are needed in order to refine some of the current ideas based on extrapolated values from relatively low pressure–temperature experiments. Similar experimental data are needed for volatile isotopic systems as well, which are even more restricted,[174,175] and for many systems, such as C,[64] H, and S,[175] they are either nonexistent or do not extend to the pressure–temperature and compositional space directly relevant for MOs.

Acknowledgments

The authors thank many former and current members of Rice University's experimental petrology team, including Han Chi, Yuan Li, Megan Duncan, Alexandra Holmes, Kyusei Tsuno, and James Eguchi, whose research results influenced the views expressed in this chapter. The authors thank Jonathan Tucker and coeditor Isabelle Daniel for useful reviews. This work was made possible by funding from NASA grants 80NSSC18K0828 and 80NSSC18K1314, the Deep Carbon Observatory, and a Packard Fellowship for Science and Engineering to RD.

Questions for the Classroom

1 How well do the chondritic meteorites (undifferentiated materials; i.e. mixtures or metals, sulfides, and silicates), derived from the main asteroid belt, represent the building blocks of Earth and other rocky planets and planetary embryos in the inner Solar System? In other words, should the plausible compositions of building blocks including their LEVE budgets remain restricted to the known classes of meteorites?

2 How could we improve our knowledge on the concentration of C, N, S, and H in the BSE? What other approaches can be used to determine the mantle inventory of LEVEs?

3 How can we improve our understanding of the relative contributions of giant impacts versus impacts of planetesimals on the acquisition of LEVEs for large, rocky planets?

4 What information from our knowledge on the origins of LEVEs on Earth can be applied to build a framework for exoplanetary chemical habitability?

References

1. Birch, F., Elasticity and constitution of the Earth's interior. *J. Geophys. Res.* **57**, 227–286 (1952).
2. Birch, F., Density and composition of mantle and core. *J. Geophys. Res.* **69**, 4377–4388 (1964).
3. Poirier, J.-P., Light elements in the Earth's outer core: a critical review. *Phys. Earth Planet. Int.* **85**, 319–337 (1994).
4. Javoy, M., Pineau, F., & Allègre, C.J., Carbon geodynamic cycle. *Nature* **300**, 171–173 (1982).
5. Sleep, N.H. & Zahnle, K., Carbon dioxide cycling and implications for climate on ancient earth. *J. Geophys. Res.* **106**, 1373–1399 (2001).
6. Hayes, J.F. & Waldbauer, J.R., The carbon cycle and associated redox processes through time. *Phil. Trans. Royal Soc. London* **B361**, 931–950 (2006).
7. Hirschmann, M.M. & Dasgupta, R., The H/C ratios of Earth's near-surface and deep reservoirs, and consequences for deep Earth volatile cycles. *Chem. Geol.* **262**, 4–16 (2009).
8. Dasgupta, R., Ingassing, storage, and outgassing of terrestrial carbon through geologic time. *Rev. Min. Geochem.* **75**, 183–229 (2013).
9. Dasgupta, R. & Hirschmann, M.M., The deep carbon cycle and melting in Earth's interior. *Earth Planet. Sci. Lett.* **298**, 1–13 (2010).
10. Bergin, E.A., Blake, G.A., Ciesla, F., Hirschmann, M.M., & Li, J., Tracing the ingredients for a habitable earth from interstellar space through planet formation. *Proc. Natl Acad. Sci.* **112**, 8965–8970 (2015).
11. Hirschmann, M.M., Constraints on the early delivery and fractionation of Earth's major volatiles from C/H, C/N, and C/S ratios. *Am. Mineral.* **101**, 540–553 (2016).

12. Halliday, A.N., The origins of volatiles in the terrestrial planets. *Geochim. Cosmochim. Acta* **105**, 146–171 (2013).
13. Marty, B., The origins and concentrations of water, carbon, nitrogen and noble gases on Earth. *Earth Planet. Sci. Lett.* **313–314**, 56–66 (2012).
14. Dasgupta, R. & Walker, D., Carbon solubility in core melts in a shallow magma ocean environment and distribution of carbon between the Earth's core and the mantle. *Geochim. Cosmochim. Acta* **72**, 4627–4641 (2008).
15. Kuramoto, K. & Matsui, T., Partitioning of H and C between the mantle and core during the core formation in the Earth: its implications for the atmospheric evolution and redox state of early mantle. *J. Geophys. Res.* **101**, 14909–14932 (1996).
16. Li, Y., Dasgupta, R., Tsuno, K., Monteleone, B., & Shimizu, N., Carbon and sulfur budget of the silicate Earth explained by accretion of differentiated planetary embryos. *Nat. Geosci.* **9**, 781–785 (2016).
17. Tsuno, K., Grewal, D.S., & Dasgupta, R., Core–mantle fractionation of carbon in Earth and Mars: the effects of sulfur. *Geochim. Cosmochim. Acta* **238**, 477–495 (2018).
18. Grewal, D.S., Dasgupta, R., Sun, C., Tsuno, K., & Costin, G., Delivery of carbon, nitrogen and sulfur to the silicate Earth by a giant impact. *Science Adv.* **5**, eaau3669 (2019).
19. Morbidelli, A., Lunine, J.I., O'Brien, D.P., Raymond, S.N., & Walsh, K.J., Building terrestrial planets. *Ann. Rev. Earth Planet. Sci.* **40**, 251–275 (2012).
20. O'Brien, D.P., Walsh, K.J., Morbidelli, A., Raymond, S.N., & Mandell, A.M., Water delivery and giant impacts in the "Grand Tack" scenario. *Icarus* **239**, 74–84 (2014).
21. Marchi, S., Canup, R.M., & Walker, R.J., Heterogeneous delivery of silicate and metal to the Earth by large planetesimals. *Nat. Geosci.* **11**, 77–81 (2018).
22. Schlichting, H.E. & Mukhopadhyay, S., Atmosphere impact losses. *Space Sci. Rev.* **214**, 34 (2018).
23. Schlichting, H.E., Sari, R.E., & Yalinewich, A., Atmospheric mass loss during planet formation: the importance of planetesimal impacts. *Icarus* **247**, 81–94 (2015).
24. Ahrens, T.J., Impact erosion of terrestrial planetary atmospheres. *Ann. Rev. Earth Planet. Sci.* **21**, 525–555 (1993).
25. Ringwood, A.E., Chemical evolution of the terrestrial planets. *Geochim. Cosmochim. Acta* **30**, 41–104 (1966).
26. Anders, E. & Ebihara, M., Solar-system abundances of the elements. *Geochim. Cosmochim. Acta* **46**, 2363–2380 (1982).
27. Allègre, C., Manhès, G., & Lewin, É., Chemical composition of the Earth and the volatility control on planetary genetics. *Earth Planet. Sci. Lett.* **185**, 49–69 (2001).
28. Clayton, R.N., Onuma, N., & Mayeda, T.K., A classification of meteorites based on oxygen isotopes. *Earth Planet. Sci. Lett.* **30**, 10–18 (1976).
29. Clayton, R.N., Oxygen isotopes in meteorites. *Ann. Rev. Earth Planet. Sci.* **21**, 115–149 (1993).
30. Young, E.D. et al., Oxygen isotopic evidence for vigorous mixing during the Moon-forming giant impact. *Science* **351**, 493–496 (2016).
31. Dauphas, N. et al., Calcium-48 isotopic anomalies in bulk chondrites and achondrites: evidence for a uniform isotopic reservoir in the inner protoplanetary disk. *Earth Planet. Sci. Lett.* **407**, 96–108 (2014).
32. Dauphas, N., The isotopic nature of the Earth's accreting material through time. *Nature* **541**, 521 (2017).
33. Fischer-Gödde, M. & Kleine, T., Ruthenium isotopic evidence for an inner Solar System origin of the late veneer. *Nature* **541**, 525 (2017).

34. Javoy, M. et al., The chemical composition of the Earth: enstatite chondrite models. *Earth Planet. Sci. Lett.* **293**, 259–268 (2010).
35. Javoy, M., The integral enstatite chondrite model of the Earth. *Geophys. Res. Lett.* **22**, 2219–2222 (1995).
36. McDonough, W.F. & Sun, S.-S., The composition of the Earth. *Chem. Geol.* **120**, 223–253 (1995).
37. Wasson, J.T. & Kallemeyn, G.W., Compositions of chondrites. *Phil. Trans. Royal Soc. A.* **325**, 535–544 (1988).
38. Fitoussi, C., Bourdon, B., & Wang, X., The building blocks of Earth and Mars: a close genetic link. *Earth Planet. Sci. Lett.* **434**, 151–160 (2016).
39. Palme, H. & O'Neill, H.S.C., Cosmochemical estimates of mantle composition, in *The Mantle and Core, Vol. 2*, ed. R.W. Carlson (Oxford: Pergamon Press, 2007), pp. 1–38.
40. Murakami, M., Ohishi, Y., Hirao, N., & Hirose, K., A perovskitic lower mantle inferred from high-pressure, high-temperature sound velocity data. *Nature* **485**, 90 (2012).
41. Fitoussi, C. & Bourdon, B., Silicon isotope evidence against an enstatite chondrite Earth. *Science* **335**, 1477–1480 (2012).
42. Shahar, A. et al., Experimentally determined Si isotope fractionation between silicate and Fe metal and implications for Earth's core formation. *Earth Planet. Sci. Lett.* **288**, 228–234 (2009).
43. Shahar, A. et al., High-temperature Si isotope fractionation between iron metal and silicate. *Geochim. Cosmochim. Acta* **75**, 7688–7697 (2011).
44. Ballmer, M.D., Houser, C., Hernlund, J.W., Wentzcovitch, R.M., & Hirose, K., Persistence of strong silica-enriched domains in the Earth's lower mantle. *Nat. Geosci.* **10**, 236 (2017).
45. Dauphas, N., Poitrasson, F., Burkhardt, C., Kobayashi, H., & Kurosawa, K., Planetary and meteoritic Mg/Si and δ^{30}Si variations inherited from solar nebula chemistry. *Earth Planet. Sci. Lett.* **427**, 236–248 (2015).
46. McNaughton, N.J., Borthwick, J., Fallick, A.E., & Pillinger, C.T., Deuterium/hydrogen ratios in unequilibrated ordinary chondrites. *Nature* **294**, 639 (1981).
47. Robert, F., Merlivat, L., & Javoy, M., Deuterium concentration in the early Solar System: hydrogen and oxygen isotope study. *Nature* **282**, 785 (1979).
48. Alexander, C.M.O.D. et al., The provenances of asteroids, and their contributions to the volatile inventories of the terrestrial planets. *Science* **337**, 721–723 (2012).
49. Pearson, V.K., Sephton, M.A., Franchi, I.A., Gibson, J.M., & Gilmour, I., Carbon and nitrogen in carbonaceous chondrites: elemental abundances and stable isotopic compositions. *Meteor. Planet. Sci.* **41**, 1899–1918 (2006).
50. Marty, B., Alexander, C.M.O.D., & Raymond, S., Primordial origins of Earth's carbon. *Rev. Mineral. Geochem.* **75**, 149–181 (2013).
51. Sarafian, A.R., Nielsen, S.G., Marschall, H.R., McCubbin, F.M., & Monteleone, B.D., Early accretion of water in the inner solar system from a carbonaceous chondrite-like source. *Science* **346**, 623–626 (2014).
52. Grady, M.M., Wright, I.P., Carr, L.P., & Pillinger, C.T., Compositional differences in enstatite chondrites based on carbon and nitrogen stable isotope measurements. *Geochim. Cosmochim. Acta* **50**, 2799–2813 (1986).
53. Alexander, C.M.O.D., Swan, P., & Prombo, C.A., Occurrence and implications of silicon nitride in enstatite chondrites. *Meteor. Planet. Sci.* **29**, 79–85 (1994).

54. Alexander, C.M.O.D., McKeegan, K.D., & Altwegg, K., Water reservoirs in small planetary bodies: meteorites, asteroids, and comets. *Space Sci. Rev.* **214**, 36 (2018).
55. Sarafian, A.R. et al., Early accretion of water and volatile elements to the inner Solar System: evidence from angrites. *Phil. Trans. Royal Soc. A.* **375**, 20160209 (2017).
56. Geiss, J. & Gloeckler, G., Isotopic composition of H, HE and NE in the protosolar cloud. *Space Sci. Rev.* **106**, 3–18 (2003).
57. Ko, H., Marc, C., Bernard, M., & Kentaro, T., Protosolar carbon isotopic composition: implications for the origin of meteoritic organics. *Astrophys. J.* **600**, 480 (2004).
58. Marty, B., Chaussidon, M., Wiens, R.C., Jurewicz, A.J.G., & Burnett, D.S., A ^{15}N-poor isotopic composition for the Solar System as shown by genesis solar wind samples. *Science* **332**, 1533–1536 (2011).
59. Marty, B. & Yokochi, R., Water in the early Earth. *Rev. Mineral. Geochem.* **62**, 421–450 (2006).
60. Wood, B.J., Li, J., & Shahar, A., Carbon in the core: its influence on the properties of core and mantle. *Rev. Mineral. Geochem.* **75**, 231–250 (2013).
61. Kerridge, J.F., Carbon, hydrogen and nitrogen in carbonaceous chondrites: abundances and isotopic compositions in bulk samples. *Geochim. Cosmochim. Acta* **49**, 1707–1714 (1985).
62. Deines, P., The carbon isotope geochemistry of mantle xenoliths. *Earth Sci. Rev.* **58**, 247–278 (2002).
63. Labidi, J., Cartigny, P., & Moreira, M., Non-chondritic sulphur isotope composition of the terrestrial mantle. *Nature* **501**, 208 (2013).
64. Satish-Kumar, M., So, H., Yoshino, T., Kato, M., & Hiroi, Y., Experimental determination of carbon isotope fractionation between iron carbide melt and carbon: ^{12}C-enriched carbon in the Earth's core? *Earth Planet. Sci. Lett.* **310**, 340–348 (2011).
65. Walsh, K.J., Morbidelli, A., Raymond, S.N., O'Brien, D.P., & Mandell, A.M., A low mass for Mars from Jupiter's early gas-driven migration. *Nature* **475**, 206–209 (2011).
66. Raymond, S.N., O'Brien, D.P., Morbidelli, A., & Kaib, N.A., Building the terrestrial planets: constrained accretion in the inner Solar System. *Icarus* **203**, 644–662 (2009).
67. Dauphas, N. & Pourmand, A., Hf–W–Th evidence for rapid growth of Mars and its status as a planetary embryo. *Nature* **473**, 489–492 (2011).
68. Lambrechts, M. & Johansen, A., Rapid growth of gas-giant cores by pebble accretion. *Astronom. Astrophys.* **544**, A32 (2012).
69. Johansen, A., Low, M.-M.M., Lacerda, P., & Bizzarro, M., Growth of asteroids, planetary embryos, and Kuiper belt objects by chondrule accretion. *Sci. Adv.* **1**, e1500109 (2015).
70. Wetzel, D.T., Hauri, E.H., Saal, A.E., & Rutherford, M.J., Carbon content and degassing history of the lunar volcanic glasses. *Nat. Geosci.* **8**, 755–758 (2015).
71. Chi, H., Dasgupta, R., Duncan, M., & Shimizu, N., Partitioning of carbon between Fe-rich alloy melt and silicate melt in a magma ocean – implications for the abundance and origin of volatiles in Earth, Mars, and the Moon. *Geochim. Cosmochim. Acta* **139**, 447–471 (2014).
72. Saal, A.E. et al., Volatile content of lunar volcanic glasses and the presence of water in the Moon's interior. *Nature* **454**, 192–195 (2008).
73. Hauri, E.H., Weinreich, T., Saal, A.E., Rutherford, M.C., & Van Orman, J.A., High pre-eruptive water contents preserved in lunar melt inclusions. *Science* **333**, 213–215 (2011).

74. Saal, A.E., Hauri, E.H., Van Orman, J.A., & Rutherford, M.J., Hydrogen isotopes in lunar volcanic glasses and melt inclusions reveal a carbonaceous chondrite heritage. *Science* **340**, 1317–1320 (2013).
75. Greenwood, R.C. et al., Oxygen isotopic evidence for accretion of Earth's water before a high-energy Moon-forming giant impact. *Sci. Adv.* **4**, eaao5928 (2018).
76. Palme, H., Lodders, K., & Jones, A., Solar system abundances of the elements, in *Treatise on Geochemistry (2nd edn.)*, eds. H.D. Holland & K.K. Turekian (Oxford: Elsevier, 2014), pp. 15–36.
77. Anders, E. & Grevesse, N., Abundances of the elements: meteoritic and solar. *Geochim. Cosmochim. Acta* **53**, 197–214 (1989).
78. Lodders, K., Solar system abundances of the elements, in *Principles and Perspectives in Cosmochemistry: Lecture Notes of the Kodai School on "Synthesis of Elements in Stars"*, eds. A. Goswami & B. Eswar Reddy (Berlin: Springer-Verlag, 2010), pp. 379–417.
79. Palme, H. & Jones, A., Solar system abundances of the elements, in *Meteorites, Comets, and Planets*, ed. A.M. Davis (Amsterdam: Elsevier Ltd., 2003), pp. 41–61.
80. Grady, M.M. & Wright, I.P., Elemental and isotopic abundances of carbon and nitrogen in meteorites. *Space Sci. Rev.* **106**, 231–248 (2003).
81. Hutson, M. & Ruzicka, A., A multi-step model for the origin of E3 (enstatite) chondrites. *Meteor. Planet. Sci.* **35**, 601–608 (2000).
82. Hirschmann, M.M., Water, melting, and the deep Earth H_2O cycle. *Ann. Rev. Earth Planet. Sci.* **34**, 629–653 (2006).
83. McDonough, W.F., Compositional model for the Earth's core, in *The Mantle and Core, Vol. 2*, ed. R.W. Carlson (Oxford: Elsevier-Pergamon, 2003), pp. 547–568.
84. Rosenthal, A., Hauri, E.H., & Hirschmann, M.M., Experimental determination of C, F, and H partitioning between mantle minerals and carbonated basalt, CO_2/Ba and CO_2/Nb systematics of partial melting, and the CO_2 contents of basaltic source regions. *Earth Planet. Sci. Lett.* **412**, 77–87 (2015).
85. Dalou, C., Hirschmann, M.M., von der Handt, A., Mosenfelder, J., & Armstrong, L.S., Nitrogen and carbon fractionation during core–mantle differentiation at shallow depth. *Earth Planet. Sci. Lett.* **458**, 141–151 (2017).
86. Nabiei, F. et al., A large planetary body inferred from diamond inclusions in a ureilite meteorite. *Nat. Commun.* **9**, 1327 (2018).
87. Kana, N., Masayuki, N., & Jun-ichi, M., Raman spectroscopic study of diamond and graphite in ureilites and the origin of diamonds. *Meteor. Planet. Sci.* **47**, 1728–1737 (2012).
88. Barrat, J.-A., Sansjofre, P., Yamaguchi, A., Greenwood, R.C., & Gillet, P., Carbon isotopic variation in ureilites: evidence for an early, volatile-rich inner Solar System. *Earth Planet. Sci. Lett.* **478**, 143–149 (2017).
89. Schiller, M., Bizzarro, M., & Fernandes, V.A., Isotopic evolution of the protoplanetary disk and the building blocks of Earth and the Moon. *Nature* **555**, 507 (2018).
90. Ding, S. & Dasgupta, R., The fate of sulfide during decompression melting of peridotite – implications for sulfur inventory of the MORB-source depleted upper mantle. *Earth Planet. Sci. Lett.* **459**, 183–195 (2017).
91. Saal, A.E., Hauri, E., Langmuir, C.H., & Perfit, M.R., Vapour undersaturation in primitive mid-ocean-ridge basalt and the volatile content of Earth's upper mantle. *Nature* **419**, 451–455 (2002).
92. Rehkämper, M. et al., Ir, Ru, Pt, and Pd in basalts and komatiites: new constraints for the geochemical behavior of the platinum-group elements in the mantle. *Geochim. Cosmochim. Acta* **63**, 3915–3934 (1999).

93. Walker, R.J., Highly siderophile elements in the Earth, Moon and Mars: update and implications for planetary accretion and differentiation. *Geochemisty* **69**, 101–125 (2009).
94. Mann, U., Frost, D.J., Rubie, D.C., Becker, H., & Audétat, A., Partitioning of Ru, Rh, Pd, Re, Ir and Pt between liquid metal and silicate at high pressures and high temperatures – implications for the origin of highly siderophile element concentrations in the Earth's mantle. *Geochim. Cosmochim. Acta* **84**, 593–613 (2012).
95. Morgan, J.W., Walker, R.J., Brandon, A.D., & Horan, M.F., Siderophile elements in Earth's upper mantle and lunar breccias: data synthesis suggests manifestations of the same late influx. *Meteor. Planet. Sci.* **36**, 1257–1275 (2001).
96. Willbold, M., Elliott, T., & Moorbath, S., The tungsten isotopic composition of the Earth's mantle before the terminal bombardment. *Nature* **477**, 195 (2011).
97. Albarede, F., Volatile accretion history of the terrestrial planets and dynamic implications. *Nature* **461**, 1227–1233 (2009).
98. Albarede, F. et al., Asteroidal impacts and the origin of terrestrial and lunar volatiles. *Icarus* **222**, 44–52 (2013).
99. Dasgupta, R., Chi, H., Shimizu, N., Buono, A., & Walker, D., Carbon solution and partitioning between metallic and silicate melts in a shallow magma ocean: implications for the origin and distribution of terrestrial carbon. *Geochim. Cosmochim. Acta* **102**, 191–212 (2013).
100. Alexander, C.M.O.D., Fogel, M., Yabuta, H., & Cody, G.D., The origin and evolution of chondrites recorded in the elemental and isotopic compositions of their macromolecular organic matter. *Geochimi. Cosmochim. Acta* **71**, 4380–4403 (2007).
101. Vdovykin, G.P., Ureilites. *Space Sci. Rev.* **10**, 483–510 (1970).
102. Grady, M.M., Wright, I.P., Swart, P.K., & Pillinger, C.T., The carbon and nitrogen isotopic composition of ureilites: implications for their genesis. *Geochim. Cosmochim. Acta* **49**, 903–915 (1985).
103. Downes, H. et al., Isotopic composition of carbon and nitrogen in ureilitic fragments of the Almahata Sitta meteorite. *Meteor. Planet. Sci.* **50**, 255–272 (2015).
104. Siebert, J., Badro, J., Antonangeli, D., & Ryerson, F.J., Metal–silicate partitioning of Ni and Co in a deep magma ocean. *Earth Planet. Sci. Lett.* **321–322**, 189–197 (2012).
105. Grewal, D.S., Dasgupta, R., Holmes, A.K., Costin, G., & Li, Y., The fate of nitrogen during core-mantle separation on Earth. *Geochim. Cosmochim. Acta* **251**, 87–115 (2019) .
106. Boujibar, A. et al., Metal–silicate partitioning of sulphur, new experimental and thermodynamic constraints on planetary accretion. *Earth Planet. Sci. Lett.* **391**, 42–54 (2014).
107. Clesi, V. et al., Low hydrogen contents in the cores of terrestrial planets. *Science Adv.* **4**, e1701876 (2018).
108. Armstrong, L.S., Hirschmann, M.M., Stanley, B.D., Falksen, E.G., & Jacobsen, S.D., Speciation and solubility of reduced C–O–H–N volatiles in mafic melt: implications for volcanism, atmospheric evolution, and deep volatile cycles in the terrestrial planets. *Geochim. Cosmochim. Acta* **171**, 283–302 (2015)
109. Roskosz, M., Bouhifd, M.A., Jephcoat, A.P., Marty, B., & Mysen, B.O., Nitrogen solubility in molten metal and silicate at high pressure and temperature. *Geochim. Cosmochim. Acta* **121**, 15–28 (2013).

110. O'Neill, H.S.C. & Mavrogenes, J.A., The sulfide capacity and the sulfur content at sulfide saturation of silicate melts at 1400°C and 1 bar. *J. Petrol.* **43**, 1049–1087 (2002).
111. Hirschmann, M.M., Withers, A.C., Ardia, P., & Foley, N.T., Solubility of molecular hydrogen in silicate melts and consequences for volatile evolution of terrestrial planets. *Earth Planet. Sci. Lett.* **345**, 38–48 (2012).
112. Rubie, D.C. et al., Highly siderophile elements were stripped from Earth's mantle by iron sulfide segregation. *Science* **353**, 1141–1144 (2016).
113. Corgne, A., Wood, B.J., & Fei, Y., C- and S-rich molten alloy immiscibility and core formation of planetesimals. *Geochim. Cosmochim. Acta* **72**, 2409–2416 (2008).
114. Dasgupta, R., Buono, A., Whelan, G., & Walker, D., High-pressure melting relations in Fe–C–S systems: implications for formation, evolution, and structure of metallic cores in planetary bodies. *Geochim. Cosmochim. Acta* **73**, 6678–6691 (2009).
115. Tsuno, K. & Dasgupta, R., Fe–Ni–Cu–C–S phase relations at high pressures and temperatures – the role of sulfur in carbon storage and diamond stability at mid- to deep-upper mantle. *Earth Planet. Sci. Lett.* **412**, 132–142 (2015).
116. Zhang, Z., Hastings, P., Von der Handt, A., & Hirschmann, M.M., Experimental determination of carbon solubility in Fe–Ni–S melts. *Geochim. Cosmochim. Acta* **225**, 66–79 (2018).
117. Li, Y., Dasgupta, R., & Tsuno, K., The effects of sulfur, silicon, water, and oxygen fugacity on carbon solubility and partitioning in Fe-rich alloy and silicate melt systems at 3 GPa and 1600 °C: implications for core–mantle differentiation and degassing of magma oceans and reduced planetary mantles. *Earth Planet. Sci. Lett.* **415**, 54–66 (2015).
118. Raymond, S.N., Quinn, T., & Lunine, J.I., High-resolution simulations of the final assembly of Earth-like planets. 2. Water delivery and planetary habitability. *Astrobiology* **7**, 66–84 (2007).
119. Hallis, L.J. et al., Evidence for primordial water in Earth's deep mantle. *Science* **350**, 795–797 (2015).
120. Tucker, J.M. & Mukhopadhyay, S., Evidence for multiple magma ocean outgassing and atmospheric loss episodes from mantle noble gases. *Earth Planet. Sci. Lett.* **393**, 254–265 (2014).
121. Kress, V.C. & Carmichael, I.S.E., The compressibility of silicate liquids containing Fe_2O_3 and the effect of composition, temperature, oxygen fugacity and pressure on their redox states. *Contrib. Mineral. Petrol.* **108**, 82–92 (1991).
122. O'Neill, H.S.C. et al., An experimental determination of the effect of pressure on the $Fe^{3+}/\sum Fe$ ratio of an anhydrous silicate melt to 3.0 GPa. *Am. Mineral.* **91**, 404–412 (2006).
123. Zhang, H.L., Hirschmann, M.M., Cottrell, E., & Withers, A.C., Effect of pressure on $Fe^{3+}/\Sigma Fe$ ratio in a mafic magma and consequences for magma ocean redox gradients. *Geochim. Cosmochim. Acta* **204**, 83–103 (2017).
124. Schaefer, L. & Elkins-Tanton, L.T., Magma oceans as a critical stage in the tectonic development of rocky planets. *Phil. Trans. Royal Soc. A.* **376**, 20180109 (2018).
125. Duncan, M.S., Dasgupta, R., & Tsuno, K., Experimental determination of CO_2 content at graphite saturation along a natural basalt–peridotite melt join: implications for the fate of carbon in terrestrial magma oceans. *Earth Planet. Sci. Lett.* **466**, 115–128 (2017).
126. Libourel, G., Marty, B., & Humbert, F., Nitrogen solubility in basaltic melt. Part I. Effect of oxygen fugacity. *Geochim. Cosmochim. Acta* **67**, 4123–4135 (2003).

127. Kadik, A.A. et al., Influence of oxygen fugacity on the solubility of nitrogen, carbon, and hydrogen in $FeO–Na_2O–SiO_2–Al_2O_3$ melts in equilibrium with metallic iron at 1.5 GPa and 1400°C. *Geochem. Int.* **49**, 429–438 (2011).
128. Holloway, J.R., Pan, V., & Gudmundsson, G., High-pressure fluid-absent melting experiments in the presence of graphite; oxygen fugacity, ferric/ferrous ratio and dissolved CO_2. *Eur. J. Mineral.* **4**, 105–114 (1992).
129. Li, Y., Dasgupta, R., & Tsuno, K., Carbon contents in reduced basalts at graphite saturation: implications for the degassing of Mars, Mercury, and the Moon. *J. Geophys. Res. Planets* **122**, doi:10.1002/2017JE005289 (2017).
130. Ding, S., Hough, T., & Dasgupta, R., New high pressure experiments on sulfide saturation of high-FeO* basalts with variable TiO_2 contents – implications for the sulfur inventory of the lunar interior. *Geochim. Cosmochim. Acta* **222**, 319–339 (2018).
131. Baker, D.R. & Moretti, R., Modeling the solubility of sulfur in magmas: a 50-year old geochemical challenge. *Rev. Mineral. Geochem.* **73**, 167–213 (2011).
132. Genda, H. & Abe, Y., Enhanced atmospheric loss on protoplanets at the giant impact phase in the presence of oceans. *Nature* **433**, 842–844 (2005).
133. Honda, M., McDougall, I., Patterson, D.B., Doulgeris, A., & Clague, D.A., Possible solar noble-gas component in Hawaiian basalts. *Nature* **349**, 149 (1991).
134. Sarda, P., Staudacher, T., & Allègre, C.J., Neon isotopes in submarine basalts. *Earth Planet. Sci. Lett.* **91**, 73–88 (1988).
135. Mizuno, H., Nakazawa, K., & Hayashi, C., Dissolution of the primordial rare gases into the molten Earth's material. *Earth Planet. Sci. Lett.* **50**, 202–210 (1980).
136. Yokochi, R. & Marty, B., A determination of the neon isotopic composition of the deep mantle. *Earth Planet. Sci. Lett.* **225**, 77–88 (2004).
137. Lynne, A.H., Disk-dispersal and planet-formation timescales. *Phys. Scipta* **2008**, 014024 (2008).
138. Yin, Q. et al., A short timescale for terrestrial planet formation from Hf–W chronometry of meteorites. *Nature* **418**, 949 (2002).
139. Tang, H. & Dauphas, N., $^{60}Fe–^{60}Ni$ chronology of core formation in Mars. *Earth Planet. Sci. Lett.* **390**, 264–274 (2014).
140. Rose-Weston, L., Brenan, J.M., Fei, Y., Secco, R.A., & Frost, D.J., Effect of pressure, temperature, and oxygen fugacity on the metal-silicate partitioning of Te, Se, and S: implications for earth differentiation. *Geochim. Cosmochim. Acta* **73**, 4598–4615 (2009).
141. Zhang, Y. & Yin, Q.-Z., Carbon and other light element contents in the Earth's core based on first-principles molecular dynamics. *Proc. Natl Acad. Sci.* **109**, 19579–19583 (2012).
142. Suer, T.-A., Siebert, J., Remusat, L., Menguy, N., & Fiquet, G., A sulfur-poor terrestrial core inferred from metal–silicate partitioning experiments. *Earth Planet. Sci. Lett.* **469**, 84–97 (2017).
143. Tsymbulov, L.B. & Tsmekhman, L.S., Solubility of carbon in sulfide melts of the system Fe–Ni-S. *Russ. J. Appl. Chem.* **74**, 925–929 (2001).
144. Stolper, E., Fine, G., Johnson, T., & Newman, S., Solubility of carbon dioxide in albitic melt. *Am. Mineral.* **72**, 1071–1085 (1987).
145. Duncan, M.S. & Dasgupta, R., Pressure and temperature dependence of CO_2 solubility in hydrous rhyolitic melt – implications for carbon transfer to mantle source of volcanic arcs via partial melt of subducting crustal lithologies. *Contrib. Mineral. Petrol.* **169**, 1–19 (2015).

146. Eguchi, J. & Dasgupta, R., A CO_2 solubility model for silicate melts from fluid saturation to graphite or diamond saturation. *Chem. Geol.* **487**, 23–38 (2018).
147. Behrens, H., Ohlhorst, S., Holtz, F., & Champenois, M., CO_2 solubility in dacitic melts equilibrated with H_2O–CO_2 fluids: implications for modeling the solubility of CO_2 in silicic melts. *Geochim. Cosmochim. Acta* **68**, 4687–4703 (2004).
148. Duncan, M.S. & Dasgupta, R., Rise of Earth's atmospheric oxygen controlled by efficient subduction of organic carbon. *Nat. Geosci.* **10**, 387–392 (2017).
149. Stolper, E. & Holloway, J.R., Experimental determination of the solubility of carbon dioxide in molten basalt at low pressure. *Earth Planet. Sci. Lett.* **87**, 397–408 (1988).
150. Pan, V., Holloway, J.R., & Hervig, R.L., The pressure and temperature dependence of carbon dioxide solubility in tholeiitic basalt melts. *Geochim. Cosmochim. Acta* **55**, 1587–1595 (1991).
151. Yoshioka, T., McCammon, C.A., Shcheka, S., & Keppler, H., The speciation of carbon monoxide in silicate melts and glasses. *Am. Mineral.* **100**, 1641–1644 (2015).
152. Wetzel, D.T., Rutherford, M.J., Jacobsen, S.D., Hauri, E.H., & Saal, A.E., Degassing of reduced carbon from planetary basalts. *Proc. Natl Acad. Sci.* **110**, 8010–8013 (2013).
153. Dasgupta, R., Hirschmann, M.M., & Smith, N.D., Partial melting experiments of peridotite + CO_2 at 3 GPa and genesis of alkalic ocean island basalts. *J. Petrol.* **48**, 2093–2124 (2007).
154. Mallik, A. & Dasgupta, R., Effect of variable CO_2 on eclogite-derived andesite-lherzolite reaction at 3 GPa – implications for mantle source characteristics of alkalic ocean island basalts. *Geochem. Geophys. Geosyst.* **15**, 1533–1557 (2014).
155. Vetere, F., Holtz, F., Behrens, H., Botcharnikov, R.E., & Fanara, S., The effect of alkalis and polymerization on the solubility of H_2O and CO_2 in alkali-rich silicate melts. *Contrib. Mineral. Petrol.* **167**, 1014 (2014).
156. Ghiorso, M.S. & Gualda, G.A.R., An H_2O–CO_2 mixed fluid saturation model compatible with rhyolite-MELTS. *Contrib. Mineral. Petrol.* **169**, 53 (2015).
157. Stanley, B.D., Hirschmann, M.M., & Withers, A.C., Solubility of C–O–H volatiles in graphite-saturated martian basalts. *Geochim. Cosmochim. Acta* **129**, 54–76 (2014).
158. Wade, J. & Wood, B.J., Core formation and the oxidation state of the Earth. *Earth Planet. Sci. Lett.* **236**, 78–95 (2005).
159. Rubie, D.C. et al., Heterogeneous accretion, composition and core–mantle differentiation of the Earth. *Earth Planet. Sci. Lett.* **301**, 31–42 (2011).
160. Badro, J., Brodholt, J.P., Piet, H., Siebert, J., & Ryerson, F.J., Core formation and core composition from coupled geochemical and geophysical constraints. *Proc. Natl Acad. Sci.* **112**, 12310–12314 (2015).
161. Okuchi, T., Hydrogen partitioning into molten iron at high pressure: implications for Earth's core. *Science* **278**, 1781–1784 (1997).
162. O'Neill, H.S.C., The origin of the moon and the early history of the earth – a chemical model. Part 2: the earth. *Geochim. Cosmochim. Acta* **55**, 1159–1172 (1991).
163. Kleine, T. et al., Hf–W chronology of the accretion and early evolution of asteroids and terrestrial planets. *Geochim. Cosmochim. Acta* **73**, 5150–5188 (2009).
164. Rubie, D.C. et al., Accretion and differentiation of the terrestrial planets with implications for the compositions of early-formed Solar System bodies and accretion of water. *Icarus* **248**, 89–108 (2015).

165. Canup, R.M. & Asphaug, E., Origin of the Moon in a giant impact near the end of the Earth's formation. *Nature* **412**, 708 (2001).
166. Ćuk, M. & Stewart, S.T., Making the Moon from a fast-spinning Earth: a giant impact followed by resonant despinning. *Science* **338**, 1047–1052 (2012).
167. Deguen, R., Olson, P., & Cardin, P., Experiments on turbulent metal–silicate mixing in a magma ocean. *Earth Planet. Sci. Lett.* **310**, 303–313 (2011).
168. Deguen, R., Landeau, M., & Olson, P., Turbulent metal–silicate mixing, fragmentation, and equilibration in magma oceans. *Earth Planet. Sci. Lett.* **391**, 274–287 (2014).
169. Wohlers, A. & Wood, B.J., A Mercury-like component of early Earth yields uranium in the core and high mantle ^{142}Nd. *Nature* **520**, 337–340 (2015).
170. Wohlers, A. & Wood, B.J., Uranium, thorium and REE partitioning into sulfide liquids: implications for reduced S-rich bodies. *Geochim. Cosmochim. Acta* **205**, 226–244 (2017).
171. Münker, C., Fonseca, R.O.C., & Schulz, T., Silicate Earth's missing niobium may have been sequestered into asteroidal cores. *Nat. Geosci.* **10**, 822 (2017).
172. Wade, J. & Wood, B.J., The oxidation state and mass of the Moon-forming impactor. *Earth Planet. Sci. Lett.* **442**, 186–193 (2016).
173. Krasnokutski, S.A. et al., Low-temperature condensation of carbon. *Astrophys. J.* **847**, 89 (2017).
174. Li, Y., Marty, B., Shcheka, S., Zimmermann, L., & Keppler, H., Nitrogen isotope fractionation during terrestrial core–mantle separation. *Geochem. Perspect. Lett.* **2**, 138–147 (2016).
175. Labidi, J. et al., Experimentally determined sulfur isotope fractionation between metal and silicate and implications for planetary differentiation. *Geochim. Cosmochim. Acta* **175**, 181–194 (2016).

3

Carbon versus Other Light Elements in Earth's Core

JIE LI, BIN CHEN, MAINAK MOOKHERJEE, AND GUILLAUME MORARD

3.1 Introduction

Carbon is a candidate light element in Earth's core.[1–3] The core consists of a liquid outer shell ranging from 2971 to 5210 km in depth and a solid inner sphere with a radius of 1220 km.[4] Without direct samples, its iron-dominant composition has been inferred from seismological, geochemical, and cosmochemical observations, together with mineral physics constraints from laboratory measurements and theoretical simulations. Both the outer and inner cores are lighter than iron or iron–nickel alloys at relevant pressure–temperature (*P–T*) values, indicating the presence of one or more elements with smaller atomic numbers than iron.[5] Candidates for the light alloying elements of the core include hydrogen (H), carbon (C), oxygen (O), silicon (Si), and sulfur (S).

Earth's core may be the largest repository for terrestrial carbon. As the fourth most abundant element in the solar photosphere, carbon occurs in carbonaceous chondrites and ordinary chondrites as a major or minor element.[6] The silicate Earth is depleted in carbon with respect to CI chondrite by more than two orders of magnitude, and by five- to ten-fold after accounting for evaporative loss to outer space during accretion.[7] Some of the missing carbon in the silicate Earth is likely found in its core, considering the large solubility of carbon in the iron-rich melt[8–10] and the strong affinity of carbon for iron metal during core–mantle differentiation.[11–14] Core sequestration can also explain the ^{13}C enrichment in silicate Earth relative to Mars, Vesta, and chondrites.[15] Cosmochemical and geochemical considerations suggest that the core may contain as much as 1 wt.% (5 at.%) carbon.[15] A lower estimate of 0.2 wt.% carbon in the core is derived by assuming that carbon depletion follows the volatility trend.[7] More details are found in Chapter 2. A core containing 1 wt.% carbon would exceed the combined budget of known carbon in the atmosphere, hydrosphere, biosphere, crust, and mantle by one order of magnitude (Figure 3.1). Even with the lowest estimate of 0.1 wt.% carbon, the core would still account for more than half of Earth's total carbon budget.

Constraining the carbon budget of the core is crucial for identifying Earth's building blocks and reconstructing its accretion history. In this chapter, we review constraints on the carbon content of the core from the phase relation, density, and sound velocities of iron–carbon alloys and compare carbon with other light elements in terms of their ability to

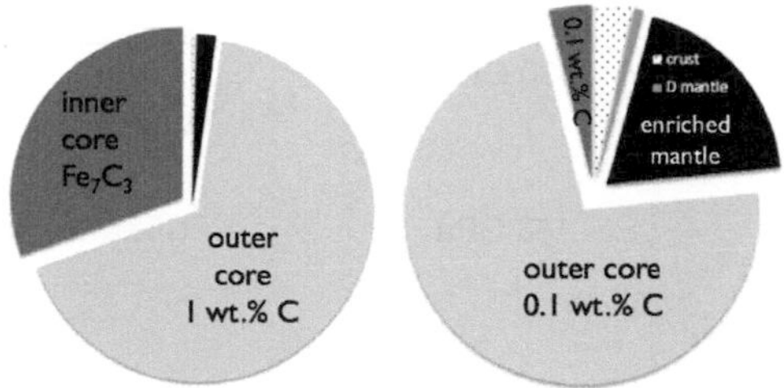

Figure 3.1 Pie diagrams showing the relative sizes of Earth's carbon reservoirs for two end-member models. The concentrations of carbon are assumed to be 0.2 wt.%, 20 ppm, and 165 ppm in the crust, depleted mantle, and enriched mantle, respectively.[16] With 100 ppm in the atmosphere, biosphere, and hydrosphere,[16] the total carbon in these reservoirs is negligible and hence not shown.

match the physical properties of the core. We will also provide a brief discussion of how carbon may have redistributed among various Earth reservoirs through geological time.

3.2 Constraints on Carbon versus Other Light Elements in Earth's Core

3.2.1 Constraints from Phase Relations of Iron–Light Element Systems

Carbon as a core component has attracted special attention through the proposal of a carbide inner core.[9] Based on long extrapolations of equation of state (EoS) data available at the time, Fe_3C with 6.67 wt.% C was predicted to be the first phase to crystallize from an Fe–S–C liquid to form the inner core, even for carbon contents below 1 wt.%.

Testing the model of a carbide inner core requires knowledge of the phase relations at core pressures. As an initial step, the simplified Fe–C binary system has been investigated through experiments and thermodynamic modeling (Figure 3.2). At 1 bar, the system has a eutectic point between iron and Fe_3C at 4.1 wt.% carbon.[17] At pressures above 10 GPa, the eutectic point lies between iron and Fe_7C_3 with 8.41 wt.% carbon,[18] hence Fe_7C_3 is expected to solidify from any composition on the carbon-rich side of the eutectic point at core pressures.

While some studies support the predicted shift of the eutectic composition toward the iron end member with increasing pressure,[19,20] others conclude that the eutectic composition contains 3 ± 1 wt.% carbon between 40 and ~100 GPa in pressure[21] and ~2 wt.% carbon at the pressure of the inner core boundary (ICB).[22] If the outer core contains less carbon than the eutectic composition, then a hexagonal close-packed (hcp) Fe incorporating carbon instead of Fe_7C_3 would be the liquidus phase to form the inner core.

The carbide inner core model can also be tested against the density increase across the ICB. Isochemical freezing of pure Fe or an Fe–light element (Fe–L) alloys produces 1.7% or 2.4% increases in density.[23,24] These are smaller than the 0.6–0.9 g/cm^3 or 4.7–7.1% observed density increases,[25] suggesting that the inner core contains less of the light elements than the outer core. In the ICB condition, a candidate Fe–L composition must reproduce the observed density contrast. For a simplified Fe–L binary, a match is possible

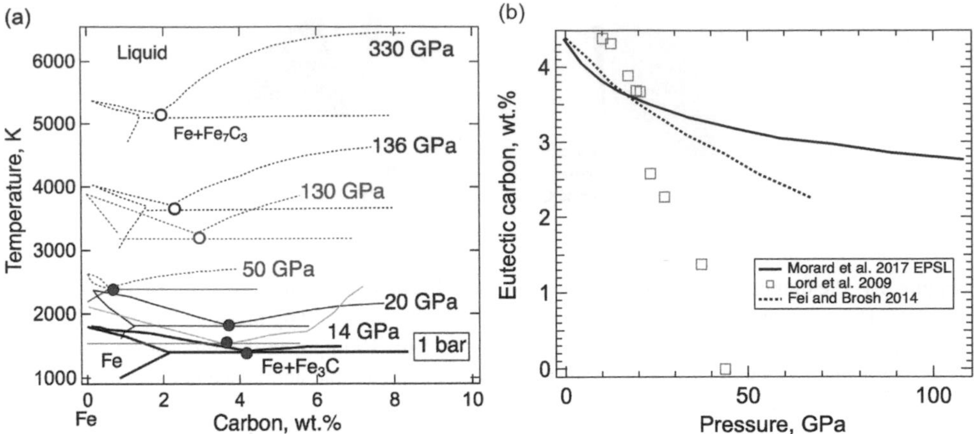

Figure 3.2 Fe–C binary system and eutectic composition. (a) Schematic phase diagram of the Fe–C binary system near the iron end member. 1 bar: thick black solid line,[17] 14 GPa: gray solid line,[18] 50 GPa and 130 GPa: red solid or dotted lines,[20] 20 GPa, 136 GPa, and 330 GPa: thick black solid or dotted lines.[22] Solid traces and filled circles are based on experimental measurements. Dotted traces and open circles are based on calculations and/or extrapolations. (b) Carbon content of the Fe–C eutectic liquid as a function of pressure.

only if the core composition is on the Fe-rich side of the eutectic point. Moreover, the light element contents of the solid and liquid must be sufficiently high and different to match the density contrast. If the eutectic composition is below 1 wt.%, it is unlikely to find a binary Fe–C composition with 5% density contrast between coexisting solid and liquid. It follows that carbon alone is unable to account for the density contrast at the ICB. The presence of sulfur and/or oxygen could help if they partition more strongly into the liquid phase. If the eutectic carbon content is as high as 3 wt.%, then a match by an Fe–C binary composition is possible (Figure 3.2).

Fe–L binary phase relations at 1 bar differ according to the nature of the light element, as is known from the metallurgy literature.[26] The phase relations at pressure and temperature conditions relevant for Earth's core are drastically different from those at 1 bar (Figure 3.3).

The Fe–S binary exhibits eutectic behavior between Fe and FeS at 1 bar and the sulfur content of the eutectic decreases with pressure (Figure 3.3). At core pressures, we may expect that a eutectic liquid containing $<$10 wt.% sulfur coexists with a solid with slightly less sulfur.[23,29,31,32] Therefore, sulfur alone cannot explain the density contrast at the ICB. At least 1–2 wt.% sulfur is likely to be present in the liquid core in addition to carbon and may enhance the stability of carbides or Fe–C alloys on the liquidus.[9]

The Fe–Si binary shows a narrow melting loop and only slight enrichment of silicon in the liquid at pressures up to 120 GPa (Figure 3.3). The eutectic composition contains 25 wt.% silicon at 21 GPa pressure[33] and $<$10 wt.% silicon at 80 GPa or higher,[34] and falls below 1.5 $\pm$ 0.1 wt.% at 127 GPa pressure.[30] Such a silicon-poor eutectic composition implies that FeSi may be a candidate for the inner core. Because Si stabilizes the

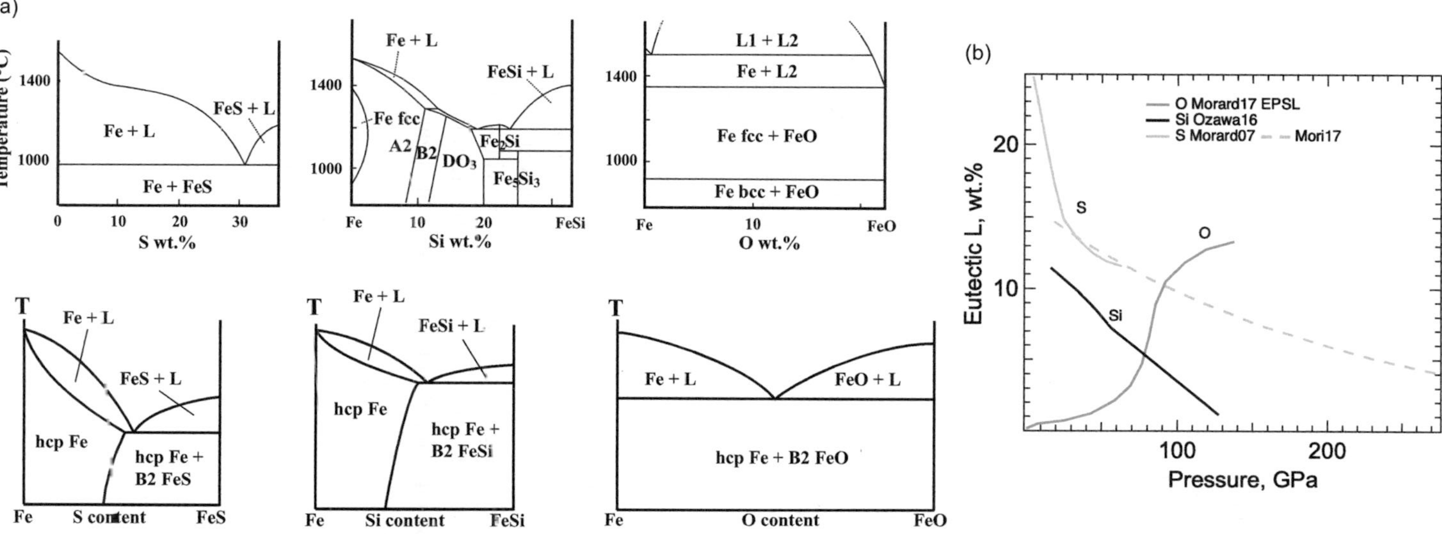

Figure 3.3 Fe–S, Fe–Si, and Fe–O binary phase diagrams and eutectic compositions. (a) Phase diagrams on the Fe-rich side of Fe–S, Fe–Si, and Fe–O systems at 1 bar (upper) and 330 GPa (lower).[27] (b) Eutectic composition as a function of pressure.

Data sources are Refs. 21 and 28–30. bcc = body-centered cubic.

body-centered cubic (bcc) structure, the inner core may be hcp Fe alloyed with Si or a mixture of a Si-rich bcc phase and a Si-poor hcp phase.[35,36] On the other hand, the silicon-poor eutectic composition and the nearly equal partitioning of silicon between solid and liquid iron at the ICB pressure[23,37] imply that silicon alone cannot explain the ICB density contrast.

While oxygen is a leading candidate for the light element in the liquid outer core, little oxygen is expected to be present in the solid inner core. At 1 bar, the Fe–O binary is characterized by a vast liquid miscibility gap.[26] At core pressures, the Fe–O system is more likely to be a eutectic with nearly pure Fe coexisting with Fe–O liquid (Figure 3.3). The eutectic oxygen content increases with pressure and exceeds 10 wt.% at >100 GPa.[21] Given its low solubility in solid Fe, the amount of oxygen in the inner core is probably negligible, but oxygen is the best candidate to explain the density difference between the solid and liquid cores.

3.2.2 Constraints from Densities of Fe–C Alloys and Compounds

The presence of light elements in Earth's core was initially inferred from comparing the observed density of the core with the measured density of iron under corresponding conditions. The pressure of the core is well constrained by geophysical and seismological data.[4] The temperature profile of the core is more uncertain and bears at least ±500 K uncertainties.[38] Compared with pure iron or iron–nickel alloys at the core *P–T* conditions,[39–42] the core is lighter than pure iron by 5–8% in the liquid outer shell and by 2–5% in the solid inner sphere.[5,43–45]

A viable composition model of the core must account for the density deficits. This is a straightforward and effective test, but requires knowledge of the phase relation and EoS of relevant Fe alloys in solid and liquid states at multi-megabar pressures and temperatures exceeding 4000 K. A wide range of mixtures of iron with C, O, Si, and S have been proposed as possible constituents of the outer core, whereas the solid inner core is most likely an iron alloy or a compound of iron with one of the light elements,[1–3] and therefore the test is somewhat simpler for the inner core.

Stimulated by the suggestion that the density of Fe_3C should be close to the observed value of the inner core,[9] measurements and calculations of the densities and elastic properties of iron carbides have been carried out (Tables 3.1 and 3.2). First-principles simulations coupled with structure search algorithms have been used to predict the iron–carbon alloys that are likely to be stable at Earth's inner core conditions. The energetically competitive stoichiometry ranges from Fe:C of 3:1 to 1:1 and includes Fe_3C, Fe_7C_3, Fe_5C_2, Fe_2C, and FeC stoichiometry.[46,47]

3.2.2.1 Fe_3C

The natural form of Fe_3C (cementite) occurs in iron meteorites and is known as cohenite. The composition of synthetic Fe_3C ranges from C deficiency with 4.2 wt.% or 17 at.% C

Table 3.1 *Elasticity parameters for solid Fe–C alloys*

Composition (wt.% L)	ρ_0 (g cm^{-3})	K_0 (GPa)	K_0'	P (GPa)	T (K)	Method	Ref.
Density							
Fe_3C							
	7.70(1)	175(4)	5.2(3)	0–73	300	PXD	121
	7.70(1)	174(6)	4.8(8)	0–30	300	PXD	61
	8.03(1)	290(13)	3.76(18)	0–187	300	PXD	50
	7.67	167	6.7	0–35	300	PXD	49
fm Fe_3C^a	7.68(1)	192(3)	4.5(1)	0–31	300–1473	PXD	63
pm Fe_3C		161(2)	5.9(2)	0–50	300	SXD	62
pm Fe_7C_3	7.68(1)	201(12)	8.0(1.4)	4–158	300	SXD	64
nm Fe_7C_3	7.75(2)	307(6)	3.2(1)	7–167	300	SXD	64
fm Fe_7C_3	7.62(1)	186(5)	6.9(2.2)	0–7	300	PXD	70
Nonlinear Fe_7C_3	7.59(2)	166(13)	4.9(1.1)	7–20	300	PXD	
pm Fe_7C_3	7.68(2)	196(9)	4.9(2)	20–66	300	PXD	
fm Fe_7C_3	7.61(1)	201(2)	4 (fixed)	0–18		PXD	
pm $Fe_7C_3{}^b$	7.70(2)	253(7)	3.6(2)	18–72	300–1973	PXD	41
	V_0 (m/s)	$V = a0 + a1 \bullet \rho$					
V_P		a0	a1				
Fe bcc	5800						89
Fe_3C	5330–5140						122
	5890	–3990	1290	0–50	300	NRIXS	55
	6103(413)	–8671	1900	0–68	300	HERIX	123
		–1138	9823	60–153	300	NRIXS	54
Fe_7C_3		2160	660	70–154	300	NRIXS	69
V_S							
Fe bcc	3000						89
Fe_3C	3010–3030						122
	3050(70)	1450	240	0–50	300	NRIXS	55
				0–50	300– 1450	NRIXS	94
		–961	4429	60–153	300	NRIXS	54
Fe_7C_3		843	242	70–154	300	NRIXS	69

^{a} $\Theta_0 = 490(120)$ K, $\gamma_0 = 2.09(4)$, q = –0.1(3).
^{b} $\Theta_0 = 920(14)$ K, $\gamma_0 = 2.57(5)$, q = 2.2(5).
HERIX = high-energy-resolution inelastic X-ray scattering; NRIXS = nuclear resonant inelastic X-ray scattering; PXD = powder X-ray diffraction; SXD = single-crystal X-ray diffraction.

(roughly Fe_5C) to C excess with 8.8 wt.% or 31 at.% C (exceeding Fe_7C_3).[48] At 1 bar and 300 K, Fe_3C has an orthorhombic structure (Figure 3.4). Although metastable at ambient conditions, the crystal structure remains unchanged to 187 GPa at 300 K[49,50] and to 25–70 GPa and 2200–3400 K.[51] Upon heating at pressures above 145 GPa, Fe_3C decomposes into a mixture of solid orthorhombic Fe_7C_3 and hcp Fe, then melts incongruently above

Table 3.2 *Elasticity parameters for liquid Fe–L alloys*

Composition (%L)		ρ_0 (g cm^{-3})	K_0 (GPa)	K_0'	P (GPa)	T (K)	Method	Ref.
wt.%	at.%							
Fe								
		7.02	109.7(7)	4.66(4)		1811	Shockwave	44
		5.19(3)	24.6(6)	6.65(4)	50–350	7000	FPMD	82
Fe–S								
10	6	5.2	48.0(2.0)	4	0–6	1770	X-ray absorption	124
10		5.5	63	4.8	0–20	1773–2123	Sink–float	125
20	12.5	4.41	35(1)	4.9	0–8	1673	Ultrasonic	96
27	17.4	4.07	25(1)	5.3	0–8	1673	Ultrasonic	
30	19.7				0–5.4	1573–1673	Ultrasonic	126
11.7	7	6.28	83.7	4.98	150–300	4000	FPMD	97
		5.43	49.6	5.08		6000		
16	9.8	5.72	64.4	4.94		4000		
		5.06	42.9	5.02		6000		
Fe–C								
0–4	0–0.9	[a]			0	1523–1823	Sessile drop	73
3.5	0.8	6.91	83.9	5.9(2)	0–4	1700	Ultrasonic	127
		6.91	100(1)	6.2(6)		1700		
		7.02(1.5)	55.3(2.5)	5.2(1.5)	2–7	1500		
2.0–4.0	0.4–0.9		65.0	6.0	42	3000	X-ray diffraction	66
	10.9(4)							
	12.1(4)							
5.7	1.3		Similar to Tera10		0–5.4		X-ray absorption	128
			Larger		5.4–7.8			
3.9	16	6.51	110(9)	5.1(3)	7–70	2500	X-ray absorption	65
6.7	1.5	6.5	54(3)	4	0–10	1973	X-ray absorption	129
Fe–Si								
17	9.3	5.88	68(1)	4	0–12	1773	Sink–float	130
17		6.33	75	4	0–5	1650	X-ray absorption	131
Fe–O								
22	7.5	5.45	128	3.85		5000	Thermo	132
Fe–H								
0.8	0.01	6.2	82.4	4.79	125–200	4000	FPMD	101
		5.63	62.9	4.76		6000		
1.2	0.02	5.88	73.1	5.02		4000		
		5.23	53.2	4.82		6000		

[a] $\rho = 7.10 - 0.0732x - (8.28 - 0.874x) \cdot 10^{-4} \cdot (T - 1823)$, x = wt.% C, T in K.

FPMD = first-principles molecular dynamics.

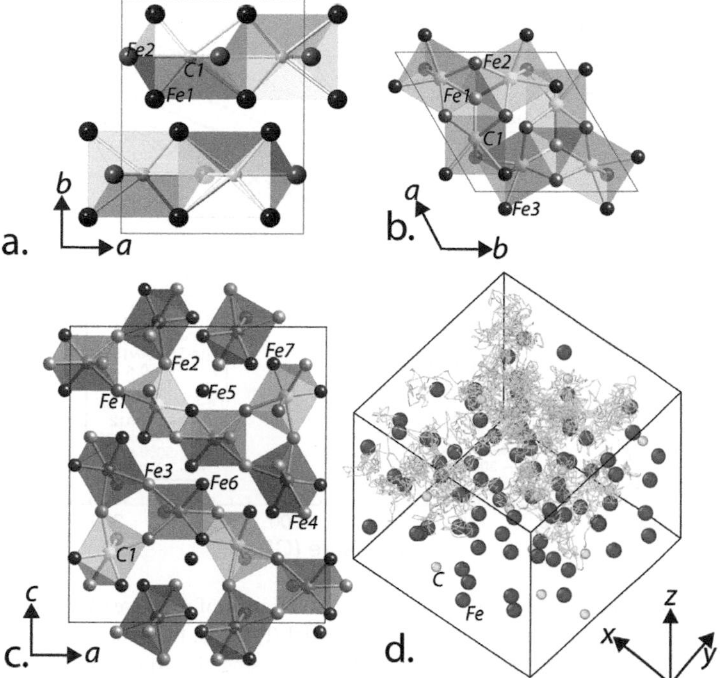

Figure 3.4 Atomic-scale structures of crystalline and molten iron carbide alloys. (a) Orthorhombic Fe_3C (space group Pnma), (b) hexagonal Fe_7C_3 (space group $P6_3mc$) and (c) orthorhombic Fe_7C_3 (space group Pbca). In both Fe_3C and Fe_7C_3 polymorphs, the fundamental building blocks are triangular prisms (CFe_6). Three such prisms are connected via shared vertices in a triangular arrangement (triads). The triads are stacked up along the c-axes for hexagonal polymorphs and along b-axes for orthorhombic polymorphs of Fe_7C_3. The carbon atoms are shown as gray spheres and the iron atoms are colored based on the distinct Wyckoff sites.[57,58] (d) A snapshot of a molten iron carbide alloy from molecular dynamics simulations. The computational supercell is shown and has orthogonal axes with $x = y = z$. The diffusion trajectory of a carbon atom is shown for reference.

3400 K.[52] Cemenite is ferromagnetic at ambient conditions and its Curie temperature is sensitive to small deviations from stoichiometry.[53] It undergoes ferromagnetic to paramagnetic transition and spin-pairing transition at high pressures.[54–56]

The density of Fe_3C at ambient conditions is 2.5% smaller than that of fictive hcp iron, corresponding to ~1.4% density reduction for 1 wt.% carbon (i.e. a compositional expansion coefficient α_c of 1.4).[59] Pressure-induced magnetic transitions lead to abrupt but small reductions in density and/or compressibility.[54,55,57,60] The calculated density of Fe_3C at the ICB pressure and 300 K is comparable to that of the inner core, but too low when thermal expansion is considered (Figure 3.5). A more appropriate test requires knowledge of the thermoelastic parameters of the non-magnetic phase.

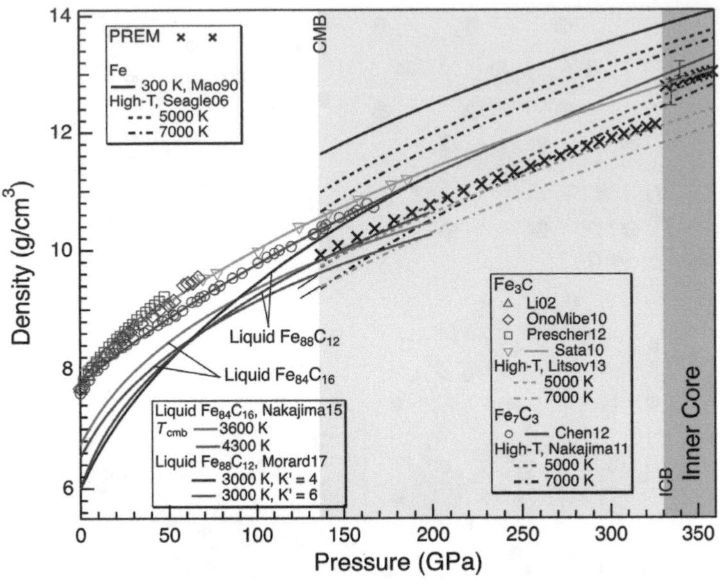

Figure 3.5 Density of Fe–C alloys and compounds as a function of pressure of iron carbides. CMB = core–mantle boundary. Preliminary reerence Earth model (PREM): black crosses;[4] hcp Fe at 300 K: black solid curve;[40] hcp Fe at 5000–7000 K calculated using the Mie–Grüneisen–Debye EoS.[42] Fe_3C at 300 K;[49,50,61,62] Fe_3C at 5000–7000 K.[63] Fe_7C_3 at 300 K;[64] Fe_7C_3 at 5000–7000 K.[41] Uncertainties are shown as error bars.[64] Liquid with $Fe_{84}C_{16}$ compoisition.[65] Liquid with $Fe_{88}C_{12}$ composition.[66]

3.2.2.2 Fe_7C_3

The metallurgical form of Fe_7C_3, known as Eckström–Adcock carbide, adopts a hexagonal structure at 1 bar and 300 K (Figure 3.3). An orthorhombic structure is also observed and may be stabilized with silicon impurities.[67] Non-stoichiometry is also observed in Fe_7C_3 and ranges from 8.0 to 10.8 wt.% (29–36 at.%) C, where the C-excess end member exceeds Fe_2C stoichiometry.[48] The crystal structure of Fe_7C_3 remains stable up to 185 GPa and 5200 K,[52,68] but it undergoes pressure-induced magnetic transitions.[18,69–71] At ambient conditions, the compositional expansion coefficients of h-Fe_7C_3 (~1.0) is smaller than that of Fe_3C (~1.4). The calculated density of the non-magnetic Fe_7C_3 is broadly consistent with that of the inner core at the relevant pressures and temperatures, thus supporting the carbide inner core model (Figure 3.5).

3.2.2.3 Fe–C Alloy Near the Iron End Member

In the simplified Fe–C model, the inner core may consist of an Fe–C alloy rather than a carbide.[22] The Fe–C alloy would contain no more than 1 wt.% carbon according to geochemical considerations and the measured solubility of carbon at pressures greater than 40 GPa.[20,21] However, 1.0–2.5 wt.% carbon may not be sufficient to reproduce the density deficit of the inner core[72] and hence would require the presence of other light elements.

3.2.2.4 Liquid Fe–C Alloy

A carbide inner core implies that the liquid outer core contains more carbon than the eutectic composition at relevant pressures (Figure 3.2). Even if the solid inner core is not made of carbides, a substantial amount of carbon may still be present in the liquid outer core, which occupies more than 90% of the core by mass or volume.

At ambient pressure, adding 1.3–2.8% carbon only reduces the density of liquid Fe by ~1% (α_c = 0.4–0.8).[73] Experimental measurements of an Fe liquid with 2.8 wt.% carbon suggest an α_c of 2–4 at the core–mantle boundary (CMB) pressure of 136 GPa and 3000 K,[66] which is in broad agreement with the calculated value of 1.3,[74] considering uncertainty and extrapolation. The larger α_c values at core pressures are consistent with Fe–C liquid being less compressible than Fe liquid.[65] Even with α_c = 2–4, 1.8–2.7 wt.% carbon is needed to explain the 5–8% density deficit in the outer core. This is higher than the upper limit from cosmochemical and geochemical considerations; hence, carbon cannot be the sole light element in the outer core.

3.2.2.5 Other Light Elements

All candidate light elements have been shown to reduce the density of solid Fe (Figure 3.6). The fitted compositional expansion coefficients of light elements in solid Fe alloys are comparable to the calculated results for liquid Fe alloys.[74] On the per wt.% basis, carbon may be slightly more efficient than O, Si, and O at reducing the density of iron, and therefore a slightly smaller amount is needed to account for the 5–8% density deficit in the outer core (Table 3.3). Combinations of light element such as that of sulfur and silicon[75] are found to satisfy the density constraints.

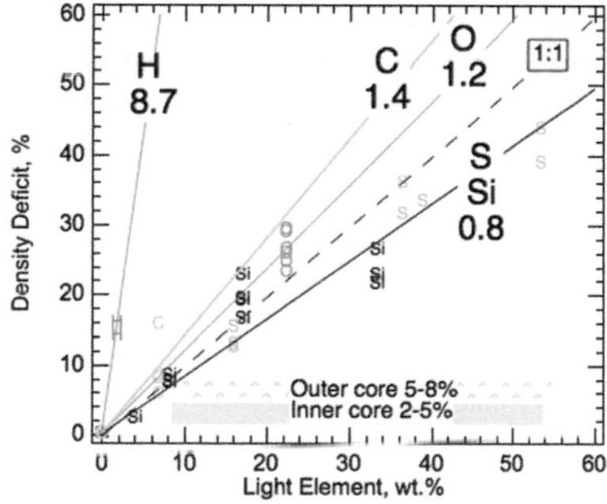

Figure 3.6 Compositional expansion coefficients of light elements in solid iron alloys. The values are derived from fits to solid Fe–L alloys and compounds.[3]

Table 3.3 *Compositional expansion coefficients*

	Solid[a]	Liquid at CMB[b]	Liquid at ICB[b]	LE, wt.%[c]	LE, wt.%[d]
H	8.7	–	–	0.6	0.9
C	1.4	1.3	1.3	4	6
O	1.2	1.1	1.0	4	7
Si	0.8	0.7	0.6	6	10
S	0.8	0.8	0.7	6	10

Compositional expansion coefficient is defined as the relative amount of density reduction per wt.% light element.[59]

[a] Li and Fei.[3]

[b] Badro et al.[74]

[c] Amount of light element needed to account for 5% density deficit in the outer core.

[d] Amount of light element needed to account for 8% density deficit in the outer core.

LE = light element.

3.2.3 Constraints from Sound Velocities of Fe–C Alloys and Compounds

Comparison between the preliminary reerence earth model (PREM) and iron reveals a prominent mismatch in the shear wave velocity, V_S, between the inner core and Fe or Fe–Ni alloys at corresponding pressures and 300 K (Figure 3.7). The discrepancy cannot be explained by the effect of temperature alone[76–78] and has been attributed to partial melting,[79] strong pre-melting effects,[80,81] and/or the presence of light elements.[55] In contrast, the compressional wave velocity, V_P, in the inner core is broadly consistent with that of hcp Fe (Figure 3.7). In the outer core, the bulk sound velocity may be comparable to or as much as 4% higher than liquid iron at corresponding conditions.[43,82] The presence of light elements, therefore, should not significantly affect the V_P of iron for this match to hold.

The sound velocities in the core increase linearly with density, following Birch's law (Figure 3.7). The velocity–density relations of solid and liquid Fe are consistent with Birth's law, but for solid Fe the V_P slope at 300 K or along a Hugoniot is steeper than that of the core. For V_S, deviation from Birch's law behavior was predicted by theory[83] and observed at high temperatures,[77] although this is not resolved in all studies.[84] A candidate Fe–L alloy must reproduce the velocity gradients in the core.

The speed of sound traversing the inner core is anisotropic by 3–4% in V_P and ~1% in V_S.[85,86] The anisotropy in sound speed may reflect convective alignment of anisotropic hcp Fe crystals[87] or an Fe–L alloy.[88] A candidate inner core phase needs to exhibit large enough elastic anisotropy to match the observations.

3.2.3.1 Fe_3C

As a candidate for the inner core phase, Fe_3C stands out in terms of its potential to account for the observed anisotropy. If the measured and calculated strong anisotropy in the sound

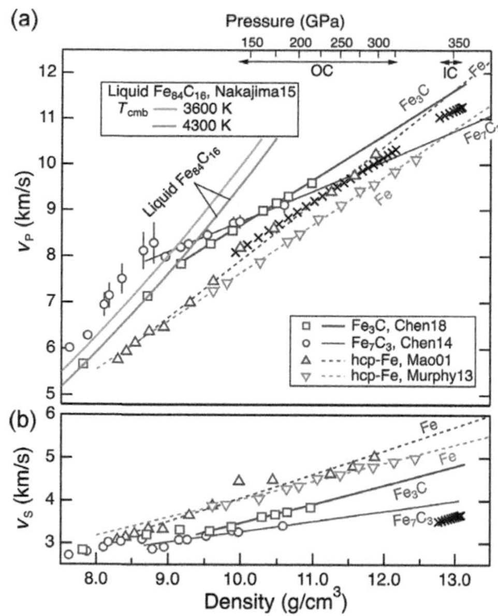

Figure 3.7 Sound velocity of Fe–C alloys and compounds. V_P and V_S of Fe carbides and liquid Fe–C as a function of density. Data are from Refs. 54, 65, 69, 89, and 90. The velocities of Fe–Ni alloys (not shown)[91] are similar to that of Fe. The top axis denotes the pressure range of the outer core (OC) and inner core (IC) according to the density–pressure relationship in PREM.

velocity of Fe_3C at ambient conditions[92,93] is applicable at core conditions, then only a small degree of alignment would be needed for Fe_3C to match the observations.

Existing data suggest that Fe_3C may provide a good match for the V_S in the inner core. At ambient conditions, the V_S of Fe_3C is similar to that of bcc Fe (Table 3.1). At 300 K, a magnetic transition near 5 GPa leads to a reduction in the V_S and its Birch's law slope so that the extrapolated V_S of Fe_3C at the inner core pressure is much smaller than that of hcp Fe and closer to the core values.[55] The high-spin to low-spin transition near 50 GPa leads to a further decrease in the Birch's law slope.[54] Moreover, at high temperatures, the V_S of Fe_3C deviates from Birch's law behavior toward the inner core values; hence, it can potentially explain the anomalously low V_S in the inner core without invoking partial melt or strong pre-melting effects.[94]

A potential match in V_P is also consistent with existing data. The range of measured V_P of Fe_3C at 1 bar and 300 K encompasses that of bcc Fe (Table 3.1). The magnetic transition to the paramagnetic phase of Fe_3C results in elastic softening and a shallower Birch's law slope of V_P, whereas the paramagnetic to non-magnetic transition does not seem to produce a visible effect.[54] At 300 K and inner core pressures, the extrapolated V_P of Fe_3C is higher than that of the inner core (Figure 3.7). A close match is possible if V_P at high temperature is lowered by a suitable amount as a result of deviation from Birch's law.

3.2.3.2 Fe_7C_3

The most compelling support for an Fe_7C_3 inner core comes from its ability to match the anomalously low V_S and high Poisson ratio, in addition to reproducing the density deficit.[58,69] While the ferro- to para-magnetic transition at 7.0–7.5 GPa does not seem to have obvious effect on sound velocities, significant shear softening accompanies the magnetic collapse at 40–50 GPa, resulting in pronounced reductions in V_P, V_S, and their Birch's law slopes (Figure 3.7). At pressures relevant to Earth's inner core, the extrapolated value of V_S of Fe_7C_3 at 300 K is only slightly higher than the observed value. There is likely a good match for V_S after considering further reduction at high temperature. It remains to be tested whether Fe_7C_3 can simultaneously match V_S, V_P, and anisotropy.

3.2.3.3 Fe–C Alloy Near the Iron End Member

First-principles calculations show that adding 1.0–2.5 wt.% carbon into the hcp Fe crystal structure increases its V_P and decreases its V_S, and this would help explain the observed anisotropy in compressional wave velocities, although there is a mismatch in shear wave anisotropy.[72]

3.2.3.4 Liquid Fe–C Alloy

Adding carbon increases the V_P of liquid iron (Table 3.1). For 1 at.% carbon, the average effect is 0.2% at 1 bar. It may increase to an estimated value of 0.8–1.2% at the core conditions, presumably because liquid Fe–C is less compressible than liquid Fe,[65] or remains at 0.2% at high pressures and high temperatures.[74] In any case, the V_P of an Fe–C alloy with <1 wt.% carbon would be consistent with the observed value in the outer core.

3.2.3.5 Other Light Elements

The sound velocities of other Fe–L alloys remain poorly constrained (Figure 3.8 and Table 3.4). The effect of sulfur on the sound velocities is not yet sufficiently understood to allow firm tests of Fe–S models for the core.[28,74,95–98] Further studies are needed to

Table 3.4 *Melting curve parameters of Fe–L alloys*

	a	c	P_0 (GPa)	T_0 (K)	T_{eut} CMB (K)	X_{eut} CMB (at.%)	dT/dx (K/at.%)
Fe–C	8.5	3.8	0	1420	2990(200)	11(5)	110(80)
Fe–O	17	3.8	0	1800	3200(200)	30(3)	33(11)
Fe–18 wt.% Si	23.6	1.89	0	1600	–	4	–
Fe–S	10.5	3	21	1260	2870(200)	15(5)	89(56)

The parameters are fitted to the Simon–Glatzel equation $(T_m/T_{m0})^c = (P_m - P_{m0})/a$. Data are from Morard et al.[75]

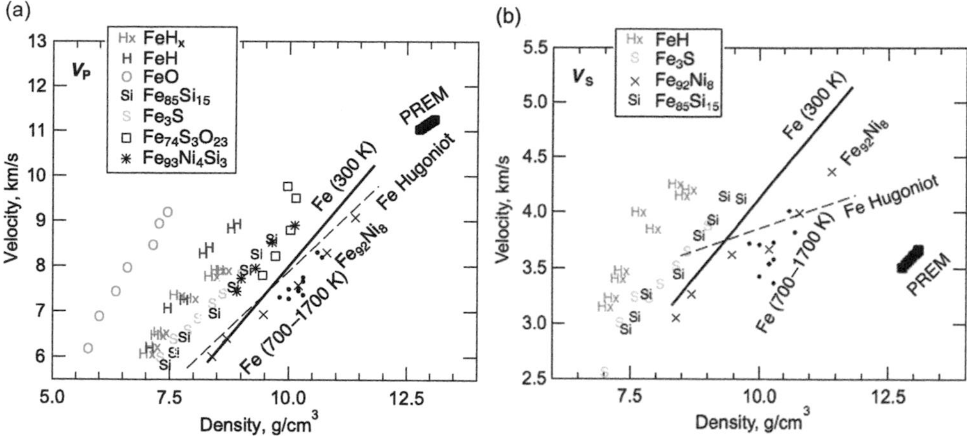

Figure 3.8 Sound velocities of Fe–H, Fe–O, Fe–S, and Fe–Si alloys and compounds. Compressional wave velocity V_P (a) and shear-wave velocity V_S (b) versus density relations. PREM;[4] hcp Fe at 300 K: solid line;[89] hcp Fe at temperatures between 700 and 1700 K: solid circles;[77] Fe from shockwave experiments: dashed line;[76] $Fe_{92}Ni_8$ at 300 K: crosses;[91] Fe_3S at 300 K;[102] $Fe_{85}Si_{15}$ at 300 K;[91] FeO at 300 K;[103] FeH_x at 300 K;[104] FeH at 300 K;[105] $Fe_{74}S_3O_{23}$ from shockwave experiments;[106] $Fe_{93}Ni_4Si_3$ at 300 K.[107]

resolve the disagreements concerning oxygen as a major light element in the core.[74,99] Computations suggest that an Fe–H alloy with 1 wt.% H can reproduce the density and V_P of the liquid outer core and therefore could be the primary alloy element, but Fe–H alloys cannot reproduce the V_S of the inner core.[100,101]

3.2.4 Constraints from Melting Temperatures of Fe–C Alloys

An independent constraint on the carbon content of the outer core can be obtained from the melting temperatures of iron alloys (Figure 3.8). The outer core is entirely molten, whereas the base of the mantle is mostly solid;[108] hence, the melting temperature of a candidate Fe–C alloy must be lower than the solidus of overlying mantle at the CMB pressure. In addition, as the geotherm is expected to follow an adiabat, which has a smaller dT/dP slope than the melting curve, the temperature at CMB is expected to be 400–900 K lower than its crystallization temperature at the ICB.[38,45]

The solidus temperature at the CMB is estimated at 4100–4200 K for peridotitic composition.[109] For comparison, core temperature profiles for pure Fe or Fe–Ni alloys would lead to a temperature at the CMB of 5400–5900 K,[38] which clearly exceeds the upper bounds on the mantle side (Figure 3.9); thus, these compositions are incompatible with a molten iron alloy and solid silicate coexisting at the CMB.

Carbon reduces the melting point of iron. Using linear interpolation between pure Fe and the eutectic liquid, the melting point reduction is estimated at >100 K per at.% carbon

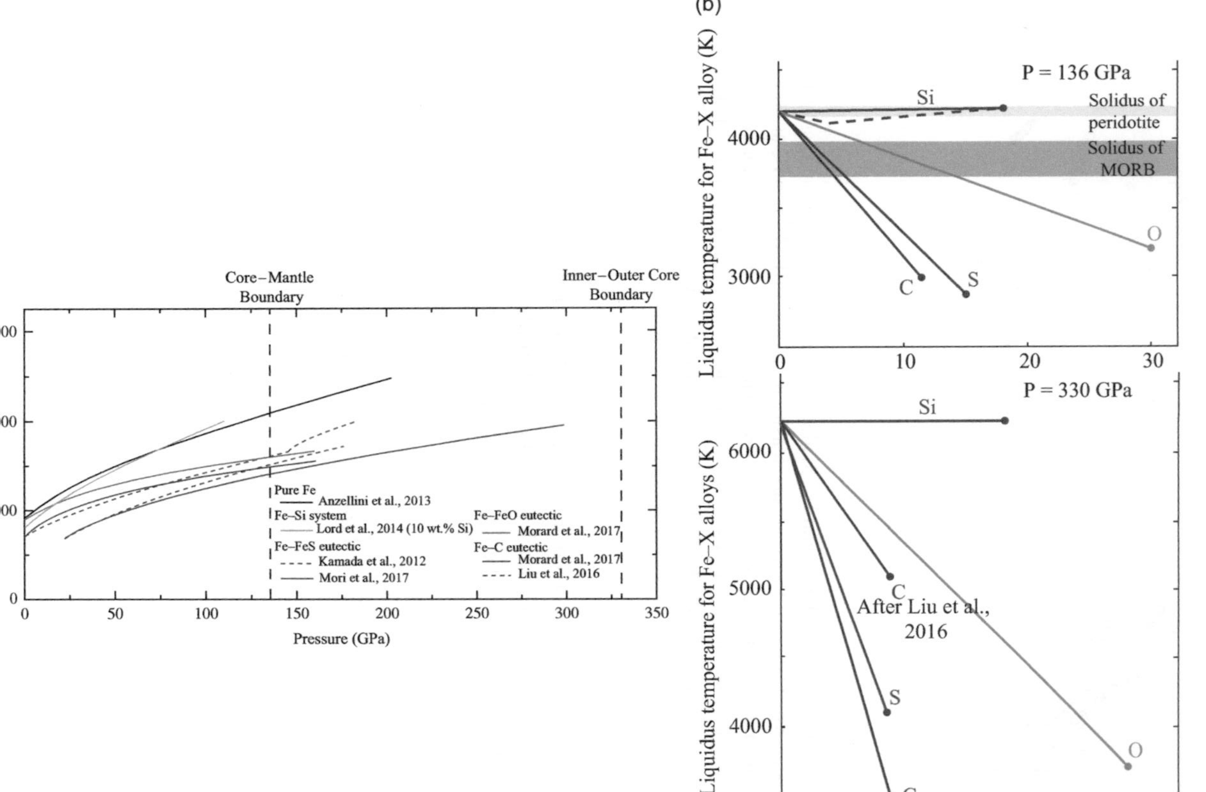

Figure 3.9 Melting temperatures of Fe-rich alloys. (a) Melting curves of pure iron,[38] and Fe–10 wt.% Si[110] and eutectic melting curves of Fe–Fe_3S (dashed line,[31] solid line[29]), Fe–FeO,[21] and Fe–Fe_3C (dashed line,[111] solid line[21]). The different melting curves are represented over the pressure range at which experiments were performed without any extrapolation. Pressures for the CMB and ICB are indicated by thick vertical dashed lines. (b, top) Liquidus temperatures in Fe–X systems compared with melting temperatures of mantle materials at the CMB (136 GPa), represented as linear interpolations between the melting point of pure Fe[38] and the eutectic compositions.[21] Solidi at CMB pressure for the peridotitic[109] and mid-ocean ridge basalt mantle[112] are represented by horizontal bands. (b, bottom) Extrapolated liquidus under ICB pressure for sulfur,[29] oxygen,[21] silicon,[110] and carbon.[21,52]

at 136 GPa.[21] At the ICB pressure, the melting point reduction effect of carbon may be similar to that at the CMB[52] or as much as 350 K/at.%.[21]

Experimentally determined eutectic melting temperatures agree within 150 K for the Fe–S, Fe–Si, and Fe–O systems.[3,21,30] Adding 1 at.% C, O, Si, and S to liquid iron reduces its melting point by 100 K for C and S, 50 K for O, and <30 K for Si at the pressure of the CMB (Figure 3.8). To pass the physical state test, a core with a single light element must contain at least 5 at.% S or C, or at least 15 at.% O.[29] The melting points of Fe–Si alloys are too high and therefore silicon cannot be the only light element in the outer core. The presence of other light elements such as carbon, oxygen, and/or sulfur are required to lower its crystallization temperature.

Compositions containing two or more lighter elements exhibit more complex behavior. While the alloying effect of oxygen on the eutectic point of the Fe–S system was found to be minor,[113] shock experiments at 100–200 GPa estimated that the presence of 8 wt.% (2.4 at.%) oxygen and 2 wt.% (1.2 at.%) sulfur would reduce the melting point of iron by 600 K.[106,114] This is more than twice the combined reductions of oxygen (120 K) and sulfur (120 K), suggesting non-ideal mixing in the ternary system.

3.3 Implications of Carbon as a Major Light Element in the Core

If the inner core consists of Fe_7C_3 with 8.41 wt.% carbon, the average concentration of carbon in the core would be at least ~0.3 wt.%, implying that the core has nearly one order of magnitude more carbon than the total amount in the surface reservoirs and silicate Earth, and hence it is by far the largest carbon reservoir in Earth (Figure 3.1). The bulk Earth would contain 0.1 wt.% carbon, higher than the estimated 0.03 wt.% for a half-mass condensation temperature of 40 K.[7,115] This result would question the validity of the volatility trend for highly volatile elements such as carbon.

Recent experiments show that Fe_7C_3 exhibits the highest electrical resistivity among all Fe–L alloys.[116] As a major element in the core, carbon may influence the thermal transport properties of the core, with implications for the evolution of the geodynamo.

3.4 Carbon in the Core Over Time

Carbon may move across the CMB over geological time if chemical disequilibrium was introduced during Earth's accretion or subsequent evolution. Earth's core may have been initially out of equilibrium with the mantle,[117] or the silicate Earth may have acquired most of its highly volatile elements through a late veneer.[118] Furthermore, chemical equilibrium at the CMB may have been perturbed as a result of secular cooling or inner core growth, which may have enriched or depleted carbon in the outer core depending on the carbon partitioning between the solid and liquid (Figure 3.2). Experiments suggest that mobility of carbon along grain boundaries may allow its transport over geologically significant length scales of 10 km over the age of Earth.[119] Facilitated by mantle convection, rapid

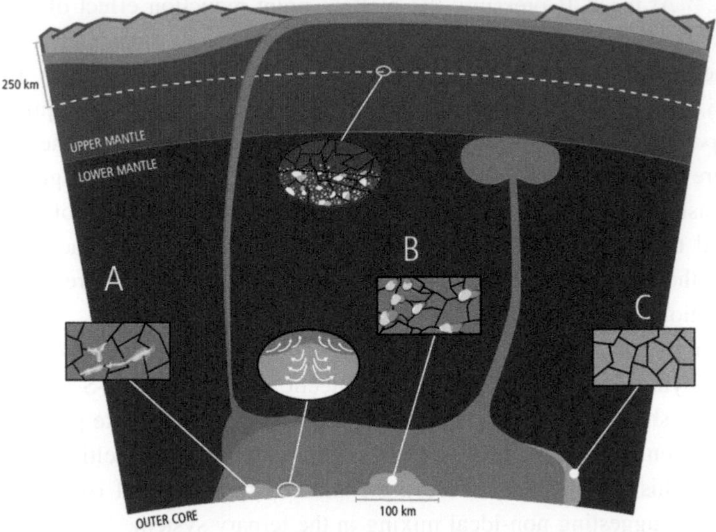

Figure 3.10 Carbon transport from subducted slabs to Earth's core. Schematic illustration of slab-derived Fe–C melt bringing carbon from Earth's surface to the core, modified after Liu et al.[111] The upper oval-shaped balloon shows elemental carbon or iron carbides (gray) associated with metallic iron (white) in the mantle at depths greater than 250 km. Three rectangular boxes represent Fe–C melts at the base of the mantle (heights are exaggerated): (a) Fe–C melt (yellow) that wets the solid silicate matrix (gray); (b) non-wetting Fe–C melt (yellow) coexisting with a small degree of silicate melt (green) in a solid silicate matrix (gray); and (c) solid phases (yellow–gray) that have become iron rich through reaction with the Fe–C melt. The lower oval-shaped balloon indicates dynamic stirring, which may prevent or slow down the draining of dense Fe–C melts to the core.

grain-boundary diffusion may have brought core-derived carbon to Earth's surface and thus connected the billion-year deep carbon cycle to the near-surface million-year shallow carbon cycle.

Ongoing carbon sequestration by the core may have resulted from subduction of the hydrothermally altered oceanic lithosphere carrying carbonates and organic matter into the deep Earth. While $CaCO_3$ in slabs may have been preserved under reducing lower-mantle conditions, the $MgCO_3$ component could have been destabilized by metallic iron-form diamonds or iron carbides.[120] Slab-derived Fe–C mixtures are expected to partially melt in the D′′ layer.[111] The melt may have accumulated near the CMB over time and episodically drained into the core (Figure 3.10).

3.5 Conclusion

We have evaluated constraints on the carbon budget of Earth's core by comparing the density, velocity, and elastic anisotropy of Fe–C alloys and compounds at core conditions with seismic observations. Existing data support the model of the inner core consisting of iron carbide Fe_7C_3, which could solidify from an Fe–C–S liquid core containing up to 1 wt.% carbon. Fe_7C_3 is unique in its ability to match the anomalous V_S and high Poisson ratio of the inner core. Its density and V_P are also broadly consistent with the PREM, but need to be further tested against the anisotropy observations. On the contrary, Fe_3C seems unstable and too light to match the inner core density. Given the upper limit of 1 wt.% carbon in the core, an Fe–C alloy is unable to generate the observed density deficit in the inner core.

The presence of 1 wt.% carbon in the outer core provides a good match to the V_P and is consistent with the coexistence of a molten iron alloy with solid silicate at the CMB. However, 1 wt.% carbon is insufficient to account for the density deficit in the outer core and cannot reproduce the density contrast at the ICB, and therefore other light elements such as H, O, S, or Si must be present in the outer core.

Earth's core remains potentially by far the largest carbon reservoir of the planet. It may participate in the long-term global carbon cycle through carbon transport across the CMB via grain-boundary diffusion, mantle convection, and sequestering slab-derived Fe–C melts.

The outer core likely contains multiple light elements. At least 1–2 wt.% sulfur is likely to be present in the outer core and would help account for its density deficit and the core's largely molten state. Oxygen may be required in the liquid outer core to explain the density contrast at the ICB, although the amount of oxygen remains uncertain. Silicon does not help explain the density contrast across the ICB or the coexistence of the liquid core with the overlying solid mantle. Existing data are insufficient to resolve the competing models of core composition because of limited data coverage in the relevant pressure–temperature–composition space and uncertainties in experimental measurements and theoretical simulations. Future studies should focus on expanding the experimental data range and investigating complex systems that contain more than one light element.

3.6 Limits to Knowledge and Unknowns

Earth's core is potentially by far the largest carbon reservoir of the planet. To assess the role of the core in Earth's deep carbon cycle, we need to test the hypothesis of iron carbide as the dominant component of the solid inner core and quantify the carbon content of the liquid outer core. In the past decade, research in deep carbon has significantly improved our knowledge of the physical properties and melting behavior of carbon-bearing iron alloys at the extreme pressure and temperature conditions in the deep Earth. Limits to our knowledge mainly stem from incomplete data coverage for the relevant pressures, temperatures, and compositions. For simplified compositions, the properties of liquid iron or iron alloys are still limited to relatively low pressures and temperatures far below the relevant ranges of the core. Investigations of complex iron alloys containing nickel and two or more light elements have only covered small subsets of the entire plausible pressure–temperature–composition space. Effects of temperature on the magnetic transitions and elasticities of solid iron alloys remain poorly constrained. Direct measurements of the densities and velocities of solids at inner core pressures are not yet available.

Acknowledgments

The authors thanks Rajdeep Dasgupta, James Badro, an anonymous reviewer, and Dave Walker for providing critical comments and constructive suggestions. JL acknowledges NSF EAR-1763189, NSF AST-1344133, and Sloan Foundation Deep Carbon Observatory Grant G-2017-9954. BC acknowledges NSF EAR-1555388 and NSF EAR-1565708. MM acknowledges XSEDE resources, NSF EAR-1634422, and NSF EAR-1753125.

Questions for the Classroom

1 How do researchers infer the presence of volatile elements such as carbon in Earth's liquid outer core?
2 As a candidate for the principal light element in Earth's core, what are the strongest arguments for and against carbon?
3 What is the plausible range of carbon content in Earth's core, and how do we know this?
4 Why was an iron carbide proposed as a candidate for the dominant component of Earth's solid inner core? How can we test this hypothesis?
5 Why is the knowledge of the eutectic composition of binary systems Fe–X, where X is an element lighter than iron such as hydrogen, carbon, oxygen, silicon, or sulfur, important for constraining Earth's core composition?
6 How do pressure and temperature affect magnetism in iron-rich alloys?
7 What are "spin-pairing" or "high-spin to low-spin" transitions in iron-rich alloys?
8 How is the elasticity of an iron alloy affected by pressure-induced magnetic transition?
9 How do light elements such as carbon affect the thermodynamic stability of iron–nickel alloys?

References

1. Jeanloz, R., The nature of the Earth's core. *Annu Rev Earth Planet Sci*, **18**, 357–386 (1990). doi:10.1146/annurev .ea.18.050190.002041
2. Poirier, J.P., Light elements in the Earth's outer core: a critical review. *Phys Earth Planet Inter*, **85**, 319–337 (1994). doi:10.1016/0031-9201(94)90120-1
3. Li, J. & Fei, Y., Experimental constraints on core composition. In *Treatise on Geochemistry, Vol*, eds. H.D. Holland & K.K. Turekian (Amsterdam: Elsevier Ltd., 2014), pp. 521–546.
4. Dziewonski, A.M. & Anderson, D.L., Preliminary reference Earth model. *Phys Earth Planet Inter*, **25**, 297–356 (1981). doi:10.1016/0031-9201(81)90046-7
5. Birch, F., Elasticity and constitution of the Earth's interior. *J Geophys Res*, **57**, 227–286 (1952). doi:10.1029/JZ057i002p00227
6. Jarosewich, E., Chemical analyses of meteorites – a compilation of stony and iron meteorite analyses. *Meteoritics*, **25**, 323–337 (1990). doi:10.1111/j.1945-5100.1990.tb00717.x
7. McDonough, W.F., Compositional model for the Earth's core, in *Treatise on Geochemistry, Vol. 3*, eds. H.D. Holland & K.K. Turekian (Oxford: Elsevier Ltd., 2014), pp. 559–576.

8. Dasgupta, R. & Walker, D., Carbon solubility in core melts in a shallow magma ocean environment and distribution of carbon between the Earth's core and the mantle. *Geochim. Cosmochim. Acta*, 72, 4627–4641 (2008). doi:10.1016/j.gca.2008.06.023
9. Wood, B.J., Carbon in the core. *Earth Planet Sci Lett*, **117**, 593–607 (1993). doi:10.1016/0012-821X(93)90105-I
10. Wang, C., Hirama, J., Nagasaka, T., & Ban-Ya, S., Phase equilibria of liquid Fe–S–C ternary system. *ISIJ Int*, **31**, 1292–1299 (1991). doi:10.2355/isijinternational.31.1292
11. Chi, H., Dasgupta, R., Duncan, M.S., & Shimizu, N., Partitioning of carbon between Fe-rich alloy melt and silicate melt in a magma ocean – implications for the abundance and origin of volatiles in Earth, Mars, and the Moon. *Geochim Cosmochim Acta*, **139**, 447–471 (2014). doi:10.1016/j.gca.2014.04.046
12. Dalou, C., Hirschmann, M.M., von der Handt, A., Mosenfelder, J., & Armstrong, L.S., Nitrogen and carbon fractionation during core–mantle differentiation at shallow depth. *Earth Planet Sci Lett*, **458**, 141–151 (2017). doi:10.1016/j.epsl.2016.10.026
13. Li, Y., Dasgupta, R., & Tsuno, K., The effects of sulfur, silicon, water, and oxygen fugacity on carbon solubility and partitioning in Fe-rich alloy and silicate melt systems at 3 GPa and 1600 C. *Earth Planet Sci Lett*, **415**, 54–66 (2015). doi:10.1016/j.epsl.2015.01.017
14. Tsuno, K., Grewal, D.S., & Dasgupta, R., Core–mantle fractionation of carbon in Earth and Mars: the effects of sulfur. *Geochim Cosmochim Acta*, **238**, 477–495 (2018). doi:10.1016/j.gca.2018.07.010
15. Wood, B.J., Li, J., & Shahar, A., Carbon in the core: its influence on the properties of core and mantle. *Rev Mineral Geochem*, **75**, 231–250 (2013). doi:10.2138/rmg.2013.75.8
16. Dasgupta, R., Ingassing, storage, and outgassing of terrestrial carbon through geological time, in *Carbon in Earth*, eds. R.M. Hazen, A.P., Jones, & J.A. Baross (Washington, DC: Mineralogical Society of America, 2013), pp. 183–229.
17. Chipman, J., Thermodynamics and phase diagram of the Fe–C system. *Metall Trans*, **3**, 55–64 (1972). doi:10.1007/BF02680585
18. Nakajima, Y., Takahashi, E., Suzuki, T., & Funakoshi, K.I., "Carbon in the core" revisited. *Phys Earth Planet Inter*, **174**, 202–211 (2009). doi:10.1016/j.pepi.2008.05.014
19. Hirayama, Y., Fujii, T., & Kei, K., The melting relation of the system iron and carbon at high pressure and its bearing on the early stage of the Earth. *Geophys Res Lett*, **20**, 2095–2098 (1993). doi:10.1029/93GL02131
20. Lord, O.T., Walter, M.J., Dasgupta, R., Walker, D., & Clark, S.M., Melting in the Fe–C system to 70 GPa. *Earth Planet Sci Lett*, **284**, 157–167 (2009). doi:10.1016/j.epsl.2009.04.017
21. Morard, G. et al., Fe–FeO and Fe–Fe–C melting relations at Earth's core–mantle boundary conditions: implications for a volatile-rich or oxygen-rich core. *Earth Planet Sci Lett*, **473**, 94–103 (2017). doi:10.1016/j.epsl.2017.05.024
22. Fei, Y. & Brosh, E., Experimental study and thermodynamic calculations of phase relations in the Fe–C system at high pressure. *Earth Planet Sci Lett*, **408**, 155–162 (2014). doi:10.1016/j.epsl.2014.09.044
23. Alfè, D., Gillan, M.J., & Price, G.D., *Ab initio* chemical potentials of solid and liquid solutions and the chemistry of the Earth's core. *J. Chem Phys*, **116**, 7127–7136 (2002). doi:10.1063/1.1464121

24. Luo, F., Cheng, Y., Chen, X.R., Cai, L.C., & Jing, F.Q., The melting curves and entropy of iron under high pressure. *J Chem Eng Data*, **56**, 2063–2070 (2011). doi:10.1021/je1010483
25. Shearer, P. & Masters, G., The density and shear velocity contrast at the inner core boundary. *Geophys J Int*, **102**, 491–408 (1990). doi:10.1111/j.1365-246X.1990.tb04481.x
26. Kubaschewski, O., *Iron-Binary Phase Diagrams* (New York: Springer-Verlag, 1982).
27. Morard, G., Andrault, D., Antonangeli, D., & Bouchet, J., Properties of iron alloys under the Earth's core conditions. *C R Geosci*, **346**, 130–139 (2014). doi:10.1016/j.crte.2014.04.007
28. Morard, G. et al., *In situ* determination of Fe–Fe3S phase diagram and liquid structural properties up to 65 GPa. *Earth Planet Sci Lett*, **272**, 620–626 (2008). doi:10.1016/j.epsl.2008.05.028
29. Mori, Y. et al., Melting experiments on Fe–Fe_3S system to 254 GPa. *Earth Planet Sci Lett*, **464**, 135–141 (2017). doi:10.1016/j.epsl.2017.02.021
30. Ozawa, H., Hirose, K., Yonemitsu, K., & Ohishi, Y., High-pressure melting experiments on Fe–Si alloys and implications for silicon as a light element in the core. *Earth Planet Sci Lett*, **456**, 47–54 (2016). doi:10.1016/j.epsl.2016.08.042
31. Kamada, S. et al., Phase relationships of the Fe–FeS system in conditions up to the Earth's outer core. *Earth Planet Sci Lett*, **294**, 94–100 (2010). doi:10.1016/j.epsl.2010.03.011
32. Li, J., Fei, Y., Mao, H.K., Hirose, K., & Shieh, S.R., Sulfur in the Earth's inner core. *Earth Planet Sci Lett*, **193**, 509–514 (2001). doi:10.1016/S0012-821X(01)00521-0
33. Kuwayama, Y. & Hirose, K., Phase relations in the system Fe–FeSi at 21 GPa. *Am Mineral*, **89**, 273–276 (2004). doi:10.2138/am-2004-2-303
34. Fischer, R.A. et al., Phase relations in the Fe–FeSi system at high pressures and temperatures. *Earth Planet Sci Lett*, **373**, 54–64 (2013). doi:10.1016/j.epsl.2013.04.035
35. Belonoshko, A.B., Rosengren, A., Burakovsky, L., Preston, D.L., & Johansson, B., Melting of Fe and Fe0.9375Si0.0625 at Earth's core pressures studied using *ab initio* molecular dynamics. *Phys Rev B*, **79**, 220102 (2009). doi:10.1103/PhysRevB.79.220102
36. Lin, J.-F. et al., Iron–nickel alloy in the Earth's core. *Geophys Res Lett*, **29**, 109–111 (2002). doi:10.1029/2002GL015089
37. Morard, G., Siebert, J., & Badro, J., Partitioning of Si and platinum group elements between liquid and solid Fe–Si alloys. *Geochim Cosmochim Acta*, **132**, 94–100 (2014). doi:10.1016/j.gca.2014.01.044
38. Anzellini, S., Dewaele, A., Mezouar, M., Loubeyre, P., & Morard, G., Melting of iron at Earth's inner core boundary based on fast X-ray diffraction. *Science*, **340**, 464–466 (2013). doi:10.1126/science.1233514
39. Dewaele, A. et al., Quasihydrostatic equation of state of iron above 2 Mbar. *Phys Rev Lett*, **97**, 29–32 (2006). doi:10.1103/PhysRevLett.97.215504
40. Mao, H.K., Wu, Y., Chen, L.C., & Shu, J.F., Static compression of iron to 300 GPa and Fe0.8Ni0.2 alloy to 260 GPa: implications for composition of the core. *J Geophys Res*, **95**, 21737–21742 (1990). doi:10.1029/JB095iB13p21737
41. Nakajima, Y. et al., Thermoelastic property and high-pressure stability of Fe7C3: implication for iron-carbide in the Earth's core. *Am Mineral*, **96**, 1158–1165 (2011). doi:10.2138/am.2011.3703

42. Seagle, C.T., Campbell, A.J., Heinz, D.L., Shen, G., & Prakapenka, V., Thermal equation of state of Fe3S and implications for sulfur in Earth's core. *J Geophys Res*, **111**, B06209 (2006). doi:10.1029/2005JB004091
43. Anderson, O.L. & Isaak, D.G., Another look at the core density deficit of Earth's outer core. *Phys Earth Planet Int*, **131**, 19–27 (2002). doi:10.1016/S0031-9201(02)00017-1
44. Anderson, W.W. & Ahrens, T.J., An equation of state for liquid iron and implications for the Earth's core. *J Geophys Res*, **99**, 4273–4284 (1994). doi:10.1029/93JB03158
45. Komabayashi, T. & Fei, Y., Internally consistent thermodynamic database for iron to the Earth's core conditions. *J Geophys Res*, **115**, B03202 (2010). doi:10.1029/2009JB006442
46. Bazhanova, Z.G., Oganov, A.R., & Gianola, O., Fe–C and Fe–H systems at pressures of the Earth's inner core. *Physics-Uspekhi*, **55**, 489–497 (2012). doi:10.3367/UFNe.0182.201205c .0521
47. Weerasinghe, G.L., Needs, R.J., & Pickard, C.J., Computational searches for iron carbide in the Earth's inner core. *Phys Rev B*, **84**, 1–7 (2011). doi:10.1103/PhysRev B.84.174110
48. Walker, D., Dasgupta, R., Li, J., & Buono, A., Nonstoichiometry and growth of some Fe carbides. *Contrib Mineral Petr*, **166**, 935–957 (2013). doi:10.1007/s00410-013-0900-7
49. Ono, S. & Mibe, K., Magnetic transition of iron carbide at high pressures. *Phys Earth Planet Int*, **180**, 1–6 (2010). doi:10.1016/j.pepi.2010.03.008
50. Sata, N. et al., Compression of FeSi, Fe3C, Fe0.95O, and FeS under the core pressures and implication for light element in the Earth's core. *J Geophys Res*, **115**, 1–13 (2010). doi:10.1029/2009JB006975
51. Rouquette, J. et al., Iron–carbon interactions at high temperatures and pressures. *Appl Phys Lett*, **12**, 121912 (2008). doi:10.1063/1.2892400
52. Liu, J., Lin, J.-F., Prakapenka, V., Prescher, C., & Yoshino, T., Phase relations of Fe–C and Fe7C3 up to 185 GPa and 5200 K: implication for the stability of iron carbide in the Earth's core. *Geophys Res Lett*, **43**, 12415–12422 (2016). doi:10.1002/2016GL071353
53. Walker, D., Li, J., Kalkan, B., & Clark, S.M., Thermal, compositional, and compressional demagnetization of cementite. *Am Mineral*, **100**, 2610–2624 (2015). doi:10.2138/am-2015-5306
54. Chen, B. et al., Experimental constraints on the sound velocities of cementite Fe3C to core pressures. *Earth Planet Sci Lett*, **494**, 164–171 (2018). doi:10.1016/j .epsl.2018.05.002
55. Gao, L. et al., Pressure-induced magnetic transition and sound velocities of Fe3C: implications for carbon in the Earth's inner core. *Geophys Res Lett*, **35**, L17306 (2008). doi:17310.11029/12008GL034817
56. Lin, J.F. et al., Magnetic transition in compressed Fe3C from X-ray emission spectroscopy. *Earth Planet Sci Lett*, **70**, 1–4 (2004). doi:10.1103/PhysRevB.70 .212405
57. Mookherjee, M., Elasticity and anisotropy of Fe3C at high pressures. *Am Mineral*, **96**, 1530–1536 (2011). doi:10.2138/am.2011.3917
58. Prescher, C. et al., High Poisson's ratio of Earth's inner core explained by carbon alloying. *Nat Geosci*, **8**, 220–223 (2015). doi:10.1038/ngeo2370
59. Roberts, P.H., Jones, C.A., & Calderwood, A., Energy fluxes and Ohmic dissipation in the Earth's core, in *Earth's Core and Lower Mantle*, eds. C.A. Jones et al. (Abingdon: Taylor & Francis, 2003), pp. 100–129.

60. Vočadlo, L. et al., The effect of ferromagnetism on the equation of state of Fe3C studied by first-principles calculations. *Earth Planet Sci Lett*, **203**, 567–575 (2002). doi:10.1016/S0012-821X(02)00839-7
61. Li, J. et al., Compression of Fe3C to 30 GPa at room temperature. *Phys Chem Mineral*, **29**, 166–169 (2002). doi:10.1007/s00269-001-0224-4
62. Prescher, C. et al., Structurally hidden magnetic transitions in Fe3C at high pressures. *Phys Rev*, **85**, 6–9 (2012). doi:10.1103/PhysRevB.85.140402
63. Litasov, K.D. et al., Thermal equation of state and thermodynamic properties of iron carbide Fe_3C to 31 GPa and 1473 K. *J Geophys Res*, **118**, 1–11 (2013). doi:10.1002/2013JB010270
64. Chen, B. et al., Magneto-elastic coupling in compressed Fe7C3 supports carbon in Earth's inner core. *Geophys Res Lett*, **39**, 2–5 (2012). doi:10.1029/2012GL052875
65. Nakajima, Y. et al., Carbon-depleted outer core revealed by sound velocity measurements of liquid iron–carbon alloy. *Nature Comm*, **6**, 8942 (2015). doi:10.1038/ncomms9942
66. Morard, G. et al., Structure and density of Fe–C liquid alloys under high pressure. *J Geophys Res*, **122**, 7813–7823 (2017). doi:10.1002/2017JB014779
67. Das, T., Chatterjee, S., Ghosh, S., & Saha-Dasgupta, T., First-principles prediction of Si-doped Fe carbide as one of the possible constituents of Earth's inner core. *Geophys Res Lett*, **44**, 8776–8784 (2017). doi:10.1002/2017GL073545
68. Raza, Z., Shulumba, N., Caffrey, N.M., Dubrovinsky, L., & Abrikosov, I.A., First-principles calculations of properties of orthorhombic iron carbide Fe7C3 at the Earth's core conditions. *Phys Rev B*, **91**, 1–7 (2015). doi:10.1103/PhysRevB.91.214112
69. Chen, B. et al., Hidden carbon in Earth's inner core revealed by shear softening in dense Fe7C3. *Proc Natl Acad Sci USA*, **111**, 17755–17758 (2014). doi:10.1073/pnas.1411154111
70. Liu, J., Li, J., & Ikuta, D., Elastic softening in Fe7C3 with implications for Earth's deep carbon reservoirs. *J Geophys Res*, **121**, 1514–1524 (2016). doi:10.1002/2015JB012701
71. Mookherjee, M. et al., High-pressure behavior of iron carbide Fe7C3 at inner core conditions. *J Geophys Res B*, **116**, 1–13 (2011). doi:10.1029/2010JB007819
72. Caracas, R., The influence of carbon on the seismic properties of solid iron. *Geophys Res Lett*, **44**, 128–134 (2017). doi:10.1002/2015GL063478
73. Jimbo, I. & Cramb, A.W., The density of liquid iron–carbon alloys. *Metall Trans B*, **24**, 5–10 (1993). doi:10.1007/BF02657866
74. Badro, J., Cote, A.S., & Brodholt, J.P., A seismologically consistent compositional model of Earth's core. *Proc Natl Acad Sci USA*, **111**, 7542–7545 (2014). doi:10.1073/pnas.1316708111
75. Morard, G. et al., The Earth's core composition from high pressure density measurements of liquid iron alloys. *Earth Planet Sci Lett*, **373**, 169–178 (2013). doi:10.1016/j.epsl.2013.04.040
76. Brown, J.M. & McQueen, G., Phase transitions, Grüneisen parameter, and elasticity. *J Geophys Res*, **91**, 7485–7494 (1986). doi:10.1029/JB091iB07p07485
77. Lin, J.-F. et al., Sound velocities of hot dense iron: Birch's law revisited. *Science*, **308**, 1892–1894 (2005). doi:10.1126/science.1111724
78. Ohtani, E. et al., Sound velocity of hexagonal close-packed iron up to core pressures. *Geophys Res Lett*, **40**, 5089–5094 (2013). doi:10.1002/grl.50992
79. Singh, S.C., Taylor, M.A.J., & Montagner, J.P., On the presence of liquid in Earth's inner core. *Science*, **287**, 2471–2474 (2000). doi:10.1126/science.287.5462.2471

80. Li, Y., Vočadlo, L., Brodholt, J., & Wood, I.G., Thermoelasticity of Fe7C3 under inner core conditions. *J Geophys Res*, **B121**, 5828–5837 (2016). doi:10.1002/2016JB013155
81. Martorell, B., Brodholt, J., Wood, I.G., & Vočadlo, L., The effect of nickel on the properties of iron at the conditions of Earth's inner core: *ab initio* calculations of seismic wave velocities of Fe–Ni alloys. *Science*, **365**, 143–151 (2013). doi:10.1016/j.epsl.2013.01.007
82. Ichikawa, H., Tschuchiya, T., & Tange, Y., The P–V–T equation of state and thermodynamic properties of liquid iron. *J Geophys Res*, **119**, 240–252 (2014). doi:10.1002/2013JB010732.Received
83. Vočadlo, L., Alfè, D., Gillan, M.J., & Price, G.D., The properties of iron under core conditions from first principles calculations. *Phys Earth Planet Int*, **140**, 101–125 (2003). doi:10.1016/j.pepi.2003.08.001
84. Kantor, A.P. et al., Sound wave velocities of fcc Fe–Ni alloy at high pressure and temperature by mean of inelastic X-ray scattering. *Phys Earth Planet Int*, **164**, 83–89 (2007). doi:10.1016/j.pepi.2007.06.006
85. Souriau, A. & Poupinet, G., The velocity profile at the base of the liquid core from PKP(BC+Cdiff) data: an argument in favor of radial inhomogeneity. *Geophys Res Lett*, **18**, 2023–2026 (1991). doi:10.1029/91GL02417
86. Wookey, J. & Helffrich, G., Inner-core shear-wave anisotropy and texture from an observation of PKJKP waves. *Nature*, **454**, 873–876 (2008). doi:10.1038/nature07131
87. Mao, H.K. et al., Elasticity and rheology of iron above 220 GPa and the nature of the Earth's inner core. *Nature*, **396**, 741–743 (1998). doi:10.1038/25506
88. Antonangeli, D. et al., Elastic anisotropy in textured hcp-iron to 112 GPa from sound wave propagation measurements. *Earth Planet Sci Lett*, **225**, 243–251 (2004). doi:10.1016/j.epsl.2004.06.004
89. Mao, H.K. et al., Phonon density of states of iron up to 153 gigapascals. *Science*, **292**, 914–916 (2001). doi:10.1126/science.1057670
90. Murphy, C.A., Jackson, J.M., & Sturhahn, W., Experimental constraints on the thermodynamics and sound velocities of hcp-Fe to core pressures. *J Geophys Res*, **118**, 1999–2016 (2013). doi:10.1002/jgrb.50166
91. Lin, J.-F. et al., Sound velocities of iron–nickel and iron–silicon alloys at high pressures. *Geophys Res Lett*, **30**, 1–4 (2003). doi:10.1029/2003GL018405
92. Gao, L. et al., Sound velocities of compressed Fe3C from simultaneous synchrotron X-ray diffraction and nuclear resonant scattering measurements. *J Synchrotron Rad*, **16**, 714–722 (2009). doi:10.1107/S0909049509033731
93. Nikolussi, M. et al., Extreme elastic anisotropy of cementite, Fe3C: first-principles calculations and experimental evidence. *Scripta Mater*, **59**, 814–817 (2008). doi:10.1016/j.scriptamat.2008.06.015
94. Gao, L. et al., Effect of temperature on sound velocities of compressed Fe3C, a candidate component of the Earth's inner core. *Earth Planet Sci Lett*, **309**, 213–220 (2011). doi:10.1016/j.epsl.2011.06.037
95. Huang, H. et al., Shock compression of Fe–FeS mixture up to 204 GPa. *Geophys Res Lett*, **40**, 687–691 (2013). doi:10.1002/grl.50180
96. Jing, Z. et al., Sound velocity of Fe–S liquids at high pressure: implications for the Moon's molten outer core. *Earth Planet Sci Lett*, **396**, 78–87 (2014). doi:10.1016/j.epsl.2014.04.015

97. Umemoto, K. et al., Liquid iron sulfur alloys at outer core conditions by first-principles calculations. *Geophys Res Lett*, **41**, 6712–6717 (2014). doi:10.1002/2014GL061233
98. Kawaguchi, S.I. et al., Sound velocity of liquid Fe–Ni–S at high pressure. *J Geophys Res*, **122**, 3624–3634 (2017). doi:10.1002/2016JB013609
99. Huang, H. et al., Evidence for an oxygen-depleted liquid outer core of the Earth. *Nature*, **479**, 513–516 (2011). doi:10.1038/nature10621
100. Caracas, R., The influence of hydrogen on the seismic properties of solid iron. *Geophys Res Lett*, **42**, 3780–3785 (2015). doi:10.1002/2015GL063478
101. Umemoto, K. & Hirose, K., Liquid iron–hydrogen alloys at outer core conditions by first-principles calculations. *Geophys Res Lett*, **42**, 7513–7520 (2015). doi:10.1002/2015GL 065899
102. Lin, J.F. et al., Magnetic transition and sound velocities of Fe3S at high pressure: implications for Earth and planetary cores. *Phy Rev B*, **226**, 33–40 (2004). doi:10.1016/j.epsl.2004.07.018
103. Badro, J. et al., Effect of light elements on the sound velocities in solid iron: implications for the composition of Earth's core. *Earth Planet Sci Lett*, **254**, 233–238 (2007). doi:10.1016/j.epsl.2006.11.025
104. Mao, W.L. et al., Nuclear resonant X-ray scattering of iron hydride at high pressure. *Geophys Res Lett*, **31**, L15618 (2004). doi:10.11029/2004GL020541
105. Shibazaki, Y. et al., Effect of hydrogen on the melting temperature of FeS at high pressure: implications for the core of Ganymede. *Earth Planet Sci Lett*, **301**, 153–158 (2013). doi:10.1016/j.epsl.2010.10.033
106. Huang, H. et al., Melting behavior of Fe–O–S at high pressure: a discussion on the melting depression induced by O and S. *J Geophys Res*, **115** (2010). doi:10.1029/2009JB006514
107. Antonangeli, D. et al., Composition of the Earth's inner core from high-pressure sound velocity measurements in Fe–Ni–Si alloys. *Earth Planet Sci Lett*, **295**, 292–296 (2010). doi:10.1016/j.epsl.2010.04.018
108. McNamara, A.K., Garnero, E.J., & Rost, S., Tracking deep mantle reservoirs with ultra-low velocity zones. *Earth Planet Sci Lett*, **299**, 1–9 (2010). doi:10.1016/j.epsl.2010.07.042
109. Fiquet, G. et al., Melting of peridotite to 140 gigapascals. *Science*, **329**, 1516–1518 (2010). doi:10.1016/j.epsl.2010.04.018
110. Lord, O.T. et al., The FeSi phase diagram to 150 GPa. *J Geophys Res*, **115**, 1–9 (2010). doi:10.1029/2009JB006528
111. Liu, J., Li, J., Hrubiak, R., & Smith, J.S., Origin of ultra-low velocity zones through mantle-derived metallic melt. *Proc Natl Acad Sci ISA*, **113**, 5547–5551 (2016). doi:10.1073/pnas.1519540113
112. Andrault, D. et al., Melting of subducted basalt at the core-mantle boundary. *Science*, **344**, 892–895 (2014). doi:10.1126/science.1250466
113. Terasaki, H. et al., Liquidus and solidus temperatures of a Fe–O–S alloy up to the pressures of the outer core: implication for the thermal structure of the Earth's core. *Earth Planet Sci Lett*, **304**, 559–564 (2011). doi:10.1016/j.epsl.2011.02.041
114. Tsuno, K. & Ohtani, E., Eutectic temperatures and melting relations in the Fe–O–S system at high pressures and temperatures. *Phys Chem Mineral*, **36**, 9–17 (2009). doi:10.1007/s00269-008-0254-2
115. Lodders, K., Solar system abundances and condensation temperatures of the elements. *Astrophys J*, **591**, 1220–1247 (2003). doi:1210.1086/375492

116. Zhang, C.W. et al., Electrical resistivity of Fe–C alloy at high pressure: effects of carbon as a light element on the thermal conductivity of the Earth's core. *J Geophys Res*, **123**, 3564–3577 (2018). doi:10.1029/2017JB015260
117. Rudge, J.F., Kleine, T., & Bourdon, B., Broad bounds on Earth's accretion and core formation constrained by geochemical models. *Nature Comms*, **3**, 439–443 (2010). doi:10.1038/ngeo872
118. Albarède, F., Volatile accretion history of the terrestrial planets and dynamic implications. *Nature*, **461**, 1227–1233 (2009). doi:10.1038/nature08477
119. Hayden, L. & Watson, B., Grain boundary mobility of carbon in Earth's mantle: a possible carbon flux from the core. *Proc Natl Acad Sci USA*, **105**, 8537–8541 (2008). doi:10.1073 pnas.0710806105
120. Dorfman, S.M. et al., Carbonate stability in the reduced lower mantle. *Earth Planet Sci Lett*, **498**, 84–91 (2018). doi:10.1016/j.epsl.2018.02.035
121. Scott, H.P., Williams, Q., & Knittle, E., Stability and equation of state of Fe_3C to 73 GPa: implications for carbon in the Earth's core. *Geophys Res Lett*, **28**, 1875–1878 (2001). doi:10.1029/2000GL012606
122. Dodd, S.P., Saunders, G.A., Cankurtaran, M., James, B., & Acet, M., Ultrasonic study of the temperature and hydrostatic-pressure dependences of the elastic properties of polycrystalline cementite (Fe3C). *Phys Status Solidi A*, **198**, 272–281 (2003). doi:10.1002/pssa.200306613
123. Fiquet, G., Badro, J., Gregoryanz, E., Fei, Y., & Occelli, F., Sound velocity in iron carbide (Fe_3C) at high pressure: implications for the carbon content of the Earth's inner core. *Phys Earth Planet Inter*, **172**, 125–129 (2009). doi:10.1016/j.pepi.2008.05.016
124. Sanloup, C. et al., Density measurements of liquid Fe–S alloys at high-pressure. *Earth Planet Sci Lett*, **27**, 811–814 (2000). doi:10.1029/1999GL008431
125. Balog, P.S., Secco, R.A., Rubie, D.C., & Frost, D.J., Equation of state of liquid Fe-10 wt % S: implications for the metallic cores of planetary bodies. *J Geophys Res*. **108**, B2 (2003). doi:10.1029/2001jb001646
126. Nishida, K. et al., Towards a consensus on the pressure and composition dependence of sound velocity in the liquid Fe–S system. *Phys Earth Planet Int*, **257**, 230–239 (2016). doi:10.1016/j.pepi.2016.06.009
127. Shimoyama, Y. et al., Thermoelastic properties of liquid Fe–C revealed by sound velocity and density measurements at high pressure. *J Geophys Res*, **121**, 7984–7995 (2016). doi:10.1002/2016JB012968
128. Sanloup, C., van Westrenen, W., Dasgupta, R., Maynard-Casely, H., & Perrillat, J.P., Compressibility change in iron-rich melt and implications for core formation models. *Earth Planet Sci Lett*, **306**, 118–122 (2011). doi:10.1016/j.epsl.2011.03.039
129. Terasaki, H. et al., Density measurement of Fe3C liquid using X-ray absorption image up to 10 GPa and effect of light elements on compressibility of liquid iron. *J Geophys Res*, **115**, 1–7 (2010). doi:10.1029/2009JB006905
130. Yu, X. & Secco, R.A., Equation of state of liquid Fe–17 wt%Si to 12 GPa. *High Pressure Res*, **28**, 19–28 (2008). doi:10.1080/08957950701882138
131. Sanloup, C., Fiquet, G., Gregoryanz, E., Morard, G., & Mezouar, M., Effect of Si on liquid Fe compressibility: implications for sound velocity in core materials. *Geophys Res Lett*, **31**, 7 (2004). doi:10.1029/2004GL019526
132. Komabayashi, T., Thermodynamicsof melting relations in the system Fe–FeO at high pressure: implications for oxygen in the Earth's core. *J Geophys Res*, **119**, 4164–4177 (2014). doi:10.1002/2014JB010980

4

Carbon-Bearing Phases throughout Earth's Interior

Evolution through Space and Time

VINCENZO STAGNO, VALERIO CERANTOLA, SONJA AULBACH, SERGEY LOBANOV, CATHERINE A. MCCAMMON, AND MARCO MERLINI

4.1 Introduction

Carbon (C) occurs in the mantle in its elemental state in the form of graphite and diamond, but also as oxidized compounds that include carbonate minerals and carbonated magmas, as reduced components such as methane and carbide, and as gaseous phases in the C–O–H chemical system. The occurrence of C-bearing phases characterized by different oxidation states reflects magmatic processes occurring in Earth's interior that link to its oxygenation through space and time.[1] Improving our understanding of the physical and chemical behavior of carbon at extreme conditions sheds light on the type and depth of possible reactions taking place in the interior of Earth and other planets over time and allows the identification of deep carbon reservoirs and mechanisms that move carbon among different reservoirs from the surface to the atmosphere, thereby affecting the total terrestrial budget of carbon ingassing and outgassing.

Carbon occurs in diverse forms depending on surrounding conditions such as pressure, temperature, oxygen fugacity (fO_2), and the availability of chemical elements that are particularly reactive with carbon to form minerals and fluids. Despite the low abundance of carbon within Earth,[2] the stability of C-rich phases in equilibrium with surrounding minerals provides an important geochemical tracer of redox evolution in Earth and other planets, as well as an important economic resource in the form of diamonds.

Knowledge of carbon cycling through the mantle requires an understanding of the stable forms of carbon-bearing phases and their abundance at pressures, temperatures, and fO_2 values that are representative of Earth's interior. Such information is necessary to identify potential carbon reservoirs and the petrogenetic processes by which carbon may be (re) cycled through the mantle over time, eventually being brought to the surface by magmas and to the atmosphere as dissolved gaseous species.

Accurate estimates of carbon abundance in Earth's interior are challenging for many reasons, such as the unknown primordial budget of carbon,[3] the low solubility of carbon in the dominant silicate minerals of the upper and lower mantle,[4–7] the low modal abundance of accessory carbon-bearing minerals and graphite/diamond in mantle xenoliths, and because magmas occurring at shallow depths are the product of igneous differentiation, magma chamber processes, and degassing. Experimental studies conducted at high pressures and high temperatures help to constrain the initial carbon budget resulting from accretion and core–mantle differentiation processes. These estimates can then be used to determine the amount of carbon in the present-day terrestrial mantle by simulating possible reactions within simplified mantle mineral assemblages that identify the roles of pressure, temperature, and bulk chemistry on carbon speciation.

This chapter addresses fundamental questions about carbon within Earth: At which depths does carbon transform into different phases? How has the redox state of the mantle influenced carbon speciation throughout geological history? Are carbonate minerals more stable than diamonds or CO_2 fluids under certain conditions? Can minerals control the distribution of carbon within the mantle? Answers to these questions reveal how carbon has evolved in the mantle through time and where it may reside in the present-day mantle.

4.2 The Abundance, Speciation, and Extraction of Carbon from the Upper Mantle over Time

Estimates of carbon abundance in Earth's interior can be made through analysis of mantle rocks sampled as peridotitic and eclogitic xenoliths carried to the surface by successively erupted lavas, combined with experimental investigations of the carbon contents of synthetic liquids and minerals at upper- and lower-mantle pressures and temperatures. Carbon abundance in upper-mantle rock-forming minerals such as olivine, orthopyroxene, clinopyroxene, and garnet in equilibrium with carbonated magmas (Na_2CO_3, Ref. 4; olivine tholeiite with ~4 wt.% CO_2, Ref. 5) were shown to be in the range of a few ppm, as also detected in synthetic minerals representative of the transition zone and lower mantle (<1 ppm; Refs. 6 and 7). These findings highlight the extreme geochemical incompatibility of carbon and the role of magmas as its main carriers in space and time.

Experimental investigation of carbon solubility in magmas has revealed that most carbon is generally present as molecular CO_2 and CO_3^{2-} molecules. Their concentrations follow a linear positive trend with pressure and alkali content ($K_2O + Na_2O$) and show a negative correlation with SiO_2 from alkali basalts to rhyolites and the water content of the melt.[8] Among the diverse suites of effusive products, basaltic lavas that have erupted along mid-ocean ridges have dissolved CO_2 concentrations that are inherited from deep mantle source rock and can therefore be used to estimate the mantle abundance of CO_2. However, due to the low solubility of CO_2 in magmas during ascent and decompression, the possibility of tracking its history from the carbonated source rock to the surface is limited to few known undegassed lavas (e.g. Siqueiros and 2πD43 popping rocks). Using the observed correlation of undegassed CO_2 versus trace elements (e.g. Rb, Ba, and Nb) from olivine-hosted melt inclusions, LeVoyer et al.[9] determined a global average CO_2 content of 137 ± 54 ppm, corresponding to an average flux out of mid-ocean ridges (60,864 km length) of 1.8×10^{12} mol·yr^{-1}. This estimate pertains to a depleted mantle source, given that the carbon content of the bulk silicate Earth, generally determined from mantle plume-like magmas, is higher, ranging from 500 to 1000 ppm.[3] In addition, the LeVoyer value is higher than the 72 ppm measured from Siqueiros melt inclusions by Saal et al.,[10] but much lower than the 1300 ± 800 ppm measured in the 1990s by Dixon et al.[11] on alkalic Atlantic lavas. Whether these carbon contents reflect those of the primordial mantle strongly depends on the origin of carbon, its initial budget at the time of Earth's accretion, and the petrological processes capable of mobilizing deep carbon.

Based on recent experimental studies, it was proposed that mantle carbon was inherited at the time of accretion from a Mercury-sized oxidized C-bearing S-rich body, which established the partitioning of carbon between the metallic core and silicate magma ocean during solidification of the terrestrial mantle.[12] An initially delivered carbon content ranging from 700 to 1000 ppm has been proposed for a reduced or S-rich impactor with a mass of 5–30% of Earth in order to match the current carbon concentration in the mantle. The mobilization of this primordial carbon, likely in the form of elemental carbon in Fe(Ni) alloys and CH_4 in the magma ocean as proposed by Li et al.,[12] links to the evolution of the mantle redox state over time. This issue remains controversial, however, and there is still an open debate on the different C species (CH_4, CO_2, and CO) that can be transported by mantle magmas (Ref. 8 and references therein).

Late addition of oxidized meteoritic bodies and enrichment of the post-core formation mantle in Fe_2O_3 consequent to the loss of Fe to the metallic core have been proposed, among a plethora of other processes, to explain the increase of fO_2 from conditions where C would be stable as diamond and in Fe(Ni) alloys ($fO_2 \approx$ iron–wüstite buffer (IW) – 2; Refs. 13 and 14) to conditions where C^{4+} is the dominant species incorporated in carbonate minerals and melts ($IW + 1 < fO_2 < IW + 5$). Geochemical tracers such as Ce in zircons believed to have crystallized in equilibrium with Hadaean magmas[15] and the V/Sc ratio of erupted continental basalts[16] have been used to argue in favor of a constant mantle redox state over the last 4.4 billion years at values where C^{4+} is stable, and therefore able to mobilize carbon as CO_2 and/or CO_3^{2-} dissolved in magmas. In contrast, based on the reconstruction of the mantle redox state over time using the V/Sc ratio of Archean spreading ridge-derived metabasalts, Aulbach and Stagno[17] found that the ambient mantle became gradually more oxidized by ~1.5 log units between ca. 3 Ga and 1.9 Ga, when fO_2 values similar to modern mid-ocean ridge basalts (MORBs) are recorded (Figure 4.1a). Figure 4.1b illustrates how a reduced asthenospheric Archaean mantle expected to host C in the form of diamond (or graphite) and CH_4 becomes more oxidized along a decompression path (Figure 4.1b, red line) as a result of the pressure effect on the equilibrium:[18]

$$4\,Fe^{3+}{}_{(bulk)} + C^0 = 4\,Fe^{2+}{}_{(bulk)} + C^{4+}{}_{(vap/melt/solid)}, \tag{4.1}$$

until the local fO_2 at which carbon turns into carbonate[19] is reached (fayalite–magnetite–quartz buffer (FMQ) – 2 log units; Figure 4.1b, blue line) at ~100 km depth. In Eq. (4.1), Fe^{3+} and Fe^{2+} refer to the ferric and ferrous iron of the bulk peridotite mantle rock, respectively. A similar equilibrium could be written to explain the distribution and fate of other important magmatic volatiles like H_2 and H^+, N^{3-} and N_2, and S^{2-} and S^{6+} in igneous systems dominated by Fe-bearing silicates, with dramatic implications for volcanic outgassing and atmospheric composition over time.

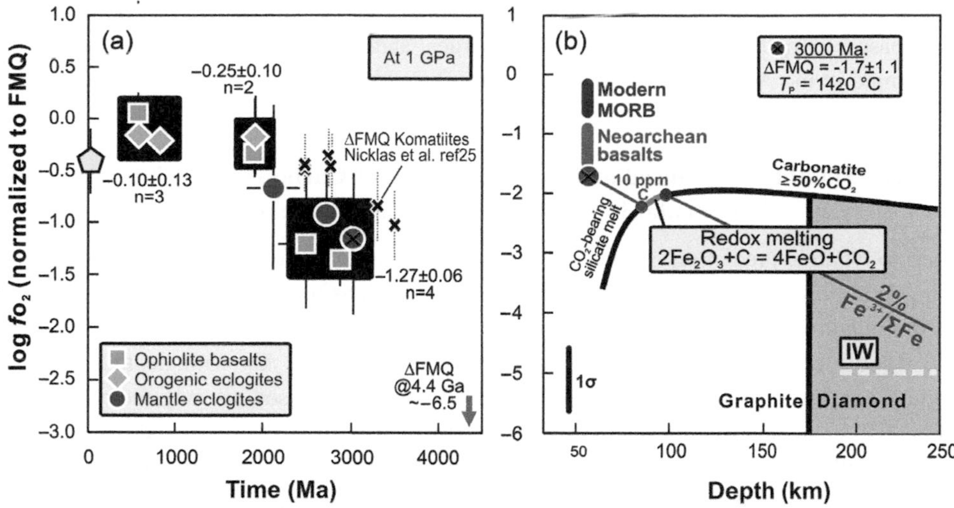

Figure 4.1 (a) fO_2 as a function of time calculated from V/Sc of mid-ocean ridge-derived basalts and metabasalts occurring in ophiolites, as orogenic eclogites, or mantle eclogites (data in Aulbach and Stagno[17]). fO_2 values were projected to 1 GPa to avoid bias due to deeper onset of partial melting in warmer ancient convecting mantle. Shown for comparison are results for komatiites from Nicklas et al.[25] (b) Thermodynamic prediction of the variation in the mantle oxidation state upon decompression, obtained using oxy-thermobarometry for garnet-bearing peridotite.[18] The graph shows the depth at which melting would occur as soon as 10 ppm C, expected in the Archean convecting mantle and hosted in diamonds, is oxidized to carbonate by the redox reaction shown. A source mantle with 2% Fe_2O_3 would generate a melt at the local temperature corresponding to the solidus of a carbonated peridotite that would have the fO_2 calculated for the igneous protolith of the Lace eclogite suite shown in (a) (after Aulbach and Stagno[17]).

At these depths, carbonate–silicate melts are likely to have formed by redox melting of graphite-bearing mantle at the near-constant fO_2 favored by hotter temperatures during the Archaean, until all carbon (~10 ppm in the case of reduced early Earth[20]) was oxidized. Upon further decompression, the Archaean mantle would have continued to become more oxidized until the oxidation state of Archean MORBs, represented by a metabasaltic eclogite suite from Lace in the Kaapvaal craton,[21] is reached below the spreading ridge. Given that redox melting – today as well as in the early mantle – requires the reduction of Fe_2O_3 (i.e. Eq. 4.1), the oxidation state of MORBs does not directly reflect the mantle oxidation state at the source, a concept introduced by Ballhaus and Frost[22] in 1994 when they proposed the oxidation of MORBs from a more reduced mantle source due to the coupling between C/carbonate reactions and Fe^{3+}/Fe^{2+} variations upon decompression as represented by Eq. (4.1). This was demonstrated in recent experiments where measured fO_2 values of synthetic MORBs appeared different from the fO_2 of the coexisting spinel-bearing peridotite rock source by up to ~3 log units.[23] Nevertheless, a gradual increase of Fe^{3+} in the bulk asthenospheric mantle over time has been corroborated by the finding of extremely low $Fe^{3+}/\sum Fe$ in mantle eclogites, which is not correlated with indices of subsequent melt depletion,[24] as well as V-based estimates of fO_2 in komatiites, picrites, and basalts of various ages showing a clear trend in the Archaean that levels out at ca. 2 Ga.[25]

4.3 The Stability of Reduced and Oxidized Forms of Carbon in the Upper Mantle: Continental Lithosphere versus Convective Mantle

Thermodynamic prediction of the stable form of carbon in the upper mantle can be made using oxy-thermobarometry applied to mantle rocks such as peridotites and eclogites based on the relation between the ferric/ferrous iron ratio of minerals such as spinel and garnet and the buffering capacity of the host rocks. The decreasing fO_2 with depth shown in Figure 4.2 links to the volume change of the reactions used as oxy-thermobarometers[18,26,27] and suggests that the Archaean continental lithospheric mantle down to ~270 km in depth might have equilibrated with elemental carbon in the form of diamond or graphite, plus a small volume of CO_2-bearing melts varying from carbonatitic (Figure 4.2, blue line) to kimberlitic (Figure 4.2, red region) in composition. The fO_2 at which diamond and CO_2-bearing melts can coexist with peridotite and eclogite depends on the effects that temperature (the local geotherm) and the presence of water have on the equilibria:[19,26]

$$\underset{\text{enstatite}}{2\,Mg_2Si_2O_6} + \underset{\text{dolomite/melt}}{CaMg(CO_3)_2} = \underset{\text{diopside}}{CaMgSi_2O_6} + \underset{\text{olivine}}{2\,Mg_2SiO_4} + \underset{\text{diamond}}{2\,C} + 2\,O_2 \tag{4.2}$$

and

$$\underset{\text{dolomite/melt}}{CaMg(CO_3)_2} + \underset{\text{coesite}}{2\,SiO_2} = \underset{\text{diopside}}{CaMgSi_2O_6} + \underset{\text{diamond}}{2\,C} + 2\,O_2. \tag{4.3}$$

A direct sampling of deep carbon phases is represented by diamantiferous peridotite and eclogite of Archaean age (1–3% diamonds in the cratonic lithosphere),[28] while the observation of carbonated magmas of mantle origin, although predicted by experimental phase equilibria studies on synthetic carbonated peridotites and eclogites, appears quite rare (about 490 deposits)[29] due to a decreasing trend over time coupled to secular mantle cooling and late-stage fractionation and assimilation processes[30,31] that make these products difficult to identify. The decrease of fO_2 with depth in the case of eclogite is expected to result in carbonate minerals (or fluids) being reduced to diamond during subduction of a carbonated oceanic crust at variable depths through oxidation of coexisting garnet and clinopyroxene mineral phases[26] depending on the Fe^{3+} of the bulk rock.[26,32]

The Fe(Ni) precipitation curve (Figure 4.2, orange line) marks the fO_2 conditions at which diamonds and Fe(Ni) alloys might coexist, therefore promoting the diffusion of carbon into adjacent metal or the formation of Fe(Ni)C intermetallic compounds depending on the local Fe(Ni)/C ratio.[33,34]

The extent to which the redox state of the continental mantle resembles that of the convective mantle is highly debated. The model proposed in Figure 4.1b is based on

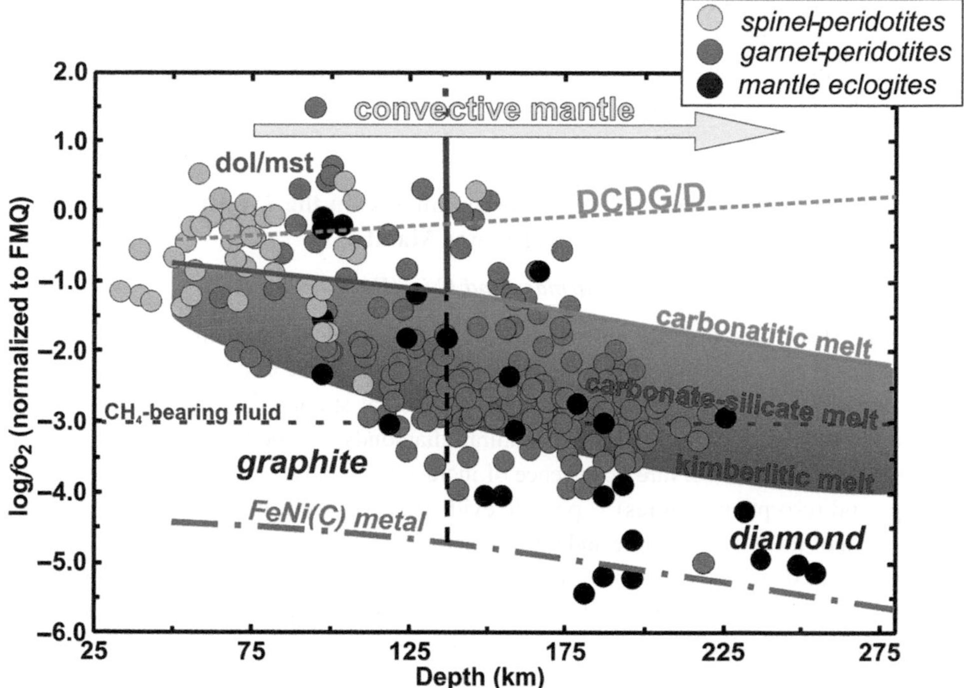

Figure 4.2 Log fO_2 (normalized to the FMQ buffer) determined for peridotitic and eclogitic xenoliths using oxy-thermobarometry for spinel/garnet peridotite and eclogite.[18,26,27] The blue curve is the fO_2 calculated for Eq. (4.2) along a cratonic geotherm of 40 mW·m^{-2} that defines the stability field between diamond (or graphite) and solid (liquid) carbonate within peridotite rocks. The orange curve indicates where Fe–Ni metallic alloys (with some C) are stable. The green line indicates the fO_2 buffered by Eq. (4.3) (see text).[26,35] CO_2-bearing silicate melts are stable at lower fO_2 values than carbonatitic melts in equilibrium with diamantiferous peridotite due to the temperature effect on Eq. (4.1).[18] The yellow arrow indicates the proposed oxidation state of a convective mantle contaminated by variable volumes of subducted carbonated lithologies.[37] DCDD/G = dolomite–coesite–diopside–(diamond/graphite).

conservative estimates in which a bulk silicate Earth composition[36] is used as an analogue for the asthenospheric mantle. For this case, the convective mantle is predicted to follow an fO_2 trend similar to the one in Figure 4.2 for natural mantle peridotites along a decompression path. This model might apply to a convective early Earth mantle not equilibrated with subducted lithologies. However, correlations between geochemical tracers such as CO_2 versus Nb and Ba measured in modern natural oceanic basalts suggest a different trend. These correlations were used to constrain mixing models of two components: a residual mantle peridotitic partial melt plus graphite-saturated partial melts of subducted lithologies (e.g. pyroxenite, MORB-eclogites, sediments) in different proportions. These models, supported by melting experiments, show that generation of CO_2-rich basalts of up to ~6 wt.% in the melt like those from Atlantic popping rocks[37] requires the contribution of subduction-related melts originating from partial melting occurring in the carbonate stability field ($fO_2 \geq$ FMQ − 1; DCDD/G in Figure 4.2) rather than graphite/diamond-saturated sources. This results in a contaminated convective present-day mantle that is more oxidized than the Archaean continental lithospheric mantle (Figure 4.2).

4.4 The Redox State and Speciation of C in the Transition Zone and Lower Mantle

4.4.1 Carbides and C in (Fe,Ni) Alloys

In contrast to the top of the upper mantle, gaining knowledge of the redox state of the deeper upper mantle (300 km and below) is more challenging due to the lack of rocks that can be sampled at the surface. Previous experimental studies,[38,39] recently supported by observations of 53 sublithospheric natural diamonds (termed CLIPPIR due to their Cullinan-like large size, rare occurrence of inclusions, relatively pure nature, and irregular shape and resorption features),[40] provide evidence of C-bearing Fe metal occurring at the pressures of the transition zone and lower mantle. This observation would limit the stability of carbon to diamond and carbide phases, which is also supported by the discovery of Fe (Ni)/C alloys trapped in diamonds.[40–43] At the base of the upper mantle, the fO_2 at which Fe-carbides (e.g. Fe_3C and Fe_7C_3) would be stable in equilibrium with mantle silicates requires either extremely reducing conditions (~7 log units below the FeNi precipitation curve; orange line in Figure 4.2) or coexistence with olivine and orthopyroxene that is more enriched in ferrous iron[44] than would be expected for peridotite. In addition, the Fe–Ni–C solidus temperature ranges between 1150 and 1250°C at 10 GPa,[34] which is low compared to generally agreed cratonic geotherms. It is likely, therefore, that Fe–Ni–C alloys are either: (1) stable as molten phases;[34] (2) limited in their formation to the grain scale as a result of interaction with surrounding C-saturated reduced fluids;[44] or (3) result from solid–solid reactions between preexisting Fe–Ni alloys and subducted carbon (carbonate or diamond) of particularly cold slabs.[45] In the deep mantle, carbides can be stable in the form of molten Fe_7C_3 along with coexisting diamonds[33] if C is locally more abundant (>~100 ppm) than Fe. Interestingly, the presence of 100–200 ppm of sulfur in the Fe–Ni–C system at high pressures and temperatures was shown to prevent the formation of carbides and promote precipitation of diamonds even in particularly C-poor (5–30 ppm) regions of the deep mantle,[46] thus explaining the occurrence of Fe–Ni–S phases trapped in diamonds.[40]

4.4.2 Carbonate Minerals in Earth's Interior

In contrast to the view that bulk mantle redox conditions are reduced at transition-zone and lower-mantle depths, inclusions hosted in super-deep (sublithospheric) diamonds have also shown the presence of solid carbonates.[47,48] These observations raise important questions: (1) Do these inclusions reflect local fO_2 heterogeneities of the deep mantle? (2) What transformations do solid carbonates experience to preserve their stability at high pressures and temperatures to act as main carbon hosts? The latter question links to the possibility that the fO_2 of the transition zone and lower mantle could be buffered by an influx of oxidized carbon in the form of solid carbonate, but questions remain as to which carbonates these might be. The speciation of carbon at conditions from the upper to the lower mantle, whether in the form of diamond or carbonate (either liquid or solid), has been a focus of investigations over the last decades. All experimental studies and theoretical calculations suggest that solid carbonates (i.e. the end members $CaCO_3$, $MgCO_3$, $FeCO_3$, and their solid solutions) are stabilized at conditions of Earth's lower mantle through a series of high-pressure phase transitions that are summarized below. This conclusion is also supported by the observations of extremely low solubility of carbon in deep mantle silicates such as wadsleyite, ringwoodite, bridgmanite, and ferropericlase in experimental studies conducted at high pressures and high temperatures.[6,7]

4.4.2.1 Dolomite and Its High-Pressure Polymorphs

It is generally accepted that dolomite, $CaMg(CO_3)_2$, is the dominant carbonate mineral up to about 4 GPa within simplified $CaO–MgO–Al_2O_3–SiO_2 + CO_2$ igneous systems, and is then replaced in dominance by magnesite at higher pressures.[49] However, both the bulk Ca/(Ca + Mg) ratio of the rock and the presence of a small amount of Fe in the structure that reduces the cation-sized misfit between Ca and Mg may play important roles in stabilizing high-pressure polymorphs at the conditions of Earth's lower mantle. Upon compression at ambient temperature, the rhombohedral dolomite structure is stable up to 17 GPa.[50] Above this pressure, a transition to dolomite-II with triclinic symmetry is observed with no observable volume change. Dolomite-II is stable upon heating to 1500 K;[51] however, it decomposes between 2000 and 2400 K below 35 GPa to form a mixture of oxides and diamond.[52] Dolomite-II is stable up to 35 GPa, whereupon a second phase transition is observed to triclinic dolomite-III.[51,52] Heating dolomite-III above 35 GPa to its melting point without evidence of decomposition demonstrates that dolomite-III can be a stable C-bearing mineral at Earth's mantle conditions. Theoretical computations indicate that dolomite transforms into a ring-carbonate structure at 115 GPa and 2500°C that features tetrahedrally coordinated carbon and threefold carbonate rings,[53] as shown in Figure 4.3.

In a recent experimental study conducted using a laser-heated diamond anvil cell, Dorfman and coauthors[54] explored chemical reactions occurring at the boundary between dolomite and Fe metal to simulate subducted carbonates penetrating the reduced lower mantle saturated in Fe metal (about 1 wt.%[38]) at ~51–113 GPa and 1800–2400 K. Their results suggest that the $MgCO_3$ component of subducted crustal dolomite would react with iron metal to form a mixture of diamond + Fe_7C_3 + ferropericlase, while $CaCO_3$ with a post-aragonite structure would be preserved. This study supports previous experimental observations on the formation of diamond + carbide by reduction of subsolidus carbonates,[45] and it demonstrates the potential role of dolomite as a deep carbon reservoir. So far,

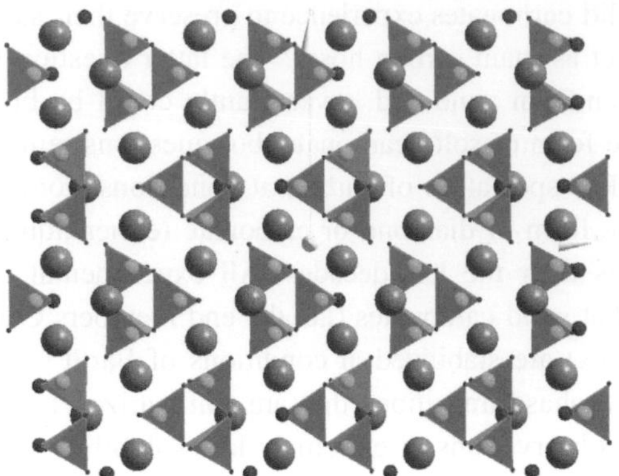

Figure 4.3 Crystal structure of dolomite-IV based on experimental single-crystal X-ray data collected in a diamond anvil cell at 115 GPa after annealing at 2500°C. The structure is based on threefold-ring carbonate units, with carbon in tetrahedral coordination.

however, its transport to the lower mantle appears limited to exotic Ca-rich, Si-poor thick portions of the oceanic slab, likely related to the subduction of sedimentary lithologies.[55]

4.4.2.2 Deep Carbon Stored as $CaCO_3$-like Phases

Calcite (the rhombohedral form of $CaCO_3$) and aragonite (orthorhombic $CaCO_3$) undergo a series of structural transitions toward denser phases with increasing pressure. Above 15 GPa, a new calcite polymorph, $CaCO_3$-IV, was discovered that shows stability at least up to 40 GPa.[56] The density of $CaCO_3$-IV is higher than aragonite (3.78 g/cm^3 at 30.4 GPa), which suggests that it might replace aragonite at intermediate mantle depths.[56] At pressures of more than ~40 GPa, $CaCO_3$ undergoes yet another phase transition into a so-called post-aragonite phase.[57] Compression to pressures of more than ~100 GPa leads to a different polymorph with a pyroxene-like structure based on CO_4 groups with sp^3-hybridized carbon.[57–60] Chemical bonding in sp^3-carbonates is different from bonding in carbonates based on sp^2-hybridized carbon; thus, sp^2–sp^3 crossover, if it occurs in mantle carbonates, may disrupt the silicate–carbonate chemical equilibrium. To date, however, the geochemical and geophysical consequences of sp^2–sp^3 transitions are not well understood. It is noteworthy, however, that calcite was found as an inclusion in a sublithospheric diamond,[47] confirming its stability in the lower mantle as a potential carbon reservoir and candidate for potential redox reactions through which diamond itself could form.

4.4.2.3 Magnesite and Fe-Bearing Solid Solutions as Deep Carbon Reservoirs

Magnesite and ferro-magnesite solid solutions have been found to occur as inclusions in diamonds that likely originated in Earth's lower mantle.[48] Despite the lack of evidence of siderite inclusions hosted in such super-deep diamonds, the role of iron in the stability of Fe-carbonates appears relevant in the mid-lower mantle, where spin crossover of Fe^{2+} from high spin (HS) to low spin (LS) changes the physical and chemical properties of these carbonates,[61–63] with important implications for their seismic detectability (discussed below).

Magnesite remains stable at up to 100 GPa and 2200 K without decomposition or associated phase transformations.[64] Congruent melting of magnesite was reported between 2100 and 2650 K at pressures between 12 and 84 GPa, whereas decomposition to MgO and diamond was observed only above ~3000 K at similar pressures.[65] Isshiki et al.[64] first reported the transformation of magnesite to its high-pressure polymorph, magnesite-II, at ~115 GPa and ~2200 K. Maeda et al.[66] discussed phase relations in the $MgCO_3$–SiO_2 system up to 152 GPa and 3100 K. They reported a reaction between $MgCO_3$ phase II and SiO_2 ($CaCl_2$-type SiO_2 or seifertite) to form diamond and $MgSiO_3$ (bridgmanite or post-perovskite) at deep lower-mantle conditions via the following possible reaction:

$$SiO_2 + MgCO_3 \rightarrow MgSiO_3 + C + O_2, \tag{4.4}$$

indicating that CO_2 dissociates above ~33 GPa and at >1700–1800 K to form diamond plus free oxygen and bridgmanite. These results describe a possible mechanism of formation of super-deep diamond in cold slabs descending into the deep lower mantle resulting from the instability of magnesite phase II in the presence of exotic assemblages.

The physical and chemical properties of Fe-bearing carbonates, such as their density, Fe content, elasticity, and optical properties, may have important geological consequences. Melting of $FeCO_3$ was investigated experimentally at high pressures and temperatures equal to or above the terrestrial geotherm. Experiments indicate that iron carbonates melt incongruently up to ~70 GPa and at >2000 K to produce a minor quenched Fe^{3+}-rich phase and CO_2 in the liquid.[67,68] At higher pressures and similar temperatures, new carbonates were observed with tetrahedral C_3O_9 rings.[69] Merlini et al.[70] reported the transformation of Mg-siderite at pressures and temperatures corresponding to the top of the D′′ layer (i.e. ~135 GPa and ~2650 K), with formation of $Mg_2Fe^{(III)}{}_2(C_4O_{13})$ and a new oxide phase, $Fe_{13}O_{19}$, thus demonstrating that self-oxidation–reduction reactions can preserve carbonates in Earth's lower mantle to 2500 km in depth.

When slabs penetrate the lower mantle, redox reactions take place between the hotter outer part of the slab and the surrounding bulk mantle in order to balance the high Fe^{3+} content of bridgmanite (~80% in equilibrium with magnesite and diamonds),[71] similar to Eq. (4.1). The origin of diamonds in the lower mantle might be therefore linked to redox reactions involving Fe^{3+}-bearing minerals, which could also account for the high proportion of ferropericlase in lower-mantle diamond inclusions.[72] At greater depths, the high degree of trivalent iron incorporation into high-pressure carbonate structures and iron oxides can influence redox processes in the deep mantle, where changes in fO_2 could be buffered by the activity of carbon (Eq. 4.5) or oxygen (Eq. 4.6) rather than being controlled by charge balance between iron cations.

$$4\,FeCO_3 = Fe_4C_3O_{12} + C, \tag{4.5}$$

$$8\,Fe_4C_3O_{12} = 6\,Fe_4C_4O_{13} + 4\,Fe_2O_3 + 3\,O_2. \tag{4.6}$$

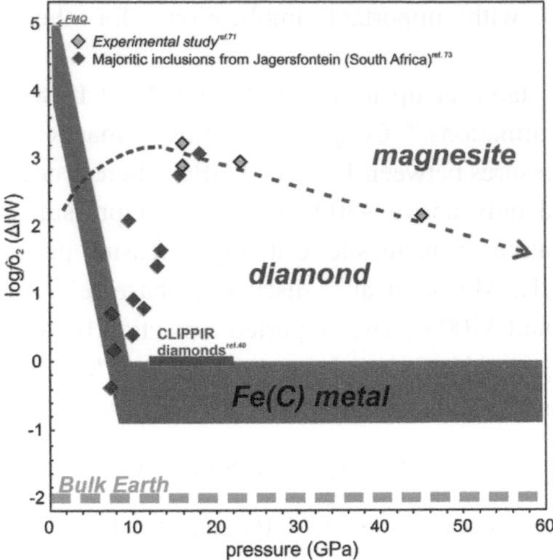

Figure 4.4 fO_2 stability of carbon phases at high pressures based on experimental studies,[71] as well as majorite inclusions in diamonds[73] and metal-bearing "CLIPPIR" diamonds.[40] The blue dashed line indicates the fO_2 values at which diamonds and carbonate (magnesite) are in equilibrium.[71]

4.4.3 Toward Oxy-Thermobarometry of the Deep Mantle and Implications for Carbon Speciation

Through the number of redox reactions that can be written to explore the mechanisms of diamond formation in the transition zone and lower mantle, a fundamental question arises: Is the deep mantle intrinsically buffered by abundant Fe-bearing minerals similarly to the upper mantle? A recent estimate of the deep-mantle redox state was proposed by Kiseeva et al.,[73] who measured the ferric iron content of majorite inclusions and developed an oxy-thermobarometer based on the self-redox capacity of majorite as a function of pressure. Figure 4.4 shows the fO_2 values calculated using Eq. (4.7):

$$\begin{aligned} &2\,Ca_3Al_2Si_3O_{12} + 4/3\,Fe_3Al_2Si_3O_{12} + 2.5\,Mg_4Si_4O_{12} + O_2 \\ &= 2\,Ca_3Fe_2Si_3O_{12} + 10/3\,Mg_3Al_2Si_3O_{12} + 4\,SiO_2, \end{aligned} \tag{4.7}$$

from majoritic inclusions in super-deep diamonds from the Jagersfontein kimberlite (South Africa).[74] The carbon/carbonate equilibrium buffer[71] refers to experimental measurements of the fO_2 at which diamond and magnesite coexist with transition-zone (wadsleyite/ringwoodite and clino-enstatite) and lower-mantle (bridgmanite and ferropericlase) mineral phases.[71] The results indicate the heterogeneity of mantle fO_2 in the transition zone, varying from about IW to IW + 3, where the high concentration of Fe^{3+} in majorite (between 8% and 30%) is linearly correlated with pressure and likely related to the oxidizing effect of CO_2-rich fluids rather than FeO disproportionation. This suggests that carbonate is, at least locally, a stable phase in Earth's deep mantle, stabilized at an fO_2 value about 2 log units above the iron–wüstite buffer. Carbonate could therefore participate in melting processes depending on the local geotherm and promote carbon mobility through the formation of carbonatitic to kimberlitic melts. At higher pressures, ferropericlase and bridgmanite are expected to play major roles in buffering redox conditions through chemical reactions that involve the Fe^{3+}-bearing mineral end members. However, the crystal chemistry of these phases has been shown to be extremely complex due to simultaneous element substitution in different sites.[75]

4.5 Seismic Detectability of Reduced and Oxidized Carbon in Earth's Mantle

The presence of carbon within the mantle may be detectable using seismic data. A recent multidisciplinary study conducted by Garber and coauthors[28] showed that diamondiferous lithologies may be responsible for elevated shear-wave velocities (V_S; ≥4.7 km/s) detected at 120–150-km depths in cratonic lithospheric mantle. Diamond and eclogite (known to contain higher concentrations of diamond compared to peridotite)[35] are the most likely high-V_S candidates that could explain the observed seismic anomalies, implying the presence of up to about 3 vol.% diamond, depending on the peridotite and eclogite compositions and the local geotherm.

The seismic detectability of carbonates in the upper mantle and transition zone is related primarily to their crystal structure. Sanchez-Valle et al.[76] measured the elastic tensor of $Mg_{1-x}Fe_xCO_3$ with four compositions extending from pure magnesite to pure siderite using single-crystal Brillouin scattering spectroscopy at ambient conditions. They found that Fe substitution has a negligible effect on the adiabatic bulk modulus, K_S, of Mg–Fe carbonates, whereas the shear modulus, μ, decreases by 34% from $MgCO_3$ to $FeCO_3$. They concluded that, based on current seismic resolution (with a threshold of 2%), detection of carbonated regions from seismic velocity contrast would require high carbonate contents (~15 wt.% CO_2) in eclogite and >20 wt.% CO_2 in peridotite. These results are supported by the work of Yang et al.,[77] who measured the full elastic stiffness tensor of $MgCO_3$ up to 14 GPa at ambient temperature and up to 750 K at ambient pressure using Brillouin scattering in the diamond anvil cell.

In Fe-bearing carbonates, the pressure-induced splitting of iron 3d energy levels into two leads to the possibility of two different spin configurations for Fe^{2+}: (1) the HS state with four unpaired and two paired electrons; and (2) the LS state with six paired electrons. The spin pairing of electrons causes a volume collapse of the iron atomic volume, followed by shrinking of the octahedral site and ultimately of the unit cell. The first experimental observation of the HS–LS transition of Fe^{2+} in siderite was reported by Mattila et al.[78] using X-ray emission spectroscopy. They observed the transition to occur at ~50 GPa and at ambient temperature in natural siderite powder. Subsequent experimental and theoretical work provided further evidence of the spin transition in Fe-carbonates at ~45 GPa, as well as insights into how the spin transition affects the physical and chemical properties of carbonates.[79–86]

The spin transition causes changes in the thermoelastic properties of carbonates and may enhance the seismic detectability of carbon in the lower mantle. Liu et al.[87] observed anomalous thermoelastic behavior in natural magnesiosiderite at high pressure and high temperature across the spin transition, including a dramatic increase in the thermal expansion coefficient and decreases in the isothermal bulk modulus and the bulk sound velocity by 75% and 50%, respectively. When compared to $MgCO_3$ at relevant pressure–temperature conditions of subducted slabs, HS magnesiosiderite with 65 mol.% $FeCO_3$ is approximately 21–23% denser and its unit-cell volume is 2–4% larger, whereas the LS state is 28–29% denser and 2% smaller than end-member magnesite. These results indicate that dense LS ferromagnesite can become more stable than HS ferromagnesite at pressures above ~50 GPa, providing a mechanism for (Mg,Fe)-bearing carbonate to be a major carbon host in the deeper lower mantle. Fu et al.[88] measured the full elastic stiffness tensor of $Mg_{0.35}Fe_{0.65}CO_3$ up to 70 GPa at ambient temperature using Brillouin light scattering and impulsive stimulated light scattering in a diamond anvil cell. They observed a dramatic softening of the C_{11}, C_{33}, C_{12}, and C_{13} moduli and stiffening of the C_{44} and C_{14} moduli across the spin transition in the mixed spin state. Outside the region of the spin transition, they observed a linear increase of all elastic moduli with pressure. Based on their work, mixed spin-state ferromagnesite is expected to exhibit abnormal elasticity in the mid-lower mantle, including a negative Poisson's ratio and a drastically reduced compressional wave velocity (V_P). Similar results were obtained by Stekiel et al.,[89] who determined the elastic stiffness moduli of $FeCO_3$ across the spin transition up to 60 GPa and at ambient temperature by inelastic X-ray scattering and density functional theory calculations. Based on calculations employing a pyrolitic mantle model and considering varying amounts of $Mg_{1-x}Fe_xCO_3$, the presence of magnesioferrite solid solutions changed the V_S more than the V_P. At least 3 vol.% of $FeCO_3$ would be needed to produce a shear velocity contrast of more than 1% due to the spin transition, and 8 vol.% of $Mg_{0.85}Fe_{0.15}CO_3$ would be needed to produce the same contrast between carbonated and non-carbonated pyrolitic mantle.

Our knowledge of the iron spin state of $FeCO_3$ is limited to near-room temperature conditions, as probing different electronic structures at combined high pressure–temperature conditions is challenging. Temperature broadens the pressure range of the HS to LS transition, but the depth range over which spin transitions take place remains uncertain (Figure 4.5). In addition, siderite is unstable at greater than ~1500 K (greater than ~40 GPa) and decomposes into iron oxides and C-bearing phases. As a result, $FeCO_3$ is unlikely to occur in the lower mantle at depths greater than ~1200 km. Instead, solid solutions involving iron oxides and sp^3-bonded novel Fe-carbonates ($Fe_4C_4O_{13}$ and $Fe_4C_3O_{12}$) may be the most common carbon-bearing phases at sufficiently oxidizing conditions.

4.6 Conclusion

In contrast to the immobile behavior of carbon in the Archaean lithospheric mantle, the deep carbon cycle in modern Earth linked with carbon transport through the convective mantle at varying pressure–temperature–fO_2 conditions over time shows evidence of large

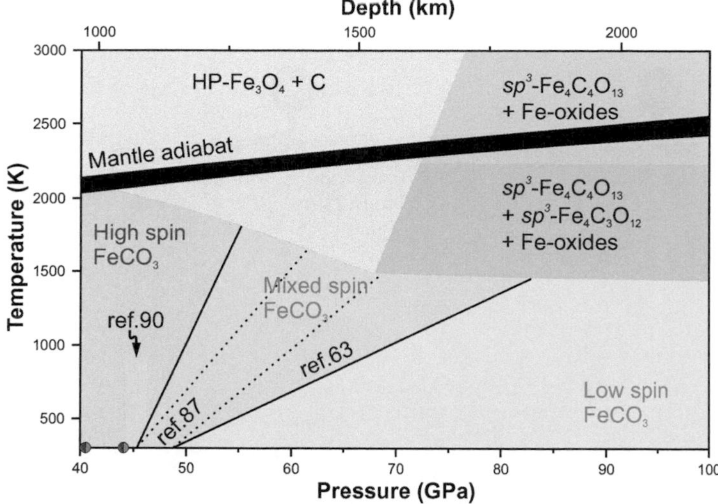

Figure 4.5 Stability and spin-state diagram of $FeCO_3$ at high pressure–temperature. Mixed blue–red circles depict the spin transition pressure in siderite at 300 K as observed in a soft pressure medium (e.g. Ne) at 40.5 GPa and in more rigid media (KCl, Ar, etc.) at ~44 GPa. Pink and blue regions are high and low spin-state regions, respectively. The pressure–temperature conditions of mixed-spin iron (0.1–0.9 LS) in $FeCO_3$ are uncertain (shown by different green-shaded areas). High-temperature transformations in the $FeCO_3$ system are shown after Cerantola et al.[67]

heterogeneities in mantle redox state, both vertically and laterally, which likely result from the contribution of subducted lithologies with varying degrees of oxidation through space and time. We conclude that the movement of carbon over geologic time is fundamentally controlled by the level of oxygenation of Earth's interior, which causes elemental carbon to oxidize to CO_2-bearing magmas, whose rheology mainly controls their transport properties.[91,92] The observation of both oxidized carbon and metal trapped as inclusions in sublithospheric diamonds testifies to the heterogeneity of mantle fO_2, thereby implying the stability of carbon in various redox states and phases (mineral, melt, or fluid), and even favored by the presence of additional elements like sulfur. The relation between iron oxidation state in mineral inclusions in diamonds and fO_2 is an important tool for determining the local redox conditions needed to model carbon speciation. Fluids in equilibrium with reduced mantle regions would be methane dominated and may precipitate diamonds by reduction of Fe, liberating water that could induce partial melting. On the other hand, the possible presence of carbonate as a source of carbon in natural diamond-forming processes is supported by experimental studies at high pressure and temperature that demonstrate the stability of diverse geologically relevant carbonates (and their solid solutions) such as magnesite, calcite/aragonite, and siderite down to the lower mantle. Redox reactions of these carbonates with Fe-bearing minerals in the bulk mantle under initially reducing conditions have been demonstrated to promote the formation of diamond as well as the crystallization of carbide phases, leading to mantle oxidation. Such

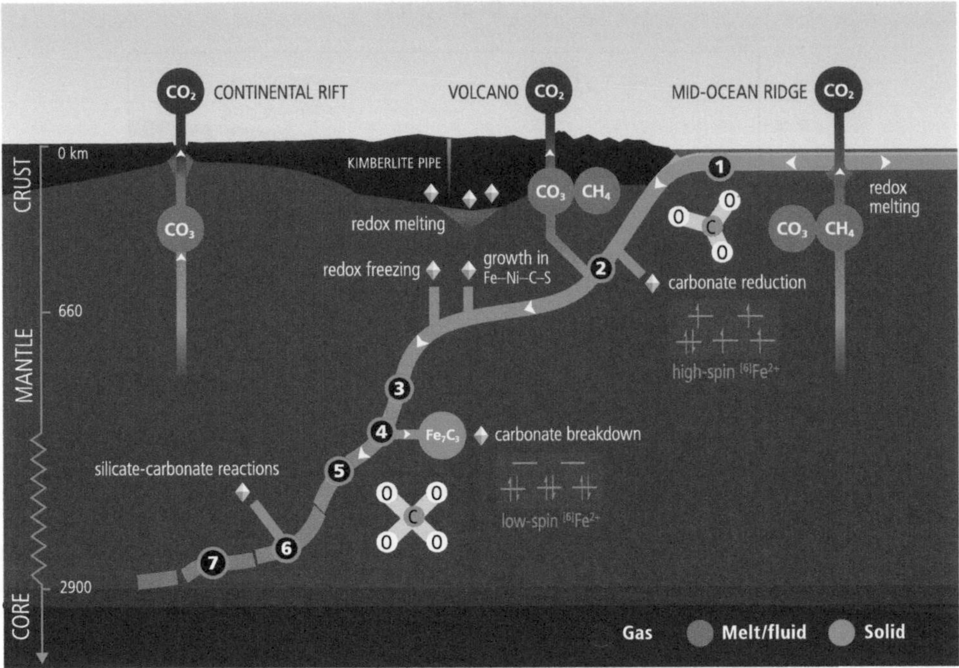

Figure 4.6 Simplified cartoon showing the distribution and forms of carbon inside Earth. Carbon has low solubility in mantle minerals;[4–7] hence, it occurs primarily in the form of gas (red circles), fluids or melts (orange circles), and accessory solid phases (green circles), including diamonds (octahedra). Reactions involving carbon include redox melting (diamond consuming) and freezing (diamond producing),[17,18] reduction of carbonate to diamond,[26,32] growth of diamond from metallic liquids,[33,34,40,46] breakdown of carbonate under reduced conditions,[45,54,66] redox reactions that produce tetracarbonates,[67,69] and reactions between carbonate and silicate.[66,71] Numbers indicate the depth at which important phase transitions in carbonates have been proposed to occur: (1) calcite to aragonite,[49] (2) dolomite to dolomite-II,[50,52] (3) aragonite to post-aragonite,[57] (4) dolomite-II to dolomite-III,[51,52] (5) Fe-carbonate to Fe-tetracarbonate,[67,69] (6) post-aragonite to $CaCO_3$ with sp^3-hybridized carbon,[57–60] and (7) magnesite to magnesite-II.[64] Ferrous iron in carbonate undergoes a HS to LS transition in the mid-mantle.[78–90] The thickness of the crust and slab is vertically exaggerated for clarity.

"conditioned" mantle regions may eventually allow carbonated melts and fluids to be mobilized along ascending (plumes) or descending (slabs) redox fronts, facilitating melt-mediated carbon transfer between different terrestrial reservoirs. Finally, elemental carbon transported upwards in upwelling mantle regions will be subjected to redox melting upon oxidation due to decompression. The resulting carbonated melts eventually erupt at the surface, where they interact with the hydrosphere and atmosphere. Figure 4.6 summarizes the stability of different carbon species through space and (indirectly) time with a particular focus on the asthenospheric mantle, the continental lithosphere, and subducted slabs.

4.7 Limits to Knowledge and Unknowns

Although great advances have been made, important limits to knowledge remain regarding the stability and mobility of carbon-bearing phases and their interaction with silicates and metals at pressure–temperature–fO_2 conditions relevant to the deep Earth. Redox melting and freezing reactions, as a function of fO_2, are critical in mediating transitions between the relatively refractory forms of carbon, such as diamond/graphite, carbide, and metal alloys, to highly mobile oxidized forms, such as carbonated melts and CO_2-bearing fluids.[18,93,94] However, the amount of carbon involved in these reactions, the participating minerals, and the fO_2 in the mantle with depth and through time remain highly uncertain. This uncertainty arises because most of our knowledge relies on indirect estimates based on mostly degassed magmas and direct estimates from mantle xenoliths and diamonds entrained in magmas that are irregularly distributed in space and time. Moreover, mantle xenoliths and diamonds are exclusively or dominantly derived from the lithospheric mantle and may sample multiply reactivated fluid and melt pathways that are not representative of the lithospheric mantle and even less so of the convective mantle.[95] The few sublithospheric diamonds available for study sample anomalous mantle domains characterized by strong compositional and redox gradients at the interface between deeply subducted slabs and ambient mantle.[93,96] Thus, although it is assumed that the fO_2 of the metal-saturated mantle below ~250–300 km in depth is effectively buffered close to the iron–wüstite redox buffer (where carbonates and carbonated melts are unstable[94]), sublithospheric diamonds appear to record much higher fO_2 values, linked to the influx of slab-derived oxidizing agents.[73] In addition, the CO_2 content of oceanic basalts has been suggested to require carbonated redox melting rather than graphite-saturated melting, implying higher fO_2 values than are inferred from lithospheric mantle xenoliths.[37] The problem of constraining the dominant fO_2 and redox equilibria with depth is only exacerbated in deep time, where debate over appropriate sample material and redox proxies continues.[15–17,98]

Progress on deciphering the speciation and distribution of carbon in modern Earth's interior will be made by combining observations from experiments and theoretical studies with increasingly sophisticated remote-sensing tools. For example, the ponded products of redox melting may be increasingly traceable by geophysics due to the increasingly well-characterized melting relations of carbon-bearing mantle[99] and constraints on the physical properties of carbonated melts (e.g. viscosity, seismic velocity, and electrical conductivity),[91,92,98] and they have indeed been linked to low-velocity zones below the oceanic lithosphere.[100] The fate of carbonates in subducting slabs, however, remains poorly constrained, in part due to the unknown variation of mantle fO_2 with depth and time. Carbonates do show rich polymorphism at high-pressure, high-temperature conditions, which renders them stable along the geotherm. Therefore, studies that explore the interaction of carbonates with silicates at well-controlled pressure, temperature, and redox conditions will provide important insights into the survival and mobility of oxidized carbon species in the mantle. Recent advances in experimental design that allow fO_2 to be controlled and monitored at deep mantle pressures[71,94] will enable the stabilities of various carbonate minerals to be mapped out as a function of depth for complex systems, combined with their predicted seismic velocities based on their crystal structure.[76] Future improvements in the sensitivity of remote-sensing tools and data processing may allow deep carbonated mantle regions to be detected. Recent studies indicate that Fe-bearing carbonates show remarkably complex chemical and physical behavior at mid-lower mantle depths. The Fe spin transition may shift the chemical equilibrium in favor of Fe-rich carbonates, which in turn would significantly affect their physical properties, with potential implications for carbonate seismic detectability. However, the relevant depth range of spin transitions is not well constrained because of the immense technological challenge of probing spin transitions at combined high-pressure, high-temperature conditions. The iron oxides, oxygen, and elemental carbon that may be produced during reactions that take place when carbonate is transported to the deep mantle may also affect redox conditions and have important (but as yet unexplored) geological consequences.

Acknowledgments

VS acknowledges financial support from the Deep Carbon Observatory and Sapienza University of Rome through "Fondi di Ateneo." SA is grateful for funding from the German Research Foundation (DFG) under grant AU 356/10. Thoughtful comments by the reviewer Arno Rohrbach and the editor Rajdeep Dasgupta helped to improve the quality of this chapter.

Questions for the Classroom

1 Describe the different ways in which carbon bonds with other elements. Which type of bonding is involved in the carbon phases discussed in this chapter? Can carbon bond with any element in the periodic table? Why or why not? Do you think there are any forms of carbon that have not yet been discovered?
2 Select one form of carbon that occurs on modern Earth's surface (e.g. dolomite) and follow its subduction journey in terms of the chemical reactions that take place involving its carbon atoms. Repeat the exercise for other carbon forms that are found on modern Earth's surface.
3 Repeat the previous exercise for the subduction of different forms of carbon in early Earth (at least 3 Ga).
4 Which of the chemical reactions in the two previous exercises are redox reactions? Describe how the cycles of other elements may be linked to the deep carbon cycle.
5 What is the estimated content of carbon in Earth as a function of depth? In which phases does carbon dominantly occur? Construct a histogram showing the variations of carbon phases and their abundance with depth. Now repeat the exercise showing only how the abundance of different oxidation states of carbon varies with depth.

6 Why is it so challenging to estimate the amount of carbon in Earth? Why does knowing how much carbon is present matter?

7 Are there any external manifestations of the deep carbon cycle on Earth's surface? Has this changed through geologic time?

8 How does the deep carbon cycle affect the habitability of modern Earth? Are there any links to climate change? Do any of the answers to these questions change if we consider early Earth?

9 Describe a high-pressure experiment that could help to answer some of the unknowns regarding carbon and its forms inside Earth.

10 Which types of natural samples could help to resolve these unknowns regarding carbon and its forms inside Earth? Note that the natural samples do not have to have been discovered yet.

References

1. Kasting, J.F., Eggler, D.H. & Raeburn, S.P. Mantle redox evolution and the oxidation state of the Archean atmosphere. *Journal of Geology*, **101**, 245–257 (1993).
2. McDonough, W.F. & Sun, S.-S. The composition of the Earth. *Chemical Geology*, **120**, 223–253 (1995).
3. Marty, B., Alexander, C.M.O. & Raymond, S. Primordial origins of Earth's carbon. *Reviews in Mineralogy and Geochemistry*, **75**, 149–181 (2013).
4. Keppler, H., Wiedenbeck, M., & Shcheka, S.S. Carbon solubility in olivine and the mode of carbon storage in the Earth's mantle. *Nature*, **424**, 414–416 (2003).
5. Rosenthal, A., Hauri, E. & Hirschmann, M. Experimental determination of C, F, and H partitioning between mantle minerals and carbonated basalt, CO_2/Ba and CO_2/Nb systematics of partial melting, and the CO_2 contents of basaltic source regions. *Earth and Planetary Science Letters*, **412**, 77–87 (2015).
6. Shcheka, S.S., Wiedenbeck, M., Frost, D.J. & Keppler, H. Carbon solubility in mantle minerals. *Earth and Planetary Science Letters*, **245**, 730–742 (2006).
7. Hayden, L.A. & Watson, E.B. Grain boundary mobility of carbon in Earth's mantle: a possible carbon flux from the core. *Proceedings of the National Academy of Sciences*, **105**, 8537–8541 (2008).
8. Ni, H. & Keppler, H. Carbon in silicate melts. *Reviews in Mineralogy and Geochemistry*, **75**, 251–287 (2013).
9. LeVoyer, M., Kelley, K., Cottrell, E. & Hauri, E. Heterogeneity in mantle carbon content from CO_2-undersaturated basalts. *Nature Communications*, **8**, 14062 (2017).
10. Saal, A.E., Hauri, E., Langmuir, C.H. & Perfit, M.R. Vapour undersaturation in primitive mid-ocean-ridge basalts and the volatile content of Earth's upper mantle. *Nature*, **419**, 451–455 (2002).
11. Dixon, J.E., Clague, D.A., Wallace, P. & Poreda, R. Volatiles in alkalic basalts from the north arc volcanic field, Hawaii: extensive degassing of deep submarine-erupted alkalic series lavas. *Journal of Petrology*, **38**, 911–939 (1997).
12. Li, Y., Dasgupta, R., Tsuno, K., Monteleone, B. & Shimizu, N. Carbon and sulfur budget of the silicate Earth explained by accretion of differentiated planetary embryos. *Nature Geoscience*, **9**, 781–785 (2016).

13. Frost, D.J., Mann, U., Asahara, Y. & Rubie, D.C. The redox state of the mantle during and just after core formation. *Philosophical Transactions of the Royal Society: Series A*, **366**, 4315–4337 (2008).
14. Scaillet, B. & Gaillard, F. Redox state of early magmas. *Nature*, **480**, 48–49 (2011).
15. Trail, D., Watson, E.B. & Tailby, N.D. The oxidation state of Hadean magmas and implications for early Earth's atmosphere. *Nature*, **480,** 79–82 (2011).
16. Li, Z.-X.A. & Lee, C.-T.A. The constancy of upper mantle fO_2 through time inferred from V/Sc ratios in basalts. *Earth and Planetary Science Letters*, **228**, 483–493 (2004).
17. Aulbach, S. & Stagno V. Evidence for a reducing Archean ambient mantle and its effects on the carbon cycle. *Geology*, **44**, 9 (2016).
18. Stagno, V., Ojwang, D.O., McCammon, C.A. & Frost, D.J. The oxidation state of the mantle and the extraction of carbon from Earth's interior. *Nature*, **493**, 84–88 (2013).
19. Stagno, V. & Frost D.J. Carbon speciation in the asthenosphere: experimental measurements of the redox conditions at which carbonate-bearing melts coexist with graphite or diamond in peridotite assemblages. *Earth and Planetary Science Letters*, **30**, 72–84 (2010).
20. Dasgupta, R. Ingassing, storage, and outgassing of terrestrial carbon through geologic time. *Reviews in Mineralogy and Geochemistry*, **75**, 183–229 (2013).
21. Aulbach, S. & Viljoen, K.S. Eclogite xenoliths from the Lace kimberlite, Kaapvaal craton: from convecting mantle source to palaeo-ocean floor and back. *Earth and Planetary Science Letters*, **431**, 274–286 (2015).
22. Ballhaus, C. & Frost, B.R. The generation of oxidized CO_2-bearing basaltic melts from reduced CH_4-bearing upper mantle sources. *Geochimica et Cosmochimica Acta*, **58**, 4931–4940 (1994).
23. Sorbadere, F., Laurenz, V., Frost, D.J., Wenz, M., Rosenthal, A., McCammon, C.A. & Rivard, C. The behaviour of ferric iron during partial melting of peridotite. *Geochimica et Cosmochimica Acta*, **239**, 235–254 (2018).
24. Aulbach, S., Jacob, D.E., Cartigny, P., Stern, R.A., Simonetti, S.S. & Viljoen, K.S. Eclogite xenoliths from Orapa: ocean crust recycling, mantle metasomatism and carbon cycling at the western Zimbabwe craton margin. *Geochimica et Cosmochimica Acta*, **213**, 574–592 (2017).
25. Nicklas, R.W., Puchtel, I.S. & Ash, R. Redox state of the Archean mantle: evidence from V partitioning in 3.5–2.4 Ga Komatiites. *Geochimica et Cosmochimica Acta*, **222**, 447–466 (2018).
26. Stagno V., Frost, D.J., McCammon, C.A., Mohseni, H. & Fei, Y. The oxygen fugacity at which graphite or diamond forms from carbonate-bearing melts in eclogitic rocks. *Contributions to Mineralogy and Petrology*, **169**, 16 (2015).
27. Ballhaus, C., Berry, R.F. & Green, D.H. High pressure experimental calibration of the olivine–orthopyroxene–spinel oxygen geobarometer: implications for the oxidation state of the upper mantle. *Contributions of Mineralogy and Petrology*, **107**, 27–40 (1991).
28. Garber, J.M. et al. Multidisciplinary constraints on the abundance of diamond and eclogite in the cratonic lithosphere. *Geochemistry, Geophysics, Geosystems*, **19**, 2062–2086 (2018).
29. Woolley, A.R. & Kjarsgaard, B.A. Carbonatite Occurrences of the World: Map and Database; *Geological Survey of Canada*, Open File 5796, 1 CD-ROM + 1 map (2008).
30. Tappe, S., Romer, R.L., Stracke, A., Steenfelt, A., Smart, K.A., Muehlenbachs, K. & Torsvik, T.H. Sources and mobility of carbonate melts beneath cratons, with

implications for deep carbon cycling, metasomatism and rift initiation. *Earth and Planetary Science Letters*, **466**, 152–167 (2017).
31. Tappe, S., Smart, K., Torsvik, T., Massuyeau, M. & de Wit, M. Geodynamics of kimberlites on a cooling Earth: clues to plate tectonic evolution and deep volatile cycles. *Earth and Planetary Science Letters*, **484**, 1–14 (2018).
32. Aulbach, S., Woodland, A.B., Vasilyev, P., Galvez, M.E. & Viljoen, K.S. Effects of low-pressure igneous processes and subduction on $Fe^{3+}/\sum Fe$ and redox state of mantle eclogites from Lace (Kaapvaal craton). *Earth and Planetary Science Letters*, **474**, 283–295 (2017).
33. Dasgupta, R. & Hirschmann, M.M. The deep carbon cycle and melting in Earth's interior. *Earth and Planetary Science Letters (Frontiers)*, **298**, 1–13 (2010).
34. Rohrbach, A., Ghosh, S., Schmidt, M.W., Wijbrans, C.H. & Klemme, S. The stability of Fe–Ni carbides in the Earth's mantle: evidence for a low Fe–Ni–C melt fraction in the deep mantle. *Earth and Planetary Science Letters*, **388**, 211 (2014).
35. Luth, R.W. Diamonds, eclogites and the oxidation state of the Earth's mantle. *Science*, **261**, 66–68 (1993).
36. McDonough, W.F. & Sun, S.-S. The composition of the Earth. *Chemical Geology*, **120**, 223–253 (1995).
37. Eguchi, J. & Dasgupta, R. Redox state of the convective mantle from CO_2-trace element systematics of oceanic basalts. *Geochemical Perspectives Letters,* **8**, 17–21 (2018).
38. Frost, D.J., Liebske, C., Langenhorst, F. & McCammon, C.A. Experimental evidence for the existence of iron-rich metal in the Earth's lower mantle. *Nature*, **428**, 409–412 (2004).
39. Rohrbach, A., Ballhaus, C., Ulmer, P., Golla-Schindler, U. & Schönbohm, D. Experimental evidence for a reduced metal-saturated upper mantle. *Journal of Petrology*, **52**, 717–731 (2011).
40. Smith, E.M., Shirey, S., Nestola, F., Bullock, E.S., Wang, J., Richardson, S. & Wang, W. Large gem diamonds from metallic liquid in Earth's deep mantle. *Science*, **354**, 1403–1405 (2016).
41. Jacob, D.E. Nature and origin of eclogite xenoliths from kimberlites. *Lithos*, **77**, 295–316 (2004).
42. Kaminsky, F.V. & Wirth, R. Iron carbide inclusions in lower-mantle diamond from Juina, Brazil, *Canadian Mineralogist*, **49**, 555–572 (2011).
43. Mikhail, S. et al. Empirical evidence for the fractionation of carbon isotopes between diamond and iron carbide from the Earth's mantle. *Geochemistry, Geophysics, Geosystems*, **15**, 855–866 (2014).
44. Schmidt, M.W., Gao, C., Golubkova, A., Rohrbach, A. & Connolly, J.A.D. Natural moissanite (SiC) – a low temperature mineral formed from highly fractionated ultra-reducing COH-fluids. *Progress in Earth and Planetary Science*, **1**, 27 (2014).
45. Palyanov, Y.N., Bataleva, Y.V., Sokol, A.G., Borzdov, Y.M., Kupriyanov, I.N., Reutsky, V.N. & Sobolev, N.V. Mantle–slab interaction and redox mechanism of diamond formation. *Proceedings of the National Academy of Sciences*, **110**, 20408–20413 (2013).
46. Tsuno, K. & Dasgupta, R. Fe–Ni–Cu–C–S phase relations at high pressures and temperatures – the role of sulfur in carbon storage and diamond stability at mid- to deep-upper mantle. *Earth and Planetary Science Letters*, **412**, 132–142 (2015).
47. Brenker, F.E. et al. Carbonates from lower part of transition zone or even the lower mantle. *Earth and Planetary Science Letters*, **260**, 1–9 (2007).

48. Kaminsky, F.V., Ryabchikov, I.D. & Wirth, R. A primary natrocarbonatitic association in the deep Earth. *Mineralogy and Petrology*, **110**, 387–398 (2016).
49. Hammouda, T. & Keshav, S. Melting in the mantle in the presence of carbon: review of experiments and discussion on the origin of carbonatites. *Chemical Geology*, **418**, 171–188 (2015).
50. Santillan, J., Williams, Q. & Knittle, E. Dolomite-II: a high-pressure polymorph of $CaMg(CO_3)_2$. *Geophysical Research Letters*, **30**, 2 (2003).
51. Mao, Z., Armentrout, M., Rainey, E., Manning, C.E., Dera, P., Prakapenka, V.B. & Kavner A. Dolomite III: a new candidate lower mantle carbonate. *Geophysical Research Letters*, **38**, L22303 (2011).
52. Merlini, M., Crichton, W., Hanfland, M., Gemmi, M., Mueller, H., Kupenko, I. & Dubrovinsky, L. Dolomite-II and dolomite-III: crystal structures and stability in the Earth's lower mantle. *Proceedings of the National Academy of Sciences*, **109**, 13509–13514 (2012b).
53. Merlini, M. et al. Dolomite-IV: candidate structure for a carbonate in the Earth's lower mantle. *American Mineralogist*, **102**, 1763–1766 (2017).
54. Dorfman, S.M., Badro, J., Nabiei, F., Prakapenka, V.B., Cantoni, M. & Gillet, P. Carbonate stability in the reduced lower mantle. *Earth and Planetary Science Letters*, **489**, 84 (2018).
55. Liu, Y. et al. First direct evidence of sedimentary carbonate recycling in subduction-related xenoliths. *Scientific Reports*, **5**, 11547 (2015).
56. Merlini, M., Hanfland, M., & Crichton, W. $CaCO_3$-III and $CaCO_3$-VI, high-pressure polymorphs of calcite: possible host structures for carbon in the Earth's mantle. *Earth and Planetary Science Letters*, **333–334**, 265–271 (2012).
57. Ono, S., Kikegawa, T. & Ohishi, Y. High-pressure transition of $CaCO_3$. *American Mineralogist*, **92**, 1246–1249 (2007).
58. Oganov, A.R., Glass, C.W., & Ono, S. High-pressure phases of $CaCO_3$: crystal structure prediction and experiment. *Earth and Planetary Science Letters*, **241**, 95–103 (2006).
59. Pickard, C.J. & Needs, R.J. Structures and stability of calcium and magnesium carbonates at mantle pressures. *Physical Review B*, **91**, 104101 (2015).
60. Lobanov, S.S. et al. Raman spectroscopy and X-ray diffraction of sp^3 $CaCO_3$ at lower mantle pressures. *Physical Review B*, **96**, 104101 (2017).
61. Cerantola, V. et al. High-pressure spectroscopic study of siderite ($FeCO_3$) with a focus on spin crossover. *American Mineralogist*, **100**, 2670–2681 (2015).
62. Lobanov, S.S., Goncharov, A.F. & Litasov, K.D. Optical properties of siderite ($FeCO_3$) across the spin transition: crossover to iron-rich carbonates in the lower mantle. *American Mineralogist*, **100**, 1059–1064 (2015).
63. Lobanov, S.S., Holtgrewe, N. & Goncharov, A.F. Reduced radiative conductivity of low spin FeO_6-octahedra in $FeCO_3$ at high pressure and temperature. *Earth and Planetary Science Letters*, **449**, 20–25 (2016).
64. Isshiki, M. et al. Stability of magnesite and its high-pressure form in the lowermost mantle. *Nature*, **427**, 60–63 (2004).
65. Solopova, N.A. et al. Melting and decomposition of $MgCO_3$ at pressures up to 84 GPa. *Physics and Chemisrty of Minerals*, **42**, 73–81 (2014).
66. Maeda, F., Ohtani, E., Kamada, S., Sakamaki, T., Hirao, N. & Ohishi, Y. Diamond formation in the deep lower mantle: A high-pressure reaction of $MgCO_3$ and SiO_2. *Scientific Reports*, **7**, 40602 (2017).

67. Cerantola, V. et al. Stability of iron-bearing carbonates in the deep Earth's interior. *Nature Communications*, **8**, 15960 (2017).
68. Kang, N. et al. Melting of siderite to 20 GPa and thermodynamic properties of $FeCO_3$-melt. *Chemical Geology*, **400**, 34–43 (2015).
69. Boulard, E. et al. New host for carbon in the deep Earth. *Proceedings of the National Academy of Sciences*, **108**, 5184–5187 (2011).
70. Merlini, M., Hanfland, M., Salamat, A., Petitgirard, S. & Müller, H. The crystal structures of $Mg_2Fe_2C_4O_{13}$, with tetrahedrally coordinated carbon, and Fe13O19, synthesized at deep mantle conditions. *American Mineralogist*, **100**, 2001–2004 (2015).
71. Stagno, V., Tange, Y., Miyajima, N., McCammon, C.A., Irifune, T. & Frost, D.J. The stability of magnesite in the transition zone and the lower mantle as function of oxygen fugacity. *Geophysical Research Letters*, **38**, L19309 (2011).
72. McCammon, C. Microscopic properties to macroscopic behavior: the influence of iron electronic states. *Journal of Mineralogical and Petrological Science*, **101**, 130–144 (2006).
73. Kiseeva, E. et al. Oxidized iron in garnets from the mantle transition zone. *Nature Geoscience*, **11**, 144–147 (2018).
74. Beyer, C. & Frost, D.J. The depth of sub-lithospheric diamond formation and the redistribution of carbon in the deep mantle. *Earth and Planetary Science Letters*, **461**, 30–39 (2017).
75. Ismailova, L. et al. Stability of Fe,Al-bearing bridgmanite in the lower mantle and synthesis of pure Fe-bridgmanite. *Science Advances*, **2**, e1600427 (2016).
76. Sanchez-Valle, C., Ghosh S. & Rosa A.D. Sound velocities of ferromagnesian carbonates and the seismic detection of carbonates in eclogites and the mantle. *Geophysical Research Letters*, **38**, L24315 (2011).
77. Yang, J., Mao, Z., Lin, J.-F. & Prakapenka, V. Single-crystal elasticity of the deep-mantle magnesite at high pressure and temperature. *Earth and Planetary Science Letters*, **392**, 292–299 (2014).
78. Mattila, A. et al. Pressure induced magnetic transition in siderite $FeCO_3$ studied by X-ray emission spectroscopy. *Journal of Physics: Condensed Matter*, **19**, 386206 (2007).
79. Shi, H., Luo, W., Johansson, B. & Ahuja, R. First-principles calculations of the electronic structure and pressure-induced magnetic transition in siderite $FeCO_3$. *Physical Review B*, **78**, 155119 (2008).
80. Lavina, B., Dera, P., Downs, R.T., Prakapenka, V., Rivers, M., Sutton, S. & Nicol, M. Siderite at lower mantle conditions and the effects of the pressure-induced spin-pairing transition. *Geophysical Research Letters*, **36**, L23306 (2009).
81. Lavina, B. et al. Structure of siderite $FeCO_3$ to 56 GPa and hysteresis of its spin-pairing transition. *Physical Review B*, **82**, 064110 (2010b).
82. Lin, J.-F., Liu, J., Jacobs, C. & Prakapenka, V.B. Vibrational and elastic properties of ferromagnesite across the electronic spin-pairing transition of iron. *American Mineralogist*, **97**, 583–591 (2012).
83. Farfan, G., Wang, S., Ma, H., Caracas, R. & Mao, W.L. Bonding and structural changes in siderite at high pressure. *American Mineralogist*, **97**, 1421–1426 (2012).
84. Cerantola, V. et al. High-pressure spectroscopic study of siderite ($FeCO_3$) with a focus on spin crossover. *American Mineralogist*, **100**, 2670–2681 (2015).

85. Lobanov, S.S., Goncharov, A.F. & Litasov, K.D. Optical properties of siderite ($FeCO_3$) across the spin transition: crossover to iron-rich carbonates in the lower mantle. *American Mineralogist*, **100**, 1059–1064 (2015).
86. Weis, C. et al. Pressure driven spin transition in siderite and magnesiosiderite single crystals. *Scientific Reports*, **7**, 16526 (2017).
87. Liu, J., Lin, J.F., Mao, Z., & Prakapenka, V.B. Thermal equation of state and spin transition of magnesiosiderite at high pressure and temperature. *American Mineralogist*, **99**, 84–93 (2014).
88. Fu, S., Yang, J. & Lin, J.F. Abnormal elasticity of single-crystal magnesiosiderite across the spin transition in Earth's lower mantle. *Physical Reviews Letters*, **118**, 036402 (2017).
89. Stekiel, M. et al. High pressure elasticity of $FeCO_3$–$MgCO_3$ carbonates. *Physics of the Earth and Planetary Interiors*, **271**, 57–63 (2017).
90. Müller, J., Efthimiopoulos, I., Jahn, S. & Koch-Müller, M. Effect of temperature on the pressure-induced spin transition in siderite and iron-bearing magnesite: a Raman spectroscopy study. *European Journal of Mineralogy*, **29**, 785–793 (2017).
91. Kono, Y. et al. Ultralow viscosity of carbonate melts at high pressure. *Nature Communications*, **5**, 5091 (2014).
92. Stagno, V., Stopponi, V., Kono, Y., Manning, C.E. & Irifune T. Experimental determination of the viscosity of Na_2CO_3 melt between 1.7 and 4.6 GPa at 1200–1700 °C: implications for the rheology of carbonatite magmas in the Earth's upper mantle. *Chemical Geology*, **501**, 19–25 (2018).
93. Frost, D.J. & McCammon, C.A. The redox state of the Earth's mantle. *Annual Review of Earth and Planetary Science*, **36**, 389–420 (2008).
94. Rohrbach, A. & Schmidt, M.W. Redox freezing and melting in the Earth's deep mantle resulting from carbon–iron redox coupling. *Nature*, **472**, 209–212 (2011).
95. O'Reilly, S.Y. & Griffin, W.L. Moho vs crust–mantle boundary: evolution of an idea. *Tectnophysics*, **609**, 535–546 (2013).
96. Foley, S.F. A reappraisal of redox melting in the Earth's mantle as a function of tectonic setting and time. *Journal of Petrology*, **52**, 1363–1391 (2011).
97. Gaillard, F., Scaillet, B., Pichavant, M. & Iacono-Marziano, G. The redox geodynamics linking basalts and their mantle sources through space and time. *Chemical Geology*, **418**, 217–233 (2015).
98. Litasov, K.D. & Shatskiy, A. Carbon-bearing magmas in the Earth's deep interior. In: Kono, Y. & Sanloup, C. (eds.), *Magmas under Pressure: Advances in High-Pressure Experiments on Structure and Properties of Melts*. Amsterdam: Elsevier, pp. 43–82 (2018).
99. Sakamaki, T. Structure and properties of silicate magmas. In: Kono, Y. & Sanloup, C. (eds.), *Magmas under Pressure: Advances in High-Pressure Experiments on Structure and Properties of Melts*. Amsterdam: Elsevier, pp. 323–341 (2018).
100. Sifre', D., Gardes, E., Massuyeau, M., Hashim, L., Hier-Majumder, S. & Gaillard, F. Electrical conductivity during incipient melting in the oceanic low-velocity zone. *Nature*, **509**, 81–85 (2014).

5

Diamonds and the Mantle Geodynamics of Carbon

Deep Mantle Carbon Evolution from the Diamond Record

STEVEN B. SHIREY, KAREN V. SMIT, D. GRAHAM PEARSON, MICHAEL J. WALTER, SONJA AULBACH, FRANK E. BRENKER, HÉLÈNE BUREAU, ANTONY D. BURNHAM, PIERRE CARTIGNY, THOMAS CHACKO, DANIEL J. FROST, ERIK H. HAURI, DORRIT E. JACOB, STEVEN D. JACOBSEN, SIMON C. KOHN, ROBERT W. LUTH, SAMI MIKHAIL, ODED NAVON, FABRIZIO NESTOLA, PAOLO NIMIS, MEDERIC PALOT, EVAN M. SMITH, THOMAS STACHEL, VINCENZO STAGNO, ANDREW STEELE, RICHARD A. STERN, EMILIE THOMASSOT, ANDREW R. THOMSON, AND YAAKOV WEISS

5.1 Introduction

The importance of diamond to carbon in Earth is due to the fact that diamond is the only mineral and especially the only carbon mineral to crystallize throughout the silicate Earth – from the crust to the lower mantle. To study diamond *is* to study deep carbon directly throughout Earth, allowing us to see the inaccessible part of the deep carbon cycle. By using the properties of diamond, including its ability to preserve included minerals, important questions relating to carbon and its role in planetary-scale geology can be addressed:

- What is the mineralogy of phases from Earth's mantle transition zone and lower mantle?
- What are the pressures and temperatures of diamond growth?
- What is the chemical speciation of recycled and deep carbon?
- What are the reactions that produce reduced carbon?
- What are the sources of carbon and its associated volatiles (H_2O, CH_4, CO_2, N_2, NH_3, and S)?
- How do these findings vary with global-scale geological processes?
- How have these processes changed over billions of years of geologic history?

Diamonds for scientific study are difficult to obtain and the nature of diamond presents special research challenges. Diamonds, whether they are lithospheric or sublithospheric (see the paragraph after next below), are xenocrysts in kimberlitic magma that travel a long path (as much as 150 to >400 km!) during eruption to Earth's surface. On strict petrologic grounds, by the time a diamond reaches Earth's surface, there is little direct evidence that it is related to any neighboring diamond. However, the suites of diamonds that occur in close spatial association at Earth's surface in a mine may have similar physical characteristics and may also record similar pressure–temperature conditions and ages. If so, these features would suggest that the host kimberlite delivered a diamond suite to the surface from a spatially restricted mantle source hundreds of kilometers distant. Kimberlite magmas can transport some diamonds in mantle blocks that apparently disaggregate near the end of their upward journey. Nonetheless, each diamond and its inclusions is a case study unto itself until its association with other diamonds can be confirmed.

The Diamonds and Mantle Geodynamics of Carbon (DMGC) consortium was conceived early in the existence of the Deep Carbon Observatory (DCO) for the specific purpose of breaking down the traditional barriers to research on single diamonds and directing research toward global carbon questions. From the outset, the DMGC consortium focused on making cross-disciplinary research tools available, sharing samples so that more definitive results could be obtained, and enhancing intellectual stimulation across research groups so that new ideas would develop. The purpose of this chapter is to showcase results from the major collaborative research areas that have emerged within the DMGC consortium: (1) geothermobarometry to allow the depth of diamond crystallization and constraints on diamond exhumation to be determined (Section 5.2); (2) diamond-forming reactions, C and N isotopic compositions, and diamond-forming fluids to understand how diamonds form in the mantle (Section 5.3); (3) the sources of carbon either from the surface or within the mantle to provide information on the way carbon and other volatiles are recycled by global processes (Section 5.4); and (4) the mineralogy, trace element, and isotopic composition of mineral inclusions and their host diamonds to relate diamond formation to geologic conditions in the lithospheric and deeper convecting mantle (Section 5.5).

A general review and summary of diamond research can be obtained by consulting previous works.[1–16] Much of this literature focuses on diamonds and the mineral inclusions that have been encapsulated when these diamonds crystallized in the subcontinental lithospheric mantle (SCLM). These so-called lithospheric diamonds can be classified as eclogitic or peridotitic (harzbugitic, lherzolitic, or websteritic) based on the composition of silicate or sulfide inclusions.[10,16] Lithospheric diamonds crystallize at depths of around 100–200 km and temperatures of around $1160 \pm 100°C$.[16] Peridotitic diamonds typically have restricted, mantle-like C isotopic compositions, whereas eclogitic diamonds have more variable and sometimes distinctly lighter C isotopic compositions.[8] Lithospheric diamonds are likely to contain appreciable nitrogen (mostly Type I; 0–3830 at. ppm, median = 91 at. ppm[16]). Their ages range from Cretaceous to Mesoarchean, but most are Proterozoic to Neoarchean.[1,7,17] The study of lithospheric diamonds has led to advances in understanding of the stabilization of the continents and their mantle keels, the onset of plate tectonics, and the nature of continental margin subduction, especially in the ancient past.

In the last two decades, attention has turned to the study of diamonds whose inclusion mineralogies and estimated pressures of origin put them at mantle depths well below the lithospheric mantle beneath continents. These "sublithospheric" or so-called super-deep diamonds can occur at any depth down to and including the top of the lower mantle (660–690 km), but a great many crystallize in the mantle transition zone (e.g. 410–660 km)[9,12,18] at higher temperatures (between 100 and 400°C higher)[19] than lithospheric diamonds. Unlike lithospheric diamonds, super-deep diamonds are not as easily classified as eclogitic or peridotitic. However, super-deep diamonds do carry mineral phases that are the high-pressure derivatives of basaltic and ultramafic precursors such as majorite or bridgmanite, respectively,[12] so a petrologic parallel exists in super-deep diamonds with the eclogitic and peridotitic lithospheric diamond designation. Super-deep diamonds are typically low nitrogen (e.g. Type IIa or IIb) and display quite variable C isotopic compositions, even extending to quite light compositions (Section 5.4.2 and Ref. 20). Age determinations on super-deep diamonds are rare, but the few that exist[21] support their being much younger than lithospheric diamonds – an expected result given the known antiquity of the continents and their attached mantle keels relative to the convecting mantle. The study of super-deep diamonds has led to advances in understanding of the deep recycling of elements from the surface (e.g. H_2O, B, C, and S), the redox structure of the mantle, and the highly heterogeneous nature of the mantle transition zone.

Much of the research described in this chapter focuses on super-deep diamonds since the study of super-deep diamonds has the greatest relevance to the deep carbon cycle.

5.2 Physical Conditions of Diamond Formation

5.2.1 Measuring the Depth of Diamond Formation

An essential question is the depth at which a diamond forms. Geobarometry of diamonds based on the stability of their included minerals has provided important constraints on the deep carbon cycle. Application of these methods has yielded the whole range of depths from 110 to 150 km, corresponding to the graphite–diamond boundary in the lithosphere, to over 660 km, lying within the lower mantle.[14,22–25] Thus, these studies have provided direct evidence for the recycling of surficial carbon to lower-mantle depths. Traditional geobarometric methods, however, have several limitations: they can only be applied to rare types of mineral inclusions; touching inclusions may re-equilibrate after diamond growth; non-touching inclusions may be incorporated under different conditions and may not be in equilibrium; and protogenetic inclusions[26,27] may not re-equilibrate completely during diamond growth.

In order to avoid some of these drawbacks, alternative approaches that are independent of chemical equilibria are increasingly being explored. Elastic geobarometry is based on the determination of the residual pressure on the inclusion, P_{inc}, which builds up on an inclusion when the diamond is exhumed to the surface as a result of the difference in the elastic properties of the inclusion and host. If these properties are known and the entrapment temperature is derived independently or its effect is demonstrably negligible, then the entrapment pressure can be calculated back from the P_{inc} determined at room conditions. The idea of using P_{inc} to calculate entrapment conditions is not new,[28] but practical methods have recently been developed that allow more robust estimates of minerals with known elastic properties.[29,30] In principle, elastic geobarometry can be applied to any inclusion in a diamond if: (1) the inclusion–diamond interaction is purely elastic, otherwise only minimum estimates can generally be obtained; (2) the geometry of the inclusion–host system is properly considered; and (3) mineral-specific calibrations are available to calculate P_{inc} from X-ray diffraction or micro-Raman spectroscopy data. Contrary to common practice, calibrations should take into account the effect of deviatoric stresses, which typically develop in inclusion–diamond systems upon exhumation. For example, Anzolini et al.[31] showed that only an accurate choice of Raman peaks could provide reliable estimates for a $CaSiO_3$–walstromite inclusion in diamond, yielding a minimum formation depth of 260 km and supporting $CaSiO_3$–walstromite as a potential indicator of sublithospheric origin. The effect of the presence of fluid films around the inclusions, which has recently been documented in some lithospheric diamonds,[32] still demands proper evaluation. In addition, the ability of diamond to deform plastically, especially under sublithospheric conditions, is well known, but methods to quantify any effects on elastic geobarometry are not available. Therefore, in many cases, only minimum estimates can be obtained. Nonetheless, we are now able to provide depth or minimum depth estimates for a number of new single-phase assemblages that would not be possible with more traditional methods. Future geobarometry of larger sets of diamonds, using both elastic and traditional approaches, will allow more comprehensive data to be gathered on the conditions for diamond-forming reactions and on the deep carbon cycle.

5.2.2 Thermal Modeling of Diamond in the Mantle from Fourier-Transform Infrared Spectroscopy Maps

The defects trapped in diamonds can be used to constrain estimates of the temperature that prevailed during the residence of a diamond in the mantle and can help constrain estimates of the return path of carbon to the surface. Pressure and temperature covary with depth in Earth, and the ability of the diamond lattice to record temperature history in its defect structure provides an additional independent constraint on estimates of mantle location. These measurements are therefore complementary to those on inclusions that can be used to determine the pressure and temperature conditions during the *trapping* of inclusions during diamond growth. The general concepts and calibration of a thermochronometer based on nitrogen defect aggregation are well established.[33] The technique is based on the kinetics of aggregation of pairs of nitrogen atoms (called A centers) into groups of four nitrogen atoms around a vacancy (called B centers) and measurement of these defect concentrations using Fourier-transform infrared (FTIR) spectroscopy. The FTIR spectroscopy method has long been used as one of the standard characterization techniques for diamonds, mostly for whole stones, but also as FTIR spectroscopy maps showing the distribution of defect concentrations across diamond plates.[34,35] Only recently has the full potential of FTIR spectroscopy for determining the thermal history of a diamond been recognized. The major recent developments have been: (1) improvements in the methods for acquiring and processing FTIR spectroscopy maps[34]; (2) a better understanding of the temperature history that is available from zoned diamonds[36]; and (3) unlocking the abundant information that is provided by the FTIR spectroscopy signal of platelets – planar defects created with B centers during nitrogen aggregation.[37,38]

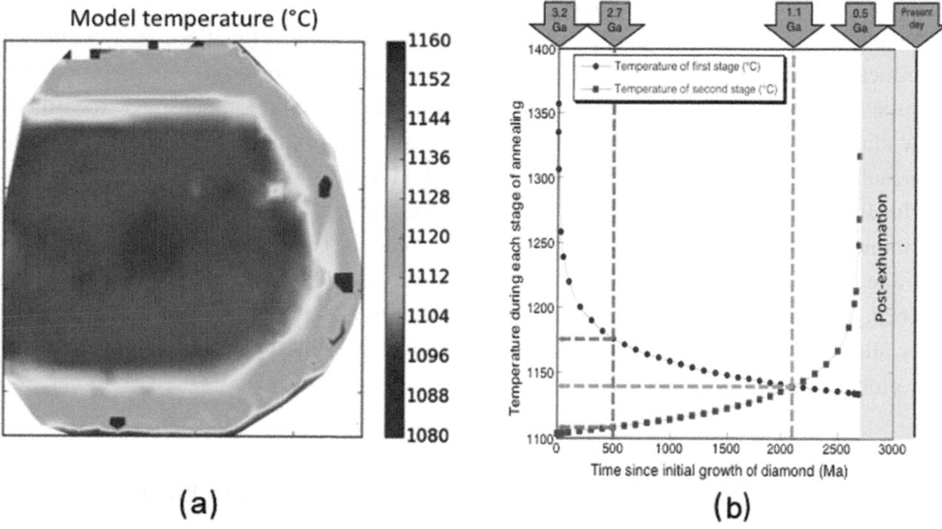

Figure 5.1 (a) Example of a map of "model temperatures" made up by automated fitting of several thousand FTIR spectra in a map of a diamond from Murowa, Zimbabwe. Model temperatures are calculated using a single assumed mantle residence time. The higher model temperatures in the core and lower model temperatures in the rim reflect a growth and annealing history with at least two stages. (b) Modeling the possible combinations of temperature and time that could explain the FTIR spectroscopy characteristics of a zoned diamond from Murowa.[36]

Figure 5.1a shows a map of "model temperatures" made up by automated fitting of several thousand FTIR spectra in a map of a diamond from Murowa, Zimbabwe. The higher temperatures in the core and lower temperatures in the rim reflect a growth and annealing history with at least two stages. The key idea is that the N aggregation in the rim only occurs during the second stage of annealing, but that N aggregation in the core occurs throughout the residence period of the diamond in the mantle (i.e. during both stages of annealing). Even if the date of rim growth is not known, there is an interplay between the temperatures of the two stages and the time of rim formation (Figure 5.1b). While these data provide a combination of time and temperature, if the dates of each stage of diamond formation are accurately known (by dating of inclusions) and the date of kimberlite eruption is known, the temperatures during the two stages can be determined. The model in Figure 5.1b assumes core growth at 3.2 Ga followed by a period of annealing, then rim growth and finally a second period of annealing. If a constant temperature prevailed throughout the history of the diamond's residence in the lithosphere, the ages of the two periods of growth are 3.2 and 1.1 Ga. If the earlier history of the diamond was hotter, the overgrowth must be older. Using this method, the mean temperature variation over very long (billion-year) timescales at a specific location in the lithosphere can be determined.

An alternative way to learn about the history of a diamond is to study the production and degradation of platelets. By comparing transmission electron microscopy and FTIR spectroscopy measurements on platelets, we now have a much better understanding of the meaning of the FTIR spectroscopy platelet peak characteristics (position, area, width, and symmetry) and how platelets evolve with thermal history.[37,38] In addition to diamonds with regular platelet behavior and irregular, platelet-degraded behavior, we have identified a new class of sub-regular diamonds with anomalously small platelets that have experienced unusually low mantle temperatures (below about 1120°C).

In summary, improvements in our understanding of the defects incorporated into diamonds will contribute to better quantitative models of carbon precipitation, carbon storage in the mantle as diamond, and carbon exhumation during continental tectonics. Indeed, advances in the correlation of spectral features with newly understood defect types may allow diamond to emerge as a prime mineral for studying the uplift and exhumation in the global tectonic cycle (e.g. Ref. 39).

5.3 Diamond-Forming Reactions, Mechanisms, and Fluids

5.3.1 Direct Observation of Reduced Mantle Volatiles in Lithospheric and Sublithospheric Diamonds

Some carbon in the deep Earth is not stored in crystalline silicates but as fluids such as highly mobile metallic and carbonatitic liquids or supercritical (also known as high-density) C-H-O fluids. As diamond is thought to crystallize from these species by different mechanisms, its study becomes a key way to understand these carbon-bearing fluids. Reduced volatiles in diamonds have only recently been identified,[40] although they have long been predicted.[41] Observations of reduced volatiles in any mantle sample are rare because they must avoid oxidation in the shallow mantle and crust on the way to Earth's surface. Direct samples of these reduced volatiles are crucial to understanding the redox speciation of mantle fluids and melts since they influence both melt composition and physical properties such as solidus temperature, viscosity, and density.[42,43]

Experimental results and thermodynamic modeling of C-H-O fluids at pressure and temperature and oxygen fugacities (fO_2) relevant to the lithospheric mantle show that methane (CH_4) is stable at 2 log units below the fayalite–magnetite–quartz buffer ($\Delta \log fO_2$ (fayalite–magnetite–quartz buffer (FMQ)) < -2) and becomes the dominant C-H-O species at around $\Delta \log fO_2$ (FMQ) 4.5.[44,45] At diamond-stable pressures, the lithosphere should typically have fO_2 between FMQ = –2 and FMQ = –4,[46] and the implication is that metasomatic and diamond-forming fluids should also generally be reduced.[47]

Due to their metasomatic origin from fluids with CO_2, CO_3, or CH_4 as the dominant carbon species,[48] diamonds are the ultimate tracers of carbon cycling into the mantle. Mantle CH_4 has recently been directly detected for the first time in diamond samples from both the lithospheric and sublithospheric mantle.[40,49] Diamond is the ideal carrier, since it shields any trapped reduced volatiles from oxygen exchange during its rapid transport to the surface. Analyses of reduced volatiles, along with any coexisting phases and the host

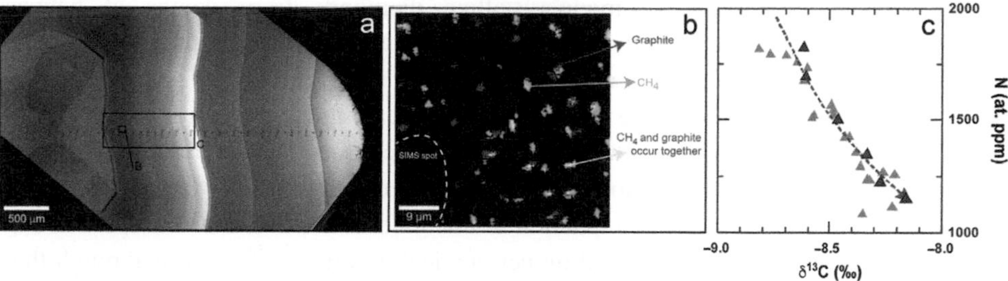

Figure 5.2 (a) Cathodoluminescence (CL) image of Marange diamond MAR06b,[40] showing core-to-rim secondary ion mass spectrometry analytical spots. (b) Raman map showing distribution of graphite and CH_4 micro-inclusions in a homogeneously grown cuboid zone. (c) Outwardly decreasing nitrogen content (atomic ppm) with increasing $\delta^{13}C$ (‰) in this same cuboid growth zone (red) and other cuboid growth zones (gray). The modeled trend (red dashed line) is for a CH_4: CO_2 ratio of 1:1 and assumes an initial $\delta^{13}C$ for the fluid of –7.4‰. For an assumed water content of 98 mol.%, the observed variation corresponds to 0.7% crystallization of the entire fluid (and 35% of the carbon in the fluid).
For details on modelling, see Stachel et al.[51]

diamonds, allow us to better understand the storage and transport of reduced volatiles into the mantle. These studies can also help us to evaluate whether methanogenesis occurs in the mantle[50] or whether mantle CH_4 has a subducted origin.

Lithospheric diamonds (depths <200 km) from Marange (Zimbabwe) are rare, mixed-habit diamonds that trap abundant inclusions in their cuboid sectors and also contain octahedral sectors that grew simultaneously.[40] Confocal Raman imaging of the faster-grown cuboid sectors shows that they contain both crystalline graphite and CH_4 inclusions (Figure 5.2). Both graphite and CH_4 are evenly distributed throughout the cuboid sectors, usually but not always occurring together, a syngenetic texture that suggests that they co-crystallized along with diamond from the same C-H-O fluid. Clear octahedral sectors never contain graphite or CH_4.

In fluid inclusion-free diamonds, core-to-rim trends in $\delta^{13}C$ and N content have been (and probably should not have been) used to infer the speciation of the diamond-forming fluid. Outwardly decreasing $\delta^{13}C$ with decreasing N content is interpreted as diamond growth from reduced fluids, whereas oxidized CO_2 or carbonate-bearing fluids should show the opposite trends. Within the CH_4-bearing sectors of Marange diamonds, however, such reduced trends are not observed. Rather, $\delta^{13}C$ increases outwardly within a homogeneously grown zone that also contains CH_4 (Figure 5.2). These contradictory observations can be explained through either mixing between CH_4- and CO_2-rich end-members of hydrous fluids[40] or through closed-system precipitation from an already mixed CH_4–CO_2–H_2O-maximum fluid.[51] The relatively low $\delta^{13}C$ value of the initial fluid (modeled at approximately –7.4‰),[51] along with positive $\delta^{15}N$ values (calculated using the diamond–fluid fractionation factor from Petts et al.[52]), suggests that the CH_4-rich Marange diamond source fluids may in part have been subduction derived.

Sublithospheric gem-quality monocrystalline diamonds from depths of between 360 and 750 km have also recently been found to contain inclusions with reduced volatile budgets.[49] Specifically, iron- and nickel-rich metallic inclusions are consistently trapped along with CH_4 (and H_2) in large gem-quality monocrystalline diamonds. This suggests that C-H-O fluids in metal-saturated regions of the deep mantle are buffered to reduced compositions, either dissolved into metallic liquids or existing as CH_4-rich fluids.

The key role of CH_4 in the precipitation of diamond and its efficient transport through Earth's mantle has long been inferred on petrological grounds,[53,54] but it is through these studies on Zimbabwe and sublithospheric diamonds that we now have the first observations of this important fluid species. Further investigation of fluid species in diamonds are needed: (1) to establish the frequency of the involvement of reduced volatiles in diamond formation; and (2) to differentiate the geological environments where transport of carbon and diamond precipitation involves CH_4-rich fluids[55] versus carbonate-rich fluids or melts.[56,57]

5.3.2 Redox-Neutral Diamond Formation and Its Unexpected Effect on Carbon Isotope Fractionation

Using models of diamond crystallization from fluids of specific composition, the variation of carbon isotopic compositions across a diamond crystal can be used to estimate fluid composition, interaction with fluid-hosting wall rock, and C source characteristics. This approach has been applied most successfully to lithospheric diamonds. Studies of garnet peridotite xenoliths[46,47] demonstrate that subcratonic lithospheric mantle typically lies on the reducing side of the fO_2 of the EMOD buffer (enstatite + magnesite → olivine + diamond: the transition from carbonate- to diamond-bearing peridotite), but well above that of the IW buffer (iron → wüstite: where native iron becomes stable). Consequently, in the deep lithospheric mantle, carbon will generally be stored as diamond rather than carbonate or carbide. From the study of mineral inclusions in diamond, we know that strongly depleted harzburgite and dunite are the principal (~56%) diamond substrates in lithospheric mantle. On this basis, we set out to address two fundamental questions: (1) in what form is carbon transported to lithospheric diamond substrates? And (2) what exactly is the process that drives the conversion of carbon-bearing fluid species into elemental carbon?

Luth and Stachel[47] modeled that <50 ppm O_2 has to be removed from or added to depleted cratonic peridotite to move its oxidation state from the EMOD to the IW buffer (or vice versa). This extremely low buffering capacity of cratonic peridotites has two important implications: (1) the redox state of subcratonic lithospheric mantle is fluid buffered and, consequently, studies of peridotite xenoliths can only reveal the redox state of the last fluid with which they interacted; and (2) redox reactions buffered by depleted cratonic peridotite cannot explain the formation of large diamonds or large quantities of diamonds (per volume unit of peridotite).

At the typical fO_2 conditions of diamond-stable cratonic peridotite ($\Delta \log fO_2$ (FMQ) = –1.5 to –3.5), C-H-O fluids will be water rich (90–99 mol.%) with minor amounts of CH_4

and CO_2. During cooling or ascent along a geotherm (cooling plus decompression), such near-water-maximum fluids precipitate diamond isochemically, without the need for oxygen exchange with their peridotitic wall rocks.[47,58] At conditions just below the EMOD buffer ($\Delta \log fO_2$ (FMQ) = –1.5 to –2.4, at 5 GPa and 1140°C), diamond precipitation can occur by the oxygen-conserving reaction:

$$CO_2 + CH_4 \rightarrow 2C + 2H_2O \tag{5.1}$$

At more reducing conditions ($\Delta \log fO_2$ (FMQ) $\leq$ –3), ascending fluids may precipitate diamond via a second redox-neutral reaction:

$$2C_2H_6 \rightarrow 3CH_4 + C \tag{5.2}$$

These modes of isochemical diamond precipitation require that fluids remain relatively pure (i.e. that progressive dilution of the fluid through addition of a melt component does not occur). For water-maximum fluids ascending along a normal cratonic geotherm (40 mW/m^2),[59] this condition is only met for harzburgite and dunite, whereas the higher melting temperatures of peridotite in the presence of more reducing fluids permits the reaction in Eq. (5.2) to occur in lherzolites as well.[58]

In such fluid-buffered systems, the fractionation of carbon isotopes during diamond growth occurs in the presence of two dominant carbon species in the fluid: either CH_4 + CO_2 or CH_4 + C_2H_6. The equations to model Rayleigh isotopic fractionation in these multi-component systems (RIFMS) were developed by Ray and Ramesh[60] and applied to the geologically likely case (based on xenolith fO_2 measurements) of diamond precipitation from ascending or cooling near-water-maximum fluids (reaction 1).[51]

Calculations revealed unexpected fundamental differences relative to diamond crystallization from a single carbon fluid species:

(1) Irrespective of which carbon species (CH_4 or CO_2) dominates the fluid, diamond crystallization from mixed CH_4–CO_2 fluids will always lead to minor (<1‰) enrichment in ^{13}C as crystallization proceeds. In contrast, diamond precipitation through wall rock-buffered redox reactions from a fluid containing only a single carbon species can result in either progressive ^{13}C enrichment (CO_2 or carbonate fluids) or ^{13}C depletion (CH_4 fluids) in diamond. These two contrasting models of diamond formation can be tested through $\delta^{13}C$–N content profiles in individual diamonds as the mixed fluid model predicts that zoning profiles should be characterized by progressive ^{13}C enrichments, whereas the single fluid redox model predicts both ^{13}C enrichments and depletions, depending on whether the fluids are oxidized or reduced. Notably, the available zoning profile data are more consistent with the mixed fluid model in that coherent trends in $\delta^{13}C$ values almost invariably involve rimward enrichments in ^{13}C and total variations within individual growth zones (i.e. zones precipitated from a single fluid pulse) are generally small (<1‰).

(2) Because all mantle-derived fluids should have mantle-like $\delta^{13}C$ values near –5‰ irrespective of their redox state, fluid speciation exerts the principal control on

diamond $\delta^{13}C$ values. For example, the $\delta^{13}C$ value of the first diamond precipitated from a relatively oxidized CO_2–CH_4 H_2O fluid in which CO_2 is the dominant fluid carbon species ($CO_2/[CO_2 + CH_4] = 0.9$) will be 3.7‰ lower than the first diamond crystallized from a reduced mixed fluid in which CH_4 is the dominant fluid carbon species ($CO_2/[CO_2 + CH_4] = 0.1$). Accordingly, the observed tight mode of peridotitic diamonds worldwide at $\delta^{13}C$ of -5 ± 1‰ requires that CH_4 generally constitutes $\geq$50% of the carbon species in peridotitic diamond-forming fluids.

The RIFMS equations were applied to CH_4-bearing Marange diamonds (Figure 5.2) to model their *in situ* stable isotope and nitrogen content data.[40] Application of Eq. (5.1) allowed us to perfectly match the observed covariations in $\delta^{13}C$–$\delta^{15}N$–N content and at the same time explain the previously counterintuitive observation of progressive ^{13}C enrichment in diamonds (Figure 5.2) that appear to have grown from a fluid with CH_4 as the dominant carbon species.

Importantly, the observation of CH_4 in Marange diamonds (Section 5.3.1) along with detailed *in situ* isotope analyses[40] have allowed us to confirm the important role of CH_4-rich fluids in worldwide peridotitic diamond formation. At this time, *in situ* data on carbon and nitrogen isotope and N content zoning profiles across diamond plates are still fairly scarce. Future research and the acquisition of many more isotopic profiles across peridotitic diamonds is needed to test whether water-maximum fluids are indeed the prevalent way for peridotitic diamonds to form.

5.3.3 Progress in Understanding Diamond-Forming Metasomatic Fluids

Gem-quality monocrystalline diamond is often devoid of fluid inclusions, and so the best samples of C-rich diamond-forming fluids are those trapped by fast-growing, so-called fibrous diamond. Here, millions of microinclusions (normally 0.2–0.5 μm in size) populate cuboid diamonds, internal fibrous zones of octahedral diamonds, or an overgrowth of a fibrous 'coat' around a preexisting diamond. The microinclusions carry a secondary mineral assemblage and a residual low-density hydrous fluid,[61] which at mantle temperatures constitute a uniform high-density fluid (HDF; either a melt or a high-density supercritical C–H–O fluid). The major elements define four compositional end-members (Figure 5.3): saline HDFs rich in Cl, K, Na, water, and carbonate; high-Mg carbonatitic HDFs rich in Mg, Ca, Fe, K, and carbonate; and a continuous array between a low-Mg carbonatitic end-member rich in Ca, Fe, K, Mg, and carbonate and a silicic end-member rich in Si, K, Al, Fe, and water.[62] The incompatible trace elements are highly enriched in all HDFs, reaching levels of ~1000 times the primordial mantle concentrations, and they are characterized by two main trace element patterns: one with alkali and high field strength element depletions and large ion lithophile element (LILE) enrichments similar to calcalkaline magmas and continental rocks; and the other with lower LILE abundances and smoother overall patterns similar to oceanic basalts.[63,64] Radiogenic isotopic studies of Sr, Nd, and Pb tracers from HDFs are very scarce; nevertheless, the

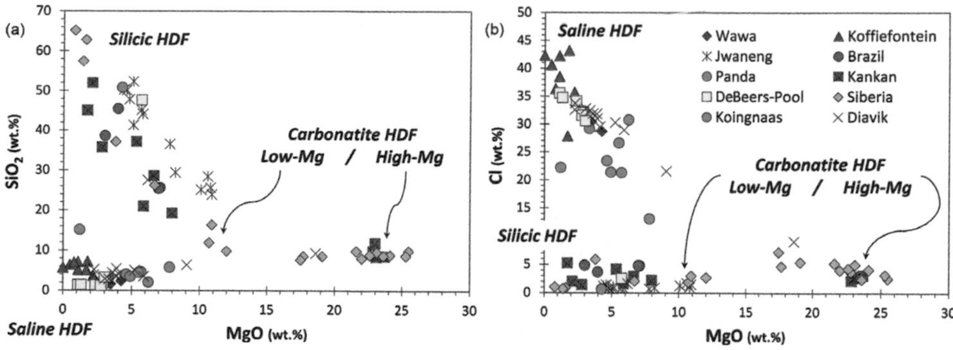

Figure 5.3 (a and b) SiO_2 and Cl versus MgO content of HDF microinclusions in 89 fibrous diamonds from different lithospheric provinces (in wt.% on a water- and carbonate-free basis). The high-Mg carbonatitic compositions are close to experimental near-solidus melts of carbonate-peridotite, while the low-Mg carbonatitic to silicic HDFs form an array that is close in composition to experimentally produced fluids/melts in the eclogite + carbonate ± water system.[62,63,65–67] The saline HDF end-members have been related to fluids derived from seawater-altered subducted slabs.[68]

Data: DeBeers-Pool, Koingnaas, and Kankan from Weiss et al.;[62,69] Koffiefontein from Izraeli et al.;[70] Brazil from Shiryaev et al.;[71] Diavik and Siberia from Klein Ben-David et al.;[72,73] Jwaneng from Schrauder and Navon;[74] Panda from Tomlinson et al.;[75] Wawa from Smith et al.[76]

available data indicate their derivation from sources varying between the "depleted" convecting mantle and ancient incompatible element-enriched lithosphere (including recycled old continental crust[65]).

As a rule, the composition of HDF microinclusions in an individual fibrous diamond is homogenous, with only a handful out of the ~250 fibrous diamonds analyzed to date showing conspicuous radial (core-to-rim) changes.[71] These diamonds revealed correlative variations in hydrous silicic fluids films around mineral inclusions[32] and saline and carbonatitic HDF microinclusions in octahedral diamonds[77] and in twinned crystals (macles[78]). Together, these similarities suggest that many lithospheric diamonds could have formed from the four carbonate-bearing HDF end-members known from fibrous diamonds. These HDF fluids provide another growth mechanism that is different from the mechanism by which lithospheric diamonds form from non-carbonate-bearing fluids (see Sections 5.3.1 and 5.3.2).

Diamond formation is a by-product of mantle metasomatism, whereby HDFs migrate through and react with different mantle reservoirs. Their entrapment in the diamond gives us a unique glance at the initial stages of melting and at the enigmatic mantle process known as metasomatism. A strong connection exists between high-Mg carbonatitic HDFs and a carbonated peridotite, indicating that the diamond grew from carbon supplied by the HDFs. Also, the origins of silicic and low-Mg carbonatitic HDFs have been related to the melting of hydrous carbonated eclogites.[65,67] On the other hand, fibrous diamonds generally make up less than a few percent of a mine's production,[2] and differences in the textures, nitrogen aggregations, diamond ages, and the range of carbon isotopic compositions between fibrous and monocrystalline gem-quality diamonds mean that further work is needed to establish whether HDFs are responsible for the formation of all types of diamonds. Accumulating evidence indicates the involvement of HDFs in the growth of many monocrystalline diamonds – the most abundant type of diamond. The age difference between monocrystalline and fibrous diamonds was bridged by finding fibrous diamonds of Archean age,[76] as well as fibrous diamonds with aggregated nitrogen (25–70% B centers[79]). The interaction of HDFs with depleted garnets was shown to closely produce sinusoidal rare earth element (REE) patterns,[62] which are one of the primary features of harzburgitic garnet inclusions in monocrystalline diamonds.[10] The deep mantle source of saline HDFs has for many years remained ambiguous, until recently, when the first conclusive trace element and Sr isotopic fingerprints indicated that they are derived from seawater-altered subducted slabs.[68] Moreover, clear chemical evolutionary trends in these Northwest Territories Canadian diamonds identify saline HDFs as parental to *in situ*-forming carbonatitic and silicic melts in the deep continental lithosphere. These advances open a new window on understanding the history of magmatism and metasomatism in the deep SCLM and their relationship to carbon and water mobility and diamond formation.

5.4 Sources of Carbon and Recycling of Volatiles

5.4.1 Atmospheric and Biotic Recycling of Sulfur into the Mantle

An important way to trace potential carbon sources in the deep carbon cycle is by using the petrogenesis of mineral inclusions in diamond. Iron–nickel–copper sulfides represent the most common type of mineral enclosed in diamonds. This overabundance of sulfide inclusions in diamonds compared with silicates suggests that a genetic link exists between sulfides and diamonds, but the exact nature of this relationship remains unclear.

Sulfide inclusion mineralogy implies at least two distinct origins for these sulfides, with Ni-rich specimens akin to peridotitic affinity and, more frequently, Ni-poor specimens that originate from a mafic crustal rock. Re/Os studies have provided robust evidence that sulfide inclusion compositions evolved with age.[17] Sulfides older than 3 Ga are all peridotitic, while eclogitic specimens prevail from the Mesoarchean until the Proterozoic. They display a discrete age distribution with at least one major age peak coeval with continental collision, which is well characterized in the Kaapvaal craton. These observations indicate irreversible changes in crust/mantle dynamics, with episodic subduction events starting ca. 3 Ga that would have driven crustal sulfur into the cratonic keel.[80]

This scenario has major implications for the global budget of volatile elements and its evolution through time. In particular, the flux linking the shallow crust (which is a major sink of volatiles) and the mantle is a key parameter because it allows the recycling of light elements (C, O, H, N, S) and thus partially controls the deep Earth budget.

Unlike carbon or oxygen, for which fractionated isotopic compositions can lead to ambiguous interpretation due to high-temperature fractionation processes[81] or perturbation by mantle metasomatism,[82] S isotopic systematics provides a unique way to assess the contribution of Archean surficial reservoirs in mantle rocks.[83] Sulfur is present in all of the external envelopes of Earth (ocean, atmosphere, and biosphere). It participates in many chemical (biotic and abiotic) and photochemical reactions. Large variations of $\delta^{34}S$ in sedimentary rocks are mainly attributed to microbe-mediated sulfur metabolisms. On the other hand, in the Archean atmosphere, photochemical reactions involving UV light induced sulfur mass-independent fractionation (MIF). Photochemical products – elemental sulfur and sulfate aerosols – carried anomalous ^{33}S-enrichment ($\Delta^{33}S > 0‰$) and depletion ($\Delta^{33}S < 0‰$), respectively. Both species have been transferred to seawater and then preserved as two independent isotopic pools in chemical sediments (banded iron formations and black shales) or hydrothermally altered oceanic crust older than 2.4 Ga. MIF sulfur anomalies ceased sharply at the Archean/Proterozoic boundary as a consequence of UV screening by ozone. Thus, multiple S isotopic systematics is a robust tracer of the Archean surficial sulfur,[83] but can also be used to track the fate of specific sedimentary pools.

Pioneering studies of S isotopes in sulfide inclusions reported departure from mantle composition and concluded that altered oceanic crust[84] or sediments[85] were recycled in the diamond growth environment. Multiple sulfur isotope measurements ($\delta^{33}S$, $\delta^{34}S$) provide a more complete assessment of the recycled sulfur pools.[83,86] In addition to a wide range of $\delta^{34}S$ ($-9‰ < \delta^{34}S < 3.4‰$), eclogitic sulfide inclusions from the Orapa and Jwaneng diamonds carry MIFs that are mostly positive ($-0.5‰ < \Delta^{33}S < 1‰$). While the most anomalous sulfur isotopic compositions match the sulfur compositional trend produced by photochemical reactions with 220-nm radiation[87] previously found in Archean sediments, some ^{34}S-depleted specimens require additional fractionation, most likely related to biologic activity (Figure 5.4). By comparison, peridotitic sulfides from the Slave and Kaapvaal cratons[86,98] do not carry significant MIF ($\Delta^{33}S$ from –0.12 to 0.19‰). These results provide compelling evidence that MIF S isotopic signatures are not produced by high-temperature processes in the mantle, but indubitably reflect an input of chemical sediments from the surface to the diamond growth region.

Geologic evidence indicates that cratonic keels have been isolated early from the convecting mantle. This is consistent with the absence of $\Delta^{33}S$ anomalies in mid-ocean ridge basalt (MORB; Figure 5.4) reported by Labidi et al.[89,90] Surprisingly, however, two independent studies on sulfides from ocean island basalt (OIB)[91] have reported small but significant negative $\Delta^{33}S$ anomalies (down to –0.8‰) correlated with strictly negative $\delta^{34}S$ (Figure 5.4). In both Pitcairn (enriched mantle reservoir 1 (EM1)) and Mangaïa (high-μ mantle reservoir (HIMU)) samples, these trends match the composition previously reported for sulfides from hydrothermal barite veins in altered oceanic crust.[93,94] Accordingly, these studies indicate that deep mantle heterogeneities inherited from seawater Archean sulfates have been preserved over billions of years. It is worth noting that negative anomalies are underrepresented in the sedimentary record. One of the most exciting hypotheses coming along with the presence of negative $\Delta^{33}S$ in some OIB is that part of the missing surficial sulfur could be stored in the deep mantle.[95]

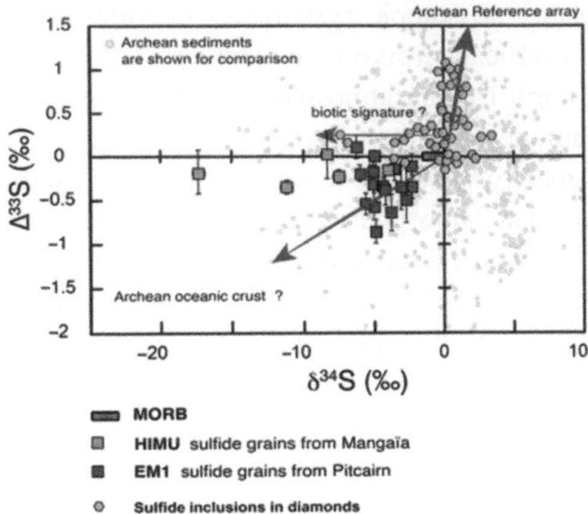

Figure 5.4 $\Delta^{33}S$ (‰) versus $\delta^{34}S$ in mid-ocean ridge basalt (MORB; blue rectangle[87,89,90]), sulfide inclusions in diamonds (yellow hexagons[83,86]), and sulfides from high-μ mantle reservoir (HIMU) ocean island basalt (OIB; Mangaïa,[91] light green squares) and enriched mantle reservoir 1 (EM1) OIB (Pitcairn,[92] dark green squares). While MORB are homogeneous and devoid of MIF, sulfides from SCLM and from some OIB contain the relict of Archean surficial sulfur. Sulfide inclusion compositions are best explained by a combination of atmospheric and biotic effect and resemble what has been previously observed in Archean chemical sediments. Sulfide in OIB carry negative $\Delta^{33}S$ together with negative $\delta^{34}S$, as previously observed in sulfides from altered oceanic crust.

The data obtained so far on mantle samples tend to indicate that the Archean surficial components recycled in the SCLM differ from those found in some OIB. Additional data are required to confirm this view. The relative abundance of the minor isotope of sulfur, ^{36}S, is also affected by both mass-dependent reactions (related to microbial cycling; e.g. Ono et al.[96]) and mass-independent atmospheric reactions that lead to variations of $\delta^{34}S$ and $\Delta^{33}S$. In the future, studying covariations of $\Delta^{33}S$ and $\Delta^{36}S$ may help to provide a more complete assessment of the recycled sulfur pools and ultimately add new constraints upon crust/mantle dynamics.

5.4.2 Carbon and Nitrogen Cycling into the Mantle Transition Zone

Studies of the carbon and nitrogen isotopic composition of diamonds represent an opportunity to examine volatile migration within the transition zone (410–660 km depth), a key region within Earth's interior that may be the main host for recycled material. Seismology and mineral physics show the tendency of subducted slabs to pond at the lower mantle transition zone boundary at ~660 km depth within a region where diamonds capture rare inclusions of majorite garnet[97] and assemblages comprising Ca-rich inclusions (CaSi-perovskite, -walstromite, -larnite, -titanite).[98] These rare diamonds principally originate from four localities: the Juína area in Brazil, Kankan in Guinea, and Monastery and Jagersfontein in South Africa.

Brazilian diamonds with majorite inclusions exhibit carbon ($\delta^{13}C = -10.3 \pm 5.5‰$, 1σ; Figure 5.5a[20,21,81,99–104]) and nitrogen isotope compositions ($\delta^{15}N = +0.4 \pm 2.9‰$, 1σ; Figure 5.5b) well outside of the current mantle range ($\delta^{13}C = -5 \pm 3‰$ and $\delta^{15}N = -5 \pm 4‰$). Most likely, these diamond isotopic compositions record subduction-related carbon and nitrogen.[101] The partial melting of former carbonated oceanic crust in the transition zone may produce carbonatitic melts, and the interaction of these melts with ambient convecting mantle may be responsible for the formation of many/most mantle transition zone diamonds and their inclusions (Figure 5.5c).[56]

Some Brazilian diamonds with super-deep Ca-rich assemblages are thought to derive from even greater depth, at the transition zone–lower mantle boundary.[18] These diamonds with distinct carbon isotope compositions ($\delta^{13}C = -5.9 \pm 3.7‰$; Figure 5.5a) that overlap with the main mantle range are suggested to originate either from homogenized subducted sediments (composed of 80% carbonate and 20% organic carbon) or mantle-related fluids from the convecting mantle (i.e. non-primordial; Figure 5.5c).

The carbon isotope compositions of Kankan diamonds with majorite- and with Ca-silicate inclusions are distinct from Brazilian samples by being ^{13}C enriched, with an overall $\delta^{13}C$ of $0.3 \pm 2.2‰$, outside the mantle range (Figure 5.5a).[107,108] This carbon isotopic signature is consistent with derivation from subducted carbonate with no/little former organic carbon involved. The identification of carbonate inclusions in ultra-deep diamonds indicates that carbonate may be efficiently transported deep into the mantle.[98] The positive $\delta^{15}N$ values of transition-zone diamonds from Kankan also strongly support a subduction origin. Modeling of the local covariations of $\delta^{13}C$–$\delta^{15}N$–N compositions within individual diamonds indicates that they grew from parental fluids involving both oxidized (majority) and reduced (minority) fluids,[107] highlighting the likely heterogeneity of the transition zone (Figure 5.5c).

Diamonds from Monastery and Jagersfontein containing majorite inclusions exhibit extremely depleted $\delta^{13}C$ values of $-16.7 \pm 1.2‰$ and $-19.7 \pm 2.1‰$, respectively, and strictly positive $\delta^{15}N$ values (Figure 5.5a), which are again consistent with subducted material.[106] The relatively low nitrogen contents (<55 at. ppm) of the host diamonds together with the positive chondrite-normalized REE slopes and high $\delta^{18}O$ values of the majorite inclusions are consistent with the formation of these diamonds within hosts that originated from hydrothermally altered basaltic protoliths. The preferred mechanism to form these diamonds is by dissolution and re-precipitation (Figure 5.5c), where subducted metastable graphite would be converted into an oxidized or reduced species during fluid-aided dissolution, before being re-precipitated as diamond.[109] In this situation, carbon remains in the subducting slab and is locally redistributed to form sublithospheric diamonds beneath the Kaapvaal Craton (Figure 5.5c).

The carbon and nitrogen isotopic signatures of transition zone diamonds worldwide indicate that they likely crystallized from fluids derived from subducted material, illustrating the deep cycling of surficial carbon and nitrogen into and through the transition zone. Carbon and nitrogen seem to be efficiently retained in the oceanic lithosphere during subduction, prior to being locally mobilized in the transition zone to form diamonds.

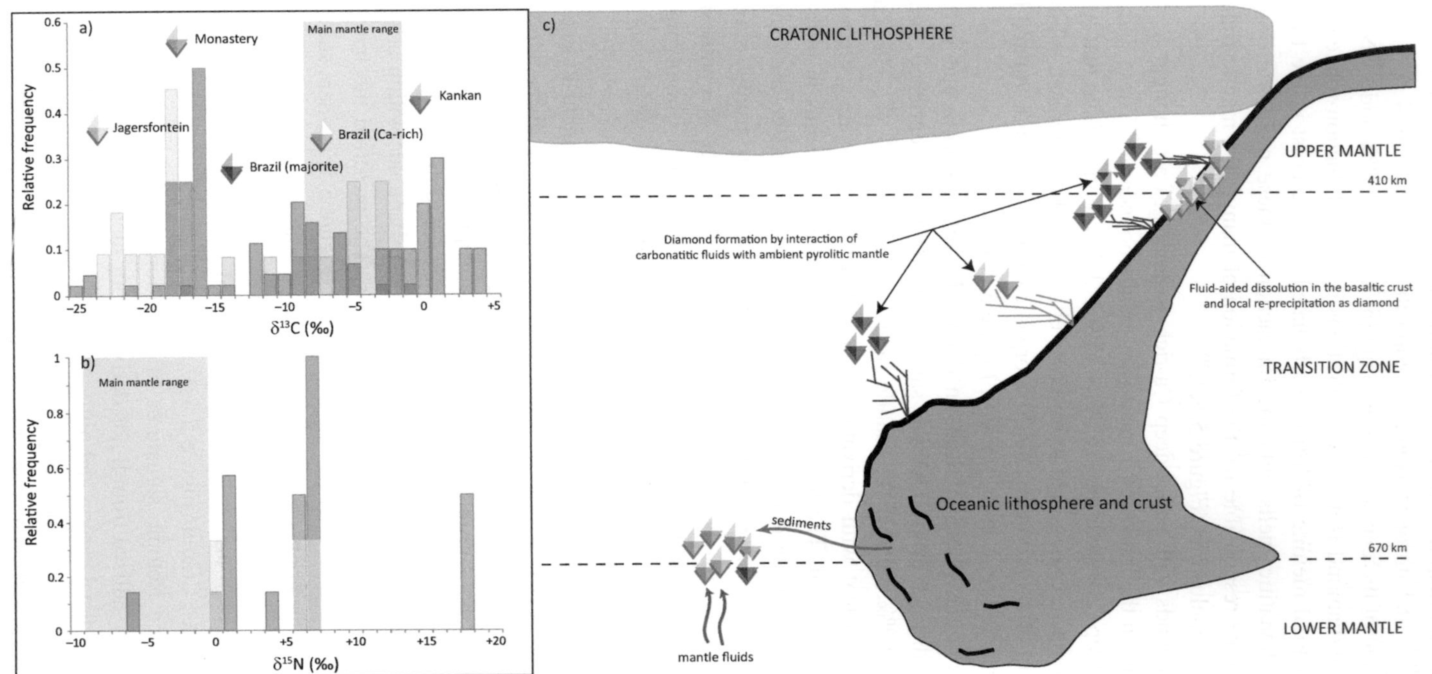

Figure 5.5 (a) Histogram of $\delta^{13}C$ values of transition-zone diamonds from Jagersfontein and Monastery (South Africa), the Juína area in Brazil (containing either majorite or Ca-rich inclusions), and Kankan (Guinea). The mantle range (gray band) is defined by the study of fibrous diamonds, mid-ocean ridge basalts, carbonatites, and kimberlites.[20,21,81,99–107] (b) Histogram of $\delta^{15}N$ values of transition zone diamonds from Jagersfontein, Monastery, Brazil, and Kankan.[99,101,106,107] (c) Schematic history of diamond formation in the transition zone, illustrating the deep recycling of surficial carbon and nitrogen in the mantle. At each locality, transition zone diamonds did not necessarily form during single subduction events.

5.4.3 Earth's Deep Water and the Carbon Cycle

Water coexists with carbon as CH_4 or CO_2 in mantle C–H–O fluids. Studies of the diamond-bearing carbon cycle also afford a chance to follow water – one of the defining components of Earth's mantle. The mantle transition zone, from ~410 to 660 km depth, was identified 30 years ago as a potentially major sink for water in Earth,[110] where H_2O could incorporate into nominally anhydrous minerals as hydroxyl species. Seismic tomography images of some subducted oceanic slabs ponding within the transition zone[111] brought into focus the potential for this geochemical reservoir to store volatiles recycled along with the oceanic slabs. Despite experimental verification of the high water storage capacity of high-pressure polymorphs of olivine – wadsleyite and ringwoodite (e.g. Kohlstedt et al.[112]) – considerable debate ensued regarding the degree of hydration of the mantle transition zone. Its state of hydration is poorly constrained at regional and global scales because of the compensating effects of temperature, bulk composition, and mineralogy in modeling geophysical observations. Diamonds have a unique role to play in illuminating this problem.

Since the discovery of super-deep diamonds,[18] their immense value in providing samples of the upper mantle, transition zone, and lower mantle has become clear.[24,25,98,113–115] Several studies discovered olivine inclusions suggested to have previously been wadsleyite or ringwoodite based on frequent spinel exsolutions[116] or their coexistence with other phases thought to be from the transition zone.[18,117,118]

During the DMGC consortium initiative on super-deep diamonds, a diamond from the Rio Aripuanã in the Juína district of Mato Grosso, Brazil, was found to contain the first terrestrial occurrence of un-retrogressed ringwoodite (Figure 5.6).[23] This ~30-μm inclusion was estimated to contain ~1.4 wt.% H_2O. Subsequent recalibration of IR absorbance for hydrous ringwoodite by absolute methods (proton–proton scattering) across the Mg_2SiO_4–Fe_2SiO_4 solid solution[119] refined this estimate to 1.43 ± 0.27 wt.% H_2O. The observed water content in the ringwoodite inclusion is close to the maximum storage capacity (~2 wt.% H_2O) observed in experiments at conditions representative of cold-slab geotherms.[120] This constraint is strong evidence that the host environment for the ringwoodite was a subducted slab carrying significant H_2O into the transition zone. With such a restricted data set, it remains to be determined how representative the Juína ringwoodite found by Pearson et al.[23] is for the mantle transition zone at regional scales; however, Nestola and Smyth[121] estimated that when this water content is applied to the whole mantle transition zone (~7% of Earth's mass), the total water content would be ~2.5 times the volume of water in Earth's oceans. Even if the single natural specimen represents a local phenomenon of water enrichment, seismological evidence of dehydration melting above and below the transition zone[122–124] and the report of the super-hydrous "Phase Egg" in a super-deep diamond by Wirth et al.,[125] along with other recently described phenomena such as the presence of brucite within ferropericlase[126] and the documentation of ice VII within diamonds originating in the transition zone and lower mantle by Tschauner et al.,[127] provide stunning evidence of linked water and carbon cycles in Earth's mantle extending down into the transition zone and possibly as deep as the top of the lower mantle.

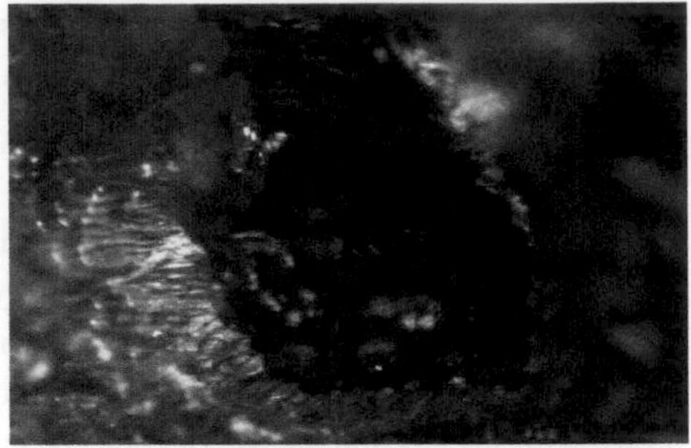

Figure 5.6 Photograph of Juína diamond JuC-29 and a magnified view of the ringwoodite inclusion (right panel, center of image) showing the characteristic indigo–blue color of ringwoodite.

The presence of water in ringwoodite, likely hosted in a cool, subducted oceanic slab along with the recycled signature of carbon and nitrogen evident in transition zone diamonds (see Section 5.4.2) indicate that the transition zone as sampled by diamonds is a reservoir that is potentially dominated by subducted slabs and their recycled volatiles. This region of Earth's mantle is therefore a key zone for the storage and ultimate return of recycled volatiles, including carbon, in Earth.

5.5 Mineral Inclusions and Diamond Types

5.5.1 Experiments to Study Diamond Formation and Inclusion Entrapment

Experiments on diamond formation and growth have two main goals: gaining a better understanding of diamond-forming reactions and simulating the mechanisms for inclusion entrapment. Both goals address an essential step in the deep carbon cycle where carbon is liberated from fluids or melts and reduced to form diamond. Synthetic diamond growth was one of the driving forces of high-pressure technology,[128] but for many years the focus remained on their industrial production, which employs metallic liquid solvents, and not on identifying the growth media of natural diamonds. Arima et al.[129] first synthesized diamond from a kimberlite composition, but since then a very wide range of plausible mantle fluid and melt compositions have yielded synthetic diamonds, including pure carbonates,[130] C–H–O fluids,[131] carbonate–silicate mixtures with water and/or chloride,[132,133] and metal sulfides.[134] Palyanov et al.[130] achieved diamond nucleation using Na_2CO_3 + C–H–O fluid mixtures at 1150°C and 5.7 GPa, overcoming the nucleation barrier exhibited by direct graphite transformation that exists throughout cratonic lithosphere conditions. Further studies have shown nucleation and growth to be greatly promoted by the presence of H_2O.[135–137] There is further evidence that NaCl may act to reduce the growth rate.[132,136] Experiments to synthesize diamonds from reduced C–H–O fluids have been less successful. Diamond formation was found to be inhibited in the presence of CH_4–H_2 fluids, for example,[131,138] although recent measurements of CH_4 imply that diamonds do indeed grow naturally with such reduced fluids.[40]

Experiments examining natural diamond growth media are generally performed with excess graphite, and in the absence of an obvious reducing agent the main formation mechanism is dissolution and precipitation along a decreasing temperature gradient.[139] Luth and Stachel[47] have argued that a thermal gradient growth process may form diamonds in the mantle because the dissolved carbon content of a C–H–O fluid in equilibrium with diamond decreases with temperature and also with pressure. Diamond growth from cooling or decompressing fluids overcomes the difficulty of dissolved carbon species being reduced or oxidized by external agents such as ferric or ferrous iron, which are not present in wall rocks at the concentrations required to account for natural macroscopic diamond growth. Calculations show that, in a C–H–O fluid, the greatest change in diamond solubility with temperature occurs at the so-called H_2O maximum, where the carbon content of the fluid is actually at a minimum.[47] This may explain why diamond growth is promoted in experiments with high H_2O concentrations.[136,137] One problem with this occurring in the mantle is that the H_2O contents of mineral inclusions such as olivine trapped in diamonds should be near H_2O saturation, which seems not to be the case for the lithospheric diamonds measured to date.[140] Reduced CH_4-rich fluids, on the other hand, experience relatively small changes in carbon solubility with temperature, but much larger changes with pressure.[47] This, coupled with problems of H_2 loss from capsules at reducing conditions, may at least partly explain the difficulties in diamond synthesis from CH_4-rich fluids.

Inclusions trapped in diamonds are among the few diagnostic tools that can constrain diamond growth media, and recent experiments have attempted to explore this link by capturing inclusions in synthetically grown diamonds. Experiments using water-rich mixtures containing carbonates and silicates have successfully produced a broad range of mineral, melt, and fluid inclusions at 6–7 GPa and 1300–1400°C (Figure 5.7).[136,137] Starting compositions were based on fluid inclusion analyses of fibrous diamonds,[61] and polycrystalline and fibrous growth textures were reproduced, which tended toward monocrystalline growth at higher temperatures. Bureau et al.[137] used observations of coexisting melt and fluid inclusions to infer the temperature where the two phases become miscible in the system examined. Bataleva et al.[141] also captured inclusions in monocrystalline diamonds grown from SiO_2–(Mg,Ca)CO_3–(Fe,Ni)S mixtures at 6.3 GPa and 1650–1750°C. The inclusions, which encompassed quenched carbonate–silicate melts, sulfide melts, and CO_2 fluid, reflected, in part, the growth medium that is generally assumed for natural diamond inclusions, but Bureau et al.[136] demonstrated that carbonate minerals are readily trapped in diamonds that are actually growing from H_2O-dominated liquids.

The alternative to diamond growth from cooling or decompressing C–H–O-rich liquids is precipitation due to redox reactions with iron or potentially sulfide species in surrounding minerals or melts. This would seem to be problematic, as mentioned above, because iron species and sulfides lack the redox capacity to reduce or oxidize macroscopic

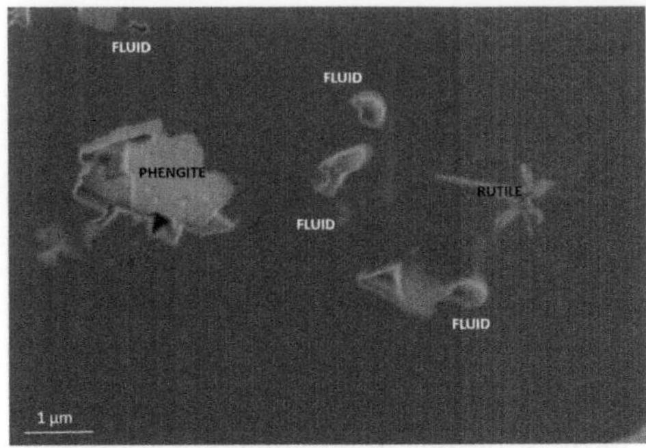

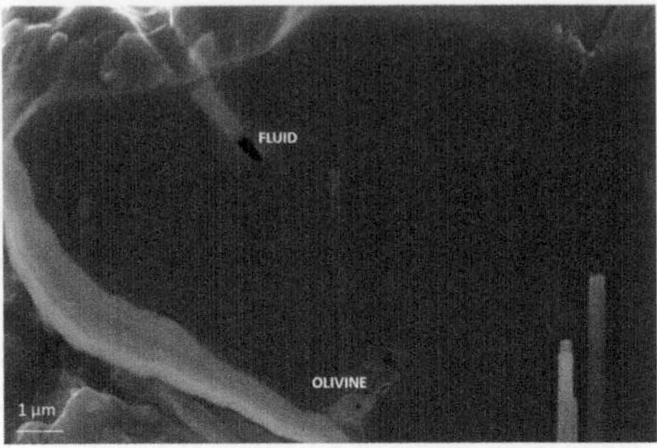

Figure 5.7 Scanning electron microscope images of monocrystalline diamond slices containing trapped inclusions, prepared by focused ion beam thinning from diamonds synthesized at 7 GPa and 1300°C for 30 hours. The fluid is lost from the inclusions once they become exposed, leaving only cavities.

diamonds from fluids.[47] With the recognition that the mantle may become metal saturated at depths >200 km, however, a number of experiments have studied a so-called redox freezing scenario[142] where carbonate melts may migrate out of subducting slabs and rise into the overlying metal-saturated sublithospheric mantle.[57,143] An intriguing result is that the oxidation of iron and reduction of carbonate in these experiments form magnesiowüstite with a wide range of Fe contents,[143] which is consistent with its occurrence in sublithospheric diamonds.

5.5.2 Nanoscale Evidence for Polycrystalline Diamond Formation

Carbon that does not occur as monocrystalline diamond but rather as a polycrystalline diamond aggregate (PDAs; "framesites,"[144] boart,[145] or diamondite[146]) represent a little-studied variety of carbon in the mantle that has the potential to reveal how carbon percolates at shallower levels of diamond stability. The PDAs can make up 20% of the diamond production in some Group I kimberlites (K. de Corte, pers. comm. 2012), but are not reported from Group II kimberlites, nor other diamondiferous volcanic rocks. Their polycrystalline nature indicates rapid precipitation from carbon-oversaturated fluids.[145] Compared to monocrystalline diamonds, individual PDAs often contain a more varied and chemically heterogeneous suite of inclusions and minerals intimately intergrown with the diamond crystals. However, these inclusions are sometimes not shielded from metasomatism and alteration, which can accompany deformation and recrystallization of the diamonds as seen in electron-backscatter diffraction images.[147] The suite of minerals found in PDAs, while derived from Earth's mantle, is unlike inclusions in diamond; websteritic and pyroxenitic parageneses dominate and olivine is absent. Individual grains commonly are chemically and structurally heterogeneous,[148,149] suggesting that many grew by reaction between mantle minerals and metasomatic fluids and/or melts. Trace element patterns of the silicates can show signatures of carbonatite metasomatism, supporting their formation from oxidized fluids. However, some samples contain very reduced inclusions, such as iron carbide.[150] Thus, a single formation process is unlikely. Like the associated websteritic and pyroxenitic silicates, the non-silicate phases such as magnetite may represent reaction products associated with diamond formation. Jacob et al.[148] documented epitactic relations between grains of iron sulfide, which oxidized to magnetite and served as nucleation templates for the host diamond crystals, underlining the role of redox reactions in diamond formation. Textural evidence for a high-pressure Fe_3O_4 precursor phase constrains the depth of formation of the PDAs to the base of the lithosphere. Hence, this variety of diamond, due to its rapid formation, provides insights into the extreme "corners" of the diamond formation process in Earth's SCLM. Such evidence is often eradicated in larger monocrystalline diamonds, which grow slowly and record time-integrated evidence within the deep carbon cycle. Several questions require answers in order for us to understand the role of PDAs and their context in the deep carbon cycle.

Applying robust constraints on absolute ages proves difficult for PDAs, and the sparse data available suggest episodic formation. Jacob et al.[149] observed trace element zonation in PDA-hosted garnets from Venetia (South Africa); modeling using known diffusion coefficients showed that the preservation of the zoning requires that these samples precipitated shortly before kimberlite eruption. The garnets show unradiogenic ε_{Nd} (–16 to –22); this implies that they contain older lithospheric material remobilized with the carbon-bearing fluid to form PDAs. However, the nitrogen aggregation states of PDAs (probably from southern Africa) showed that these samples are not exclusively young, but formed in several distinct events over a long time span, possibly more than millions of years.[151]

Carbon isotope values ($\delta^{13}C$) in PDAs span a range from ca. –1‰ to ca. –30‰ (n = 115) with peaks at –5‰ and at –19‰.[152,153] The first peak is the typical mantle,[8] but the second peak at –19‰ is typical only for PDAs[8] and, to date, remains unexplained. Nitrogen concentrations and $\delta^{15}N$ in PDAs cover large ranges: 4–3635 at. ppm and –6.1‰ to +22.6‰,[152,154,155] but both are independent of $\delta^{13}C$ values, arguing against broad-scale fractionation processes.[5,155,156] Some authors have argued for Rayleigh fractionation of a mantle-derived fluid,[5,156] while others inferred a subducted fluid origin.[151,152] A role for a subducted component is supported by the high nitrogen concentrations in some of the diamonds, combined with heavy nitrogen isotopic compositions – typical for material from Earth's surface.[151,152]

Collectively, these data imply that PDAs are the product of small-scale reactions with fluids related to both ambient mantle carbon, remobilized SCLM material, and subducted crustal carbon. The formation of PDAs also requires interaction of melts and rocks within the subcratonic lithosphere, and these events are episodic. These fluid-driven, rapid reactions serve to "freeze" carbon as diamond in the subcratonic lithosphere, and they may preserve some of the best evidence for small-scale chemical heterogeneity at the site of formation. Obtaining accurate information on the age and depth of polycrystalline diamond formation is the next step to addressing their role in the deep carbon cycle, since they may represent the shallowest form of diamond-forming fluid.

5.5.3 Proterozoic Lherzolitic Diamond Formation: A Deep and Early Precursor to Kimberlite Magmatism

Lithospheric diamonds are ancient and form during episodes of fluid infiltration into the lithospheric mantle keel in response to large-scale geotectonic processes. Studies of lithospheric diamonds provide the main way to look back at carbon cycling over the past 3.5 billion years. The continental lithosphere has been recognized as an important carbon reservoir, which, after initial melt extraction imposing highly refractory and reducing conditions, was gradually re-enriched and re-oxidized through infiltration of volatile-bearing fluids and melts from episodically impinging plumes and subducting slabs.[157] Kimberlite eruptions, which increased in frequency through time (Figure 5.8), sporadically connect this deep lithospheric carbon cycle to surface reservoirs, but require conditions that are favorable to their formation and extraction from a carbonated mantle source.[158–160] The cryptic part of the lithospheric carbon cycle involving diamonds can be illuminated through a combination of age dating and chemical characterization of the inclusions in the diamonds, which reveal the nature of their source rocks and their formation conditions.[58] Diamond formation through time reveals a trend, from harzburgitic sources prior to 3 Ga, to the first appearance of eclogitic diamonds ca. 3 Ga[17] and of lherzolitic diamonds and renewed diamond growth in the Proterozoic recording ages from ca. 2.1 to 0.7 Ga (Figure 5.8). Thus, Proterozoic lherzolitic diamond formation is a widespread phenomenon, corresponding to ~12% of inclusion-bearing peridotitic lithospheric diamonds.[10] Refertilization (lherzolitization) of the initially depleted lithosphere and minor associated diamond growth likely occurred in the presence of small-volume melts, producing garnet inclusions with characteristic, generally mildly sinusoidal REE patterns.[58]

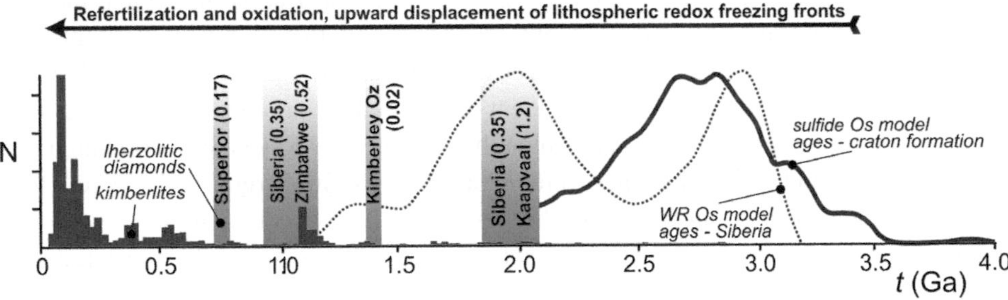

Figure 5.8 Lherzolitic diamond formation through time: ca. 2.1 to 1.8 Ga, diamonds from Premier (Kaapvaal craton) and 23rd Party Congress/Udachnaya (Siberian craton); 1.4 Ga, diamonds from Ellendale (Western Australia);[161] 1.1 to 1.0 Ga, diamonds from 23rd Party Congress/Mir (Siberian craton)[162] and Venetia (Zimbabwe craton);[163] and 0.72 Ga, diamonds from Attawapiskat (Superior craton).[164] Numbers in parentheses give host kimberlite eruption ages (in giga-years) to illustrate the delay between lherzolitization and kimberlite magmatism. Shown for comparison is the age distribution of kimberlites from Tappe et al.,[160] Os model ages of mantle sulfides from Griffin et al.,[165] which predominantly reflect the time of Archaean craton formation and the creation of strongly refractory and reducing mantle lithosphere, and for xenoliths from the Siberian craton, which show a major Paleoproterozoic lithospheric mantle formation event.[166]

The successful eruption of kimberlites has been suggested to require, *inter alia*, metasomatic oxidative "preconditioning" of deep lithospheric pathways in the presence of diamond[167] and of shallower lithospheric levels through the precipitation of hydrous and Ti minerals from kimberlite precursors.[168] Although harzburgitic source rocks are too Fe^{3+} depleted to permit substantial diamond formation by wall rock-buffered redox reactions,[47] lherzolitization accompanied by (ferrous) iron introduction may have restored sufficient redox buffering capacity to precipitate some diamonds by reduction of carbonate from small-volume melts through simultaneous oxidation of Fe^{2+} to Fe^{3+} of the host rock. Indeed, very depleted initial source compositions inferred from average FeO contents in refractory garnet peridotites from various cratons (5.3–6.8 wt.%) give way to higher FeO contents (6.3–7.4 wt.%) when metasomatized lherzolites are included (data in Aulbach[169]). Some metasomatism was likely accompanied by oxidation and an increase in $Fe^{3+}/\sum Fe$, as indicated by the higher fO_2 determined for enriched compared to depleted mantle xenoliths.[167,170,171] Pervasive interaction of the deep mantle column with (proto)kimberlite melts is confirmed by isotopic and trace element studies of mantle xenoliths.[172,173] Moreover, (hydrous) carbonated melts diluted with silicate components, similar to kimberlites, can be stable at lower fO_2 than pure carbonatite.[46] Notably, lherzolitic diamond formation is temporally and genetically dissociated from the later-emplaced kimberlite hosts of these diamonds (Figure 5.8). Within this framework, Proterozoic lherzolitic diamond formation has recently been suggested to represent the deep and early component of the refertilization and reoxidation of the lithosphere required for successful kimberlite eruption.[164] Given the recent advances in experimental and thermodynamic groundwork, the task is now to quantitatively delineate carbon speciation in the cratonic lithosphere through space and time, including the potential role of lherzolitic diamond formation as part of the upward displacement of redox freezing fronts that culminate in kimberlite eruption.

5.5.4 Diamond Growth by Redox Freezing from Carbonated Melts in the Deep Mantle

The bulk composition and trace element distributions in mineral inclusions in super-deep diamonds provide information about the conversion of carbonate to diamond in the deep upper mantle, transition zone, and lower mantle. Of the wide range of inclusions found in super-deep diamonds, those with bulk compositions consistent with Ca-rich majorite garnet and Ca-rich silicate perovskite have provided the most compelling evidence for the role of carbonated, subducted oceanic crustal materials in their origin (e.g. Refs. 18, 20, 21, 49, 56, 98, 100, 102, 105, 118, 174–176).

Inclusions interpreted as former majorite garnet or Ca-perovskite often exhibit composite mineralogy that is interpreted to have formed by unmixing from primary precursor minerals exhibiting solid solution (e.g. Refs. 12, 20, 21, 56, 102, 175, 177). Inclusions formed as majorite solid solutions typically unmix upon uplift in the mantle to a mixture of pyrope-rich garnet and clinopyroxene (e.g. Refs. 12, 21, 177), and on the basis of majorite

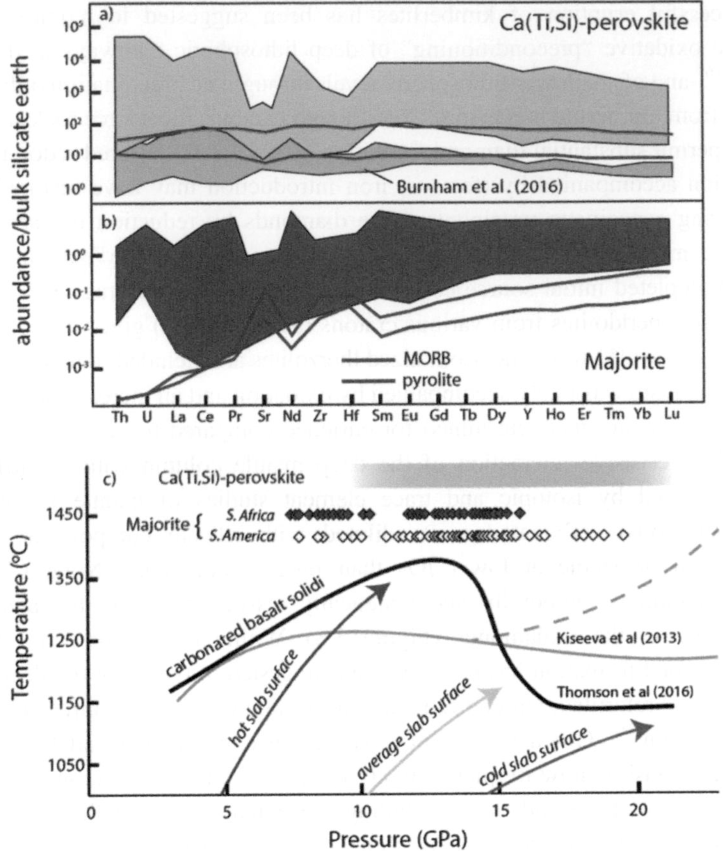

Figure 5.9 Bulk silicate earth normalized trace element composition of (a) "calcium silicate perovskite" and (b) majoritic garnet inclusions compared with models for these phases in subsolidus peridotite (blue) and MORB (red) at transition-zone conditions, as described in Thomson et al.[57] Inclusion compositions are from Davies et al.,[180] Stachel et al.,[98,178] Kaminsky et al.,[100] Tappert et al.,[105] Bulanova et al.,[21] Hutchison,[99,181] Moore et al.,[182] and Burnham et al.[103] (c) Pressure–temperature plot showing the solidi of model carbonated MORB with 2.5 wt.% CO_2[176] and ~4.5% CO_2[179] relative to model geotherms for slab surface temperature at modern subduction zones (Syracuse et al.[183]). The solidi create a depth interval over which most slab surface temperatures intersect the melting curves, producing a region of carbonated melt generation. Also shown are calculated pressures of majoritic garnet inclusions in diamonds from South America and South Africa,[21,57,100,102,105,177,178,181,184,185] calculated from the barometer of Beyer et al. [186]

geobarometry, these inclusions originated at pressures of ~7–20 GPa (Figure 5.9), with most inclusions forming in the range of ~10–17 GPa; that is, in the deep upper mantle and shallow transition zone. Super-deep majorite inclusions are almost exclusively high Ca and low Cr, indicating petrogenesis involving mafic protoliths (e.g. Refs. 12, 21, 178, 179).

Inclusions interpreted to have originated as Ca-rich perovskite come in two varieties: Ti rich and Ti poor. Ti-rich inclusions typically unmix to nearly phase-pure

$CaTiO_3$-perovskite plus walstromite, larnite, titanite, or wollastonite, whereas $CaSiO_3$ varieties are typically walstromite. An important feature of nearly all Ca-perovskite inclusions is their very low MgO content, typically <0.5 wt.%, which effectively precludes equilibration in the lower mantle with bridgmanite at mantle temperatures.[187] This observation, together with phase relations in the $CaSiO_3$–$CaTiO_3$ system,[188] constrains the original Ti-rich perovskites to have formed between ~10 and 25 GPa, similar to the majorite inclusions. The exception are some rare occurrences of Ti-poor $CaSiO_3$ inclusions that have moderate MgO contents, and these may have formed in a lower mantle, peridotitic protolith.[103]

A feature common to most Ca-perovskite and majorite inclusions is extreme enrichment in trace elements.[20,21,56,174,178,189] Figure 5.9 shows normalized trace element abundances in inclusions compared to the modeled abundances for these phases in either mantle peridotite or subducted basalt at transition-zone pressures. These inclusions are not fragments of ambient solid mantle or subducted basalt, and in many cases the most incompatible elements are four to five orders of magnitude more abundant than expected for minerals in these lithologies. Wang et al.[174] suggested a role for carbonated melt in the origin of Ca-perovskite inclusions, and the distinct trace element patterns in Ca-perovskite and majorite inclusions from Juína, Brazil, indicate that the inclusions crystallized from carbonated melt derived from subducted oceanic crust.[20,21,56,176,189]

The involvement of recycled crustal components in the origin of super-deep diamonds is supported by observations that the diamond hosts show a wide range of carbon isotope compositions extending to very light values (e.g. ~0% to –25%)[20,21] and that majorite garnet and other silicate inclusions have isotopically heavy oxygen isotope compositions.[104,109] The high and variable ferric iron content of majorite inclusions is also consistent with an oxidized carbonate component in their origin.[190]

The solidus of carbonated basalt is notably depressed at deep upper-mantle and transition zone pressures (e.g. Refs. 57, 179), and in model MORB composition with 2.5% CO_2, a deep a solidus ledge occurs at pressures of 10–15 GPa (Figure 5.9).[57] Model geotherms for hot and average surface slab temperatures[183] intersect the carbonated basalt solidi, and only the coldest slab surface geotherms can escape melting. Carbonated basalt will melt in the deep upper mantle and transition zone and produce alkali-rich silico-carbonate melts.

Mantle peridotite is reducing and expected to be saturated in metal at transition zone depths.[142] Upon infiltration into the mantle, oxidized carbonated melts will reduce to diamond plus oxygen in a process called "redox freezing."[142] Experiments mimicking this process show that, upon reaction with peridotite at transition zone pressures, such melts can crystallize calcic majorite and Ca-perovskite with compositions matching those found as inclusions.[57] This melt metasomatism can also produce both Fe- and Mg rich periclase,[57] which are common minerals in super-deep assemblages (e.g. Refs. 12, 191). Some large, gem-quality super-deep diamonds show evidence of growing directly from carbon-saturated metallic melts,[49] testifying to a wide range of redox conditions in deeply subducted material.

Future studies will concentrate on determining the role of low-degree, volatile-rich melts in a wider array of inclusions in super-deep diamonds, elucidate the role of water, model the reactive transport processes by which this melt metasomatism occurs, and determine its role in modifying mantle elemental and isotopic compositions.

5.5.5 Evidence for Carbon-Reducing Regions of the Convecting Mantle

The existence, extent, and scale of oxidized versus reduced regions of the convecting mantle are critically important to understanding how carbon moves around at depth. Large and relatively pure diamonds have recently been shown to contain key physical evidence of metallic iron from the deep sublithospheric mantle,[49,192] suggesting that metallic iron may be one of the principal reservoirs of mantle carbon in this region. As a family, diamonds like the historic 3106 carat Cullinan diamond tend to be large, inclusion poor, relatively pure (usually Type IIa), and in their rough state they are irregularly shaped and significantly resorbed.[49,193,194] These characteristics are combined into the acronym "CLIPPIR" (Cullinan-like, Large, Inclusion Poor, Pure, Irregular, and Resorbed) to label this genetically distinct diamond variety.

Out of 81 inclusion-bearing CLIPPIR diamonds, the most common inclusion encountered is a composite, metallic iron–nickel–carbon–sulfur mixture (Figure 5.10). In fact, 60 of the 81 samples contain only this inclusion, which attests to its predominance in the CLIPPIR variety.[192] The metallic inclusions are a made up of cohenite ($(Fe,Ni)_3C$), interstitial Fe–Ni alloy, segregations of Fe-sulfide, and minor occasional Fe–Cr-oxide, Fe-oxide, and Fe-phosphate.[192] A thin fluid layer of CH_4 and lesser H_2 is trapped at the interface between the solid inclusion and surrounding diamond. The mixture is interpreted to have been trapped as a molten metallic liquid.

Other inclusions found in CLIPPIR diamonds represent high-pressure silicates, which provide a constraint for the depth of diamond formation. The second most abundant

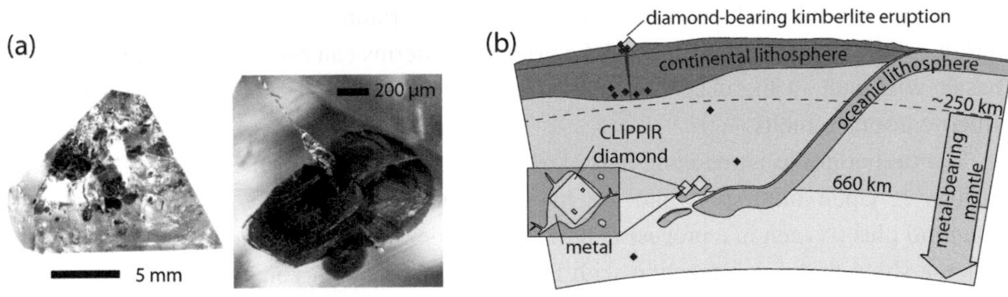

Figure 5.10 (a) Metallic inclusions in a 9.56 carat CLIPPIR diamond with an enlargement of one of the inclusions. These metallic inclusions sometimes have a needle-like tail and typically have large, graphitic decompression cracks around them. (b) Depth constraints place the origin of these diamonds within 360–750-km depths in the mantle, where they are associated with subducted lithologies. The metallic inclusions are evidence for reduced, metal-bearing regions of the deep mantle below a depth of approximately 250 km.

inclusion, after the metallic Fe–Ni–C–S, is a mix of calcium silicates interpreted as retrogressed $CaSiO_3$-perovskite (CaPv), which is a high-pressure mineral stable at depths beyond about 360 km.[25,31,195] An additional inclusion phase found was low-Cr majoritic garnet, which provides a maximum depth bracket, since it is not stable deeper than about 750 km.[196] Silicate inclusion phases therefore bracket the depth to 360–750 km, overlapping the mantle transition zone.

The Fe–Ni–C–S inclusions in CLIPPIR diamonds are physical samples of an Fe-rich metallic liquid from the deep mantle, which help to confirm the fundamental process of Fe^{2+} disproportionation at depth.[49] Disproportionation is driven by the progressive increase in the capacity for silicate minerals to host Fe^{3+} preferentially over Fe^{2+} with increasing pressures, promoting the following reaction: $3Fe^{2+} \rightarrow 2Fe^{3+} + Fe^0$. The "oxidized" Fe^{3+} is partitioned into the silicates, but the "reduced" Fe^0 separates into its own metallic phase, which is thought to generate up to about 1 wt.% metal in the lower mantle.[142,197–199] This metal budget should regulate fO_2, keeping it near the iron–wüstite buffer below ~250 km and establishing a carbon-reducing environment in much of the deep mantle.[142] The solubility of carbon in metallic Fe liquid is very high, up to ~6 wt.%, and it is ~2% in solid Fe metal at lower temperatures. Thus, in regions of the transition zone and lower mantle where metallic Fe exists, the mantle's entire budget of carbon might be dissolved in Fe metal and diamond would not be a stable phase.

In order to confirm the experiments and theory that the bulk of Earth's mantle below ~250 km is likely saturated with metallic iron,[198–200] obtaining direct samples is important because the behavior of carbon in the mantle is so strongly affected by oxygen fugacity. In a metal-saturated mantle, carbon is expected to be efficiently reduced and dissolved into the metal phase, or potentially precipitated as carbide or diamond,[142] and this has large-scale implications for the behavior of carbon in the mantle over geologic time. In CLIPPIR diamonds, variably light carbon isotopic signatures as well as the composition of majoritic garnet inclusions suggest the involvement of subducted materials.[49] This additional observation implies that deeply recycled carbon can enter into carbon-reducing regions of the mantle and become dissolved into metallic phases. Thus, dispersed Fe-rich metal in the deeper, convecting mantle may contain both primordial and recycled carbon, whose proportions may have changed with time. Further research into the influence of metallic iron on carbon in the mantle will explore the evolution of storage and cycling, from core formation to the onset of modern-style plate tectonics.

5.6 Limits to Knowledge and Questions for the Future

Limits to knowledge in diamond research have been the typical occurrence of each diamond as a single isolated xenocryst, its pure crystal structure, the rarity of inclusions, and the availability of samples (see Section 5.1). The DMGC consortium was conceived early in the existence of the DCO to help overcome these and other obstacles to reaching the following major goals: (1) a research focus on the most widely occurring carbon mineral on Earth; (2) the use of diamond's special ability to preserve mineral inclusions, fluid species, and pathways from great depths and ages to provide a view of the otherwise inaccessible deep carbon cycle; and (3) application of cross-disciplinary research tools and sharing of samples necessary to go beyond a single investigator or research group.

Future directions in diamond research in relation to the bigger picture of carbon in the deep Earth are best summed up with a list of questions:

- What is the source of carbon and other volatiles returned by diamond from the deep mantle?
- How can the findings from diamond studies be extrapolated to the bulk mantle?
- What is the mineralogy and composition of the major deep mantle minerals?
- What is the capacity for carbon and water storage in deep mantle minerals?
- How and why do contrasting (carbonatitic versus metallic versus supercritical C–H–O fluids) diamond-forming environments exist?
- What are the most effective mechanisms for diamond formation in these different mantle environments?
- Do the C and N isotopic compositions of diamonds vary with geologic time, and if so, do they record major geodynamic changes?
- What is the nature of diamond-forming fluids in the lithosphere, what are the mechanisms for their movement through the lithosphere, and what is their relation to geologic events?
- How are accurate entrapment pressures and temperatures determined for mineral inclusions?
- How do the fluids around mineral inclusions in gem diamonds compare with the hydrous–silicic fluids in fibrous diamonds?
- What are the phase transformations in subducting slabs that allow some carbon, water, and other volatiles to be carried to the transition zone?

Taken in total, the studies made by the DMGC consortium will provide new insights into how carbon behaves and resides in both the lithosphere and the deeper convecting mantle. Moreover, through diamond's remarkable attributes, diamond studies will allow us to go beyond the study of carbon alone, to make fundamental discoveries on the nature of the deep Earth that is inaccessible in any other way, and to understand the spectrum of geological processes that govern how carbon gets into the mantle and the form in which it resides.

Acknowledgments

The authors thank the following institutions and individuals: our home institutions; SBS and EHH for support from the US National Science Foundation (EAR-104992); FN and PN for support from the European Research Council Starting Grant (#307322); Wuyi Wang and Tom Moses of the Gemological Institute of America (GIA) for the support of the research projects undertaken by KVS and EMS; and SCK for the support of De Beers Technologies. This is contribution 1168 from the ARC Centre of Excellence for Core to

Crust Fluid Systems (www.ccfs.mq.edu.au) and 1130 in the Geochemical Evolution and Metallogeny of the Continents Key Centre (www.gemoc.mq.edu.au). Over the years, we have benefitted from discussions and collaborations with other DMGC consortium researchers who are not coauthors of this chapter: Galina Bulanova, Jeffrey Harris, Dan Howell, Shantanu Keshav, Stephen Richardson, Chris Smith, Nikolai Sobolev, and Laura Speich. During the preparation of this chapter, our coauthor and leader within the DCO, Erik H. Hauri, died tragically. We remember him as a champion of diamond research and thank him posthumously for his strong support.

References

1. Gurney, J. J., Helmstaedt, H. H., Richardson, S. H. & Shirey, S. B. Diamonds through time. *Economic Geology* **105**, 689–712 (2010).
2. Harris, J. W. Diamond geology. In *The properties of natural and synthetic diamond* (ed. J. E. Field) 345–394 (Academic Press, 1992).
3. Harris, J. W. The recognition of diamond inclusions. Part I: Syngenetic inclusions. *Industrial Diamond Review* **28**, 402–410 (1968).
4. Harris, J. W. & Gurney, J. J. Inclusions in diamond. In *The properties of diamond* (ed. J. E. Field) 555–591 (Academic Press, 1979).
5. Deines, P. The carbon isotopic composition of diamonds: Relationship to diamond shape, color, occurrence and vapor composition. *Geochimica et Cosmochimica Acta* **44**, 943–962 (1980).
6. Pearson, D. G., Canil, D. & Shirey, S. B. Mantle samples included in volcanic rocks: Xenoliths and diamonds. In *Teatise on geochmistry: Vol. 2, the mantle* (ed. R. W. Carlson) 171–277 (Elsevier, 2003).
7. Pearson, D. G. & Shirey, S. B. Isotopic dating of diamonds. In *Application of radiogenic isotopes to ore deposit research and exploration. Vol. 12, reviews in economic geology* (eds. D. Lambert David & Joaquin Ruiz) 143–171 (Society of Economic Geologists, 1999).
8. Cartigny, P. Stable isotopes and the origin of diamond. *Elements* **1**, 79–84 (2005).
9. Stachel, T., Brey, G. P. & Harris, J. W. Inclusions in sublithospheric diamonds: Glimpses of deep earth. *Elements* **1**, 73–87 (2005).
10. Stachel, T. & Harris, J. W. The origin of cratonic diamonds – Constraints from mineral inclusions: The genesis of gem deposits. *Ore Geology Reviews* **34**, 5–32 (2008).
11. Spetsius, Z. V. & Taylor, L. A. *Diamonds of Yakutia: Photographic evidence for their origin* (Tranquility Base Press, 2008).
12. Harte, B. Diamond formation in the deep mantle: The record of mineral inclusions and their distribution in relation to mantle dehydration zones. *Mineralogical Magazine* **74**, 189–215 (2010).
13. Tappert, R. & Tappert, M. C. *Diamonds in nature: A guide to rough diamonds* (Springer Verlag, 2011).
14. Shirey, S. B. et al. Diamonds and the geology of mantle carbon. *Reviews in Mineralogy and Geochemistry* **75**, 355–421 (2013).
15. Shirey, S. B. & Shigley, J. E. Recent advances in understanding the geology of diamonds. *Gems and Gemology* **49**, 188–222 (2013).

16. Stachel, T. & Harris, J. Formation of diamond in the Earth's mantle. *Journal of Physics: Condensed Matter* **21**, 364206 (2009).
17. Shirey, S. B. & Richardson, S. H. Start of the Wilson cycle at 3 Ga shown by diamonds from subcontinental mantle. *Science* **333**, 434–436 (2011).
18. Harte, B., Harris, J. W., Hutchison, M. T., Watt, G. R. & Wilding, M. C. Lower mantle mineral associations in diamonds from Sao Luiz, Brazil. In *Mantle petrology; field observations and high-pressure experimentation: A tribute to Francis R. (Joe) Boyd. Vol. 6, special publication – Geochemical Society* (eds. Y. Fei, C. M. Bertka & B. O. Mysen) 125–153 (Geochemical Society, 1999).
19. Anzolini, C. et al. Depth of formation of $CaSiO_3$-walstromite included in super-deep diamonds. *Lithos* **265**, 138–147 (2016).
20. Thomson, A. R. et al. Origin of sub-lithospheric diamonds from the Juina-5 kimberlite (Brazil): Constraints from carbon isotopes and inclusion compositions. *Contributions to Mineralogy and Petrology* **168**, 1081 (2014).
21. Bulanova, G. P. et al. Mineral inclusions in sublithospheric diamonds from Collier 4 kimberlite pipe, Juina, Brazil: Subducted protoliths, carbonated melts and primary kimberlite magmatism. *Contributions to Mineralogy and Petrology* **160**, 489–510 (2010).
22. Beyer, C. & Frost, D. J. The depth of sub-lithospheric diamond formation and the redistribution of carbon in the deep mantle. *Earth and Planetary Science Letters* **461**, 30–39 (2017).
23. Pearson, D. G. et al. Hydrous mantle transition zone indicated by ringwoodite included within diamond. *Nature* **507**, 221–224 (2014).
24. Walter, M. J. et al. Deep mantle cycling of oceanic crust: Evidence from diamonds and their mineral inclusions. *Science* **334**, 54–57 (2011).
25. Nestola, F. et al. $CaSiO_3$ perovskite in diamond indicates the recycling of oceanic crust into the lower mantle. *Nature* **555**, 237–241 (2018).
26. Nestola, F. et al. Olivine with diamond-imposed morphology included in diamonds: Syngenesis or protogenesis? *International Geology Review* **56**, 1658–1667 (2014).
27. Milani, S., Nestola, F., Angel, R. J., Nimis, P. & Harris, J. W. Crystallographic orientations of olivine inclusions in diamonds. *Lithos* **265**, 312–316 (2016).
28. Nimis, P. Trapped minerals under stress. *Geology* **46**, 287–288 (2018).
29. Angel, R. J., Mazzucchelli, M. L., Alvaro, M. & Nestola, F. EosFit-Pinc: A simple GUI for host-inclusion elastic thermobarometry. *American Mineralogist* **102**, 1957–1960 (2017).
30. Mazzucchelli, M. L. et al. Elastic geothermobarometry: Corrections for the geometry of the host-inclusion system. *Geology* **46**, 231–234 (2018).
31. Anzolini, C. et al. Depth of formation of super-deep diamonds: Raman barometry of $CaSiO_3$-walstromite inclusions. *American Mineralogist* **103**, 69–74 (2018).
32. Nimis, P. et al. First evidence of hydrous silicic fluid films around solid inclusions in gem quality diamonds. *Lithos* **260**, 384–389 (2016).
33. Taylor, W. R., Jaques, A. L. & Ridd, M. Nitrogen-defect aggregation characteristics of some Australasian diamonds: Time–temperature constraints on the source regions of pipe and alluvial diamonds. *American Mineralogist* **75**, 1290–1310 (1990).
34. Howell, D. et al. μ-FTIR mapping: Distribution of impurities in different types of diamond growth. *Diamond and Related Materials* **29**, 29–36 (2012).
35. Spetsius, Z. V., Bogush, I. N. & Kovalchuk, O. E. FTIR mapping of diamond plates of eclogitic and peridtotitc xenoliths from the Nyurinskaya pipe, Yakutia: Genetic implications. *Russian Geology and Geophysics* **56**, 344–353 (2015).

36. Kohn, S. C., Speich, L., Smith, C. B. & Bulanova, G. P. FTIR thermochronometry of natural diamonds: A closer look. *Lithos* **265**, 148–158 (2016).
37. Speich, L., Kohn, S. C., Wirth, R., Bulanova, G. P. & Smith, C. B. The relationship between platelet size and the B′ infrared peak of natural diamonds revisited. *Lithos* **278–281**, 419–426 (2017).
38. Speich, L., Kohn, S. C., Bulanova, G. P. & Smith, C. B. The behaviour of platelets in natural diamonds and the development of a new mantle thermometer. *Contributions to Mineralogy and Petrology* **173**, 39 (2018).
39. Smit, K. V., D'Haenens-Johansson, U. F. S., Howell, D., Loudin, L. C. & Wang, W. Deformation-related spectroscopic features in natural Type Ib–IaA diamonds from Zimmi (West African craton). *Mineralogy and Petrology* **69**, 1–15 (2018).
40. Smit, K. V., Shirey, S. B., Stern, R. A., Steele, A. & Wang, W. Diamond growth from C-H-N-O recycled fluids in the lithosphere; evidence from CH_4 micro-inclusions and delta ^{13}C-delta ^{15}N-N content in Marange mixed-habit diamonds. *Lithos* **265**, 68–81 (2016).
41. Frost, D. J. & McCammon, C. A. The redox state of Earth's mantle. *Annual Reviews of Earth Planetaary Science* **36**, 389–420 (2008).
42. Eggler, D. H. & Baker, D. R. Reduced volatiles in the system C O H: Implications to mantle melting, fluid formation, and diamond genesis. In *Advances in Earth and planetary sciences: Vol. 12, high pressure research in geophysics* (eds. S. Akimoto & M. H. Manghnani) 237–250 (Kluwer, 1982).
43. Green, D. H., Falloon, T. J. & Taylor, W. R. Mantle-derived magmas – Roles of variable source peridotite and variable C–H–O fluid compositions. In *Magmatic processes: Physiochemical principles* (ed. B. O. Mysen) 139–154 (The Geochemical Society, 1987).
44. Matveev, S., Ballhaus, C., Fricke, K., Truckenbrodt, J. & Ziegenben, D. Volatiles in the Earth's mantle: I. Synthesis of CHO fluids at 1273 K and 2.4 GPa. *Geochimica et Cosmchimica Acta* **61**, 3081–3088 (1997).
45. Zhang, C. & Duan, Z. A model for C–O–H fluid in the Earth's mantle. *Geochimica et Cosmochimica Acta* **73**, 2089–2102 (2009).
46. Stagno, V., Ojwang, D. O., McCammon, C. A. & Frost, D. J. The oxidation state of the mantle and the extraction of carbon from Earth's interior. *Nature* **493**, 84–88 (2013).
47. Luth, R. W. & Stachel, T. The buffering capacity of lithospheric mantle; implications for diamond formation. *Contributions to Mineralogy and Petrology* **168**, 1–12 (2014).
48. Stachel, T., Harris, J. W. & Muehlenbachs, K. Sources of carbon in inclusion bearing diamonds. *Lithos* **112**, 625–637 (2009).
49. Smith, E. M. et al. Large gem diamonds from metallic liquid in Earth's deep mantle. *Science* **354**, 1403–1405 (2016).
50. Scott, H. P. et al. Generation of methane in the Earth's mantle: *In situ*, high pressure–temperature measurements of carbonate reduction. *Proceedings of the National Academy of Sciences* **101**, 14023–14026 (2004).
51. Stachel, T., Chacko, T. & Luth, R. W. Carbon isotope fractionation during diamond growth in depleted peridotite; counterintuitive insights from modelling water-maximum CHO fluids as multi-component systems. *Earth and Planetary Science Letters* **473**, 44–51 (2017).
52. Petts, D. C., Chacko, T., Stachel, T., Stern, R. A. & Heaman, L. M. A nitrogen isotope fractionation factor between diamond and its parental fluid derived from

detailed SIMS analysis of a gem diamond and theoretical calculations. *Chemical Geology* **410**, 188–200 (2015).
53. Wyllie, P. J. Metasomatism and fluid generation in mantle xenoliths. In *Mantle xenoliths* (ed. P. H. Nixon) 609–621 (John Wiley and Sons, 1987).
54. Taylor, W. R. & Green, D. H. Measurement of reduced peridotite-C–O–H solidus and implications for redox melting of the mantle. *Nature* **332**, 349–352 (1988).
55. Thomassot, E., Cartigny, P., Harris, J. W. & Viljoen, K. S. Methane-related diamond crystallization in the Earth's mantle; stable isotope evidences from a single diamond-bearing xenolith. *Earth and Planetary Science Letters* **257**, 362–371 (2007).
56. Walter, M. J. et al. Primary carbonatite melt from deeply subducted oceanic crust. *Nature* **454**, 622–625 (2008).
57. Thomson, A. R., Walter, M. J., Kohn, S. C. & Brooker, R. A. Slab melting as a barrier to deep carbon subduction. *Nature* **529**, 76–79 (2016).
58. Stachel, T. & Luth, R. W. Diamond formation; where, when and how? *Lithos* **220–223**, 200–220 (2015).
59. Hasterok, D. & Chapman, D. S. Heat production and geotherms for the continental lithosphere. *Earth and Planetary Science Letters* **307**, 59–70 (2011).
60. Ray, J. S. & Ramesh, R. Rayleigh fractionation of stable isotopes from a multi-component source. *Geochimica et Cosmochimica Acta* **64**, 299–306 (2000).
61. Navon, O., Hutcheon, I. D., Rossman, G. R. & Wasserburg, G. J. Mantle-derived fluids in diamond micro-inclusions. *Nature* **335**, 784–789 (1988).
62. Weiss, Y. et al. A new model for the evolution of diamond-forming fluids; evidence from microinclusion-bearing diamonds from Kankan, Guinea. *Lithos* **112**, 660–674 (2009).
63. Tomlinson, E. L. & Mueller, W. A snapshot of mantle metasomatism; trace element analysis of coexisting fluid (LA-ICP-MS) and silicate (SIMS) inclusions in fibrous diamonds. *Earth and Planetary Science Letters* **279**, 362–372 (2009).
64. Weiss, Y., Griffin, W. L. & Navon, O. Diamond-forming fluids in fibrous diamonds; The trace-element perspective. *Earth and Planetary Science Letters* **376**, 110–125 (2013).
65. Klein-Bendavid, O. et al. The sources and time-integrated evolution of diamond-forming fluids – Trace elements and isotopic evidence. *Geochimica et Cosmchimica Acta* **125**, 146–169 (2014).
66. Weiss, Y., Griffin, W. L., Bell, D. R. & Navon, O. High-Mg carbonatitic melts in diamonds, kimberlites and the sub-continental lithosphere. *Earth and Planetary Science Letters* **309**, 337–347 (2011).
67. Elazar, O., Navon, O. & Kessel, R. Melting of hydrous-carbonated eclogite at 4–6 GPa and 900–1200°C: Implications for the sources of diamond-forming fluids. *11th International Kimberlite Conference Extended Abstract*, No. 11IKC-4471 (2017).
68. Weiss, Y., McNeill, J., Pearson, D. G., Nowell, G. M. & Ottley, C. J. Highly saline fluids form a subducting slab as the source for fluid-rich diamonds. *Nature* **524**, 339–342 (2015).
69. Weiss, Y., Navon, O., Goldstein, S. L. & Harris, J. W. Inclusions in diamonds constrain thermo-chemical conditions during Mesozoic metasomatism of the Kaapvaal cratonic mantle. *Earth and Planetary Science Letters* **491**, 134–147 (2018).
70. Izraeli, E. S., Harris, J. W. & Navon, O. Brine inclusions in diamonds; a new upper mantle fluid. *Earth and Planetary Science Letters* **187**, 323–332 (2001).
71. Shiryaev, A. A. et al. Chemical, optical and isotopic investigation of fibrous diamonds from Brazil. *Russian Geology and Geophysics* **46**, 1185–1201 (2005).

72. Klein-BenDavid, O., Pearson, D. G., Nowell, G. M., Ottley, C. & Cartigny, P. Origins of diamond forming fluids; constraints from a coupled Sr–Nd–Pb isotope and trace element approach. *Abstracts – Israel Geological Society, Annual Meeting* 2009, 75 (2009).
73. Klein-BenDavid, O., Wirth, R. & Navon, O. Micrometer-scale cavities in fibrous and cloudy diamonds; a glance into diamond dissolution events. *Earth and Planetary Science Letters* **264**, 89–103 (2007).
74. Schrauder, M. & Navon, O. Solid carbon dioxide in a natural diamond. *Nature* **365**, 42–44 (1993).
75. Tomlinson, E. L., Jones, A. P. & Harris, J. W. Co-existing fluid and silicate inclusions in mantle diamond. *Earth and Planetary Science Letters* **250**, 581–595 (2006).
76. Smith, E. M., Kopylova, M. G., Nowell, G. M., Pearson, D. G. & Ryder, J. Archean mantle fluids preserved in fibrous diamonds from Wawa, Superior Craton. *Geology* **40**, 1071–1074 (2012).
77. Weiss, Y., Kiflawi, I., Davies, N. & Navon, O. High density fluids and the growth of monocrystalline diamonds. *Geochimica et Cosmochimica Acta* **141**, 145–159 (2014).
78. Jablon, B. M. & Navon, O. Most diamonds were created equal. *Earth and Planetary Science Letters* **443**, 41–47 (2016).
79. Zedgenizov, D. A. et al. Directional chemical variations in diamonds showing octahedral following cuboid growth. *Contributions to Mineralogy and Petrology* **151**, 45–57 (2006).
80. Shirey, S. B., Richardson, S. H. & Harris, J. W. Integrated models of diamond formation and craton evolution. *Lithos* **77**, 923–944 (2004).
81. Cartigny, P., Palot, M., Thomassot, E. & Harris, J. W. Diamond formation; a stable isotope perspective. *Annual Review of Earth and Planetary Sciences* **42**, 699–732 (2014).
82. Huang, J. et al. Magnesium and oxygen isotopes in Roberts Victor eclogites. *Chemical Geology* **438**, 73–83 (2016).
83. Farquhar, J. et al. Mass-independent sulfur of inclusions in diamond and sulfur recycling on early Earth. *Science* **298**, 2369–2372 (2002).
84. Chaussidon, M., Albarede, F. & Sheppard, S. M. F. Sulfur isotope heterogeneity in the mantle from ion microprobe measurements of sulphide inclusions in diamonds. *Nature* **330**, 242–244 (1987).
85. Eldridge, C. S., Compston, W., Williams, I. S., Harris, J. W. & Bristow, J. W. Isotope evidence for the involvement of recycled sediments in diamond formation. *Nature* **353**, 649–653 (1991).
86. Thomassot, E. et al. Metasomatic diamond growth; a multi-isotope study (^{13}C, ^{15}N, ^{33}S, ^{34}S) of sulphide inclusions and their host diamonds from Jwaneng (Botswana). *Earth and Planetary Science Letters* **282**, 79–90 (2009).
87. Thomassot, E., O'Neil, J., Francis, D., Cartigny, P. & Wing, B. A. Atmospheric record in the Hadean Eon from multiple sulfur isotope measurements in Nuvvuagittuq Greenstone Belt (Nunavik, Quebec). *Proceedings of the National Academy of Science* **112**, 707–712 (2015).
88. Cartigny, P. et al. A mantle origin for Paleoarchean peridotitic diamonds from the Panda kimberlite, Slave Craton; evidence from ^{13}C, ^{15}N and $^{33,34}S$ stable isotope systematics. *Lithos* **112**, 852–864 (2009).
89. Labidi, J., Cartigny, P., Hamelin, C., Moreira, M. & Dosso, L. Sulfur isotope budget (^{32}S, ^{33}S, ^{34}S and ^{36}S) in Pacific–Antarctic ridge basalts: A record of mantle source heterogeneity and hydrothermal sulfide assimilation. *Geochimica et Cosmchimica Acta* **133**, 47–67 (2014).

90. Labidi, J., Cartigny, P. & Moreira, M. Non-chondritic sulphur isotope composition of the terrestrial mantle. *Nature* **501**, 208–211 (2013).
91. Cabral, R. A. et al. Anomalous sulphur isotopes in plume lavas reveal deep mantle storage of Archaean crust. *Nature* **496**, 490–493 (2013).
92. Delavault, H., Chauvel, C., Thomassot, E., Devey, C. W. & Dazas, B. Sulfur and lead isotopic evidence of relic Archean sediments in the Pitcairn mantle plume. *Proceedings of the National Academy of Sciences* **113**, 12952–12956 (2016).
93. Ueno, Y., Ono, S., Rumble, D. & Maruyama, S. Quadruple sulfur isotope analysis of ca. 3.5 Ga Dresser Formation: New evidence for microbial sulfate reduction in the early Archean. *Geochimica et Cosmochimica Acta* **72**, 5675–5691 (2008).
94. Roerdink, D. L., Mason, P. R. D., Whitehouse, M. J. & Brouwer, F. M. Reworking of atmospheric sulfur in a Paleoarchean hydrothermal system at Londozi, Barberton Greenstone Belt, Swaziland. *Precambrian Research* **280**, 195–204 (2016).
95. Farquhar, J. & Jackson, M. Missing Archean sulfur returned from the mantle. *Proceedings of the National Academy of Science* **113**, 12893–12895 (2016).
96. Ono, S., Wing, B., Johnston, D. & Farquhar, J. Mass-dependent fractionation of quadruple stable sulfur isotope system as a new tracer of sulfur biogeochemical cycles. *Geochimica et Cosmchimica Acta* **70**, 2238–2252 (2006).
97. Moore, R. O. & Gurney, J. J. Pyroxene solid solution in garnets included in diamond. *Nature* **318**, 553–555 (1985).
98. Stachel, T., Harris, J. W., Brey, G. P. & Joswig, W. Kankan diamonds (Guinea) II: Lower mantle inclusion parageneses. *Contributions to Mineralogy and Petrology* **140**, 16–27 (2000).
99. Hutchison, M. T., Cartigny, P. & Harris, J. W. Carbon and nitrogen compositions and physical characteristics of transition zone and lower mantle diamonds from Sao Luiz, Brazil. In *The J. B. Dawson volume; Proceedings of the VIIth International Kimberlite Conference; Vol. 1.* (eds. J. J. Gurney, J. L. Gurney, M. D. Pascoe & S. H. Richardson) 372–382 (1999).
100. Kaminsky, F. V. et al. Superdeep diamonds form the Juina area, Mato Grosso State, Brazil. *Contributions to Mineralogy and Petrology* **140**, 734–753 (2001).
101. Palot, M., Cartigny, P., Harris, J. W., Kaminsky, F. V. & Stachel, T. Evidence for deep mantle convection and primordial heterogeneity from nitrogen and carbon stable isotopes in diamond. *Earth and Planetary Science Letters* **357–358**, 179–193 (2012).
102. Zedgenizov, D. A., Kagi, H., Shatsky, V. S. & Ragozin, A. L. Local variations of carbon isotope composition in diamonds from Sao Luis (Brazil); evidence for heterogenous carbon reservoir in sublithospheric mantle. *Chemical Geology* **363**, 114–124 (2014).
103. Burnham, A. D. et al. Diamonds from the Machado River alluvial deposit, Rondonia, Brazil, derived from both lithospheric and sublithospheric mantle. *Lithos* **265**, 199–213 (2016).
104. Burnham, A. D. et al. Stable isotope evidence for crustal recycling as recorded by superdeep diamonds. *Earth and Planetary Science Letters* **432**, 374–380 (2015).
105. Tappert, R. et al. Diamonds from Jagersfontein (South Africa); messengers from the sublithospheric mantle. *Contributions to Mineralogy and Petrology* **150**, 505–522 (2005).
106. Palot, M. et al. The transition zone as a host for recycled volatiles: Evidence from nitrogen and carbon isotopes in ultra-deep diamonds from Monastery and Jagersfontein (South Africa). *Chemical Geology* **466**, 733–749 (2017).

107. Palot, M., Pearson, D. G., Stern, R. A., Stachel, T. & Harris, J. W. Isotopic constraints on the nature and circulation of deep mantle C–H–O–N fluids; carbon and nitrogen systematics within ultra-deep diamonds from Kankan (Guinea). *Geochimica et Cosmochimica Acta* **139**, 26–46 (2014).
108. Stachel, T., Harris, J. W., Aulbach, S. & Deines, P. Kankan diamonds (Guinea) III; delta ^{13}C and nitrogen characteristics of deep diamonds. *Contributions to Mineralogy and Petrology* **142**, 465–475 (2002).
109. Ickert, R. B., Stachel, T., Stern, R. A. & Harris, J. W. Extreme ^{18}O-enrichment in majorite constrains a crustal origin of transition zone diamonds. *Geochemical Perspectives Letters* **1**, 65–74 (2015).
110. Smyth, J. R. β-Mg_2SiO_4: A potential host for water in the mantle? *American Mineralogist* **72**, 1051–1055 (1987).
111. Van der Hilst, R. D., Widiyantoro, S. & Engdahl, E. R. Evidence for deep mantle circulation from global tomography. *Nature* **386**, 578–584 (1997).
112. Kohlstedt, D. L., Keppler, H. & Rubie, D. C. Solubility of water in the α, β and γ phases of $(Mg,Fe)_2SiO_4$. *Contributions to Mineralogy and Petrology* **123**, 345–357 (1996).
113. Brenker, F. E. et al. Detection of a Ca-rich lithology in the Earth's deep (>300 km) convecting mantle. *Earth and Planetary Science Letters* **236**, 579–587 (2005).
114. Harte, B. & Hudson, N. F. C. Mineral associations in diamonds from the lowermost upper mantle and uppermost lower mantle. In *Proceedings of 10th International Kimberlite Conference* (ed. D. G. Pearson) 235–253 (Springer, 2013).
115. Nestola, F. Inclusions in super-deep diamonds: Windows on the very deep Earth. *Rendiconti Lincei* **28**, 595–604 (2017).
116. Brenker, F. E., Stachel, T. & Harris, J. W. Exhumation of lower mantle inclusions in diamond; ATEM investigation of retrograde phase transitions, reactions and exsolution. *Earth and Planetary Science Letters* **198**, 1–9 (2002).
117. Brey, G. P., Bulatov, V., Girnis, A., Harris, J. W. & Stachel, T. Ferropericlase – A lower mantle phase in the upper mantle. *Lithos* **77**, 655–663 (2004).
118. Hayman, P. C., Kopylova, M. G. & Kaminsky, F. V. Lower mantle diamonds from Rio Soriso (Juina area, Mato Grosso, Brazil). *Contributions to Mineralogy and Petrology* **149**, 430–445 (2005).
119. Thomas, S.-M. et al. Quantification of water in hydrous ringwoodite. *Frontiers in Earth Science* **2**, 38 (2015).
120. Litasov, K. D. & Ohtani, E. Effect of water on the phase relations in Earth's mantle and deep water cycle. In *Advances in high pressure mineralogy* (ed. E. Ohtani) 115–156 (Geological Society of America, 2007).
121. Nestola, F. & Smyth, J. R. Diamonds and water in the deep Earth; a new scenario. *International Geology Review* **58**, 263–276 (2016).
122. Schmandt, B., Jacobsen, S. D., Becker, T. W., Liu, Z. & Dueker, K. G. Dehydration melting at the top of the lower mantle. *Science* **344**, 1265 (2014).
123. Liu, Z., Park, J. & Karato, S.-I. Seismic evidence for water transport out of the mantle transition zone beneath the European Alps. *Earth and Planetary Science Letters* **482**, 93–104 (2018).
124. Zhang, Z., Dueker, K. G. & Huang, H.-H. Ps mantle transition zone imaging beneath the Colorado Rocky Mountains: Evidence for an upwelling hydrous mantle. *Earth and Planetary Science Letters* **492**, 197–205 (2018).
125. Wirth, R., Vollmer, C., Brenker, F., Matsyuk, S. & Kaminsky, F. Inclusions of nanocrystalline hydrous aluminium silicate "phase egg" in superdeep diamonds from

Juina (Mato Grosso State, Brazil). *Earth and Planetary Science Letters* **259**, 384–399 (2007).

126. Palot, M. et al. Evidence for H_2O-bearing fluids in the lower mantle from diamond inclusion. *Lithos* **265**, 237–243 (2016).
127. Tschauner, O. et al. Ice-VII inclusions in diamonds: Evidence for aqueous fluid in Earth's deep mantle. *Science* **359**, 1136–1139 (2018).
128. Bundy, F. P., Hall, H. T., Strong, H. M. & Wentorf Jun, R. H. Man-made diamonds. *Nature* **176**, 51–55 (1955).
129. Arima, M., Nakayama, K., Akaish, M., Yamaoka, S. & Kanda, H. Crystallization of diamond from a silicate melt of kimberlite composition in high-pressure and high-temperature experiments. *Geology* **21**, 968–970 (1993).
130. Palyanov, Y. N., Sokol, A. G., Borzdov, Y. M., Khokhryakov, A. F. & Sobolev, N. V. Diamond formation from mantle carbonate fluids. *Nature* **400**, 417–418 (1999).
131. Sokol, A. G., Borzdov, Y. M., Pal'yanov, Y. N., Khokhryakov, A. F. & Sobolev, N. V. An experimental demonstration of diamond formation in the dolomite–carbon and dolomite–fluid–carbon systems. *European Journal of Mineralogy* **13**, 893–900 (2001).
132. Fagan, A. J. & Luth, R. W. Growth of diamond in hydrous silicate melts. *Contributions to Mineralogy and Petrology* **161**, 229–236 (2011).
133. Palyanov, Y. N., Sokol, A. G., Khokhryakov, A. F. & Kruk, A. N. Conditions of diamond crystallization in kimblerlite melt: Experimental data. *Russian Geology and Geophysics* **36**, 196–210 (2015).
134. Litvin, Y. A. & Butvina, V. G. Diamond-forming media in the system eclogite–carbonatite–sulfide–carbon: Experiments at 6.0–8.5 GPa. *Petrology* **12**, 377–387 (2004).
135. Sokol, A. G. & Palyanov, Y. N. Diamond formation in the system $MgO–SiO_2-H_2O–C$ at 7.5 GPa and 1,600°C. *Contributions to Mineralogy and Petrology* **155**, 33–43 (2008).
136. Bureau, H. et al. Diamond growth in mantle fluids. *Lithos* **265**, 4–15 (2016).
137. Bureau, H. et al. The growth of fibrous, cloudy and polycrystalline diamonds. *Geochimica et Cosmochimica Acta* **77**, 202–214 (2012).
138. Sokol, A. G., Palyanova, G. A., Palyanov, Y. N., Tomilenko, A. A. & Melenevskiy, V. N. Fluid regime and diamond formation in the reduced mantle; experimental constraints. *Geochimica et Cosmochimica Acta* **73**, 5820–5834 (2009).
139. Palyanov, Y. N. & Sokol, A. G. The effect of composition of mantle fluids/melts on diamond formation processes. *Lithos* **112**, 690–700 (2009).
140. Novella, D., Bolfan-Casanova, N., Nestola, F. & Harris, J. W. H_2O in olivine and garnet inclusions still trapped in diamonds from the Siberian Craton; implications for the water content of cratonic lithosphere peridotites. *Lithos* **230**, 180–183 (2015).
141. Bataleva, Y. V., Palyanov, Y. N., Borzdov, Y. M., Kupriyanov, I. N. & Sokol, A. G. Synthesis of diamonds with mineral, fluid and melt inclusions. *Lithos* **265**, 292–303 (2016).
142. Rohrbach, A. & Schmidt, M. W. Redox freezing and melting in the Earth's deep mantle resulting from carbon-iron redox coupling. *Nature* **472**, 209–214 (2011).
143. Palyanov, Y. N. et al. Mantle–slab interaction and redox mechanism of diamond formation. *Proceedings of the National Academy of Sciences* **110**, 20408–20413 (2013).

144. Gurney, J. J., Harris, J. W. & Rickard, R. S. Silicate and oxide inclusions in diamonds from the Orapa Mine, Botswana. In *Developments in petrology, Vol. 11* (ed. J. Kornprobst) 3–9 (Elsevier, 1984).
145. Orlov, Y. L. *The mineralogy of the diamond* (John Wiley and Sons, 1977).
146. Dobosi, G. & Kurat, G. On the origin of silicate-bearing diamondites. *Mineralogy and Petrology* **99**, 29–42 (2010).
147. Rubanova, E. V., Piazolo, S., Griffin, W. L. & O'Reilly, S. Y. Deformation microstructures reveal a complex mantle history for polycrystalline diamond. *Geochemistry Geophysics Geosystems* **13**, Q10010 (2012).
148. Jacob, D. E., Piazolo, S., Schreiber, A. & Trimby, P. Redox-freezing and nucleation of diamond via magnetite formation in the Earth's mantle. *Nature Communications* **21**, 7 (2016).
149. Jacob, D. E., Viljoen, K. S., Grassineau, N. & Jagoutz, E. Remobilization in the cratonic lithosphere recorded in polycrystalline diamond. *Science* **289**, 1182–1185 (2000).
150. Jacob, D. E., Kronz, A. & Viljoen, K. S. Cohenite, native iron and troilite inclusions in garnets from polycrystalline diamond aggregates. *Contributions to Mineralogy and Petrology* **146**, 566–576 (2004).
151. Mikhail, S. et al. Constraining the internal variability of the stable isotopes of carbon and nitrogen within mantle diamonds. *Chemical Geology* **366**, 14–23 (2014).
152. Mikhail, S., Dobosi, G., Verchovsky, A. B., Kurat, G. & Jones, A. P. Peridotitic and websteritic diamondites provide new information regarding mantle melting and metasomatism induced through the subduction of crustal volatiles. *Geochimica et Cosmchimica Acta* **107**, 1–11 (2013).
153. Jacob, D. E., Dobrzhinetskaya, L. & Wirth, R. New insight into polycrystalline diamond genesis from modern nanoanalytical techniques. *Earth-Science Reviews* **136**, 21–35 (2014).
154. Gautheron, C., Cartigny, P., Moreira, M., Harris, J. W. & Allegre, C. J. Evidence for a mantle component shown by rare gases, C and N isotopes in polycrystalline diamonds from Orapa (Botswana). *Earth and Planetary Science Letters* **240**, 559–572 (2005).
155. Mikhail, S., Howell, D. & McCubbin, F. M. Evidence for multiple diamondite-forming events in the mantle. *American Mineralogist* **99**, 1537–1543 (2014).
156. Maruoka, T., Kurat, G., Dobosi, G. & Koeberl, C. Isotopic composition of carbon in diamonds of diamondites: Record of mass fractionation in the upper mantle. *Geochimica et Cosmchimica Acta* **68**, 1635–1644 (2004).
157. Foley, S. F. & Fischer, T. P. An essential role for continental rifts and lithosphere in the deep carbon cycle. *Nature Geoscience* **10**, 897–902 (2017).
158. Stagno, V. & Frost, D. J. Carbon speciation in the asthenosphere: Experimental measurements of the redox conditions at which carbonate-bearing melts coexist with graphite or diamond in peridotite assemblages. *Earth and Planetary Science Letters* **300**, 72–84 (2010).
159. Kjarsgaard, B. A., Heaman, L. M., Sarkar, C. & Pearson, D. G. The North America mid-Cretaceous kimberlite corridor: Wet, edge-driven decompression melting of an OIB-type deep mantle source. *Geochemistry Geophysics Geosystems* **18**, 2727–2747 (2017).
160. Tappe, S., Smart, K., Torsvik, T., Massuyeau, M. & de Wit, M. Geodynamics of kimberlites on a cooling Earth: Clues to plate tectonic evolution and deep volatile cycles. *Earth and Planetary Science Letters* **484**, 1–14 (2018).

161. Smit, K. V., Shirey, S. B., Richardson, S. H., le Roex, A. P. & Gurney, J. J. Re–Os isotopic composition of peridotitic sulphide inclusions in diamonds from Ellendale, Australia: Age constraints on Kimberley cratonic lithosphere. *Geochimica et Cosmochimica Acta* **74**, 3292-3306, (2010).
162. de Vries, D. F. W. et al. Re–Os dating of sulphide inclusions zonally distributed in single Yakutian diamonds: Evidence for multiple episodes of Proterozoic formation and protracted timescales of diamond growth. *Geochimica et Cosmochimica Acta* **120**, 363–394 (2013).
163. Koornneef, J. M. et al. Archaean and Proterozoic diamond growth from contrasting styles of large-scale magmatism. *Nature Communications* **8**, 648 (2017).
164. Aulbach, S. et al. Diamond ages from Victor (Superior Craton): Intra-mantle cycling of volatiles (C, N, S) during supercontinent reorganisation. *Earth and Planetary Science Letters* **490**, 77–87 (2018).
165. Griffin, W. L. et al. The world turns over: Hadean–Archean crust–mantle evolution. *Lithos* **189**, 2–15 (2014).
166. Ionov, D. A., Carlson, R. W., Doucet, L. S., Golovin, A. V. & Oleinikov, O. B. The age and history of the lithospheric mantle of the Siberian craton: Re–Os and PGE study of peridotite xenoliths from the Obnazhennaya kimberlite. *Earth and Planetary Science Letters* **428**, 108–119 (2015).
167. Yaxley, G. M., Berry, A. J., Rosenthal, A., Woodland, A. B. & Paterson, D. Redox preconditioning deep cratonic lithosphere for kimberlite genesis - evidence from the central Slave Craton. *Scientific Reports* **7**, 30 (2017).
168. Giuliani, A. et al. Petrogenesis of mantle polymict breccias: Insights into mantle processes coeval with kimberlite magmatism. *Journal of Petrology* **55**, 831–858 (2014).
169. Aulbach, S. Craton nucleation and formation of thick lithospheric roots. *Lithos* **149**, 16–30 (2012).
170. McCammon, C. A., Griffin, W. L., Shee, S. R. & O'Neill, H. S. C. Oxidation during metasomatism in ultramafic xenoliths from the Wesselton kimberlite, South Africa: Implications for the survival of diamond. *Contributions to Mineralogy and Petrology* **141**, 287–296 (2001).
171. Creighton, S. et al. Oxidation of the Kaapvaal lithospheric mantle driven by metasomatism. *Contributions to Mineralogy and Petrology* **157**, 491–504 (2009).
172. Aulbach, S., Griffin, W. L., Pearson, N. J. & O'Reilly, S. Y. Nature and timing of metasomatism in the stratified mantle lithosphere beneath the central Slave Craton (Canada). *Chemical Geology* **352**, 153–169 (2013).
173. Simon, N. S. C., Irvine, G. J., Davies, G. R., Pearson, D. G. & Carlson, R. W. The origin of garnet and clinopyroxene in "depleted" Kaapvaal peridotites. *Lithos* **71**, 289–322 (2003).
174. Wang, W., Gasparik, T. & Rapp, R. P. Partitioning of rare earth elements between $CaSiO_3$ perovskite and coexisting phases; constraints on the formation of $CaSiO_3$ inclusions in diamonds. *Earth and Planetary Science Letters* **181**, 291–300 (2000).
175. Brenker, F. E. et al. Carbonates from the lower part of transition zone or even the lower mantle. *Earth and Planetary Science Letters* **260**, 1–9 (2007).
176. Thomson, A. R. et al. Trace element composition of silicate inclusions in sub-lithospheric diamonds from the Juina-5 kimberlite; evidence for diamond growth from slab melts. *Lithos* **265**, 108–124 (2016).
177. Harte, B. & Cayzer, N. Decompression and unmixing of crystals included in diamonds from the mantle transition zone. *Physics and Chemistry of Minerals* **34**, 647–656 (2007).

178. Stachel, T., Brey, G. P. & Harris, J. W. Kankan diamonds (Guinea) I: From the lithosphere down to the transition zone. *Contributions to Mineralogy and Petrology* **140**, 1–15 (2000).
179. Kiseeva, E. S., Litasov, K. D., Yaxley, G. M., Ohtani, E. & Kamenetsky, V. S. Melting and phase relations of carbonated eclogite at 9–21 GPa and the petrogenesis of alkali-rich melts in the deep mantle. *Journal of Petrology* **54**, 1555–1583 (2013).
180. Davies, R. M., Griffin, W. L., O'Reilly, S. Y. & McCandless, T. E. Inclusions in diamonds from the K14 and K10 kimberlites, Buffalo Hills, Alberta, Canada; diamond growth in a plume? *Lithos* **77**, 99–111 (2004).
181. Hutchison, M. T. *Constitution of the deep transition zone and lower mantle shown by diamonds and their inclusions* (University of Edinburgh, 1997).
182. Moore, R. O., Gurney, J. J., Griffin, W. L. & Shimizu, N. Ultra-high pressure garnet inclusions in Monastery diamonds: Trace element abundance patterns and conditions of origin. *European Journal of Mineralogy* **3**, 213–230 (1991).
183. Syracuse, E. M. et al. The global range of subduction zone thermal models. *Physics of the Earth and Planetary Interiors* **183**, 73–90 (2010).
184. Moore, R. O. & Gurney, J. J. Mineral inclusions in diamond from the Monastry kimberlite, South Africa. In *Kimberlites and related rocks, Vol. 14* (eds. J. Ross et al.) 1029–1041 (Geological Society of Australia, 1989).
185. Wilding, M., Harte, B. & Harris, J. W. Carbon isotope variation in a zoned Bultfontein diamond determined by SIMS. In *Seventh International Conference on Geochronology, Cosmochronology and Isotope Geology; abstracts volume, Vol. 27* (ed. W. Compston) 112 (Geological Society of Australia, 1990).
186. Beyer, C., Frost, D. J. & Miyajimas, N. Experimental calibration of a garnet-clinopyroxene geobarometer for mantle eclogites. *Contributions to Mineralogy and Petrology* **169**, 18 (2015).
187. Armstrong, L. S. et al. Perovskite phase relations in the system $CaO–MgO–TiO_2–SiO_2$ and implications for deep mantle lithologies. *Journal of Petrology* **53**, 611–635 (2012).
188. Kubo, A., Suzuki, T. & Akaogi, M. High pressure phase equilibria in the system $CaTiO_3–CaSiO_3$: Stability of perovskite solid solutions. *Physics and Chemistry of Minerals* **24**, 488–494 (1997).
189. Zedgenizov, D. A., Ragozin, A. L., Kalinina, V. V. & Kagi, H. The mineralogy of Ca-rich inclusions in sublithospheric diamonds. *Geochemistry International* **54**, 890–900 (2016).
190. Kiseeva, E. S., Wood, B. J., Ghosh, S. & Stachel, T. The pyroxenite–diamond connection. *Geochemical Perspectives Letters* **2**, 1–9 (2016).
191. Kaminsky, F. V. et al. Oxidation potential in the Earth's lower mantle as recorded by ferropericlase inclusions in diamond. *Earth and Planetary Science Letters* **417**, 49–56 (2015).
192. Smith, E. M., Shirey, S. B. & Wang, W. The very deep origin of the world's biggest diamonds. *Gems and Gemology* **53**, 388–403 (2017).
193. Moore, A. E. The origin of large irregular gem-quality Type II diamonds and the rarity of blue Type IIB varieties. *South African Journal of Geology* **117**, 219–236 (2014).
194. Bowen, D. C., Ferraris, R. D., Palmer, C. E. & Ward, J. D. On the unusual characteristics of the diamonds from Letseng-la-Terae kimberlites, Lesotho. *Lithos* **112**, 767–774 (2009).

195. Brenker, F. et al. Detection of a Ca-rich lithology in the Earth's deep (>300 km) convecting mantle. *Earth and Planetary Science Letters* **236**, 579–587 (2005).
196. Wijbrans, C. H., Rohrbach, A. & Klemme, S. An experimental investigation of the stability of majoritic garnet in the Earth's mantle and an improved majorite geobarometer. *Contributions to Mineralogy and Petrology* **171**, 50 (2016).
197. Bell, P., Mao, H., Weeks, R. & Valkenburg, A. High pressure disproportionation study of iron in synthetic basalt glass. *Carnegie Institution of Washington Yearbook* **75**, 515–520 (1976).
198. Frost, D. J. et al. Experimental evidence for the existence of iron-rich metal in the Earth's lower mantle. *Nature* **428**, 409–412 (2004).
199. Rohrbach, A. et al. Metal saturation in the upper mantle. *Nature* **449**, 456–458 (2007).
200. Ballhaus, C. Is the upper mantle metal-saturated? *Earth and Planetary Science Letters* **132**, 75–86 (1995).

6

CO_2-Rich Melts in Earth

GREGORY M. YAXLEY, SUJOY GHOSH, EKATERINA S. KISEEVA, ANANYA MALLIK, CARL SPANDLER, ANDREW R. THOMSON, AND MICHAEL J. WALTER

6.1 Introduction

Carbonate-rich magmas in Earth play a critical role in Earth's deep carbon cycle. They have been emplaced into or erupted onto the crust as carbonatites (i.e. igneous rocks composed of >50% carbonate minerals, with SiO_2 contents <20 wt.%) for the last 2.5 Ga of geological history, and one volcano (Oldoinyo Lengai, Tanzania) has erupted sodic carbonatite lavas since 1960.[1]

Carbonate melts are inferred to exist in the upper mantle, largely on the basis of high-pressure experimental studies.[2–6] Their existence has also been inferred from the mineralogy and geochemistry of some suites of peridotite xenoliths recovered from alkali basalts,[7–9] and they have been observed directly in some inclusions in diamonds[10] and minerals in sheared garnet peridotite xenoliths.[11] They may also be present in the mantle transition zone or lower mantle in association with deeply subducted, carbonate-bearing slabs[12–14] and as inclusions in lower-mantle diamonds.[15]

Because of their low density, low viscosity, and ability to wet the surfaces of silicate minerals in the mantle,[16–18] carbonate melts are able to migrate upwards from their source regions rapidly and at extremely low melt fractions. They are able to transport significant amounts of incompatible trace and minor elements, volatile elements (H_2O, halogens, sulfur), and major components such as C, Mg, Ca, Fe, Na, and K. This renders them highly effective metasomatic agents and potentially major contributors to fluxes of carbon between reservoirs in the deep and shallow Earth. They are also of particular economic importance as hosts or sources of many critical metals, including the rare earth elements (REEs) Nb, Ta, P, and others.

In this chapter, we review the current understanding of the occurrence, stability, and role of carbonatites emplaced into or onto Earth's crust and carbonate melts in the deep Earth, from lower mantle to crust. We first outline constraints from high-pressure experimental petrology and thermodynamic considerations on their stability, as functions of variables such as pressure (P), temperature (T), and oxygen fugacity (fO_2). These constraints are then used in the context of different tectonic settings in Earth to infer the presence and nature of carbonate melts in those various locations.

Carbonate melts and carbonatite magmas are also often proposed to be genetically linked to some CO_2-bearing silicate melts (melts with >20 wt.% SiO_2 and dissolved, oxidized carbon, such as kimberlites, intraplate basalts, continental alkali basalts, etc.) through processes such as carbonate–silicate liquid immiscibility, crystal fractionation, and oxidation of diamond or graphite. Genetic relationships between carbonate melts and CO_2-bearing silicate melts in appropriate settings are also considered in this chapter.

6.2 Constraints on Carbonate Stability in Earth's Mantle

Critical to the stability, distribution, movement, and capacity for mass transport of carbonate melts at mantle pressure and temperature conditions is the fO_2 of the ambient mantle. This intensive variable exerts very strong control over the speciation of C in the mantle, which can range from highly reduced metal carbides, to methane fluids, to crystalline graphite or diamond, to oxidized carbon species such as CO, CO_2, or CO_3^{2-} in fluids or melts.[19]

The various species of carbon in the peridotite upper mantle are further limited by P–T–fO_2 conditions relative to: (1) the univariant graphite/diamond phase transition; (2) redox-dependent reactions such as enstatite–magnesite–olivine–diamond (EMOD) and enstatite–magnesite–forsterite–diopside–diamond (EMFDD; see below); (3) the carbonate peridotite (or eclogite) solidus; (4) decarbonation reactions involving carbonate minerals or carbonate in melts, and silicate phases, such as

$$\begin{array}{ccccccccc} 4MgSiO_3 & + & CaMg(CO_3)_2 & = & 2Mg_2SiO_4 & + & CaMgSi_2O_6 & + & 2CO_2 \\ \text{orthopyroxene} & + & \text{dolomite} & = & \text{olivine} & + & \text{clinopyroxene} & + & CO_2 \end{array} \tag{6.1}$$

and (5) fluid absent equilibria such as

$$\begin{array}{ccccccc} 2MgSiO_3 & + & CaMg(CO_3)_2 & = & 2MgCO_3 & + & CaMgSi_2O_6 \\ \text{orthopyroxene} & + & \text{dolomite} & = & \text{magnesite} & + & \text{clinopyroxene} \end{array} \tag{6.2}$$

For example, the stability of carbonate phases versus diamond in a melt- or fluid-free, Ca-poor, magnesite harzburgite assemblage is limited in P–fO_2 space by the "EMOG/D" reaction (6.2) (enstatite–magnesite–olivine–graphite/diamond):[20,21]

$$\begin{array}{lll} MgSiO_3 & + MgCO_3 & = Mg_2SiO_4 + C + O_2 \\ \text{orthopyroxene} & + \text{magnesite} & = \text{olivine} \end{array} \tag{6.3}$$

At depths from around 40–240 km on a typical cratonic geotherm, the univariant reaction will increase slightly in fO_2 from about –1.2 log units below fayalite–magnetite–quartz buffer (FMQ) to about –0.5 log units below FMQ.[21] At fO_2 values above this reaction, CO_2-fluid, dolomite, or magnesite will be stable depending on pressure relative to reactions (6.1) and (6.2). At fO_2 values below this reaction, graphite or diamond will be stable, depending on pressure relative to the univariant graphite–diamond reaction,[22] which lies at a pressure corresponding to about 150 km depth on a cratonic geotherm. The majority of kimberlite-borne garnet peridotite xenoliths for which fO_2 has been determined lie below EMOG/D, consistent with the sampling of diamond by deeply sourced kimberlites passing through the cratonic lithosphere. This also means that carbonate melts are unlikely to be stable at depths throughout most of the cratonic mantle lithosphere, except in minor volumes (perhaps adjacent to conduits for kimberlites) where metasomatic enrichment processes may have locally oxidized the wall-rock significantly.[23]

In carbon-bearing lherzolite assemblages, the EMFDD reaction (6.4) limits carbonate stability in P–T–fO_2 space:

$$\begin{array}{llllll} 3MgSiO_3 & + CaCO_3 & = Mg_2SiO_4 & + CaMgSi_2O_6 & + C + O_2 \\ \text{orthopyroxene} & + \text{carbonate} & = \text{olivine} & + \text{clinopyroxene} & + \text{diamond/graphite} \end{array} \quad (6.4)$$

At 5 GPa, this reaction lies at about –1.2 log units at 500°C, decreasing to –1.5 log units at 1300°C, for realistic activities of the various components.[20]

Melting of carbon-bearing peridotite to form melts with high activities of carbonate therefore will be restricted to those regions of the peridotitic upper mantle where the oxidation state is consistent with carbonate stability (i.e. where fO_2 lies above the appropriate limiting reaction at given pressure and temperature).

The magnitude and variation of fO_2 in the mantle have been the subjects of many studies over recent decades. These have used measurements by a variety of techniques (wet chemical methods, Mössbauer spectroscopy, flank method, Fe K-edge micro-X-ray absorption near-edge structure spectroscopy) of $Fe^{3+}/\sum Fe$ in phases in peridotite xenoliths (spinel, garnet, pyroxenes) from the upper mantle, synchrotron Mössbauer measurements of Fe^{3+} in majoritic garnets from the sublithospheric upper mantle or mantle transition zone,[24] and $Fe^{3+}/\sum Fe$ measurements on primitive mid-ocean ridge basalt (MORB) glasses[25,26] coupled with high-pressure experimental and thermodynamic calibrations of relevant redox-controlling reactions.[27–29] These studies indicate that it is likely that carbonate stability in peridotite is generally limited to relatively shallow parts of the continental lithosphere (i.e. depths <~100 km; see Section 6.6.2 for more details). In the next section, we review high-pressure experimental constraints on the melting of carbonate peridotite.

6.3 Experimental Constraints on the Melting of Carbonate Peridotite in the Mantle

Many high-pressure experimental studies have investigated the phase and partial melting relations of (oxidized) carbonate-bearing mantle lithologies of peridotite at upper-mantle pressures.[3,5,6,30–41] The stable species of carbonate in the subsolidus peridotite upper mantle is controlled by some key carbonate–silicate reactions. These include reactions (6.1), (6.2), and (6.5):

$$\begin{array}{ccccccc} Mg_2SiO_4 & + & CO_2 & = & MgSiO_3 & + & MgCO_3 \\ \text{olivine} & + & CO_2 & = & \text{orthopyroxene} & + & \text{magnesite} \end{array} \quad (6.5)$$

These reactions have been delineated using high-pressure experiments in the CaO–MgO–SiO_2 ± H_2O (CMS ± H) system.[42–49]

In more complex natural systems, particularly those containing Fe and alkali metal components, reaction (6.4) intersects the peridotite + CO_2 system at the carbonate solidus at about 2.1 GPa and 1030°C (Hawaiian pyrolite + 5 wt.% dolomite[30]), dividing the shallow subsolidus lithospheric mantle into a shallower zone in which crystalline carbonate is unstable at the expense of CO_2 fluid and a deeper zone in which dolomite crystallizes as part of a spinel or garnet lherzolite assemblage. At the solidus, this reaction forms an approximately isobaric solidus ledge. At pressures greater than the ledge is a near-solidus field of sodic dolomitic carbonate melt in equilibrium with lherzolite residue. At pressures below the ledge, CO_2-rich fluid coexists with spinel lherzolite.

This reaction may act as a barrier to the migration of carbonate melts formed at higher pressures than reaction (6.1) to shallower depths in continental settings where geotherms are likely to intersect it. Dolomitic carbonate melts are predicted to react according to reaction (6.1), and this may lead to elimination of the melt and crystallization of secondary clinopyroxene and olivine at the expense of orthopyroxene and, in extreme cases, conversion of harzburgite or lherzolite mantle to orthopyroxene-free, clinopyroxene-rich wehrlite along with liberation of a CO_2-rich fluid. Such a process was inferred to have occurred in some spinel wehrlite xenoliths hosted in the Newer Volcanics of Victoria, southeastern Australia,[8,50] and in the Olmani Cinder cone, northern Tanzania.[51] It has been suggested that only in the circumstances where magma conduits become armored with orthopyroxene-free wehrlite are dolomitic carbonatites able to ascend to pressures less than the solidus ledge, potentially entering the crust, evolving to more calcic compositions, and, in some cases, becoming emplaced in the crust or erupted.[52]

In oceanic settings, convective geotherms are at higher temperatures than conductive geotherms in the continental lithosphere and are not expected to intersect the solidus ledge or by reaction (6.1). At higher pressures and temperatures (3.4 GPa, 1080°C),[30] the vapor-absent reaction (6.2) intersects the peridotite solidus, dividing the subsolidus regime into a lower-pressure field of dolomite garnet lherzolite and a higher pressure field of magnesite garnet lherzolite. Low-degree partial melts in experimentally investigated peridotite–$CO_2 \pm H_2O$ systems with natural compositions are generally broadly alkali rich and calcio-dolomitic to dolomitic in composition.[3] Solidus temperatures are considerably lower than those of volatile-free systems, and although experimental studies that include CO_2 *and* H_2O are relatively rare, the available evidence suggests solidus temperatures are even lower.[3,53]

In the following sections, we apply the experimental and other constraints to infer the existence and behavior of carbonate melts in different tectonic settings.

6.4 Carbonate Melts Associated with Subduction Zones

The mantle is believed to have played a key role in controlling the long-term carbon budget in the exosphere through cycling of carbon from the surface to the mantle and back again.[6,54,55] The deep carbon cycle is regulated at the surface by the quantity of carbon subducted into the mantle at convergent margins and by volcanic degassing of mantle-derived melts releasing carbon into the exosphere at mid-ocean ridges, ocean islands, and arc volcanoes. A substantial proportion of the mass of carbon drawn into the mantle at subduction zones (anywhere between 20% and 100%) is recycled back to the surface via fore-arc degassing and arc magmatism,[56] as discussed in detail in Chapter 10 of this volume. Estimates of the net annual flux of carbon ingassing and outgassing via these processes are difficult to constrain with certainty, and range from negligible values to ~60 Mt/year net recycling.[6,56]

Carbon enters the mantle at subduction zones in sediments, altered oceanic crust, and mantle lithosphere, with the total input flux in the range of ~50–100 Mt/year at modern subduction zones.[6,56] Sedimentary carbon includes both biogenic organic carbon and carbonate. In more than a third of studied modern subduction zones, carbonate-rich materials make up a substantial fraction of the downgoing sediment, whereas in others it is absent altogether (e.g. Refs. 57, 58). However, organic carbon is expected to be at least a minor component in pelagic sediments and turbidites.[56]

Carbon is deposited during hydrothermal processes at mid-ocean ridges where carbonate (calcite and aragonite) forms during alteration of oceanic crust due to its reaction with CO_2 in seawater; biotic organic carbon in oceanic crust is minor relative to abiotic organic compounds and inorganic carbonates. The top few hundred meters of oceanic crust contain an average of ~2.5 wt.% CO_2, and at deeper levels the carbon content, mostly in the form of organic hydrocarbon species, drops below 0.2 wt.% throughout the remainder of the crustal section. Altered lithospheric mantle that is exposed to alteration by seawater also carries carbonate, although likely at an overall fraction that is much less than that of oceanic crust.[56,59]

Downgoing slab materials never reach temperatures high enough for "dry" partial melting during blueschist and eclogite facies metamorphism at fore-arc depths (up to ~80 km),[60] and portions of the slab that do not experience pervasive dehydration can effectively transport carbon to greater mantle depths. Nevertheless, at pressures above ~0.5 GPa, carbonate solubility in aqueous fluids increases with temperature,[61] so fluids produced by metamorphic devolatilization of the slab can be effective at dissolving carbonate minerals and mobilizing carbon.[62] Much of this carbon may be redistributed within the slab[63] or sequestered into serpentinized mantle rock that overlies the slab,[64–66] some of which in turn is dragged down with the descending slab.

At sub-arc depths (80–200 km), slab surface temperatures reach between 600 and 1000°C,[67] and carbonate mineral dissolution becomes much more efficient due to higher solubilities in hydrous fluids[68] and silicate melts[69] at these depths. In some cases, carbonatite liquids may also form via fluid-flux melting of carbonate-rich metasedimentary rocks[70] or carbonate-bearing metagabbros.[71] As a consequence, CO_2 ($\pm$ CO_3^{2-})-rich fluid phases migrating from the downgoing slab or from buoyantly upwelling slab diapirs[72,73] can introduce significant carbon flux from the slab to the overlying mantle wedge. Evidence in support of C-rich fluid phases at the slab surface comes from garnet and clinopyroxene inclusions in diamonds from Dachine, South America, which have major and trace element characteristics indicating growth at the surface of a subducting slab at ~200 km depth, possibly in metalliferous metasediment.[74]

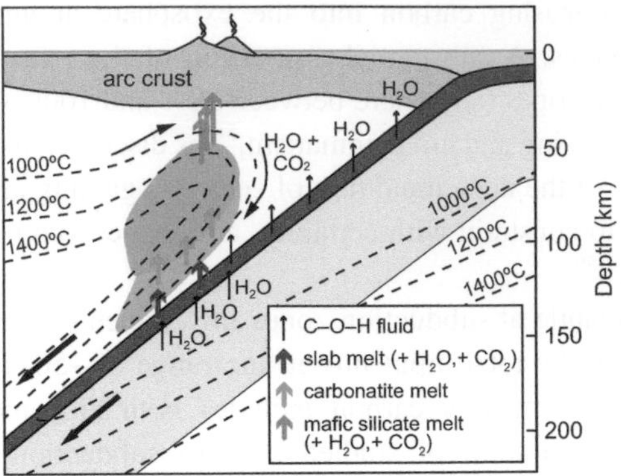

Figure 6.1 Schematic cross-section of a subduction zone (modified after Green[75]) depicting progressive slab devolatilization during subduction and zones of the mantle wedge containing carbonatitic and silicate-rich partial melts (green and orange fields, respectively).

Slab-derived fluids migrating into the overlying mantle will experience progressive heating as they ascend through the inverted temperature gradient of the mantle wedge (Figure 6.1).[75] The introduction of C–O–H fluids or melts results in a significant lowering of the wedge peridotite solidus, such that carbonatite liquids can be produced at temperatures below 950°C.[75] Such carbonatitic liquids (+ H_2O) are expected to be highly mobile, but will react with the wedge upon ascent and heating to produce carbonated hydrous silicate melts (>1020°C; Refs. 73, 75). Further ascent will favor further peridotite melting, increasing melt fractions, and diluting dissolved volatile contents. Upwelling from the hot core of the wedge may impart retrograde melt-rock or fluid-rock reactions, locking some carbonate (+ H_2O) phases in the mantle lithosphere and lower crust,[56] but the most volatile flux from the slab is expected ultimately to be delivered to the upper-arc crust via fractionating and degassing arc magmas.

The unique petrophysical evolution of magmas traversing the mantle wedge means that carbonatitic liquids are unable to be tapped from the mantle to the surface, which explains the lack of carbonatites found in supra-subduction zone settings.[76] Further, carbonated sub-arc mantle lithosphere and arc lower crust may eventually be preserved as subcontinental lithosphere through the reaction of volatile-rich melts with the mantle[77] or recycled into the convecting mantle, possibly thereafter to undergo melting to produce carbonatites in intraplate settings.

6.5 Melting of Subducted, Carbonated Sediment and Ocean Crust in the Deep Upper Mantle and Transition Zone

While subduction to ~200 km depth can remove a significant fraction of the initial down-welling carbon flux to the mantle wedge, experimental evidence of carbonate stability, modeling of phase equilibria, and slab devolatilization indicate that, in some subduction zones, a substantial portion of subducted carbon or carbonate may make it past the dehydration zone and into the deeper mantle.[78,79] Experiments also suggest that carbonate may be reduced to elemental carbon (graphite or diamond) at depths shallower than 250 km in more reducing eclogitic assemblages, although for oxidation states typical of MORB (e.g. Ref. 80), eclogitic assemblages should remain in the carbonate stability field to at least 250 km.[28]

Inclusions of carbonate minerals in superdeep diamonds provide the strongest direct evidence for a carbonate component subducted past the volcanic front and at least to transition-zone depths.[81–84] In addition, the distinctive major and trace element compositions of silicate inclusions in many superdeep diamonds (e.g. majorite garnet, Ca- and Ti-rich perovskite) have been interpreted to preserve a direct record in their origin of a low-degree carbonated melt derived from subducted oceanic crust.[82,85,86] Both the carbon isotopic composition of the diamonds and the oxygen isotope composition of the inclusions provide further evidence for a key role of subducted crustal components in the origin of many superdeep diamonds and their inclusions.[87,88] Thus, it seems that melting of carbonated sediment and oceanic crust in the deep upper mantle and transition zone may play a key role in the deep carbon cycle.

Figure 6.2 compares the solidus determinations from published studies on the melting behavior of carbonated pelitic sediment and carbonated oceanic crust at upper-mantle and transition-zone conditions. The solidus of carbonated pelitic sediment was determined by Tsuno and Dasgupta[89] at 2.5–3.0 GPa and by Grassi and Schmidt[90] at 8–13 GPa, with melts ranging from granitic at low pressures to K-rich carbonatitic at higher pressures. On the basis of solidus determinations in these studies, only in the hottest subduction zones would melting of anhydrous carbonated sediments occur in the sub-arc region. The addition of water reduces the solidus of pelitic sediments, but unless sediments are water saturated, perhaps by fluxing of water-rich fluids from below, carbonated slab sediments may reach the deep upper mantle and transition zone.[91] At higher pressures approaching the transition zone, the experiments of Grassi and Schmidt[90] indicate that carbonated sediments can melt along warm and hot slab-top geotherms, but colder slabs could transport sedimentary carbonate into the deeper mantle (Figure 6.2).

There have been many experimental studies of melting carbonated basaltic compositions, both in simplified[92,93] and natural systems.[12,13,35,94–97] The small number of phases in basaltic compositions hampers studies in simplified compositions, whereas subtle compositional dependencies hamper studies in natural compositions. Indeed, subtle variations in bulk compositions between studies (Table 6.1) likely cause the significant variations in position and shape of the carbonated basalt solidus (e.g. Refs. 13, 98).

In the lower pressure range (e.g. $<$3–5 GPa), a carbonate phase is not always stable at the solidus depending on the bulk CO_2 and SiO_2 contents. In those with higher SiO_2/CO_2 ratios, CO_2 and/or silicate melts containing dissolved CO_2 define the solidus,[13,95,97] whereas compositions with lower SiO_2/CO_2 ratios showed carbonate stability and carbonate melt production along the solidus from low pressure.[13,35,98,99] Additionally, near-solidus melts have been identified with a wide range of compositions from mafic to

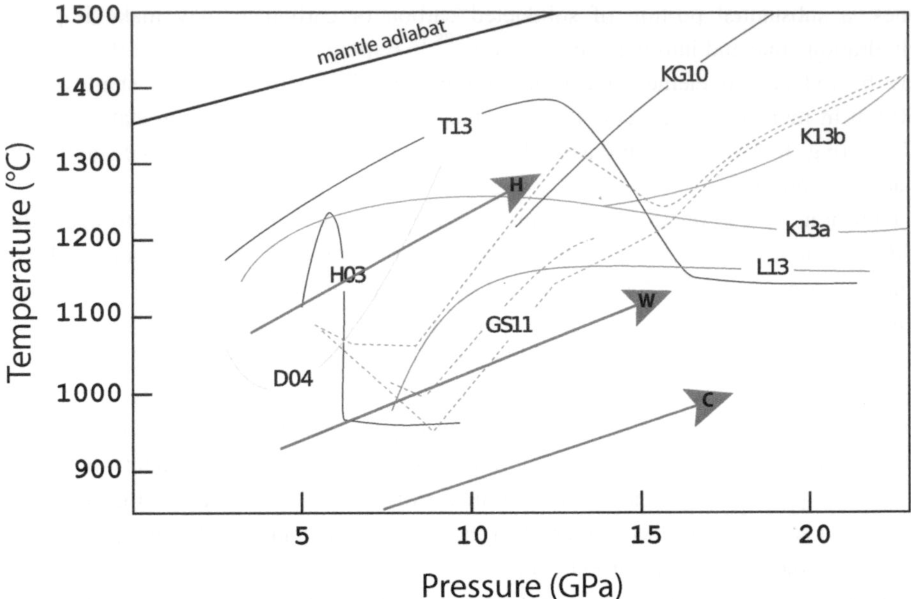

Figure 6.2 Summary of the experimentally determined solidus curves for carbonated pelitic sediment and basaltic compositions. Solidus curves for carbonated basalt are shown as solid colored curves and are from: Hammouda,[95] H03; Yaxley and Brey,[148] YB04; Dasgupta et al.,[94] D04; Keshav and Gudfinnsson,[186] KG10; Kiseeva et al.,[12] K13a,b; and Thomson et al.,[13] T13. Labels are keyed to the compositions listed in Table 6.1. Solidus curves for carbonated sediment are shown as dashed curves and are from: Tsuno and Dasgupta,[89] TD11; and Grassi and Schmidt,[187] GS11. Also shown is the solidus of alkaline carbonatite (Litasov et al.,[101] L13). The mantle adiabat is from Katsura et al.[188] Model geotherms at the top of a subducting slab are extrapolations shown for examples of hot, warm, and cold slabs.[67]

silicic. In some cases silicate melts are observed at the solidus before carbonate melts, whereas in other cases this relationship is reversed, and both kinds of melts have been interpreted to coexist together as immiscible liquids.[12,99,100] The observation of immiscible melts may reflect the maximum CO_2 solubility in silicate melts, and the appearance of liquid immiscibility will therefore depend on the CO_2 content of the bulk composition.

At higher pressures of the deep upper mantle and transition zone, starting compositions again show remarkable control of melting behavior (Figure 6.2). When comparing the carbonated starting compositions used in the various studies, there are considerable differences, perhaps most notably in SiO_2 and CO_2 contents and Ca#. Thomson et al.[13] observed that the Ca# has an important controlling effect on the stable carbonate phase at the solidus beyond the pressure of dolomite breakdown ($>\sim 10$ GPa). Phase relations apparently preclude coexistence of magnesite and aragonite in majorite- and clinopyroxene-bearing assemblages, and which of these carbonate phases is stable has an important controlling effect on the solidus shape and melt compositions.

Table 6.1 *Bulk compositions (as wt.%) of anhydrous MORB and carbonated MORB used in melting studies*

MORB examples[a]	G12 All MORB	Y94 MAN-7	W99 JB1	H03 OTBC[b]	D04 SLEC1	K13a GA1cc	K13b Volgacc	T16 ATCM1
SiO_2	50.47	47.71	53.53	47.23	41.21	45.32	42.22	50.35
TiO_2	1.68	1.71	1.44	–	2.16	1.34	1.43	1.33
Al_2O_3	14.7	15.68	14.85	15.35	10.89	14.88	15.91	13.66
FeO	10.43	9.36	7.92	8.93	12.83	8.85	9.46	11.35
MnO	0.18	0.18	0.16	–	0.12	0.15	0.14	0.21
MgO	7.58	8.43	7.64	6.24	12.87	7.15	7.64	7.15
CaO	11.39	11.73	9.12	14.77	13.09	14.24	14.85	10.8
Na_2O	2.79	2.76	2.64	2.91	1.63	3.14	3.36	2.48
K_2O	0.16	0.23	1.31	0.02	0.11	0.4	0.42	0.06
P_2O_5	0.18	0.02	–	–	–	0.14	0.15	0.1
CO_2	–	–	–	4.43	5.0	4.4	4.4	2.52
Total	99.57	99.81	98.61	99.98	99.91	100.01	99.98	100
Ca#[c]	0.38	0.38	0.38	0.48	0.32	0.46	0.45	0.36
Mg#[c]	0.57	0.62	0.62	0.56	0.64	0.59	0.59	0.53

[a] G12 = Gale et al.;[190] Y94 = Yasuda et al.;[191] W99 = Wang and Takahashi;[192] D04 = Dasgupta et al.;[94] H03 = Hammouda;[95] K13 = Kiseeva et al.[12]

[b] OTBC additionally contains 1200 ppm H_2O.

[c] Ca number = Ca/(Ca + Mg + Fe); Mg number = Mg/(Mg + Fe).

In Ca-rich carbonated basalt bulk compositions, the stable phase is aragonite, which is also a host for sodium.[12] However, in lower Ca# bulk compositions, magnesite is stable, and because sodium is relatively insoluble in magnesite, another Na-rich carbonate stabilizes in the subsolidus assemblage at pressures greater than ~15 GPa. The appearance of a Na-carbonate phase in the subsolidus produces a dramatic lowering of the melting temperature and a deep trough along the solidus between ~10 and 15 GPa that is not observed where aragonite is stable (Figure 6.2). When Na-carbonate occurs on the solidus, it is observed that the melting temperature at pressures >~15 GPa (~1150°C) is indistinguishable from that observed for a simplified Na-carbonate-rich bulk composition at these conditions (Figure 6.2).[101] As the majority of natural MORB compositions, fresh and altered, fall to the Mg-rich side of the majorite–clinopyroxene join, they should have magnesite and not aragonite as the stable carbonate at high pressures and experience the lowered solidus at transition-zone conditions.[13]

Figure 6.3 shows melt compositions projected onto a plane differentiating major and minor components of sediment and basalt-derived carbonated melts, demonstrating a generally continuous evolution with increasing pressure and temperature. The lowest-degree melts of carbonated basalt are highly calcic, even when the subsolidus carbonate is magnesite, because magnesite is the liquidus phase in most carbonate systems.[101] Carbonated melts are enriched in

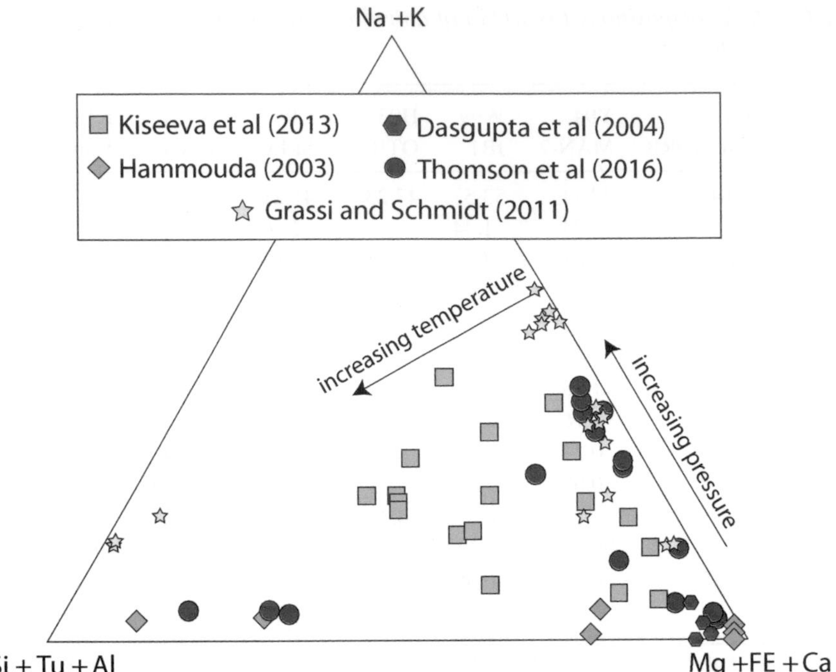

Figure 6.3 The compositions of partial melts from carbonated natural basalt and sediment from the studies of Hammouda,[95] Dasgupta et al.,[94] Kiseeva et al.,[12] Thomson et al.,[13] and Grassi and Schmidt.[90] The generalized effects of increasing pressure and temperature are also shown.

incompatible elements and have high concentrations of TiO_2, P_2O_5, and alkalis (Na_2O and K_2O), the relative abundances of which will be controlled by the bulk composition of the protolith. Sediment melts are dominated by potassium, whereas melts from carbonated basalt have alkali contents that are dominated by Na_2O. Melt Na_2O content increases systematically with pressure in basaltic compositions, resulting from the decreasing compatibility of Na_2O in the coexisting residual phase assemblage. Thomson et al.[13] observed that melt compositions of carbonated basalt in the transition zone remain approximately constant over a wide temperature interval of >300°C above the solidus, and only when temperature exceeds ~1,500°C does the silica content of the melt increase. This behavior is reminiscent of the melting behavior observed in carbonated peridotite assemblages at lower pressures.[102]

6.6 Carbonate Melts and Kimberlites in the Cratonic Lithospheric Mantle

The cratonic mantle lithosphere underlies ancient, continental blocks that have been geologically stable for billions of years. It is chemically depleted, thick, and buoyant. Unlike other tectonic settings, any surface expression of cratonic magmatism is manifested by emplacement of small-volume, rare, exotic, volatile-rich, alkali- and carbonate-rich magmas, such as carbonatites, kimberlites, lamproites, various lamprophyres, and other highly silica-undersaturated magmas,[103,104] all of which have a deep mantle origin. Describing the detailed petrology of these rocks is beyond the scope of this chapter, and the reader is referred to the work of Jones et al.[76] Below, we give a short overview of an important carbonate-bearing silicate volcanic rock found in cratonic settings: kimberlites.

6.6.1 Kimberlites

Kimberlites are volatile (chiefly CO_2)-rich, silica-poor alkaline, ultrabasic magmas generally believed to be derived from a depth of ≥150–250 km and almost exclusively emplaced into cratonic crust. Kimberlite magmas are economically significant because of their association with diamonds. They have been subdivided into two major groups on the basis of petrographic and geochemical characteristics: Group 1 kimberlites and Group 2 kimberlites.[105–107] Group 1 kimberlites are CO_2-rich, potassic, ultrabasic rocks with a typical porphyritic texture, with large phenocrysts of most commonly olivine surrounded by a fine-grained matrix consisting of olivine, phlogopite, spinel, ilmenite, monticellite, calcite, apatite, perovskite, and other phases.[108] Group 2 kimberlites, or orangeites, are restricted to southern Africa. However, similar rocks have also been identified in Australia, India, Russia, and Finland. They are texturally similar to the Group 1 kimberlites, but have distinct compositional and isotopic differences, manifested mainly by the presence of phlogopite, K–Ba–V titanites, and Zr-bearing minerals, such as kimzeytic garnets,[108] lower ε_{Hf} and ε_{Nd} values, and high radiogenic ^{87}Sr isotope compositions.[109]

Similar to carbonatites, kimberlites are volumetrically very minor, but they are widespread throughout the cratonic parts of continents, with the latest discoveries extending the Gondwanan Cretaceous kimberlite province to Antarctica.[110] The main difference with carbonatites is in their origin in the subcratonic lithospheric or asthenospheric mantle enriched by an ocean island basalt source in the case of Group 1 kimberlites, whereas Group 2 kimberlites are derived from metasomatized lithospheric mantle,[107,111] and their eruption is predominantly through stable parts of cratons.[112] To our knowledge, there are no reported occurrences of kimberlites in the oceanic crust.[112,113]

The accurate determination of the kimberlite parental magma composition and its origin is hampered by the ubiquitous presence of a large fraction of foreign materials (mantle and crustal xenocrysts and xenoliths), in particular olivine crystals, collected from lithospheric mantle during their ascent to the surface, and by low-temperature alteration during and after emplacement. Petrological studies of the unaltered Udachnaya kimberlite pipe in the Siberian Craton showed high enrichment in CO_2, halogens, and alkalis, resulting in a hypothesis in which genetic links between kimberlites and carbonatites were important.[114,115] Later, experimental studies concluded that magmas potentially parental to kimberlites originate as dolomitic carbonate liquids in metasomatized, oxidized zones in the deep cratonic lithospheric mantle[23] that segregate and become progressively more silicate-rich due to the assimilation of silicate material (mostly orthopyroxene) during ascent through the refractory lithosphere.[116]

Alternative models are based on high-pressure experimental studies of estimated compositions of melts parental to kimberlites and have attempted to identify pressure, temperature, and volatile conditions at which the melts are multiply saturated in garnet peridotite phases.[117,118] Such studies have indicated that the most likely source for kimberlite parental melts is hydrous, carbonate-bearing garnet harzburgite in the deep cratonic lithospheric mantle or sublithospheric asthenosphere.

In either case, kimberlite genesis clearly requires a sufficiently oxidized mantle for crystalline carbonate (Ca-magnesite) to be stable in the peridotitic source,[23] and the source is constrained to be greater than ~150 km depth in the cratonic lithospheric mantle because of the presence of xenocrystic diamonds.

6.6.2 Redox Constraints on Carbonate Stability in the Cratonic Lithospheric Mantle

In any volume of the cratonic mantle, in the absence of externally derived melts or fluids, the local fO_2 is controlled by silicate or oxide mineral exchange equilibria such as reactions (6.6)–(6.9), which involve oxidation of Fe^{2+}-bearing components and reduction of Fe^{3+}-bearing components.

$$\underset{\text{orthopyroxene}}{6FeSiO_3} + \underset{\text{spinel}}{2Fe_3O_4} = \underset{\text{olivine}}{6Fe_2SiO_4} + O_2 \tag{6.6}$$

$$\underset{\text{spinel}}{2Fe_3O_4} + \underset{\text{orthopyroxene/olivine}}{3SiO_2} = \underset{\text{olivine}}{3Fe_2SiO_4} + O_2 \tag{6.7}$$

$$\underset{\text{garnet}}{2Fe^{2+}{}_3Fe^{3+}{}_2Si_3O_{12}} = \underset{\text{olivine}}{4Fe_2SiO_4} + \underset{\text{orthopyroxene}}{2FeSiO_3} + O_2 \tag{6.8}$$

$$\underset{\text{garnet}}{2Ca_3Fe_2Si_3O_{12}} + \underset{\text{garnet}}{2Mg_3Al_2Si_3O_{12}} + \underset{\text{orthopyroxene}}{4FeSiO_3} = \underset{\text{garnet}}{2Ca_3Al_2Si_3O_{12}} + \underset{\text{olivine}}{4Fe_2SiO_4} + \underset{\text{opx}}{6MgSiO_3} + O_2 \tag{6.9}$$

For example, (6.6)–(6.9) have been calibrated experimentally,[21,27,119–121] meaning that, in principle, the fO_2 of a spinel or garnet peridotite from the upper mantle can be determined if the activities of the Fe-bearing components in the minerals that contain Fe^{2+} and Fe^{3+} can be determined, along with other mineral component activities, pressure, and temperature.

These experimental calibrations can be combined with conventional thermobarometry to calculate the variation in fO_2 as a function of depth in the peridotite lithospheric upper mantle, as recorded by peridotite xenoliths from kimberlites in the case of cratonic mantle lithosphere, for example. The uppermost cratonic mantle lithosphere based on spinel peridotite xenoliths sampled by kimberlites has fO_2 between 0 and –1 log units relative to FMQ for primitive ($\Delta logfO_2^{FMQ}$),[122] well within the carbonate stability field. In the garnet peridotite facies, fO_2 decreases systematically with increasing pressure (depth) through the cratonic lithosphere (Figure 6.4),[23,122,123] although the xenolith record exhibits complications associated with oxidative overprinting associated with metasomatism, particularly at depths >150 km. Decreasing fO_2 with increasing depth is expected on a thermodynamic basis, because of the molar volume changes of reactions such as (6.8) and (6.9).[27,124] It is therefore likely that melts with high carbonate activities are unstable relative to graphite or diamond at depths greater than ~90–120 km in the cratonic lithospheric mantle.

At depths of ~250–300 km, the fO_2–P path is expected intersect the Ni precipitation curve, an FeNi alloy will exsolve from peridotite, and fO_2 will be buffered near the iron–wüstite (IW) buffer deeper into the upper mantle. In the presence of metallic FeNi alloy, some reduced carbon will be accommodated as (Fe,Ni) carbides. Rohrbach et al.[125] have

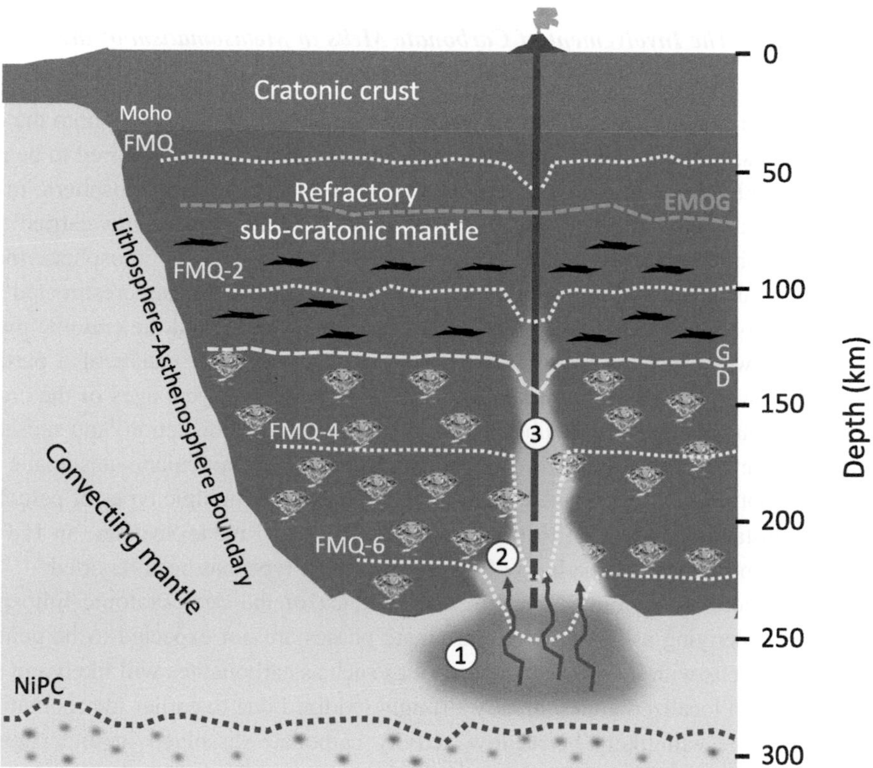

Figure 6.4 Highly schematic representation of a section through Earth's upper mantle modified from figure 2c of Foley and Fischer,[112] showing fO_2 as a function of depth. The dashed white contours are approximate contours of fO_2 variation in the cratonic lithosphere based on studies of garnet peridotite xenoliths (references in the text). The dashed yellow line is the graphite–diamond transition for a typical cratonic geotherm (38 mW m^{-2}). The red line (3) represents the path of an erupting kimberlite and the blue zone (2) near the base of the lithosphere represents a local mantle volume metasomatized to higher fO_2 by asthenospheric melts derived from the brown field labeled (1).

calculated that, assuming the mantle contains 50–700 ppm FeNi and 50–500 ppm C at 300 km depth, it would contain an assemblage of $(Fe,Ni)_3C$ + FeNi alloy + diamond or an Fe–Ni–S–C melt.[126]

Although the cratonic mantle becomes more reduced with increasing depth, local redox heterogeneities may be introduced, particularly at depths greater than 150 km, by metasomatic fluids such as carbonate-bearing silicate melts[23,127] with low carbonate activities,[21] possibly derived from the asthenosphere or from oxidized crustal material recycled via subduction. In some cases, the degree of oxidation in the lower part of the cratonic lithospheric upper mantle could be sufficient to stabilize crystallization of carbonates (magnesite in peridotite), which, in the presence of H_2O, could melt and produce magmas parental to kimberlites.[117,118]

6.6.3 The Involvement of Carbonate Melts in Metasomatism of the Deep Cratonic Lithospheric Mantle

Apart from carbonatites and kimberlites that were emplaced into or erupted onto the crust, CO_2-rich fluids, carbonate melts, or carbonated silicate melts are often inferred to be major agents of mantle metasomatism and trace element enrichment in the lithospheric mantle. Unlike, for example, alpine massifs, most peridotite and eclogite xenoliths carried to the surface by kimberlites are chemically enriched relative to the depleted lithospheric mantle. It may be that the degree of this enrichment was previously severely underestimated[112,128] and that CO_2-rich metasomatism is a widespread process throughout the cratonic mantle.

It is not always easy to distinguish and characterize the exact nature of a particular metasomatic overprint observed in natural rocks. Given the giga-year ages of the cratonic lithospheric mantle, there is a likelihood of multiple melting (i.e. depletion) and subsequent re-enrichment events. Moreover, the challenge of deciphering a particular metasomatic event in a given mantle xenolith is exacerbated by the possibility of multiple types of percolating fluids. For instance, in addition to the carbonatitic CO_2-rich metasomatism, an H_2O-rich alkali silicate metasomatism, perhaps of a proto-lamproite type, has been reported.[129–131]

At the reduced conditions likely to exist in much of the deep cratonic lithospheric mantle and underlying asthenosphere, carbonate phases are not expected to be generally stable and so melts with high carbonate activities such as carbonatites will likely not form, except locally in localized zones already strongly oxidized due to earlier metasomatism.[23] Carbon transport will likely be as low activity carbonate dissolved in undersaturated silicate melts[21] or as $CH_4 + H_2O$ fluids.[128,132,133] However, it should also be noted that the depth–fO_2 profile of the asthenospheric mantle is not necessarily the same as that observed in xenolith studies from the cratonic mantle or inferred from thermodynamic calculations. Based on CO_2–Ba–Nb systematics of oceanic basalts, the convecting mantle may in fact be considerably more oxidized at pressures greater than about 3 GPa.[134]

If undersaturated silicate melts with dissolved carbonate or CO_2 at low activities form as a result of adiabatic upwelling in plumes or rifts, segregate from their asthenospheric sources, and percolate upwards into parts of the cooler, deep cratonic lithosphere, they will freeze into the lithospheric mantle as the appropriate peridotite + volatile solidus is locally crossed. Oxidized carbon species will be exsolved from the melt during crystallization and would reduce to C or CH_4 ($CO_2 = C + O_2$; $CO_2 + 2H_2O = CH_4 + 2O_2$) because of the low ambient fO_2 in the deep cratonic lithosphere. Fe^{2+} in the melt and the wall rock will oxidize to Fe^{3+} ($2FeO + ½O_2 = Fe_2O_3$) and be incorporated into garnet and pyroxenes, leading to the observed increase in lithospheric fO_2 associated with metasomatism. The increased activity of H_2O may cause a decrease in the solidus temperatures of peridotite locally and lead to partial melting in a process known as hydrous redox melting. Only after sufficient oxidation by this type of metasomatic process could the fO_2 of the deep cratonic lithospheric mantle be raised sufficiently to allow carbonate stability and the formation of carbonatites, kimberlites, and related rocks at these depths.

6.7 Carbonate Melts beneath Ocean Islands in Intraplate Settings

Observations from natural samples indicate the role of CO_2, specifically CO_2-rich silicate melts, in the metasomatism of the upper mantle beneath ocean islands in intraplate settings. Carbonate phases, interpreted to be quenched carbonate liquids, have been identified in metasomatized harzburgite xenoliths in the Kerguelen and Canary Islands.[135,136] These carbonate inclusions are texturally associated with silicate glass inclusions, either as globules within a silicate matrix or as intimate associations with silicate inclusions. Such textural associations between the carbonate-rich phases and the silicate glasses have led to the conclusion that the carbonate-rich phases are products of immiscibility of a single carbonated silicate melt phase that exists at depth, possibly at upper-mantle conditions. Geochemical modeling of rejuvenated Hawaiian lavas also indicates the presence of carbonate-rich liquids in the source of these lavas.[137]

6.7.1 How Do CO_2-Rich Silicate Melts Form in the Upper Mantle? Can These CO_2-Rich Melts Explain the Chemistry of Erupted Magmas in Intraplate Ocean Islands?

Previous experimental studies have demonstrated that CO_2-rich silicate melts can be produced in upper-mantle conditions by the following mechanism. Carbon may exist in its oxidized form as carbonate mineral or as liquid in the mantle at fO_2 values that are about 2 log units higher than that of the IW buffer (given by equilibrium (6.10)):

$$\underset{\text{iron}}{2Fe + O_2} = \underset{\text{wüstite}}{2FeO} \tag{6.10}$$

which corresponds to depths above 150–250 km[21,138] in the oceanic lithosphere. Carbonate may be present deeper in the mantle in locally oxidized regions.[139,140] Due to cryoscopic depression of the freezing point, carbonate-bearing lithologies (peridotite and recycled oceanic crust or eclogite) have lower solidus temperatures than nominally anhydrous or carbonate-free lithologies, as demonstrated by experimental studies.[5,33,47,94–96,99,141–148] This implies that in an upwelling mantle partial melting of carbonated lithologies is initiated deeper than in the surrounding carbonate-free lithologies. Near-solidus partial melting of carbonated peridotite and/or carbonated recycled oceanic crust (eclogite) produces carbonatitic liquids (>25 wt.% CO_2, <20 wt.% SiO_2). These carbonatitic liquids tend to be very mobile owing to their low viscosities and low dihedral angles,[149] which lead to their escape from the site of generation. These rising liquids can cause flux-based partial melting of peridotite and eclogite in an adiabatically upwelling plume mantle in intraplate settings, which produces CO_2-bearing silicate melts (Figure 6.5a).

Peridotite is the dominant lithology of Earth's mantle. This increases the likelihood of partial melts of CO_2-bearing peridotite being the best candidates for explaining the geochemistry of ocean island basalts that commonly erupt in ocean islands in intraplate settings. The next likely candidate for ocean island basalts would be partial melts of CO_2-bearing eclogite, given recycled oceanic crust forms the dominant chemical heterogeneity in Earth's mantle.[150] Comparisons of the major element chemistry of partial melts of CO_2-bearing peridotite and eclogite with natural, near-primary ocean island basalts indicate that carbonated peridotite-derived partial melts are too TiO_2 poor,[151] while carbonated eclogite-derived partial melts are too depleted in MgO.[99] Also, the peridotite-derived partial melts can explain the MgO content very well for the ocean island basalts, while the eclogite-derived partial melts can explain the TiO_2 contents. This implies that the source of ocean island basalts would be best explained by a hybrid source involving contributions from both peridotite (high MgO) and eclogite (high TiO_2). A previous study indicated a couple of geodynamic scenarios involving CO_2-bearing eclogite and peridotite as the source of ocean island basalts (Figure 6.5b).[152]

Deep carbonatites are formed by very-low-degree partial melting of carbonate-bearing eclogites that are present in deep, locally oxidized upper-mantle domains. These carbonatites rise up and cause flux melting of overlying volatile-free eclogites. The flux melting produces eclogite-derived carbonated silicate melts (ranging from basanites to andesites) that are out of chemical equilibrium with their surrounding peridotite and undergo reactive infiltration. The metasomatic process or melt-rock reaction associated with such reactive infiltration forms ocean island basalts.

Deep carbonatites can also be generated by very-low-degree partial melting of graphite/diamond-bearing peridotite or eclogite at the redox front (150–250 km depth, where reduced carbon in the form of diamond/graphite is oxidized to a carbonatitic melt). These carbonatites can cause fluxed melting of volatile-free eclogite and can undergo subsequent reactive infiltration through the peridotite, as described above.

A question arises as to whether the involvement of CO_2 in the source is required for all ocean island basalts. Studies have shown that basanitic ocean island basalts can be explained by partial melting of volatile-free peridotite ± eclogite in the source; however, more silica-undersaturated ocean island basalts such as nephelinites and melilitites require the involvement of CO_2.[152,153]

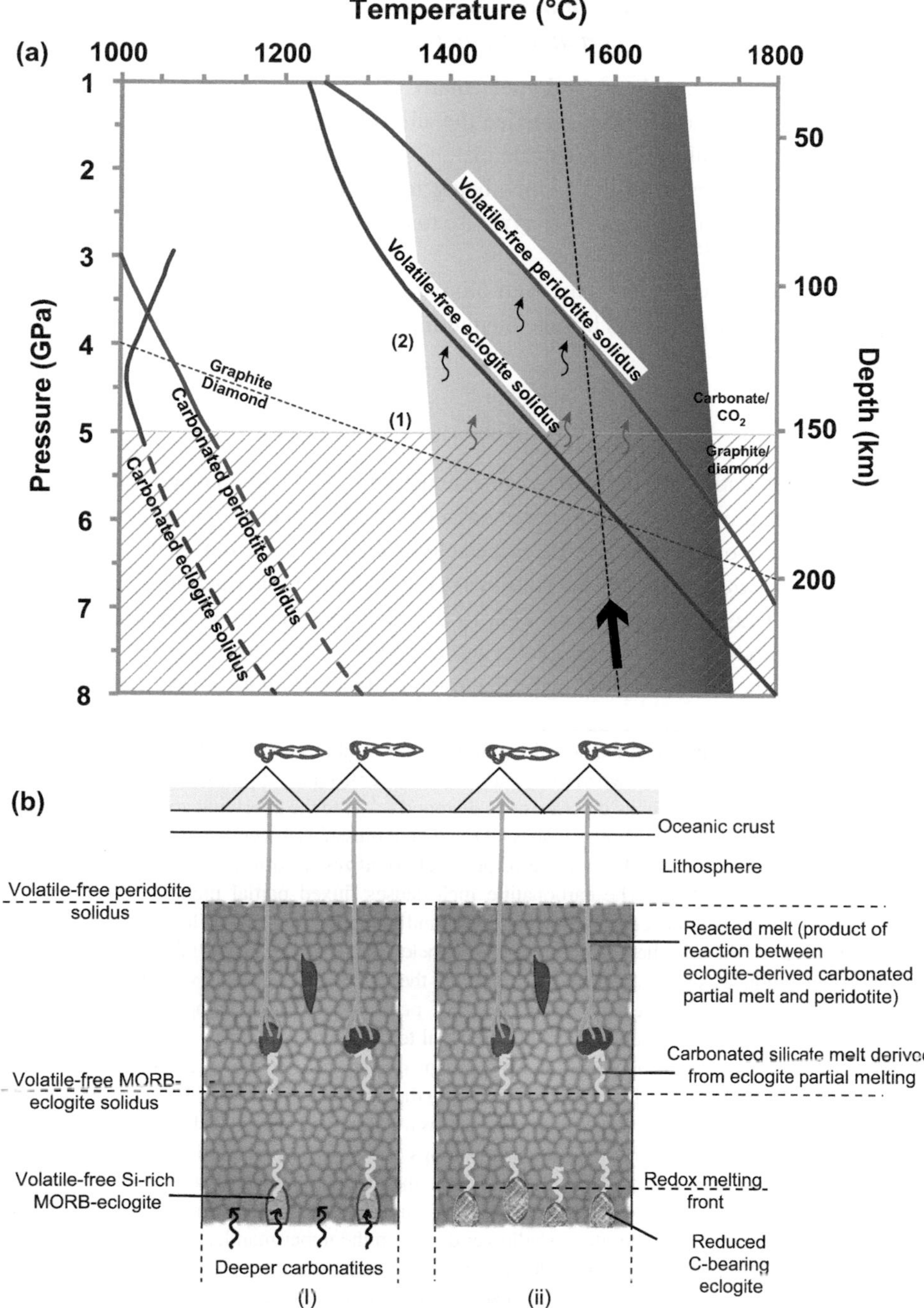

Figure 6.5 (a) Pressure–temperature plot showing the generation of carbonated silicate melt in Earth's upper mantle. A parcel of carbon-bearing mantle peridotite with pods of eclogite upwells

6.7.2 Effect of CO_2 on the Reaction between Eclogite-Derived Partial Melts and Peridotite

CO_2 dissolves in silicate melts as both molecular CO_2 and CO_3^{2-} anions. When network-modifying cations (those that depolymerize the silicate network) such as Na^+, K^+, Ca^{2+}, Mg^{2+}, and Fe^{2+} are available in the silicate melt, CO_3^{2-} bonds with these cations.[154] These carbonate complexes result in the cations being removed from their network-modifying roles. Thus, these carbonate complexes give rise to pockets of polymerized networks in the silicate melt structure.[155]

When eclogite-derived melts react with olivine- and orthopyroxene-bearing peridotite, the following equilibrium (6.11) buffers the thermodynamic activity of SiO_2 (a_{SiO2}) in the reacted melt:

$$\underset{\text{olivine}}{Mg_2SiO_4} + \underset{\text{melt}}{SiO_2} = \underset{\text{orthopyroxene}}{Mg_2Si_2O_6} \tag{6.11}$$

The localized separation of polymerized silicate networks and carbonate complexes in the silicate melt structure results in the requirement for excess free energy of mixing between the two structural components. This increases the activity coefficient of SiO_2 in the melt (γ_{SiO2}). Higher γ_{SiO2} when a_{SiO2} is buffered results in a decrease in the mole fraction of SiO_2 (X_{SiO2}) in the melt.[151] Thus, involvement of CO_2 in the melt-rock reaction decreases the SiO_2 content of the product melt, driving eclogite-derived basanites and andesites to produce nephelinites and melilitites (depending on the amount of CO_2 available in the system).[152,153]

Figure 6.5 (*cont.*) beneath ocean islands along an adiabat (bold arrow along the dotted line). As it upwells, redox melting takes place when the mantle crosses over from the stippled zone where the carbon is present in reduced form (either as graphite or as diamond[21,138]) to the overlying zone where carbon is present in its oxidized form (as carbonates or as CO_2 vapor at pressures lower than 2 GPa[3]). At the redox front, reduced carbon present in the eclogite or peridotite oxidizes to form trace to minor amounts of carbonatitic melt (labeled (1)). The carbonatitic melt causes fluxed partial melting of eclogite and peridotite because the carbonated solidi of peridotite and eclogite are at much lower temperatures at the pressure range of the upwelled parcel of mantle (labeled (2)). The fluxed partial melting of eclogite and peridotite produces carbonated silicate melts in the upper mantle. The volatile-free solidi of peridotite[189] and eclogite,[191] the carbonated solidi of peridotite[33,142] and eclogite,[148] the graphite–diamond transition,[193] and the range of mantle potential temperatures in Earth's upper mantle (orange shaded area) from ridges[194] to plumes[195] are plotted for reference. The broken curves for carbonated solidi are only applicable for locally oxidized domains in the mantle where carbonates can exist at depths below the redox-melting front. (b) Geodynamic scenarios involving carbonated eclogite and peridotite in the source of ocean island volcanism in intraplate settings (modified from a previous study[152]). (i) This scenario is applicable for locally oxidized domains in the mantle where deeper carbonatites generated by very-low-degree partial melting of carbonated peridotite and/or eclogite rise upwards and cause fluxed partial melting of volatile-free eclogite at shallower depths in the upper mantle. The carbonated partial melt of eclogite, a basanite, or andesite (a lower eclogite to carbonate ratio produces basanite, while a higher ratio produces andesite) reacts with subsolidus volatile-free peridotite. The melt-rock reaction produces product melts that display similarity in composition with ocean island basalts. (ii) Reduced carbon-bearing eclogite upwells and produces carbonated silicate partial melts of eclogite after crossing the redox-melting front. The carbonated silicate partial melt of eclogite reacts with the surrounding subsolidus volatile-free peridotite, and a similar melt-rock reaction as proposed in (i) takes place.

A higher γ_{SiO2} and an increased degree of polymerization of the silicate domain of the melt network also imply that saturation of orthopyroxene is preferred over saturation of olivine in the residue of melt-rock reaction. Thus, the stability field of orthopyroxene is enhanced over that of olivine.[47,151–153] Also, cations such as Ca^{2+}, Mg^{2+}, Na^{+}, and K^{+} prefer to enter the melt structure to form carbonate complexes in the presence of CO_2. This enhances the calcium, magnesium, and alkali contents of the product melts.

6.8 Carbonate Melts under Mid-ocean Ridges

Along mid-ocean ridges, where divergent tectonic plates move apart from each other, the rising asthenospheric mantle decompresses and partially melts to create basaltic magma that buoyantly rises to Earth's surface and solidifies to form new oceanic crust. During decompression melting, volatile-free mantle lithologies melt at ~60 km, producing silicate melts.[156] Beneath ridges at shallow depths, carbonate-bearing silicate melts (basalts) are stable.[5,102,142] Melting in carbonated mantle lithologies may take place at depths of ~300 km[142] due to depression of the mantle solidus temperature.[5,29,157]

Figure 6.6 shows the stability fields of carbonates and carbonate and silicate melts in an adiabatically upwelling mantle and is compared with the fO_2 of the asthenospheric mantle. The fO_2 determined using reaction (6.10) suggests that carbonatitic melts are unstable at depths greater than 300 km due to the increased activity of Fe^{3+}-rich components in mantle minerals.[138,158,159] At the lower end of carbon concentrations in the mantle (30 ppm C for a MORB source mantle[54] and an initial $Fe^{3+}/\sum Fe$ content of 4%), graphite would transform to diamond at depths of ~160 km (Figure 6.2) and diamond would be the stable phase instead of carbonates or carbonatitic melts.[29,122,123] If the mantle is reduced below ~250 km depth and becomes metal saturated, then carbonatitic melts would be stable only to a depth of 150 km. During adiabatic upwelling of the upper mantle, reduction of Fe^{3+} locked in mantle silicates to Fe^{2+} results in concomitant oxidation of diamond or graphite and carbonatitic melting.[138] This redox process can be explained by reaction (6.12):

$$\underset{\text{graphite/diamond}}{C} + \underset{\text{silicate}}{2Fe_2O_3} = \underset{\text{carbonatitic melt}}{CO_2} + \underset{\text{silicate}}{4FeO} \tag{6.12}$$

As soon as 30 ppm of graphite in the upper mantle is oxidized, the $Fe^{3+}/\sum Fe$ content of the remaining mantle drops from 4% to 3%. After such redox melting, the fO_2 of the upper mantle increases again and finally reaches the level of mid-ocean ridge basalts at around FMQ (Figure 6.6).

6.9 Crustally Emplaced Carbonatites

Natural carbonate melts represent a tiny proportion of the magmatic rocks of Earth's crust, but nonetheless provide important insights into mantle-to-crust carbon transfer over geological time. Carbonate melts primarily refer to carbonatites, but could also include kimberlites and various other alkaline mafic–ultramafic silicate magma series (e.g.

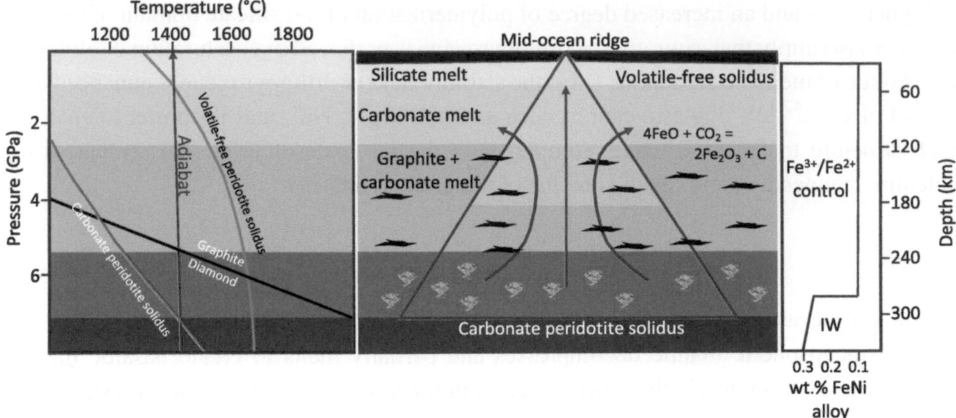

Figure 6.6 A simplified model for the extraction of carbonatitic/carbonated silicate melts from the mantle based on Stagno et al.[28] If carbonates are present in the adiabatically upwelling mantle beneath mid-ocean ridges, they will lower the melting temperature of the mantle rocks and melting will take place at ~300 km depth if the adiabat has a potential temperature of 1400°C. The left-hand panel shows that the solidus of peridotite and eclogite is depressed by several hundred degrees compared to the volatile-free solidus. The central panel shows the focusing of deeply derived carbonate melts (red triangle) formed by redox melting at about 1.5 GPa lower pressures than the carbonate peridotite solidus for oxidized mantle. The right-hand panel shows estimated $Fe^{3+}/\sum Fe$ in peridotite buffered at about 0.1 at depths below about 290 km, and buffered by FeNi alloy (≈IW) to higher values at greater depths.

lamprophyres). All of these magma types have indisputable mantle origins, as revealed by stable and radiogenic isotope signatures, as well as a common association with mantle-derived alkaline silicate magmas.[76]

Carbonatites (magmatic rocks with >50% carbonate minerals) are variably classified into a number of groups based on their chemistry, with most researchers agreeing on at least five groups; calcio-carbonatite, dolomite carbonatite, ferro-carbonatite, natro- (or alkali-) carbonatite, and rare earth carbonatites.[76] The Ca–Mg–Fe-rich carbonatites are primarily found as intrusive complexes that may to some extent represent crystal cumulates from carbonatite magma,[160] or as melt inclusions in igneous and mantle minerals.[161,162] Extrusive carbonatites are relatively rare in the rock record[163] and include the only known active carbonatite volcano, Oldoinyo Lengai (Tanzania), which erupts natrocarbonatite. Other occurrences of natrocarbonatite are found only very rarely as melt inclusions within igneous minerals from alkaline rocks[161,164] and kimberlites.[11] The lack of natrocarbonatite from the rock record is often ascribed to its extreme instability under atmospheric and most geological conditions.[165] Rare earth carbonatites (with >1 wt.% REE) also tend to be enriched in other incompatible trace elements (e.g. U, Th, Nb, Ta) and are regarded to be products of fractional crystallization of parental carbonatite magma.[166] This carbonatite class is of significant economic importance as a source of the critical metals needed for many technological applications that support modern societies.[167]

Despite consensus regarding their mantle origins, there are differing opinions on carbonatite evolution. Some researchers assert that carbonatites are primary magmas derived directly by low-degree melting of CO_2-rich mantle sources,[168] whereas experimental work by Watkinson and Wyllie[169] demonstrated that carbonate liquids could be formed as residua of fractionation from carbonated peralkaline silicate magmas; this latter origin would be consistent with the classification as "carbothermal residua" of Mitchell.[166] Supporting a primary mantle origin are studies of mantle xenoliths of distinctive petrological and geochemical character formed via metasomatism by carbonatitic melts.[8,76,170] Carbonatitic melt inclusions within diamond represent further unequivocal evidence of a mantle origin.[76,171]

A third origin proposed for carbonatite genesis is via liquid immiscibility from evolving CO_2-bearing alkaline silicate melts at crustal pressures.[76] Support for this model comes not only from experimental evidence and the common field association between carbonatite and alkaline silicate magmatic rocks, but crucially from a growing number of textural, petrological, and geochemical studies of carbonatites.[164,172–174] Given the compelling evidence for their primary mantle and immiscibility origins, it is likely that carbonatite magmas may be formed in a range of geological environments, from shallow crustal down to lower-mantle conditions.

Carbonatites and other mantle-derived CO_2-rich magmas, such as kimberlites, are found across all continents, although over half are in Africa.[175] They are most commonly found within Precambrian cratons despite the fact that most are of Phanerozoic age.[176,177] These settings would be consistent with formation via low-degree (and hence relatively low-temperature) melting of fertile mantle sources, such as thick mantle lithosphere beneath stable cratons. There tends to be an increasing frequency of their formation toward younger ages, which is not easily explained as an artefact of preservation,[177] but rather may reflect more favorable conditions for the production of mantle-derived alkaline magmatism in the modern Earth. This would be consistent with more effective recycling of crustal material into the mantle via modern-style subduction processes, allowing the production of the re-fertilized mantle domains that are requisite sources of these magmas.

There are diverse views on the tectonic settings in which carbonatites and other CO_2-bearing magmas form and are emplaced into the crust. Carbonatite magmatism is a distinctive feature of the recent igneous activity of the East African Rift, demonstrating a link to intracontinental rifting, and potentially mantle plumes.[178] Many older carbonatite complexes have also been linked to continental rifting, mantle plumes, or large igneous provinces.[179–181] However, Woolley and Bailey[177] argue that ephemeral mantle controls, such as mantle plumes, do not account for the episodic and repeated nature of carbonatitic (and kimberlitic) magmatism within restricted areas, sometimes over billions of years. Instead, these authors argue for reactivation of lithospheric-scale structures or lineaments due to far-field plate reorganization (e.g. due to rift initiation or continental assembly[182]), which allows for low-degree melting of the fertile cratonic mantle lithosphere to produce carbonatites or kimberlites, as well as efficient tapping of the melts to the surface.

Carbonatites in oceanic settings have only been described from the hot spot volcanos of Cape Verde and the Canary Islands.[183,184] In these settings, the degree of mantle melting is expected to be too high to produce carbonatites directly; rather, the carbonatites are interpreted to be products of fractionation and unmixing from alkali-rich primitive basanite melts.[185] In this case, the rarity of oceanic carbonatites may reflect a lack of preservation and exposure rather than unsuitable petrogenesis conditions.

6.10 Concluding Remarks

The existence of carbonate-rich melts in Earth is likely limited in P–T–fO_2 space to the crust, to the uppermost part of the upper mantle including oceanic and continental lithosphere, to the mantle wedge above subducting slabs, to locally strongly metasomatized regions of the deep cratonic lithospheric mantle and asthenosphere, and to local regions of the deep upper mantle, mantle transition zone, and uppermost lower mantle associated with deep subduction of the carbonate-bearing oceanic lithosphere. Much of the deep Earth is too reduced for carbonate or carbonate-rich melts to be stable, where carbon will exist in reduced forms, such as diamond, methane, and other alkanes or volatile organic molecules, as well as Fe-carbides.

Despite this, carbonate melts are of tremendous importance as agents of mass transport and metasomatic enrichment in Earth's interior. They are also the primary source of almost all commercially viable REE resources, and they may contain other critical metals in extractable quantities as well. Although crustally emplaced carbonatites are small in volume, they are likely to be much more abundant as broadly alkali-rich calcio-dolomitic to dolomitic melts in the upper mantle and mantle transition zone, where they contribute significantly to the carbon fluxes to the surface.

6.11 Limits to Knowledge and Unknowns

Key areas relating to carbonate or CO_2-bearing melts in Earth in which current knowledge is limited include the solubility of CO_2 in hydrous fluids and partial melts derived from subducting oceanic crust, particularly in the sub-arc environment. These melts may play a critically important role in metasomatizing the mantle wedge and flux melting, ultimately leading to arc volcanism and continental crust formation. They are agents by which subducted carbon is recycled on relatively short timescales of tens of Ma from the subducted slab in the sub-arc environment (blueschist or eclogite facies conditions) back to the atmosphere and hydrosphere. We cannot quantify carbon fluxes in this relatively shallow part of Earth's deep carbon cycle unless we understand the magnitude of volatile fluxes from slab to arc volcanism. This will require high-pressure experimental studies as well as field studies of high-grade, carbonated metamorphic rocks in orogenic zones.

This in turn bears on estimates of carbon fluxes via subduction into the deeper mantle, and ultimately to the mantle transition zone or lower mantle. These estimates currently vary from 0 to 15 Mt/year.[56] It appears likely that some carbon does escape relatively shallow recycling during subduction to sub-arc conditions based on studies of sublithospheric diamonds and their inclusions. If so, it may then be transported as part of the oceanic lithosphere to the deeper mantle. However, it remains unclear how the fO_2 of the oceanic crustal components being subducted as part of the lithospheric package varies along the prograde path. This will be a critical control on the ultimate fate of carbon in the oceanic crust. For example, as has been shown recently by Kiseeva et al.[12] and Thomson et al.,[13] carbonate in the mafic crust is unlikely to be subducted beyond the mantle transition zone, as it will melt to form carbonate liquids, which will segregate from their source and undergo redox freezing on contact with reduced peridotite wall-rock.[138] However, if carbonate is reduced to diamond on the prograde path, melting will not occur, and in those cases where slabs enter the lower mantle, carbon will likely also be transported beyond the mantle transition zone.

Finally, we currently lack understanding of aspects of the return cycle of carbon from the deep upper mantle, mantle transition zone, or lower mantle to the surface. In particular, some sublithospheric diamonds originate from these very deep parts of the upper and lower mantle based on their mineral inclusions. However, it is unclear exactly how and where the kimberlites formed that host these diamonds and transported them to the surface. Did they form at deep upper-mantle, transition-zone, or lower-mantle pressures, or were the sublithospheric diamonds first transported from great depths during mantle convection and then entrained in kimberlites formed at depths just below the cratonic lithospheric mantle? Answering these questions will require careful high-pressure experimental studies in order to deepen understanding of the role of C–O–H volatiles in causing melting at depths in and around the transition zone.

Acknowledgments

We gratefully thank the editorial team for this book, Beth Orcutt, Isabelle Daniel, and Raj Dasgupta, for their encouragement and immense patience in the development of this chapter and for their expert editorial handling. We also thank Raj Dasgupta and an anonymous reviewer for two constructive formal reviews.

Questions for the Classroom

1 Describe the different forms in which carbon is believed to exist in Earth's mantle. In which parts of the mantle are these different carbon-bearing species located and what is the evidence for their presence?

2 Explain how pressure, temperature, and fO_2 in Earth's mantle influence the species present in different parts of the mantle (lower mantle, mantle transition zone, upper mantle) and different tectonic settings.

3 What are some of the reasons that carbonate melts inferred to exist in parts of the subcontinental lithosphere are considered to be highly effective metasomatic agents?

4 Carbonate is a widespread component of oceanic crust hydrothermally altered near mid-ocean ridges. When transported by plate tectonics to subduction zones, this carbonate may be recycled back into the mantle. Describe some of the processes that may influence the fate and redistribution of this subducted carbon in the fore-arc, sub-arc, and deeper environments.

5 How does fO_2 vary within Earth's mantle? What are the major controls on this variation and why is it important in understanding Earth's deep carbon cycle?
6 Sublithospheric diamonds are believed to have formed at depths greater than the lithosphere–asthenosphere boundary (i.e. they did not form within the cratonic mantle lithosphere like the majority of kimberlite-borne diamonds). From where are they derived and how are they likely to have formed? What is the evidence for this?
7 The only currently active carbonatite volcano on Earth is Oldoinyo Lengai in Tanzania. How are the carbonatites erupted by this volcano different from all other known crustally emplaced carbonatites? Do some literature research to understand some of the hypotheses for the formation of Oldoinyo Lengai's lavas and the possible reasons that they are so anomalous relative to other older carbonatites.
8 What are the likely reasons for the absence of crustally emplaced carbonatite melts in subduction zones?
9 Describe ways in which CO_2-rich silicate melts can form in Earth's upper mantle. In what sorts of tectonic settings are these melts likely to erupt?
10 Draw a cross-section of Earth showing the main tectonic settings (e.g. mid-ocean ridge, subduction zone, intraplate volcanic setting, continental and cratonic setting, etc.) and illustrate the locations of major carbon reservoirs and the nature of the carbon-bearing species that may be present in each. Discuss processes by which carbon may move between these reservoirs.

References

1. Dawson, J. B. Sodium carbonate lavas from Oldoinyo Lengai, Tanganyika. *Nature* **195**, 1075 (1962).
2. Wyllie, P. J., Baker, M. B. & White, B. S. Experimental boundaries for the origin and evolution of carbonatites. *Lithos* **26**, 3–19 (1990).
3. Wallace, M. E. & Green, D. H. An experimental-determination of primary carbonatite magma composition. *Nature* **335**, 343–346 (1988).
4. Yaxley, G. & Brey, G. Phase relations of carbonate-bearing eclogite assemblages from 2.5 to 5.5 GPa: implications for petrogenesis of carbonatites. *Contrib Mineral Petrol* **146**, 606–619 (2004).
5. Dasgupta, R. & Hirschmann, M. M. Melting in the Earth's deep upper mantle caused by carbon dioxide. *Nature* **440**, 659–662 (2006).
6. Dasgupta, R. & Hirschmann, M. M. The deep carbon cycle and melting in Earth's interior. *Earth Planet Sci Lett* **298**, 1–13 (2010).
7. Yaxley, G., Crawford, A. & Green, D. Evidence for carbonatite metasomatism in spinel peridotite xenoliths from western Victoria, Australia. *Earth Planet Sci Lett* **107**, 305–317 (1991).
8. Yaxley, G., Green, D. & Kamenetsky, V. Carbonatite metasomatism in the southeastern Australian lithosphere. *J Petrol* **39**, 1917–1930 (1998).
9. Rudnick, R. L., Mcdonough, W. F. & Chappell, B. W. Carbonatite metasomatism in the northern Tanzanian mantle – petrographic and geochemical characteristics. *Earth Planet Sci Lett* **114**, 463–475 (1993).

10. Weiss, Y., McNeill, J., Pearson, D. G., Nowell, G. M. & Ottley, C. J. Highly saline fluids from a subducting slab as the source for fluid-rich diamonds. *Nature* **524**, 339–342 (2015).
11. Golovin, A. V., Sharygin, I. S., Kamenetsky, V. S., Korsakov, A. V. & Yaxley, G. M. Alkali-carbonate melts from the base of cratonic lithospheric mantle: links to kimberlites. *Chem Geol* **483**, 261–274 (2018).
12. Kiseeva, E. S., Litasov, K. D., Yaxley, G. M., Ohtani, E. & Kamenetsky, V. S. Melting and phase relations of carbonated eclogite at 9-21 GPa and the petrogenesis of alkali-rich melts in the deep mantle. *J Petrol* **54**, 1555–1583 (2013).
13. Thomson, A. R., Walter, M. J., Kohn, S. C. & Brooker, R. A. Slab melting as a barrier to deep carbon subduction. *Nature* **529**, 76–79 (2016).
14. Litasov, K. & Ohtani, E. Solidus of carbonate eclogite in the system $CaO–Al_2O_3–MgO–SiO_2–Na_2O–CO_2$ to 32 GPa and carbonatite liquid in the deep mantle. *Earth Planet Sci Lett* **295,** 115–126 (2010).
15. Kaminsky, F. Mineralogy of the lower mantle: a review of "super-deep" mineral inclusions in diamond. *Earth Sci Rev* **110**, 127–147 (2012).
16. Hunter, R. H. & Mckenzie, D. The equilibrium geometry of carbonate melts in rocks of mantle composition. *Earth Planet Sci Lett* **92**, 347–356 (1989).
17. Hammouda, T. & Laporte, D. Ultrafast mantle impregnation by carbonatite melts. *Geology* **28**, 283–285 (2000).
18. Shatskiy, A. et al. Silicate diffusion in alkali-carbonatite and hydrous melts at 16.5 and 24 GPa: implication for the melt transport by dissolution–precipitation in the transition zone and uppermost lower mantle. *Phys Earth Planet Int* **225**, 1–11 (2013).
19. Taylor, W. R. & Green, D. H. The petrogenic role of methane: effect on liquidus phase relations and the solubility mechanism of reduced C–H volatiles. *The Geochemical Society* **Special Publication No. 1**, 121–137 (1987).
20. Luth, R. W. Diamonds, eclogites, and the oxidation-state of the Earth's mantle. *Science* **261**, 66–68 (1993).
21. Stagno, V., Ojwang, D. O., McCammon, C. A. & Frost, D. J. The oxidation state of the mantle and the extraction of carbon from Earth's interior. *Nature* **493**, 84–88 (2013).
22. Kennedy, C. S. & Kennedy, G. C. The equilibrium boundary between graphite and diamond. *J Geophys Res* **81**, 2467–2470 (1976).
23. Yaxley, G. M., Berry, A. J., Rosenthal, A., Woodland, A. B. & Paterson, D. Redox preconditioning deep cratonic lithopshere for kimberlite genesis – evidence from the central Slave Craton. *Nat Sci Rep* **7,** 30 (2017).
24. Kiseeva, E. S. et al. Oxidized iron in garnets from the mantle transition zone. *Nat Geosci* **11,** 144–147 (2018).
25. Cottrell, E. & Kelley, K. A. The oxidation state of Fe in MORB glasses and the oxygen fugacity of the upper mantle. *Earth Planet Sci Lett* **305**, 270–282, (2011).
26. Berry, A. J., Stewart, G. A., O'Neill, H. S. C., Mallmann, G. & Mosselmans, J. F. W. A re-assessment of the oxidation state of iron in MORB glasses. *Earth Planet Sci Lett* **483**, 114–123 (2018).
27. Gudmundsson, G. & Wood, B. J. Experimental tests of garnet peridotite oxygen barometry. *Contrib Mineral Petrol* **119**, 56–67 (1995).
28. Stagno, V., Frost, D. J., McCammon, C. A., Mohseni, H. & Fei, Y. The oxygen fugacity at which graphite or diamond forms from carbonate-bearing melts in eclogitic rocks. *Contrib Mineral Petrol* **169**, 16 (2015).
29. Stagno, V. & Frost, D. J. Carbon speciation in the asthenosphere: experimental measurements of the redox conditions at which carbonate-bearing melts coexist with

graphite or diamond in peridotite assemblages. *Earth Planet Sci Lett* **300**, 72–84 (2010).

30. Falloon, T. J. & Green, D. H. The solidus of carbonated, fertile peridotite. *Earth Planet Sci Lett* **94**, 364–370 (1989).
31. Falloon, T. J. & Green, D. H. Solidus of carbonated fertile peridotite under fluid-saturated conditions. *Geology* **18**, 195–199 (1990).
32. Dasgupta, R. & Hirschmann, M. M. Effect of variable carbonate concentration on the solidus of mantle peridotite. *Am Mineral* **92**, 370–379 (2007).
33. Dasgupta, R., Hirschmann, M. M. & Smith, N. D. Partial melting experiments of peridotite + CO_2 at 3 GPa and genesis of alkalic ocean island basalts. *J. Petrol* **48**, 2093–2124 (2007).
34. Dasgupta, R., Hirschmann, M. M. & Smith, N. D. Water follows carbon: CO_2 incites deep silicate melting and dehydration beneath mid-ocean ridges. *Geology* **35**, 135–138 (2007).
35. Dasgupta, R., Stalker, K., Withers, A. C. & Hirschmann, M. M. The transition from carbonate-rich to silicate-rich melts in eclogite: partial melting experiments of carbonated eclogite at 3 GPa. *Lithos* **73**, S23 (2004).
36. Lee, W. J. & Wyllie, P. J. Petrogenesis of carbonatite magmas from mantle to crust, constrained by the system CaO–(MgO+FeO*)(Na_2O+K_2O)–(SiO_2+Al_2O_3+TiO_2)–CO_2. *J Petrol* **39**, 495–517 (1998).
37. Ryabchikov, I. D., Edgar, A. D. & Wyllie, P. J. Partial melting in the systen carbonate phosphate peridotite at 30 kbar. *Geokhimiya* **2**, 163–168 (1991).
38. White, B. S. & Wyllie, P. J. Solidus reactions in synthetic lherzolite–H_2O–CO_2 from 20–30kbar, with applications to melting and metasomatism. *J Volcanol Geoth Res* **50**, 117–130 (1992).
39. Wyllie, P. J. Peridotite–CO_2–H_2O and carbonatitic liquids in upper asthenosphere. *Nature* **266**, 45–47 (1977).
40. Wyllie, P. J. Mantle fluid compositions buffered in peridotite–CO_2–H_2O by carbonates, amphibole, and phlogopite. *J Geol* **86**, 687–713 (1978).
41. Wyllie, P. J. Fusion of peridotite–H_2O–CO_2 in system peridotite–H–C–O at upper mantle pressures. *Eos Trans Am Geophysl Union* **59**, 398–398 (1978).
42. Newton, R. C. & Sharp, W. E. Stability of forsterite + CO_2 and its bearing on the role of CO_2 in the mantle. *Earth Planet Sci Lett* **26**, 239–244 (1975).
43. Haselton, H. T., Sharp, W. E. & Newton, R. C. CO_2 fugacity at high temeratures and pressures from experimental decarbonation reactions. *Geophys Res Lett* **5**, 753 (1978).
44. Eggler, D. H., Kushiro, J. & Holloway, J. R. Free energies of decarbonation reactions at mantle pressures, I. Stability of the assemblage forsterite-enstatite-magnesite in the system MgO–SiO_2–CO_2–H_2O to 60 kbar. *Am Mineral* **64**, 288 (1979).
45. Johannes, W. An experimental investigation of the system MgO–SiO_2–H_2O–CO_2. *Am J Sci* **267**, 1083 (1969).
46. Wyllie, P. J. Magmass and volatile components. *Am Mineral* **64**, 469–500 (1979).
47. Eggler, D. H. The effect of CO_2 on partial melting of peridotite in the system Na_2O–CaO–Al_2O_3–MgO–SiO_2–CO_2 to 35 kbar, with an analysis of melting in a peridotite–H_2O–CO_2 system. *Am J Sci* **278**, 305 (1978).
48. Kushiro, I. Carbonate–silicate reactions at high presures and possible presence of dolomite and magnesite in the upper mantle. *Earth Planet Sci Lett* **28**, 116–120(1975).
49. Brey, G. et al. Pyroxene–carbonate reactions in the upper mantle. *Earth Planet Sci Lett* **62**, 63–74 (1983).

50. Yaxley, G. M., Crawford, A. J. & Green, D. H. Evidence for carbonatite metasomatism in spinel peridotite xenoliths from Western Victoria, Australia. *Earth Planet Sci Lett* **107**, 305–317 (1991).
51. Rudnick, R. L., McDonough, W. F. & Chappell, B. W. Carbonatite metasomatism in the northern Tanzanian mantle: petrographic and geochemical characteristics. *Earth Planet Sci Lett* **114**, 463–475 (1993).
52. Dalton, J. A. & Wood, B. J. The compositions of primary carbonate melts and their evolution through wallrock reaction in the mantle. *Earth Planet Sci Lett* **119**, 511–525 (1993).
53. Foley, S. F. et al. The composition of near-solidus melts of peridotite in the presence of CO_2 and H_2O between 40 and 60 kbar. *Lithos* **112**, 274–283 (2009).
54. Marty, B. The origins and concentrations of water, carbon, nitrogen and noble gases on Earth. *Earth Planet Sci Lett* **313–314**, 56–66 (2012).
55. Sleep, N. H. & Zahnle, K. Carbon dioxide cycling and implications for climate on ancient Earth. *J Geophys Res Planets* **106**, 1373–1399 (2001).
56. Kelemen, P. B. & Manning, C. E. Reevaluating carbon fluxes in subduction zones, what goes down, mostly comes up. *Proc Natl Acad Sci* **112**, E3997–E4006 (2015).
57. Plank, T. & Langmuir, C. H. The chemical composition of subducting sediment and its consequences for the crust and mantle. *Chem Geol* **145**, 325–394 (1998).
58. Plank, T. The chemical composition of subducting sediments. *Treatise Geochem* **4**, 607–629 (2014).
59. Alt, J. C. & Teagle, D. A. H. The uptake of carbon during alteration of ocean crust. *Geochim Cosmochim Acta* **63**, 1527–1535 (1999).
60. Connolly, J. A. D. Computation of phase equilibria by linear programming; a tool for geodynamic modelling and its application to subduction zone decarbonation. *Earth Planet Sci Lett* **236**, 524–541 (2005).
61. Manning, C. E., Shock, E. L. & Sverjensky, D. A. The chemistry of carbon in aqueous fluids at crustal and upper-mantle conditions: experimental and theoretical constraints. *Rev Mineral Geochem* **75**, 109–148 (2013).
62. Ague, J. J. & Nicolescu, S. Carbon dioxide released from subduction zones by fluid-mediated reactions. *Nat Geosci* **7**, 355 (2014).
63. Piccoli, F. et al. Carbonation by fluid–rock interactions at high-pressure conditions: implications for carbon cycling in subduction zones. *Earth Planet Sci Lett* **445**, 146–159 (2016).
64. Spandler, C., Hermann, J., Faure, K., Mavrogenes, J. A. & Arculus, R. J. The importance of talc and chlorite "hybrid" rocks for volatile recycling through subduction zones; evidence from the high-pressure subduction mélange of New Caledonia. *Contrib Mineral Petrol* **155**, 181–198 (2008).
65. Scambelluri, M. et al. Carbonation of subduction-zone serpentinite (high-pressure ophicarbonate; Ligurian Western Alps) and implications for the deep carbon cycling. *Earth Planet Sci Lett* **441**, 155–166 (2016).
66. Sieber, M. J., Hermann, J. & Yaxley, G. M. An experimental investigation of C–O–H fluid-driven carbonation of serpentinites under forearc conditions. *Earth Planet Sci Lett* **496**, 178–188 (2018).
67. Syracuse, E. M., van Keken, P. E. & Abers, G. A. The global range of subduction zone thermal models. *Phys Earth Planet Interiors* **183**, 73–90 (2010).
68. Frezzotti, M. L., Selverstone, J., Sharp, Z. D. & Compagnoni, R. Carbonate dissolution during subduction revealed by diamond-bearing rocks from the Alps. *Nat Geosci* **4**, 703 (2011).

69. Schmidt, M. W. Melting of pelitic sediments at subarc depths: 2. Melt chemistry, viscosities and a parameterization of melt composition. *Chem Geol* **404**, 168–182 (2015).
70. Korsakov, A. V. & Hermann, J. Silicate and carbonate melt inclusions associated with diamonds in deeply subducted carbonate rocks. *Earth Planet Sci Lett* **241**, 104–118 (2006).
71. Poli, S. Carbon mobilized at shallow depths in subduction zones by carbonatitic liquids. *Nat Geosci* **8**, 633 (2015).
72. Marschall, H. R. & Schumacher, J. C. Arc magmas sourced from melange diapirs in subduction zones. *Nat Geosci* **5**, 862–867 (2012).
73. Tumiati, S., Fumagalli, P., Tiraboschi, C. & Poli, S. An experimental study on COH-bearing peridotite up to 3.2 GPa and implications for crust–mantle recycling. *J Petrol* **54**, 453–479 (2013).
74. Smith, C. B. et al. Diamonds from Dachine, French Guiana: a unique record of early Proterozoic subduction. *Lithos* **265**, 82–95 (2016).
75. Green, D. H. Experimental petrology of peridotites, including effects of water and carbon on melting in the Earth's upper mantle. *Phys Chem Mineral* **42**, 95–122 (2015).
76. Jones, A. P., Genge, M. & Carmody, L. Carbonate melts and carbonatites. *Rev Mineral Geochem* **75**, 289–322 (2013).
77. Saha, S., Dasgupta, R. & Tsuno, K. High pressure phase relations of a depleted peridotite fluxed by CO_2–H_2O-bearing siliceous melts and the origin of Mid-Lithospheric Discontinuity. *Geochem Geophys Geosyst* **19**, 595–620 (2018).
78. Gorman, P. J., Kerrick, D. M. & Connolly, J. A. D. Modeling open system metamorphic decarbonation of subducting slabs. *Geochem Geophys Geosyst* **7**, 1–21 (2006).
79. Molina, J. F. & Poli, S. Carbonate stability and fluid composition in subducted oceanic crust: an experimental study on H_2O–CO_2-bearing basalts. *Earth Planet Sci Lett* **176**, 295–310 (2000).
80. Kelley, K. A. & Cottrell, E. The influence of magmatic differentiation on the oxidation state of Fe in a basaltic arc magma. *Earth Planet Sci Lett* **329–330**, 109–121 (2012).
81. Brenker, F. E. et al. Carbonates from the lower part of transition zone or even the lower mantle. *Earth Planet Sci Lett* **260**, 1–9 (2007).
82. Bulanova, G. P. et al. Mineral inclusions in sublithospheric diamonds from Collier 4 kimberlite pipe, Juina, Brazil: subducted protoliths, carbonated melts and primary kimberlite magmatism. *Contrib Mineral Petrol* **160**, 489–510 (2010).
83. Thomson, A. R. et al. Origin of sub-lithospheric diamonds from the Juina-5 kimberlite (Brazil): constraints from carbon isotopes and inclusion compositions. *Contrib Mineral Petrol* **168**, 1081 (2014).
84. Zedgenizov, D. A., Kagi, H., Shatsky, V. S. & Ragozin, A. L. Local variations of carbon isotope composition in diamonds from Sao-Luis (Brazil): evidence for heterogenous carbon reservoir in sublithospheric mantle. *Chem Geol* **363**, 114–124 (2010).
85. Thomson, A. R. et al. Trace element composition of silicate inclusions in sub-lithospheric diamonds from the Juina-5 kimberlite: evidence for diamond growth from slab melts. *Lithos* **265**, 108–124 (2016).
86. Walter, M. J. et al. Primary carbonatite melt from deeply subducted oceanic crust. *Nature* **454**, 622 (2008).
87. Burnham, A. D. et al. Stable isotope evidence for crustal recycling as recorded by superdeep diamonds. *Earth Planet Sci Lett* **432**, 374–380 (2015).

88. Ickert, R. B., Stachel, T., Stern, R. A. & Harris, J. W. Diamond from recycled crustal carbon documented by coupled $\delta^{18}O$–$\delta^{13}C$ measurements of diamonds and theirinclusions. *Earth Planet Sci Lett* **364**, 85–97 (2013).
89. Tsuno, K. & Dasgupta, R. Melting phase relation of nominally anhydrous, carbonated pelitic-eclogite at 2.5–3.0 GPa and deep cycling of sedimentary carbon. *Contrib Mineral Petrol* **161**, 743–763 (2011).
90. Grassi, D. & Schmidt, M. W. Melting of carbonated pelites at 8-13 GPa: generating K-rich carbonatites for mantle metasomatism. *Contrib Mineral Petrol* **162**, 169–191 (2011).
91. Mann, U. & Schmidt, M. W. Melting of pelitic sediments at subarc depths: 1. Flux vs. fluid-absent melting and a parameterization of melt productivity. *Chem Geol* **404**, 150–167 (2015).
92. Keshav, S. & Gudfinnsson Gudmundur, H. Experimentally dictated stability of carbonated oceanic crust to moderately great depths in the Earth: results from the solidus determination in the system CaO–MgO–Al_2O_3–SiO_2–CO_2. *J Geophys Res Solid Earth* **115**, B05205 (2010).
93. Litasov, K. & Ohtani, E. The solidus of carbonated eclogite in the system CaO–Al_2O_3–MgO–SiO_2–Na_2O–CO_2 to 32GPa and carbonatite liquid in the deep mantle. *Earth Planet Sci Lett* **295**, 115–126 (2010).
94. Dasgupta, R., Hirschmann, M. M. & Withers, A. C. Deep global cycling of carbon constrained by the solidus of anhydrous, carbonated eclogite under upper mantle conditions. *Earth Planet Sci Lett* **227**, 73–85 (2004).
95. Hammouda, T. High-pressure melting of carbonated eclogite and experimental constraints on carbon recycling and storage in the mantle. *Earth Planet Sci Lett* **214**, 357–368 (2003).
96. Kiseeva, E. S. et al. An experimental study of carbonated eclogite at 3.5–5.5 GPa – implications for silicate and carbonate metasomatism in the cratonic mantle. *J Petrol* **53**, 727–759 (2012).
97. Yaxley, G. M. & Green, D. H. Experimental demonstration of refractory carbonate-bearing eclogite and siliceous melt in the subduction regime. *Earth Planet Sci Lett* **128**, 313–325 (1994).
98. Dasgupta, R., Hirschmann, M. M. & Dellas, N. The effect of bulk composition on the solidus of carbonated eclogite from partial melting experiments at 3 GPa. *Contrib Mineral Petrol* **149**, 288–305 (2005).
99. Gerbode, C. & Dasgupta, R. Carbonate-fluxed melting of MORB-like pyroxenite at 2.9 GPa and genesis of HIMU ocean island basalts. *J Petrol* **51**, 2067–2088 (2010).
100. Dasgupta, R., Hirschmann, M. M. & Stalker, K. Immiscible transition from carbonate-rich to silicate-rich melts in the 3 GPa melting interval of eclogite + CO_2 and genesis of silica-undersaturated ocean island lavas. *J Petrol* **47**, 647–671 (2006).
101. Litasov, K. D., Shatskiy, A., Ohtani, E. & Yaxley, G. M. Solidus of alkaline carbonatite in the deep mantle. *Geology* **41**, 79–82 (2013).
102. Gudfinnsson, G. H. & Presnall, D. C. Continuous gradations among primary carbonatitic, kimberlitic, melilititic, basaltic, picritic, and komatiitic melts in equilibrium with garnet lherzolite at 3–8GPa. *J Petrol* **46**, 1645–1659 (2005).
103. Tappe, S., Foley, S. F., Jenner, G. A. & Kjarsgaard, B. A. Integrating ultramafic lamprophyres into the IUGS classification of igneous rocks: rationale and implications. *J Petrol* **46**, 1893–1900 (2005).

104. Tappe, S. et al. Between carbonatite and lamproite – diamondiferous Torngat ultramafic lamprophyres formed by carbonate-fluxed melting of cratonic MARID-type metasomes. *Geochim Cosmochim Acta* **72**, 3258–3286 (2008).
105. Sparks, R. S. J. Kimberlite volcanism. *Ann Rev Earth Planet Sci* **41**, 497–528 (2013).
106. Mitchell, R. H. *Kimberlites, Orangeites, and Related Pocks* (Plenum Press, 1995).
107. Becker, M. & Le Roex, A. P. Geochemistry of South African on- and off-craton, Group I and Group II kimberlites: petrogenesis and source region evolution. *J Petrol* **47**, 673–703 (2006).
108. Mitchell, R. H. Kimberlites and lamproites – primary sources of diamond. *Geosci Canada* **18**, 1–16 (1991).
109. Nowell, G. M. et al. Hf isotope systematics of kimberlites and their megacrysts: new constraints on their source regions. *J Petrol* **45**, 1583–1612 (2004).
110. Yaxley, G. M. et al. The discovery of kimberlites in Antarctica extends the vast Gondwanan Cretaceous province. *Nat Commun* **4**, 2921 (2013).
111. Novella, D. & Frost, D. J. The composition of hydrous partial melts of garnet peridotite at 6 GPa: implications for the origin of Group II kimberlites. *J Petrol* **55**, 2097–2124 (2014).
112. Foley, S. F. & Fischer, T. P. An essential role for continental rifts and lithosphere in the deep carbon cycle. *Nat Geosci* **10**, 897–902 (2017).
113. Patterson, M., Francis, D. & McCandless, T. Kimberlites: magmas or mixtures? *Lithos* **112**, 191–200 (2009).
114. Kamenetsky, M. B. et al. Kimberlite melts rich in alkali chlorides and carbonates: a potent metasomatic agent in the mantle. *Geology* **32**, 845–848 (2004).
115. Kamenetsky, V. S. et al. Olivine in the Udachnaya-East kimberlite (Yakutia, Russia): types, compositions and origins. *J Petrol* **49**, 823–839 (2008).
116. Russell, J. K., Porritt, L. A., Lavallee, Y. & Dingwell, D. B. Kimberlite ascent by assimilation-fuelled buoyancy. *Nature* **481**, 352–356 (2012).
117. Girnis, A. V., Brey, G. P. & Ryabchikov, I. D. Origin of Group 1A kimberlites: fluid-saturated melting experiments at 45–55 kbar. *Earth Planet Sci Lett* **134**, 283–296 (1995).
118. Girnis, A. V., Bulatov, V. K. & Brey, G. P. Formation of primary kimberlite melts – constraints from experiments at 6–12GPa and variable CO_2/H_2O. *Lithos* **127**, 401–413 (2011).
119. Nell, J. & Wood, B. J. High-temperature electrical measurements and thermodynamic properties of Fe_3O_4–$FeCr_2O_4$–$MgCr_2O_4$–$FeAl_2O_4$ spinels. *Am Mineral* **76**, 405–426 (1991).
120. Ballhaus, C., Berry, R. F. & Green, D. H. High-pressure experimental calibration of the olivine–ortho-pyroxene–spinel oxygen geobarometer – implications for the oxidation-state of the upper mantle. *Contrib Mineral Petrol* **107**, 27–40 (1991).
121. O'Neill, H. S. C. & Wall, V. J. The olivine–orthopyroxene–spinel oxygen geobarometer, the nickel precipitation curve, and the oxygen fugacity of the Earth's upper mantle. *J Petrol* **28**, 1169–1191 (1987).
122. Woodland, A. B. & Koch, M. Variation in oxygen fugacity with depth in the upper mantle beneath the Kaapvaal craton, Southern Africa. *Earth Planet Sci Lett* **214**, 295–310 (2003).
123. Yaxley, G. M., Berry, A. J., Kamenetsky, V. S., Woodland, A. B. & Golovin, A. V. An oxygen fugacity profile through the Siberian Craton–Fe K-edge XANES determinations of $Fe^{3+}/\sum Fe$ in garnets in peridotite xenoliths from the Udachnaya East kimberlite. *Lithos* **140**, 142–151 (2012).

124. Frost, D. J. & McCammon, C. A. The redox state of Earth's mantle. *Ann Rev Earth Planet Sci* **36**, 389–420 (2008).
125. Rohrbach, A., Ghosh, S., Schmidt, M. W., Wijbrans, C. H. & Klemme, S. The stability of Fe–Ni carbides in the Earth's mantle: evidence for a low Fe–Ni–C melt fraction in the deep mantle. *Earth Planet Sci Lett* **388**, 211–221 (2014).
126. Tsuno, K. & Dasgupta, R. Fe–Ni–Cu–C–S phase relations at high pressures and temperatures – the role of sulfur in carbon storage and diamond stability at mid- to deep-upper mantle. *Earth Planet Sci Lett* **412**, 132–142 (2015).
127. Yaxley, G. M., Berry, A. J., Kamenetsky, V. S., Woodland, A. B. & Golovin, A. V. An oxygen fugacity profile through the Siberian Craton; Fe K-edge XANES determinations of $Fe^{3+}/\sum Fe$ in garnets in peridotite xenoliths from the Udachnaya East kimberlite. *Lithos* **140–141**, 142–151 (2012).
128. Foley, S. F. Rejuvenation and erosion of the cratonic lithosphere. *Nat Geosci* **1**, 503–510 (2008).
129. Coltorti, M., Beccaluva, L., Bonadiman, C., Salvini, L. & Siena, F. Glasses in mantle xenoliths as geochemical indicators of metasomatic agents. *Earth Planet Sci Lett* **183**, 303–320 (2000).
130. Delpech, G. et al. Feldspar from carbonate-rich silicate metasomatism in the shallow oceanic mantle under Kerguelen Islands (South Indian Ocean). *Lithos* **75**, 209–237 (2004).
131. Misra, K. C., Anand, M., Taylor, L. A. & Sobolev, N. V. Multi-stage metasomatism of diamondiferous eclogite xenoliths from the Udachnaya kimberlite pipe, Yakutia, Siberia. *Contrib Mineral Petrol* **146**, 696–714 (2004).
132. Taylor, W. R. & Green, D. H. Measurement of reduced peridotite–C–O–H solidus and implications for redox melting of the mantle. *Nature* **332**, 349–352 (1988).
133. Litasov, K. D., Shatskiy, A. & Ohtani, E. Melting and subsolidus phase relations in peridotite and eclogite systems with reduced COH fluid at 3–16 GPa. *Earth Planet Sci Lett* **391**, 87–99 (2014).
134. Eguchi, J. & Dasgupta, R. Redox state of the convective mantle from CO_2-trace element systematics of oceanic basalts. *Geochem Perspect Lett* **8**, 7–21 (2018).
135. Kogarko, L. N., Henderson, C. M. B. & Pacheco, H. Primary Ca-rich carbonatite magma and carbonate–silicate–sulfide liquid immiscibility in the upper-mantle. *Contrib Mineral Petrol* **121**, 267–274 (1995).
136. Schiano, P., Clocchiatti, R., Shimizu, N., Wcis, D. & Mattielli, N. Cogenetic silica-rich and carbonate-rich melts trapped in mantle minerals in Kerguelen ultramafic xenoliths – implications for metasomatism in the oceanic upper-mantle. *Earth Planet Sci Lett* **123**, 167–178 (1994).
137. Dixon, J., Clague David, A., Cousens, B., Monsalve Maria, L. & Uhl, J. Carbonatite and silicate melt metasomatism of the mantle surrounding the Hawaiian plume: evidence from volatiles, trace elements, and radiogenic isotopes in rejuvenated-stage lavas from Niihau, Hawaii. *Geochem Geophys Geosyst* **9**, 9 (2008).
138. Rohrbach, A. & Schmidt, M. W. Redox freezing and melting in the Earth's deep mantle resulting from carbon-iron redox coupling. *Nature* **472**, 209–212, (2011).
139. Dasgupta, R. Volatile-bearing partial melts beneath oceans and continents – where, how much and of what compositions. *Am J Sci* **318**, 141–165 (2018).
140. Eguchi, J. & Dasgupta, R. CO_2 content of andesitic melts at graphite-saturated upper mantle conditions with implications for redox state of oceanic basalt source regions and remobilization of reduced carbon from subducted eclogite. *Contrib Mineral Petrol* **172**, 12 (2017).

141. Dalton, J. A. & Presnall, D. C. Carbonatitic melts along the solidus of model lherzolite in the system CaO–MgO–Al_2O_3–SiO_2–CO_2 from 3 to 7 GPa. *Contrib Mineral Petrol* **131**, 123–135 (1998).
142. Dasgupta, R. et al. Carbon-dioxide-rich silicate melt in the Earth's upper mantle. *Nature* **493**, 211–215 (2013).
143. Green, D. H. & Wallace, M. E. Mantle metasomatism by ephemeral carbonatite melts. *Nature* **336**, 459–462 (1988).
144. Keshav, S. & Gudfinnsson, G. H. Melting phase equilibria of model carbonated peridotite from 8 to 12 GPa in the system CaO–MgO–Al_2O_3–SiO_2–CO_2 and kimberlitic liquids in the Earth's upper mantle. *Am Mineral* **99**, 1119–1126 (2014).
145. Mysen, B. O. & Boettcher, A. L. Melting of a hydrous mantle: I. Phase relations of natural peridotite at high pressures and temperatures with controlled activities of water, carbon dioxide, and hydrogen. *J Petrol* **16**, 520–548 (1975).
146. Wendlandt, R. F. & Mysen, B. O. Melting relations of natural peridotite+CO_2 as a function of degree of partial melting at 15 and 30 kbar. *Am Mineral* **65**, 37–44 (1980).
147. Wyllie, P. J. & Huang, W. L. Carbonation and melting reactions in system CaO–MgO–SiO_2–CO_2 at mantle pressures with geophysical and petrological applications. *Contrib Mineral Petrol* **54**, 79–107 (1976).
148. Yaxley, G. M. & Brey, G. P. Phase relations of carbonate-bearing eclogite assemblages from 2.5 to 5.5 GPa: implications for petrogenesis of carbonatites. *Contrib Mineral Petrol* **146**, 606–619 (2004).
149. Minarik, W. G. & Watson, E. B. Interconnectivity of carbonate melt at low melt fraction. *Earth Planet Sci Lett* **133**, 423–437 (1995).
150. Helffrich, G. R. & Wood, B. J. The Earth's mantle. *Nature* **412**, 501 (2001).
151. Dasgupta, R., Hirschmann, M. M. & Smith, N. D. Partial melting experiments of peridotite + CO_2 at 3 GPa and genesis of alkalic ocean island basalts. *J Petrol* **48**, 2093–2124 (2007).
152. Mallik, A. & Dasgupta, R. Reactive infiltration of MORB-eclogite-derived carbonated silicate melt into fertile peridotite at 3 GPa and genesis of alkalic magmas. *J Petrol* **54**, 2267–2300 (2013).
153. Mallik, A. & Dasgupta, R. Effect of variable CO_2 on eclogite-derived andesite and lherzolite reaction at 3 GPa – implications for mantle source characteristics of alkalic ocean island basalts. *Geochem Geophys Geosyst* **15**, 1533–1557 (2014).
154. Guillot, B. & Sator, N. Carbon dioxide in silicate melts: a molecular dynamics simulation study. *Geochim Cosmochim Acta* **75**, 1829–1857 (2011).
155. Morizet, Y. et al. Towards the reconciliation of viscosity change and CO_2-induced polymerization in silicate melts. *Chem Geol* **458**, 38–47 (2017).
156. Langmuir, C., Klein, E. M. & Plank, T. Petrological systematics of mid-ocean ridge basalts: contraints on melt generation beneath ocean ridges. In: *Mantle Flow and Melt Generation at Mid-ocean Ridges, Geophysical Monograph Series* (eds. J. Phipps-Morgan, D. K. Blackman, & J. M. Sinton), 183–280 (American Geophysical Union, 1992).
157. Ghosh, S., Ohtani, E., Litasov, K. D. & Terasaki, H. Solidus of carbonated peridotite from 10 to 20 GPa and origin of magnesiocarbonatite melt in the Earth's deep mantle. *Chem Geol* **262**, 17–28 (2009).
158. Frost, D. J. et al. Experimental evidence for the existence of iron-rich metal in the Earth's lower mantle. *Nature* **428**, 409–412 (2004).
159. Rohrbach, A. et al. Metal saturation in the upper mantle. *Nature* **449**, 456–458 (2007).

160. Harmer, R. E. & Gittins, J. The origin of dolomitic carbonatites: field and experimental constraints. *J African Earth Sci* **25**, 5–28 (1997).
161. Nielsen, T. F. D., Solovova, I. P. & Veksler, I. V. Parental melts of melilitolite and origin of alkaline carbonatite: evidence from crystallised melt inclusions, Gardiner complex. *Contrib Mineral Petrol* **126**, 331–344 (1997).
162. Guzmics, T. et al. Primary carbonatite melt inclusions in apatite and in K-feldspar of clinopyroxene-rich mantle xenoliths hosted in lamprophyre dikes (Hungary). *Mineral Petrol* **94**, 225 (2008).
163. Woolley, A. R. & Church, A. A. Extrusive carbonatites: a brief review. *Lithos* **85**, 1–14 (2005).
164. Guzmics, T. et al. Carbonatite melt inclusions in coexisting magnetite, apatite and monticellite in Kerimasi calciocarbonatite, Tanzania: melt evolution and petrogenesis. *Contrib Mineral Petrol* **161**, 177–196 (2011).
165. Zaitsev, A. N. & Keller, J. Mineralogical and chemical transformation of Oldoinyo Lengai natrocarbonatites, Tanzania. *Lithos* **91**, 191–207 (2006).
166. Mitchell, R. H. Carbonatites and carbonatites and carbonatites. *Can Mineral* **43**, 2049–2068 (2005).
167. Weng, Z., Jowitt, S. M., Mudd, G. M. & Haque, N. A detailed assessment of global rare earth element resources: opportunities and challenges. *Econ Geol* **110**, 1925–1952 (2015).
168. Harmer, R. E. & Gittins, J. The case for primary, mantle-derived carbonatite magma. *J Petrol* **39**, 1895–1903 (1998).
169. Watkinson, D. H. & Wyllie, P. J. Experimental study of the composition join $NaAlSiO_4$–$CaCO_3$–H_2O and the genesis of alkalic rock–carbonatite complexes. *J Petrol* **12**, 357–378 (1971).
170. Neumann, E. R., Wulff-Pedersen, E., Pearson, N. J. & Spencer, E. A. Mantle xenoliths from Tenerife (Canary Islands): evidence for reactions between mantle peridotites and silicic carbonatite melts inducing Ca metasomatism. *J Petrol* **43**, 825–857 (2002).
171. Klein-BenDavid, O. et al. High-Mg carbonatitic microinclusions in some Yakutian diamonds – a new type of diamond-forming fluid. *Lithos* **112**, 648–659 (2009).
172. Kjarsgaard, B. & Peterson, T. Nephelinite–carbonatite liquid immiscibility at Shombole volcano, East Africa: petrographic and experimental evidence. *Miner Petrol* **43**, 293–314 (1991).
173. Guzmics, T., Zajacz, Z., Mitchell, R. H., Szabó, C. & Wälle, M. The role of liquid–liquid immiscibility and crystal fractionation in the genesis of carbonatite magmas: insights from Kerimasi melt inclusions. *Contrib Mineral Petrol* **169**, 17 (2015).
174. Weidendorfer, D., Schmidt, M. W. & Mattsson, H. B. Fractional crystallization of Si-undersaturated alkaline magmas leading to unmixing of carbonatites on Brava Island (Cape Verde) and a general model of carbonatite genesis in alkaline magma suites. *Contrib Mineral Petrol* **171**, 43 (2016).
175. Woolley, A. R. & Kjarsgaard, B. *Carbonatite Occurrences of the World: Map and Database* (Geological Survey of Canada, 2008).
176. Jelsma, H., Barnett, W., Richards, S. & Lister, G. Tectonic setting of kimberlites. *Lithos* **112**, 155–165 (2009).
177. Woolley, A. R. & Bailey, D. K. The crucial role of lithospheric structure in the generation and release of carbonatites: geological evidence. *Mineral Mag* **76**, 259–270 (2012).

178. Bell, K. & Tilton, G. R. Nd, Pb and Sr isotopic compositions of east African carbonatites: evidence for mantle mixing and plume inhomogeneity. *J Petrol* **42**, 1927–1945 (2001).
179. Bailey, D. K. Episodic alkaline igneous activity across Africa: implications for the causes of continental break-up. *Geol Soc Lond Spec Publ* **68**, 91–98 (1992).
180. Bell, K. Carbonatites: relationships to mantle-plume activity. *Geol Soc Am Special Paper*, **352**, 267–290 (2001).
181. Ernst, R. E. & Bell, K. Large igneous provinces (LIPSs) and carbonatites. *Mineral Petrol* **98**, 55–76 (2010).
182. Jelsma, H., Barnett, W., Richards, S. & Lister, G. Tectonic setting of kimberlites. *Lithos* **112**, 155–165 (2009).
183. Hoernle, K., Tilton, G., Le Bas, M. J., Duggen, S. & Garbe-Schonberg, D. Geochemistry of oceanic carbonatites compared with continental carbonatites: mantle recycling of oceanic crustal carbonate. *Contrib Mineral Petrol* **142**, 520–542 (2002).
184. Holm, P. M. et al. Sampling the cape verde mantle plume: evolution of melt compositions on Santo Antao, Cape Verde Islands. *J Petrol* **47**, 145–189 (2006).
185. Schmidt, M. W. & Weidendorfer, D. Carbonatites in oceanic hotspots. *Geology* **46**, 435–438 (2018).
186. Keshav, S. & Gudfinnsson, G. H. Experimentally dictated stability of carbonated oceanic crust to moderately great depths in the Earth: results from the solidus determination in the system CaO–MgO–Al_2O_3–SiO_2–CO_2. *J Geophys Res B Solid Earth* **115**, B05205 (2010).
187. Grassi, D. & Schmidt, M. W. The melting of carbonated pelites from 70 to 700 km depth. *J Petrol* **52**, 765–789 (2011).
188. Katsura, T., Yoneda, A., Yamazaki, D., Yoshino, T. & Ito, E. Adiabatic temperature profile in the mantle. *Phys Earth Planet Interiors* **183**, 212–218 (2010).
189. Hirschmann, M. M. Mantle solidus: experimental constraints and the effects of peridotite composition. *Geochem Geophys Geosyst* **24**, 1–26 (2000).
190. Gale, A., Dalton, C. A., Langmuir, C. H., Su, Y. & Schilling, J.-G. The mean composition of ocean ridge basalts. *Geochem Geophys Geosyst* **14**, 489–518 (2012).
191. Yasuda, A., Fujii, T. & Kurita, K. Melting relations of an anhydrous mid-ocean ridge basalt from 3 to 20 GPa: implications for the behviour of subducted oceanic crust in the mantle. *J Geophys Res* **99**, 9401–9414 (1994).
192. Wang, W. & Takahashi, E. Subsolidus and melting experiments of a K-rich basaltic composition to 27 GPa: implication for the behaviour of potassium in the mantle. *Am Mineral* **84**, 357–361 (1999).
193. Day, H. W. A revised diamond–graphite transition curve. *Am Mineral* **97**, 52–62 (2012).
194. Herzberg, C. & Gazel, E. Petrological evidence for secular cooling in mantle plumes. *Nature* **458**, 619 (2009).
195. Herzberg, C. et al. Temperatures in ambient mantle and plumes: constraints from basalts, picrites, and komatiites. *Geochem Geophys Geosyst* **8**, 2 (2007).

7

The Link between the Physical and Chemical Properties of Carbon-Bearing Melts and Their Application for Geophysical Imaging of Earth's Mantle

FABRICE GAILLARD, NICOLAS SATOR, EMMANUEL GARDÉS, BERTRAND GUILLOT, MALCOLM MASSUYEAU, DAVID SIFRÉ, TAHAR HAMMOUDA, AND GUILLAUME RICHARD

7.1 Introduction: Toward a Geophysical Definition of Incipient Melting and Mantle Metasomatism

Geochemical observations on mantle xenoliths and experiments at pressure and temperature on CO_2- and H_2O-bearing mantle rocks have provided the widely accepted picture that melts and fluids are flowing and reacting within the solid mantle.[1–7] Whether this must be seen as a transient and local process or a broad and planetary-scale mantle dynamic is unknown.[2,4,7] Understanding this could establish whether these melt advection processes explain some remote geophysical observations and could help with clarifying the geodynamic roles played by these melting dynamics. The question is rendered difficult since these melts may not be easily linked to the volcanic products reaching Earth's surface; somehow, most of the mantle melting processes may produce melts that never leave the mantle and therefore remain inaccessible.

The fingerprints of such deep melts have been historically characterized by geochemical means: trace element abundances and some isotopic ratios of mantle xenoliths are modified by the reactive passage of these melts.[2,4,6,8,9] Major element abundances and the modal proportions of minerals can also be significantly affected.[6,8] All of this is named mantle metasomatism,[2,8,9] and this process may explain some geophysical observations.[2,10] Notably, mantle metasomatism has been characterized on lithospheric samples only, therefore representing the shallowest part of a deeper melting dynamic that will be presented hereafter.

The melt causing such modifications is usually not observable in mantle rocks and is also not generally found in most volcanic exposures (except for the enigmatic petit-spot volcanoes[11]). Experimental petrology has therefore been used to reconstruct the chemical compositions of the parental melt coexisting at equilibrium with the solid mantle assemblage.[3,5,7,12,13] Experiments at upper-mantle conditions have shown the key role of volatile species (i.e. H_2O and CO_2) in stabilizing CO_2-rich melts or fluid versus SiO_2-rich melts.[1,3,5,7,12,13] The take-home message of such experimental approaches is that, in the presence of volatiles, mantle melting can occur in most of the upper mantle;[7] melting regions are commonly limited by redox process reactions favoring diamonds in the deep upper mantle and decarbonation reactions in the shallowest part of the mantle.[5,7,14]

The particularity of the mantle melting regime due the presence of H_2O and CO_2 is that it produces small amounts of melt (i.e. $<1\%$) embedded within the solid mantle matrix. These small amounts of melt, named incipient melt, are directly linked to small amounts of H_2O and CO_2 (i.e. tens to thousands of part per millions (ppm)) being present in the mantle.[15] These incipient melts are very CO_2 and H_2O rich. Their volatile-rich nature imparts physical properties to these melts that are at odds with the conventional basaltic products that reach Earth's surface. A large part of this chapter will review the state of the art on the unconventional properties of these CO_2–H_2O-rich melts.

What are the origins of these melts? They are certainly diverse and linked to large-scale recycling processes, but broadly speaking, mantle convection (large or small scale[16–18]) causes decompression melting in many mantle regions. In the upper mantle, convection occurs in the asthenosphere – the convective mantle. Upwelling regions undergo decompression and produce incipient melts. The asthenosphere remains enigmatic as we do not have any mantle xenolith samples from it. Mantle xenoliths are lithospheric, being part of the plates. The mantle metasomatic processes captured from mantle xenoliths therefore describe what happens to these melts when they reach the top of the convection limbs and meet the base of tectonic plates.

The repetitive passage of small amounts of melts and their freezing and reactivity with solids can result in major geodynamic modifications. Seminal examples of reactive transports of melts within an ancient lithosphere achieving a sort of completion are reported as cases of rejuvenation.[2,6] By this process, lithospheres, being thick, cold, depleted, and rigid lids, can become the warm, enriched, and soft asthenosphere. This means that the boundary between the nonconvective and the convective mantle, the so-called lithosphere–asthenosphere boundary (LAB), must be partly controlled by melt advection–reaction processes.[2,4] The LAB must be, at a geological timescale, a movable boundary.[4,6] Notably, though this is not further discussed here, in addition to mass transfer processes, melt advection at the LAB also conveys heat (including latent heat[19]).

A critical issue is whether such small melt fractions can be detected using remote geophysical probing of the electrical conductivity (EC) and seismic properties of Earth's mantle.[20,21] This requires the long-studied geochemical processes of mantle melting and metasomatism to be converted into the physical numbers that are addressed by geophysical probing. This also requires issues regarding the fluid mechanics of melt advection in the mantle to be tackled (i.e. how fast the incipient melt moves with respect to conventional convection rates and plate velocities).

Here, we address the stability domains, the atomic structures, and the physical properties of these incipient melts together with their connectivity at small-volume fractions in mantle aggregates. The objective of this assessment is to highlight the possible links between mantle melting–metasomatism and the geophysical observations that supposedly mark the LAB.

These geophysical observations indicate bright contrasts between resistive mantle lids also featuring high seismic wave velocities (V_P and V_S) and the underlying conductive mantle[22,23] also featuring low V_S.[23–25] The seismic discontinuities have been named as the

Gutenberg discontinuity (G)[25] and the underlying region is the low-velocity zone (LVZ). This broad description is not a rule, as there are specific settings where such bright contrasts are not clearly observed, such as beneath cratons[26] and in the enigmatic NoMelt area,[27] a setting in the Pacific Ocean where no geodynamic perturbations seem to operate. We also note that the depth of this geophysical discontinuity varies from being shallow beneath oceanic plates (50–90 km) to being deep (and elusive) under cratons (>200 km). The cause of low V_S and high conductivity remain debated; it does not have to be a unique cause, but a common explanation would be an elegant simplification. Whether these discontinuities reveal the ponding of melts is an increasingly accepted though still debated concept.[23,25] Here, we will focus on the melting processes that are able to cause anomalously high ECs because the effect of melting on the genesis of the LVZ remains elusive.[25]

7.2 CO_2-Rich Melts in the Mantle: Stability, Composition, and Structure

7.2.1 Partial Melting in the Presence of CO_2 and H_2O: Incipient Melting

The fluxing effect of volatiles on silicate melting has been long recognized.[1,3,5,7,12–14] In the case of mantle melting, the effects of H_2O and CO_2, whether separately[3,12,14,28] or mixed,[1,3,7,13,29,30] have been investigated. It is clear that peridotite systems equilibrated with H_2O–CO_2 mixtures have been poorly investigated in comparison to the (nominally) CO_2-only bearing system.[5]

Figure 7.1 summarizes the main features for the case of melting at undersaturated CO_2–H_2O fluid conditions. In the temperature region below the fluid-free high-pressure solidus, the amount of melt is controlled by the low amount of available fluid. This melting regime is incipient melting. In the incipient melting regime, several cases must be highlighted.

For pressures less than 2 GPa (corresponding to depths of less than 60 km), the solidus is weakly depressed. In this pressure range, the solubility of CO_2 in silicate melts is low, most CO_2 is in the fluid, and the solidus is controlled by the availability of water.[28]

At ~2 GPa, the CO_2 of the fluid reacts with the silicates, yielding carbonate mineral formation. This carbonation reaction has the effect of strongly depressing the solidus temperature, and the carbonate ledge (i.e. nearly isobaric melting curve) is developed in the phase diagram. Melts at the solidus are carbonatitic.[1,3,7,29] Away from the solidus, hydrated silicate melting takes places and the melt compositions shift from carbonatitic to carbo-hydrous silicates.[3,7,29] As long as the temperature remains below that of the fluid-free solidus, the amount of produced melt is controlled by the fluid availability. Major melting only happens as the fluid-free solidus is crossed (temperature >1350°C at 2 GPa).

As pressure is further increased, the CO_2–H_2O fluid may be reduced by interaction with mantle silicates.[3,5,13,14] Along the mantle geotherm, carbonate reduction following this reaction would occur at about 120–250 km depth.[5,14] In the presence of hydrogen, a mixed CH_4–H_2O fluid is formed.[13] In this fluid, water activity is lowered, resulting in increasing solidus as depth (and therefore reduction) increases. The solidus evolution caused by fluid reduction at high pressure is indicated by the dashed line with arrows in Figure 7.1.

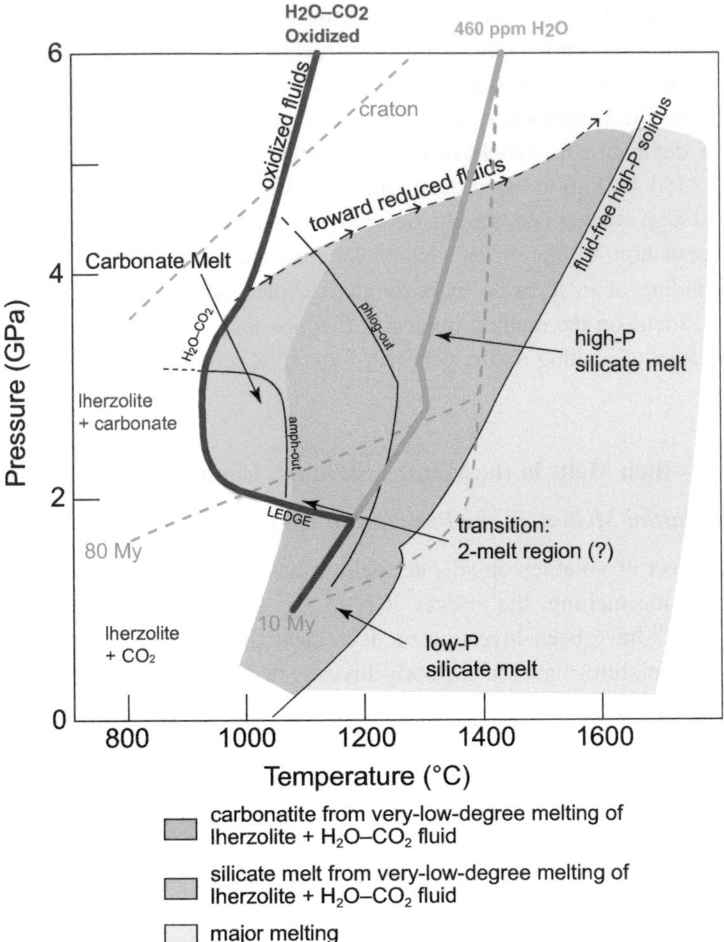

Figure 7.1 Pressure–temperature plot showing the stability fields for different types of mantle melt as a function of the volatile contents. Different geotherms (10 My, 80 My, cratons) are superimposed. The CO_2-bearing hydrous peridotite solidus is calculated from the combination of solidus temperatures of Ref. 1 from 0 to ~4 GPa and Ref. 13 at higher pressures. We connected the melting curve of CO_2-bearing hydrous peridotite to that of the dehydration solidus of nominally anhydrous peridotite[28] (considering peridotite with 460 ppm H_2O) at low pressures.

7.2.2 Carbonate to Silicate Melts in Various Geodynamic Settings

Following this broad picture of the process of incipient melting, we present here the range of melt compositions produced in various geodynamic settings. Two end-member cases are illustrated: adiabatic conditions showing regions where volcanism occurs at Earth's surface versus intraplate conditions showing mantle melting without volcanism.

Figure 7.2 illustrates melting along an adiabatic mantle (i.e. convective mantle) involving various H_2O- and CO_2-enrichments[15] and a potential temperature of 1360°C. The

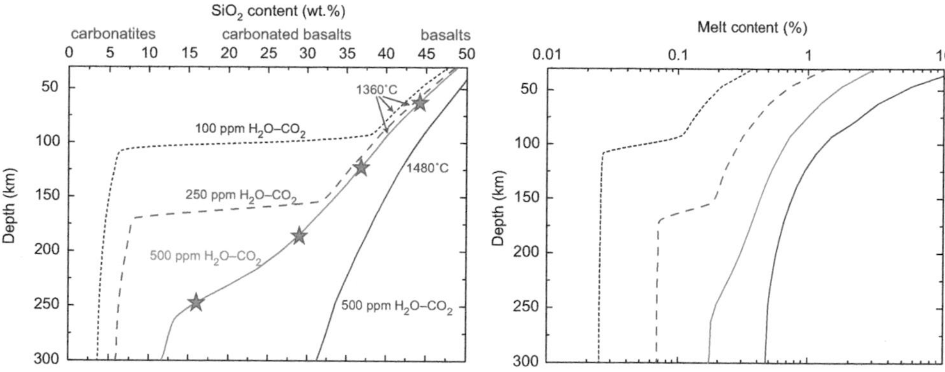

Figure 7.2 Melt composition (left) and melt fraction (right) produced during adiabatic mantle melting (temperature = 1360°C or 1480°C when specified). Under most of the pressure–temperature–volatile content conditions shown here, incipient melting occurs. Incipient melting of depleted (100 ppm H_2O–100 ppm CO_2) to enriched (500 ppm) mantle sources is considered. The stars show compositions for which viscosities were determined by molecular dynamics (see Section 7.3).

calculated melts[29] define an array of compositions ranging from low to high silica content in response to decompression melting: incipient melts evolving from carbonatite (i.e. <15 wt.% SiO_2), to carbonated basalts (15–50 wt.% SiO_2), to alkali basalts (SiO_2 >40 wt.%). Hereafter in the chapter, some of these compositions are used as reference compositions for which density, viscosity, and EC are defined (stars in Figure 7.2; see Section 7.3). Decreasing the bulk volatile contents decreases the absolute depth and the depth interval at which the equilibrium melt composition changes from carbonate to silicate melts: while the volatile-enriched mantle can produce silicate-bearing liquids down to 250 km, the volatile-depleted case shows an abrupt shift in composition at ~100 km depth within a ~10–20-km interval. This indicates that dry systems must have a greater tendency for carbonate–silicate immiscibility. Notably, Figure 7.2 does not consider the possibility of redox melting at ~250 km depth.

Figure 7.3 illustrates the nature of equilibrium partial melts formed in an intraplate thermal regime considering both CO_2–H_2O depleted and enriched mantles. A young plate (10 Ma, red), an intermediate plate (80 Ma, green), and a craton (gray) are illustrated.[31] These lithospheres of variable ages, and therefore variable thicknesses, can host melts with contrasting and strongly depth-dependent compositions. These lithospheres overlay a convective mantle having an age-independent pressure–temperature–melt composition pattern. The lithospheric and convective mantles are, by definition, separated by a thermal boundary layer, the LAB, separating the diffusive (lid) and the adiabatic (convective) mantles.[31] To what extent discontinuities in ECs[22,23] and seismic wave velocities[25,32] map the depth of this thermal LAB is a matter of a great debate. The depth range of the putative seismic G as broadly compiled from various studies and for various geodynamic environments is shown in Figure 7.3.[22,23] Notice that G values are reported at depths of ~60–80 km for oceanic

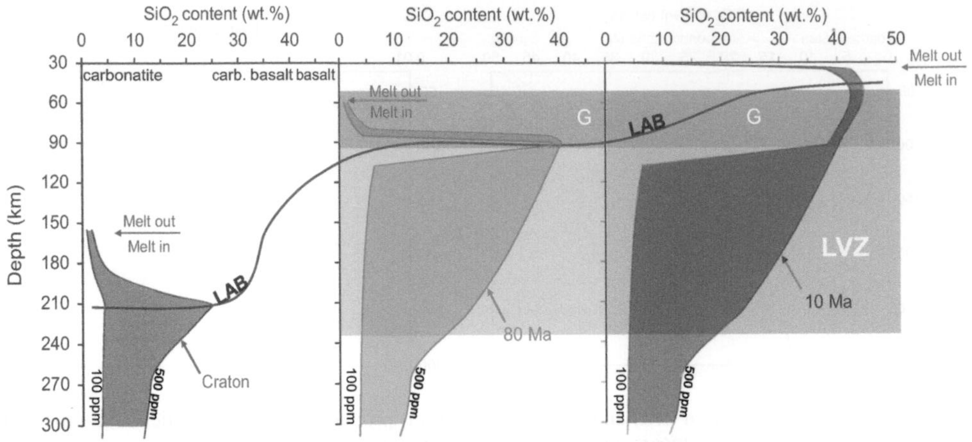

Figure 7.3 Profiles of melt compositions in intraplate geodynamic settings of variable ages (cratons, 80 Ma, and 10 Ma from left to right) compared to the depth range of the Gutenberg discontinuity (G) and the LVZ. The curve labeled "LAB" corresponds to the thermal lithosphere–asthenosphere boundary. The LAB displays a depth that changes with the age of the plate.[31] One sees that the degree of H_2O–CO_2 enrichment moderately affects the type of melt composition formed at lithospheric depth. It is essentially the temperature change with depth that controls the melt composition. This narrow range of lithospheric melt compositions is at odds with the large range of melt compositions that are formed in the convective mantle, which strongly depends on the degree of H_2O–CO_2 enrichment, as shown in Figure 7.2.

lithospheres of variable ages (0–120 Ma),[23,25] while the conventional LAB depths for similar lithospheric ages vary over a range of 0–110 km.[31] We must also notice that the seismic signal of G may be manifold and feature variable depth–age relationships.[32]

The broad correspondence between the stability field of incipient melting and the geophysical signals supposed to mark the LAB has long been known.[3,4,7,20,30,33] These observations must be advanced by considering the physical properties of incipient melts and how they impact the mantle rock properties near to the LAB.

7.2.3 Structural Differences between Silicate and Carbonate Melts

From a structural viewpoint, silicate and carbonate melts are opposites. However, they are end members of a continuum going from network-forming iono-covalent silicate liquids (e.g. silica) to ionic carbonate liquids (e.g. molten salts). The former are characterized by their degree of polymerization[34] (estimated from the NBO/T ratio, with "NBO" being the number of nonbridging oxygens and "T" the number of tetracoordinated cations, Si and Al), whereas the latter are fully depolymerized, with the liquid structure being controlled by the size ratio between anions and cations and by the ion valence state.[35] The microscopic structure of silicate melts is also quantified using structural indicators such as the Q_n species (with n = 0–4, where "n" is the number of TO_4 tetrahedra sharing a common

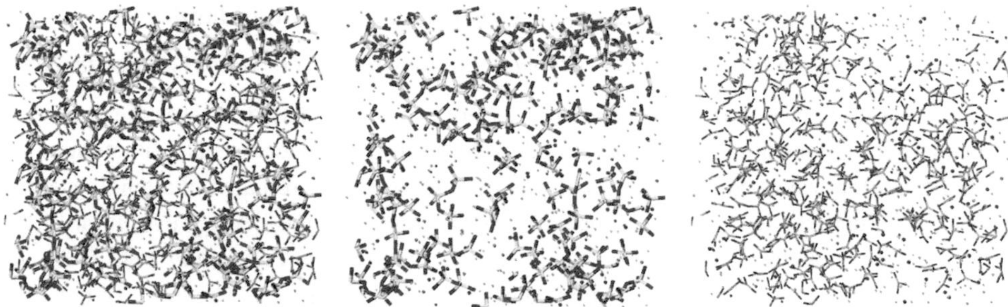

Figure 7.4 MD-generated snapshots of a carbo-silicate melt (17 wt.% SiO_2 and 28 wt.% CO_2) at 8 GPa and 1727 K. In the left panel, all atoms are depicted (SiO_4 in yellow and red, CO_3 in cyan and red, Mg in green, Ca in cyan, Na in blue, K in pink, and Fe in purple, with Al and Ti not being represented for clarity reasons). In the middle panel, the carbonate ions are not depicted in order to show the silicate network, whereas in the right panel, the SiO_4 units are not depicted in order to better visualize the arrangement of the carbonated component of the melt. It is clear that the SiO_4 and CO_3 ions do not mix well and form two subnetworks. A movie of the MD simulation may be found in the supplementary online material.

oxygen) and the coordination numbers associated with the network-forming and network-modifying cations (e.g. $Nc \approx 4$ for Si and Al, and $Nc \approx 5 - 8$ for Fe, Mg, Ca, and Na).

In carbonate melts, experimental and simulation studies[35–38] have shown that the number of carbonate ions (playing the role of O^{2-} in silicate melts) around Mg and Ca is about five to six, with one oxygen of each carbonate ion pointing preferentially toward the cation (with $d_{Mg-O} = 2.0$ Å and $d_{Ca-O} = 2.35$ Å, distances similar to those found in silicate melts), whereas the number of carbonate ions around each carbonate is about 12, a value similar to the oxygen coordination number in depolymerized silicate melts.[39] The self-diffusion coefficients (D) of Mg^{2+}, Ca^{2+}, and $CO_3{}^{2-}$ are of the same order of magnitude, contrary to silicate melts in which the D values of network-forming ions (e.g. Si and O) are much smaller than those of network-modifying ions (e.g. Mg and Ca).[40] In carbonate melts, cations and anions exchange with each other at the same rate, preventing network formation, which is at odds with silicate melts where network formation occurs.

As a silicate component is added to a carbonate melt, recent spectroscopic studies and molecular dynamics (MD) simulations[38,41,42] indicate that the carbonate ions are preferentially linked to alkaline earth cations,[41] as in carbonate melts (Figure 7.4). The silicate-forming network therefore mixes poorly with the carbonate units. This atomic configuration reveals a two-subnetwork structure in carbonated basalts, reflecting a tendency to immiscibility, though this has been observed in samples quenched into clear glassy structures that do not show immiscibility.[41,42] The link between a two-subnetwork atomic structure and macroscopic separation in a two-liquid system remains unclear despite its major geochemical importance.[43,44] Finally, the transport properties of such a melt are difficult to predict *a priori*, as the silicate component implies high viscosity while the carbonate component implies low viscosity.[45]

7.3 Physical Properties of CO_2-Rich Melts in the Mantle

7.3.1 Evolution of the Melt Density with Composition, CO_2, and H_2O Contents

It is well documented that the incorporation of both CO_2 and H_2O decreases the density of silicate melt. Establishing quantitative models describing this decrease as a function of the volatile content, melt composition, and pressure–temperature remains challenging, however. For illustration, we have evaluated by MD simulation the density evolution of the CO_2-bearing melt (H_2O free for technical reason, since melts containing both CO_2 and H_2O cannot be run) along the adiabatic geotherm used in Figure 7.2. The melt composition and its CO_2 content change along this adiabatic path (from a CO_2-rich kimberlitic composition at 8 GPa to a CO_2-poor basaltic composition at 2 GPa). The resulting density–pressure path is shown in Figure 7.5 ("melt + CO_2"). For the sake of comparison, the density evolutions of two CO_2-free compositions – a basalt and a kimberlite – are also reported in Figure 7.5. It is clear that, at 2 GPa, the addition of 4 wt.% CO_2 (to the basaltic composition) has a small effect on the melt density, whereas at 8 GPa, addition of 28 wt.% CO_2 causes a large density drop (notice the huge density contrast between the CO_2-free and CO_2-bearing kimberlitic melts at 8 GPa). However, it is noteworthy that along the chosen thermodynamic path, the melt also contains a significant amount of H_2O (from 8–9

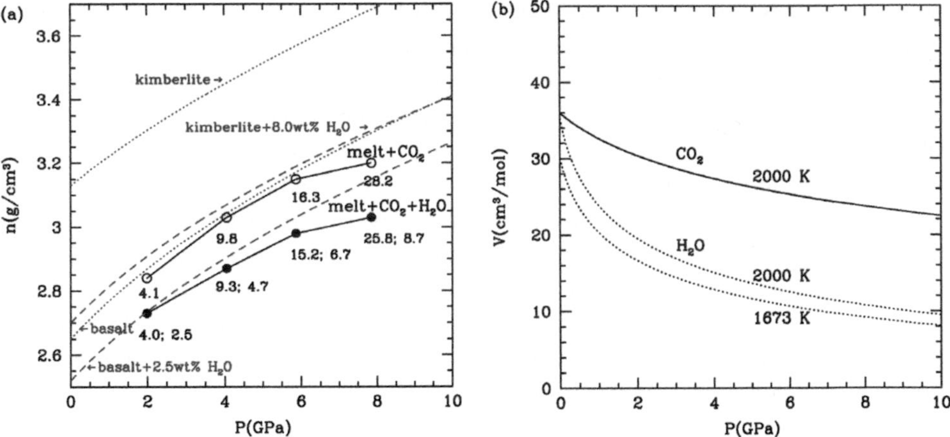

Figure 7.5 Effects of H_2O and CO_2 on the melt density curve as a function of pressure. (a) Calculations of melt density are performed for the melts produced along the adiabatic path of Figure 7.2. The black dots correspond to the chemical compositions marked by stars in Figure 7.2. Open symbols are H_2O free and full symbols contain both CO_2 and H_2O. Values along the line "melt + CO_2" indicate the CO_2 content in wt.%, and those along the line "melt + CO_2 + H_2O" indicate the CO_2 content (first number) in wt.% and the H_2O content (second number) in wt.%. The effects of H_2O content on the compressibility curve of a basaltic melt (red) and a CO_2-free kimberlitic melt (blue) are given for comparison. (b) The evolution of the partial molar volumes of CO_2 and H_2O as a function of pressure at 2000 K as given by the Vinet equation of state for CO_2[51] and for H_2O.[49] Notice that these partial molar volumes are independent of the melt composition. The temperature dependence of VCO_2 is negligible in the range 1673–2000 K.

wt.% H_2O at 8 GPa to 2–3 wt.% at 2 GPa). We have reported in Figure 7.5 the density curve of a basaltic melt with 2.5 wt.% H_2O and of a (CO_2-free) kimberlitic melt incorporating 8 wt.% H_2O. Briefly, along the chosen geotherm, the melts must be increasingly less dense with increasing pressure in comparison to conventional basaltic liquids. This yields an apparent lesser compressibility due to CO_2 incorporation.

To generalize these density curves, we will assume ideal mixing between CO_2, H_2O, and the silicate melt:

$$V(T,P) = x_{H_2O}V_{H_2O}{}^{(T,P)} + x_{CO_2}V_{CO_2}{}^{(T,P)} + (1 - x_{H_2O} - x_{CO_2})V_{sm}{}^{(T,P)}, \tag{7.1}$$

where $V(T,P)$ is the molar volume of the volatile bearing melt, $V_{H_2O}(T,P)$ is the partial molar volume of H_2O in the melt, $V_{CO_2}(T,P)$ is the partial molar volume of CO_2 in the melt, and $V_{sm}(T,P)$ is the molar volume of the volatile-free silicate melt. It has been shown[40,46,47] that the assumption of ideal mixing is very accurate when only one volatile species is considered, and we will assume that this assumption still holds in the presence of both H_2O and CO_2.

$V_{sm}(T,P)$ is given with a very good accuracy ($\pm 1\%$) by a third-order Birch–Murnaghan equation of state parametrized either from density measurements[48] or using MD calculations.[39] For $V_{H_2O}(T,P)$, we adopt the Vinet equation of state,[49] which is based on the partial molar volume data of water in various melts (78 wt.% SiO_2 to 35 wt.% SiO_2, 1473–2573 K, and 1–20 GPa). Astonishingly, experimental and simulation studies[39,47] show that $V_{H_2O}(T,P)$ is independent of water concentration and depends very little on the melt composition. For this reason, we can describe for any melt composition the evolution of $V_{H_2O}(T,P)$ with pressure and temperature (see Figure 7.5b). Similarly, empirical measurements and MD simulation studies[38,40,50–53] show that the value for $V_{CO_2}(T,P)$ depends very little on the melt composition, while in all these studies CO_2 is found to be much less compressible than H_2O in silicate melts. Changes in $V_{CO_2}(T,P)$ along the geotherm were calculated using the Vinet equation of state.[51]

The density curve of the H_2O–CO_2-bearing melts is shown in Figure 7.5 (see the curve entitled "melt + CO_2 + H_2O"). As expected, the influence of H_2O content on the density curve of the carbonated silicate melt is significant (compare the two curves with and without H_2O). We also note that the volatile-bearing melt becomes more buoyant as pressure increases.

7.3.2 Transport Properties: Viscosity–Diffusion

That carbonatite melts have unconventional transport properties in comparison to mantle silicate melts has long been clear to the research community, but acquiring robust quantitative data at the relevant pressure and temperature conditions has been challenging.[35] The first hints were indirectly found using analyses of molten salts,[35] which are well studied at atmospheric pressures by material scientists. MD have then provided insights on viscosity and diffusion properties[54] at various pressures and temperatures, and Dobson et al.[55]

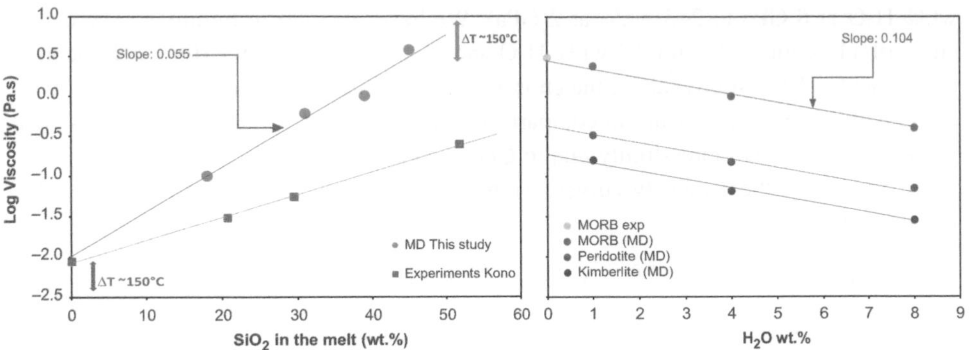

Figure 7.6 Viscosity changes as a function of melt silica content in dry carbonated melts (left) and as a function of H_2O content in CO_2-free melts (right). The conditions of calculation are 1400–1450°C and 2–8 GPa. Left: the melt compositions change from carbonatite (0 wt.% SiO_2) to basalt (45 wt.% SiO_2); both MD calculations (this work) and experimental measurements[56] are shown. Right: the effect of water on the viscosity of silicate melts; basaltic, peridotitic, and kimberlitic melts are similarly affected by water. MORB = mid-ocean ridge basalt.

provided the first *in situ* viscosity measurements at various pressures and temperatures. A significant step in our understanding of the transport properties of molten carbonates was realized by MD calculations,[37] which, together with high-precision viscosity measurements,[56] recently provided an internally consistent model. All of this indicates that carbonate melts have viscosity values of ~0.01 Pa/s with small to negligible temperature and melt chemical composition dependences. As these are ionic liquids in which all ionic groups move at similar rates, viscosity and ionic diffusion properties are closely linked.[37] This has been made clear experimentally,[57] with a remarkably simple relationship existing between the viscosity and EC of carbonate melts in the system Na–K–Ca–Mg–CO_3: $\log \eta = -\log \sigma$. As EC corresponds to the charge transfers due to the transport of all ionic groups, this essentially means that the ionic groups constituting carbonate melts have diffusion coefficients of ~10^{-9} m^2/s. Notably, this also matches diffusion-limited processes such as those from melt infiltration experiments in olivine aggregates.[58] This relatively well-established model of the transport properties of molten carbonate contrasts with our poor understanding of the physical properties of mixed carbonated basalts (e.g. kimberlites).

These findings have motivated a series of MD simulations determining the viscosity of hydrated carbonated silicate melts (Figure 7.6). The investigated compositions are those shown in Figure 7.2, and they globally correspond to a temperature range 1360–1440°C (pressure = 2–8 GPa; Figure 7.6). From carbonate to basalt melts, the viscosity increases by 2.5 log units, and it follows a logarithmic relationship with the melt SiO_2 content. Such variations due to changes in melt composition clearly overwhelm the expected changes due to pressure and temperature (ΔT of 150°C implies a change in basalt viscosity of ~0.5 log units). We noticed that the calculated viscosity change from carbonate to basalt melts exceeds that which has been experimentally determined.[56] This is not surprising since Kono et al.[56] considered $CaSiO_3$ melts as silicate melt end members instead of basalts (i.e. Al-bearing systems) in our case. Interestingly, we notice that the logarithmic relationship still holds, but the slope differs. The effect of water has been addressed for three different CO_2-free compositions (basalt, peridotite, and kimberlite) and it also shows a logarithmic relationship. The magnitude of the viscosity change due to water incorporation in the melt is much less than that described upon changing from carbonate to basalt, but an important point is that the effect of water is independent of the melt chemical composition: addition of 1 wt.% H_2O decreases the melt viscosities by ~0.1 log units for all melt compositions. This implies that the following relationship can describe the viscosity changes of hydrated melts having a composition between carbonate (0 wt.% SiO_2) and basalts (45 wt.% SiO_2):

$$\text{Log}\,\eta\ (\pm 0.2) = [\,0.055 \times SiO_2(\text{wt\%})\,] - 2.00 - [\,0.104 \times H_2O\ (\text{wt\%})\,]. \qquad (7.2)$$

In this relationship, notice that the effects of pressure and temperature are neglected. These effects were determined in basalts,[59] but the above equation quantifies an effect of melt chemical compositions being much greater than the expected pressure–temperature effects (see arrows in Figure 7.6).

7.3.3 Electrical Conductivity

The EC of mantle melts is an important geophysical property as it may guide the interpretation of magnetotelluric data. Conductivities in specific regions of Earth's mantle reaching or exceeding 0.1 S m^{-1} have long been considered as anomalously high. Several interpretations are possible, including water in olivine[60] (which is, however, a controversial issue[61]), hydrated basalts,[62] and carbonatite.[21] Carbonatite melts and hydrated basalts are not singular geological objects, but they constitute end-member products of incipient melting.[30] Carbonated basalts, constituting intermediate compositions between carbonatites and hydrated basalts, form the dominant melt compositions featuring incipient melting. Incipient melting could therefore be mapped from mantle EC, offering the exciting prospect of a direct visualization of the deep carbon and water cycles using geophysical data. This has motivated several recent experimental and theoretical surveys.[33,36,37,57,62–65] A summary of these surveys is shown in Figure 7.7, delineating the conductivity–temperature domains for dry and hydrated basalts, kimberlites, and carbonatites. Carbonatites at mantle pressure and temperature have conductivities exceeding 200 S m^{-1}, while carbonated basalts (e.g. kimberlites) are in the range 30–200 S m^{-1}. Sifré et al.[33] proposed a model describing the changes in conductivity as the melt chemical composition evolves from carbonate to basalt melts:

$$\sigma_{model} = \left(\sigma_0^{\text{basalt}} \times \exp\left(\frac{-E_a^{\text{basalt}}}{RT}\right)\right) + \left(\sigma_0^{CO_2} \times \exp\left(\frac{-E_a^{CO_2}}{RT}\right)\right). \qquad (7.3)$$

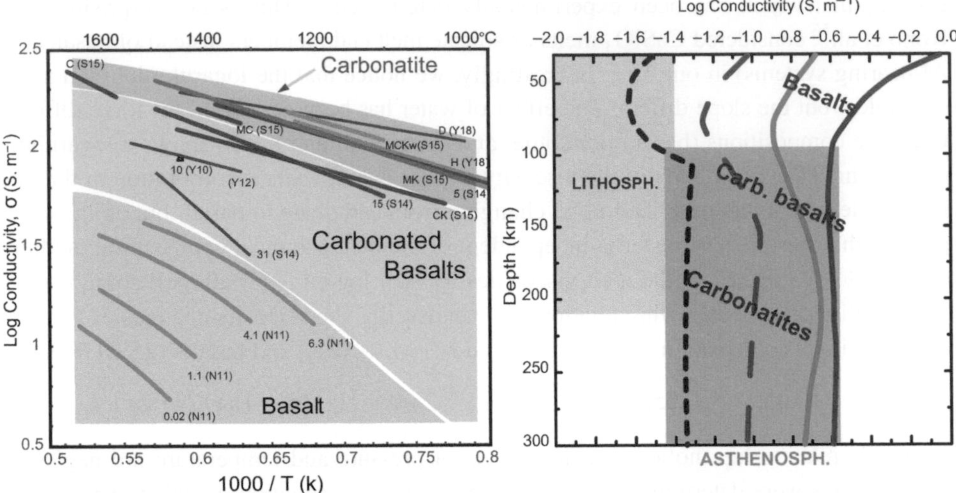

Figure 7.7 The EC of incipient melts either pure (left) or embedded into a olivine matrix (right). (Left) Basalts, kimberlites, and carbonatites are shown. The basalts are labeled in terms of water contents.[62] The kimberlite are labeled in terms of CO_2.[33,63,64] Carbonatite melts are compiled from Refs. 33, 57, 65. (Right) The conductivity of the mantle during incipient melting. The four cases from Figure 7.2 are converted here into EC versus depth signals. The solid mantle is approximated by hydrated olivine using the model of hydrated olivine.[85] The melt conductivity is calculated from Ref. 33. Values of 0.1 S m^{-1} are identified by the magnetotelluric community as anomalies.

Figure 7.7 shows the effect of incipient melting on the conductivity of mantle rocks along the adiabatic path of Figure 7.2. It shows that the depleted mantle produces moderately high ECs (<0.1 S m^{-1}), while the enriched mantle produces conductivities that match or exceed most geophysical assessments (>0.1 S m^{-1}). Given that the mantle source of mid-ocean ridge basalts (MORBs) varies in CO_2 content from 20 to 1200 ppm, this implies that mantle ECs may vary by greater than an order of magnitude.

7.4 Interconnection of CO_2-Rich Melts in the Mantle

Carbonate melt–olivine wetting angles θ were found to range narrowly around 30° (23–36°) over various experimental conditions (1200–1400°C, 0.5–3.0 GPa, $CaMg(CO_3)_2$, Na_2CO_3, K_2CO_3, and $CaCO_3 \pm H_2O$ carbonate compositions).[66,67] Thus, similarly to silicate melts, carbonate melts have <60° wetting angles. This implies that they should form interconnected networks at any melt fractions according to theories of melt equilibrium distribution.[68] This has consequences for transport properties in mantle rocks.

One of the most employed and powerful tools for investigating melt interconnectivity is *in situ* EC measurement in high-pressure and high-temperature apparatus. EC measurements provide bulk electrical responses of partially molten samples at mantle conditions. *In situ* measurements are also of particular importance for carbonated melts since they are

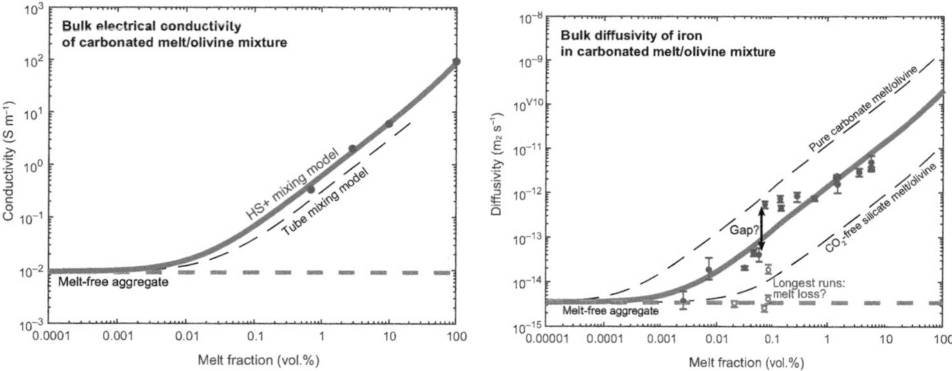

Figure 7.8 Evidence for interconnectivity at small melt fractions within olivine aggregates. (Left) Bulk EC of a carbonated melt/olivine mixture, evidencing melt interconnection over the range of investigated melt fractions down to 0.7 vol.%. Experimental data[63] (blue circles) were collected at 1377°C and 3 GPa. The HS+ mixing model (7.4) does reproduce the data very well, while the tube mixing model (7.5) appears inappropriate. Models were calculated (no adjustments) using $\sigma_{melt} = 8.91 \times 10^{1}$ S m^{-1} (conductivity at 100% melt) and $\sigma_{solid} = 9.23 \times 10^{-3}$ S m^{-1} (olivine conductivity at 1377°C).[85] (Right) Bulk diffusivity of iron in a carbonated melt/olivine mixture evidencing melt interconnection down to very small melt fractions <0.01 vol.%. Experimental data[70] were obtained at 1300°C and 1 GPa. Runs with longest durations (95–127 hours, empty blue circles) are differentiated from the other runs (5–49 hours, full blue circles), since they were possibly affected by melt loss (see text). Data roughly follow a trend that is intermediate between the bulk diffusivities of pure molten carbonate/olivine mixture and of a CO_2-free molten silicate/olivine mixture (dashed curves). The gap at ~0.07 vol.% melt, if significant, might mark a transition from tube to HS+ types of interconnection. The dashed curves are HS+ mixing laws with $D_{melt} = 1.8 \times 10^{-9}$ m^2 s^{-1} for pure a molten carbonate/olivine mixture[37] (diffusivity of Ca in $CaCO_3$ at 1300°C and 1 GPa) and $D_{melt} = 1.3 \times 10^{-11}$ m^2 s^{-1} for a CO_2-free molten silicate/olivine mixture (diffusivity of Fe in basaltic melt at 1300°C and 1 atm), where the diffusivity in a melt-free solid matrix is given by runs with no added carbonate,[70] $D_{solid} = 3.45 \times 10^{-15}$ m^2 s^{-1}. The blue curve is an adjustment of the experimental data with a HS+ mixing law. The added carbonate weight fractions reported by Ref. 70 were converted to the melt volume fractions using conversion factors of ~1.5.

hardly quenchable, and thus *post mortem* characterizations may be misleading. *In situ* EC measurements were measured at 1377°C and 3 GPa on mixtures of olivine and dolomitic melts with ~10–20 wt.% SiO_2.[63] Measured bulk conductivities are significantly enhanced compared to melt-free solid conductivities over the range of investigated melt fractions (down to 0.7 vol.%). As illustrated in Figure 7.8, the data are remarkably well reproduced by the Hashin–Shtrikman upper-bound (HS+) model:

$$\sigma_{bulk} = \sigma_{melt} + (1-\varphi)\left(\frac{1}{\sigma_{solid}-\sigma_{melt}} + \frac{\varphi}{3\sigma_{melt}}\right)^{-1}, \tag{7.4}$$

where φ is the melt volume fraction, σ_{melt} and σ_{solid} are the conductivities of melt and solid end members, respectively, and σ_{bulk} is the conductivity of the mixture.

This result is rather unexpected since the HS+ model considers a liquid–solid system where the liquid completely wets the matrix grains (no solid–solid contact), which is for $\theta = 0°$ according to melt equilibrium distribution theories. For systems with $0° < \theta < 60°$ (i.e. where interconnections occur via tubules along grain edges), the tube mixing model is expected:

$$\sigma_{bulk} = \frac{1}{3}\varphi\sigma_{melt} + (1 - \varphi)\sigma_{solid}. \tag{7.5}$$

However, this model does not reproduce the experimental data (Figure 7.8).[63,64] This highlights at least three points: (1) the wetting angles could have been overestimated and grain boundary wetting overlooked;[69] (2) some of the simplifying assumptions of melt equilibrium distribution theories (e.g. equally sized spherical grains, no anisotropy) might significantly depart from actual systems; and (3) the tube model might underestimate the conductivity of actual melt–solid geometries and the HS+ model might also, though fortuitously, apply to tubular melt distributions.

Evidence for the interconnection of CO_2-bearing melts down to very small fractions has been provided using "diffusion-sink" experiments in the Na_2CO_3–olivine system at 1300°C and 1 GPa with carbonate additions as low as ~0.001 wt.%.[70] In these experiments, iron is lost from olivine and diffuses via the molten intergranular medium into a platinum sink placed at the top of the sample as a proxy for melt interconnectivity. Figure 7.8 illustrates the iron bulk diffusivities as a function of melt fractions. In Figure 7.8, we converted added Na_2CO_3 weight content to the melt volume fraction by taking into account that the melt dissolves ~30 wt.% olivine (~14 wt.% SiO_2; weight to volume fraction conversion factors of ~1.5).[70] The striking point is that significant diffusivity enhancement occurs down to below 0.01 vol.% melt. We have to highlight here that our interpretation differs from that of Minarik and Watson,[70] who concluded that carbonated melt interconnectivity stops at 0.03–0.07 wt.% added carbonate (~0.04–0.10 vol.% melt). They based their conclusion on the fact that the longest experiments (95–127 hours) do not yield increases in bulk diffusivities (Figure 7.8), while they are the most likely to reach textural equilibrium. However, millimetric melt migrations can be achieved in less than 1 hour in the Na_2CO_3–forsterite system at 1300°C and 1 GPa, with a melt distribution along grain edges consistent with textural equilibrium.[58] Thus, the shortest experiments (5 hours) of Minarik and Watson[70] were very likely all texturally equilibrated. Conversely, it is possible that some melt was lost from the charges into the surrounding graphite medium in the longest experiments. When ignoring the four longest runs (95–127 hours), the remaining runs (5–49 hours) roughly follow a trend that is intermediate between the bulk diffusivities of pure molten carbonate/olivine mixing and that of CO_2-free molten silicate/olivine mixing. There might be a gap at ~0.07 vol.% melt, which requires corroborations, but it is not an interconnection stoppage, since bulk diffusivity enhancement is still observed down to 0.007 vol.%.

To sum up, wetting angle measurements indicate that CO_2-rich melts should form interconnected networks at any melt fraction in mantle rocks according to melt equilibrium distribution theories. This is validated by experiments down to below 0.01 vol.% melt. No interconnection stoppage can be clearly evidenced. The lowest melt fraction so far investigated in silicate melt/mantle rock systems (i.e. 0.15 vol.%) also reveals conductivity enhancement.[71] From 100 down to ~1 vol.% melt, the bulk transport properties of carbonated melt-bearing mantle rocks (e.g. EC, mass transfer, etc.) can be calculated using the HS+ mixing model. Whether the HS+ model still applies down to very small melt fractions (<0.01%) remains to be elucidated. These results stem from experiments using high-pressure apparatus where grain size is typically in the order of ~10 μm. Both melt interconnectivity and melt wetness increase as a function of grain size,[70,72] and thus should be strongly enhanced by millimetric grain sizes in the mantle.

7.5 Mobility and Geophysical Imaging of Incipient Melts in the Upper Mantle

7.5.1 Melt Mobility as a Function of Melt Composition

Knowledge of melt compositions and their viscosities (Sections 7.2 and 7.3) and evidence for their interconnection at very small melt fractions (Section 7.4) allow us to estimate the mobility of small fractions of CO_2-rich melts in the upper mantle. The mobility of a fluid embedded in a solid matrix is commonly addressed using the well-known Darcy's law. According to Darcy's law, the melt mobility is due to buoyancy ($\delta \rho g$) and is favored by low viscosity of the fluid (η_f) and high permeability ($k(\Phi)$):

$$V = \frac{k(\Phi)}{\eta_f} \delta \rho g. \tag{7.6}$$

In the case of mantle incipient melting and metasomatism, permeability goes down to very small values. Viscosity of the liquids increases by more than two orders of magnitude as the melt changes from carbonatite to basalt (Figure 7.7), while density variations are moderate (at 2 GPa, density changes from 2.4 to 2.8 g cm^{-3} from carbonatite[36] to basalt[34,47,49,59]).

Permeability is related to the degree of melting (porosity) as:

$$k(\Phi) = \frac{d^2}{C} \Phi^n, \tag{7.7}$$

where d is the grain size, Φ is the melt fraction (variable), and the parameters C and n are empirical values.

Here, we address the mobility of three types of incipient melts: carbonatites, carbonated basalts, and hydrated basalts. We assume a mantle grain size of 1 mm and we use an experimental permeability law[73] that is based on observations of samples with basaltic melt contents in the range of 1.5–18.0 vol.% (C = 58 and n = 2.6). We therefore operate an extrapolation toward much smaller melt fractions, which is justified in Section 7.4.

As is shown in Figure 7.9, carbonatites are very mobile, reaching velocities in the order of centimeters per year, even at melt fractions as low as 0.1 vol.% (typical of mantle

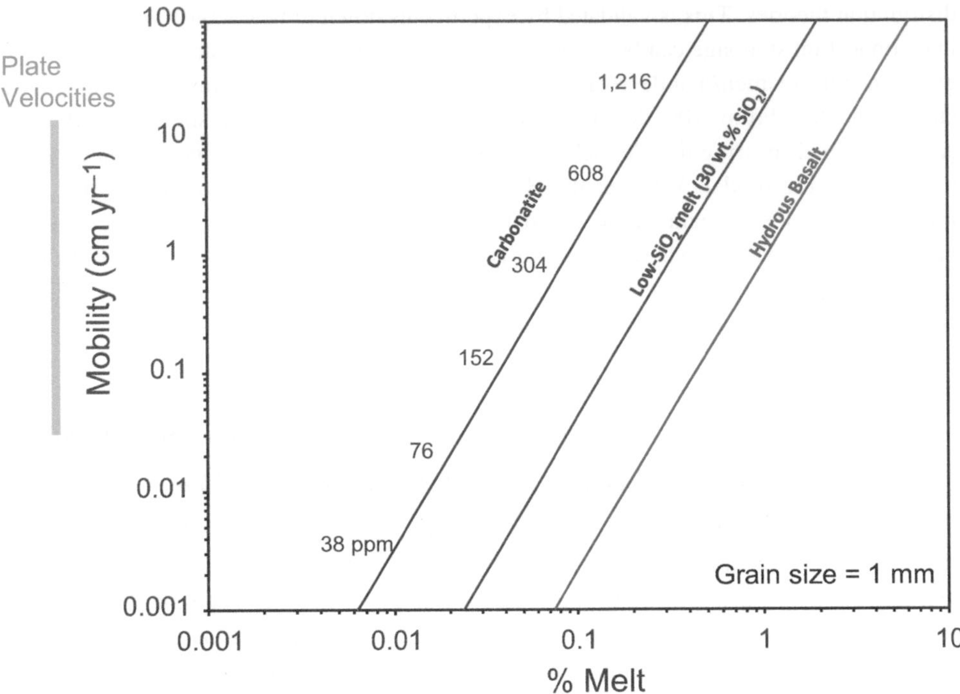

Figure 7.9 The melt vertical velocity at mantle depth versus melt fractions during incipient melting. Carbonatites, carbonated basalts, and hydrated basalts are shown. Carbonatites (containing 40 wt.% CO_2) are labeled in equivalent ppm CO_2 contents in the bulk rock. This concentration range covers depleted to enriched MORB sources.[15] The basalt contains 2 wt.% H_2O and 0.2 wt.% CO_2, while the carbonated basalt contains 2 wt.% H_2O and 15 wt.% CO_2.

metasomatism[2,4,6–8]). Hydrated basalts can move as fast as carbonatites when they have at ten times greater melt fractions (i.e. 1 vol.% hydrated basalt moves as fast as 0.1 vol.% carbonatites). A velocity of 1 cm yr^{-1} (0.1% of carbonatite melt) is high, but this only corresponds to 10 km of ascent within 1 Myrs. Nevertheless, ten times greater velocities are reached at 0.2–0.3% carbonatite. Furthermore, compaction processes, which are not considered here, must enhance the melt velocities.[19] This simple analysis implies that carbonatites can be efficient metasomatic agents operating over long distances at geological timescales, as has been suggested many times.[2,4,6,7,9,58] Hydrated basalts can have similar metasomatic roles if present at melt fractions exceeding 2–3%.

In the most depleted mantle sources containing <100 ppm CO_2, incipient melts are not mobile due to too small melt fractions (i.e. <0.02 vol.% carbonatite). In contrast, carbonatite melts formed in the most enriched mantle are very mobile (>10 cm yr^{-1}). Such a high mobility implies that these melts would rapidly migrate and would hardly be preserved if in physical contact with their source. This also implies unavoidable mixing processes in the column of melting where deep incipient melts rise fast and mix with upper-mantle regions (Figures 7.2 and 7.3), where melts produced at 200 km depth would rise fast and mix with the melts produced at shallower levels. The radioactive disequilibria found in MORBs have already been interpreted considering these mixing processes,[74] but how they impact the highly variable CO_2/Nb ratios of MORBs[15] remains unaddressed.

7.5.2 EC versus Mobility of Incipient Melts

The identification of the exact nature of the melts responsible for high EC anomalies in the mantle is critical. It would allow us to decipher whether these electrical anomalies reflect lithospheric processes, being cold and involving carbonatites, or asthenospheric processes, being warm and involving (hydrated) basalts. If the electrical anomalies result from carbonated basalt (low-SiO_2) melts, they may indicate both asthenospheric and lithospheric processes, since these melts can be stable in both domains (see Figure 7.3).

We calculate that incipient melting of a peridotite can produce an electrical anomaly (~0.1 S m^{-1})[22,23,27,75] if it contains 0.1 ± 0.04% of carbonatite melts, 0.3 ± 0.15% of carbonated basalts, or 1–6% of hydrated basalts at 1350°C and 3 GPa (Figure 7.10). Note that the calculation is only weakly dependent on temperature given the low temperature dependence of the EC of incipient melts (Figure 7.7).

The comparison of Figures 7.3 and 7.10 indicates that electrical anomalies in a young plate (<10 Ma) at ~80 km depth[23] can reasonably be matched with hydrated basalts,[62] while deeper anomalies observed beneath older plates (i.e. >50 Ma)[23,75] may match any kind of CO_2-rich melt.[33]

The mobility of melts produced during incipient melting conditions matching a conductivity of 0.1 S m^{-1} are reported in Figure 7.10. Both carbonatite and the hydrated basalt cases yield the highest velocities, favoring melt extraction, while carbonated basalts (low-SiO_2 melts) yield minimum velocities, favoring melt stability. Whether melts can be stabilized and detected by geophysical means therefore depends on the convection velocity of the mantle sourcing the melts. Highly mobile melts such as >0.1 vol.% carbonatite or 3 vol.% basalts can be detected in settings with mantle velocities of ~10 cm yr^{-1} or more (e.g. the center of mantle plumes feeding volcanic hotspots such as Hawai[76] or the East Pacific Rise[23,77]).

Away from these extreme geodynamic settings, many types of convections can occur in the mantle at rates in the range 0.1–1.0 cm yr^{-1}.[16,17,19] There, the most likely type of incipient melts producing high conductivity are carbonated basalts. These melts contain 30–40 wt.% SiO_2, implying that their viscosity is close to that of basalt but with enhanced EC, as their CO_2 contents are ~10–20 wt.%. About 0.3 ± 0.15% of such melts are required to produce high EC. This corresponds to 180–440 ppm CO_2 in the mantle. Notably, these melts resemble petit-spot volcanism.[11,78] These CO_2-rich melts are believed to remain in the mantle beneath plates and to be extracted when particular stress regimes just ahead of subduction zones trigger diking at a lithospheric scale.[11]

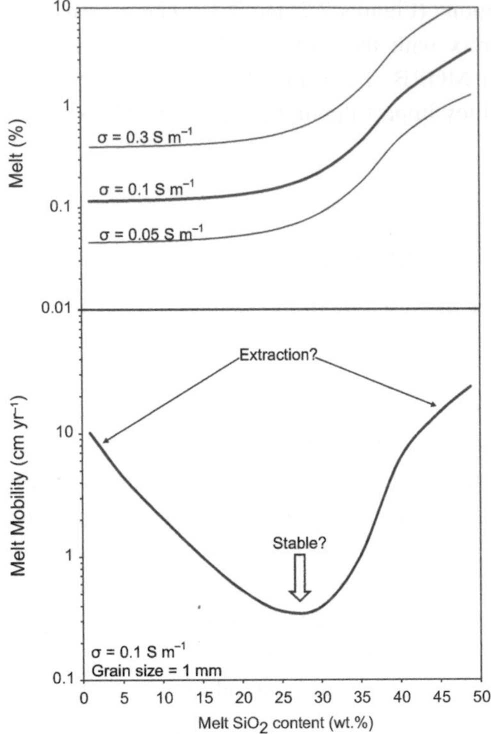

Figure 7.10 Incipient melting conditions producing high EC and the corresponding melt mobility. (Top) The range of melt fraction–melt compositions producing high ECs; three curves corresponding to three values of conductivity are shown. (Bottom) The mobility of incipient melt at the melt content required to produce high EC. A minimum in melt mobility appears in the case of carbonate basalts, while carbonatites and hydrated basalts are very mobile.

7.6 Conclusions

Due to the presence of small amounts of CO_2 and H_2O in Earth's mantle, incipient melting can occur in the regions close to the LAB. The produced melts have unconventional physical properties such as low densities, low viscosities, high diffusion rates, and high ECs, but there are strong interplays between the chemical compositions and physical properties of these melts. In a melting column (e.g., in ridges (Figure 7.2) or intraplates (Figure 7.3)), the structural composition of the produced incipient melts greatly changes from ionic to polymerized as a function of depth. This must cause a series of extraction–accumulation processes, so far unidentified, which are critical in our interpretation the geophysical and geochemical fingerprints of magmatism. Notably, these incipient melts remain interconnected even at very small melt fractions. This implies that a broad distribution of interconnected incipient melts can impact geophysical observations. On the other hand, interconnected low-viscosity melts also mean high melt mobility, which speaks against their stability within their source regions. In particular, for the incipient melting conditions capable of producing geophysical anomalies, melt extraction must occur rapidly at a geological scale. Production–extraction of incipient melts must therefore be considered altogether within the convective mantle.

7.6.1 LAB versus Geophysical Discontinuities

The LAB is a concept assuming that Earth's upper mantle is composed of two layers: the lower one, the asthenosphere, being adiabatic (convection controls heat transport); and the upper one, the lithosphere, being diffusive (diffusion dissipates heat). The electrical and seismic discontinuities mapped worldwide supposedly mark these boundaries, but the magmatic processes at the LAB and their geophysical visibility remain debated. The analysis provided in this chapter shows that several types of incipient melting can produce high mantle ECs. These melting processes can be lithospheric or asthenospheric, and therefore geophysical discontinuities may not image the LAB, but rather illuminate the dynamics of melting and melt transfers in the region of the LAB. Furthermore, geochemical observations have long defined the LAB as a movable boundary; that is to say, melt productions, melt infiltrations, and mantle metasomatism can cause major modifications of lithospheric roots.

Anomalous EC and regions with low seismic velocities have been mapped in many mantle region.[22,23,77,79,80] A broadly distributed LVZ beneath oceanic domains is deduced from large-scale surveys,[23,25] but the magnitudes of the velocity decreases locally vary. Beneath cratons, no such LVZ is observed.[26] After the Mantle ELectromagnetic and Tomography (MELT) experiment that imaged the mantle beneath the ultrafast East Pacific Rise,[77] several surveys have investigated quieter geodynamic settings hoping for more conventional geophysical signals, but high conductivities and low seismic velocities have often been observed.[23] The recently investigated NoMelt area[27] (70 My old oceanic plates) provides the first geophysical survey identifying no anomalous conductivity and a weak LVZ. If mantle melting is one of the ingredients causing geophysical anomalies, why does such an area exist?

7.6.2 Manifold Types of Mantle Convection Fuel Incipient Melting

The driving force of melt production is mantle convection, which produces decompression melting. There are sound geodynamic reasons to argue that decompression melting does not only occur where hot spot volcanoes pierce the surface: the inward mass transfers associated with slab sinking must be compensated by an upward flow being broadly distributed and probably more pronounced beneath oceanic basins.[16] The rate at which this upward mantle flow occurs should not exceed the melt velocities described in Figure 7.10 (i.e. ≤ 1 cm yr^{-1}), except in large plumes such as Hawaii.[76] Clearly, convections, decompressions, incipient melting, and melt extractions must occur in many regions of Earth's mantle. Convection-related decompression melting must fuel the LAB where upwelling occurs. Where mantle upwelling does not occur, a source of incipient melts simply does not exist. It is not completely clear whether this vision can explain the distribution of mantle geophysical anomalies.[23] Furthermore, if high EC can be explained by incipient melts, it remains unclear whether the same process could explain the low S-wave velocities in the asthenosphere.

7.7 Limits to Knowledge and Unknowns

What is the role of partial melting in the LVZ? Grain size and temperature distributions in the solid mantle may be accounted for by the LVZ,[81] but a recent experimental survey suggests a key role is played by incipient melting.[82] Melting is also an attractive model to account for the radial seismic anisotropy[23] of the LVZ. Yet the classical theories of melt equilibrium distribution[68] have been developed that predict that several volume percentage points of melt are required to significantly reduce seismic wave velocities.[83] This is at odds with recent experimental measurements[82] revealing that $<1\%$ melts drastically reduce S- and P-wave velocities.

Can the LVZ be a low-viscosity zone and can it play a role in the development of plate tectonics? The question is asked by Holtzman,[69] who tentatively responded by a "yes it can." If minute amounts of melt can wet the mantle grain boundaries and impact large-scale geophysical properties such as EC and seismic wave velocities, they may well affect the viscosity of the mantle. If diffusion creep is the mechanism of deformation in such systems, the tremendous diffusion properties of CO_2-rich melts demand an assessment of their impact on mantle viscosity. If the LVZ is indeed a low-viscosity zone, it certainly facilitates the motion of plates as suggested by geodynamic models,[84] and the conjunction of there being no observed LVZ beneath cratons and the relatively slow motion of cratons in comparison to younger plates speaks to a link between the magnitude of the LVZ and the velocity of the overlying plate.

Is there a continuous process from the metasomatic rejuvenation of cratonic roots to the deployment of the LVZ?[2] The chain of processes, broadly named rejuvenation, involve a combination of mechanical, thermal, chemical, and mineralogical processes.[2] How does melt ascend? Certainly via a combination of dikes and porous flows, but can we fit this within a proper petrological framework? The first numerical attempts were recently conducted.[19] Additional efforts are expected.

Acknowledgments

The authors acknowledge funding from the European Research Council (ERC project #279790), the French National Research Agency (ANR project #2010 BLAN62101), and the South African DST/NRF Research Chairs Initiative (Geometallurgy; Fanus Viljoen #64779). This is Laboratory of Excellence ClerVolc contribution 325.

Questions for the Classroom

1 What is the difference between incipient melting and partial melting of the mantle?
2 What is the nature of the produced melts and what are their physical properties?
3 What is the link between the atomic structure and the physical properties of these incipient melts?
4 Which effects could other volatile elements like S (very redox sensitive), F, Cl, and B have on the chemical–physical properties of Earth's mantle and on the dynamics of incipient melting?
5 Why can incipient melts barely rise above 60 km, which is the depth at which the pressure is ~2 GPa?
6 What is the LAB? Can we observe it by geophysical means? What are the geophysical observables?
7 Why can or cannot the LVZ be straightforwardly attributed to partial melting?
8 Is there a unique solution to account for by high EC layers in the mantle?
9 Why can incipient melting be accounted for by EC?
10 Why are some considerations of melt mobility needed in order to interpret high mantle conductivity?
11 What is the geodynamic process that causes incipient melting?
12 Where should we then (not) observe high EC near the LAB?

References

1. Wallace, M.E. & Green, D.H., An experimental determination of primary carbonatite magma composition. *Nature* 335, 343–346 (1988).
2. Aulbach, S., Massuyeau, M., & Gaillard, F., Origins of cratonic mantle discontinuities: a view from petrology, geochemistry and thermodynamic models. *Lithos* 268–271, 364–382 (2017).
3. Dasgupta, R. et al., Carbon-dioxide-rich silicate melt in the Earth's upper mantle. *Nature* 493, 211–215 (2013).
4. O'Reilly, S.Y. & Griffin, W.L., The continental lithosphere–asthenosphere boundary: can we sample it? *Lithos* 120, 1–13 (2010).
5. Hammouda, T. & Keshav S., Melting in the mantle in the presence of carbon: review of experiments and discussion on the origin of carbonatites. *Chem. Geol.* 418, 171–188 (2015).
6. Tappe, S. et al., Craton reactivation on the Labrador Sea margins: $^{40}Ar/^{39}Ar$ age and Sr–Nd–Hf–Pb isotope constraints from alkaline and carbonatite intrusives. *Earth Planet. Sci. Lett.* 256, 433–454 (2007).
7. Dasgupta, R., Volatile bearing partial melts beneath oceans and continents – where, how much, and of what compositions? *Am. J. Sci.* 318, 141–165 (2018).
8. Coltorti, M.C. et al., Carbonatite metasomatism of the oceanic upper mantle: evidence from clinopyroxenes and glasses in ultramafic xenoliths of Grande Comore, Indian Ocean. *J. Petrol.* 40, 133–165 (1999).
9. Pilet, S., Baker, M.B., & Stolper, E.M., Metasomatized lithosphere and the origin of alkaline lavas. *Science* 320, 916–919 (2008).

10. Pinto, L.G.R. et al. Magnetotelluric deep soundings, gravity and geoid in the south São Francisco craton: geophysical indicators of cratonic lithosphere rejuvenation and crustal underplating. *Earth Planet. Sci. Lett.* 297, 423–434 (2010).
11. Hirano, N. et al., Volcanism in response to plate flexure. *Science* 313, 1426–1428 (2006).
12. Wyllie, P.J. & Huang, W.L., Carbonation and melting reactions in the system CaO–MgO–SiO_2–CO_2 at mantle pressures with geophysical and petrological applications. *Contrib. Mineral. Petrol.* 54, 79–107 (1976).
13. Taylor, W.R. & Green, D.H., Measurement of reduced peridotite–C–O–H solidus and implications for redox melting of the mantle. *Nature* 332, 349–352 (1988).
14. Stagno, V., Ojwang, D.O., McCammon, C.A., & Frost, D.J., The oxidation state of the mantle and the extraction of carbon from Earth's interior. *Nature* 493, 84–88 (2013).
15. Le Voyer, M., Kelley, K.A., Cottrell, E., & Hauri, E.H., Heterogeneity in mantle carbon content from CO_2-undersaturated basalts. *Nat. Commun.* 8, 14062 (2017).
16. Morency, C., Doin, M.-P., & Dumoulin, C., Three-dimensional numerical simulations of mantle flow beneath mid-ocean ridges. *J. Geophys. Res. Solid Earth* 110, 344–356 (2005).
17. Ballmer, M.D., van Hunen, J., Ito, G., Tackley, P.J., & Bianco, T.A., Non-hotspot volcano chains originating from small-scale sublithospheric convection. *Geophys. Res. Lett.* 34, L23310 (2007).
18. French, S., Lekic, V., & Romanowicz, B., Waveform tomography reveals channeled flow at the base of the oceanic asthenosphere. *Science* 342, 227–230 (2013).
19. Keller, T. & Katz, R.F., The role of volatiles in reactive melt transport in the asthenosphere. *J. Petrol.* 57, 1073–1108 (2016).
20. Eggler, D.H., Does CO_2 cause partial melting in the low-velocity layer of the mantle? *Geology* 4, 69–72 (1976)
21. Gaillard, F. et al., Carbonatite melts and electrical conductivity in the asthenosphere. *Science* 322, 1363–1365 (2008).
22. Naif, S., Key, K., Constable, S., & Evans R.L., Melt-rich channel observed at the lithosphere–asthenosphere boundary. *Nature* 495, 356–359 (2013).
23. Kawakatsu, H. & Utada, H., Seismic and electrical signatures of the lithosphere–asthenosphere system of the normal oceanic mantle. *Ann. Rev. Earth Planet. Sci.* 45, 139–167 (2017).
24. Rychert, C.A., Laske, G., Harmon, N., & Shearer, P.M., Seismic imaging of melt in a displaced Hawaiian plume. *Nat. Geosci.* 6, 657–660 (2013)
25. Schmerr, N., The Gutenberg discontinuity: melt at the lithosphere–asthenosphere boundary. *Science* 380, 1480–1483 (2012).
26. Eaton, D.W. et al., The elusive lithosphere–asthenosphere boundary (LAB) beneath cratons. *Lithos* 109, 1–22 (2009).
27. Sarafian, E. et al., The electrical structure of the central Pacific upper mantle constrained by the NoMelt experiment. *Geochem. Geophys. Geosyst.* 16, 1115–1132 (2015).
28. Hirschmann, M., Tenner, T., Aubaud, C., & Withers, A.C., Dehydration melting of nominally anhydrous mantle: the primacy of partitioning. *Phys. Earth Planet. Inter.* 176, 54–68 (2009).
29. Massuyeau, M., Gardés, E., Morizet, Y., & Gaillard, F., A model for the activity of silica along the carbonatite–kimberlite–mellilitite–basanite melt compositional joint. *Chem. Geol.* 418, 206–216 (2015).
30. Hirschmann, M.M., Partial melt in the oceanic low velocity zone. *Phys. Earth Planet. Inter.* 179, 60–71 (2010).

31. McKenzie, D., Jackson, J., & Priestley, K., Thermal structure of oceanic and continental lithosphere. *Earth Planet. Sci. Lett.* 233, 337–349 (2005).
32. Burgos, G. et al., Oceanic lithosphere–asthenosphere boundary from surface wave dispersion data. *J. Geophys. Res. Solid Earth* 119, 1079–1093 (2014).
33. Sifré, D. et al., Electrical conductivity during incipient melting in the oceanic low-velocity zone. *Nature* 509, 81-85 (2014).
34. Mysen, B.O. & Richet, P., *Silicate Glasses and Melts: Properties and Structure*. Elsevier, Amsterdam (2005).
35. Jones, A.P., Genge, M., & Carmody, L., Carbonate melts and carbonatites. *Rev. Mineral. Geochem.* 75, 289–322 (2013).
36. Desmaele, E., Physico-chemical properties of molten carbonates by atomistic simulations. PhD thesis, Sorbonne University (2017).
37. Vuilleumier, R., Seitsonen, A., Sator, N., & Guillot, B., Structure, equation of state and transport properties of molten calcium carbonate ($CaCO_3$) by atomistic simulations. *Geochim. Cosmochim. Acta* 141, 547–566 (2014).
38. Vuilleumier, R., Seitsonen, A.P., Sator, N., & Guillot, B., Carbon dioxide in silicate melts at upper mantle conditions: insights from atomistic simulations. *Chem. Geol.* 418, 77–88 (2015).
39. Dufils, T., Sator, N., & Guillot, B., 2018. Properties of planetary melts by molecular dynamics simulation. *Chem. Geol.* 493, 298–315 (2018).
40. Guillot, B. & Sator, N., Carbon dioxide in silicate melts: a molecular dynamics simulation study. *Geochim. Cosmochim. Acta* 75, 1829–1857 (2011).
41. Moussallam, Y. et al., The molecular structure of melts along the carbonatite–kimberlite–basalt compositional joint: CO_2 and polymerisation. *Earth Planet. Sci. Lett.* 434, 129–140 (2016).
42. Morizet, Y., Florian, P., Paris, M., & Gaillard, F., ^{17}O NMR evidence of free ionic clusters Mn^+ CO_3^{2-} in silicate glasses: precursors for carbonate–silicate liquids immiscibility. *Am. Mineral.* 102, 1561–1564 (2017)
43. Brooker, R.A. & Kjarsgaard, B.A., Silicate–carbonate liquid immiscibility and phase relations in the system SiO_2–Na_2O–Al_2O_3–CaO–CO_2 at 0.1–2.5 GPa with applications to carbonatite genesis. *J. Petrol.* 52, 1281–1305 (2011).
44. Novella, D. et al. Melting phase relations of model carbonated peridotite from 2 to 3 GPa in the system CaO–MgO–Al_2O_3–SiO_2–CO_2 and further indication of possible unmixing between carbonatite and silicate liquids. *J. Geophys. Res. Solid Earth* 119, 2780–2800 (2014).
45. Morizet, Y. et al., Towards the reconciliation of viscosity change and CO_2-induced polymerization in silicate melts. *Chem. Geol.* 458, 38–47 (2017).
46. Liu, Q. & Lange, R.A., New density measurements on carbonate liquids and the partial molar volume of the $CaCO_3$ component. *Contrib. Mineral. Petrol.* 146, 370–381 (2003).
47. Bouhifd, M.A., Whittington, A.G., & Richet, P., Densities and volumes of hydrous silicate melts: new measurements and predictions. *Chem. Geol.* 418, 40–50 (2015).
48. Jing, Z. & Karato, S., Compositional effect on the pressure derivatives of bulk modulus of silicate melts. *Earth Planet. Sci. Lett.* 272, 429–436 (2008).
49. Sakamaki, T., Density of hydrous magma. *Chem. Geol.* 475, 135 139 (2017).
50. Ghosh, S. et al., Stability of carbonated magmas at the base of the Earth's upper mantle. *Geophys. Res. Lett.* 34, L22312 (2007).
51. Sakamaki, T. et al., Density of carbonated peridotite magma at high pressure using an X-ray absorption method. *Am. Mineral.* 96, 553–557 (2011).

52. Duncan, M.S. & Agee, C.B., The partial molar volume of carbon dioxide in peridotite partial melt at high pressure; *Earth Planet. Sci. Lett.* 312, 429–436 (2011).
53. Ghosh, D.B., Bajgain, S.K., Mookherjee, M., & Karki, B.B. Carbon-bearing silicate melt at deep mantle conditions. *Sci. Rep.* 7, 1–5 (2017).
54. Genge, M.J., Price, G.D., & Jones, A.P. Molecular dynamics simulations of $CaCO_3$ melts to mantle pressures and temperatures: implications for carbonatite magmas. *Earth Planet. Sci. Lett.* 131, 225–238 (1995).
55. Dobson, D.P. et al., *In-situ* measurement of viscosity and density of carbonate melts at high pressure. *Earth Planet. Sci. Lett.* 143, 207–215 (1996).
56. Kono, Y. et al., Ultralow viscosity of carbonate melts at high pressures. *Nature Commun.* 5, 5091 (2014).
57. Sifré, D., Hashim, L., & Gaillard, F., Effects of temperature, pressure and chemical compositions on the electrical conductivity of carbonated melts and its relationship with viscosity. *Chem. Geol.* 418, 189–197 (2015).
58. Hammouda, T. & Laporte, D., Ultrafast mantle impregnation by carbonatite melts. *Geology* 28, 283–285 (2000).
59. Sakamaki, T. et al., Ponded melt at the boundary between the lithosphere and asthenosphere. *Nat. Geosci.* 6, 1041–1044 (2013).
60. Karato, S., The role of hydrogen in the electrical conductivity of the upper mantle. *Nature* 347, 272–273 (1990).
61. Gardés, E., Gaillard, F., & Tarits, P., Comment to "High and highly anisotropic electrical conductivity of the asthenosphere due to hydrogen diffusion in olivine" by Dai and Karato [*Earth Planet. Sci. Lett.* 408 (2014) 79–86]. *Earth Planet. Sci. Lett.* 427, 296–299, (2015).
62. Ni, H., Keppler, H., & Behrens, H., Electrical conductivity of hydrous basaltic melts: implications for partial melting in the upper mantle. *Contrib. Mineral. Petrol.* 162, 637–650 (2011).
63. Yoshino, T., Laumonier, M., McIsaac, E., & Katsura, T., Electrical conductivity of basaltic and carbonatite melt-bearing peridotites at high pressures: implications for melt distribution and melt fraction in the upper mantle. *Earth Planet. Sci. Lett.* 295, 593–602 (2010)
64. Yoshino, T. et al., Electrical conductivity of partial molten carbonate peridotite. *Phys. Earth Planet. Inter.* 194–195, 1–9 (2012).
65. Yoshino, T., Gruber, B., & Reinier, C., Effects of pressure and water on electrical conductivity of carbonate melt with implications for conductivity anomaly in continental mantle lithosphere. *Phys. Earth Planet. Inter.* 281, 8–16 (2018)
66. Hunter, R.H. & McKenzie, D., The equilibrium geometry of carbonate melts in rocks of mantle composition. *Earth Planet. Sci. Lett.* 92, 347–356 (1989).
67. Watson, E.B., Brenan, J.M., & Baker, D.R., Distribution of fluids in the continental mantle, in Menzies, M.A., ed., *Continental Mantle: Oxford Monographs on Geology and Geophysics*. Oxford University Press, Oxford, 111–125 (1990).
68. von Bargen, N. & Waff, H.S., Permeabilities, interfacial areas and curvatures of partially molten systems: results of numerical computations of equilibrium microstructures. *J. Geophys. Res.* 91, 9261–9276 (1986).
69. Holtzman, B.K., Questions on the existence, persistence, and mechanical effects of a very small melt fraction in the asthenosphere, *Geochem. Geophys. Geosyst.* 17, 470–484 (2016).
70. Minarik, W.G. & Watson, E.B., Interconnectivity of carbonate melt at low melt fraction. *Earth Planet. Sci. Lett.* 133, 423–437 (1995).

71. Laumonier, M. et al., Experimental determination of melt interconnectivity and electrical conductivity in the upper mantle. *Earth Planet. Sci. Lett.* 463, 286–297 (2017).
72. Mu, S. & Faul, U.H., Grain boundary wetness of partially molten dunite. *Contrib. Mineral. Petrol.* 171, 40 (2016).
73. Miller, K.J., Montési, L.G.J., & Zhu, W-l., Estimates of olivine–basaltic melt electrical conductivity using a digital rock physics approach. *Earth Planet. Sci. Lett.* 432, 332–341 (2015).
74. Faul, U., Melt retention and segregation beneath mid-ocean ridges. *Nature* 410, 920–923 (2001).
75. Tada, N. et al., Electromagnetic evidence for volatile-rich upwelling beneath the society hotspot, French Polynesia. *Geophys. Res. Lett.* 43, 12021–12026 (2016).
76. Ballmer, M.D., Ito, G., van Hunen, J., & Tackley, P.J., Spatial and temporal variability in Hawaiian hotspot volcanism induced by small-scale convection. *Nat. Geosci.* 4, 457–460 (2011).
77. Evans, R.L. et al., Geophysical controls from the MELT area for compositional controls on oceanic plates, *Nature* 437, 249–252 (2005).
78. Okumura, S. & Hirano, N., Carbon dioxide emission to Earth's surface by deep-sea volcanism. *Geology* 41, 1167–1170 (2013).
79. Tada, N., Tarits, P., Baba, K., Utada, H., Kasaya, T., & Suetsugu, D., Electromagnetic evidence for volatile-rich upwelling beneath the society hotspot, French Polynesia. *Geophys. Res. Lett.* 43, 12021–12026 (2016).
80. Baba, K., Utada, H., Goto, T.-N., Kasaya, T., Shimizu, H., & Tada, N., Electrical conductivity imaging of the Philippine Sea upper mantle using seafloor magnetotelluric data. *Phys. Earth Planet. Inter.* 183, 44–62 (2010).
81. Jackson, I. & Faul, U.H., Grain size-sensitive viscoelastic relaxation in olivine: towards a robust laboratory-based model for seismological application. *Phys. Earth Planet. Inter.* 183, 151–163 (2010).
82. Chantel, J. et al., Experimental evidence supports mantle partial melting in the asthenosphere. *Sci. Adv.* 2, e1600246 (2016).
83. Wimert, J. & Hier-Majumder, S., A three-dimensional microgeodynamic model of melt geometry in the Earth's deep interior. *J. Geophys. Res.* 117, B04203 (2012).
84. Höink, T., Jellinek, A.M., & Lenardic, A. Viscous coupling at the lithosphere–asthenosphere boundary. *Geochem. Geophys. Geosyst.* 12, Q0AK02 (2011).
85. Gardés, E., Gaillard, F., & Tarits, P., Toward a unified hydrous olivine electrical conductivity law. *Geochem. Geophys. Geosyst.* 15, 4984–5000 (2014).

8

Carbon Dioxide Emissions from Subaerial Volcanic Regions

Two Decades in Review

CYNTHIA WERNER, TOBIAS P. FISCHER, ALESSANDRO AIUPPA, MARIE EDMONDS, CARLO CARDELLINI, SIMON CARN, GIOVANNI CHIODINI, ELIZABETH COTTRELL, MIKE BURTON, HIROSHI SHINOHARA, AND PATRICK ALLARD

8.1 Introduction

Volcanism and metamorphism are the principal geologic processes that drive carbon transfer from the interior of Earth to the surface reservoir.[1–4] Input of carbon to the surface reservoir through volcanic degassing is balanced by removal through silicate weathering and the subduction of carbon-bearing marine deposits over million-year timescales. The magnitude of the volcanic carbon flux is thus of fundamental importance for stabilization of atmospheric CO_2 and for long-term climate. It is likely that the "deep" carbon reservoir far exceeds the size of the surface reservoir in terms of mass;[5,6] more than 99% of Earth's carbon may reside in the core, mantle, and crust. The relatively high flux of volcanic carbon to the surface reservoir, combined with the reservoir's small size, results in a short residence time for carbon in the ocean–atmosphere–biosphere system (~200 ka).[7] The implication is that changes in the flux of volcanic carbon can affect the climate and ultimately the habitability of the planet on geologic timescales. In order to understand this delicate balance, we must first quantify the current volcanic flux of carbon to the atmosphere and understand the factors that control this flux.

The three most abundant magmatic volatiles are water (H_2O), carbon dioxide (CO_2), and sulfur (S), with CO_2 being the least soluble in silicate melts.[8] For this reason, it is not only Earth's active volcanoes that are a source of magmatic CO_2, but also numerous inactive volcanoes with magma bodies present at depth in the crust that contribute to the carbon emissions (Figure 8.1). Emissions from active volcanoes are released through crater fumaroles and open vents to form visible volcanic plumes, but diffuse degassing and degassing through springs on the volcano flanks also contribute to the total flux of carbon from a volcano. Plume gas emissions typically dominate over flank gas emissions and are highest during periods of eruptive activity.[9] Due to the hazard associated with eruptions and the value of volcanic gas monitoring to aid in eruption forecasting, much of our knowledge about the degassing of volcanic systems comes from active volcanoes, and typically during periods of unrest.

At less active and dormant (i.e. inactive) volcanoes, magmatic emissions of CO_2 are less obvious. CO_2 emissions are typically highest in thermal areas where gases are emitted through small fumaroles, soils, and fractures as diffuse degassing and through hot and cold

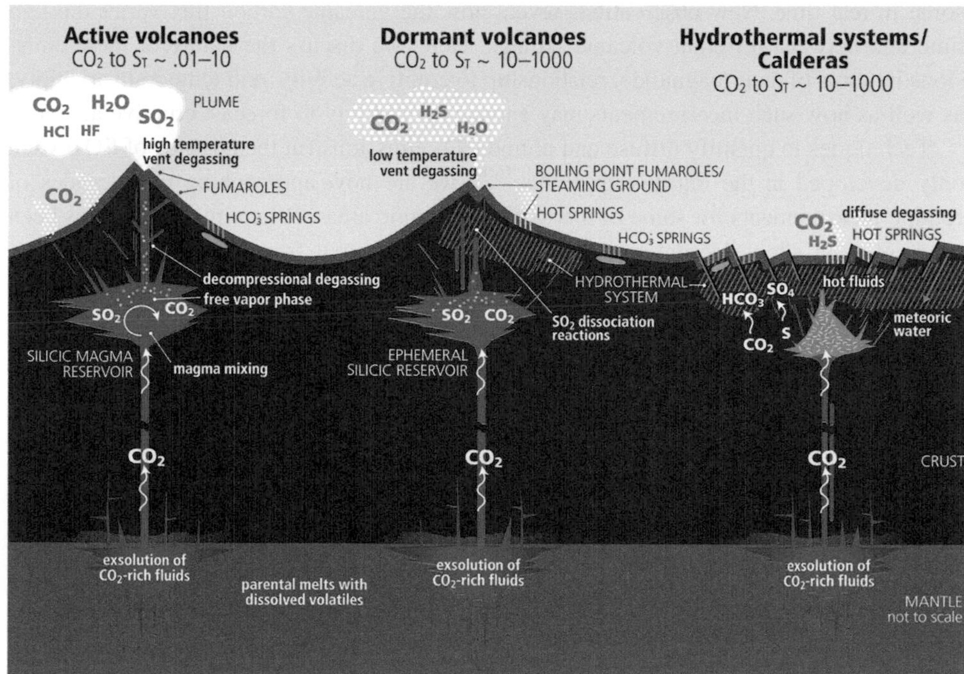

Figure 8.1 Conceptual models showing typical CO_2 emission patterns from volcanic and magmatic systems. CO_2 may be sourced from magma bodies deep in the crust, whereas other volatiles may remain largely dissolved in magma until much shallower depths. Visible plumes are typical for active volcanoes, whereas CO_2 degassing from dormant/inactive volcanoes and hydrothermal systems is less obvious. Low-temperature degassing may or may not result in a visible plume even when CO_2 is present. Significant quantities of CO_2 are emitted from areas of diffuse degassing, and CO_2 also reacts with groundwater.

springs. Occasionally, older volcanic areas can also exhibit cold degassing of CO_2.[10] CO_2 is denser than air, and therefore an asphyxiation hazard can exist in low-lying areas. Visual indicators of CO_2 release include thermally perturbed or bare ground and the odor of H_2S. Atmospheric plumes of CO_2 can also form in such areas, even if a region is not thermal,[11] and these emissions may not be visible. Additional contributions of volcanic carbon can be found in groundwaters,[12,13] but globally this contribution is less well studied compared to gas emissions.

In this chapter, we review recent advances in our understanding of the flux of CO_2 emitted in subaerial volcanic areas and how these emissions vary in space and time. Carbon released through mid-ocean ridges (MORs) and other oceanic environments is reviewed in Chapter 9. Through the focused efforts funded by the Deep Carbon Observatory and the Deep Earth CArbon DEgassing (DECADE) research initiative,[14] there is now greater global coverage of subaerial volcanic areas emitting CO_2 compared to previous efforts.[15] CO_2 emission rates have now been quantified for many of the most active volcanoes, and some in real time. New observations reveal how the volcanic carbon flux varies through time and between different volcanic settings. Here, we discuss the nature of these emissions in terms of their magnitude, relationship to eruptive activity, and temporal variability, as well as how such measurements may enhance our ability to forecast eruptive activity.

Techniques to quantify diffuse and plume CO_2 emissions (in the absence of SO_2) were only developed in the mid-to-late 1990s;[16,17] we are now approaching two decades of routine measurements for some of the world's volcanic areas. Where monitoring has been frequent, data allow decadal-scale evaluations of the output from a number of the most prolific carbon-emitting volcanic regions. We discuss the magnitude of emissions from some of the largest diffuse degassing regions and the challenges in extrapolating diffuse measurements globally.

We review the advances in understanding the sources of carbon outgassing from volcanoes, showing how the isotopic signature of carbon and other species has allowed distinction of the carbon contribution from subducting slabs, the crust, and mantle sources on arc scales. These insights into magmatic–tectonic controls on carbon outgassing then allow us to begin to link the modern volcanic carbon outgassing picture to that which might have existed in the geological past.

8.2 Methods for Measuring Volcanic CO_2: Established Techniques and Recent Advances

The principal challenges in the measurement of CO_2 from volcanic regions are related to the detection of volcanic CO_2 over the atmospheric background, logistical difficulties associated with accessing gas plumes, and technical issues that accompany deployment of instruments in the field. Techniques to measure CO_2 emission rates from different types of sources and the related uncertainties in these measurements have been reviewed previously.[15] Here, we expand on aspects of these methods as they pertain to information presented here and review emerging techniques and measurement biases.

8.2.1 Measurements of CO_2 Emissions in Volcanic Plumes

Volcanoes that are most active display persistent gas emissions during either frequent eruptions or as "passive" degassing of shallow magma bodies.[18] These volcanoes typically produce a volcanic plume, measurable with either direct or indirect techniques. The *indirect* or *ratio* technique underpins much of the recent progress in the measurement of volcanic CO_2 emission rates. Here, the SO_2 emission rate is measured using ultraviolet (UV) spectroscopy either from the ground,[19] airborne,[20,21] or space-based platforms,[22] and then multiplied by the C/S mass ratio determined by fumarole sampling,[8] Fourier-transform infrared (FTIR) spectroscopy,[23] or Multi-GAS measurements[24,25] and the plume speed. Indirect techniques rely on reliable and representative SO_2 emissions and C/S data. Uncertainties in the SO_2 flux (e.g. due to in-plume light scattering) can produce a bias to lower SO_2 fluxes by a factor of two or more.[26,27] Uncertainties in the C/S ratio of the gas arise from calibration of the Multi-GAS at a different altitude from the measurements, variable sensor response times to CO_2 and SO_2,[28–30] low plume concentrations (close to detection limits), and poorly mixed plumes.[31]

Direct measurements of CO_2 plumes use an airborne platform to measure the vertical profile of CO_2 concentration in the atmosphere downwind of the volcanic vent. The volcanic CO_2 (in excess of atmospheric background) is multiplied by the plume speed to derive an emission rate.[17,32,33] Uncertainties in plume speed affect both measurement types and vary greatly depending on whether the speed is estimated from radiosonde or weather models or is measured on site. Direct CO_2 measurement is the only option for the quantification of plume emissions where SO_2 is not present.[34] In-plume concentrations of 2–5 ppm above background are typically needed, depending on the CO_2 analyzer used, and ~10-50 t/d is a reasonable detection limit for airborne measurements depending on the aircraft, plume speed, and distance from the vent. Emissions less than this range or in areas where airborne measurements are not feasible are challenging to quantify and represent a recognized gap in current budgets. Future approaches to such challenging field measurements will include use of miniaturized IR (and other) sensors on drones[35–37] and ground-based light detection and ranging (LIDAR).[38,39]

8.2.2 *Diffuse CO_2 Emissions and Groundwater Contributions*

Many volcanic systems support areas of diffuse degassing often associated with hydrothermal activity due to magmatic intrusions at depth. A common method of quantifying diffuse emissions through soils is the accumulation chamber technique,[16,40] where a chamber is set on the ground and the concentration of the accumulated CO_2 is measured with time. Here, point measurements of the flux of CO_2 are made over an area of interest and total emissions are quantified by applying geostatistical techniques (see Refs. 16, 41 and references therein). The same chamber technique can be applied to lakes.[42,43] Eddy covariance (EC) is an aboveground technique that has been used successfully to measure the flux of CO_2 derived from diffuse, fumarole, and pool sources in regions with relatively low topographic relief.[44–47] The EC footprint (the source area on the ground contributing to the measured EC CO_2 flux) varies with atmospheric conditions such as wind speed and direction and is typically smaller than most degassing regions. Thus, to determine the CO_2 emission rate from a region of interest requires assumptions about the representativeness of the average EC flux to the larger area.[48] Alternatively, months-long deployments and inverse modeling have also been used to determine emission rates,[47,49] though such models also have inherent uncertainty. While promising for long-term hazard monitoring, more work is needed for utilizing the full potential of EC for determining emission rates.

In volcanic areas, CO_2 also dissolves into groundwaters and can emit through springs as a dissolved constituent. This flux can be quantified through chemical sampling and stream gauging,[12,50] or through mass balance of the aquifer (i.e. coupling hydrogeological and hydrogeochemical data). For instance, using this technique, the amount of CO_2 transported by Vesuvio groundwaters was estimated at about 150 t/d, or in the same order of magnitude as the diffuse emission of CO_2 in the crater area.[13]

8.2.3 Significant Recent Advances: Continuous and Remote Techniques

One major advance toward producing robust long-term records of volcanic CO_2 emissions has arisen from the advent of autonomous Multi-GAS instruments.[51] When combined with independent SO_2 flux time series,[52] measurements from these instruments have refined the CO_2 output for several volcanoes, characterizing the variability of emissions on temporal scales of days to years for the first time.[53–57] Multi-GAS stations are being adapted for deployment at high-latitude volcanoes[28,30,58] and can perform automated calibrations for improved accuracy.[28,48] Multi-GAS has recently been used on manned airborne missions and on unmanned aerial vehicles (UAVs).[35,37,59] Overall, these measurements are fundamentally changing the way volcanic hazard is evaluated at active volcanoes.

Techniques to quantify CO_2 remotely and from smaller features have also developed in recent years. Tunable diode laser spectrometer[60–62] measurements have shown that the CO_2 output from fumaroles is significant (hundreds of t/d) at some volcanoes, illustrating a nontrivial contribution to the volcanic CO_2 flux from this largely unquantified source at a global scale. In addition, LIDAR, specifically differential absorption LIDAR,[38,63] the smaller CO_2 differential absorption LIDAR,[39] and miniaturized light laser sensing spectrometers,[64] have been used to determine path-integrated CO_2 concentrations over kilometer scales. While these studies offer new prospects for quantifying CO_2 flux, further work is required to standardize and widen their potential applications.

Advances have been made with satellite remote sensing of CO_2, although CO_2 is among the most challenging volcanic gases to detect due to high atmospheric concentrations (~400 ppm and rising due to anthropogenic contributions) that dominate the signal of column-average measurements. Even at some of the strongest volcanic gas sources (e.g. Etna, Italy), the volcanic CO_2 signal may be only up to tens of ppm above background,[21,65] requiring high precision and accuracy for detection from space. NASA's Orbiting Carbon Observatory 2 (OCO-2), with a small footprint size (1.3×2.3 km) and $<0.2\%$ accuracy, permitted the first reported satellite detection of volcanic CO_2 emissions at Yasur volcano in 2015.[66] However, neither OCO-2 nor the Japanese Greenhouse Gases Observing Satellite (GOSAT) provide sufficient temporal resolution or spatial coverage to be effective volcano monitoring tools. The future of volcanic CO_2 monitoring from space is inextricably linked to the politics of greenhouse gas measurements and climate change. Several planned or proposed satellite missions (NASA's OCO-3, JAXA's GOSAT-2, and ESA's CarbonSat) offer the potential for volcanic CO_2 detection, but it is unlikely to ever become as routine as volcanic SO_2 measurements, and will likely be restricted to "spot" measurements of the strongest persistent volcanic CO_2 sources.

8.3 Estimating Global Emission Rates of CO_2

Quantifying global emissions of volcanic CO_2 is an area of ongoing research that will continue to evolve as more measurements become available. Some of the first estimates of global volcanic CO_2 degassing, published in the 1990s, were based on only seven to nine measurements of passive CO_2 degassing;[67,68] our understanding of CO_2 degassing in volcanic areas has progressed greatly since then. Here, we review methodologies from recent studies quantifying global volcanic CO_2 and how our new understanding may allow us to reduce some of the uncertainties in these estimates.

Until recently, most estimates of global CO_2 emissions were determined by proxy where volcanic CO_2 was scaled globally by a tracer (e.g. SO_2 or ^3He). Work focused on determining the C/S ratio of fumaroles based on level of activity[69] or on an arc-wide basis[70] and combining these data with global SO_2 emission rate compilations.[71,72] While seemingly straightforward, numerous uncertainties exist in these methods. First, C/S is not constant in time, and it is challenging to discern whether variations in C/S reflect changing mixtures of sources (magmatic vs hydrothermal) or progressive degassing of a single magmatic source due to gas loss or decompression (Figure 8.1).[73,74] Second, SO_2 emission rate data are skewed toward easily accessible locations and volcanoes experiencing unrest. Progress has been made with satellite remote sensing data that, when averaged over long time periods, are sensitive enough to measure lower emission rates of SO_2, thereby reducing some measurement bias. For example, recent work used Ozone Monitoring Instrument (OMI) satellite data[75] to calculate a global passive volcanic SO_2 flux of 23 ± 2 Tg SO_2/yr during the decade 2005–2015 from 91 volcanoes, half of which also have the C/S ratio measured. However, 91 volcanoes only represents 16% of the 570 volcanoes active in historic time, and 6% of the volcanoes active in the Holocene,[76] many of which might be passively degassing CO_2, but not emitting SO_2 over the satellite detection limit. Some previous global estimates of CO_2 emissions have assumed that the strongest emitters of SO_2[57] also produce the most CO_2, but the time frame of measurement is important to consider, as is the number of degassing systems. The temporal distribution of volcanic CO_2 outgassing could be very different from that of SO_2,[77] which is dominated by a relatively small number of erupting and persistently degassing volcanoes. Further clarity on this issue may be provided in the near future with the recently launched Tropospheric Ozone Monitoring Instrument (TROPOMI) sensor (www.tropomi.eu), which has 12-times higher spatial resolution than the earlier OMI sensor, and may reveal weaker plumes.

Global ^3He fluxes have also been used to estimate global arc CO_2 fluxes.[78,79] The estimated ^3He flux from arcs is based on the well-constrained ^3He flux from MORs and the assumption that 80% of volcanic activity on Earth is associated with MORs and the remainder mainly from volcanic arcs.[80] Intra-oceanic arc magma fluxes were revised[81] and show a factor of approximately two times higher rates compared to the early studies.[80,82] While MOR ^3He fluxes appear to be quite well constrained within a factor of approximately two,[83] work on global arc-magma production rates is still sparse, and therefore arc ^3He fluxes are likely associated with uncertainties that remain challenging to quantify. The most recent volcanic CO_2 flux from arcs is estimated to be 22×10^{11} mol/yr or 95 Tg CO_2/yr,[84] based on the $CO_2/^3He$ ratio of volcanic gases with outlet temperatures of >200°C, although variability in the mantle $CO_2/^3He$ adds considerable uncertainty to such calculations.

A third approach has been to extrapolate CO_2 data based on actual measurements.[15,57] The latest budget calculation[15] separated emissions based on the source (plume degassing, diffuse degassing from historically active volcanoes, hydrothermal and inactive areas, volcanic lakes, and MORs) and extrapolated them based on the number of similar systems globally. Roughly 50% (271 Tg CO_2/yr) of the total global subaerial emission of CO_2 (540 Tg CO_2/yr) was estimated to come from ~150 passively degassing volcanoes, based on the average CO_2 emissions measured at 33 active volcanic systems. An additional 20% was estimated by extrapolating observed diffuse emissions to the ~550 historically active volcanoes. CO_2 emissions from hydrothermal systems were treated separately, as were volcanic lakes and MORs.

The two main quantification challenges in extrapolating empirical data include the determination of a representative flux and the estimation of the total number of degassing volcanoes. The Global Volcanism Program (GVP) Volcanoes of the World catalog has been used to assess the number of volcanoes degassing globally, but it is important to note that the catalog quantifies the number of degassing volcanoes (i.e. "fumarolic" volcanoes) only where there has been "no (other) explicit evidence for Holocene eruptive activity." Thus, of the 1545 volcanoes with known or inferred eruptive activity in the Holocene, it is unclear how many are degassing other than those defined as "fumarolic" or "solfataric." In the latest volume, this category has been reduced from over 100 to 64 as more systems now have other data indicating Holocene activity.

Burton et al.[15] suggested ~150 volcanoes were degassing today, or 10% of the ~1500 volcanoes active in the Holocene.[85] Satellite measurements show[16] instead that 91 systems have emitted significant amounts of SO_2 (and thus CO_2) between 2005 and 2015, yet these data are representative of eruptive periods only, as higher-altitude plumes are more readily detected from space. As satellite surveillance of SO_2 emissions improves, the number of degassing sources measurable from space will likely increase. Our compilation shows there are now 201 Holocene volcanic systems associated with some form of CO_2 degassing observations (Supplemental Tables 8.1 and 8.2) and an additional 22 where the last eruptive activity was in the Pleistocene (Supplemental Table 8.3). Thus, future attempts to estimate global CO_2 degassing from volcanic areas should not assume that only historically active or Holocene volcanoes are actively degassing, but also consider the 1325 Pleistocene volcanoes.[76]

8.4 Current State of Knowledge of CO_2 Degassing from Volcanoes

8.4.1 CO_2 Emissions from Earth's Most Active Volcanoes

Earth's most active volcanoes are those that are best studied due to the hazards they pose. Over decadal timescales, many of the most active volcanoes alternate between periods of strong degassing, typically associated with periods of eruptive activity, and phases of reduced (or arrested) degassing, with the former preferentially targeted by observations. Global CO_2 compilations calculate average emissions based on all published estimates of CO_2 flux,[15,67] yet many of these are spot measurements acquired during periods of heightened activity that may span decades. It is therefore likely that combining sparse measurements collected over several decades may lead to an overestimation of the real time-averaged global volcanic CO_2 output.

In an attempt to reduce the above uncertainty, the average CO_2 fluxes for some volcanoes in Burton et al.[15] were revised (Supplemental Table 8.1) using more recent observations that have been obtained in the last decade (2005–2017) where available, and including both eruptive and quiescent periods whenever possible. Our revised average fluxes are lower than previously published[15] for all of the major volcanic CO_2 sources (Supplemental Table 8.1). For example, recent observations of CO_2 emissions from Nyiragongo volcano are lower by approximately a factor of ten (i.e. ~9300 t/d,[86,87] compared to ~95,000 t/d collected during the 1950s–1970s[15,88]). Similarly, we report a new time-averaged CO_2 flux for Miyakejima volcano in Japan (1070 t/d[57]) based on nearly two decades of systematic observations. This longer data set yields one order of magnitude lower CO_2 emissions than implied by the intense degassing unrest of early 2000.[89,90] Likewise, emission rates from Augustine and Mount Spurr volcanoes are considerably lower than previously estimated when quiescent periods are considered as well as the unrest/eruptive periods that occurred between 2005 and 2015 (Supplemental Table 8.1).[65,91–93]

Our compilation also includes new results for more than 50 volcanoes whose volcanic CO_2 fluxes have been quantified for the first time since 2013 (Supplemental Table 8.1), mostly due to the DECADE initiative.[14] While the number of volcanoes with a measured CO_2 plume has more than tripled since 2013 (33[15] vs. 102), the total (cumulative) CO_2 emitted is roughly two-thirds of the previous estimate (44 Mt CO_2/yr, or Tg/yr, vs 59.7;[15] Supplemental Table 8.1), largely due to the diminished estimates for the top volcanic CO_2 emitters by including inter-eruptive periods.

Given that our data set, by necessity, includes a number of spot measurements, the relative contribution of the most active volcanoes might continue to diminish as longer records are obtained at more volcanoes. As a first-order test, we compare the data set from direct measurements (many of which are spot measurements) with the CO_2 flux estimated from global compilations of the most active volcanoes that represent longer time frames. We utilize the 2005–2015 OMI volcanic SO_2 flux measurements[75] and combine these with CO_2/SO_2 ratios from Aiuppa et al.[73] and elsewhere where available to estimate CO_2 emissions from these sources (Supplemental Table 8.1). At the time of writing, 49 of the 91 volcanoes in the OMI data set[75] have their volcanic gas CO_2/SO_2 ratio signatures characterized (Supplemental Table 8.1), leaving a sizable gap in our knowledge. If only the CO_2 emissions from these 49 volcanoes are summed, the OMI-based data result in a total CO_2 emission of only 27 Tg/yr, compared to the 44 Tg/yr CO_2 by direct measurements (Supplemental Table 8.1). Overall, a reasonable correlation exists between the two data

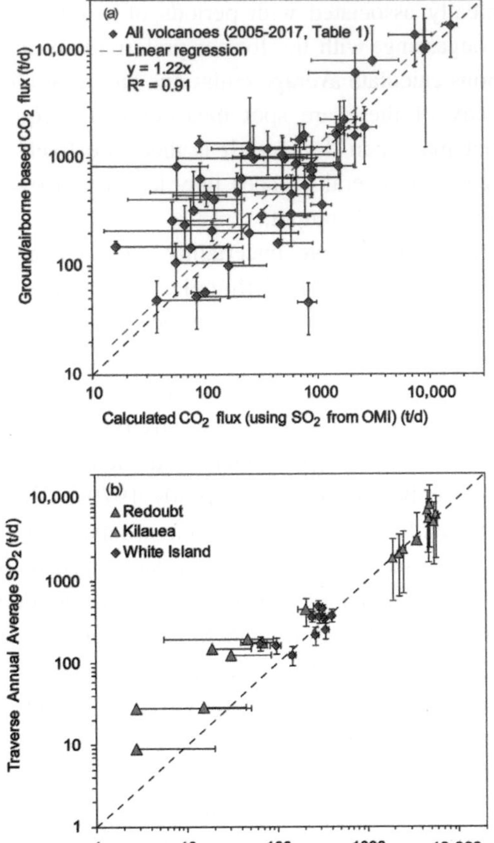

Figure 8.2 (a) Decadal average of CO_2 emissions from direct measurements for the period 2005–2017 (as available; see Supplemental Table 8.1) compared to that calculated from decadal average SO_2 emission from OMI and C/S ratio estimates. (b) Annual average SO_2 measurements from direct measurements for the years 2005–2015 when available (Redoubt,[65,91,93] White Island,[33,94,95] and Kilauea[96] compared to annual OMI estimates[75]). Error bars show one standard deviation over the period of observations in both (a) and (b), and uncertainties in OMI-derived values are propagated. Note that annual variability is low at open-system volcanoes.

sets, with scatter about the 1:1 line (Figure 8.2a), but each data set is associated with significant variability over the decadal period. The variability in the decadal average CO_2 flux is mirrored by the spread of the SO_2 annual averages over a decadal period (Figure 8.2b), pointing to inherently large temporal variability in the emission rates from active volcanoes over multiyear periods (see also Section 8.6.1) and suggesting that the spread in CO_2 data could be related to different observation periods. This lies in contrast to variability in SO_2 emission on an annual basis from the open-vent volcanoes; where frequent measurements have been made, variability is low and data cluster around the

1:1 line. This suggests that OMI-derived estimates of CO_2 emissions are accurate when the C/S ratios of volcanic gases are measured frequently. Future work should focus on a rigorous, systematic inter-comparison study between satellite and ground-based SO_2 flux data sets and on capturing the temporal variability in volcanic gas chemistry.

Given the importance and potential further use of the long-term OMI SO_2 data sets[75,77] for estimating global CO_2 emission rates, it is important to note that many of the "passive" degassing measurements[75] in fact represent eruptive periods. For instance, the OMI decadal data set omits eruptive emissions using a threshold SO_2 amount,[75] which excludes large-scale explosive emissions, but here we show that the data set includes emissions from eruptive and inter-eruptive periods. If we consider passive degassing to be degassing in the absence of eruption, we can compare the GVP volcanic eruption database[97] with the OMI SO_2 degassing data set. On an annual basis, 10 of the 91 volcanoes (11%) reflect true passive degassing such that the volcanoes did not experience an eruption between 2005 and 2015. Furthermore, 24% of the volcanoes erupted at least once every year; and, in any given year, at least half of the volcanoes experienced an eruption during this decade (the minimum number was 45 volcanoes erupted in a given year). For comparison, on average of 83 ($\pm$1.6) volcanoes experienced a non-zero SO_2 flux in a given year, suggesting that roughly half of the volcanoes might be considered to be passively degassing on an annual basis.

It is also important to note that the OMI SO_2 data set almost exclusively represents volcanoes with predominantly basaltic or basaltic–andesite compositions and is thus not globally representative of Earth's more silicic systems. Basaltic systems have been shown to have the shortest repose periods (averaging $<$1 year), whereas basaltic–andesite systems can show much longer periods of repose (averaging roughly 20 years),[98] with the latter approaching the length of time that the volcanic gas community has been making CO_2 emission rate measurements. Thus, we suggest that future work should also focus on analyzing time series data that span eruptive cycles at the more silicic of this set of dominantly mafic volcanoes to understand how emissions vary over multi-decadal time periods that include both periods of repose and open-vent degassing.

8.4.2 CO_2 Emissions during Explosive Eruptions

Our present knowledge of CO_2 emissions from large, explosive eruptions is limited owing to both proximal hazards and instrumental challenges in measuring volcanic gases during such events. Direct assessment of CO_2 emission rates during explosive eruptions has been achieved on rare occasions, either from airborne plume measurements (e.g. during the 2009 Redoubt eruption,[65] though here the most explosive events were not captured) or by coupling real-time FTIR spectroscopy of CO_2/SO_2 ratios in eruptive gases with the UV-sensed SO_2 flux (e.g. during the 2010 Eyjafjallajökull eruption[99]). In other cases, bulk CO_2 emissions from explosive eruptions have been estimated by combining CO_2/SO_2 data from *in situ* measurements with satellite-based SO_2 data or by modeling the pre-eruptive vapor-phase composition.[100–102] Such techniques yield, for instance, estimates of ~10 and ~50 Mt CO_2 for the 1980 Mount St Helens and 1991 Pinatubo eruptions, respectively.[103] These estimates imply CO_2/SO_2 mass ratios of ~3–10 in the eruptive emissions, although higher ratios cannot be excluded.[100] Indeed, mass budgets for these and other explosive eruptions of silicic magmas strongly suggest pre-eruptive accumulation of a CO_2-rich vapor phase.[100,104] Gas accumulation in silicic magma reservoirs between eruptions can result from the second boiling of vapor-saturated crystallizing magma and/or volatile supply from basalt underplating.[105] Magmatic vapor may migrate to the roof zones of reservoirs via gas transport through channelized flow in crystal-rich mush.[106] Because of its low solubility, CO_2 becomes preferentially enriched in the accumulating vapor phase. Therefore, high CO_2/SO_2 ratios[107] and high CO_2 and SO_2 fluxes[108,109] can be expected during the initial phases of explosive eruptions that tap the gas-rich upper levels of magma reservoirs.

Measuring volcanic CO_2 emissions during explosive eruptions will continue to be challenging regardless of whether one is using *in situ* or satellite techniques. *In situ* measurements of explosive eruption plumes are hampered by proximal volcanic hazards and high atmospheric ash loadings,[101] and spaceborne CO_2 measurements will also be hindered by volcanic ash. However, UAV (or drone) technology and improved satellite SO_2 instruments (e.g. TROPOMI) hold great promise to improve measurements of explosive volcanic CO_2 emissions in the coming decade.

8.4.3 CO_2 Emissions from Dormant Volcanoes

It has been recognized for some time that volcanoes that are dormant (defined here as not erupting but likely to erupt again) emit significant amounts of CO_2.[110,111] These volcanoes may support smaller CO_2 plumes that may or may not contain SO_2 derived from fumarolic emissions, or they may host large regions of diffuse degassing (Supplemental Table 8.3) related to silicic volcanism that have long repose times typical of caldera settings. Below we review each source separately.

8.4.3.1 Small Volcanic Plumes: Fumarolic Contributions

Volcanoes that produce small plumes, or CO_2 plumes in the absence of significant SO_2 emission, are more difficult to characterize for their CO_2 emission rate than those that have strong SO_2 plumes. Roughly 40%[44] of the 102 direct CO_2 flux measurements listed in Supplemental Table 8.1 are from volcanoes where the volcanic plume was not detected by OMI[75] and thus fall in this category (we refer to these as "small plumes," although some do not have low CO_2 emissions). The CO_2 emissions associated with these volcanoes range from 13 to nearly 1500 t/d, with an average of 300 t/d ($1\sigma = \pm 360$) and a median of 147 t/d, excluding the large emission from Oldoinyo Lengai (Tanzania). Some of the largest CO_2 emissions are from active volcanoes that host crater lakes (e.g. Taal and Pinatubo, Philippines; Ruapehu, New Zealand; Supplemental Table 8.1) and from better-studied sections of arcs in the United States (Cascades and Alaska), Central and South America, and Indonesia (Supplemental Table 8.1). Where airborne methods and easy access have allowed for measurements, the data show that such emissions are common and are likely widespread in many arcs globally (Figure 8.3a).

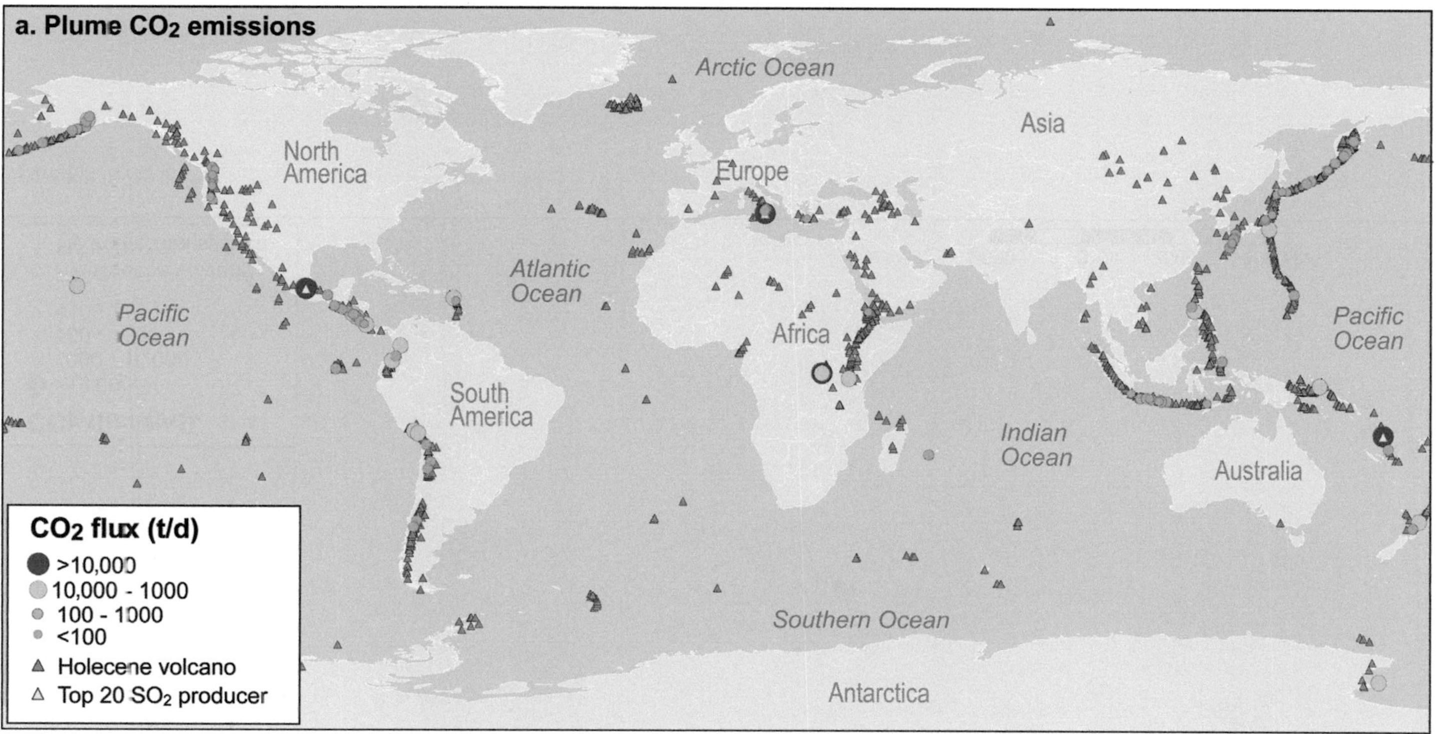

Figure 8.3 Measured CO_2 emissions from (a) active volcanoes (plume emissions; Supplemental Table 8.1) and (b) diffuse degassing sources (Supplemental Table 8.2). All plume emissions are from Holocene volcanoes. Diffuse emissions are from volcanic sources with a broader period of activity; hydrothermal locations are often colocated with active volcanoes (i.e. Holocene volcanoes). Volcano locations from Ref. 97, top 20 SO_2 producers from Ref. 75, and hydrothermal system locations from a modified version of the database from Ref. 113.

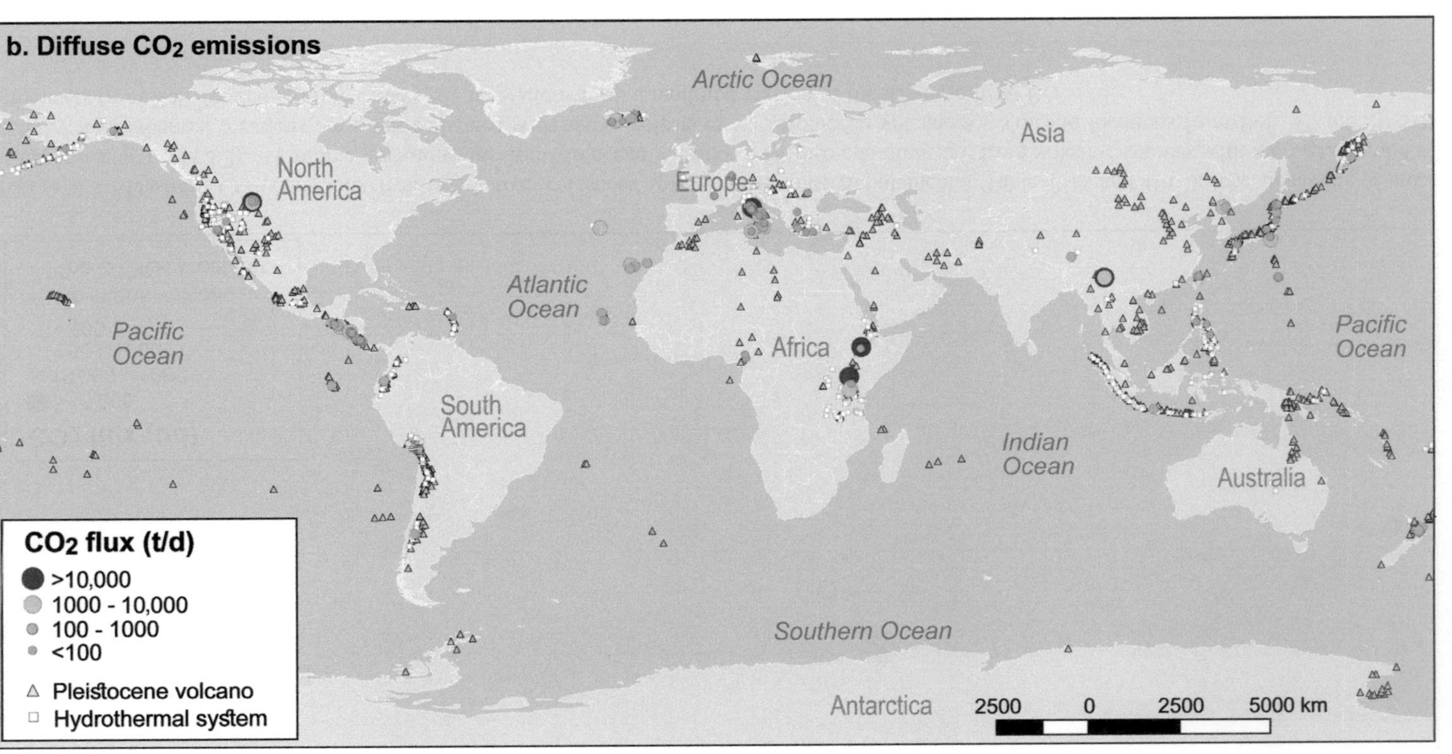

Figure 8.3 (*cont.*)

If we simply sum the CO_2 emissions from these "small" volcanic plumes, they amount to only 5 Tg CO_2/yr of the ~44 Tg/year in Supplemental Table 8.1, demonstrating the dominance of strongly emitting volcanoes in the data set. However, if we take the volcanoes that have erupted in the last 100 years (n = 407) minus the 83 volcanoes that are shown to be degassing each year (Section 8.4.1), resulting in 324 volcanoes globally, and assume that each outputs on average 300 t/d, this equates to ~35 Tg CO_2/yr. If we use the median (147 t/d) instead of the average, this results in 17 Tg CO_2/yr. This exercise suggests that the emissions from volcanoes with small plumes (or in the absence of SO_2 emission) could potentially emit a similar order of magnitude of CO_2 globally as volcanoes whose SO_2 plumes were detected by satellite. This result, if robust, would potentially stand in contrast to recent regional studies in Japan[57,112] that suggest that the global volatile budget is dominated by the high SO_2-emitting volcanoes, although this study[57] recognized that the data set lacked comprehensive measurements for the less active, diffusely degassing volcanoes. While challenging, more work is needed to verify the global contribution of CO_2 emissions from volcanoes that do not emit satellite-detectable SO_2.

8.4.3.2 Diffuse Emission of CO_2: Hydrothermal Systems, Calderas, and Continental Rifts

Our understanding of the magnitude of the diffuse CO_2 flux from volcanic and magmatically active regions on Earth continues to evolve with each year of new measurements, and we now understand this to be a significant outgassing source. What we show here is that diffuse CO_2 outgassing from calderas and dormant volcanic regions can rival outgassing from actively erupting volcanoes (Figures 8.3, 8.4, and 8.5). Quantification of such fluxes on global scales, however, remains a great challenge. Available flux data for diffuse gas emissions have been gathered, together with data from active volcanoes, into a database (the MaGa web database: www.magadb.net[114]). The data show that there are large regions where measurements have not yet been made (e.g. South America, Kamchatka, and Southeast Asia; Figure 8.3b). As new discoveries of large emission sources have been made in the last 10 years in areas with large magmatic intrusions and concentrations of hydrothermal systems (e.g. the East African Rift (EAR) and the Technong volcanic province, China), we expect that additional important areas will be located in the future.

If we compare the distribution of measured diffuse emissions of CO_2 (Figure 8.5a and Supplemental Table 8.3) with plume emissions from active volcanoes (Supplemental Table 8.1), we find significant overlap and similarity in the emission rates, especially at higher rates. The diffuse emission data tend to be bimodal, with a larger population at low emission rates (Figure 8.5a), but the lack of measurements at low CO_2 emissions for active volcanoes may simply reflect a sampling bias due to method limitations (e.g. fumarolic contributions and plumes below SO_2 satellite detection limits). Volcanic systems that have diffuse emission rates between 100 and 500 t/d are most common, representing 30% of the data, with an additional 20% falling between 500 and 5000 t/d. The highest CO_2 emission rates are for large magmatic systems (e.g. Yellowstone in the United States, the Tengchong Volcanic Field in China, the Tuscan Roman degassing structure (TRDS) and Campanian degassing structure (CDS), and the EAR system; Supplemental Table 8.3). Although

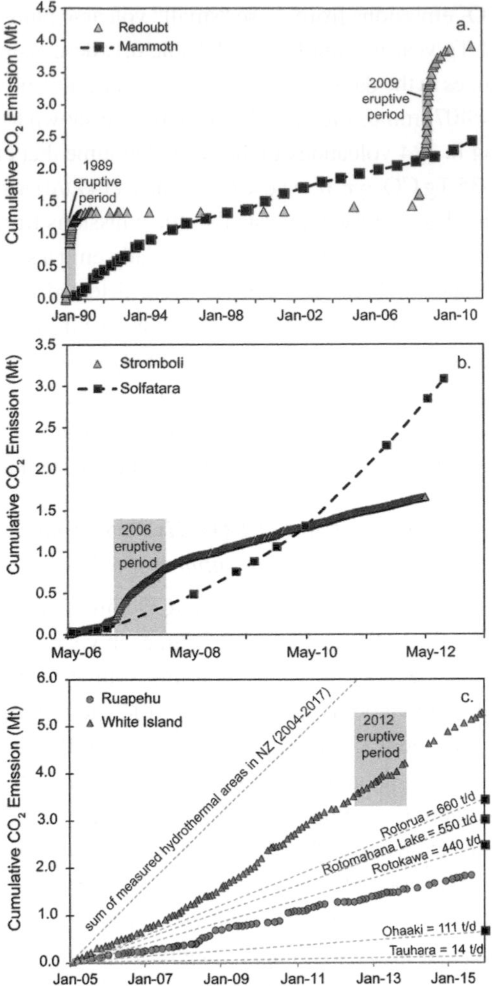

Figure 8.4 Cumulative CO_2 emissions for some of the best-studied volcanoes in the world showing the comparison of vent emissions (triangles) to diffuse emissions (squares and dotted lines). (a) Emissions from Redoubt volcano and those of Mammoth mountain are roughly equal over 20-year time frames. Redoubt data from Refs. 65, 91, 93, and 115 and Mammoth data from Ref. 116. (b) Solfatara data from Ref. 117, Stromboli data from Ref. 118. (c) White Island data from Refs. 33 and 119, Ruapehu data from Refs. 32, 94, and 95, Taupo Volcanic Zone diffuse degassing data from Refs. 120–123.

estimates for these systems have large uncertainties, high emission rates are consistent with high heat fluxes and voluminous magmatism. As run-up time, or period of unrest prior to an eruption, is positively correlated with the repose period between eruptions,[98] it should not be surprising that some of the largest and longest-lived volcanic systems (e.g. silicic calderas systems) can produce some of the largest CO_2 emissions globally (e.g. Yellowstone, Campi Flegrei, and Rotorua; Supplemental Table 8.3).

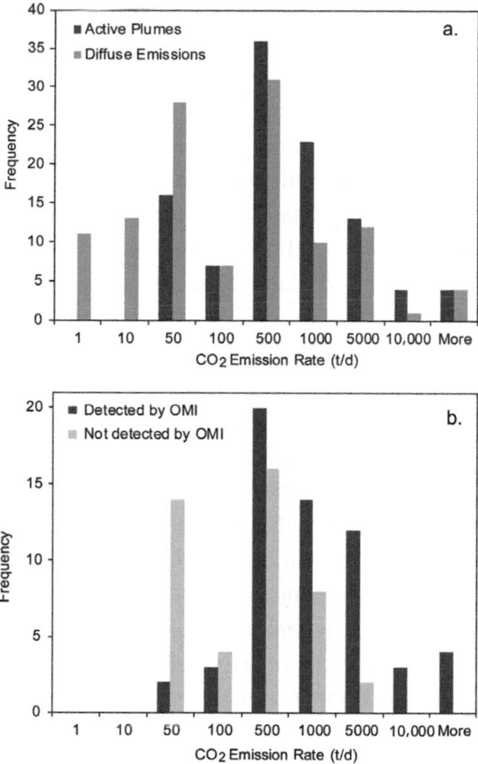

Figure 8.5 Distribution of CO_2 emission rate data for (a) active volcanic plumes (Supplemental Table 8.1) and diffuse emissions (Supplemental Table 8.2) and (b) the active volcano plumes that were detected for passive degassing by OMI[75] and those volcanoes that were not.

Unrest in caldera systems is common,[124,125] and thus using unrest catalogs may allow us to understand and constrain this likely significant CO_2 source better. Globally, there are 446 caldera systems, of which 225 have erupted in the Quaternary and 97 in the Holocene.[124,125] At caldera systems, unrest is understood to be driven by the influx of mafic, volatile-rich magma to the base of relatively shallow reservoirs containing vapor-saturated magma (Figure 8.1). In a recent study of the best-monitored caldera systems in the world, such episodes of magma intrusion were found to be the fundamental driver of unrest,[124] and 71% exhibited changes in degassing with unrest. At mafic calderas, unrest often proceeded to eruption, whereas felsic systems were thought to have a higher capacity to accommodate magmatic intrusions without leading to an eruption.[124] The hydrothermal systems and gas reservoirs that often lie above such intrusions act as buffers, such that changes in gas emission at the surface are often delayed by some time (sometimes years) from the time when fresh magma is intruded.[116,124,126] However, the time-averaged release of CO_2 from these systems is roughly similar to that of many active volcanoes (Figure 8.4). The main difference may then be the contrast in available pathways for gas release: active (and often mafic) volcanoes maintain open conduits with high permeability (Supplemental Table 8.1) and dormant (often silicic) volcanoes and caldera systems release gas along faults and fracture networks with lower permeability than open conduits, resulting in regions of diffuse emissions (Supplemental Table 8.3).

Here, for the first time, we show that net release of CO_2 over time (decades) from areas of diffuse degassing for several well-studied systems that have not erupted in recent history (e.g. Mammoth Mountain, USA; Solfatara, Italy; Rotorua, New Zealand) can rival that of active volcanoes that have experienced an eruption recently (Etna and Stromboli, Italy; Redoubt, USA; White Island, New Zealand). The active volcanoes are all on the list of top SO_2-producing volcanoes in the world.[75] For example, the CO_2 emission from Mammoth Mountain (last eruption 70 ka[116]) over ~20-year timescales is similar to that of Redoubt Volcano (Figure 8.4a). In this case, the long-term average emission is fundamentally controlled by the period of observation (e.g. note that the 2005–2015 average emission is an order of magnitude higher in Supplemental Table 8.1 than the average over two decades portrayed in Figure 8.4a). Another example is the long-term cumulative emission from Solfatara in Campi Flegrei, Italy, which exceeds that of Stromboli Volcano (Figure 8.4b). And finally, the emission from White Island is dwarfed by the sum of the cumulative CO_2 emissions from dormant volcanoes in New Zealand (Figure 8.4c).

It is not known how many hydrothermal systems exist on Earth, but the majority are associated with areas of either present or past volcanism. Our current summation of the extent of diffuse degassing from dormant volcanoes is ~64 Tg/yr (Supplemental Table 8.3), which is similar to that published previously,[15] but our estimate does not include large-scale extrapolated values for Indonesia–Philippines and the Subaerial Pacific Rim.[127] Our data highlight the importance of several large regions of localized hydrothermal activity. While work in several areas has already begun and has yielded valuable initial data (Yellowstone, USA;[128,129] Campi Flegrei, Italy;[117,130] the TRDS and CDS, central Italy;[131,132] the EAR;[133,134] and the Taupo Volcanic Zone (TVZ), New Zealand[120,122,123]), we expect several other areas will also be globally important for their CO_2 emissions. Guidance for where these areas might be located can be gleaned from global assessments of geothermal energy reserve. In a 2016 review by the World Energy Council,[135] the five nations with the highest potential geothermal generating capacity were the USA, the Philippines, Indonesia, Mexico, and New Zealand. Several of these countries have had very few CO_2 surveys to date. On the other hand, countries such as Italy and Japan (ranked 6th and 10th, respectively, on the list of top nations) have had considerably more studies.

Here, we attempt to estimate the number of hydrothermal areas worldwide by building on a list of geothermal systems capable of power production,[113] adding in hydrothermal areas located in Alaska, Kamchatka, and Peru. This results in ~670 hydrothermal regions worldwide (Figure 8.2). The average of all diffuse emissions from localities that have not experienced eruptive activity since 2000 is 340 ± 628 t/d, demonstrating a positive skew in the population. We omitted hydrothermal areas on volcanoes with eruptions since 2000 because we did not want to include anomalous data due to recent volcanism. This average also does not include large-scale magmatic emission estimates (Supplemental

Table 8.4) because these areas are also anomalous on a global scale and are not representative of individual hydrothermal regions. While admittedly simplistic, applying this average to the 670 hydrothermal regions worldwide would result in 83 Tg CO_2/year, or an additional 30% over the current summation of the diffuse data. We consider it likely that this estimate is conservative given: (1) that our data only represent ~135 of 670 localities; (2) these data largely do not include groundwater contributions that may be similar in magnitude to diffuse emissions (see Ref. 57 and references therein); and (3) the discovery of other large systems globally (such as Yellowstone) could add significantly to the budget (currently, large magmatic provinces sum to 75 Tg CO_2/yr; Supplemental Tables 8.3 and 8.4).

The TVZ is a region that deserves extra attention given its unique tectonic setting and high heat flow and because arc-scale extrapolations based on studies from the TVZ[127] continue to be used for global compilations and comparisons of CO_2.[15,134] The TVZ is an intra-arc rift zone that hosts over 20 separate hydrothermal regions with heat flux greater than 20 MW,[136] and many of these regions are exploited for geothermal energy. Previous estimates of the CO_2 output of the TVZ as a whole have been extrapolated based on the heat flux and the CO_2 content of upwelling fluids. However, diffuse CO_2 flux at the surface[120,122,123] for many of these systems greatly exceeds previous CO_2 emission estimates for these areas, without including fumarolic contributions. As an example, the emission rate estimated for the Rotorua hydrothermal system alone (estimated at ~1000 t/d, including sub-lacustrine degassing;[123] Supplemental Table 8.3) is nearly equal to that previously estimated for the whole TVZ (~1200 t/d[127]). To date, 7 of the 18 hydrothermal systems have been measured for CO_2; together, they equal a total of ~2000 t/d. Further investigation is required to determine why the previous estimates for the TVZ hydrothermal systems were low, but likely this results from underestimating the CO_2 content of the deep hydrothermal fluids or degassing from gas reservoirs at depth.[117,137] In any case, arc-scale estimates for other regions on Earth should not be based on TVZ data,[127] and continued effort should be placed on measuring the total diffuse CO_2 output from typical arcs and high-heat-flow regions around the world.

The EAR, also deserving of extra attention due to its high global output, is a series of rift valleys that extend 4000 km from the Afar region in the north to Beira in Mozambique in the south.[138] The system is split into an eastern branch, which hosts the Main Ethiopian Rift (MER) in the north and the Kenyan rift in the south. In these two sectors alone, there exist 36 volcanoes and 28 hydrothermal areas. The western branch of the EAR is characterized by a lack of recent volcanism relative to the eastern branch, but still hosts a number of geothermal prospects as well as Nyiragongo, a major emitter of CO_2. Several recent studies attempted to estimate the diffuse CO_2 flux from the EAR. One study focused on the centers of volcanic activity in the MER and extrapolated that to between 3.9 and 33 Tg CO_2/yr for the EAR.[134] A second study focused on tectonic degassing away from active volcanic centers and estimated 38–104 Tg CO_2/yr for the EAR, not including focused degassing through the active centers.[133] For our estimate, we use the midpoint of the range presented by Hunt et al.[134] because our aim is to estimate volcanic degassing; this value is one-third of the total CO_2 emissions from all diffuse sources in our compilation (Supplemental Table 8.3). While the estimates of EAR fault-related degassing[133] are not volcanic per se, isotopic evidence suggests there exists a significant flux of mantle-derived CO_2 to the atmosphere through these structures, and using these data would increase significantly the global contribution of the EAR. We caution that both studies found relatively few measurements of modest to high CO_2 flux in faulted or hydrothermal areas and that these results were then extrapolated over extensive regions. Significant uncertainty is associated with such large-scale extrapolations, particularly when diffuse CO_2 flux can vary on meter scales. However, it is clear the EAR is a very important region for global CO_2 emissions, and more work is needed to quantify the flux of CO_2 from this and other areas of continental rifting/extension that support volcanism and hydrothermal activity, such as the Rio Grande Rift in New Mexico and the Rhine Graben and the Eger Rift in Central Europe. Such areas could potentially add 30–40 Tg CO_2/yr (or 8–11 Tg C/yr) to global budgets, and potentially be on the same order as global arc fluxes.

8.5 The Next Iteration of Global Volcanic CO_2 Emissions

As our understanding of the distribution and magnitude of volcanic and magmatic CO_2 degassing evolves, so will our ability to estimate accurately the present-day global CO_2 emission from these areas. As a culmination of the DECADE program, scientists came together in May 2018 to constrain better the total global CO_2 flux from volcanic regions, as well as corresponding uncertainties. Here, we follow simple methods based on the extrapolation of measured data to determine a global subaerial volcanic CO_2 budget (Supplemental Table 8.4). Our methods are similar – and thus comparable – to previous studies,[15,57] but future work should focus on a rigorous statistical analysis of the data and more complex extrapolation procedures that lie beyond the scope of this chapter.

We break down the subaerial volcanic budget into three main categories: (1) passive degassing for active volcanoes, dividing these into those that have been detected by OMI and those that have not; (2) diffuse emissions from both active and dormant volcanoes, with groundwater contributions (not estimated) and large degassing provinces as separate categories; and (3) eruptive emissions. We first calculate the average of the measurements of CO_2 flux from the volcanoes that have been measured using ground-based or airborne techniques (i.e. measured directly) that were detected by OMI[75] (58 of 91 volcanoes in Supplemental Table 8.1) and apply the average emission from these volcanoes (1730 ± 440 t/d, mean and standard error) to the 83 volcanoes (Section 8.4.1) that were degassing and detected by OMI globally on an annual basis between 2005 and 2015, yielding 52 ± 13 Tg CO_2/yr (Supplemental Table 8.4) for this set of volcanoes. Multiplying the average of the remaining 33 volcanoes from Supplemental Table 8.1 (i.e. those that do not emit SO_2 in large enough quantities to be detected by OMI – the average CO_2 output of these volcanoes is 300 ± 68 t/d; Supplemental Table 8.4) to 324 volcanoes results in 35 ± 8 Tg CO_2/yr. The sum of these results in 88 ± 21 for passive degassing from active volcanoes (Supplemental Table 8.4). How the value of 324 volcanoes was determined is discussed in

Section 8.4.3. The uncertainty in this number is difficult to quantify without global arc-wide assessments of the numbers of expected degassing volcanoes that have not been detected by OMI. To put this number in context, there are 169 active volcanoes in the United States, and 81 (47%) have notable degassing as determined through visual surveys.[139] Of these, 28 (17%) are thought to have plumes large enough for airborne measurements, but only 8 (4%) were detected by OMI.[75] Thus, 90% of the US volcanoes that emit CO_2 and 71% of the US volcanoes that have plumes large enough for airborne surveying were not detected for passive degassing of SO_2 by OMI. Similarly, of the 19 persistently degassing volcanoes in Japan[112] during the period of OMI measurements, only 7 were detected by OMI[75] (i.e. 63% were not detected by OMI). If this relationship were to hold globally (i.e. that 63–71% of volcanoes with significant CO_2 emissions are not detected by OMI), this would suggest that 245–313 volcanoes with notable plume emissions worldwide remain undetected by OMI, which is similar to the value we used. We suggest that completing a global assessment of which volcanoes are degassing, and the nature of that degassing, based on visual assessment and documented activity in the GVP database would lead to a much more accurate estimate of the number of degassing volcanoes globally than the methods used here. However, it must be considered that additional invisible or nearly invisible emissions of CO_2 may also exist.[140–142]

As discussed above, the CO_2 contribution from explosive eruptions is poorly constrained. Here, we estimate eruptive emissions of CO_2 (Supplemental Table 8.2) by combining recent decadal-scale (2005–2018) SO_2 fluxes derived from satellite measurements of eruptions[143] with the most representative CO_2/SO_2 ratios measured at corresponding volcanoes, and separate these data into explosive and effusive events (Supplemental Table 8.2). In previous estimates, CO_2/SO_2 ratios were assumed to have uniformly high values of 10 and 7 in the pre-eruptive vapors of silicic and basaltic magmas, respectively.[57] Such an assumption is reasonable for initial phases, but is not necessarily valid for the whole eruption length. The figure we obtain by using measured ratios (0.6 Tg/yr; Supplemental Table 8.2) is much lower than previously estimated based on theoretical ratios (7 Tg/yr[57]). We anticipate that our value is likely underestimated and that the true answer may lie between these two values. Regardless, the estimates show that explosive emissions are minimal compared to the passive degassing estimates. CO_2 emissions from effusive eruptions in the same period (Supplemental Table 8.2) are inferred by subtracting explosive CO_2 emissions from the total CO_2 load from all eruptions. In this case, the calculated contribution is similar to previous estimates (1.3 compared to 1 Tg/yr CO_2 for effusive eruptions[57]), and again the contribution is a small fraction of the total subaerial budget.

Our calculations suggest that diffuse degassing of CO_2 from volcanoes is only slightly lower than that from active volcanic vent emissions, with diffuse emissions estimated at 83 ± 15 Tg/yr. However, combined contribution of 170 Tg/yr CO_2 from diffuse degassing from all volcanic–hydrothermal systems, including groundwater contributions and degassing related to large regions of intrusive magmatic activity is higher (Supplemental Table 8.4). This value is likely an underestimate given that fumarolic contributions (focused venting/small fumaroles) in regions of diffuse degassing are often not quantified as part of the estimation of degassing across such regions, and because groundwater contributions are largely unquantified. Furthermore, we anticipate the discovery of additional large emission sources as many of the countries with the highest potential for geothermal power generation have few measurements.

In total, we conservatively estimate global subaerial volcanic CO_2 emissions to lie between ~220 and 300 Tg CO_2/yr, and between ~280 and 360 Tg CO_2/yr including the contribution of MOR (Supplemental Table 8.4), based on currently known sources. Our estimates are lower than those published by Burton et al.,[15] but higher than previous estimates for global subaerial sources.[111,144]

8.6 Temporal Variability of Volcanic Degassing

8.6.1 Comparison of the Temporal Variability of CO_2 Emission from Active and Less Active Volcanoes

Most of what we now know about the temporal evolution of CO_2 emissions from volcanoes has been learned in the last two decades. Advances in instrumental techniques now permit continuous, real-time monitoring.[51,56,145] The trends emerging from these data show that emissions vary dramatically with volcanic setting (Figure 8.1) and that timescales of observation are important for understanding the relative contributions from different systems. Active volcanoes with open-vent degassing such as Stromboli and Mount Etna (Italy) show orders of magnitude variability over very short time frames (Figure 8.6) correlating with magma supply and eruptive activity.[53,54,146] In such active, often mafic systems, volatiles reach the atmosphere via magma convection, permeable gas flow, or bubble rise through low-viscosity melts.[18,147] Despite short-term variability, the long-term average output at these volcanoes stays relatively constant over multiyear periods (Figure 8.4). In some cases, paroxysmal-type activity will increase emissions for over a year before returning to the long-term average (see Stromboli, 2006; Figures 8.4b and 8.6). Minor eruptive activity, on the other hand, can be difficult to discern in long-term

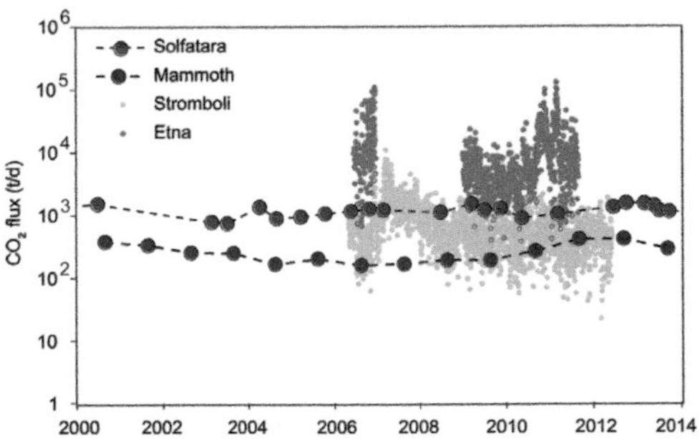

Figure 8.6 Temporal variability of CO_2 emissions from some of the best-studied volcanoes in the world. Emissions from open-vent volcanoes vary dramatically in time, whereas diffuse emissions are buffered and show less variability. Solfatara data from Ref. 117, Mammoth data from Ref. 116, Stromboli data from Ref. 118, and Etna data from Ref. 148.

trends. For instance, White Island and Ruapehu (New Zealand) demonstrate steady emissions over decadal periods during which eruptive activity is barely detectable (Figure 8.4c).

Closed-system volcanoes, or volcanoes that oscillate between closed- and open-vent degassing, can show dramatic variability in emissions over periods of years related to magma intrusion and variations in conduit permeability. Increases in CO_2 emission rates are typically associated with eruptive activity (Redoubt; Figure 8.4a) and sometimes when intrusions occur without eruption.[92] Periods of unrest can last months to years and are often accompanied by increases in emissions that then decrease exponentially following eruptive activity. Such behavior has been better documented for SO_2 emissions,[89,149] but is mirrored by CO_2 emissions where measured (e.g. Redoubt in 1989 and 2009;[65] Figure 8.4a).

Over an entire arc, the dominant volcanic CO_2 producers may vary over decadal timescales, with some volcanoes becoming more or less active. A recent compilation of data for the Central American Arc estimated an arc-scale CO_2 output one order of magnitude higher (22,500 $\pm$ 4900 t/d[56]) than previous estimates, owing to the reactivation of Turrialba Volcano, as well as an increase in CO_2 flux from Momotombo and Masaya volcanoes over the previous decade. Other arc segments have had similar changes to the overall degassing budget due to the reactivation of particular volcanoes (e.g. Miyakajima in Japan[112]).

Finally, large-caldera systems are thought to be underlain by silicic magma bodies, and in turn underplated by mafic magma.[109] The CO_2 emissions from such volcanoes, often modulated by large hydrothermal systems, show much less variability over annual or even decadal scales than emissions from active volcanoes (Figures 8.4 and 8.6). Where long-term measurements are available, small variations in the CO_2 output in these systems often follow a geophysical manifestation of magma movement at depth[116,150] whereby the transport of the gas to the surface is buffered by the overlying crust.

8.6.2 Using the Temporal Variability of CO_2/SO_2 in Volcanic Gas for Eruption Forecasting

It has been shown that the relative proportions of C and S change prior to and during eruptive activity.[151,152] Owing to the low solubility of CO_2 in silicate melts,[153] the magmatic vapor phase typically has a high molar C/S ratio at depths of $>\sim5$ km in the crust,[102] and then C/S decreases with magma ascent as more S exsolves from the magma.[141,146] With more frequent monitoring of pre-eruptive volcanic gas using Multi-GAS, we now know that an elevated C/S in gas emissions is common prior to the onset of eruptions.[51,55,154] In Figure 8.7a, we show variability of C/S over various timescales for 12 episodes at 7 volcanoes. One can observe that C/S ratios increased to between 15 and 43 in the months to hours preceding eruption at five well-monitored basaltic volcanoes, whereas the long-term C/S signature of shallow degassing at these volcanoes typically lies between 2 and 7 (Figure 8.7a to j). Such trends are often interpreted as the migration of deeply sourced gas bubbles prior to magma ascent.[53,155]

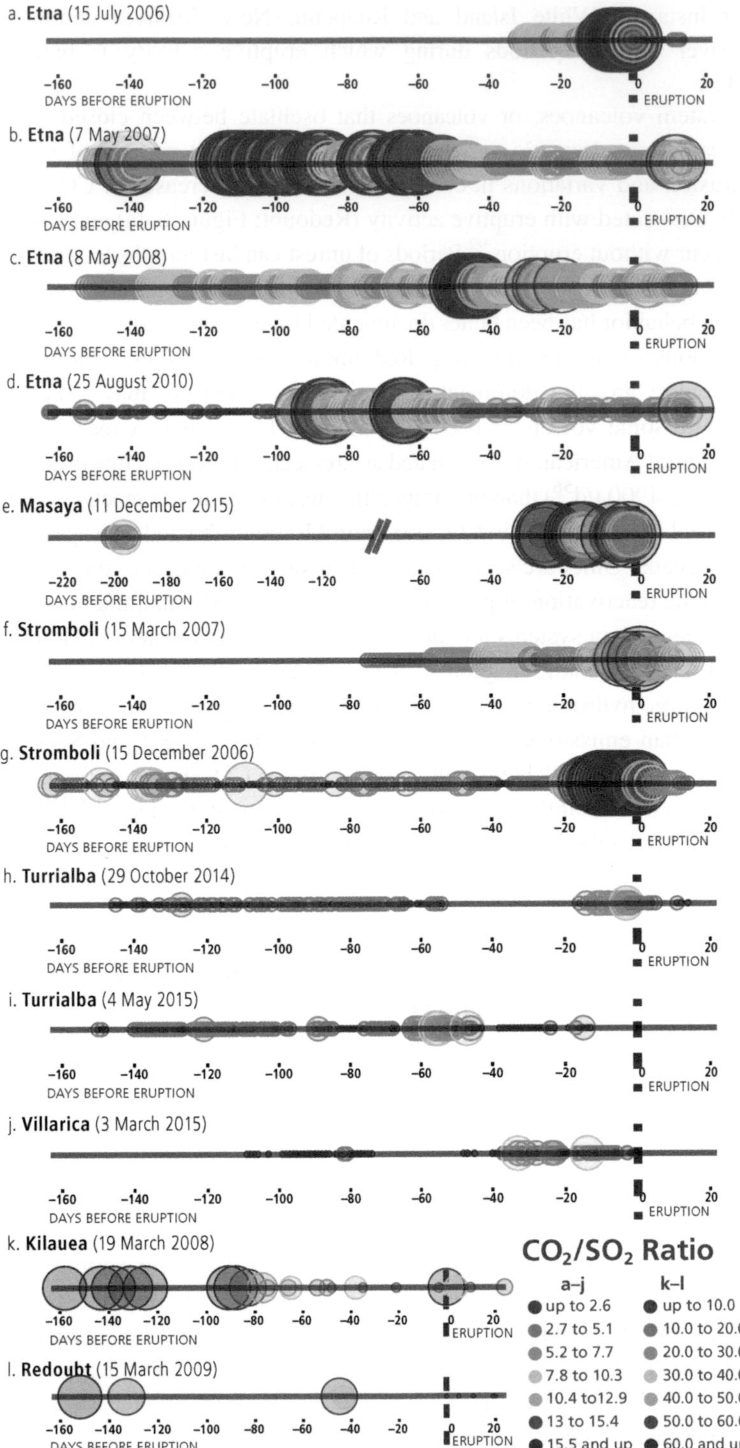

Figure 8.7 (a to j) Trends in C/S ratio observed at basaltic volcanoes monitored with Multi-GAS instruments, with elevated C/S documented in the months to hours prior to eruption. (k and l)

A second set of C/S ratio observations has been made over somewhat longer timescales and is related to degassing of deep-seated magmas prior to ascent (Figures 8.7k and l). Months- to years-long trends in C/S were observed at both Redoubt[141] and Kilauea[140] volcanoes prior to eruption (Figure 8.7b). In both cases, the C/S ratio was very high compared to the data observed at the other volcanoes, reaching values between 80 and 200, and both were associated with no visible plume (Figure 8.8). Documentation of this type of degassing is rare, but the occurrence is likely not rare. In the case of Redoubt, the pre-eruptive degassing of CO_2 only amounted to roughly 15% of the total budget,[65] but for Kilauea nearly 30 Mt of gas escaped prior to eruption. Similar patterns of pre-eruptive gas release might be inferred, for instance, from the Holuhraun/Bárðarbunga eruption, where a total of 9.6 Mt SO_2 and 5.1 Mt CO_2 was emitted over the course of the eruption.[157] The low bulk C/S ratio (0.7) compared to most high-temperature volcanic gases[73] might imply that substantial amounts of CO_2 degassed before monitoring began. In fact, a new study discovered significant degassing from glacially covered Katla volcano (~37 kt/d CO_2) in the absence of a visible plume, unusual geophysical unrest, or S gas emission.[142] These studies highlight a gap in our ability to detect CO_2 degassing from volcanoes without dedicated airborne surveys downwind of potentially degassing volcanoes.

8.7 Sources of Carbon Outgassed from Volcanoes

Carbon outgassed at subduction zone volcanoes is sourced from the mantle, the subducted slab, and the overlying lithosphere (including the crust),[158–162] whereas CO_2 released from MORs and hot spots is dominated by mantle carbon.[144,161] The carbon isotopic composition ($\delta^{13}C$) of the depleted MOR mantle (DMM) is -5 ± 1[163,164] and that of plumes is documented as -3.1 ± 1.9 (high-temperature fluids from Iceland[165]) and –3.4‰ (Kilauea summit gas[166]), whereas the subcontinental lithospheric mantle likely contains carbon of composition between –3.5‰[133,167] and –6‰.[168] Research on CO_2 sources in subduction-zone volcanic gases has emphasized the role of carbon release from subducted sediments and carbonates and has shown that the type of material subducted imprints a carbon isotopic and $C/^3He$ signature on the discharging gases. This approach, combined with CO_2 fluxes from volcanoes, led to the development of volatile budgets in subduction zones and implies that more carbon is subducted than what is currently released by volcanoes, leading to the transfer of carbon into the deeper mantle and beyond the zones of arc magma generation.[70,79,169,170] Accumulation of subducted carbon below the arc crust or

Figure 8.7 (*cont.*) Observations of elevated high C/S ratios in the years prior to eruption. Decreasing trends in C/S ratios were observed in the last 100 days prior to eruption. Both the symbol size and color scale with C/S ratio, with larger and warmer symbols relating to higher C/S ratios. Data from Refs. 118, 140, 141, 148, 155, and 156. At Masaya, the volcano (e) did not erupt, but rather experienced the opening of a new lava lake. At Redoubt, three values in excess of 80 related to a period of SO_2 scrubbing in the month prior to eruption were removed – see Ref. 141 for details.

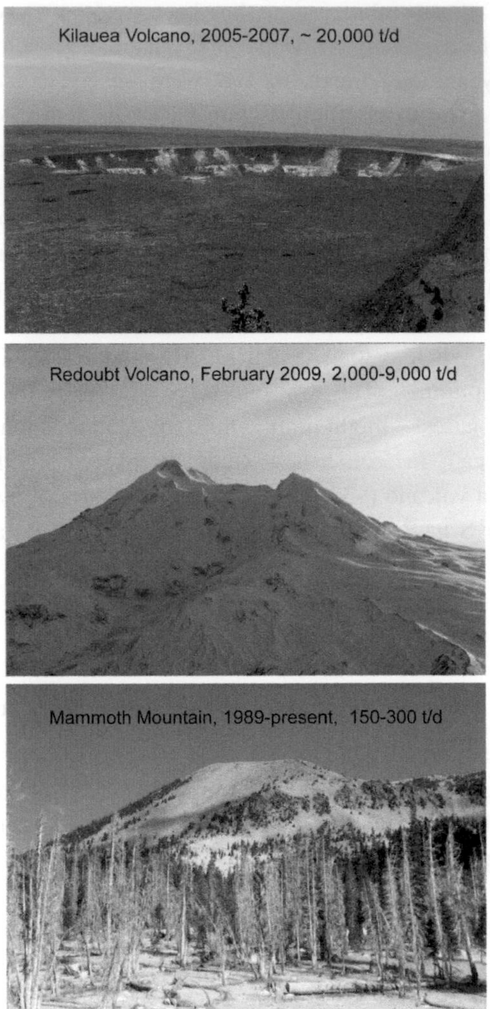

Figure 8.8 Images of volcanoes emitting significant quantities of volcanic CO_2 in the absence of a visible plume.

continental lithosphere has been suggested as a mechanism for long-term carbon storage, removing the requirement for carbon subduction into the deeper mantle to balance the input versus output budgets at arcs.[171] The extent to which this process occurs globally is poorly constrained, but it could significantly affect the carbon isotope composition of the mantle wedge and therefore the mantle component sampled by arc volcanic gases. Recently, researchers have highlighted, in addition to the subducted carbon source, the potential significance of carbon assimilation from the overlying crust in continental arc magmas as a major source of CO_2 degassing from volcanoes, both today and in the geologic past.[4,172,173] In particular, this crustally sourced carbon can have profound effects on the generation of arc magmas,[174] the explosive activity of a volcano,[173,175] and long-term climate change resulting from CO_2 release into the atmosphere.[172]

The approach most commonly taken when assessing the contribution of volatiles from the crust is to use helium isotopes, which range widely in volcanic gases from values that approach pure crustal sources of ~0.02 Ra, where Ra is the $^3He/^4He$ ratio of air at 1.4×10^{-6}, to 29 Ra in fluids discharged from hot spot hydrothermal systems and volcanoes.[176] In subduction-zone settings, $^3He/^4He$ ratios of gas discharges range from the crustal value to 10 Ra,[176] with an unweighted average of 5.4 Ra.[70] A recent compilation of maximum $^3He/^4He$ ratios of arc gases shows a global average of 7.4 ± 1.5 Ra,[176] overlapping with the mid-ocean ridge basalt (MORB) value (Figure 8.9). The main process that lowers $^3He/^4He$ in arc gases is the contribution of ^4He from crustal sources through either magma assimilation of crustal rocks or interaction of magmatic fluids with crustal fluids at shallow depths.[176] Such processes may also affect the carbon isotopic signature. Plotting the $\delta^{13}C$ and $^3He/^4He$ of arc gases shows that: (1) very few samples plot in the DMM range for both helium and carbon, implying that subducted and/or crustal carbon affects the isotopic composition and the amount of CO_2 at arc volcanoes; (2) samples where $^3He/^4He$ is 7 ± 1 Ra have $\delta^{13}C$ values ranging from +2‰ to −12‰, implying that if the source of CO_2 is from the subducted slab, it is sourced from both carbonates and organic carbon. Alternatively, the wide $\delta^{13}C$ range for samples with Ra > 7 could be the result of modification of the mantle beneath arc volcanoes due to prior subduction events that affected carbon, but not helium; and (3) gases with $^3He/^4He$ values <7 Ra show an equally wide distribution of $\delta^{13}C$ as those >7 Ra, implying that both carbonate and organic carbon derived from the overlying crust (as implied by low $^3He/^4He$) may contribute to the degassing CO_2.

Most helium and carbon isotope data are from low-temperature (<100°C) bubbling springs and fumaroles, which can be affected by low-temperature carbon isotope fractionation in the crust and shallow hydrothermal systems.[177,178] If we only consider >200°C gases (Figure 8.9b), which are more likely to reflect their source, the range in $\delta^{13}C$ remains from −12‰ to 0‰ for both gases with $^3He/^4He$ values >7 and <7 Ra, with the same implications as stated above.

Recently, a different approach has been used to evaluate the sources of carbon in volcanic emissions, using the C/S ratio of volcanic gases in crater plumes and high-temperature (>450°C) fumaroles.[73] Selection of only high-temperature samples ensures that secondary hydrothermal processes do not affect the data set. The advantage of this approach is that many more data are available for C/S ratios than for isotope systematics, allowing for a more complete global coverage. Correlations with petrologic indicators of slab-derived fluids such as the Ba/La ratios of erupted materials allows distinction between emissions that have predominantly crustally derived CO_2 and emissions that show a strong subducted slab carbon component.[73] This global data set further reveals that only some gases with high C/S ratios (>4) are from locations where volcanoes sit on upper-plate carbonates. The data set further shows that volcanoes with low C/S ratios (<2) are in locations where the subducting sediments contain only <10% CO_2. This

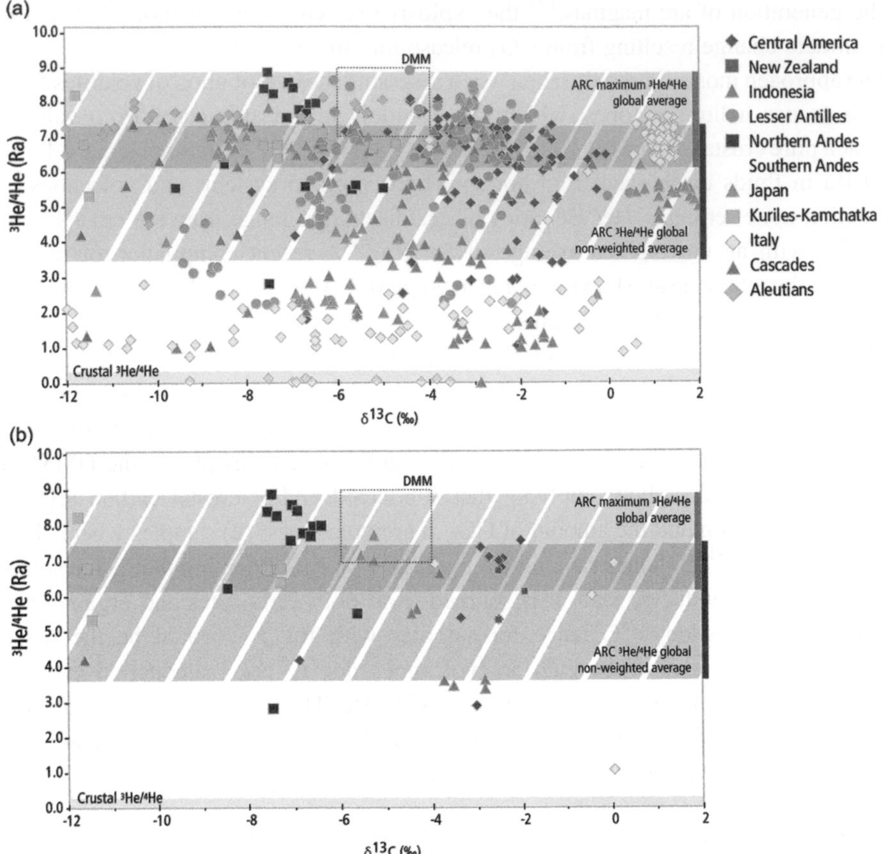

Figure 8.9 (a) Helium and carbon isotope signatures for volcanic and hydrothermal gas discharges and (b) data from discharges with vent temperatures >200°C. Data from Refs. 70 and 179–182. The field for DMM is from Refs. 162 and 163. The global data for arcs are from Ref. 70 and represent a non-weighted global average. The maximum global average for arcs is from Ref. 176 and represents the average of the maximum $^3He/^4He$ ratios measured at a given locality.

study shows that the carbon contribution from assimilation in the overriding crust may be significant in some localities (Italy, Indonesia, Central America, Lesser Antilles), but an important factor affecting the C/S ratio and CO_2 source in arc volcanoes is the subducting slab. As with the isotope approach, more work is needed since the C/S ratio in volcanic gas discharges is also significantly affected by volcanic activity, the presence of accumulated carbon-rich exsolved vapor in magma reservoirs,[107] and degassing processes. Long-term records are thus needed to constrain the average and representative C/S ratio of a particular volcano.

8.8 Volcanic Release of CO_2 over Geologic Time

The modern-day volcanic carbon flux is a snapshot in time. We have shown that the modern-day flux of volcanic carbon could be dominated by the diffuse degassing of volcanic regions and large calderas (Figure 8.3b). Equally important for the global volcanic flux are a number of large volcanic point sources that represent Earth's most active volcanoes (Figure 8.3a); these volcanoes are in a range of geologic settings (arc, ocean island/mantle plume, continental rift).

Over 1-Ma timescales, the flux of volcanic carbon to Earth's surface is counteracted by the drawdown of CO_2 by silicate weathering and the associated precipitation of marine carbonates, as well as the burial of organic carbon. Perturbations to carbon outgassing are compensated for by changes in the rate of silicate weathering (which is enhanced under conditions of high pCO_2 and atmosphere temperature), keeping the surface reservoir approximately in balance. There have been periods through Earth history, however, when volcanism has been enhanced, causing perturbations to atmospheric CO_2 that have persisted for a range of timescales. Although our study of modern volcanic carbon fluxes tells us little about the carbon cycle response to such perturbations in Earth's past, there are insights to be gained from modern observations of the magnitude of the flux from individual volcanoes and from larger regions, the nature of the flux (direct or diffuse), and its variability with time.

The processes of subduction, deep carbonated eclogite melting,[5] and fluid addition to the subcontinental lithosphere from the convecting asthenosphere over time have led to the subcontinental lithosphere becoming a large carbon sink.[183] Over Earth's past, supercontinents have accreted and broken apart. It has been recognized that periods of greenhouse climate correlate with supercontinents in the geological record,[133,184] and this has been attributed to the prevalence of high-flux continental arc volcanoes, which may capture carbon not only from the downgoing slabs, but also from the devolatilization of carbonate platforms in the overlying continental crust. In the Cretaceous, for example, Gondwana's breakup led to the closure of the Tethys Ocean, accretion and subduction of marine carbonate platforms and the formation of a long subduction zone that may have been an important source of global volcanic CO_2[4,172,185] (reviewed further in Chapter 11).

The results of our modern-day volcanic CO_2 studies have shown, however, that the continental rifts themselves may also be important sources of carbon outgassing, which may in fact be larger than the surrounding continental arcs, certainly enhancing CO_2 output and warming during the initial stages of continental breakup and providing a way for short-term tectonics and volcanism to impact climate.[183]

8.9 Synthesis

Considerable progress has been made in the last decade in quantifying CO_2 emissions from volcanic areas worldwide. Technological advances, including widespread use of miniturized UV spectometer systems and Multi-GAS instruments, have resulted in a greater number of CO_2 measurements. Volcanic CO_2 emissions have been measured for a few decades, which is short in terms of the eruptive cycles at many of Earth's volcanoes. Measurements are heavily biased toward eruptive periods, and average emissions for some active volcanoes are decreasing as longer records become available. Global satellite studies of decadal-scale SO_2 emission from Earth's most active volcanoes, when combined with a complete C/S data set, will allow for an accurate estimate of persistent CO_2 degassing in the near future. We suggest the highest 10–20 SO_2-emitting volcanoes be prioritized for measurement as these volcanoes may dominate the total CO_2 output from active volcanoes. CO_2 emissions from volcanoes with SO_2 output below satellite detection limits are not as well quantified, yet the emissions from these sources could be significant at a global scale. More work is needed to determine both the magnitude of these emissions and how widespread these volcanoes are globally.

Our knowledge of diffuse degassing at active and dormant volcanoes continues to improve. The distribution of diffuse fluxes is similar to the distribution of CO_2 emissions from active volcanoes, and there are several areas worldwide with large, intrusive magma bodies where the diffuse fluxes are globally significant. More effort is needed to quantify emissions in these vast regions, as well as in the biggest 10–20 hydrothermal areas globally.

We summarize two decades of measurements at some well-studied volcanic systems that demonstrate that the slow release of CO_2 from inactive or dormant volcanoes rivals that of active volcanoes when considered over long timescales. Regional volcanic CO_2 fluxes are heavily influenced by individual volcanoes becoming more or less active, and thus measurements must be maintained over decadal scales to assess this variability quantitatively.

Vent emissions from active volcanoes vary by orders of magnitude over short (days to years) timescales, whereas diffuse emissions are largely buffered and show steadier rates through time. This variability is largely controlled by the plumbing of the volcanic systems. Active mafic volcanoes often host an open vent (with a free magma surface or lava lake) and volatiles are delivered rapidly to the surface, allowing for rapid variability. Larger silicic magmatic systems, sustained by the underplating of basaltic magmas, are characterized by steady diffuse outgassing over time.

Continuous Multi-GAS monitoring has improved eruption forecasting potential by showing that eruptions at mafic volcanoes are often preceded by an increase in the C/S gas ratio in the days to months prior to eruption. Long-term monitoring of some volcanoes shows months-long changes in the C/S ratio prior to eruption, and sometimes in the absence of eruption, which may accompany decompression of magma as it migrates from the lower to the upper crust. These later cases were often accompanied by the lack of a visible plume or SO_2 emission, and thus more work is needed to identify when such plumes exist.

A fundamental challenge in carbon science is to constrain the deep global carbon budget and how much of the surface-derived carbon is recycled back into the mantle. In arc volcanoes, the source of volcanic CO_2 is a combination of mantle and slab-derived C with a potentially significant crustal component, at least in some locations. A more complete understanding of the carbon balance at subduction zones requires quantification of the amount of carbon in the subducted lithologies. The pathways and fate of subducted carbon beyond the zones of arc magma generation determine where this carbon is ultimately stored and how it could potentially affect processes in different tectonic settings through time. While great progress has been made in terms of the quantification of carbon emissions from volcanic regions, more work is needed to constrain our understanding of the balance between surface and deep carbon through geologic time.

8.10 Limits to Knowledge of Volcanic Carbon

For all of the new data and understanding, significant gaps remain in our knowledge. Some of these gaps may persist, limited by technology or the logistics of measurement. The first gap in knowledge is caused by the lack of data for key volcanic systems. These include important point sources of volcanic SO_2, such as Bagana, Tavurvur, and Manam (Papua New Guinea), as well as Aoba/Ambae (Vanuatu). Longer records are needed at most volcanoes globally to assess variability over decadal scales in relation to eruptive cycles and periods of repose. Better quantification of hydrothermal diffuse degassing is needed in areas already identified to be large CO_2 emitters, such as Yellowstone (USA), the EAR (Africa), and the TVZ (New Zealand), and more measurements are needed in the vast regions of Southeast Asia and South America, where large hydrothermal systems exist. Further measurements are also needed in rifts such as the Eger Rift (Germany) and the Basin and Range (USA). Many of these targets are accessible, but require dedicated efforts over many years, and they would benefit from further technological development and improved measurement strategies. In this category we include global MORs and submarine back-arcs, for which only limited data exist. These measurements are logistically challenging, and it is unlikely that significant progress can be made without considerable effort and expense. More tractable approaches use primary melt geochemistry (Chapter 9) and geodynamic models to reconstruct CO_2 budgets of submarine regions.

A fraction of the CO_2 released from the degassing of silicate melts and directly from the mantle or crust may be dissolved into groundwaters and transported through aquifers, delivered to the surface via cold springs. This source of CO_2 has not been quantified in most volcanic regions. Studies from central Italy have shown that significant quantities of inorganic carbon are dissolved in aquifers, derived from a mixture of biological sources, carbonate dissolution, and deep carbon sources.

Using the DECADE results thus far, it is possible to relate slab inputs at arc volcanoes to volcanic gas C/S signatures and to identify arcs and individual volcanoes where CO_2-rich crustal fluids play a significant role. However, much work remains to be done in linking magma geochemistry to the composition of outgassing fluids and for understanding the fate of devolatilized carbon in the mantle wedge and the behavior and dynamics of magma- and crust-derived fluids in vertically protracted, complex magma reservoirs in the crust. These studies require a range of approaches, including thermodynamic and analog modeling and building detailed databases of magma geochemistry and volcanic gas composition for detailed empirical comparisons. Understanding the amount of carbon returned to the deep mantle is of fundamental importance to the carbon budget of Earth through time. Linking our understanding of the present-day volcanic carbon budget to studies of plate tectonic reconstructions is an aim for the future and is explored further by Lee et al. in Chapter 11 of this book.

Acknowledgments

The work presented here has only been possible thanks to the enormous contributions made by the scientists and technicians in the volcano gas community in recent years. Funding provided by the Alfred P. Sloan Foundation through the Deep Carbon Observatory has enabled a significant amount of new research from many regions around the world. Funding for this chapter was provided by the US Geological Survey Volcano Hazards Program. The authors thank Josh Wood for his contribution in drafting Figures 8.1 and 8.7 and Jake Lowenstern and one anonymous reviewer for valuable reviews. The authors thank the following people for their assistance in obtaining publicly available CO_2 emission rate data, hydrothermal location data, and helpful suggestions and refinements of the text: Robin Campion, Bruce Christenson, Maarten deMoor, Tamar Elias, Kohei Kazahaya, Ryunosuke Kazahaya, Peter Kelly, Christoph Kern, Taryn Lopez, Jennifer Lewicki, Agnes Mazot, Chris Newhall, Mitch Owen, Vicentina Cruz Pauccara, Jim Stimac, Jeff Sutton, and Yuri Taran. The research leading to some of the results some of the results presented received funding from the European Research Council under the European Union's Seventh Framework Programme (FP/2007–2013)/ERC Grant Agreement 279802, and RCUK NERC funding through the DisEqm project (NE/N018575/1).

Questions for the Classroom

1 How can we improve our estimates of volcanic CO_2 emissions?
2 How can we quantify the CO_2 contribution to the atmosphere from magmatic intrusions that do not lead to eruption and how can we better combine geophysical and geochemical studies to identify when such emissions occur?
3 What methods can advance our estimation of hydrothermal CO_2 emissions?
4 How can we improve our knowledge of how many volcanoes are emitting CO_2 in the absence of significant quantities of SO_2?
5 How significant is CO_2 degassing from magmas located at the base of the crust and how would we distinguish this from shallower magma degassing?
6 Can CO_2 degas from the mantle without the presence of magma?

List of Online Resources

- Eruptions, Earthquakes & Emissions Application Citation: Global Volcanism Program, 2016. Eruptions, Earthquakes & Emissions, v. 1.0 (Mobile application software). Smithsonian Institution. Accessed March 2018, retrieved from http://volcano.si.axismaps.io
- Carn, S. A., Fioletov, V. E., McLinden, C. A., Li, C. & Krotkov, N. A. A decade of global volcanic SO_2 emissions measured from space. *Scientific Reports* 7, 44095 (2017). Direct link: www.nature.com/articles/srep44095#supplementary-information
- MaGa database: www.magadb.net
- EarthChem Library: www.earthchem.org/library

References

1. Kump, L. R. & Barley, M. E. Increased subaerial volcanism and the rise of atmospheric oxygen 2.5 billion years ago. *Nature* **448**, 1033–1036 (2007).
2. Berner, R. A. *The Phanerozoic Carbon Cycle: CO_2 and O_2* (Oxford University Press, 2004).
3. Dasgupta, R. Ingassing, storage, and outgassing of terrestrial carbon through geologic time. In: *Carbon in Earth, Vol. 75: Reviews in Mineralogy & Geochemistry* (eds. R. M. Hazen, A. P. Jones & J. A. Baross), 183–229 (Mineralogical Society of America, 2013).
4. Mason, E., Edmonds, M. & Turchyn, A. V. Remobilization of crustal carbon may dominate volcanic arc emissions. *Science* **357**, 290–294 (2017).
5. Dasgupta, R. & Hirschmann, M. M. The deep carbon cycle and melting in Earth's interior. *Earth and Planetary Science Letters* **298**, 1–13 (2010).
6. Wood, B. J. Carbon in the core. *Earth and Planetary Science Letters* **117**, 593–607 (1993).
7. Berner, R. A. Atmospheric carbon dioxide levels over Phanerozoic time. *Science* **249**, 1382–1386 (1990).
8. Symonds, R. B., Rose, W. I., Bluth, G. S. & Gerlach, T. M. Volcanic gas studies: methods, results, and applications. In: *Volatiles in Magmas, Vol. 30* (eds. M. R. Carroll and J. R. Holloway), 1–60 (Mineralogical Society of America, 1994).
9. McCormick, B. T., Edmonds, M., Mather, T. A. & Carn, S. A. First synoptic analysis of volcanic degassing in Papua New Guinea. *Geochemistry, Geophysics, Geosystems* **13**, 3 (2012).
10. Gal, F., Leconte, S. & Gadalia, A. The "Escarot" gas seep, French Massif Central: CO_2 discharge from a quiescent volcanic system – characterization and quantification of gas emissions. *Journal of Volcanology and Geothermal Research* **353**, 68–82 (2018).
11. Gerlach, T. M., Doukas, M. P., McGee, K. A. & Kessler, R. Airborne detection of diffuse carbon dioxide emissions at Mammoth Mountains, California. *Geophysical Research Letters* **26**, 3661–3664 (1999).
12. James, E. R., Manga, M. & Rose, T. P. CO_2 degassing in the Oregon cascades. *Geology* 27, 823–826 (1999).
13. Caliro, S., Chiodini, G., Avino, R., Cardellini, C. & Frondini, F. Volcanic degassing at Somma-Vesuvio (Italy) inferred by chemical and isotopic signatures of groundwater. *Applied Geochemistry* **20**, 1060–1076 (2005).

14. Fischer, T. P. DEep CArbon DEgassing: the Deep Carbon Observatory DECADE Initiative. *Mineralogical Magazine* **77**, 1089 (2013).
15. Burton, M., Sawyer, G. M. & Granieri, D. Deep carbon emissions from volcanoes. In: *Carbon in Earth. Reviews in Mineralogy and Geochemistry* (eds. R. M. Hazen, A. P. Jones & J. A. Baross), 323–354 (Mineralogical Society of America, 2013).
16. Chiodini, G., Cioni, R., Guidi, M., Raco, B. & Marini, L. Soil CO_2 flux measurements in volcanic and geothermal areas. *Applied Geochemistry* **13**, 543–552 (1998).
17. Gerlach, T. M. et al. Application of the LI-COR CO_2 analyzer to volcanic plumes: a case study, volcan Popocatepetl, Mexico, June 7 and 10, 1995. *Journal of Geophysical Research* **102**, 8005–8019 (1997).
18. Shinohara, H. Excess degassing from volcanoes and its role on eruptive and intrusive activity. *Reviews of Geophysics* **46**, RG4005 (2008).
19. McGonigle, A. J. S., Oppenheimer, C., Galle, B., Mather, T. A. & Pyle, D. M. Walking traverse and scanning DOAS measurements of volcanic gas emission rates. *Geophysical Research Letters* **29**, 1985 (2002).
20. Galle, B. et al. A miniaturised ultraviolet spectrometer for remote sensing of SO_2 fluxes: a new tool for volcano surveillance. *Journal of Volcanology and Geothermal Research* **119**, 241–254 (2003).
21. Allard, P. et al. Eruptive and diffuse emissions of CO_2 from Mount Etna. *Nature* 351, 387–391 (1991).
22. Carn, S. A. & Bluth, G. J. S. Prodigious sulfur dioxide emissions from Nyamuragira volcano, DR Congo. *Geophysical Research Letters* **30**, 2211 (2003).
23. Allard, P., Burton, M., Sawyer, G. & Bani, P. Degassing dynamics of basaltic lava lake at a top-ranking volatile emitter: Ambrym volcano, Vanuatu arc. *Earth and Planetary Science Letters* **448**, 69–80 (2016).
24. Shinohara, H. A new technique to estimate volcanic gas composition; plume measurements with a portable multi-sensor system. *Journal of Volcanology and Geothermal Research* **143**, 319–333 (2005).
25. Aiuppa, A., Federico, C., Giudice, G. & Gurrieri, S. Chemical mapping of a fumarolic field: La Fossa Crater, Vulcano Island (Aeolian Islands, Italy). *Geophysical Research Letters* **32**, 4 (2005).
26. Kern, C., Werner, C., Elias, T., Sutton, A. J. & Lübcke, P. Applying UV cameras for SO_2 detection to distant or optically thick volcanic plumes. *Journal of Volcanology and Geothermal Research* **262**, 80–89 (2013).
27. Mori, T. et al. Effect of UV scattering on SO_2 emission rate measurements. *Geophysical Research Letters* **33**, L17315 (2006).
28. Kelly, P. J. et al. Long-term autonomous volcanic gas monitoring with Multi-GAS at Mount St. Helens, Washington, and Augustine Volcano. *Alaska AGU Fall Meeting Abstracts*, V23B-3095 (2015).
29. Kelly, P. J. et al. Rapid chemical evolution of tropospheric volcanic emissions from Redoubt Volcano, Alaska, based on observations of ozone and halogen-containing gases. *Journal of Volcanology and Geothermal Research* **259**, 317–333 (2013).
30. Moussallam, Y. et al. Hydrogen emissions from Erebus volcano, Antarctica. *Bulletin of Volcanology* **74**, 2109–2120 (2012).
31. Gerlach, T. M., McGee, K. A., Elias, T., Sutton, A. J. & Doukas, M. P. Carbon dioxide emission rate of Kilauea Volcano: implications for primary magma and the summit reservoir. *Journal of Geophysical Research: Solid Earth* **107**, 2189 (2002).
32. Werner, C., Christenson, B. W., Hagerty, M. & Britten, K. Variability of volcanic gas emissions during a crater lake heating cycle at Ruapehu Volcano, New Zealand. *Journal of Volcanology and Geothermal Research* **154**, 291–302 (2006).

33. Werner, C. et al. Variability of passive gas emissions, seismicity, and deformation during crater lake growth at White Island Volcano, New Zealand, 2002–2006. *Journal of Geophysical Research: Solid Earth* **113**, B01204 (2008).
34. Werner, C., Evans, W. C., Poland, M., Tucker, D. S. & Doukas, M. P. Long-term changes in quiescent degassing at Mount Baker Volcano, Washington, USA; evidence for a stalled intrusion in 1975 and connection to a deep magma source. *Journal of Volcanology and Geothermal Research* **186**, 379–386 (2009).
35. McGonigle, A. J. S. et al. Unmanned aerial vehicle measurements of volcanic carbon dioxide fluxes. *Geophysical Research Letters* **35**, L06303 (2008).
36. Mori, T. et al. Volcanic plume measurements using a UAV for the 2014 Mt. Ontake eruption. *Earth Planets and Space* **68**, 49 (2016).
37. Shinohara, H. Composition of volcanic gases emitted during repeating Vulcanian eruption stage of Shinmoedake, Kirishima volcano, Japan. *Earth Planets and Space* **65**, 667–675 (2013).
38. Aiuppa, A. et al. New ground-based lidar enables volcanic CO_2 flux measurements. *Scientific Reports* **5**, 13614 (2015).
39. Queisser, M., Granieri, D. & Burton, M. A new frontier in CO_2 flux measurements using a highly portable DIAL laser system. *Scientific Reports* **6**, 33834 (2016).
40. Werner, C., Brantley, S. L. & Boomer, K. CO_2 Emissions related to the Yellowstone volcanic system 2. Statistical sampling, total degassing, and transport mechanisms. *Journal of Geophysical Research* **105**, 10831-10846 (2000).
41. Cardellini, C., Chiodini, G. & Frondini, F. Application of stochastic simulation to CO_2 flux from soil: mapping and quantification of gas release. *Journal of Geophysical Research: Solid Earth* **108**, 2425 (2003).
42. Mazot, A. & Bernard, A. CO_2 degassing from volcanic lakes. In: *Volcanic Lakes* (eds. D. Rouwet, B. Christenson, F. Tassi & J. Vandemeulebrouck), 341–354 (Springer, 2015).
43. Pérez, N. M. et al. Global CO_2 emission from volcanic lakes. *Geology* **39**, 235–238 (2011).
44. Werner, C. et al. Monitoring volcanic hazard using eddy covariance at Solfatara volcano, Naples, Italy. *Earth and Planetary Science Letters* **210**, 561–577 (2003).
45. Werner, C., Wyngaard, J. C. & Brantley, S. Eddy-correlation measurement of hydrothermal gases. *Geophysical Research Letters* **27**, 2925–2929 (2000).
46. Lewicki, J., Fischer, M. L. & Hilley, G. E. Six-week time series of eddy covariance CO_2 flux at Mammoth Mountain, California: performance evaluation and role of meteorological forcing. *Journal of Volcanology and Geothermal Research* **171**, 178–190 (2008).
47. Lewicki, J. L., Hilley, G. E., Dobeck, L. & Marino, B. D. V. Eddy covariance imaging of diffuse volcanic CO_2 emissions at Mammoth Mountain, CA, USA. *Bulletin of Volcanology* **74**, 135–141 (2012).
48. Lewicki, J. L., Kelly, P. J., Bergfeld, D., Vaughan, R. G. & Lowenstern, J. B. Monitoring gas and heat emissions at Norris Geyser Basin, Yellowstone National Park, USA based on a combined eddy covariance and Multi-GAS approach. *Journal of Volcanology and Geothermal Research* **347**, 312–326 (2017).
49. Lewicki, J. L. & Hilley, G. E. Multi-scale observations of the variability of magmatic CO_2 emissions, Mammoth Mountain, CA, USA. *Journal of Volcanology and Geothermal Research* **284**, 1–15 (2014).
50. Rose, T. P., Lee Davisson, M. & Criss, R. E. Isotope hydrology of voluminous cold springs in fractured rock from an active volcanic region, northeastern California. *Journal of Hydrology* **179**, 207–236 (1996).

51. Aiuppa, A. et al. Forecasting Etna eruptions by real-time observation of volcanic gas composition. *Geology* **35**, 1115–1118 (2007).
52. Galle, B. et al. Network for Observation of Volcanic and Atmospheric Change (NOVAC) – a global network for volcanic gas monitoring: network layout and instrument description. *Journal of Geophysical Research –Atmospheres* **115**, D05304 (2010).
53. Aiuppa, A. et al. Unusually large magmatic CO_2 gas emissions prior to a basaltic paroxysm. *Geophysical Research Letters* **37**, L17303 (2010).
54. Aiuppa, A. et al. Total volatile flux from Mount Etna. *Geophysical Research Letters* **35**, L24302 (2008).
55. de Moor, J. M. et al. Short-period volcanic gas precursors to phreatic eruptions: insights from Poás Volcano, Costa Rica. *Earth and Planetary Science Letters* **442**, 218–227 (2016).
56. de Moor, J. M. et al. A new sulfur and carbon degassing inventory for the Southern Central American volcanic arc: the importance of accurate time-series datasets and possible tectonic processes responsible for temporal variations in arc-scale volatile emissions. *Geochemistry, Geophysics, Geosystems* **18**, 4437–4468 (2017).
57. Shinohara, H. Volatile flux from subduction zone volcanoes: Insights from a detailed evaluation of the fluxes from volcanoes in Japan. *Journal of Volcanology and Geothermal Research* **268**, 46–63 (2013).
58. Ilyinskaya, E. et al. Degassing regime of Hekla volcano 2012–2013. *Geochimica et Cosmochimica Acta* **159**, 80–99 (2015).
59. Werner, C. et al. Magmatic degassing, lava dome extrusion, and explosions from Mount Cleveland volcano, Alaska, 2011–2015: insight into the continuous nature of volcanic activity over multi-year timescales. *Journal of Volcanology and Geothermal Research* **337**, 98–110 (2017).
60. Pedone, M. et al. Volcanic CO_2 flux measurement at Campi Flegrei by tunable diode laser absorption spectroscopy. *Bulletin of Volcanology* **76**, 13 (2014).
61. Pedone, M. et al. Tunable diode laser measurements of hydrothermal/volcanic CO_2 and implications for the global CO_2 budget. *Solid Earth* **5**, 1209–1221 (2014).
62. Pedone, M. et al. Total (fumarolic plus diffuse soil) CO_2 output from Furnas volcano. *Earth Planets and Space* **67**, 12 (2015).
63. Fiorani, L. et al. Early detection of volcanic hazard by lidar measurement of carbon dioxide. *Natural Hazards* **83**, S21–S29 (2016).
64. Queisser, M., Burton, M., Allan, G. R. & Chiarugi, A. Portable laser spectrometer for airborne and ground-based remote sensing of geological CO_2 emissions. *Optics Letters* **42**, 2782–2785 (2017).
65. Werner, C. et al. Degassing of CO_2, SO_2, and H_2S associated with the 2009 eruption of Redoubt Volcano, Alaska. *Journal of Volcanology and Geothermal Research* **259**, 270–284 (2013).
66. Schwandner, F. M. et al. Spaceborne detection of localized carbon dioxide sources. *Science* **358**, eaam5782 (2017).
67. Gerlach, T. M. Present-day carbon dioxide emissions from volcanos. *Earth in Space* **4**, 5 (1991).
68. Brantley, S. L. & Koepenick, K. W. Measured carbon dioxide emissions from Oldoinyo Lengai and the skewed distribution of passive volcanic fluxes. *Geology* **23**, 933–936 (1995).
69. Williams, S. N., Schaefer, S. J., Calvache, V. M. L. & Lopez, D. Global carbon dioxide emission to the atmosphere by volcanoes. *Geochimica et Cosmochimica Acta* **56**, 1765–1770 (1992).

70. Hilton, D. R., Fischer, T. & Marty, B. Noble gases and volatile recycling at subduction zones. *Reviews in Mineralogy and Geochemistry* **47**, 319–370 (2002).
71. Andres, R. J. et al. A summary of sulfur-dioxide emission rate measurements from Guatemalan volcanos. *Bulletin of Volcanology* 55, 379–388 (1993).
72. Stoiber, R. E., Williams, S. N. & Huebert, B. Annual contribution of sulfur dioxide to the atmosphere by volcanoes. *Journal of Volcanology and Geothermal Research* **33**, 1–8 (1987).
73. Aiuppa, A., Fischer, T. P., Plank, T., Robidoux, P. & Di Napoli, R. Along-arc, inter-arc and arc-to-arc variations in volcanic gas CO_2/S-T ratios reveal dual source of carbon in arc volcanism. *Earth Science Reviews* **168**, 24–47 (2017).
74. Fischer, T. P. Fluxes of volatiles (H_2O, CO_2, N_2, Cl, F) from arc volcanoes. *Geochemical Journal* **42**, 21–38 (2008).
75. Carn, S. A., Fioletov, V. E., McLinden, C. A., Li, C. & Krotkov, N. A. A decade of global volcanic SO_2 emissions measured from space. *Scientific Reports* **7**, 44095 (2017).
76. Siebert, L., Simkin, T. & Kimberly, P. *Volcanoes of the World*, 3rd edn. (University of California Press, 2010). https://doi.org/10.5479/si.GVP.VOTW4-2013.
77. Carn, S. A., Clarisse, L. & Prata, A. J. Multi-decadal satellite measurements of global volcanic degassing. *Journal of Volcanology and Geothermal Research* **311**, 99–134 (2016).
78. Torgersen, T. Terrestrial helium degassing fluxes and the atmospheric helium budget: Implications with respect to the degassing processes of continental crust. *Chemical Geology: Isotope Geoscience Section* **79**, 1–14 (1989).
79. Sano, Y. & Williams, S. N. Fluxes of mantle and subducted carbon along convergent plate boundaries. *Geophysical Research Letters* **23**, 2749–2752 (1996).
80. Crisp, J. A. Rates of magma emplacement and volcanic output. *Journal of Volcanology and Geothermal Research* **20**, 177–211 (1984).
81. Dimalanta, C., Taira, A., Yumul, G. P., Tokuyama, H. & Mochizuki, K. New rates of western Pacific island arc magmatism from seismic and gravity data. *Earth and Planetary Science Letters* **202**, 105–115 (2002).
82. Reymer, A. & Schubert, G. Phanerozoic addition rates to the continental crust and crustal growth. *Tectonics* **3**, 63–77 (1984).
83. Bianchi, D. et al. Low helium flux from the mantle inferred from simulations of oceanic helium isotope data. *Earth and Planetary Science Letters* **297**, 379–386 (2010).
84. Kagoshima, T. et al. Sulphur geodynamic cycle. *Scientific Reports* **5**, 8330 (2015).
85. Siebert, L. & Simkin, T. Volcanoes of the world: an illustrated catalog of holocene volcanoes and their eruptions. *Smithsonian Institution. Global volcanism program digital information series, GVP-3* (https://doi.org/10.5479/si.GVP.VOTW4-2013) (2002).
86. Bobrowski, N. et al. Multi-component gas emission measurements of the active lava lake of Nyiragongo, DR Congo. *Journal of African Earth Sciences* **134**, 856–865 (2017).
87. Sawyer, G. M., Carn, S. A., Tsanev, V. I., Oppenheimer, C. & Burton, M. Investigation into magma degassing at Nyiragongo volcano, Democratic Republic of the Congo. *Geochemistry, Geophysics, Geosystems* **9**, Q02017 (2008).
88. Le Guern, F. Mechanism of energy-transfer in the lava lake of Niragongo (Zaire), 1959–1977. *Journal of Volcanology and Geothermal Research* **31**, 17–31 (1987).
89. Kazahaya, K. et al. Gigantic SO_2 emission from Miyakejima volcano, Japan, caused by caldera collapse. *Geology* **32**, 425–428 (2004).

90. Shinohara, H., Kazahaya, K., Saito, G., Fukui, K. & Odai, M. Variation of CO_2–SO_2 ratio in volcanic plumes of Miyakejima; stable degassing deduced from heliborne measurements. *Geophysical Research Letters* **30**, 4 (2003).
91. Doukas, M. P. & McGee, K. A. *A Compilation of Gas Emission-Rate Data from Volcanoes of Cook Inlet (Spurr, Crater Peak, Redoubt, Iliamna, and Augustine) and Alaska Peninsula (Douglas, Fourpeaked, Griggs, Mageik, Martin, Peulik, Ukinrek Maars, and Veniaminof), Alaska, from 1995–2006*. Open File Report, report number 2007-1400 (US Geological Survey, 2007).
92. Werner, C. A., Doukas, M. P. & Kelly, P. J. Gas emissions from failed and actual eruptions from Cook Inlet Volcanoes, Alaska, 1989–2006. *Bulletin of Volcanology* **73**, 155–173 (2011).
93. Kelly, P., Werner, C., Kem, C., Clor, L. E. & Doukas, M. A compilation of airborne gas emissions data from Alaskan volcanoes. *USGS Data Release* (2019).
94. Christenson, B. W. et al. Cyclic processes and factors leading to phreatic eruption events: insights from the 25 September 2007 eruption through Ruapehu Crater Lake, New Zealand. *Journal of Volcanology and Geothermal Research* **191**, 15–32 (2010).
95. Kilgour, G. N. et al. Timescales of magmatic processes at Ruapehu volcano from diffusion chronometry and their comparison to monitoring data. *Journal of Volcanology and Geothermal Research* **288**, 62–75 (2014).
96. Elias, T. & Sutton, A. J. *Volcanic Air Pollution Hazards in Hawaii. Fact Sheet No. 2017-3017* (US Geological Survey, 2017).
97. Global Volcanism Program. *Volcanoes of the World, v. 4.7.7* (Smithsonian Institution, 2013). https://doi.org/10.5479/si.GVP.VOTW4-2013.
98. Passarelli, L. & Brodsky, E. E. The correlation between run-up and repose times of volcanic eruptions. *Geophysical Journal International* **188**, 1025–1045 (2012).
99. Allard, P., Burton, M., Oskarsson, N., Michel, A. & Polacci, M. Chemistry and fluxes of magmatic gases driving the explosive trachyandesitic phase of Eyjafjallajökull 2010 eruption: constraints on degassing magma volumes and processes. In: *AGU Fall Meeting*, V53F-07 (AGU, 2010).
100. Gerlach, T. M., Westrich, H. R. & Symonds, R. B. Preeruption vapor in magma of the climactic Mount Pinatubo eruption: source of the giant stratospheric sulfur dioxide cloud. In: *Fire and Mud: Eruptions and Lahars of Mount Pinatubo, Philippines* (eds. C. G. Newhall & R. S. Punongbayan), 415–434 (University of Washington Press, 1996).
101. Hobbs, P. V., Tuell, J. P., Hegg, D. A., Radke, L. F. & Eltgroth, M. W. Particles and gases in the emissions from the 1980–1981 volcanic eruptions of Mt. St. Helens. *Journal of Geophysical Research* **87**, 11062–11086 (1982).
102. Scaillet, B. & Pichavant, M. Experimental constraints on volatile abundances in arc magmas and their implications for degassing processes. *Geological Society Special Publications* **213**, 23–52 (2003).
103. Gerlach, T. M. Volcanic versus anthropogenic carbon dioxide. *EOS Transactions* **92**, 201–208 (2011).
104. Holloway, J. R. Fluids in the evolution of granitic magma: consequences of finite CO_2 solubility. *Geological Society of America Bulletin* **87**, 1513–1518 (1976).
105. Christopher, T. E. et al. Crustal-scale degassing due to magma system destabilization and magma-gas decoupling at Soufriere Hills Volcano, Montserrat. *Geochemistry, Geophysics, Geosystems* **16**, 2797–2811 (2015).
106. Parmigiani, A., Faroughi, S., Huber, C., Bachmann, O. & Su, Y. Bubble accumulation and its role in the evolution of magma reservoirs in the upper crust. *Nature* **532**, 492–495 (2016).

107. Burgisser, A., Alletti, M. & Scaillet, B. Simulating the behavior of volatiles belonging to the C–O–H–S system in silicate melts under magmatic conditions with the software D-Compress. *Computers & Geosciences* **79**, 1–14 (2015).
108. Kilbride, B. M., Edmonds, M. & Biggs, J. Observing eruptions of gas-rich compressible magmas from space. *Nature Communications* **7**, 13744 (2016).
109. Wallace, P. J. Volcanic SO_2 emissions and the abundance and distribution of exsolved gas in magma bodies. *Journal of Volcanology and Geothermal Research* **108**, 85–106 (2001).
110. Kerrick, D. M. Present and past non-anthropogenic CO_2 degassing from the solid Earth. *Reviews of Geophysics* **39**, 565–585 (2001).
111. Morner, N. A. & Etiope, G. Carbon degassing from the lithosphere. *Global and Planetary Change* **33**, 185–203 (2002).
112. Mori, T. et al. Time-averaged SO_2 fluxes of subduction-zone volcanoes: example of a 32-year exhaustive survey for Japanese volcanoes. *Journal of Geophysical Research – Atmospheres* **118**, 8662–8674 (2013).
113. Stimac, J., Goff, F. & Goff, C. J. Intrusion-related geothermal systems. In: *The Encyclopedia of Volcanoes*, 2nd edn. (eds. H. Sigurdsson, B. Houghton, S. McNutt, H. Rymer & J. Stix), 799–822 (Academic Press, 2015).
114. Cardellini, C. et al. MAGA, a new database of gas natural emissions: a collaborative webenvironment for collecting data. *Geophysical Research Abstracts* **16**, EGU2014-13715 (2014).
115. Casadevall, T. J., Doukas, M. P., Neal, C. A., McGimsey, R. G. & Gardner, C. A. Emission rates of sulfur dioxide and carbon dioxide from Redoubt Volcano, Alaska during the 1989-1990 eruptions. *Journal of Volcanology and Geothermal Research* **62**, 519–530 (1994).
116. Werner, C. et al. Decadal-scale variability of diffuse CO_2 emissions and seismicity revealed from long-term monitoring (1995-2013) at Mammoth Mountain, California, USA. *Journal of Volcanology and Geothermal Research* **289**, 51–63 (2014).
117. Cardellini, C. et al. Monitoring diffuse volcanic degassing during volcanic unrests: the case of Campi Flegrei (Italy). *Scientific Reports* **7**, 15 (2017).
118. Aiuppa, A. et al. Volcanic gas plume data from Stromboli Volcano (Italy). Interdisciplinary Earth Data Alliance (IEDA). doi:10.1594/IEDA/100643.
119. Christenson, B. W., White, S., Britten, K. & Scott, B. J. Hydrological evolution and chemical structure of a hyper-acidic spring-lake system on Whakaari/White Island, NZ. *Journal of Volcanology and Geothermal Research* **346**, 180–211 (2017).
120. Bloomberg, S. et al. Soil CO_2 emissions as a proxy for heat and mass flow assessment, Taupō Volcanic Zone, New Zealand. *Geochemistry, Geophysics, Geosystems* **15**, 4885–4904 (2014).
121. Mazot, A. et al. CO_2 discharge from the bottom of volcanic Lake Rotomahana, New Zealand. *Geochemistry, Geophysics, Geosystems* **15**, 577–588 (2014).
122. Rissmann, C. et al. Surface heat flow and CO_2 emissions within the Ohaaki hydrothermal field, Taupo Volcanic Zone, New Zealand. *Applied Geochemistry* **27**, 223–239 (2012).
123. Werner, C. & Cardellini, C. Comparison of carbon dioxide emissions with fluid upflow, chemistry, and geologic structures at the Rotorua geothermal system, New Zealand. *Geothermics* **35**, 221–238 (2006).
124. Acocella, V., Di Lorenzo, R., Newhall, C. & Scandone, R. An overview of recent (1988 to 2014) caldera unrest: knowledge and perspectives. *Reviews of Geophysics* **53**, 896–955 (2015).

125. Newhall, C. G. & Dzurisin, D. *Historical Unrest at Large Calderas of the World. USGS Bulletin No. 1855* (USGS, 1988).
126. Chiodini, G. et al. Magmas near the critical degassing pressure drive volcanic unrest towards a critical state. *Nature Communications* **7**, 13712 (2016).
127. Seward, T. M. & Kerrick, D. M. Hydrothermal CO_2 emission from the Taupo volcanic zone, New Zealand. *Earth and Planetary Science Letters* **139**, 105–113 (1996).
128. Werner, C. & Brantley, S. CO_2 emissions from the Yellowstone volcanic system. *Geochemistry, Geophysics, Geosystems* **4**, 1061 (2003).
129. Hurwitz, S. & Lowenstern, J. B. Dynamics of the Yellowstone hydrothermal system. *Reviews of Geophysics* **52**, 375–411 (2014).
130. Chiodini, G. et al. CO_2 degassing and energy release at Solfatara volcano, Campi Flegrei, Italy. *Journal of Geophysical Research: Solid Earth* **106**, 16213–16221 (2001).
131. Chiodini, G. et al. Carbon dioxide Earth degassing and seismogenesis in central and southern Italy. *Geophysical Research Letters* **31**, L07615 (2004).
132. Frondini, F. et al. Measurement of CO_2 fluxes at regional scale: the case of Apennines, Italy. *Journal of the Geological Society* **176**, 408–416 (2019).
133. Lee, H. et al. Massive and prolonged deep carbon emissions associated with continental rifting. *Nature Geoscience* **9**, 145–149 (2016).
134. Hunt, J. A., Zafu, A., Mather, T. A., Pyle, D. M. & Barry, P. H. Spatially variable CO_2 degassing in the Main Ethiopian Rift: implications for magma storage, volatile transport, and rift-related emissions. *Geochemistry, Geophysics, Geosystems* **18**, 3714–3737 (2017).
135. World Energy Council. World Energy Resources 2016. www.worldenergy.org/wp-content/uploads/2016/10/World-Energy-Resources-Full-report-2016.10.03.pdf (2016).
136. Bibby, H., Caldwell, T. G., Davey, F. J. & Webb, T. H. Geophysical evidence on the structure of the Taupo Volcanic Zone and its hydrothermal circulation. *Journal of Volcanology and Geothermal Research* **68**, 29–58 (1995).
137. Chiodini, G. et al. Carbon dioxide diffuse emission and thermal energy release from hydrothermal systems at Copahue–Caviahue Volcanic Complex (Argentina). *Journal of Volcanology and Geothermal Research* **304**, 294–303 (2015).
138. Omenda, P. A. The geothermal activity of the East African Rift, in *Short Course IV on Exploration for Geothermal Resources, organized by UNU-GTP, KenGen and GDC* (ed. UNU-GTP). https://orkustofnun.is/gogn/unu-gtp-sc/UNU-GTP-SC-10-0204.pdf.
139. Moran, S. C. et al. *Instrumentation Recommendations for Volcano Monitoring at U.S. Volcanoes Under the National Volcano Early Warning System. US Geological Survey Scientific Investigations Report No. 2008-5114* (USGS, 2008).
140. Poland, M. P., Miklius, A., Jeff Sutton, A. & Thornber, C. R. A mantle-driven surge in magma supply to Kīlauea Volcano during 2003–2007. *Nature Geoscience* **5**, 295 (2012).
141. Werner, C. et al. Deep magmatic degassing versus scrubbing: elevated CO_2 emissions and C/S in the lead-up to the 2009 eruption of Redoubt Volcano, Alaska. *Geochemistry, Geophysics, Geosystems* **13**, Q03015 (2012).
142. Ilyinskaya, E. et al. Globally significant CO_2 emissions From Katla, a subglacial volcano in Iceland. *Geophysical Research Letters* **45**, 10332–10341 (2018).
143. Carn, S. A. Multi-Satellite Volcanic Sulfur Dioxide L4 Long-Term Global Database V2. doi:10.5067/MEASURES/SO2/DATA402.

144. Marty, B. & Tolstikhin, I. N. CO_2 fluxes from mid-ocean ridges, arcs and plumes. *Chemical Geology* 145, 233–248 (1998).
145. Aiuppa, A. et al. The 2007 eruption of Stromboli volcano: Insights from real-time measurement of the volcanic gas plume CO_2/SO_2 ratio. *Journal of Volcanology and Geothermal Research* **182**, 221–230 (2009).
146. Aiuppa, A. et al. A model of degassing for Stromboli Volcano. *Earth and Planetary Science Letters* **295**, 195–204 (2010).
147. Kazahaya, K., Shinohara, H. & Saito, G. Excessive degassing of Izu-Oshima Volcano; magma convection in a conduit. *Bulletin of Volcanology* **56**, 207–216 (1994).
148. Aiuppa, A. et al. Volcanic gas plume data from Etna Volcano (Italy). Interdisciplinary Earth Data Alliance (IEDA). doi:10.1594/IEDA/100643.
149. McGee, K. A. The structure, dynamics, and chemical composition of noneruptive plumes from Mount St. Helens, 1980–1988. *Journal of Volcanology and Geothermal Research* **51**, 269–282 (1992).
150. Chiodini, G. et al. Magma degassing as a trigger of bradyseismic events: the case of Phlegrean Fields (Italy). *Geophysical Research Letters* **30**, 1434 (2003).
151. Giggenbach, W. Variations in the carbon, sulfur and chlorine contents of volcanic gas discharges from White Island, New Zealand. *Bulletin of Volcanology* **39**, 15–27 (1975).
152. Fischer, T. P., Arehart, G. B., Sturchio, N. C. & Williams, S. N. The relationship between fumarole gas composition and eruptive activity at Galeras volcano, Colombia. *Geology* **24**, 531–534 (1996).
153. Dixon, J. E. & Stolper, E. M. An experimental study of water and carbon dioxide solubilities in mid-ocean ridge basaltic liquids. 2. Applications to degassing. *Journal of Petrology* **36**, 1633–1646 (1995).
154. Aiuppa, A. et al. Tracking Formation of a Lava Lake From Ground and Space: Masaya Volcano (Nicaragua), 2014–2017. *Geochemistry, Geophysics, Geosystems* **19**, 496–515 (2018).
155. Aiuppa, A. et al. A CO_2-gas precursor to the March 2015 Villarrica volcano eruption. *Geochemistry, Geophysics, Geosystems* **18**, 2120–2132 (2017).
156. de Moor, J. M. et al. Turmoil at Turrialba Volcano (Costa Rica): degassing and eruptive processes inferred from high-frequency gas monitoring. *Journal of Geophysical Research: Solid Earth* **121**, 5761–5775 (2016).
157. Pfeffer, M. et al. Ground-Based Measurements of the 2014–2015 Holuhraun Volcanic Cloud (Iceland). *Geosciences* **8**, 29 (2018).
158. Allard, P. Composition isotopique du carbonne dans les gas d'un volcan d'arc: le Momotombo (Nicaragua). *Comptes rendus de l'Académie des Sciences* **290**, 1525–1528 (1980).
159. Allard, P. Stable isotope composition of hydrogen, carbon and sulphur in magmatic gases from rift and island arc volcanoes. *Bulletin of Volcanology* **45**, 269–271 (1982).
160. Allard, P. The origin of hydrogen, carbon, sulfur, nitrogen and rare gases in volcanic exhalations: evidence from isotope geochemistry. In: *Forecasting Volcanic Events*, Vol. 1 (eds. H. Tazieff & J. Sabroux), 337–386 (Elsevier, 1983).
161. Marty, B. & Jambon, A. $C/^3He$ in volatile fluxes from the solid Earth; implications for carbon geodynamics. *Earth and Planetary Science Letters* **83**, 16–26 (1987).
162. Marty, B., Jambon, A. & Sano, Y. Helium isotopes and CO_2 in volcanic gases of Japan. *Chemical Geology* **76**, 25-40 (1989).
163. Cartigny, P., Jendrzejewski, N., Pineau, F., Petit, E. & Javoy, M. Volatile (C, N, Ar) variability in MORB and the respective roles of mantle source heterogeneity and

degassing: the case of the Southwest Indian Ridge. *Earth and Planetary Science Letters* **194**, 241–257 (2001).
164. Marty, B. & Zimmermann, L. Volatiles (He, C, N, Ar) in mid-ocean ridge basalts: assesment of shallow-level fractionation and characterization of source composition. *Geochimica et Cosmochimica Acta* **63**, 3619–3633 (1999).
165. Barry, P. H., Hilton, D. R., Füri, E., Halldórsson, S. A. & Grönvold, K. Carbon isotope and abundance systematics of Icelandic geothermal gases, fluids and subglacial basalts with implications for mantle plume-related CO_2 fluxes. *Geochimica et Cosmochimica Acta* **134**, 74–99 (2014).
166. Gerlach, T. M. & Taylor, B. E. Carbon isotope constraints on degassing of carbon dioxide from Kilauea Volcano. *Geochemica et Cosmochimica Acta* **54**, 2051–2058 (1990).
167. Fischer, T. P. et al. Upper-mantle volatile chemistry at Oldoinyo Lengai volcano and the origin of carbonatites. *Nature* **459**, 77–80 (2009).
168. Barry, P. H. et al. Helium and carbon isotope systematics of cold "mazuku" CO_2 vents and hydrothermal gases and fluids from Rungwe Volcanic Province, southern Tanzania. *Chemical Geology* **339**, 141–156 (2013).
169. de Leeuw, G. A. M., Hilton, D. R., Fischer, T. P. & Walker, J. A. The He–CO_2 isotope and relative abundance characteristics of geothermal fluids in El Salvador and Honduras: new constraints on volatile mass balance of the Central American Volcanic Arc. *Earth and Planetary Science Letters* **258**, 132–146 (2007).
170. Shaw, A. M., Hilton, D. R., Fischer, T. P., Walker, J. A. & Alvarado, G. E. Contrasting He–C relationships in Nicaragua and Costa Rica: insights into C cycling through subduction zones. *Earth and Planetary Science Letters* **214**, 499–513 (2003).
171. Kelemen, P. B. & Manning, C. E. Reevaluating carbon fluxes in subduction zones, what goes down, mostly comes up. *Proceedings of the National Academy of Sciences of the United States of America* **112**, E3997–E4006 (2015).
172. Lee, C. T. A. et al. Continental arc-island arc fluctuations, growth of crustal carbonates, and long-term climate change. *Geosphere* **9**, 21–36 (2013).
173. Troll, V. R. et al. Crustal CO_2 liberation during the 2006 eruption and earthquake events at Merapi volcano, Indonesia. *Geophysical Research Letters* **39**, L11302 (2012).
174. Carter, L. B. & Dasgupta, R. Effect of melt composition on crustal carbonate assimilation: implications for the transition from calcite consumption to skarnification and associated CO_2 degassing. *Geochemistry, Geophysics, Geosystems* **17**, 3893–3916 (2016).
175. Deegan, F. M. et al. Magma-carbonate interaction processes and associated CO_2 release at Merapi volcano, Indonesia: insights from experimental petrology. *Journal of Petrology* **51**, 1027–1051 (2010).
176. Sano, Y. & Fischer, T. P. The analysis and interpretation of noble gases in modern hydrothermal systems. In: *Noble Gases as Geochemical Tracers* (ed. P. Burnard), 249–317 (Springer Verlag, 2013).
177. Ray, M. C., Hilton, D. R., Munoz, J., Fischer, T. P. & Shaw, A. M. The effects of volatile recycling, degassing and crustal contamination on the helium and carbon geochemistry of hydrothermal fluids from the Southern Volcanic Zone of Chile. *Chemical Geology* **266**, 38–49 (2009).
178. van Soest, M. C., Hilton, D. R. & Kreulen, R. Tracing crustal and slab contributions to arc magmatism in the Lesser Antilles island arc using helium and carbon relationships in geothermal fluids. *Geochim.Cosmochim. Acta* **62**, 3323–3335 (1998).

179. Evans, W. C. et al. *Aleutian Arc Geothermal Fluids: Chemical Analyses of Waters and Gases. US Geological Survey Data release* (USGS, 2015).
180. Motyka, R. J., Liss, S. A., Nye, C. J. & Moorman, M. A. *Geothermal Resources of the Aleutian Arc* (Alaska Division of Geological & Geophysical Surveys, 1994).
181. Oppenheimer, C., Fischer, T. P. & Scaillet, B. Volcanic degassing: process and impact. In: *Treatise on Geochemistry*, 2nd edn. (ed. K. K. Turekian), 111–179 (Elsevier, 2014).
182. Symonds, R. B. et al. *Scrubbing Masks Magmatic Degassing during Repose at Cascade-Range and Aleutian-Arc Volcanoes. Open-File Report 2003-435* (USGS, 2003).
183. Foley, S. F. & Fischer, T. P. An essential role for continental rifts and lithosphere in the deep carbon cycle. *Nature Geoscience* **10**, 897–902 (2017).
184. McKenzie, N. R. et al. Continental arc volcanism as the principal driver of icehouse-greenhouse variability. *Science* **352**, 444–447 (2016).
185. Lee, C. T. A., Thurner, S., Paterson, S. & Cao, W. R. The rise and fall of continental arcs: interplays between magmatism, uplift, weathering, and climate. *Earth and Planetary Science Letters* **425**, 105–119 (2015).
186. Allard, P. et al. Prodigious emission rates and magma degassing budget of major, trace and radioactive volatile species from Ambrym basaltic volcano, Vanuatu island Arc. *Journal of Volcanology and Geothermal Research* **322**, 119–143 (2016).
187. Mailik, N. Temperature and gas composition of the Avachinsky volcano fumaroles (Kamchatka) in 2013–2017. In: *13th CCVG-IAVCEI Gas Workshop* (CCVG-IAV-CEI, 2017).
188. Burton, M. et al. New constraints on volcanic CO_2 emissions from Java, Indonesia. *Geophysical Research Abstracts* **20**, EGU2018-15195 (2018).
189. Taran, Y. et al. Gas emissions from volcanoes of the Kuril Island Arc (NW Pacific): geochemistry and fluxes. *Geochemistry, Geophysics, Geosystems* **19**, 1859–1880 (2018).
190. Aiuppa, A. et al. First determination of magma-derived gas emissions from Bromo volcano, eastern Java (Indonesia). *Journal of Volcanology and Geothermal Research* **304**, 206–213 (2015).
191. Tamburello, G. et al. Intense magmatic degassing through the lake of Copahue volcano, 2013–2014. *Journal of Geophysical Research: Solid Earth* **120**, 6071–6084 (2015).
192. Bani, P. et al. Dukono, the predominant source of volcanic degassing in Indonesia, sustained by a depleted Indian-MORB. *Bulletin of Volcanology* **80**, 5 (2017).
193. Battaglia, A. et al. The magmatic gas signature of Pacaya volcano, with implications for the volcanic CO_2 flux from Guatemala. *Geochemistry, Geophysics, Geosystems* **19**, 667–692 (2018).
194. Gerlach, T. M., McGee, K. A. & Doukas, M. P. Emission rates of CO_2, SO_2, and H_2S, scrubbing, and preeruption excess volatiles at Mount St. Helens, 2004–2005. In: *A Volcano Rekindled: The Renewed Eruption of Mount St. Helens, 2004–2006* (eds. W. E. Scott, D. R. Sherrod & P. H. Stauffer), 554–571 (2008).
195. Aiuppa, A. et al. First volatile inventory for Gorely volcano, Kamchatka. *Geophysical Research Letters* **39**, L06307 (2012).
196. Gunawan, H. et al. New insights into Kawah Ijen's volcanic system from the wet volcano workshop experiment. *Geological Society, London, Special Publications* **437**, 35 (2016).
197. Sutton, A. J. & Elias, T. One hundred volatile years of volcanic gas studies at the Hawaiian Volcano Observatory. In: *Characteristics of Hawaiian Volcanoes. US*

Geological Survey Professional Paper 1801 (eds. M. P. Poland, T. J. Takahashi & C. M. Landowski), 295–320 (USGS, 2014).

198. Allard, P. *Isotope Geochemistry and Origins of Water, Carbon and Sulfur in Volcanic Gases: Rift Zones, Continental Margins and Island Arcs* (Paris VII University, 1986).
199. Bani, P. et al. First measurement of the volcanic gas output from Anak Krakatau, Indonesia. *Journal of Volcanology and Geothermal Research* **302**, 237–241 (2015).
200. Lopez, T. et al. Geochemical constraints on volatile sources and subsurface conditions at Mount Martin, Mount Mageik, and Trident Volcanoes, Katmai Volcanic Cluster, Alaska. *Journal of Volcanology and Geothermal Research* **347**, 64–81 (2017).
201. Burton, M. R., Oppenheimer, C., Horrocks, L. A. & Francis, P. W. Remote sensing of CO_2 and H_2O emission rates from Masaya volcano, Nicaragua. *Geology* **28**, 915–918 (2000).
202. Martin, R. S. et al. A total volatile inventory for Masaya Volcano, Nicaragua. *Journal of Geophysical Research* **115**, 1–12 (2010).
203. Allard, P., Metrich, N. & Sabroux, J. C. Volatile and magma supply to standard eruptive activity at Merapi volcano, Indonesia. In: *EGU General Assembly 2011*, Vol. 13, EGU2011-13522 (EGU, 2011).
204. Dzurisin, D. et al. The 2004–2008 dome-building eruption at Mount St. Helens, Washington: epilogue. *Bulletin of Volcanology* **77**, 17 (2015).
205. Taran, Y. A. Geochemistry of volcanic and hydrothermal fluids and volatile budget of the Kamchatka–Kuril subduction zone. *Geochimica et Cosmochimica Acta* **73**, 1067–1094 (2009).
206. Lages J. et al. Volcanic CO_2 and SO_2 emissions along the Colombia Arc Segment (Northern Volcanic Zone). In: *Geophysical Research Abstracts*, Vol. 20, EGU2018-1301 (2018).
207. Bobrowski, N. et al. Plume composition and volatile flux of Nyamulagira volcano, Democratic Republic of Congo, during birth and evolution of the lava lake, 2014–2015. *Bulletin of Volcanology* **79**, 90 (2017).
208. Lyons, J. J. et al. Long period seismicity and very long period infrasound driven by shallow magmatic degassing at Mount Pagan, Mariana Islands. *Journal of Geophysical Research: Solid Earth* **121**, 188–209 (2016).
209. Tulet, P. et al. First results of the Piton de la Fournaise STRAP 2015 experiment: multidisciplinary tracking of a volcanic gas and aerosol plume. *Atmospheric Chemistry and Physics* **17**, 5355–5378 (2017).
210. Aiuppa, A. et al. Gas measurements from the Costa Rica–Nicaragua volcanic segment suggest possible along-arc variations in volcanic gas chemistry. *Earth and Planetary Science Letters* **407**, 134-147 (2014).
211. Maldonado, L. F. M., Inguaggiato, S., Jaramillo, M. T., Valencia, G. G. & Mazot, A. Volatiles and energy released by Puracé volcano. *Bulletin of Volcanology* **79**, 84 (2017).
212. D'Aleo, R. et al. Preliminary results of a multi-parametric characterization of gas manifestations from volcanoes in west Papua New Guinea. Presented at: *Conferenze Nationale Rittmann Giovani Ricercatori*, Bari, Italy, 2016.
213. Moussallam, Y. et al. Volcanic gas emissions and degassing dynamics at Ubinas and Sabancaya volcanoes; implications for the volatile budget of the central volcanic zone. *Journal of Volcanology and Geothermal Research* **343**, 181–191 (2017).
214. Granieri, D. et al. Emission of gas and atmospheric dispersion of SO_2 during the December 2013 eruption at San Miguel volcano (El Salvador, Central America). *Geophysical Research Letters* **42**, 5847–5854 (2015).

215. Bani, P. et al. First study of the heat and gas budget for Sirung volcano, Indonesia. *Bulletin of Volcanology* **79**, 60 (2017).
216. Edmonds, M. et al. Excess volatiles supplied by mingling of mafic magma at an andesite arc volcano. *Geochemistry, Geophysics, Geosystems* **11**, Q04005 (2010).
217. Allard, P. et al. Steam and gas emission rate from La Soufriere volcano, Guadeloupe (Lesser Antilles): implications for the magmatic supply during degassing unrest. *Chemical Geology* **384**, 76–93 (2014).
218. GeoNet Database. www.geonet.org.nz (2018).
219. Campion, R. et al. Space- and ground-based measurements of sulphur dioxide emissions from Turrialba Volcano (Costa Rica). *Bulletin of Volcanology* **74**, 1757–1770 (2012).
220. Conde, V. et al. SO_2 degassing from Turrialba Volcano linked to seismic signatures during the period 2008–2012. *International Journal of Earth Sciences* **103**, 1983–1998 (2014).
221. Epiard, M. et al. Relationship between diffuse CO_2 degassing and volcanic activity. Case study of the Poás, Irazú, and Turrialba Volcanoes, Costa Rica. *Frontiers in Earth Science* **5** (2017).
222. Martini, F. et al. Geophysical, geochemical and geodetical signals of reawakening at Turrialba volcano (Costa Rica) after almost 150 years of quiescence. *Journal of Volcanology and Geothermal Research* **198**, 416–432 (2010).
223. Vaselli, O. et al. Evolution of fluid geochemistry at the Turrialba volcano (Costa Rica) from 1998 to 2008. *Bulletin of Volcanology* **72**, 397–410 (2010).
224. McGee, K. A., Doukas, M. P., McGimsey, R. G., Wessels, R. L. & Neal, C. A. Gas emissions from Augustine Volcano, Alaska 1995–2006. *EOS Transactions* **87**, 1687 (2006).
225. López, T. et al. Constraints on magma processes, subsurface conditions, and total volatile flux at Bezymianny Volcano in 2007–2010 from direct and remote volcanic gas measurements. *Journal of Volcanology and Geothermal Research* **263**, 92–107 (2013).
226. Varley, N. R. & Taran, Y. Degassing processes of Popocatepetl and Volcan de Colima, Mexico. *Geological Society Special Publications* **213**, 263–280 (2003).
227. Hidalgo, S. et al. Evolution of the 2015 Cotopaxi eruption revealed by combined geochemical and seismic observations. *Geochemistry Geophysics Geosystems* **19**, 2087-2108 (2018).
228. Fischer, T. P. et al. The chemical and isotopic composition of fumarolic gases and spring discharges from Galeras Volcano, Colombia. *Journal of Volcanology and Geothermal Research* **77**, 229–253 (1997).
229. Tamburello, G., Hansteen, T. H., Bredemeyer, S., Aiuppa, A. & Tassi, F. Gas emissions from five volcanoes in northern Chile and implications for the volatiles budget of the Central Volcanic Zone. *Geophysical Research Letters* **41**, 4961–4969 (2014).
230. Menyailov, I. A., Nikitina, L. P., Shapar, V. N. & Pilipenko, V. P. Temperature increase and chemical change of fumarolic gases at Momotombo Volcano, Nicaragua, in 1982–1985 – are these indicators of possible eruption *Journal of Geophysical Research – Solid Earth and Planets* **91**, 2199–2214 (1986).
231. Hammouya, G. et al. Pre- and syn-eruptive geochemistry of volcanic gases from Soufriere Hills of Montserrat, West Indies. *Geophysical Research Letters* **25**, 3685–3688 (1998).
232. Kusakabe, M. et al. Evolution of CO_2 in Lakes Monoun and Nyos, Cameroon, before and during controlled degassing. *Geochemical Journal* **42**, 93–118 (2008).

233. Dionis, S. M. et al. Diffuse CO_2 degassing and volcanic activity at Cape Verde islands, West Africa. *Earth, Planets and Space* **67**, 48 (2015).
234. Zhang, L., Guo, Z., Zhang, M. & Cheng, Z. Study on soil micro-seepage gas flux in the high temperature geothermal area: an example from the Yangbajing geothermal field, South Tibet. *Acta Petrologica Sinica* **30**, 3612–3626 (2014).
235. Zhang, M. L. et al. Magma-derived CO_2 emissions in the Tengchong volcanic field, SE Tibet: implications for deep carbon cycle at intra-continent subduction zone. *Journal of Asian Earth Sciences* **127**, 76–90 (2016).
236. Cheng, Z., Guo, Z., Zhang, M. & Zhang, L. CO_2 flux estimations of hot springs in the Tengchong Cenozoic volcanic field, Yunnan Province, SW China. *Acta Petrologica Sinica* **28**, 1217–1224 (2012).
237. Cheng, Z., Guo, Z., Zhang, M. & Zhang, L. Carbon dioxide emissions from Tengchong Cenozoic volcanic field, Yunnan Province, SW China. *Acta Petrologica Sinica* **30**, 3657–3670 (2014).
238. Guo, Z., Zhang, M., Cheng, Z., Zhang, L. & Liu, J. Fluxes and genesis of greenhouse gases emissions from typical volcanic fields in China. *Acta Petrologica Sinica* **30**, 3467-3480 (2014).
239. Zhang, M. et al. Greenhouse gases flux estimation of hot springs in Changbaishan volcanic field, NE China. *Acta Petrologica Sinica* **24**, 2898–2904 (2011).
240. Sun, Y. & Guo, Z. CO_2 diffuse emission from maar lake: an example in Changbai volcanic field, NE China. *Journal of Volcanology and Geothermal Research* **349**, 146–162 (2017).
241. Galindo, I. et al. Emision difusa de dioxido de carbono en el volcan Irazu, Costa Rica. Carbon dioxide emissions at Irazu Volcano, Costa Rica. *Revista Geologica de America Central* **30**, 157–165 (2004).
242. Liegler, A. *Diffuse CO_2 Degassing and the Origin of Volcabic Gas Variability from Rincon de la Vieja, Miravalles and Tenorio Volcanoes*, Master of Science in Geology thesis, Michigan Technological University (2016).
243. Melián, G. V. et al. Emisión difusa de CO_2 y actividad volcánica en el volcán Poás, Costa Rica. *Revista Geológica de América Central* **43**, 147–170 (2010).
244. Padron, E. et al. Diffuse CO_2 emission rate from Pululahua and the lake-filled Cuicocha calderas, Ecuador. *Journal of Volcanology and Geothermal Research* **176**, 163–169 (2008).
245. Padron, E. et al. Fumarole/plume and diffuse CO_2 emission from Sierra Negra *caldera, Galapagos archipelago. Bulletin of Volcanology* **74**, 1509–1519 (2012).
246. Salazar, J. M. L. et al. Spatial and temporal variations of diffuse CO_2 degassing at the Santa Ana-Izalco-Coatepeque volcanic complex, *El Salvador, Central America. Special Paper – Geological Society of America* **375**, 135–146 (2004).
247. López, D. L., Ransom, L., Pérez, N. M., Hernández, P. A. & Monterrosa, J. Dynamics of diffuse degassing at Ilopango Caldera, El Salvador. In: *Special Paper of the Geological Society of America*, Vol. 375 (eds. W. I. Rose et al.), 191–202 (Geological Society of America, 2004).
248. Hutchison, W., Mather, T. A., Pyle, D. M., Biggs, J. & Yirgu, G. Structural controls on fluid pathways in an active rift system: a case study of the Aluto volcanic complex. *Geosphere* **11**, 542–562 (2015).
249. Brombach, T., Hunziker, J. C., Chiodini, G., Cardellini, C. & Marini, L. Soil diffuse degassing and thermal energy fluxes from the southern Lakki plain, Nisyros (Greece). *Geophysical Research Letters* **28**, 69–72 (2001).

250. Caliro, S. et al. Recent activity of Nisyros volcano (Greece) inferred from structural, geochemical and seismological data. *Bulletin of Volcanology* **67**, 358–369 (2005).
251. D'Alessandro, W. et al. Diffuse and focused carbon dioxide and methane emissions from the Sousaki geothermal system, Greece. *Geophysical Research Letters* **33**, 5 (2006).
252. Parks, M. M. et al. Distinguishing contributions to diffuse CO_2 emissions in volcanic areas from magmatic degassing and thermal decarbonation using soil gas Rn-222-delta C-13 systematics: application to Santorini volcano, Greece. *Earth and Planetary Science Letters* **377**, 180–190 (2013).
253. D'Alessandro, W., Brusca, L., Kyriakopouios, K., Michas, G. & Papadakis, G. Methana, the westernmost active volcanic system of the south Aegean arc (Greece): insight from fluids geochemistry. *Journal of Volcanology and Geothermal Research* **178**, 818–828 (2008).
254. Fridriksson, T. et al. CO_2 emissions and heat flow through soil, fumaroles, and steam heated mud pools at the Reykjanes geothermal area, SW Iceland. *Applied Geochemistry* **21**, 1551–1569 (2006).
255. Hernandez, P. et al. Diffuse volcanic degassing and thermal energy release from Hengill volcanic system, Iceland. *Bulletin of Volcanology* **74**, 2435–2448 (2012).
256. Toutain, J. P. et al. Structure and CO_2 budget of Merapi volcano during inter-eruptive periods. *Bulletin of Volcanology* **71**, 815–826 (2009).
257. Carapezza, M. L. et al. Diffuse CO_2 soil degassing and CO_2 and H_2S concentrations in air and related hazards at Vulcano Island (Aeolian arc, Italy). *Journal of Volcanology and Geothermal Research* **207**, 130–144 (2011).
258. Chiodini, G., Frondini, F. & Raco, B. Diffuse emission of CO_2 from the Fossa Crater, Vulcano Island (Italy). *Bulletin of Volcanology* **58**, 41–50 (1996).
259. Inguaggiato, S. et al. Total CO_2 output from Vulcano island (Aeolian Islands, Italy). *Geochemistry, Geophysics, Geosystems* **13**, Q02012 (2012).
260. Granieri, D. et al. Correlated increase in CO_2 fumarolic content and diffuse emission from La Fossa crater (Vulcano, Italy): evidence of volcanic unrest or increasing gas release from a stationary deep magma body? *Geophysical Research Letters* **33**, L13316 (2006).
261. D' Alessandro, W. et al. Chemical and isotopic characterization of the gases of Mount Etna (Italy). *Journal of Volcanology and Geothermal Research* **78**, 65–76 (1997).
262. Giammanco, S., Bellotti, F., Groppelli, G. & Pinton, A. Statistical analysis reveals spatial and temporal anomalies of soil CO_2 efflux on Mount Etna volcano (Italy). *Journal of Volcanology and Geothermal Research* **194**, 1–14 (2010).
263. Camarda, M., De Gregorio, S. & Gurrieri, S. Magma-ascent processes during 2005–2009 at Mt Etna inferred by soil CO_2 emissions in peripheral areas of the volcano. *Chemical Geology* **330**, 218–227 (2012).
264. De Gregorio, S. & Camarda, M. A novel approach to estimate the eruptive potential and probability in open conduit volcanoes. *Scientific Reports* **6**, 30471 (2016).
265. Melian, G. et al. Diffuse and visible emission of CO_2 from Etna Volcano, Italy. American Geophysical Union, Fall Meeting #V21D-202 (2009).
266. Inguaggiato, S. et al. CO_2 output discharged from Stromboli Island (Italy). *Chemical Geology* **339**, 52–60 (2013).
267. Granieri, D., Chiodini, G., Avino, R. & Caliro, S. Carbon dioxide emission and heat release estimation for Pantelleria Island (Sicily, Italy). *Journal of Volcanology and Geothermal Research* **275**, 22–33 (2014).

268. Favara, R., Giammanco, S., Inguaggiato, S. & Pecoraino, G. Preliminary estimate of CO_2 output from Pantelleria Island volcano (Sicily, Italy): evidence of active mantle degassing. *Applied Geochemistry* **16**, 883–894 (2001).
269. Frondini, F. et al. Diffuse CO_2 degassing at Vesuvio, Italy. *Bulletin of Volcanology* **66**, 642–651 (2004).
270. Granieri, D. et al. Level of carbon dioxide diffuse degassing from the ground of Vesuvio: comparison between extensive surveys and inferences on the gas source. *Annals of Geophysics* **56**, doi:10.4401/ag-6455 (2013).
271. Aiuppa, A. et al. First observations of the fumarolic gas output from a restless caldera: implications for the current period of unrest (2005–2013) at Campi Flegrei. *Geochemistry Geophysics Geosystems* **14**, 4153–4169 (2013).
272. Pecoraino, G. et al. Total CO_2 output from Ischia Island volcano (Italy). *Geochemical Journal* **39**, 451–458 (2005).
273. Chiodini, G. et al. Fumarolic and diffuse soil degassing west of Mount Epomeo, Ischia, Italy. *Journal of Volcanology and Geothermal Research* **133**, 291–309 (2004).
274. Rogie, J. D., Kerrick, D. M., Chiodini, G. & Frondini, F. Flux measurements of non-volcanic CO2 emission from some vents in central Italy. *Journal of Geohysical Research: Solid Earth* **105**, 8435–8445 (2000).
275. Chiodini, G. et al. Geochemical evidence for and characterization of CO_2 rich gas sources in the epicentral area of the Abruzzo 2009 earthquakes. *Earth and Planetary Science Letters* **304**, 389–398 (2011).
276. Frondini, F., Caliro, S., Cardellini, C., Chiodini, G. & Morgantini, N. Carbon dioxide degassing and thermal energy release in the Monte Amiata volcanic-geothermal area (Italy). *Applied Geochemistry* **24**, 860–875 (2009).
277. Costa, A. et al. A shallow-layer model for heavy gas dispersion from natural sources: application and hazard assessment at Caldara di Manziana, Italy. *Geochemistry, Geophysics, Geosystems* **9**, Q03002 (2008).
278. Quattrocchi, F. et al. Continuous/discrete geochemical monitoring of CO_2 natural analogues and of diffuse degassing structures (DDS): hints for CO_2 storage sites geochemical monitoring protocol. In: *Greenhouse Gas Control Technologies 9, Vol. 1 Energy Procedia* (eds. J. Gale, H. Herzog & J. Braitsch), 2135–2142 (Elsevier, 2009).
279. Hernandez, P. A. et al. Diffuse emission of CO_2 from Miyakejima volcano, Japan. *Chemical Geology* **177**, 175–185 (2001).
280. Hernandez, P. A. et al. Diffuse emission of CO_2 from Showa-Shinzan, Hokkaido, Japan; a sign of volcanic dome degassing. *Pure and Applied Geophysics* **163**, 869–881 (2006).
281. Hernandez, P. A. et al. Carbon dioxide degassing by advective flow from Usu volcano, Japan. *Science* **292**, 83–86 (2001).
282. Hirabayashi, J. & Mizuhashi, M. The discharge rate of volatiles from Kusatsu-Shirane volcano, Japan. Report of 4th Joint Observation of Kusatsu-Shirane Volcano, 167–174 (2004).
283. Shimoike, Y., Kazahaya, K. & Shinohara, H. Soil gas emission of volcanic CO_2 at Satsuma-Iwojima Volcano, Japan. *Earth, Planets and Space* **54**, 239–247 (2002).
284. Saito, M., Matsushima, T., Matsuwo, N. & Shimizu, H. Observation SO_2 and CO_2 fluxes in and around the active crater of Aso Nakadake Volcano. In: *Science Reports of the Kyushu University, Department of Earth and Planetary Sciences*, Vol. 22 (2007).
285. Notsu, K., Mori, T., Do Vale, S. C., Kagi, H. & Ito, T. Monitoring quiescent volcanoes by diffuse CO_2 degassing; case study of Mt. Fuji, Japan. *Pure and Applied Geophysics* **163**, 825–835 (2006).

286. Hernandez, P. A., Mori, T., Padron, E., Sumino, H. & Perez, N. Carbon dioxide emission from Katanuma volcanic lake, Japan. *Earth, Planets and Space* **63**, 1151–1156 (2011).
287. Notsu, K. et al. Diffuse CO_2 efflux from Iwojima Volcano, Izu-Ogasawara Arc, Japan. *Journal of Volcanology and Geothermal Research* **139**, 147–161 (2005).
288. Robertson, E. et al. Diffuse degassing at Longonot volcano, Kenya: implications for CO_2 flux in continental rifts. *Journal of Volcanology and Geothermal Research* **327**, 208–222 (2016).
289. Werner, C., Christenson, B., Scott, B., Britten, K. & Kilgour, G. Monitoring CO2 emissions at White Island volcano, New Zealand: evidence for total decreases in magmatic mass and heat output. In: *Water Rock Interaction, Eleventh Symposium* (eds. R. Wanty, & R. R. Seal), 223–226 (A.A. Balkema Publishers, 2004).
290. Harvey, M. C. et al. Heat flux from magmatic hydrothermal systems related to availability of fluid recharge. *Journal of Volcanology and Geothermal Research* **302**, 225–236 (2015).
291. Salazar, J. M. L. et al. Diffuse emission of carbon dioxide from Cerro Negro Volcano, Nicaragua, Central America. *Geophysical Research Letters* **28**, 4275–4278 (2001).
292. Harvey, M. C., White, P. J., Kenzie, K. M. & Lovelock, B. G. Results from soil CO_2 flux and shallow temeperature survey at the San Jacinto-Tizate geothermal power project. In: *Nicaragua in New Zealand Geothermal Workshop 2011*, 1–8 (University of Auckland, 2011).
293. Lewicki, J. L. et al. Comparative soil CO_2 flux measurements and geostatistical estimation methods on Masaya volcano, Nicaragua. *Bulletin of Volcanology* **68**, 76–90 (2005).
294. Arpa, M. C. et al. Geochemical evidence of magma intrusion inferred from diffuse CO_2 emissions and fumarole plume chemistry: the 2010-2011 volcanic unrest at Taal Volcano, Philippines. *Bulletin of Volcanology* **75**, 747 (2013).
295. Viveiros, F. et al. Soil CO_2 emissions at Furnas volcano, Sao Miguel Island, Azores archipelago: Volcano monitoring perspectives, geomorphologic studies, and land use planning application. *Journal of Geophysical Research: Solid Earth* **115**, B12208 (2010).
296. Andrade, C., Viveiros, F., Cruz, J. V., Coutinho, R. & Silva, C. Estimation of the CO_2 flux from Furnas volcanic Lake (São Miguel, Azores). *Journal of Volcanology and Geothermal Research* **315**, 51–64 (2016).
297. Frunzeti, N. *Geogenic Emissions of Greenhouse Gases in the Southern Part of the Eastern Carpathians*. PhD thesis (Babcş-Bolyai University, 2013).
298. Frunzeti, N. & Baciu, C. Diffuse CO_2 emission at Santa Ana lake-filled crater (Eastern Carpathians, Romania). *Procedia Environmental Sciences,* **14**, 188–194 (2012).
299. Inguaggiato, S., Cardellini, C., Taran, Y. & Kalacheva, E. The CO_2 flux from hydrothermal systems of the Karymsky volcanic Centre, Kamchatka. *Journal of Volcanology and Geothermal Research* **346**, 1–9 (2017).
300. Hernandez, P. A. et al. Analysis of long- and short-term temporal variations of the diffuse CO_2 emission from Timanfaya volcano, Lanzarote, Canary Islands. *Applied Geochemistry* **27**, 2486–2499 (2012).
301. Perez, P. H. et al. Carbon dioxide emissions from soils at Hakkoda, north Japan. *Journal of Geophysical Research* **108**, 9 (2003)
302. Melian, G. et al. A magmatic source for fumaroles and diffuse degassing from the summit crater of Teide Volcano (Tenerife, Canary Islands): a geochemical evidence for the 2004–2005 seismic-volcanic crisis. *Bulletin of Volcanology* **74**, 1465–1483 (2012).

303. Hernandez, P. A. et al. Geochemical evidences of seismo-volcanic unrests at the NW rift zone of Tenerife, Canary Islands, inferred from diffuse CO_2 emission. *Bulletin of Volcanology* **79**, 30 (2017).
304. Melian, G. et al. Spatial and temporal variations of diffuse CO_2 degassing at El Hierro volcanic system: relation to the 2011-2012 submarine eruption. *Journal of Geophysical Research: Solid Earth* **119**, 6976–6991 (2014).
305. Padron, E. et al. Dynamics of diffuse carbon dioxide emissions from Cumbre Vieja volcano, La Palma, Canary Islands. *Bulletin of Volcanology* **77**, 28 (2015).
306. Lan, T. F. et al. Compositions and flux of soil gas in Liu-Huang-Ku hydrothermal area, northern Taiwan. *Journal of Volcanology and Geothermal Research* **165**, 32–45 (2007).
307. Wen, H.-Y. et al. Soil CO_2 flux in hydrothermal areas of the Tatun Volcano Group, Northern Taiwan. *Journal of Volcanology and Geothermal Research* **321**, 114-124 (2016).
308. Nisi, B., Vaselli, O., Marchev, P. & Tassi, F. Diffuse CO_2 soil flux measurements at the youngest volcanic system in Bulgaria: the 12.2 Ma old Kozhuh cryptodome. *Acta Vulcanologica* **25**, 169–178 (2013).
309. Bergfeld, D., Goff, F. & Janik, C. J. Elevated carbon dioxide flux at the Dixie Valley geothermal field, Nevada; relations between surface phenomena and the geothermal reservoir. *Chemical Geology* **177**, 43–66 (2001).
310. Bergfeld, D., Evans, W. C., Howle, J. F. & Farrar, C. D. Carbon dioxide emissions from vegetation-kill zones around the resurgent dome of Long Valley caldera, eastern California, USA. *Journal of Volcanology and Geothermal Research* **152**, 140–156 (2006).
311. Bergfeld, D., Evans, W. C., Howle, J. F. & Hunt, A. G. Magmatic gas emissions at Holocene volcanic features near Mono Lake, California, and their relation to regional magmatism. *Journal of Volcanology and Geothermal Research* **292**, 70–83 (2015).
312. Evans, W. C., Bergfeld, D., McGimsey, R. G. & Hunt, A. G. Diffuse gas emissions at the Ukinrek Maars, Alaska: implications for magmatic degassing and volcanic monitoring. *Applied Geochemistry* **24**, 527–535 (2009).
313. Sorey, M. L., Werner, C., McGimsey, R. G. & Evans, W. C. *Hydrothermal Activity and Carbon Dioxide Discharge at Shrub and Upper Klawasi Mud Volcanoes, Wrangell Mountains, Alaska. Water-Resources Investigations Report 00-4207* (USGS, 2000).
314. Werner, C. et al. Volatile emissions and gas geochemistry of Hot Spring Basin, Yellowstone National Park, USA. *Journal of Volcanology and Geothermal Research* **178**, 751–762 (2008).
315. Bergfeld, D., Evans, W. C., Lowenstern, J. B. & Hurwitz, S. Carbon dioxide and hydrogen sulfide degassing and cryptic thermal input to Brimstone Basin, Yellowstone National Park, Wyoming. *Chemical Geology* **330**, 233–243 (2012).
316. Lin, P., Deering, C. D., Werner, C. & Torres, C. Origin and quantification of diffuse CO_2 and H_2S emissions at Crater Hills, Yellowstone National Park. *Journal of Volcanology and Geothermal Research*, **377**, 117–130 (2019).
317. Marty, B., Alexander, C. M. O. & Raymond, S. N. Primordial origins of Earth's carbon. In: *Reviews in Mineralogy and Geochemistry, Vol. 75*, 149–181 (Mineralogical Society of America, 2013).

9

Carbon in the Convecting Mantle

ERIK H. HAURI, ELIZABETH COTTRELL, KATHERINE A. KELLEY, JONATHAN M. TUCKER, KEI SHIMIZU, MARION LE VOYER, JARED MARSKE, AND ALBERTO E. SAAL

9.1 Introduction

The flux of magmatic CO_2 into the oceans and atmosphere contributes to the global surface carbon cycle, and changes in basaltic magma production and associated degassing have been proposed as one of several important forcing mechanisms that have influenced past global climate variations.[1–4] Despite this degassing, the vast majority of Earth's carbon is present not at the surface but in Earth's convecting mantle.[5–10] The convecting mantle is the part of the mantle that lies beneath tectonic plates, and its motions over geologic timescales are driven by the sinking of cold oceanic plates into the interior and hot active upwellings originating from great depth. The convecting upper mantle rises, melts, and erupts mid-ocean ridge basalt (MORB) due to seafloor spreading along the 56,000-km length of the global mid-ocean ridge system. Active upwellings, driven by mantle buoyancy, produce intraplate and near-ridge hot spots (e.g. Hawaii, Iceland) that erupt ocean island basalt (OIB) with a smaller overall flux.[11] Occasionally, however, active upwellings produce large igneous provinces (LIPs; e.g. Ontong–Java Plateau, Caribbean–Columbian Plateau, Manihiki Plateau, Kerguelen Plateau) characterized by enormous outputs of magma erupted over very short timescales. These flood basalt eruptions have often coincided in time with global climate crises and mass species extinctions.[12–16] Ridges, hot spots, and LIPs represent the main volcanic expressions of the delivery of heat to Earth's surface by the ongoing motions of the convecting mantle.

Seafloor dredging and core recovery have produced tens of thousands of individual submarine glass and rock samples. Study of these samples has demonstrated that variations in the major and trace element chemistry of MORBs and OIBs are primarily the result of crystal fractionation following polybaric melting of mantle sources at different depths and variable temperatures.[17–21] At the same time, isotopic differences between MORBs and OIBs show that the mantle sources of hot spots are much more diverse and heterogeneous than those of ridges, reflecting their derivation from distinct reservoirs within the convecting mantle consisting of numerous geochemically distinct components with varied origins and evolutions.[22–25] Some of these reservoirs have retained anomalies in short-lived isotope systems that were active only in the first tens of millions of years of Earth's history, and thus appear to have been isolated from each other – and from Earth's surface – for a large fraction of the age of Earth.[26–33] It is also well established that, from segment to segment, the concentrations of highly incompatible trace elements do not correlate with crustal thickness (a proxy for potential temperature) or spreading rate (a proxy for melting dynamics). Instead, highly incompatible trace elements (e.g. Th concentrations, Ba/Nb ratios) covary with radiogenic isotopes (e.g. $^{87}Sr/^{86}Sr$) within segments,[34] approaching hot spots,[35] and globally;[36] these covariations indicate heterogeneity in the mantle source. Here, we rely on this fundamental observation of MORB and OIB geochemistry to infer mantle carbon concentrations (Section 9.3.2).

Carbon is a trace element in the mantle, and it is present in erupted magmas as CO_2-rich vesicles and as carbonate ions dissolved in the melt.[37,38] The solubility of CO_2 in basaltic magma is low at pressures of eruption on land or on the seafloor,[39–42] and as a result nearly all magmas have degassed much of their original pre-eruptive budget of carbon during eruption. Degassing has thus complicated the use of MORBs and OIBs for determining the carbon content of their mantle sources and the flux of CO_2 from basaltic eruptions. To attempt to circumvent this complication, researchers have used a variety of microbeam analysis methods to study the composition and volatile content of melt inclusions, which are small pockets of melt (tens to hundreds of microns in diameter) that become trapped in phenocryst minerals that grow in the magma prior to eruption at the surface (Figure 9.1) and are consequently less degassed than their host magmas. Fourier-transform infrared spectroscopy (FTIR) and secondary ion mass spectrometery (SIMS) have both been used to make accurate and precise measurements of CO_2 and H_2O in melt inclusions, providing a means to interrogate collections of MORB and OIB samples to study the origin and evolution of carbon in the convecting mantle. Most melt inclusions (with some exceptions; see below) contain a vapor bubble formed at vapor saturation after inclusion entrapment, during differential cooling of the melt inside the olivine crystals (Figure 9.1). These vapor bubbles typically contain most of the CO_2 within the inclusion, but they cannot be measured by SIMS or FTIR.[43,44]

In this chapter, we review the types of samples that are best studied to assessing the primary CO_2 content of magmas from mid-ocean ridges and hot spots, estimate volcanic fluxes, and estimate the CO_2 content of mantle sources present in the convecting mantle. While we are unable to assess the primary CO_2 content of LIP magmas, we review CO_2 flux estimates from the literature. The volcanism that we consider here does not include subduction zones or continental intraplate volcanism, which are covered in Chapter 8. We also do not consider petit-spot volcanism – small, young edifices discovered outboard of the Japan trench.[45] These lavas may originate from below the lithosphere–asthenosphere boundary and erupt on the seafloor due to plate flexure. They are highly alkalic and highly vesicular, such that they may carry significant CO_2, although it is unclear whether this CO_2 originated from the mantle or from crustal contamination. Their occurrence may be restricted to a few edifices outboard of the Japan trench, or could be ubiquitous. Their contribution to CO_2 outgassing is presently highly unconstrained.

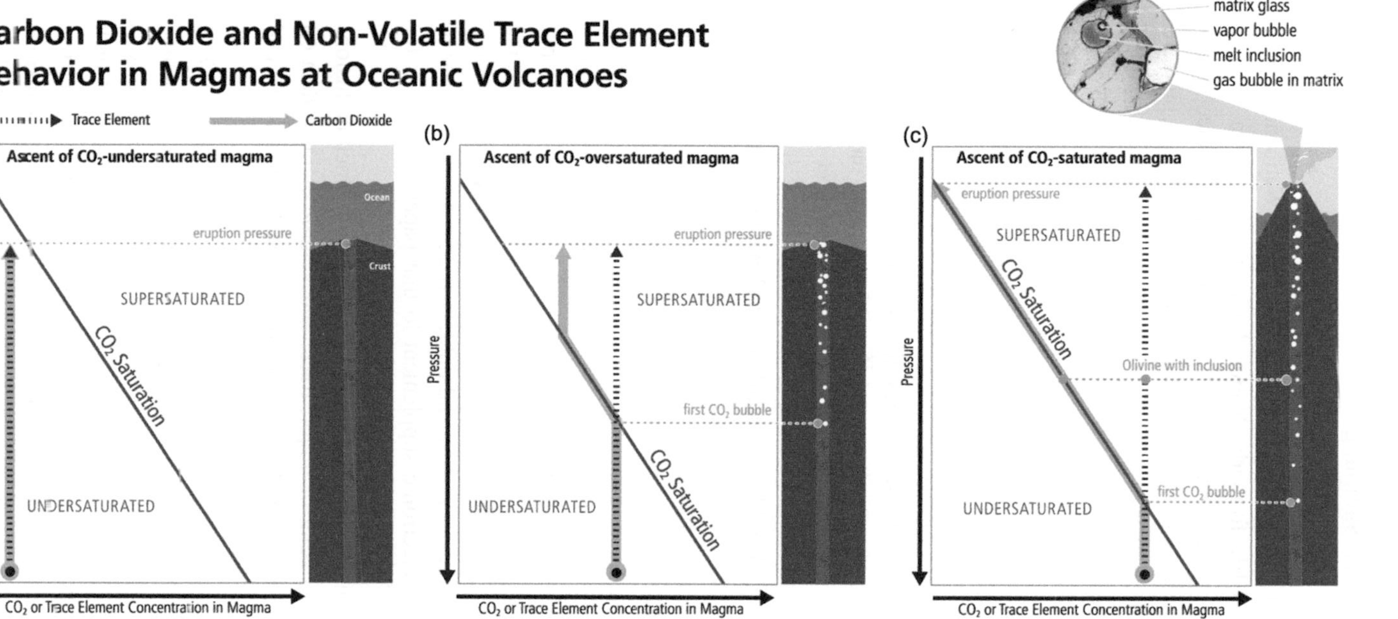

Figure 9.1 Illustration of carbon dioxide and nonvolatile trace element behavior in magmas at oceanic volcanoes. Each panel tracks the concentration of CO_2 and a nonvolatile trace element during the ascent of magma for three hypothetical cases. (a) A CO_2-undersaturated magma that begins with low volatile and trace element concentrations. In this case, the magma ascends and erupts on the seafloor without experiencing CO_2 degassing, such that the starting CO_2 concentration and that of a nonvolatile trace element remain roughly constant during ascent and eruption. (b) A magma that begins with higher volatile and trace element concentrations and experiences CO_2 saturation during ascent to the seafloor. At conditions of saturation, bubbles form as the magma moves upwards to lower pressure, removing CO_2 from the magma. The concentration of a nonvolatile trace element in this system, however, remains roughly constant. In the last stages prior to eruption, the magma may move faster than CO_2 is able to diffuse toward bubbles, resulting in an erupted magma on the seafloor that contains more CO_2 than would be present for equilibrium saturation (i.e. it is supersaturated). This is the most common condition for basalts erupted at mid-ocean ridges. (c) A volatile- and trace element-enriched magma that ascends beneath an oceanic island and experiences CO_2 saturation at greater depth. Because of the lower eruption pressure above sea level, the magma loses virtually all of its CO_2 to degassing, whereas the concentration of a nonvolatile trace element remains roughly constant. In all cases, olivine-hosted melt inclusions (see inset, sample NMNH 116111-5 melt inclusion is 100 μm in diameter) may trap melts during various stages of ascent. If trapped at pressures higher than the CO_2 saturation curve, melt inclusions may preserve the undegassed magma composition even if the external magma has experienced degassing.

9.2 Sampling

Major collections of seafloor samples (Figure 9.2) are discoverable and searchable through the Index to Marine and Lacustrine Geological Samples (IMLGS) at the National Centers for Environmental Information (NCEI; formerly National Geophysical Data Center, NGDC) at the US National Oceanographic and Atmospheric Association.[46] Once identified, samples can be requested from the individual sample repositories. In this chapter, we rely heavily on the MORB study of Le Voyer et al.[47] (and references therein), who reported new data on 753 geographically distributed MORB samples, combined with a quality-controlled database of published data from an additional 2446 samples, describing CO_2 fluxes and estimated primary magma CO_2 contents for 387 of the 458 mid-ocean ridge segments with major element data (out of a total of 711 geophysically defined MORB ridge segments).[17,36] Among these MORB samples, there exist samples with sufficiently low pre-eruptive carbon contents and high eruptive pressures that they are thought to have avoided CO_2 degassing (e.g. Figure 9.1a); these samples display correlations of CO_2 with nonvolatile trace elements, as described by Cartigny et al.[48] and Javoy and Pineau[38] (CO_2-rich samples), and by Saal et al.,[49] Michael and Graham[50], and Shimizu et al.[51] (undegassed samples; Figure 9.3). Additionally, three suites of melt inclusions from MORB sources show similar correlations of CO_2 with nonvolatile trace elements characteristic of vapor undersaturation, which have been described from the Siqueiros Transform on the East Pacific Rise,[49] the equatorial Mid-Atlantic Ridge,[52] and northern Iceland (Figure 9.3).[53] Importantly, these three melt inclusion suites are dominated by inclusions that have not formed post-entrapment vapor bubbles, such that the pre-eruptive magmatic CO_2 contents are preserved in the glass. These data are supplemented by an additional 113 melt inclusions that have been measured for CO_2 but show evidence of degassing; these come from the Gakkel Ridge,[54,55] the Juan de Fuca Ridge,[56] the Lucky Strike segment of the Mid-Atlantic Ridge,[57] and selected melt inclusions from Axial Seamount on the Juan de Fuca Ridge.[58] This entire set of new and published data, from the paper of Le Voyer et al.,[47] is available as an electronic supplement to that paper, as well as for download at the EarthChem Library.[59]

For hot spots, we rely heavily on an extensive new data set for Hawaiian melt inclusions reported by Marske and Hauri[60] consisting of major, trace, and volatile element data on 915 olivine-hosted melt inclusions in 29 samples of submarine-erupted pillow lavas and subaerial tephra deposits from the shield-building stage of five Hawaiian volcanoes (Figure 9.4). Tucker et al.[61] estimated bulk (dissolved + bubble) CO_2 contents in 437 of the Hawaiian melt inclusions studied by Marske and Hauri.[60] These data are supplemented by melt inclusion data from other hot spot localities where shrinkage of bubble dimensions are reported, namely the Azores,[62] Canaries,[63] and Iceland.[64,65] The compiled OIB melt inclusion dataset may be downloaded from the EarthChem Library.[66] We emphasize, however, that none of the microbeam and statistical methods applied to melt inclusions can account for CO_2 lost from magmas prior to melt inclusion entrapment in their crystal hosts.

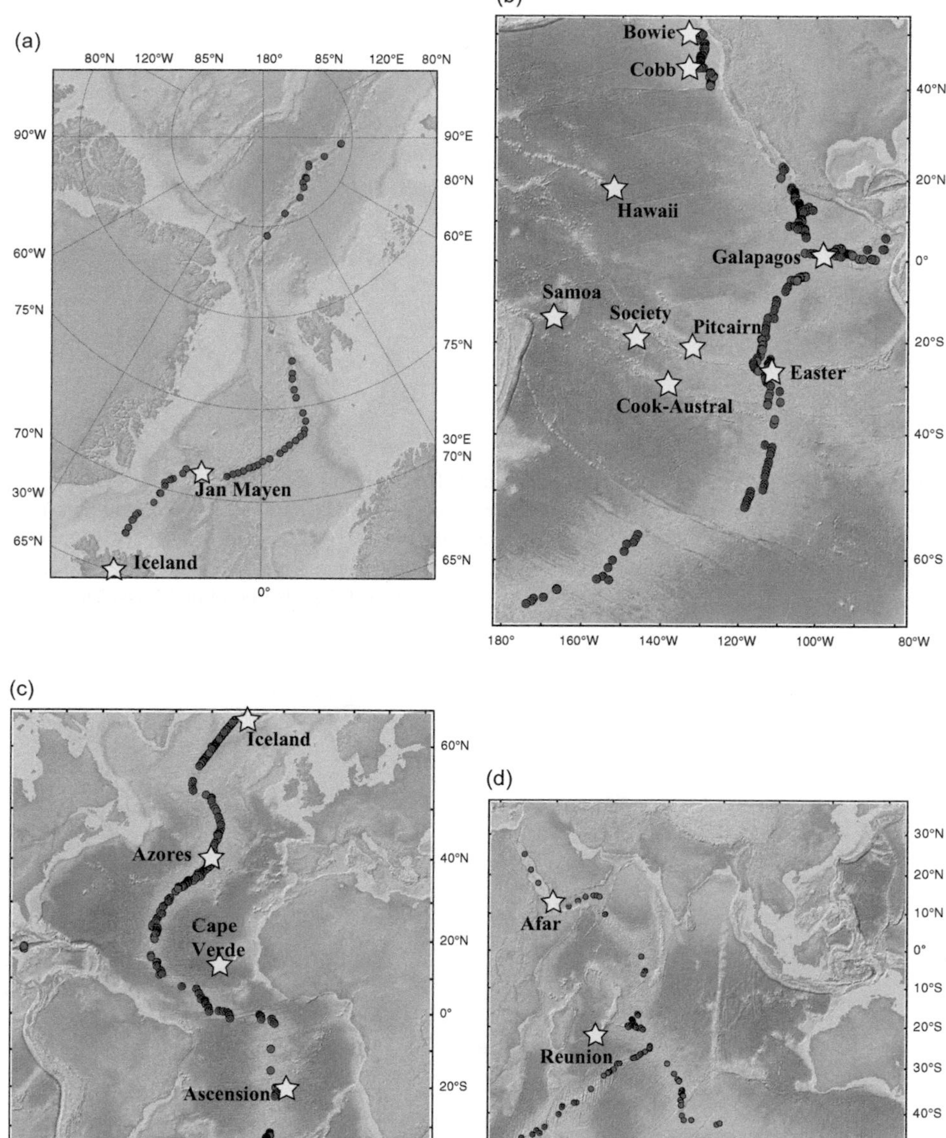

Figure 9.2 GeoMapApp figures showing MORB sample locations from the study of Le Voyer et al.[47] (red circles) and locations of previously published MORB samples (gray circles). Hot spots with estimated CO_2 fluxes are shown as yellow stars. (a) Arctic Ocean basin; (b) Pacific Ocean basin; (c) Atlantic Ocean basin; (d) Indian Ocean basin.

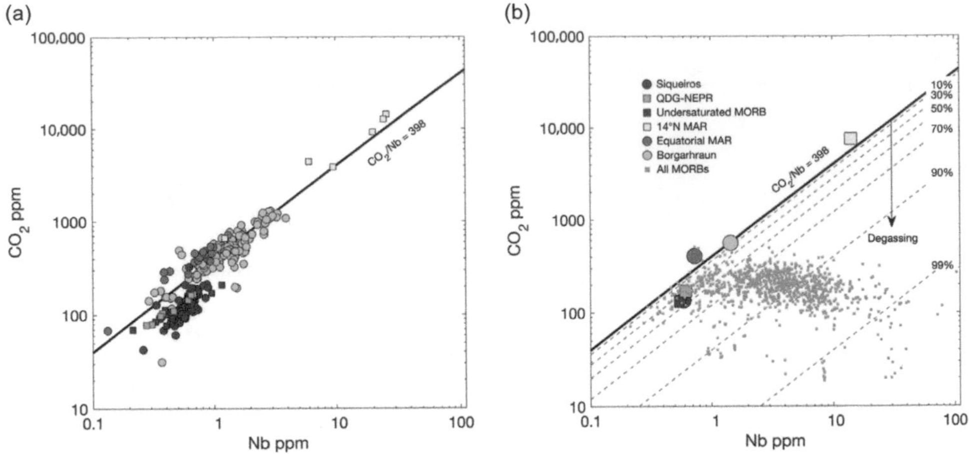

Figure 9.3 CO_2–Nb data for (a) vapor-undersaturated sample suites of submarine glasses and melt inclusions and (b) individual MORBs, with undersaturated sample suites from (a) represented by average compositions (boxes). The CO_2/Nb = 398 line represents the average of the individual average CO_2/Nb ratio for each of the undersaturated samples suites. For any given sample, one can gain a sense of the extent of CO_2 degassing by comparing the sample's CO_2/Nb ratio to the average CO_2/Nb ratio of undegassed samples.

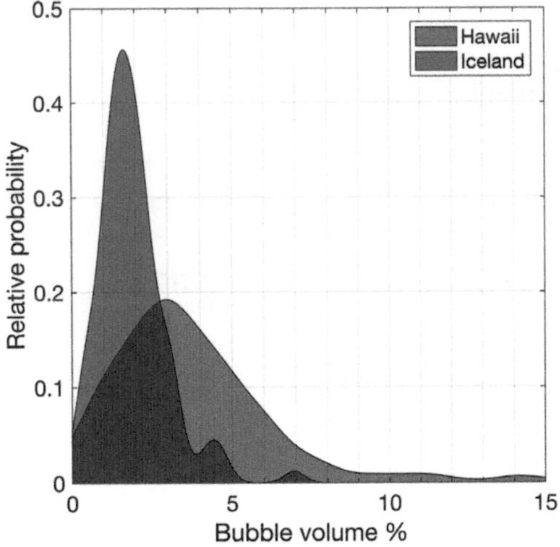

Figure 9.4 Probability distribution of the ratio of vapor bubble volume to total melt inclusion volume for vapor bubbles in melt inclusions from Hawaii[61] and Iceland.[64,65] Bubbles in Iceland melt inclusions strongly peak near 1.5 vol.%, whereas bubbles in Hawaiian melt inclusions are more variable and peak around 3 vol.%.

9.3 Fluxes of CO_2 from the Global Mid-Ocean Ridge System

The samples compiled by Le Voyer et al.[47] were assigned to individual ridge segments based on a comparison of their sampling coordinates with the global spreading center segment catalog of Gale et al.[36] (Figure 9.2). Segment-average volatile concentrations were calculated for all segments that contained an on-axis sample analyzed for H_2O, CO_2, F, S, and/or Cl. Segment averages for the volatile data were corrected to MgO = 8 wt.% and to compositions in equilibrium with Fo_{90} olivine in order to compare with the fractionation-normalized database of Gale et al.[17] We note that correction for crystal fractionation cannot account for the irreversible loss of CO_2 by degassing, although we estimate primary magma CO_2 contents using Ba and Rb as proxies, as explained below.

MORBs are erupted under hydrostatic pressure on the seafloor, determined by the water depth at the site of MORB collection. The solubility of CO_2 in MORBs is strongly dependent on pressure,[37,41] and so dissolved CO_2 concentrations in MORBs correlate with the pressure of eruption on the seafloor.[67] However, many MORBs have dissolved CO_2 contents in excess of the equilibrium solubility of CO_2 predicted at the eruption pressure. This oversaturation is due to rapid melt ascent during eruption, during which diffusion of carbon in the melt is too slow to maintain equilibrium between the dissolved CO_3^{2-} in the melt and the CO_2 vapor within the vesicles (Figure 9.1b).[37,38,67–69]

The CO_2–H_2O solubility model of Dixon et al.[39] has typically been used to constrain vapor saturation pressure (P_{sat}) in MORBs because they determined CO_2 and H_2O solubility specifically on MORB melt compositions. When we use measured CO_2 and H_2O contents in MORBs to calculate P_{sat} using the Dixon et al.[39] model and compare these with the pressures of MORB sample collection (P_{sample}), a subset of the data scatter around the 1:1 line (line in Figure 9.5) where vapor saturation pressures are close to those corresponding to the sampling location. A large number of samples are also demonstrably oversaturated in a CO_2–H_2O vapor and thus plot to high values of P_{sat} (above line in Figure 9.5), in good agreement with the presence of vapor bubbles in almost all MORB samples.[69]

Two other solubility models are often used as alternatives to the Dixon et al.[39] model: Papale et al.[70] and Ghiorso and Gualda.[71] The model of Papale et al.[70] (green field in Figure 9.5) produces P_{sat} values that are 60% lower on average compared with Dixon et al.,[39] which suggests that most MORBs are vapor undersaturated. Vesicle abundances observed in MORBs, however, contradict this prediction[69] because samples that are vapor undersaturated should not have vesicles.[49,50,52] The Papale et al.[70] model thus likely underestimates vapor saturation pressures in MORBs. The Papale et al.[70] model was calibrated over a very wide range in melt compositions, whereas the Dixon et al.[39] model was calibrated specifically on MORB compositions, so disagreement between these models is perhaps not surprising. More recently, the CO_2–H_2O solubility model of Ghiorso and Gualda[71] (pink field in Figure 9.5) results in P_{sat} values that are 20% lower on average compared with Dixon et al.,[39] producing good agreement with P_{sample} values. The Ghiorso and Gualda[71] model thus also predicts more samples than Dixon et al.[39] to be vapor undersaturated. Both models provide reasonable estimates of vapor saturation pressures, but we prefer the model of Dixon et al.[39] because it is calibrated specifically for MORBs and it predicts that fewer MORBs are vapor undersaturated, which is consistent with the presence of vesicles in the vast majority of MORBs.

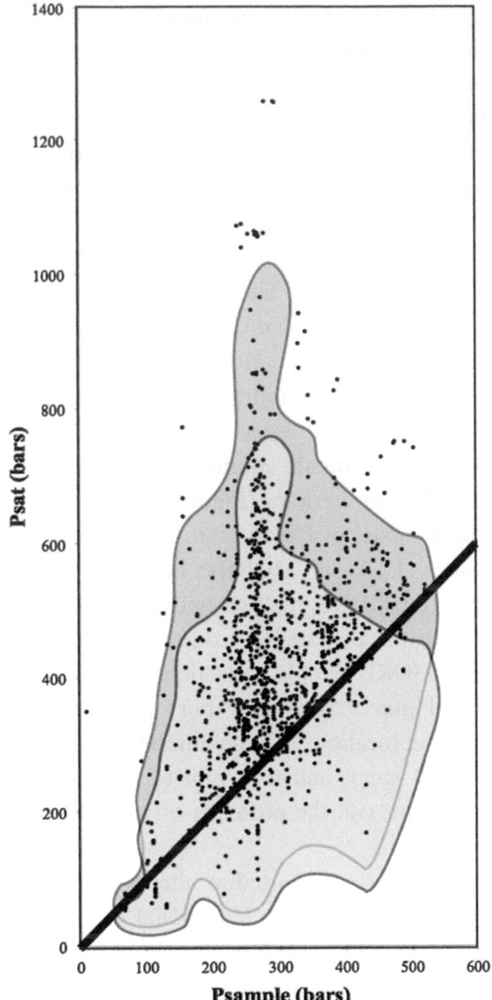

Figure 9.5 Pressure of MORB sample collection on the seafloor (P_{sample}) versus pressure of vapor saturation (P_{sat}) calculated from the CO_2 and H_2O contents of MORB samples using the vapor solubility model of Dixon et al.[39] The 1:1 line ($P_{sat} = P_{sample}$) corresponds to equilibrium of CO_2–H_2O contents at seafloor pressures. The pink field encloses the P_{sat} values calculated from the solubility model of Ghiorso and Gualda,[71] which produces P_{sat} estimates 20% lower than those of Dixon et al.;[39] the green field encloses the P_{sat} values calculated from the solubility model of Papale et al.,[70] which are 60% lower than those of Dixon et al.[39]

9.3.1 MORB Melt Inclusions and the Usefulness of Volatile/Nonvolatile Element Ratios

Because CO_2 has degassed irreversibly from most MORBs, the flux of CO_2 from the MORB mantle is often determined by establishing the mantle ratio of CO_2 to another element whose flux is well constrained. An element often used this way is helium, as the upper mantle ^3He flux is determined from oceanographic measurements[72–74] or measurements in MORBs.[75] $CO_2/^3He$ ratios measured in MORBs must be corrected for fractionation during magmatic degassing, and an equilibrium degassing model is often used for this purpose.[10,76,77] However, Tucker et al.[75] showed that MORB degassing is influenced by kinetic effects not accounted for in equilibrium degassing models. Using a disequilibrium degassing model, Tucker et al.[75] estimated a mantle CO_2 flux of 59 Tg/yr, although the extent of CO_2 fractionation from other volatile elements during degassing and the total mantle CO_2 flux strongly depend on poorly constrained diffusivities of carbon in basaltic melt.

To avoid reliance on degassing models and uncertainties in their critical parameters, we seek undegassed samples to develop nonvolatile proxies for the CO_2 contents of primary mantle melts (melts in equilibrium with mantle olivine of Fo_{90}). We therefore turn to melt inclusions. By virtue of being encapsulated in rigid host minerals, melt inclusions gain additional protection from degassing, and their volatile concentrations are not constrained by eruption depth as described in the preceding section. Indeed, some suites of melt inclusions lack vapor bubbles and are well established as being vapor (and hence carbon) undersaturated. For a proxy, we turn to highly incompatible nonvolatile trace elements with similar behavior to carbon during partial melting of the mantle (Figure 9.1). Analyses of rare undegassed melt inclusions,[49,52,53] rare undegassed MORBs,[50,51] and experimental studies[78] establish that the nonvolatile elements Rb, Ba, Nb, and Th behave similarly to CO_2 during mantle melting. In undegassed samples, CO_2 concentrations correlate with Rb, Ba, Nb, and Th. Moreover, the CO_2/Ba and CO_2/Rb ratios of undegassed mantle-derived melts are approximately constant.[53] Two fundamental assumptions follow from these observations and underpin this contribution: (1) mantle processes do not fractionate carbon from Ba and Rb; and (2) MORB Ba and Rb concentrations can be used as proxies for CO_2 in partially degassed MORBs and OIBs. Furthermore, we suggest that loss of CO_2 by degassing can be detected by the presence (or lack thereof) of a correlation of CO_2 with incompatible trace elements.

The first such correlation was observed for MORB melt inclusions from the Siqueiros Fracture Zone on the East Pacific Rise, where depleted MORB (D-MORB) is erupted at great water depth, inhibiting CO_2 loss even from the erupted magmas.[49] Subsequent studies of MORB melt inclusions, from less-depleted ridge segments and containing shrinkage bubbles, generally observed degassed magmas with no correlations of CO_2 with other trace elements.[54,55] Recently, two other suites of melt inclusions, showing CO_2 correlating with Rb, Ba, Nb, Th, and other trace elements, have been reported from the Mid-Atlantic Ridge[52] and northern Iceland.[53] These recently reported melt inclusion suites were free of vapor bubbles within the melt inclusions, with the exception of the most enriched of the Iceland melt inclusions, for which bubble measurements were made to account for the influence of their CO_2 budget on the overall CO_2 content of the inclusions.[53] Additionally, Michael and Graham[50] and Shimizu et al.[51] identified limited suites of deeply erupted D-MORB glasses that also show correlations of CO_2 with incompatible trace elements. Hauri et al.[53] showed that vapor-undersaturated MORBs have average CO_2/Rb (991 ± 26%) and CO_2/Ba (81.3 ± 56%) ratios that are more uniform than CO_2/Nb or CO_2/Th ratios. The ratio of the measured CO_2/Ba ratio to the average CO_2/Ba ratio for undegassed MORBs indicates that a majority of MORB samples have degassed >80% of their initial CO_2 (Figure 9.3).

Although vapor-undersaturated MORBs with relatively uniform CO_2/Ba, CO_2/Rb, CO_2/Th, and CO_2/Nb ratios have mostly been interpreted to be undegassed, alternative interpretations have been explored that bear on how accurately these ratios record those in the mantle sources of MORB. Rosenthal et al.,[78] argued that scattering in observed CO_2/Ba and CO_2/Nb ratios in vapor-undersaturated MORBs could be due to undetected degassing of CO_2 in amounts small enough to preserve CO_2–Ba–Nb correlations. Matthews et al.[79] modeled this undetected degassing as a partial degassing and mixing process, where near-fractional melts of the mantle are partially degassed at a specified depth and partially mixed. Despite being partially degassed, modeled melt mixtures were shown to retain positive correlations between CO_2 and nonvolatile trace elements while delineating average CO_2/nonvolatile trace element ratios that underestimated those of the mantle source. A defining characteristic of the model results from Matthews et al.[79] is the prediction that sample suites that partially degas will show a gradual as opposed to a sharp transition from positive to negative Pearson correlation coefficients between CO_2/El and 1/El ratios (where El is a nonvolatile trace element) with increasing compatibility of the El. Applying their model results to natural melt inclusion populations, Matthews et al.[79] argued that Pearson correlation coefficients in the equatorial Mid-Atlantic Ridge melt inclusions reported by Le Voyer et al.[52] show a gradual transition from positive to negative correlation coefficients, and that the melt inclusions were partially degassed. They also argued that other MORB data sets are consistent with partial degassing, and that rather than the average CO_2/Ba ratio in CO_2-undersaturated MORBs, the maximum CO_2/Ba ratio of ~140 provides a more reliable estimate of the CO_2/Ba ratio of the mantle source.

The study of Shimizu et al.[80] evaluated whether the partial degassing and mixing process described by Matthews et al.[79] can reproduce the data for the olivine-hosted MORB melt inclusions from the Garrett and Siqueiros transform faults on the East Pacific Rise and found that it is difficult to fully explain their Pearson correlation coefficients between the CO_2/El and 1/El ratios (Figure 9.6). To explain the observed coefficients, Shimizu et al.[80] considered the effect of analytical uncertainties in the model by adding noise to the modeled concentrations of CO_2 and EI, simulating the analytical uncertainties during measurements of the Siqueiros and Garrett melt inclusions. They demonstrated that when analytical uncertainties are considered, the correlations between the CO_2/El and 1/El ratios of an undegassed model show a gradual transition from positive to negative similar to a partially degassed model (Figure 9.6). They argued that although partial degassing cannot be ruled out, it may not be required to explain the Pearson correlation coefficients in the Siqueiros and Garrett melt inclusions (Figure 9.6).

Isotopic heterogeneity within melt inclusion populations derived from individual samples has been reported by several studies,[81–84] which can be explained only if the melt

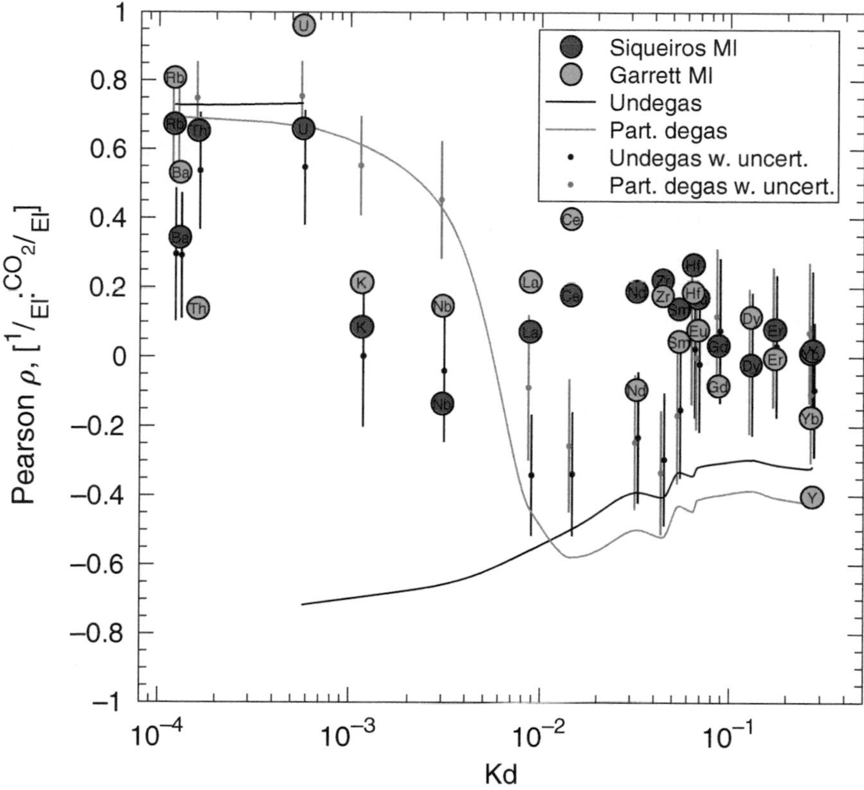

Figure 9.6 Pearson correlation coefficients for 1/El versus CO_2/El (where El is an incompatible trace element) for Siqueiros and Garrett melt inclusions (MIs) plotted against the bulk partition coefficients of the El.[80] Garrett MIs have been filtered for anomalously depleted MIs with high analytical uncertainties. Bulk partition coefficients are from Rosenthal et al.[78] for C and Kelemen et al.[173] for nonvolatile trace elements. Solid gray and black lines show correlation coefficients generated by the model of Matthews et al.[79] conducted using the pHMELTS model[174–176] to generate near-fractional melts of the DDMM from Workman and Hart,[177] which were then mixed using the model of Rudge et al.[178] The black and gray lines are the undegassed and partially degassed models, respectively, both of which do not include the analytical uncertainties. The black and gray dots are mean and 2σ uncertainties of Pearson correlation coefficients generated by the undegassed and partially degassed models ran 1000 times, both of which include analytical uncertainties ranging from 6% to 32% (2σ) depending on the element. In the partial degassing model, partial degassing is done using the CO_2 solubility model of Shishkina et al.[179] at 7 km in the oceanic crust underneath 4 km of seawater.

inclusion suite represents mixtures of melts from a heterogeneous source. Indeed, chemical and isotopic heterogeneity has been observed in MORBs by numerous studies, suggesting generation through melting of a heterogeneous mantle source[25,35,85–87] (and see expanded discussion in Section 9.3.2). This heterogeneity typically takes the form of mixing between melts from depleted and enriched sources with low and high concentrations of trace elements, respectively. In this mixing scenario, trace element ratios like CO_2/Ba and

CO_2/Nb are dominated by melts from the more enriched component with higher trace element abundances, while melts from the depleted source act to dilute the signature of melts from the enriched source. As a result, the population average of a ratio of two highly incompatible elements (such as CO_2/Ba) might also be modified by mixing of melts from heterogeneous sources, having an important and unexplored effect on the modeling results of Matthews et al.[79] While it is beyond the scope of this chapter to provide detailed discussion of all of the processes that lead to compositional heterogeneity at ridges and hot spots, we note that the Pearson correlation coefficients between the CO_2/El and 1/El ratios for each of the vapor-undersaturated sample suites is not completely explained by either partial degassing and mixing or melting of a single mantle source (e.g. La, Ce, Nd, and Hf in Figure 9.6). Given this and the result that gradual changes in Pearson correlation coefficients between CO_2/El and 1/El ratios from positive to negative may not require partial degassing (Figure 9.6), we postulate that average CO_2/El ratios observed in vapor-undersaturated melt inclusions reflect those of the mantle source(s) involved in melting, and that these ratios apply to the entire global ridge system.

9.3.2 Variations in Primary MORB CO_2 Contents and CO_2 Fluxes

We describe here the central methodology of this contribution. Trace elements vary among MORB segments by a factor of 20, and this cannot be explained by variations in the extent of melting,[17,20,21] but instead requires mantle heterogeneity. At the segment,[34] basin,[35] and global scales,[36] trace element ratios correlate with some, but not all, radiogenic isotopes. For example, segment average Ba/Nb ratios correlate with $^{87}Sr/^{86}Sr$.[36] While correlations between trace elements and isotopes can be complex because isotopes represent time-integrated histories and trace elements are highly sensitive to recent fractionation events, trace elements are highly correlated among themselves. Carbon – another trace element – has been shown to correlate with the highly incompatible trace elements and varies on an order that exceeds that which can be produced by ridge processes. By analogy with trace elements, we consequently infer large variations in primary mantle CO_2 concentrations due to mantle heterogeneity, which we calculate on the segment scale below.

The restricted ranges in CO_2/Ba and CO_2/Rb in undegassed MORBs permit an estimation of the initial primary CO_2 concentrations of MORB magmas, given an estimate of their primary trace element abundances. Gale et al.[17,36] provide such estimates based on the compositions of MORB magmas averaged by ridge segment, and corrected for shallow-level crystal fractionation to MgO = 8 wt.% and also further corrected to be in equilibrium with Fo_{90} olivine. Le Voyer et al.[47] estimated primary MORB magma CO_2 for 387 segments that contain data for Ba and/or Rb out of the 711 MORB spreading ridge segments identified by Gale et al.[36] These segment-average concentrations were then multiplied by the undegassed-average ratios from Hauri et al.[53] of CO_2/Rb (991) and CO_2/Ba (81.3) to calculate segment-averaged primary CO_2 concentrations. Unsurprisingly, the segment-average primary CO_2 contents (as derived from Ba and Rb) correlate well with segment-averaged primary Th and K_2O (Figure 9.7). These CO_2 abundances range widely, from a low of 104 ppm CO_2 at a

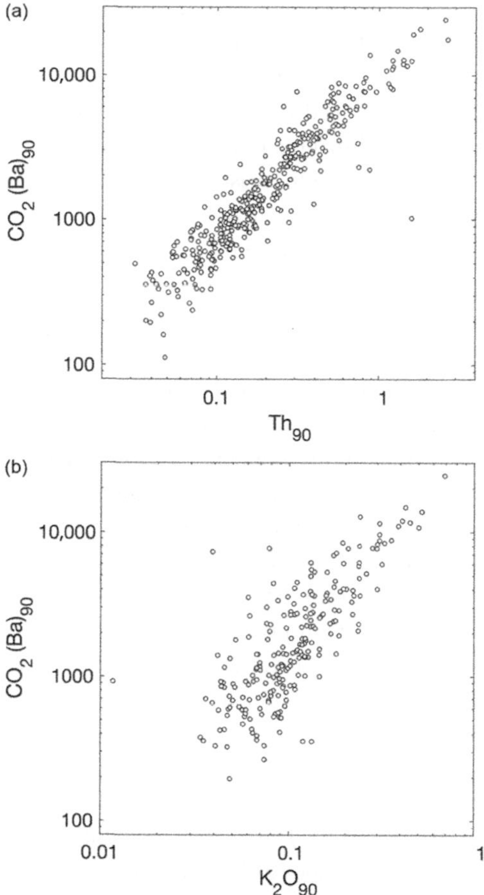

Figure 9.7 Segment-averaged primary MORB compositions corrected for low-pressure fractionation to equilibrium with Fo_{90} olivine. $CO_2(Ba)_{90}$ versus (a) Th_{90} and (b) K_2O_{90} from the study of Le Voyer et al.,[47] who estimated primary magma CO_2 contents from the average MORB CO_2/Ba ratio (81.3) and segment-average Ba_{90} concentrations.

depleted segment in the Galapagos Spreading Center to a high of 1.90 wt.% CO_2 at a highly enriched segment of the Juan de Fuca Ridge. The distribution of primary melt CO_2 contents is strongly skewed with a near log-normal distribution. The median CO_2 content is 1107 ppm and the mean is 2139 ppm. To determine the mode ("most likely" CO_2 content), we applied a kernel density analysis with an optimal bandwidth of 318 (Bowman & Azzalini, 1997[182]), and obtained 621 ppm. This preferred value for "typical primary MORB" (621 ppm CO_2) is similar to the analytical mode (486 ppm), which assumes a true log-normal distribution. (Note that the original printing of Le Voyer et al. (2019)[59] incorrectly referred to 2100 ppm as median primary CO_2. This contribution revises that error.)

Nearly five decades of deep-water oceanographic research, coupled with ocean drilling by the Deep Sea Drilling Program (DSDP) and the International Ocean Drilling Program (IODP), have demonstrated a limited range of crustal thickness of 6 ± 1 km,[20,88,89] and at variable spreading rates that are well determined globally.[90] Given the relationship between crustal thickness and seafloor depth,[20,88,91] magma production rates at mid-ocean ridges are readily calculated from the product of ridge length, crustal thickness, and spreading rate.[11] Behn and Grove[88] and Van Avendonk et al.[92] summarized seismically determined measurements of crustal thickness at ridge axes, which varies from 2 km at the slow-spreading Gakkel Ridge to 30 km at central Iceland and is a product of the local magma production rate and spreading rate. The crustal thickness of ridge segments correlates with ridge depth (reflecting isostatic compensation of MORB crust at the ridge axis). With this, Le Voyer et al.[47] estimated the crustal thickness at all of the 711 MORB segments defined by Gale et al.[36] based on the axial depth. Le Voyer et al.[47] then determined a segment-average magma production rate by multiplying the crustal thickness by the segment spreading rate, which was then used to calculate the fluxes of CO_2 (and other elements) on a segment-by-segment basis.

CO_2 fluxes, normalized by ridge length, vary by a factor of >100 because they are the product of local primary magma CO_2 content and local magma production rate, which are uncorrelated. Fluxes of CO_2 vary from 1.52×10^6 to 4.74×10^8 mol/yr/km. High CO_2 fluxes are typically correlated with high primary CO_2 in the magma (Figure 9.8), reinforcing the observations made above that variations in the MORB primary magma CO_2 content are mostly driven by variations in mantle composition rather than magma production dynamics or mantle temperature.[47]

Integrated over the length of the global mid-ocean ridge system, the global MORB magma production rate is 16.5 km^3/yr (lower than the estimate of 21 km^3/yr from Crisp[11]), and the total flux of CO_2 is 1.32×10^{12} mol/yr.[47] This estimate is significantly higher than that of Saal et al.,[49] which is based on depleted Siqueiros Fracture Zone melt inclusions. Our estimate is lower than the estimate of Cartigny et al.,[48] which is based on enriched MORBs from 14°N and 34°N on the Mid-Atlantic Ridge. Marty and Tolstikhin,[10] Michael and Graham,[50] and Hauri et al.[53] give CO_2 fluxes that are nearly a factor of two higher than the integrated value summed over the detailed fluxes determined on a segment-by-segment basis by Le Voyer et al.[47]

The MORB flux of CO_2 degassed at the ridge axis is ultimately dissolved into the overlying water column, either instantly by eruptive degassing (~80%) or later stripped out by hydrothermal circulation, where it becomes part of the marine bicarbonate cycle. Short-term global effects of large variations in the ridge CO_2 flux are thus muted by the ~100,000 year residence time of carbon dissolved in seawater,[93] and thus variations in the ridge CO_2 flux are unlikely to influence short-term surface climate conditions while the ridge is submerged. Only when the mid-ocean ridge system is exposed above sea level, such as at Iceland today, will there be a significant delivery of CO_2 from the mid-ocean ridge mantle directly into the atmosphere. On longer timescales, the ridge CO_2 flux is dwarfed by the deposition of carbonate sediments and organic carbon on the seafloor and of carbonate alteration veins within the crustal and mantle sections of the oceanic lithosphere, which make the seafloor a net sink for carbon by a large factor compared with its initial magmatic budget.[8,94–96]

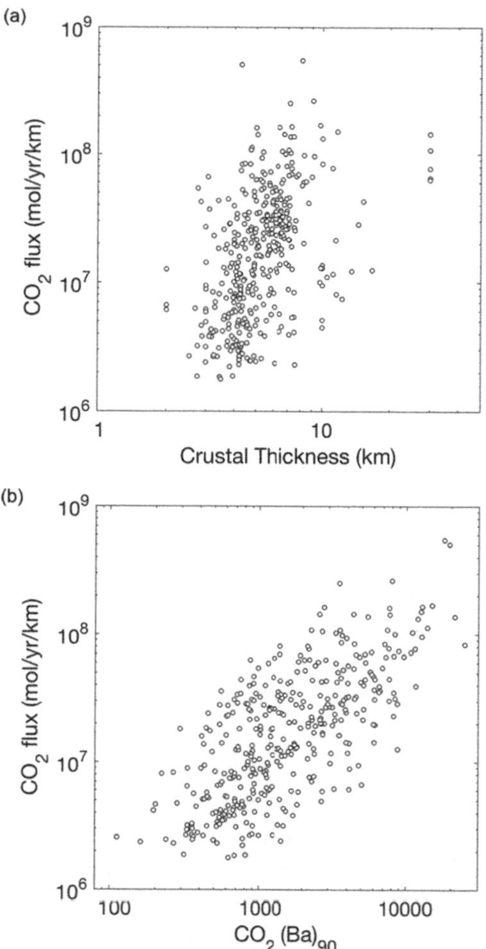

Figure 9.8 MORB segment-average CO_2 flux normalized by ridge length (mol/yr/km) versus (a) crustal thickness and (b) primary CO_2 concentration estimated from the average MORB CO_2/Ba ratio (81.3) and segment-average Ba_{90} concentrations.[47] Estimated CO_2 fluxes are more strongly correlated with mantle composition than with crustal thickness.

9.4 Fluxes of CO_2 from Mantle Plumes

The next most important flux of CO_2 from the convecting mantle occurs at intraplate hot spots, where deeper sources of carbon penetrate through the upper mantle and oceanic lithosphere to produce seamounts and volcanic chains (Figure 9.1c). The flux of carbon from the deep mantle is of critical importance to mantle geochemistry and the study of planetary volatiles. Whereas mid-ocean ridges sample the uppermost mantle, some ocean intraplate volcanics sample the deeper mantle via upwelling plumes,[97,98] which could represent a significant reservoir of terrestrial carbon and other volatile elements. Hot spot magma fluxes are similarly calculated from the volumes of the individual volcanoes in the chain; however, weathering of the volcanic edifice and poorer chronologic information on the duration of volcanism result in greater relative uncertainties for these fluxes. The flux of CO_2 from oceanic intraplate volcanics is generally considered to be much smaller than the mid-ocean ridge flux.[7,10,99] However, many oceanic intraplate volcanics erupt subaerially, so their CO_2 emissions are injected directly into the atmosphere. Furthermore, many modern-day plumes are linked to LIPs,[100] and gas emissions from eruption of these OIB progenitors can have profound and devastating environmental effects.[101]

Modern-day volatile fluxes can be directly measured at active volcanic centers by analysis of volcanic gas, and this subject is discussed in depth in Chapter 8. Here, we focus on quantification of long-term integrated fluxes from oceanic intraplate volcanics. These are most typically quantified by combining estimates of primary magmatic CO_2 concentrations with magmatic fluxes. Two broad strategies are used to estimate pre-degassing CO_2 concentrations: using maximum measured CO_2 concentrations; and using a degassing model to calculate pre-degassing concentrations. The disadvantage of using degassing models is that they require additional constraints such as noble gas or carbon isotopic measurements to quantify the extent of gas loss, and the results may be sensitive to the assumed degassing model and its physical parameters. Using maximum measured values estimated from melt inclusions does not require assumptions about degassing processes, but can only provide lower limits on pre-degassing concentrations.

The heat flux produced by oceanic intraplate hot spots amounts to only 6–10% of the total surface heat flux;[102,103] however, OIB magmas produced at hot spots are known to have enriched abundances of volatile elements[104–107] that may contribute to a CO_2 flux in excess of that inferred from their relative heat fluxes. However, enriched CO_2 contents can also result in deeper degassing of CO_2 and thus many submarine-erupted OIBs have likely lost a greater fraction of their initial CO_2 compared with MORBs erupted under the same submarine water pressures.[108] As a result, like the situation for MORBs, submarine glasses at hot spots have lost the majority of their CO_2.

Melt inclusions once again provide the means to examine the CO_2 content of OIB magmas. The recent study of Marske and Hauri[60] provides a definitive data set of volatiles in Hawaiian melt inclusions, adding to the smaller datasets of Refs. 109–111. However, for CO_2, these prior studies were all complicated by the presence of vapor bubbles within melt inclusions, which were not accounted for in determining inclusion CO_2 concentrations. Tucker et al.[61] examined hundreds of bubbles in Hawaiian melt inclusions and quantified the uncertainties associated with reconstructing total (bubble + dissolved) CO_2 in the melt inclusions. Given the proportion of bubble/inclusion volume (which strongly peaks at 3.3% for Hawaiian inclusions; Figure 9.4), an estimate of the bubble formation pressure determined from the dissolved H_2O and CO_2 contents of the glass phase, a bubble closure temperature, and an equation of state for CO_2 vapor, it is possible to reconstruct the total CO_2 content of the inclusion. At Hawaii, these CO_2 contents range over hundreds of ppm to several wt.% CO_2; however, much of this range is likely due to degassing, and only the highest CO_2 contents are likely to be representative of primary (undegassed) CO_2 contents. Tucker et al.[61] estimated an average primary Hawaiian magma CO_2 content of 7000 ppm

Table 9.1 *Melt inclusion-based estimates for primary magma CO_2 (ppm) and CO_2 fluxes from selected oceanic intraplate hot spots*

Locality	Iceland	Azores	Canary	Hawaii	Society	Pitcairn	Total
Number of MIs	104	26	29	407	N/A	N/A	
Primary CO_2 (ppm)	3200	8200	10,600	7000	14,000	5100	
Magma supply rate (g/yr)	5.40E+13	2.67E+12	2.70E+12	5.40E+14	6.75E+12	1.62E+13	
CO_2 flux (g/yr)	1.73E+11	2.19E+10	2.86E+10	3.78E+12	9.45E+10	8.26E+10	4.18E+12
C flux (Tg/yr)	4.71E–02	5.97E–03	7.81E–03	1.03E–00	2.58E–02	2.25E–02	1.14E+00
CO_2 flux (mol/yr)	3.93E+09	4.98E+08	6.5E+08	8.59E+10	2.15E+09	1.88E+09	9.50E+10
Refs.	64, 65, 180	62, 180	63, 180	61, 181	117	116	

MI = melt inclusion; N/A = not applicable.

(Table 9.1) and a modern CO_2 flux for Hawaii of 8.6×10^{10} mol/yr. This flux represents approximately 5% of the MORB CO_2 flux,[47] while the magma flux at Hawaii (0.2 km^3/yr) represents only 1% of the MORB magma flux, a comparison that highlights the enriched nature of CO_2 in Hawaiian magmas.

The melt inclusion-based estimates compare well to prior published estimates of Hawaiian magma primary CO_2 contents and volcanic fluxes. By combining dissolved and vesicular CO_2 concentrations, Dixon and Clague[104] found maximum CO_2 concentrations of up to 6300 ppm in submarine Loihi basalts. Using a noble gas-based disequilibrium degassing model, Gonnermann and Mukhopadhyay[112] estimated pre-degassing concentrations of between 1100 and 6300 ppm in basalts from Loihi and Kilauea. However, those results may be sensitive to model parameters such as the magmatic carbon diffusivity. The long-term rate from melt inclusions is similar to modern-day emission rates measured from the Kilauea summit.[113–115]

Given knowledge of the H_2O and CO_2 contents of glassy melt inclusions and vapor bubble sizes, the same methodology as applied to the Hawaiian inclusions can also be applied to published data sets from other hot spots in order to calculate their primary magma CO_2 contents and CO_2 fluxes. Melt inclusion volatile and bubble size data are available from the Azores,[62] Canary,[63] and Iceland[64,65] hot spots; primary magmatic CO_2 contents of these hot spots are conservatively estimated as the 95th percentile of CO_2 contents from the data sets. Additionally, we include estimates based on analyses of submarine glasses of the Society and Pitcairn hot spots.[116,117] The primary CO_2 contents of these hot spot OIBs range from 3200 ppm at Iceland to 10,600 ppm CO_2 at the Canaries, suggesting a large degree of heterogeneity in primary CO_2 contents between hot spots (Table 9.1). However, each of the individual CO_2 fluxes of the smaller hot spots is more than an order of magnitude lower than the CO_2 flux from Hawaii.

In their bathymetric studies of Pacific seamounts and islands, Batiza,[118] Wessel and Lyons,[119] and Kim and Wessel[120] estimated the total volume of oceanic intraplate magmatism to be 1–2 × 10^7 km^3, which equates to an extra 100–200 m (1.5–3.0%) of crustal thickness across the entirety of the Pacific plate. If this relationship holds for all of the ocean basins, then even with a primary magma CO_2 content of 1 wt.% (similar to the highest OIB primary CO_2 contents; Table 9.1), this would represent an additional 12–25% atop the ridge flux, or 0.17–0.35 × 10^{12} mol/yr (or two to four times the output of Hawaii; Table 9.1). Thus, while the overall intraplate CO_2 output is not insignificant compared with the ridge flux, long-term global variations in the CO_2 flux from ridges alone are no doubt of a similar magnitude given the spreading rates and total ridge length changes during plate tectonic evolution. And because most of the intraplate volcanism in the oceans is below sea level, the CO_2 outgassed at all but the largest hot spots is fated for dissolution into seawater, and, like the MORB flux, it is ultimately precipitated as carbonate sediments and carbonate vein fillings on the seafloor.

LIPs represent an exceptional source of hot spot-related magmatism due to their punctuated eruptions of enormous volumes over very short periods of time (~1 Myr) that approach the global MORB production rate and the well-known correlation of ages of LIP eruption events with the timing of mass species extinctions.[121–122] Such eruptions have great potential to influence global climate on short timescales due to the likelihood that their volatile inventory was delivered directly to the atmosphere; this is particularly acute for the emission of SO_2 during LIP volcanism, as the atmospheric abundance of SO_2 is typically low and its ability to absorb or reflect UV radiation is very high. Thus, LIP SO_2 emissions (of the order 10^{13} mol/yr over decadal timescales) are enormous compared with the SO_2 budget of the atmosphere (of the order 10^{10} mol/yr) and can result in severe short-term cooling effects on Earth's surface.[16,121] If flood basalts contain primary CO_2 contents similar to OIBs, the emission of CO_2 in a single decade-long eruption potentially represents a magnitude of the order 10^{13} mol/yr,[16] approximately ten times the modern emission rate from mid-ocean ridges[47] and representing an instantaneous decadal addition to atmospheric CO_2 on the order of 1000 ppm. The greenhouse effect of such LIP CO_2 emissions would be concurrent with LIP SO_2 emissions. Increased SO_2 would push conditions toward global cooling, but would have a shorter residence time in the atmosphere than CO_2. Thus, the overall impact of LIP volcanism might be expected to strongly drop global temperatures while the volcanism is active and to strongly increase global temperatures following the cessation of eruptions. The evolution of end-member "hothouse" or "icehouse" climate conditions, triggered by LIP eruptions, could depend on whether the atmospheric system was already close to a tipping point in one direction or the other.[16,122,126] Nevertheless, both SO_2 and CO_2 emissions from LIPs would ultimately lead to intense ocean acidification that could be an additional driver in triggering mass extinctions associated with LIP volcanism.

9.5 Carbon Content of Convecting Mantle Sources

The large range of primary MORB and OIB magma CO_2 contents at the segment scale is due to a combination of variations in degree of melting, as well as variations in mantle CO_2

abundance. The mean degree of melting at a given ridge segment can be evaluated by consideration of crustal thickness. For the standard model of passive upwelling at oceanic spreading centers, Klein and Langmuir[20] describe the relationships between crustal thickness (D_c), mean degree of melting ($\overline{F}$),mean pressure of melting ($\overline{P}$), and final pressure of melting (P_f). The mean degree of melting can be calculated from the crustal thickness using the following relation:

$$\overline{F} = 0.006\sqrt{D_c \rho_c / 0.0612}, \tag{9.1}$$

where crustal thickness D_c has units of kilometers and ρ_c is the density of the oceanic crust in units of Tg/km^3 (equivalent to g/cm^3). With estimates of crustal thickness for each ridge segment derived from its relationship with ridge depth, we can use (9.1) to calculate the degree of melting at each segment and then invert the primary CO_2 content for mantle CO_2 using the batch melting equation:

$$C_{mantle} = C_{melt}\left[D_{carbon}\left(1 - \overline{F}\right) + \overline{F}\right], \tag{9.2}$$

where C_{mantle} is the mantle concentration, C_{melt} is the melt concentration (in equilibrium with Fo_{90} olivine), and D_{carbon} is the carbon bulk solid/melt partition coefficient. Because D_{carbon} is much lower than $\overline{F}$, (9.2) reduces to a simple relationship between the degree of melting and mantle-melt concentrations:

$$C_{mantle} = C_{melt}\overline{F}. \tag{9.3}$$

Bearing in mind that the accuracy of the primary CO_2 in MORB contents relies on our assumption that CO_2/Ba and CO_2/Rb are relatively invariant, we can use (9.3) to calculate mantle source concentrations of CO_2 at each ridge segment. The calculations, carried out in the dataset of Cottrell et al. (2019),[183] predict that MORB mantle CO_2 content ranges over two orders of magnitude from 10 ppm (segment GALA16) to 1980 ppm (segment JUAN1; Figure 9.9) and follows a log-normal distribution. For the 387 segments with trace element constraints, we estimate that the mean mantle CO_2 content is 205 ppm, the median is 99 ppm, and the mode (estimated using a kernel density analysis as above), is 73 ppm. Segment-specific values are tabulated in Cottrell et al. (2019).[183] If we assume that the segments without trace element constraints have "typical" primary melt CO_2 contents, constrained by either the distribution mode (621 ppm) or the median (1107 ppm), then the mode of mantle CO_2 distribution is 50 or 104 ppm, respectively. The Arctic ridges are noteworthy in having uniformly high estimates of mantle CO_2. There are additional isolated ridge segments with high estimated mantle CO_2 concentrations that are far from hot spots, notably just north of the equatorial Mid-Atlantic Ridge (MARR168), much of the American–Antarctic Ridge, several slow-spreading segments of the Southwest Indian Ridge, and the northernmost Red Sea. Apart from these scattered locations, there is a strong geographic correlation of high mantle CO_2 at ridge segments that occur near hot spots (Iceland, Azores, South Atlantic hot spots, Reunion, Afar, Easter Island, Galapagos). All of the ridge segments with $>$700 ppm mantle CO_2 are located within 1000 km of the nearest hot spot, which may indicate that hot

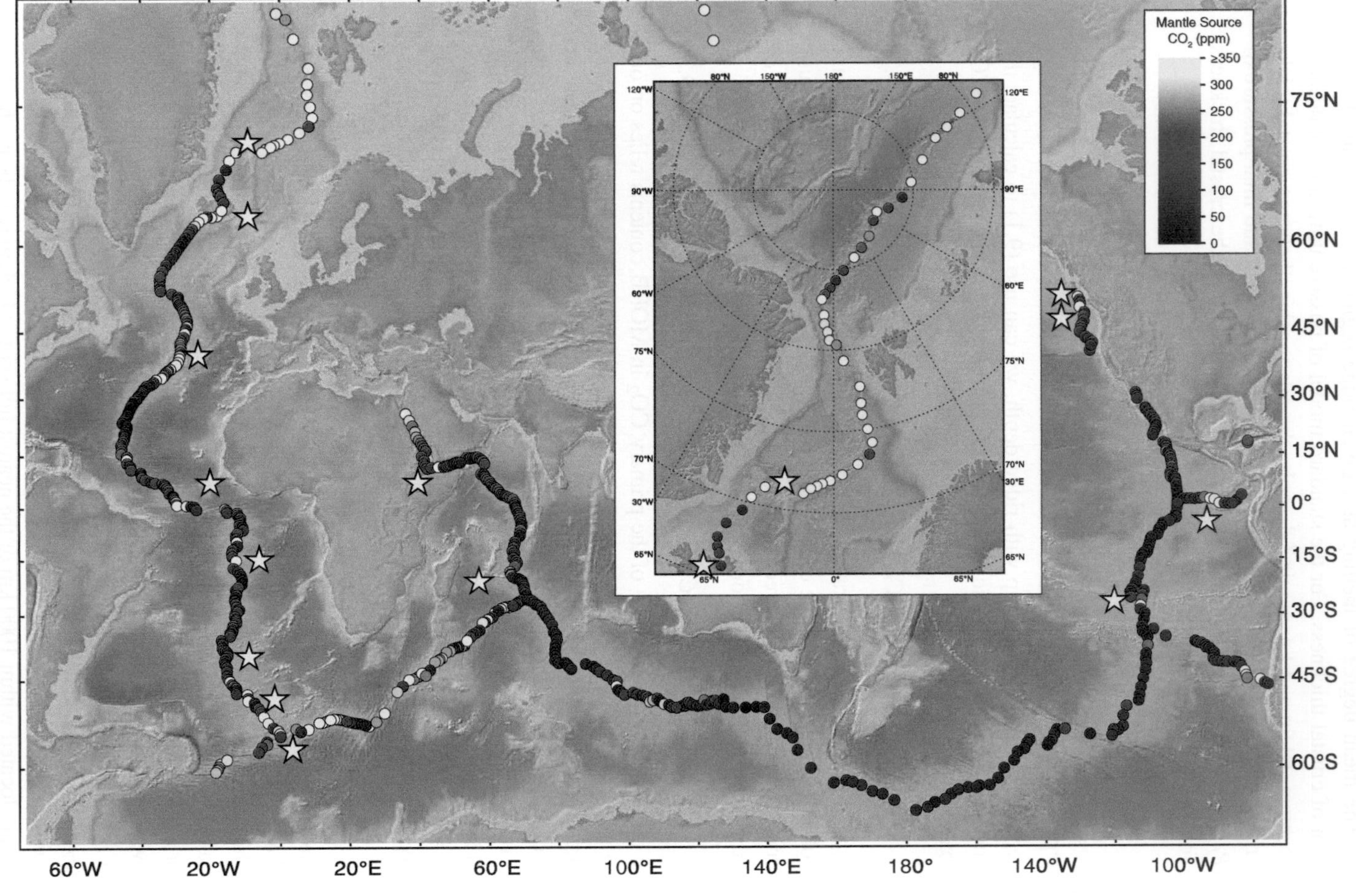

Figure 9.9 Map of segment-average mantle source CO_2 concentrations along the global mid-ocean ridge system. Mantle source CO_2 concentrations were derived from segment-average primary CO_2 abundances by estimating the degree of melting at each ridge segment and assuming batch melting (see Eqs. (9.1)–(9.3) in the text). High concentrations are observed near hot spots (yellow stars show approximate localities but are off-set so as to not obscure data), but also at isolated sections of ridge far from hot spots, particularly the Arctic ridges and the southern Indian Ocean ridges.

spots have elevated concentrations of CO_2 in their mantle sources. This inference is predicated on our underlying assumption that the mantle has uniform CO_2/Ba and CO_2/Rb ratios.[47,53] If hot pots sample recycled mantle that has had carbon added to it or stripped from it[7,8] relative to other trace elements, our assumption would not hold and our inferences about primary CO_2 at hot spots would not be valid. While this is an important topic for future inquiry, at this point there is no evidence to suggest that the CO_2/Ba or CO_2/Rb ratios at hot spots are significantly different from those of the MORB mantle.[53]

At hot spots, no simple relationship translates between volcano volume and degree of melting, as hot spots have widely different temperatures and buoyancy fluxes[127,128] and erupt through oceanic lithosphere of widely variable age and thickness. Thus, the degree of melting must be determined in a different way from MORBs. Estimates of the degree of melting for the major Hawaiian volcanoes range from ~4% to 10%,[129,130] which, when combined with estimates of primary CO_2 contents of individual volcanoes, yield mantle CO_2 concentrations ranging from ~400 to 500 ppm.[61]

Other hot spots appear to produce magmas that have high primary CO_2 concentrations similar to Hawaii. Barry et al.[131] estimated a CO_2 concentration for the Iceland plume of 530 ppm, which is similar to the 590 ppm CO_2 estimated by Le Voyer et al.[47] for the most CO_2-rich ridge segment at Iceland. Similarly, high mantle CO_2 contents are observed at hot spot-proximal ridge segments near Jan Mayen (1800 ppm), the Azores (1490 ppm), Cape Verde (830 ppm), Ascension (415 ppm), Discovery/Meteor (910 ppm), Shona (480 ppm), Bouvet (1160 ppm), Reunion (415 ppm), Afar (1420 ppm), Easter Island (350 ppm), Galapagos (850 ppm), and the Bowie hot spot on the Juan de Fuca Ridge (1980 ppm). At intraplate hot spots, mantle CO_2 concentrations in the Society and Pitcairn hot spots have been estimated to be 700 ppm[117] and between 260 and 510 ppm,[116] respectively. Despite the highly variable methods used, these hot spot mantle CO_2 concentrations are distinctly higher than mantle concentrations at nearly all MORB ridge segments located $>$1000 km away from hot spots.[47]

9.6 Carbon and Mantle Melting

Partial melting of the convecting mantle beneath ridges and hot spots is the key process and pathway by which carbon moves from Earth's interior to the surface.[7,8,132,133] Thus far in this chapter, we have addressed the quantity and dynamic range of carbon present in the mantle and how it partitions between the crust and hydrosphere. We now explore carbon's role in mantle melting and how it affects the flux of carbon reaching the ridge axis.

Decompression melting above the anhydrous solidus dominates the production of melt and oceanic crust at ridges.[134] However, the volume of mantle that experiences low degrees of partial melting far exceeds the volume above the anhydrous solidus and depends on the power of volatiles, primarily H_2O and CO_2, to flux mantle melting.[135–137] Dasgupta[138] provides an in-depth review and further exploration of near-solidus mantle melting in the presence of H_2O and CO_2. Here, we keep the focus on carbon and how its dynamic range in the mantle source relates to the flux of carbon reaching the ridge axis.

Three fundamental properties of carbon affect its fate during partial melting: (1) carbon is highly incompatible in mantle silicates;[78,139,140] (2) carbon is likely to exist in both oxidized and reduced forms within Earth's mantle;[7,141–147] and (3) oxidized carbon stabilizes melts, influencing the partial melting process itself.[132,133,137,148–151] In mantle where oxygen fugacity (fO_2) is low enough to stabilize graphite, carbon concentration in silicate melts, and hence the flux of carbon to the surface, will vary as a function of fO_2, temperature, and pressure.[145,152–154] Depending on the concentration of carbon in the source and degree of melting, graphite may be exhausted from the source during decompression melting under reduced conditions at quite low melt fractions.[155] Such conditions may be relevant on reduced terrestrial bodies, such as Mars, but on Earth, basalts[156–159,184] and their residues[160] record fO_2 near the quartz–fayalite–magnetite (QFM) buffer and are too oxidized to stabilize graphite in their residues.

In a uniformly oxidized mantle, the stable form of carbon is carbonate. The solidus of carbonated peridotite is depressed by hundreds of degrees relative to the anhydrous solidus,[132] is nearly independent of carbon concentration,[161] and is sufficiently steep to allow carbonatitic melts to form all the way to the transition zone along convecting mantle geotherms.[162] The flux of carbon from the mantle to the surface in this scenario depends on the concentration of carbon in the source, with carbonatitic (~40 wt.% CO_2) melts forming at melt fractions proportional to carbon concentration. The melt fraction across the carbonated solidus for a MORB mantle with median CO_2 (~100 ppm) would be 0.03%, but could range from 0.0025% to 0.5% spatially, consistent with the dynamic range of mantle CO_2 contents predicted here (10–2000 ppm). The addition of water to the system may stabilize carbonated silicate melts over carbonatitic melts to hundreds of kilometers;[137,138] however, consideration of silica activity–composition relationships at greater depths predicts an abrupt transition from carbonatitic to carbonated silicate melts as shallow as 75–100 km along a 1350°C geotherm.[151] Regardless of the melt composition, in a uniformly oxidized mantle, deep carbon-rich melts would be widespread, but heterogeneously distributed at depth if mantle carbon concentrations are as variable as suggested by Le Voyer et al.[47]

A uniformly oxidized mantle is unlikely on Earth, however. Theoretical and experimental considerations and observational evidence from continental xenoliths all suggest that the mantle becomes more reducing with depth relative to the QFM buffer.[142,146,163–167] When fO_2 becomes low enough to stabilize graphite/diamond over carbonate, carbonate-fluxed melting cannot proceed. The depth of the carbonate–graphite/diamond transition is uncertain, but may lie as shallow as 120 km due to increasing stabilization of the ferric iron component in garnet with pressure,[167] or perhaps at depths closer to 250 km due to increasing modes of majoritic garnet and stabilization of Fe metal.[166] It should be noted that the xenolith record of iron redox comes entirely from continental lithospheric xenoliths rather than oceanic samples; thus, our limited knowledge of the redox-depth profile of the convecting mantle stands as a major limitation to our ability to predict the extent and consequences of carbon-fluxed melting at depth.

If we assume that the mantle becomes reduced quickly with depth and that the carbonate to graphite transition takes place at 120 km depth, then this depth will correspond to the depth of incipient melting. The depth at which melts are not only present, but interconnected and detectable with geophysical probes, will depend on carbon concentration. If we assume a threshold for melt interconnectivity and detection of 0.05 wt.% melt,[168] then Le Voyer et al.[52] showed how minimum CO_2 concentrations of ~85 ppm are required at 120 km depth to generate detectable melts in anhydrous mantle with a potential temperature of 1345°C. The depth of interconnected, detectable melting will decrease as carbon concentration decreases. As the depth of the carbonate–graphite/diamond transition deepens, the importance of carbon concentration to the volume of mantle interconnected by melts increases. However, so long as mantle carbon concentrations are above 220 ppm, melts will be interconnected and detectable at all depths and carbon will be extracted as soon as the mantle becomes oxidized enough to stabilize carbonate, provided those melts can reach the ridge axis (see below). Thus, the depth range over which carbon-fluxed melting creates important geophysical signatures and actively increases the flux of carbon to the surface by enabling melting itself may be narrow. Indeed, in a mantle with only 140 ppm water, there may be no depth range over which carbon concentration strongly influences such signatures because of water's ability to contribute to melt stabilizations.[138,169]

While the fO_2 recorded by ridge lithologies suggests that the MORB source is too oxidized to retain graphite as a stable residual phase, the high concentration of CO_2 dissolved in undegassed MORB provides further independent evidence of this. Figure 9.10 shows the extent of melting required to exhaust graphite during peridotite melting in the graphite stability field as a function of fO_2 and the carbon content of the mantle (100 ppm CO_2 corresponds to 27 ppm C). The fO_2 range for graphite/diamond stability in the mantle corresponds to depths greater than 100 km and is taken from Stagno et al.[167] For typical mantle with ~50–100 ppm CO_2 (= 14–27 ppm C), graphite would be exhausted at reasonable melt fractions for MORBs (6–20%). In this scenario, we would expect carbon to behave as a highly incompatible element, just as we observe in some submarine glass and melt inclusion suites that are undersaturated in CO_2–H_2O vapor.[49–53] However, at higher mantle C contents, such as are evidenced by Le Voyer et al.[47] and presented here, melt fractions far in excess of 20% would be required to exhaust C from the residue and produce incompatible behavior for carbon. Such high melt fractions are unreasonable for MORB. Lower melt fractions would leave graphite (or diamond) present in the residue, and carbon would appear to behave as a moderately incompatible element, quite unlike the highly incompatible behavior seen in vapor-undersaturated MORB samples. Although there is a lack of detailed understanding of the fO_2 with depth in the convecting mantle, the discovery of vapor-undersaturated samples at higher total carbon contents might illuminate whether or not graphite/diamond-saturated melting is an important process to consider at the deepest parts of the melting regime beneath ridges and hot spots.

Thus far, we have considered uniform partial melting, melt interconnectivity, and melt extraction. Keller et al.[133] show how melt focusing is only effective within and below the width defined by the anhydrous melting triangle, and that this may impact the fraction of mantle carbon in the melting regime that reaches the ridge axis. They show that deep volatile-fluxed melting may not influence melt focusing to the extent previously

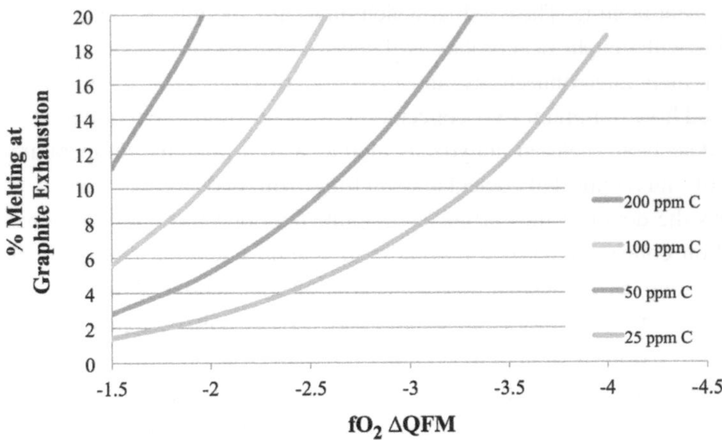

Figure 9.10 Extent of melting required to exhaust graphite from the mantle as a function of the fO_2 and the C content of the mantle. The solubility of total C in graphite-saturated silicate melt was taken from experimental studies.[145,153,154] At low mantle C contents, only small degrees of melting would be required to exhaust graphite, and C would behave similarly to highly incompatible elements (D ~0.02–0.04). At higher mantle C contents, C would appear only mildly incompatible (D ~0.1–0.2). For typical MORB melt fractions, we observe that C behaves as a highly incompatible element in vapor-undersaturated melt inclusions. This provides independent evidence that the MORB source is not graphite saturated; however, these curves could be relevant when considering more reduced planetary bodies, such as Mars.

assumed,[136] but instead introduces spatial and temporal variability in the volatile flux reaching the ridge on timescales of 10,000 years. Keller et al.[133] predict that the concentrations of carbon (or any incompatible element) in primary MORB melts could vary as a function of time for any given source concentration due to reactive channelization. Furthermore, melts generated in the "wings" of the melt triangle collect along the base of the lithosphere and ultimately freeze, and up to half the melt volume produced within the melting regime may never make it to the surface at the ridge axis.[133] Because such melts do not contribute to oceanic crust production, their contribution does not factor into estimates of CO_2 fluxes derived from geochemical and geophysical observations. The important implication for what we have presented here is that much of the heterogeneity we observe in incompatible trace elements on the dredge scale (hundreds of meters) within MORB may be generated by the melting process, rather than very-fine-scale variations in mantle source.[170] Thus, when interpreting our segment-scale variations in mantle source globally, we must keep in mind how representative the concentrations presented here may be for poorly sampled segments and recognize that within segments it can be difficult to distinguish time-dependent variations in process from small-scale variations in composition.

Keller et al.[133] and Le Voyer et al.[47] provide the only estimates of segment-specific carbon flux. MORB mantle from Le Voyer et al.[47] has ~73 ppm (mode), 99 ppm (median)

CO_2 in the sources.[183] For such sources, the model of Keller et al.[133] returns global fluxes of 9.4×10^6 to 1.3×10^7 mol/yr/km (normalized to ridge length), which is comparable to the 1.4×10^7 mol/yr/km estimated in Le Voyer et al.[47] and this study. These length-normalized ridge fluxes are in excellent agreement despite the facts that Keller et al.[133] use a constant-source CO_2 content and that the two models calculate crustal thicknesses differently. The average crust is 0.83 km thicker in the model output of Keller et al.,[133] who modeled magma production at variable potential temperatures constrained from a global-scale seismic tomography model,[171] producing a mean crustal thickness of 6.67 km. Le Voyer et al.[47] used the relationship of crustal thickness versus ridge depth observed at ridges with seismically determined crustal thickness measurements and extrapolated this relationship to calculate crustal thickness at all ridge segments from their axial depth, producing an average crustal thickness of 5.75 km despite calculating larger crustal thicknesses at hot spots compared with Keller et al.,[133] who capped their temperature input values at 1450°C.

When integrated over the entire length of ridges, we[183] calculate a global CO_2 flux of 1.3 to 1.5×10^{12} mol/yr, depending on whether the global mode or median CO_2 content is applied to segments unconstrained by trace element data. Keller et al.[133] calculate a global flux of 1.20×10^{12} mol/yr. Despite the differences in methodology, these two studies are in excellent agreement on the integrated ridge CO_2 flux, and this agreement emphasizes the depth of our understanding of CO_2 fluxes at ridges.

At the same time, mantle heterogeneity must be considered at the more granular scale of individual ridge segments. Figure 9.11 highlights the geographic distribution of differences in segment-scale CO_2 fluxes between our estimates, with variable-source CO_2, and the Keller et al.[133] estimates. Observations of Sr, Nd, Pb, and Hf isotopic heterogeneity within individual ridge segments (length scales of tens to hundreds of kilometers) are often observed where measurements are sufficiently dense (see data sets of Gale et al.[36] and Class and Lehnert[172]). It has long been recognized that mantle heterogeneity is the primary source of segment-scale variations in isotopes and highly incompatible trace elements (e.g. Langmuir et al.[21]). Keller et al.[133] show that, for a homogeneous mantle, spreading rate and potential temperature dominate the flux of carbon at ridges, as these parameters dominate the production rate of magma itself. But the ridge CO_2 flux is more sensitive to mantle carbon concentration than to mantle temperature, spreading rate, compaction length, or mantle fertility. The segment-scale CO_2 fluxes that we have calculated (normalized by segment length) exhibit a variability that is twice that observed in the model of Keller et al.;[133] in our calculations of crustal thickness, we have implicitly included the effects of mantle temperature and magma production rate, since they also influence crustal thickness. In this rendering, it is clear that consideration of variability in mantle source CO_2 concentrations results in much higher carbon fluxes near hot spots and lower carbon fluxes for the D-MORB mantle than those of the Keller et al.[133] model with a single-source composition. It is also clear that mantle source concentration has more influence on CO_2 flux at the ridge than any other parameter under consideration.

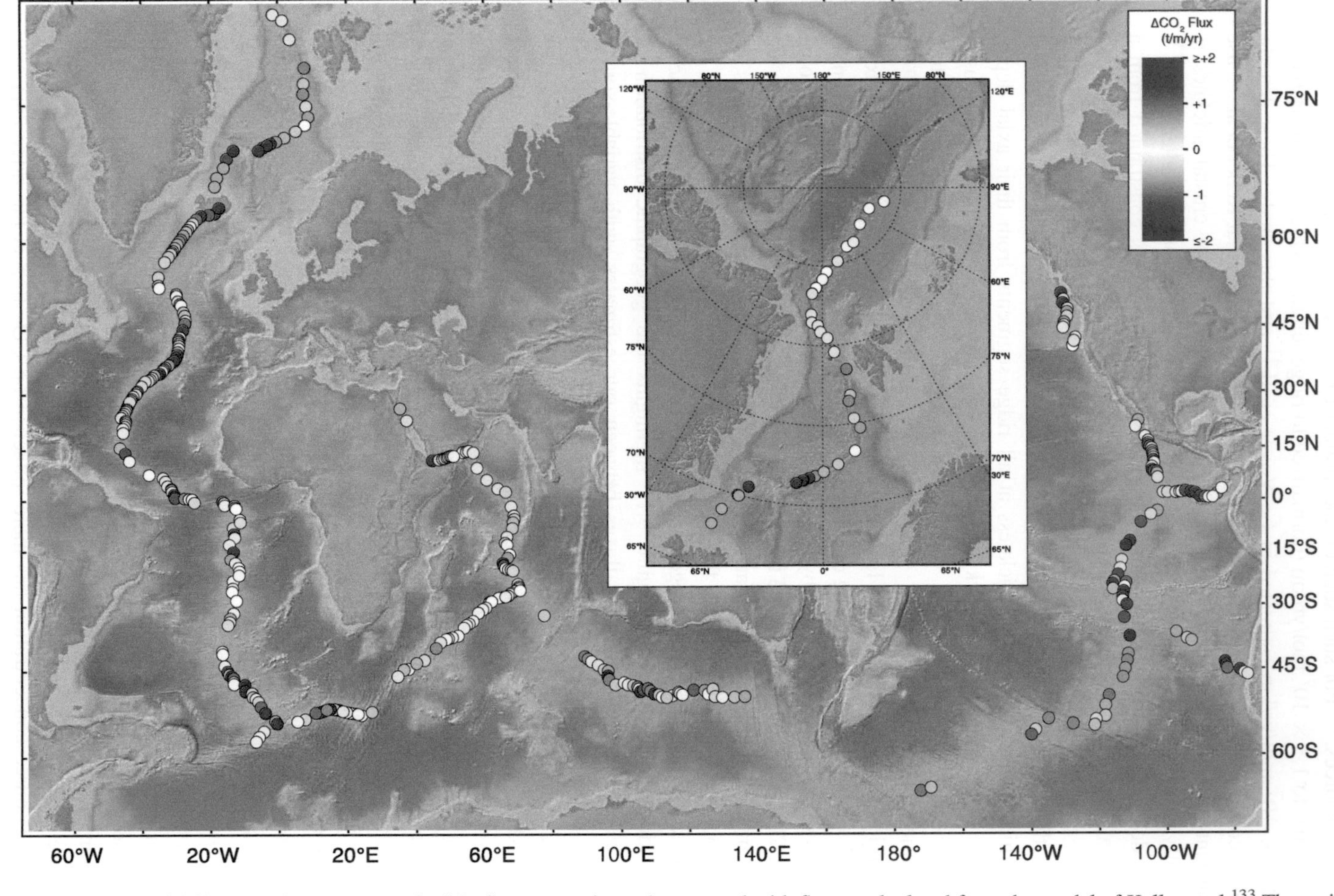

Figure 9.11 Map of differences in segment-scale CO_2 fluxes as estimated compared with fluxes calculated from the model of Keller et al.[133] The main differences are that Keller et al.[133] use a uniform mantle CO_2 content of 100 ppm; while we[183] estimate mantle CO_2 contents by calculating primary magma CO_2 and degree of melting at each segment, resulting in variations of mantle CO_2 of more than three orders of magnitude (Figure 9.8).

9.7 Conclusions

The convecting mantle holds the vast majority of the carbon in the silicate Earth, and melting of this mantle at ridges and hot spots provides a means to assess the behavior of carbon in Earth's interior. Studies of large numbers of samples of submarine volcanic glasses and melt inclusions utilizing accurate and precise high-throughput microbeam methods have uncovered suites of MORB samples that are undersaturated in CO_2 and display correlated variations of CO_2 with nonvolatile incompatible trace elements, with limited variations in CO_2/Ba and CO_2/Rb. These valuable sample suites have provided a means to assess the primary CO_2 concentrations in MORBs at each ridge segment and to quantify, with stated uncertainties, the segment-scale and total flux of CO_2 from mid-ocean ridges. We suggest that CO_2 fluxes from the mantle to the surface vary by over three orders of magnitude at the segment scale.

9.8 Limits of Knowledge and Unknowns

A key advance in the future will be the discovery of more populations of CO_2-undersaturated samples so that the assumed limited variability in CO_2/Ba and CO_2/Rb ratios and the origin of the variability of CO_2/Nb and CO_2/Th ratios can be critically assessed. It will likewise be advantageous to conduct rare gas and carbon isotope measurements on CO_2-undersaturated sample suites in order to assess the degree of possible carbon isotope heterogeneity in the convecting upper mantle. While the knowledge of MORB mantle CO_2 is now quite mature, we still lack the number of demonstrably vapor-undersaturated samples with which to assess one of our primary assumptions regarding MORB magmatism, namely the assumed homogeneity of MORB CO_2/Ba and CO_2/Rb ratios.

A critical examination of melt inclusions at Hawaii and other enriched hot spots requires knowledge of the CO_2 content of ubiquitous vapor bubbles within these inclusions. We estimate that hot spot CO_2 fluxes comprise 10–50% of the total MORB CO_2 flux. The hot spot flux is large compared to the additional 1.5–3.0% magma flux from hot spot volcanism, emphasizing the significant enrichment of CO_2 in hot spot mantle sources compared with the MORB mantle. Our understanding of hot spot CO_2 is currently limited by the lack of vapor-undersaturated melt inclusion suites; it is not yet apparent whether melts that were undersaturated at the time of entrapment will be recognized once we can precisely account for the CO_2 in melt inclusion shrinkage bubbles. However, this is presently a very active area of investigation not only at hot spots, but also at arcs, and thus conclusions (one way or the other) are likely not far behind the publication of this chapter.

LIPs can deliver enormous volatile fluxes over short time periods, although their climate impact is driven by the relative balance of SO_2 over shorter timescales (years to decades) versus CO_2 in the longer term. The pace of volcanism and relative fluxes of volatile species from LIPs remain largely unknown. So, too, does our knowledge of mantle CO_2 fluxes going back in time.

Mantle carbon can, in its oxidized form, lower the solidus of silicates and generate mantle melts. At great depths in the mantle, fO_2 is likely low enough to stabilize reduced forms of carbon that do not facilitate silicate melting. The shallow reaches of the upper mantle, however, are everywhere sufficiently oxidizing that carbonate is the stable phase. Thus, at some point during adiabatic ascent, the oxidation of carbon will initiate melting, and large variations in the CO_2 content of the mantle will translate into variations in the melt fraction that may be observable by geophysical methods. In the future, large-scale geophysical experiments involving both seismic and electromagnetic studies at CO_2-depleted and CO_2-enriched ridge segments could reveal whether a redox-melting boundary exists in the sub-ridge mantle. At the same time, improved understanding of the variation of fO_2 with depth in the convecting mantle should provide further tests of the behavior of the upper mantle in regions where melting is expected.

Acknowledgments

Erik Hauri worked relentlessly on this contribution, and many associated contributions, in the months, and even in the days, before he passed. We, his coauthors, dedicate this chapter to Erik – to his scientific vision, his insights, his magnificent data, his leadership, his spirit of discovery, his mentorship, and his friendship. We miss him terribly.

Many thanks to Jianhua Wang for expert assistance negotiating with cranky instruments in the Carnegie SIMS lab, to the Smithsonian Department of Mineral Sciences for assistance with sample preparation, to Mike Garcia and Aaron Pietruszka for provision of submarine Hawaii samples, and to Frank Trusdell and Jim Kauahikaua of the Hawaii Volcano Observatory (HVO) for going above and beyond the call of duty in making possible our sampling of Hawaiian cinder cones. We thank the geochemistry group at IPGP, University of Paris, for sharing data and samples from the Pacific–Antarctic Ridge, and the captain, crew, and scientific parties of the PACANTARCTIC 1 and 2 cruises (PAR), the GIMNAUT, CD127, and KNOX11RR cruises (CIR), the PROTEA-5, MD34, AG22, and AG53 cruises (SWIR), and the CHEPR and PANORAMA-1 cruises (EPR). We also acknowledge support from the University of California Ship Funds, and NSF grant OCE-0726573 (DRH). Conversations with Mark Behn, Pierre Cartigny, Raj Dasgupta, Marc Hirschmann, Peter van Keken, Peter Kelemen, Charles Langmuir, Terry Plank, Tobias Keller, and Richard Katz have all led to improvements in this manuscript. We thank Josh Wood for drafting Figure 9.1 and Glenn Macpherson for the Kilauea Iki melt inclusion photograph. We benefited from thorough and constructive reviews from Simon Matthews and Cyril Aubaud and from Raj Dasgupta's editorial handling. This is a contribution to the Deep Carbon Observatory.

Questions for the Classroom

1 This chapter puts forth a methodology for quantifying global fluxes based on the assumption that CO_2/Ba and CO_2/Rb ratios are constant in Earth's mantle. What

processes might fractionate these ratios? How might the hypothesis that these ratios are relatively invariant be tested?

2 Other than identifying sample suites with constant CO_2/Ba or CO_2/Rb ratios, are there other ways to test whether samples have experienced pre-eruptive degassing?

3 How could new, promising samples to determine mantle carbon concentrations be discovered?

4 What factors limit our ability quantify the amount of carbon stored in the mantle?

5 What factors limit our ability to quantify the flux of carbon out of the mantle?

6 From the current mid-ocean ridge outgassing rate, how long is required to accumulate the mass of carbon in the crust, ocean, and atmosphere (~1×10^{23} g C)? How would this number change if carbon is returned to the mantle in subduction zones?

7 How might the carbon outgassing rate from mid-ocean ridges and hot spots have changed throughout Earth's history?

8 Use the batch melting equation (9.2) to calculate the CO_2 and Ba concentrations and CO_2/Ba ratio of partial melts formed by melting a mantle with 100 ppm CO_2 and 1 ppm Ba (using peridotite-melt partition coefficients of 0.00055 and 0.00012 for CO_2 and Ba, respectively, and varying the degree of melting from 0% to 20%). At what degree of melting does the CO_2/Ba ratio of the partial melt equal to that of the mantle source (100)? What can this tell us about the usefulness of CO_2/Ba ratios measured in mid-ocean ridges and OIBs?

9 Look at the variation of other volatile species included in this chapter (H_2O, Cl, F, or S). How do they vary as a function of major elements, as a function of incompatible trace elements, and as a function of location along the ridge system? How do these trends differ from those of CO_2? How is their solubility in basalt different from CO_2?

10 Given the potential for carbon to precipitate on melt inclusion bubble walls and thus be missed by quantification with Raman spectroscopy, what is the future promise of quantifying CO_2 in melt inclusions that have nucleated bubbles?

References

1. Zhong, Y. et al., Centennial-scale climate change from decadally-paced explosive volcanism: a coupled sea ice-ocean mechanism. *Clim. Dyn.* **37**, 2373–2387 (2010).
2. Robock, A., Volcanic eruptions and climate. *Rev. Geophys.* **38**, 191–219 (2000).
3. Shindell, D.T., Volcanic and solar forcing of climate change during the preindustrial era. *J. Climate* **16**, 4094–4107 (2003).
4. Huybers, P. & Langmuir, C.H., Delayed CO– emissions from mid-ocean ridge volcanism as a possible cause of late-Pleistocene glacial cycles. *Earth Planet. Sci. Lett.* **457**, 238–249 (2017).
5. Halliday, A.N., The origins of volatiles in the terrestrial planets. *Geochim. Cosmochim. Acta* **105**, 146–171 (2013).
6. Hirschmann, M.M., Comparative deep Earth volatile cycles: the case for C recycling from exosphere/mantle fractionation of major (H_2O, C, N) volatiles and from H_2O/

Ce, CO_2/Ba, and CO_2/Nb exosphere ratios. *Earth Planet. Sci. Lett.* **502**, 262–273 (2018).

7. Dasgupta, R. & Hirschmann, M.M., The deep carbon cycle and melting in Earth's interior. *Earth Planet. Sci. Lett.* **298**, 1–13 (2010).
8. Kelemen, P.B. & Manning, C.E., Reevaluating carbon fluxes in subduction zones, what goes down, mostly comes up. *Proc. Natl Acad. Sci.* **112**, E3997–E4006 (2015).
9. Marty, B., Alexander, C.M.O.D., & Raymond, S.N., Primordial origins of Earth's carbon. *Rev. Mineral. Geochem.* **75**, 149–181 (2013).
10. Marty, B. & Tolstikhin, I.N., CO_2 fluxes from mid-ocean ridges, arcs and plumes. *Chem. Geol.* **145**, 233–248 (1998).
11. Crisp, J.A., Rates of magma emplacement and volcanic output. *J. Volcan. Geotherm. Res.* **20**, 177–211 (1984).
12. Schoene, B., Guex, J., Bartolini, A., Schaltegger, U., & Blackburn, T.J., Correlating the end-Triassic mass extinction and flood basalt volcanism at the 100 ka level. *Geology* **38**, 387–390 (2010).
13. Chung, S., Jahn, B.M., Wu, G.Y., Lo, C.H., & Cong, B.L., The Emeishan flood basant in SW China: a mantle plume initiation model and its connection with continental breakup and mass extinction at the Permian–Triassic boundary, in *Mantle Dynamics and Plate Interactions in East Asia: American Geophysical Union Geodynamics Monograph 27*, eds. M.F.J. Flower, S.-L. Chung, C.-H. Lo, & T.-Y. Lee (American Geophysical Union, Washington, DC, 1998), pp. 47–58.
14. Kamo, S.L. et al., Rapid eruption of Siberian flood-volcanic rocks and evidence for coincidence with the Permian–Triassic boundary and mass extinction at 251 Ma. *Earth Planet. Sci. Lett.* **214**, 75–91 (2003).
15. Marzoli, A. et al., Extensive 200-million-year-old continental flood basalts of the Central Atlantic Magmatic Province. *Science* **284**, 616–618 (1999).
16. Self, S., Widdowson, M., Thordarson, T., & Jay, A.E., Volatile fluxes during flood basalt eruptions and potential effects on the global environment: a Deccan perspective. *Earth Planet. Sci. Lett.* **248**, 518–532 (2006).
17. Gale, A., Langmuir, C.H., & Dalton, C.A., The global systematics of ocean ridge basalts and their origin. *J. Petrol.* **55**, 1051–1082 (2014).
18. Kelley, K.A., Kingsley, R., & Schilling, J.-G., Composition of plume-influenced mid-ocean ridge lavas and glasses from the Mid-Atlantic Ridge, East Pacific Rise, Galapagos Spreading Center, and Gulf of Aden. *Geochem. Geophys. Geosyst.* **14**, 223–242 (2013).
19. Kinzler, R.J. & Grove, T.L., Primary magmas of mid-ocean ridge basalts 1. Experiments and methods. *J. Geophys. Res. Solid Earth* **97**, 6885–6906 (1992).
20. Klein, E.M. & Langmuir, C.H., Global correlations of ocean ridge basalt chemistry with axial depth and crustal thickness. *J. Geophys. Res. Solid Earth* **92**, 8089–8115 (1987).
21. Langmuir, C.H., Klein, E.M., & Plank, T., Petrological systematics of mid-ocean ridge basalts: constraints on melt generation beneath ocean ridges, in *Mantle Flow and Melt Generation at Mid-Ocean Ridges*, eds. J.P. Morgan, D.K. Blackman, & J.M. Sinton (American Geophysical Union, Washington, DC, 1992), pp. 183–280.
22. Hart, S.R., A large-scale isotope anomaly in the Southern Hemisphere mantle. *Nature* **309**, 753–757 (1984).
23. Hofmann, A.W., Chemical differentiation of the Earth: the relationship between mantle, continental crust, and oceanic crust. *Earth Planet. Sci. Lett.* **90**, 297–314 (1988).

24. Stracke, A., Hofmann, A.W., & Hart, S.R., FOZO, HIMU, and the rest of the mantle zoo. *Geochem. Geophys. Geosyst.* **6**, Q05007 (2005).
25. Zindler, A. & Hart, S., Chemical geodynamics. *Ann. Rev. Earth Planet. Sci.* **14**, 493–571 (1986).
26. Horan, M.F. et al., Tracking Hadean processes in modern basalts with 142-neodymium. *Earth Planet. Sci. Lett.* **484**, 184–191 (2018).
27. Mukhopadhyay, S., Early differentiation and volatile accretion recorded in deep-mantle neon and xenon. *Nature* **486**, 101–104 (2012).
28. Mundl, A. et al., Tungsten-182 heterogeneity in modern ocean island basalts. *Science* **356**, 66–69 (2017).
29. Parai, R., Mukhopadhyay, S., & Standish, J.J., Heterogeneous upper mantle Ne, Ar and Xe isotopic compositions and a possibble Dupal noble gas signature recorded in basalts from the Southwest Indian Ridge. *Earth Planet. Sci. Lett.* **359–360**, 227–239 (2012).
30. Peters, B.J., Carlson, R.W., Day, J.M.D., & Horan, M.F., Hadean silicate differentiation preserved by anomalous $^{142}Nd/^{144}Nd$ ratios in the Réunion hotspot source. *Nature* **555**, 89–93 (2018).
31. Peto, M.K., Mukhopadhyay, S., & Kelley, K.A., Heterogeneities from the first 100 million years recorded in deep mantle noble gases from the Northern Lau Back-arc Basin. *Earth Planet. Sci. Lett.* **369**, 13–23 (2013).
32. Rizo, H. et al., Preservation of Earth-forming events in the tungsten isotopic composition of modern flood basalts. *Science* **352**, 809–812 (2016).
33. Tucker, J.M., Mukhopadhyay, S., & Schilling, J.-G., The heavy noble gas composition of the depleted MORB mantle (DMM) and its implications for the preservation of heterogeneities in the mantle. *Earth Planet. Sci. Lett.* **355–356**, 244–254 (2012).
34. Donnelly, K.E., Goldstein, S.L., Langmuir, C.H., & Spiegelman, M., Origin of enriched ocean ridge basalts and implications for mantle dynamics. *Earth Planet. Sci. Lett.* **226**, 347–366 (2004).
35. Schilling, J.G. et al., Petrologic and geochemical variations along the Mid-Atlantic Ridge from 29°N to 73°N. *Am. J. Sci.* **283**, 510–586 (1983).
36. Gale, A., Dalton, C.A., Langmuir, C.H., Su, Y., & Schilling, J.-G., The mean composition of ocean ridge basalts. *Geochem. Geophys. Geosyst.* **14**, 489–518 (2013).
37. Dixon, J.E. & Stolper, E.M., An experimental study of water and carbon dioxide solubilities in mid-ocean ridge basaltic liquids. Part II: applications to degassing. *J. Petrol.* **36**, 1633–1646 (1995).
38. Javoy, M. & Pineau, F., The volatiles record of a popping rock from the Mid-Atlantic Ridge at 14°N: chemical and isotopic composition of gas trapped in the vesicles. *Earth Planet. Sci. Lett.* 107, 598–611 (1991).
39. Dixon, J.E., Stolper, E.M., & Holloway, J.R., An experimental study of water and carbon dioxide solubilities in mid-ocean ridge basaltic liquids. Part I: calibration and solubility models. *J. Petrol.* **36**, 1607–1631 (1995).
40. Jendrzejewski, N., Trull, T.W., Pineau, F., & Javoy, M., Carbon solubility in mid-ocean ridge basaltic melt at low pressures (250–1950 bar). *Chem. Geol.* **138**, 81–92 (1997).
41. Pan, V., Holloway, J.R., & Hervig, R.L., The temperature and pressure dependence of carbon dioxide solubility in tholeiitic basalt. *Geochim. Cosmochim. Acta* **55**, 1587–1595 (1991).

42. Shishkina, T.A., Botcharnikov, R.E., Holtz, F., Almeev, R.R., & Portnyagin, M.V., Solubility of H_2O- and CO_2-bearing fluids in tholeiitic basalts at pressures up to 500MPa. *Chem. Geol.* **277**, 115–125 (2010).
43. Aster, E.M. et al., Reconstructing CO_2 concentrations in basaltic melt inclusions using Raman analysis of vapor bubbles. *J. Volcan. Geotherm. Res.* **323**, 148–162 (2016).
44. Moore, L.R. et al., Bubbles matter: an assessment of the contribution of vapor bubbles to melt inclusion volatile budgets. *Am. Mineral.* **100**, 806–823 (2015).
45. Hirano, N. et al., Volcanism in response to plate flexure. *Science* **313**, 1426–1428 (2006).
46. Curators of Marine and Lacustrine Geological Samples Consortium. The Index to Marine and Lacustrine Geological Samples (IMLGS). *NOAA National Centers for Environmental Information.* https://data.nodc.noaa.gov/cgi-bin/iso?id=gov.noaa.ngdc.mgg.geology:G00028.
47. Le Voyer, M. et al., Carbon fluxes and primary magma CO_2 contents along the global mid-ocean ridge system. *Geochem. Geophys. Geosyst.* **20**, 1387–1424 (2019).
48. Cartigny, P., Pineau, F., Aubaud, C., & Javoy, M., Towards a consistent mantle carbon flux estimate: insights from volatile systematics (H_2O/Ce, dD, CO_2/Nb) in the North Atlantic mantle (14°N and 34°N). *Earth Planet. Sci. Lett.* **265**, 672–685 (2008).
49. Saal, A.E., Hauri, E.H., Langmuir, C.H., & Perfit, M.R., Vapour undersaturation in primitive mid-ocean-ridge basalt and the volatile content of Earth's upper mantle. *Nature* **419**, 451–455 (2002).
50. Michael, P.J. & Graham, D.W., The behavior and concentration of CO_2 in the suboceanic mantle: inferences from undegassed ocean ridge and ocean island basalts. *Lithos* **236–237**, 338–351 (2015).
51. Shimizu, K. et al., Two-component mantle melting-mixing model for the generation of mid-ocean ridge basalts: implications for the volatile content of the Pacific upper mantle. *Geochim. Cosmochim. Acta* **176**, 44–80 (2016).
52. Le Voyer, M., Kelley, K.A., Cottrell, E., & Hauri, E.H., Heterogeneity in mantle carbon content from CO_2-undersaturated basalts. *Nat. Commun.* **8**, 14062 (2017).
53. Hauri, E.H. et al., CO_2 content beneath northern Iceland and the variability of mantle carbon. *Geology* **46**, 55–58 (2017).
54. Shaw, A.M., Behn, M.D., Humphris, S.E., Sohn, R.A., & Gregg, P.M., Deep pooling of low degree melts and volatile fluxes at the 85°E segment of the Gakkel Ridge: evidence from olivine-hosted melt inclusions and glasses. *Earth Planet. Sci. Lett.* **289**, 311–322 (2010).
55. Wanless, V.D., Behn, M.D., Shaw, A.M., & Plank, T., Variations in melting dynamics and mantle compositions along the Eastern Volcanic Zone of the Gakkel Ridge: insights from olivine-hosted melt inclusions. *Contrib. Mineral. Petrol.* **167**, 1005–1027 (2014).
56. Wanless, V.D. & Shaw, A.M., Lower crustal crystallization and melt evolution at mid-ocean ridges. *Nat. Geosci.* **5**, 651–655 (2012).
57. Wanless, V.D. et al., Magmatic plumbing at Lucky Strike volcano based on olivine-hosted melt inclusion compositions. *Geochem. Geophys. Geosyst.* **16**, 126–147 (2015).
58. Helo, C., Longpré, M.-A., Shimizu, N., Clague, D.A., & Stix, J., Explosive eruptions at mid-ocean ridges driven by CO_2-rich magmas. *Nat. Geosci.* **4**, 260–263 (2011).

59. Le Voyer, M. et al., Volatile abundances in MORB magmas along the global mid-ocean ridge system (2018). Interdisciplinary Earth Data Alliance. doi:10.1594/IEDA/111195.
60. Marske, J.P. & Hauri, E.H., Major- and trace-element compositions of 915 melt inclusions and host olivines from Hawaiian shield volcanoes. (2019). Interdisciplinary Earth Data Alliance. doi:10.1594/IEDA/111193.
61. Tucker, J.M. et al., A high carbon content of the Hawaiian mantle from olivine-hosted melt inclusions. *Geochim. Cosmochim. Acta* **254**, 156–172 (2019).
62. Métrich, N. et al., Is the "Azores Hotspot" a wetspot? Insights from the geochemistry of fluid and melt inclusions in olivine of Pico basalts. *J. Petrol.* **55**, 377–393 (2014).
63. Longpré, M.-A., Stix, J., Klügel, A., & Shimizu, N., Mantle to surface degassing of carbon- and sulphur-rich alkaline magma at El Hierro, Canary Islands. *Earth Planet. Sci. Lett.* **460**, 268–280 (2017).
64. Hartley, M.E., Maclennan, J., Edmonds, M., & Thordarson, T., Reconstructing the deep CO_2 degassing behaviour of large basaltic fissure eruptions. *Earth Planet. Sci. Lett.* **393**, 120–131 (2014).
65. Neave, D.A., Maclennan, J., Edmonds, M., & Thordarson, T., Melt mixing causes negative correlation of trace element enrichment and CO_2 content prior to an Icelandic eruption. *Earth Planet. Sci. Lett.* **400**, 272–283 (2014).
66. Tucker, J.M., Shrinkage bubble CO_2 reconstructions from ocean island basalt melt inclusions. *Interdiscip. Earth Data Alliance* (2019). doi:10.1594/IEDA/111230.
67. Dixon, J.E., Stolper, E.M., & Delaney, J.R., Infrared spectroscopic measurements of CO_2 and H_2O in Juan de Fuca Ridge basaltic glasses. *Earth Planet. Sci. Lett.* **90**, 87–104 (1988).
68. Le Roux, P.J., Shirey, S.B., Hauri, E.H., & Perfit, M.R., The effects of variable sources, processes and contaminants on the composition of northern EPR MORB (8–10°N and 12–14°N): evidence from volatiles (H_2O, CO_2, S) and halogens (F, Cl). *Earth Planet. Sci. Lett.* **251**, 209–231 (2006).
69. Chavrit, D., Humler, E., Morizet, Y., & Laporte, D., Influence of magma ascent rate on carbon dioxide degassing at oceanic ridges: message in a bubble. *Earth Planet. Sci. Lett.* **357–358**, 376–385 (2012).
70. Papale, P., Moretti, R., & Barbato, D., The compositional dependence of the saturation surface of $H_2O + CO_2$ fluids in silicate melts. *Chem. Geol.* **229**, 78–95 (2006).
71. Ghiorso, M.S. & Gualda, G.A.R., An H_2O–CO_2 mixed fluid saturation model compatible with rhyolite-MELTS. *Contrib. Mineral. Petrol.* **169**, 53 (2015).
72. Craig, H., Clarke, W.B., & Beg, M.A., Excess ^3He in deep water on the East Pacific Rise. *Earth Planet. Sci. Lett.* **26**, 125–132 (1975).
73. Schlitzer, R., Quantifying He fluxes from the mantle using multi-tracer data assimilation. *Philos. Trans. R. Soc. A* **374**, 20150288 (2016).
74. Holzer, M. et al., Objective estimates of mantle ^3He in the ocean and implications for constraining the deep ocean circulation. *Earth Planet. Sci. Lett.* **485**, 305–314 (2017).
75. Tucker, J.M., Mukhopadhyay, S., & Gonnermann, H.M., Reconstructing mantle carbon and noble gas contents from degassed mid-ocean ridge basalts. *Earth Planet. Sci. Lett.* **496**, 108–119 (2018).
76. Burnard, P., Correction for volatile fractionation in ascending magmas: noble gas abundances in primary mantle melts. *Geochim. Cosmochim. Acta* **65**, 2605–2614 (2001).

77. Colin, A., Burnard, P., & Marty, B., Mechanisms of magma degassing at mid-oceanic ridges and the local volatile composition (^4He–$^{40}Ar^*$–CO_2) of the mantle by laser ablation analysis of individual MORB vesicles. *Earth Planet. Sci. Lett.* **361**, 183–194 (2013).
78. Rosenthal, A., Hauri, E., & Hirschmann, M., Experimental determination of C, F, and H partitioning between mantle minerals and carbonated basalt, CO_2/Ba and CO_2/Nb systematics of partial melting, and the CO_2 contents of basaltic source regions. *Earth Planet. Sci. Lett.* **412**, 77–87 (2015).
79. Matthews, S., Shorttle, O., Rudge, J.F., & Maclennan, J., Constraining mantle carbon: CO_2–trace element systematics in basalts and the roles of magma mixing and degassing. *Earth Planet. Sci. Lett.* **480**, 1–14 (2017).
80. Shimizu, K., Saal, A.E., Hauri, E.H., Perfit, M.R., & Hékinian, R., Evaluating the roles of melt-rock interaction and partial degassing on the CO_2/Ba ratios of MORB: implications for the CO_2 budget in Earth's depleted upper mantle, *Geochim. Cosmochim. Acta.* (2019) (submitted).
81. Koornneef, J.M. et al., TIMS analysis of Sr and Nd isotopes in melt inclusions from Italian potassium-rich lavas using prototype 10^{13} Ω amplifiers. *Chem. Geol.* **397**, 14–23 (2015).
82. Maclennan, J., Lead isotope variability in olivine-hosted melt inclusions from Iceland. *Geochim. Cosmochim. Acta* **72**, 4159–4176 (2008).
83. Paul, B. et al., Melt inclusion Pb-isotope analysis by LA–MC-ICPMS: assessment of analytical performance and application to OIB genesis. *Chem. Geol.* **289**, 210–223 (2011).
84. Saal, A.E. et al., Pb isotopic variability in melt inclusions from the EMI–EMII–HIMU mantle end-members and the role of the oceanic lithosphere. *Earth Planet. Sci. Lett.* **240**, 605–620 (2005).
85. Dupré, B. & Allégre, C.J., Pb–Sr isotope variation in Indian Ocean basalts and mixing phenomena. *Nature* **303**, 142 (1983).
86. Niu, Y.L., Collerson, K.D., Batiza, R., Wendt, J.L., & Regelous, M., Origin of enriched-type mid-ocean ridge basalts at ridges far from mantle plumes: the East Pacific Rise at 11°20' N. *J. Geophys. Res.* **104**, 7067–7087 (1999).
87. Arevalo Jr, R. & McDonough, W.F., Chemical variations and regional diversity observed in MORB. *Chem. Geol.* **271**, 70–85 (2010).
88. Behn, M.D. & Grove, T.L., Melting systematics in mid-ocean ridge basalts: application of a plagioclase-spinel melting model to global variations in major element chemistry and crustal thickness. *J. Geophys. Res. Solid Earth* **120**, 4863–4886 (2015).
89. White, R.S., McKenzie, D., & O'Nions, R.K., Oceanic crustal thickness from seismic measurements and rare earth element inversions. *J. Geophys. Res. Solid Earth* **97**, 19683–19715 (1992).
90. DeMets, C., Gordon, R.G., Argus, D.F., & Stein, S., Effect of recent revisions to the geomagnetic reversal time scale on estimates of current plate motions. *J. Geophys. Res.* **21**, 2191–2194 (1994).
91. Asimow, P.D., Hirschmann, M.M., & Stolper, E.M., Calculation of peridotite partial melting from thermodynamic models of minerals and melts, IV. Adiabatic decompression and the composition and mean properties of mid-ocean ridge basalts. *J. Petrol.* **42**, 963–998 (2001).

92. Van Avendonk, H.J.A., Davis, J.K., Harding, J.L., & Lawver, L.A., Decrease in oceanic crustal thickness since the breakup of Pangaea. *Nat. Geosci.* **10**, 58–61 (2016).
93. Broecker, W.S. & Peng, T.-H., Carbon cycle: 1985 glacial to interglacial changes in the operation of the global carbon cycle. *Radiocarbon* **28**, 309–327 (1986).
94. Alt, J.C. & Shanks, W.C., Serpentinization of abyssal peridotites from the MARK area, Mid-Atlantic Ridge: sulfur geochemistry and reaction modeling. *Geochim. Cosmochim. Acta* **67**, 641–653 (2003).
95. Alt, J.C. & Teagle, D., The uptake of carbon during alteration of ocean crust. *Geochim. Cosmochim. Acta* **63**, 1527–1535 (1999).
96. Hayes, J.M. & Waldbauer, J.R., The carbon cycle and associated redox processes through time. *Philos. Trans. R. Soc. Lond. B Biol. Sci.* **361**, 931–950 (2006).
97. French, S.W. & Romanowicz, B., Broad plumes rooted at the base of the Earth's mantle beneath major hotspots. *Nature* **525**, 95–99 (2015).
98. Montelli, R., Nolet, G., Dahlen, F.A., & Masters, G., A catalogue of deep mantle plumes: new results from finite-frequency tomography. *Geochem. Geophys. Geosyst.* **7**, Q11007 (2006).
99. Burton, M.R., Sawyer, G.M., & Granieri, D., Deep carbon emissions from volcanoes. *Rev. Mineral. Geochem.* **75**, 323–354 (2013).
100. Richards, M.A., Duncan, R.A., & Courtillot, V.E., Flood basalts and hot-spot tracks: plume heads and tails. *Science* **246**, 103–107 (1989).
101. Kiehl, J.T. & Shields, C.A., Climate simulation of the latest Permian: implications for mass extinction. *Geology* **33**, 757–760 (2005).
102. Sleep, N.H., Hotspots and mantle plumes: some phenomenology. *J. Geophys. Res.* **95**, 6715–6736 (1990).
103. Davies, G.F., Ocean bathymetry and mantle convection: 1. Large-scale flow and hotspots. *J. Geophys. Res.* **93**, 10467–10480 (1988).
104. Dixon, J.E. & Clague, D.A., Volatiles in basaltic glasses from Loihi seamount, Hawaii: evidence for a relatively dry plume component. *J. Petrol.* **42**, 627–654 (2001).
105. Dixon, J.E., Clague, D.A., Wallace, P., & Poreda, R., Volatiles in alkalic basalts from the North Arch volcanic field, Hawaii: extensive degassing of deep submarine-erupted alkalic series lavas. *J. Petrol.* **38**, 911–939 (1997).
106. Simons, K., Dixon, J.E., Schilling, J.-G., Kingsley, R., & Poreda, R., Volatiles in basaltic glasses from the Easter-Salas y Gomez Seamount Chain and Easter Microplate: implications for geochemical cycling of volatile elements. *Geochem. Geophys. Geosyst.* **3**, 1039 (2002).
107. Workman, R.K., Hauri, E., Hart, S.R., Wang, J., & Blusztajn, J., Volatile and trace elements in basaltic glasses from Samoa: implications for water distribution in the mantle. *Earth Planet. Sci. Lett.* **241**, 932–951 (2006).
108. Hilton, D.R., Thirlwall, M.F., Taylor, R.N., Murton, B.J., & Nichols, A., Controls on magmatic degassing along the Reykjanes Ridge with implications for the helium paradox. *Earth Planet. Sci. Lett.* **183**, 43–50 (2000).
109. Hauri, E., SIMS analysis of volatiles in silicate glasses, 2: isotopes and abundances in Hawaiian melt inclusions. *Chem. Geol.* **183**, 115–141 (2002).
110. Edmonds, M. et al., Magma storage, transport and degassing during the 2008–10 summit eruption at Kilauea Volcano, Hawai'i. *Geochim. Cosmochim. Acta* **123**, 284–301 (2013).

111. Sides, I.R., Edmonds, M., MacLennan, J., Swanson, D.A., & Houghton, B.F., Eruption style at Kilauea Volcano in Hawai`i linked to primary melt composition. *Nat. Geosci.* **7**, 1–6 (2014).
112. Gonnermann, H.M. & Mukhopadhyay, S., Non-equilibrium degassing and a primordial source for helium in ocean-island basalts. *Nature* **449**, 1037–1040 (2007).
113. Anderson, K.R. & Poland, M.P., Abundant carbon in the mantle beneath Hawai'i. *Nat. Geosci.* **10**, 704–708 (2017).
114. Gerlach, T.M., McGee, K.A., Elias, T., Sutton, A.J., & Doukas, M.P., Carbon dioxide emission rate of Kilauea Volcano: implications for primary magma and the summit reservoir. *J. Geophys. Res.* **107**, ECV 3-1–ECV 3-15 (2002).
115. Poland, M.P., Miklius, A., Sutton, A.J., & Thornber, C.R., A mantle-driven surge in magma supply to Kīlauea Volcano during 2003–2007. *Nat. Geosci.* **5**, 295–300 (2012).
116. Aubaud, C., Pineau, F., Hekinian, R., & Javoy, M., Carbon and hydrogen isotope constraints on degassing of CO_2 and H_2O in submarine lavas from the Pitcairn hotspot (South Pacific). *Geophys. Res. Lett.* **33**, L02308 (2006).
117. Aubaud, C., Pineau, F., Hekinian, R., & Javoy, M., Degassing of CO_2 and H_2O in submarine lavas from the Society hotspot. *Earth Planet. Sci. Lett.* **235**, 511–527 (2005).
118. Batiza, R., Abundances, distribution and sizes of volcanoes in the Pacific Ocean and implications for the origin of non-hotspot volcanoes. *Earth Planet. Sci. Lett.* **60**, 195–206 (1982).
119. Wessel, P. & Lyons, S., Distribution of large Pacific seamounts from Geosat/ERS-1: implications for the history of intraplate volcanism. *J. Geophys. Res. Solid Earth* **102**, 22459–22475 (1997).
120. Kim, S.-S. & Wessel, P., New global seamount census from altimetry-derived gravity data. *Geophys. J. Int.* **186**, 615–631 (2011).
121. Black, B.A., Hauri, E.H., Elkins-Tanton, L.T., & Brown, S.M., Sulfur isotopic evidence for sources of volatiles in Siberian Traps magmas. *Earth Planet. Sci. Lett.* **394**, 58–69 (2014).
122. Black, B.A. & Manga, M., Volatiles and the tempo of flood basalt magmatism. *Earth Planet. Sci. Lett.* **458**, 130–140 (2017).
123. Coffin, M.F. & Eldholm, O., Scratching the surface: estimating dimensions of large igneous provinces. *Geology* **21**, 515–518 (1993).
124. Courtillot, V.E. & Renne, P.R., On the ages of flood basalt events. *Comp. Rend. Geosci.* **335**, 113–140 (2003).
125. Ernst, R.E., Large mafic magmatic events through time and links to mantle-plume heads, in *Mantle Plumes: Their Identification through Time*, eds. R.E. Ernst & K.L. Buchan (Geological Society of America Special Paper, Boulder, CO, 2001), p. 483.
126. Ward, P.L., Sulfur dioxide initiates global climate change in four ways. *Thin Solid Films* **517**, 3188–3203 (2009).
127. Sleep, N.H., Hotspots and mantle plumes. *Ann. Rev. Earth Planet. Sci.* **20**, 19–43 (1992).
128. Davies, G.F. & Richards, M.A., Mantle convection. *J. Geol.* **100**, 151–206 (1992).
129. Norman, M.D. & Garcia, M.O., Primitive magmas and source characteristics of the Hawaiian plume: petrology and geochemistry of shield picrites. *Earth Planet. Sci. Lett.* **168**, 27–44 (1999).
130. Pietruszka, A.J., Norman, M.D., Garcia, M.O., Marske, J.P., & Burns, D.H., Chemical heterogeneity in the Hawaiian mantle plume from the alteration and dehydration of recycled oceanic crust. *Earth Planet. Sci. Lett.* **361**, 198–309 (2013).

131. Barry, P.H., Hilton, D.R., Füri, E., Halldórsson, S.A., & Grönvold, K., Carbon isotope and abundance systematics, and CO_2 fluxes from Icelandic geothermal gases, fluids and subglacial basalts. *Geochim. Cosmochim. Acta* **134**, 74–99 (2014).
132. Dasgupta, R. & Hirschmann, M.M., Melting in the Earth's deep upper mantle caused by carbon dioxide. *Nature* **440**, 659–662 (2006).
133. Keller, T., Katz, R.F., & Hirschmann, M.M., Volatiles beneath mid-ocean ridges: deep melting, channelised transport, focusing, and matasomatism. *Earth Planet. Sci. Lett.* 464, 55–68 (2017).
134. Hirschmann, M.M., Mantle solidus: experimental constraints and the effects of peridotite composition. *Geochem. Geophys. Geosyst.* **1**, 2000GC000070 (2000).
135. Dalton, J.A. & Presnall, D.C., Carbonatitic melts along the solidus of model lherzolite in the system $CaO–MgO–Al_2O_3–SiO_2–CO_2$ from 3 to 7 GPa. *Contrib. Mineral. Petrol.* **131**, 123–135 (1998).
136. Asimow, P.D. & Langmuire, C.H., The importance of water to oceanic mantle melting regimes. *Nature* **421**, 815–820 (2003).
137. Dasgupta, R. et al., Carbon-dioxide-rich silicate melt in the Earth's upper mantle. *Nature* **493**, 211–216 (2013).
138. Dasgupta, R., Volatile-bearing partial melts beneath oceans and continents – where, how much, and of what compositions. *Am. J. Sci.* **318**, 141–165 (2018).
139. Shcheka, S.S., Wiedenbeck, M., Frost, D.J., & Keppler, H., Carbon solubility in mantle minerals. *Earth Planet. Sci. Lett.* **245**, 730–742 (2006).
140. Keppler, H., Wiedenbeck, M., & Shcheka, S.S., Carbon solubility in olivine and the mode of carbon storage in the Earth's mantle. *Nature* **424**, 414–416 (2003).
141. Stagno, V. & Frost, D.J., Carbon speciation in the asthenosphere: experimental measurements of the redox conditions at which carbonate-bearing melts coexist with graphite or diamond in peridotite assemblages. *Earth Planet. Sci. Lett.* **300**, 72–84 (2010).
142. Frost, D.J. & McCammon, C.A., The redox state of Earth's mantle. *Ann. Rev. Earth Planet. Sci.* **36**, 389–420 (2008).
143. Luth, R.W., Carbon and carbonates in the mantle, in *Mantle Petrology: Field Observations and High Pressure Experimentation: A Tribute to Francis R. (Joe) Boyd*, eds. Y. Fei, C.M. Bertka, & B.O. Mysen (Geochemical Society, Houston, TX, 1999), pp. 297–316.
144. Zhang, Z., Hastings, P., Von der Handt, A., & Hirschmann, M.M., Experimental determination of carbon solubility in Fe–Ni–S melts. *Geochim. Cosmochim. Acta* **225**, 66–79 (2018).
145. Armstrong, L.S., Hirschmann, M.M., Stanley, B.D., Falksen, E.G., & Jacobsen, S.D., Speciation and solubility of reduced C–O–H–N volatiles in mafic melt: implications for volcanism, atmospheric evolution, and deep volatile cycles in the terrestrial planets. *Geochim. Cosmochim. Acta* **171**, 283–302 (2015).
146. Rohrbach, A. & Schmidt, M.W., Redox freezing and melting in the Earth's deep mantle resulting from carbon-iron redox coupling. *Nature* **472**, 209–212 (2011).
147. Tsuno, K. & Dasgupta, R., Fe–Ni–Cu–C–S phase relations at high pressures and temperatures – the role of sulfur in carbon storage and diamond stability at mid- to deep-upper mantle. *Earth Planet. Sci. Lett.* **412**, 132–142 (2015).
148. Eggler, D.H. & Baker, D.R., Reduced volatiles in system C–O–H: implications to mantle melting, fluid formation and diamond genesis, in *High-Pressure Research in Geophysics*, eds. S. Akimoto & M.H. Manghnani (Springer, New York, 1982), pp. 237–250.

149. Eggler, D.H., Does CO_2 cause partial melting in low-velocity layer of mantle. *Geology* **4**, 69–72 (1976).
150. Dasgupta, R., Hirschmann, M.M., & Withers, A.C., Deep global cycling of carbon constrained by the solidus of anhydrous, carbonated eclogite under upper mantle conditions. *Earth Planet. Sci. Lett.* **227**, 73–85 (2004).
151. Massuyeau, M., Gardés, E., Morizet, Y., & Gaillard, F., A model for the activity of silica along the carbonatite–kimberlite–mellilitite–basanite melt compositional joint. *Chem. Geol.* **418**, 206–216 (2015).
152. Holloway, J.R., Pan, V., & Gudmundsson, G., High-pressure fluid-absent melting experiments in the presence of graphite; oxygen fugacity, ferric/ferrous ratio and dissolved CO_2. *Eur. J. Mineral.* **4**, 105–114 (1992).
153. Stanley, B.D., Hirschmann, M.M., & Withers, A.C., Solubility of COH volatiles in graphite-saturated Martian basalts. *Geochim. Cosmochim. Acta* **129**, 54–76 (2014).
154. Eguchi, J. & Dasgupta, R., A CO_2 solubility model for silicate melts from fluid saturation to graphite or diamond saturation. *Chem. Geol.* **487**, 23–38 (2018).
155. Balhaus, C., Redox states of lithospheric and asthenospheric upper mantle. *Contrib. Mineral. Petrol.* **114**, 331–348 (1993).
156. Haggerty, S.E., The redox state of planetary basalts. *Geophys. Res. Lett.* **5**, 443–446 (1978).
157. Bezos, A. & Humler, E., The $Fe^{3+}/\Sigma Fe$ ratios of MORB glasses and their implications for mantle melting. *Geochim. Cosmochim. Acta* **69**, 711–725 (2005).
158. Cottrell, E. & Kelley, K.A., The oxidation state of Fe in MORB glasses and the oxygen fugacity of the upper mantle. *Earth Planet. Sci. Lett.* **305**, 270–282 (2011).
159. Zhang, H.L., Cottrell, E., Solheid, P.A., Kelley, K.A., & Hirschmann, M.M., Determination of $Fe^{3+}/\Sigma Fe$ of XANES basaltic glass standards by Mössbauer spectroscopy and its application to the oxidation state of iron in MORB. *Chem. Geol.* **479**, 166–175 (2018).
160. Birner, S.K., Cottrell, E., Warren, J.M., Kelley, K.A., & Davis, F.A., Peridotites and basalts reveal broad congruence between two independent records of mantle fO_2 despite local redox heterogeneity. *Earth Planet. Sci. Lett.* **494**, 172–189 (2018).
161. Dasgupta, R. & Hirschmann, M.M., Effect of variable carbonate concentration on the solidus of mantle peridotite. *Am. Mineral.* **92**, 370–379 (2007).
162. Ghosh, S., Litasov, K., & Ohtani, E., Phase relations and melting of carbonated peridotite between 10 and 20 GPa: a proxy for alkali- and CO_2-rich silicate melts in the deep mantle. *Contrib. Mineral. Petrol.* **167**, 964 (2014).
163. Gudmundsson, G. & Wood, B.J., Experimental tests of garnet peridotite oxygen barometry. *Contrib. Mineral. Petrol.* **119**, 56–67 (1995).
164. O'Neill, H.S. et al., Ferric iron in the upper mantle and in transition zone assemblages: implications for relative oxygen fugacities in the mantle, in *Evolution of the Earth and Planets*, eds. E. Takahashi, R. Jeanloz, & D. Rubie (American Geophysical Union, Washington, DC, 1993), Vol. 74, pp. 73–88.
165. Balhaus, C., Is the upper mantle metal saturated? *Earth Planet. Sci. Lett.* **132**, 75–86 (1995).
166. Rohrbach, A. et al., Metal saturation in the upper mantle. *Nature* **449**, 456–458 (2007).
167. Stagno, V., Ojwang, D.O., McCammon, C.A., & Frost, D.J., The oxidation state of the mantle and the extraction of carbon from Earth's interior. *Nature* **493**, 84–88 (2013).

168. Minarik, W.G. & Watson, E.B., Interconnectivity of carbonate melt at low melt fraction. *Earth Planet. Sci. Lett.* **133**, 423–437 (1995).
169. Sarafian, E., Gaetani, G.A., Hauri, E.H., & Sarafian, A., Experimental constraints on the damp peridotite solidus and oceanic mantle potential temperature. *Science* **355**, 942–945 (2017).
170. Keller, T. & Katz, R.F., The role of volatiles in reactive melt transport in the asthenosphere. *J. Petrol.* **57**, 1073–1108 (2016).
171. Dalton, C.A., Langmuir, C.H., & Gale, A., Geophysical and geochemical evidence for deep temperature variations beneath mid-ocean ridges. *Science* **344**, 80–83 (2014).
172. Class, C. & Lehnert, K., PetDB expert MORB (mid-ocean ridge basalt) compilation. *EarthChem Library* (2012). doi:10.1594/IEDA/100060
173. Kelemen, P.B., Yogodzinski, G.M., & Scholl, D.W., Along-strike variation in the Aleutian island arc: genesis of high Mg# andesite and implications for continental crust, in *Inside the Subduction Factory. American Geophysical Union Geophysical Monograph*, ed. J. Eiler (American Geophysical Union, Washington DC, 2003), Vol. 138, pp. 223–276.
174. Ghiorso, M.S., Hirschmann, M.M., Reiners, P.W., & Kress, V.C., The pMELTS: a revision of MELTS for improved calculation of phase relations and major element partitioning related to partial melting of the mantle to 3 GPa. *Geochem. Geophys. Geosyst.* **3**, 1–35 (2002).
175. Asimow, P.D., Dixon, J.E., & Langmuir, C.H., A hydrous melting and fractionation model for mid ocean ridge basalts: application to the Mid Atlantic Ridge near the Azores. *Geochem. Geophys. Geosyst.* **5**, Q01E16 (2004).
176. Smith, P.M. & Asimow, P.D., Adiabat_1ph: a new public front-end to the MELTS, pMELTS, and pHMELTS models. *Geochem. Geophys. Geosyst.* **6**, Q02004 (2005).
177. Workman, R.K. & Hart, S.R., Major and trace element composition of the depleted MORB mantle (DMM). *Earth Planet. Sci. Lett.* **231**, 53–72 (2005).
178. Rudge, J.F., Maclennan, J., & Stracke, A., The geochemical consequences of mixing melts from a heterogeneous mantle. *Geochim. Cosmochim. Acta* **144**, 112–143 (2013).
179. Shishkina, T.A. et al., Compositional and pressure effects on the solubility of H_2O and CO_2 in mafic melts. *Chem. Geol.* **388**, 112–129 (2014).
180. White, S.M., Crisp, J.A., & Spera, F.J., Long-term volumetric eruption rates and magma budgets. *Geochem. Geophys. Geosyst.* **7**, Q03010 (2006).
181. Poland, M.P., Miklius, A., & Montgomery-Brown, E.K., Magma supply, storage, and transport at shield-stage Hawaiian volcanoes. *US Geol. Surv. Prof. Pap.* **1801**, 179–234 (2014).
182. Bowman, A.W. & Azzalini, A., *Applied Smoothing Techniques for Data Analysis*, (New York: Oxford University Press Inc., 1997).
183. Cottrell, E., Kelley, K.A., Hauri, E.H., & Le Voyer, M., Mantle Carbon Contents for Mid-Ocean Ridge Segments (2019). Interdisciplinary Earth Data Alliance. doi:10.1594/IEDA/111333.
184. O'Neill, H.S., Berry, A.J. & Mallmann, G., The oxidation state of iron in Mid-Ocean Ridge Basaltic (MORB) glasses: Implications for their petrogenesis and oxygen fugacities. *Earth Planet. Sci. Lett.* **504**, 152–162 (2018).

10

How Do Subduction Zones Regulate the Carbon Cycle?

MATTHIEU EMMANUEL GALVEZ AND MANUEL PUBELLIER

10.1 Carbon Distribution on Earth

The core, mantle, and crust contain more than 99% of Earth's carbon stocks.[1] The remaining 1% is in the fluid Earth, split between the biosphere, atmosphere, and oceans. But this distribution must be considered as a snapshot in time, not a fixed property of the Earth system. Continuous exchange of carbon between fluid (ocean, atmosphere, and biosphere) and solid Earth (mainly mantle and crust) has modified the size of the fluid and solid carbon reservoirs[2] over geological time, regulating atmospheric composition and climate.[3,4] The subduction zone, where converging tectonic plates sink below one another or collide, is the main pathway for this exchange. It will be the focus of this chapter.

Geologists believe that a long-term shift in regime of subduction carbon cycling is underway. Following an ecological innovation – the evolution of open-ocean calcifiers (e.g. coccolithophores and foraminifera) in the Mesozoic, marine regression and other changes[5] – it is thought that the accumulation of carbonates on the seafloor (pelagic) has increased over the Cenozoic to reach about 50–60% of the global rate today (Table 10.1).[5–7] Most of the carbonate that has accumulated over the last 100 Myr has not subducted yet (Table 10.1) and should do so sometime in the coming hundreds of millions of years. But when this will happen is unknown because there is no direct link between the precipitation of carbon on the seafloor and the birth of a subduction zone. Irrespective of when it happens, because the fates of shelf and deep-sea carbon materials differ, it has been proposed that intensification of deep-ocean carbonate deposition may eventually affect the prevailing regime of geological carbon cycling.[8,9]

To understand the link between oceanic carbon deposition centers and modes of long-term carbon cycling, we need to consider the fate of sedimentary carbon. Shelf and oceanic island carbon mostly escapes subduction and is accreted to continents during continental subduction and collision. While a fraction of pelagic carbon can also be thrusted within accretionary wedges and accreted, most[10] is bound to be subducted, dissolved, or molten at various depths (Figure 10.1) within the sinking plate, before being released in the fore-arc,[11] arc,[12] or back-arc regions,[13] or mechanically incorporated deeper into the mantle. The contrasted fate distinguishes two principal modes of tectonic carbon cycling: the shallow accretionary carbon cycle and the relatively deeper subduction zone carbon cycle (Figure 10.1). What is not clear yet is how fast those cycles operate and how they interact.

Table 10.1 *Estimates of total annual carbon flux, reservoir sizes, and residence times for various components of the carbon cycle.*

	C flux Mt/yr	Reservoir[a] ($\times 10^9$ MtC)	Residence time,[b] τ (Myr)	Refs.
Sedimentary carbonate				
Total deposition	**210–240**	–	–	[5]
Continental[c]	60–120	64 ± 14	~500–1100	[6]
Pelagic	120–150	9.3 ± 4.3	~60–75	[5,6]
Subduction[d]	40–60	–	–	[26]
Organic matter (OM)				
Total deposition[e]	**160–200**	–	–	[43,145]
Accretion	130–185	11 ± 3	~70	[43,145,146]
Pelagic	15–30	1.6 ± 0.8	~53–160	[29]
Subduction	~10	–	–	[26]
Hydrothermal carbonate total	**>26–45?**	–	–	
Subduction (AOC + UM)[f]	26–45	9.3 ± 4.3	–	[18]
Arc outgassing	**~18–43**	–	–	[21]

[a] Reservoir size from Ref. 83.

[b] Residence time is estimated by assuming a homogeneous reservoir and by dividing reservoir size by flux. Because geological reservoirs are heterogeneous, the mean age of C in those reservoirs should greatly exceed their theoretical residence times.

[c] Obtained from the difference between total deposition flux and total pelagic deposition flux provided by Ref. 5.

[d] Note that the Cenozoic average estimated by Ref. 26, in part based on Ref. 25, exceeds the estimate from Ref. 21. The pelagic deposition flux exceeds the carbonate subduction flux, which suggests net accumulation of carbonate in the pelagic reservoir.

[e] This total contains a fraction of partially graphitic petrogenic materials, estimated by Ref. 33 to be around 40 Mt/yr. The large spread of estimates of organic carbon (OC) burial (as in Ref. 29) is in part due to the variety of depositional environments considered and to the different definitions of what "burial" of biospheric OM means for different authors. For example, the estimates of Ref. 43 (on which the re-evaluations of Ref. 145 are based) accounts for a global 20 wt.% diagenetic loss by decarboxylation before OC is effectively "buried." Loss of hydrocarbons during catagenesis in accretionary wedges may reduce by another 20% or more (>25 Mt/yr) the amount of OC that eventually subducts.

[f] AOC = altered oceanic crust (22–28 Mt/yr); UM = hydrothermally altered upper mantle (4–15 Mt/yr), after Ref. 21.

The residence time of carbon in all geological reservoirs is important because it controls the response of the lithosphere–climate system to *perturbations* in carbon fluxes,[14] and long-term changes in the partitioning of oceanic carbonate may cause such perturbation. The key is that the fluid reservoirs (atmosphere, ocean, biosphere) contain so little carbon

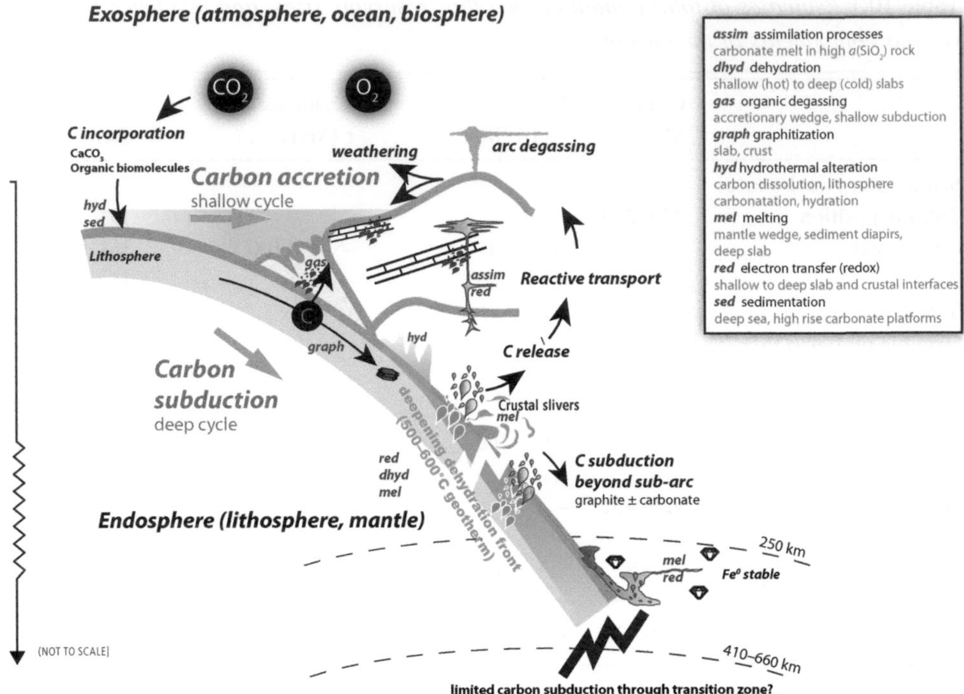

Figure 10.1 The major carbon (organic and inorganic) transformation pathways in subduction zones (layout adapted from Ref. 144). Processes that mediate these transformations are hydrothermal alteration – including reverse weathering – of the oceanic crust (seafloor), slab and mantle wedge (infiltration), sedimentation, diagenetic (CO_2) and catagenetic (e.g. CH_4) degassing of kerogen, graphitization by pressure, temperature, and deformation, dehydration of slabs, electron transfers (redox) between Fe-, C-, and S-bearing mineral and liquids, melting, and reactive transport (assimilation/deposition) of C-bearing fluids and melts from slabs to the exosphere through mantle wedge and continental crust. Dehydration of slabs and partial melting are indicated by blue and red droplets, respectively. A potential limit to deep C subduction in the transition zone is indicated.[84]

that even slight perturbations of subduction or degassing fluxes may change the size of the surface carbon reservoir and impact climate and the diversity of surface habitats over timescales that are instants in the context of geological time. To the first order, the residence time of carbon in the subduction zone depends on the efficacy of shallow subduction processes (<~150 km) to return carbon back to the atmosphere and ocean, and this can occur within timescales of ~10 Myr (Figure 10.1). The residence time of carbon in continents depends on the interaction/assimilation of continental carbon materials by arc magmas[15,16] and on orogenic processes of continental C degassing.[17,18]

This contribution reviews the processes that transform and transport carbon materials in the subduction environment. We first introduce how surface and deep processes control the fluxes of carbon in and out of the solid Earth through subduction zones. Those fluxes define what we call the *pace* of the carbon cycle. Fast reorganization of Earth's separating and amalgamating tectonic blocks is responsible for episodic modifications of carbon

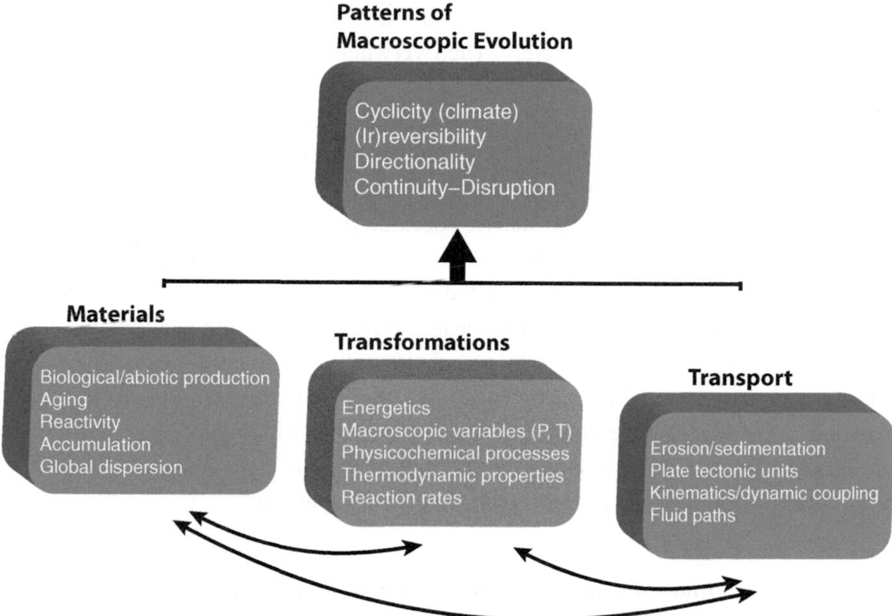

Figure 10.2 The tectonic carbon cycle is a hierarchical structure that must be studied at all scales of organization. The materials that comprise the carbon cycle and their transformations affect the higher levels of organization that transport carbon through the surface and deep Earth. In turn, the tectonic evolution of our planet influences the transformation of carbon-based geobiomaterials and the nature of the reactions that mediate those transformations. As a structure coproduced by biological and geological evolution, the subduction carbon cycle is the ideal research target to assess the link between the heterogeneity of Earth's materials, their reactivity, and the patterns of macroscopic evolution such as cyclicity, irreversibility, continuity, and disruption.

fluxes. We call these fluctuations the *pulse*, or *cadence*, of the carbon cycle. The carbon isotope record and other proxies[19] suggest that the geological carbon cycle has been uninterrupted for more than 3 Gyr. We call this last property its *longevity*. Many recent advances illuminate the origin and controls of each of these central carbon cycle properties. Our work shows that only a cross-scale understanding in space and time will illuminate the links between the microscopic processes of the carbon cycle and patterns of macroscopic evolution, providing a toolkit to decipher the meaning of atmospheric signatures on Earth and other planets (Figure 10.2).

10.2 How Do Surface Processes Control the Subduction Carbon Cycle?

10.2.1 Sources to Sinks and Back

Subduction zones act both as a source and a sink of carbon for the exosphere. Over geological timescales, the source is the volcanic flux of carbon degassed in continental and oceanic arcs, while the sink is the carbon subducted in both reduced (organic C) and oxidized (carbonate) forms. But subduction fluxes do not depend on tectonic processes and mantle convection only; they are also controlled by surface processes driven by solar energy and tied to the water cycle.[20] The overwhelming majority of carbon subducted is allochthonous,[21] which means that most of it is added to marine sediments and the oceanic lithosphere from external sources – mainly the oceans and continents. The fate of carbon in subduction zones is therefore under tight oceanographic, biological, and geomorphic control (Figure 10.1), and it is especially sensitive to the partitioning of carbon between deep (*pelagic*) and shallow (e.g. shelves, oceanic plateaus) oceanic domains.[22]

There are three main carbon sinks, or *pumps*, that mediate the transfer of allochotonous carbon from oceans to sediment and the oceanic lithosphere: the *hydrothermal carbon pump*, which stores CO_2 within the oceanic lithosphere during seafloor weathering; the *soft-tissue pump*, which leads to accumulation of soft organic tissues (of terrestrial or marine origins) in seafloor sediments; and the *carbonate pump*, which controls the topology of the carbonate compensation depth above which carbonate sedimentation on the seafloor may occur and below which carbonates are undersaturated and may lead to dissolution of old $CaCO_3$ already deposited on the seafloor.[23,24]

10.2.2 Heterogeneity of Sedimentary Carbonate Subduction (Carbonate Pump)

Knowledge of carbonate heterogeneities on the seafloor and of carbon subduction rates mostly comes from recent decades of deep-sea drilling efforts.[25] Continuous core sampling of sedimentary covers and upper oceanic crust offers a direct window onto what rocks may eventually subduct and degas. Today, most marine carbonates form and accumulate in the Atlantic and Indian Oceans, where few subduction zones have formed yet, and to a lesser degree on top of the ridges of the southeast Pacific. Estimates of total annual carbonate subduction range between ~66 and 103 MtC/yr (Table 10.1), with about 60% present in sediments (~40–60 MtC/yr[10,25,26]) and 40% in the altered oceanic lithosphere (~26–43 MtC/yr in basalts and serpentinites[21]). Carbonate subduction fluxes have been less than total carbonate deposition rates over the Cenozoic (Table 10.1), which suggests net accumulation of pelagic carbonate.

While efforts are underway to reduce these large uncertainties, the most important limitation for past and future predictions of carbon cycle dynamics is the heterogeneity of the flux (Figure 10.3): 25% of the global carbon subduction flux of the last 10 Myr[26] has occurred in the Sunda trench. The importance of the Indonesian hot spot, as well as those of the Aegean and Makran (Figure 10.3), contrasts with the western Pacific, where subduction and degassing fluxes are generally low. But the southern Pacific New Hebrides–Solomon–Vanuatu segment does not conform to this rule. Unlike most settings of the western Ring of Fire, the slab is young and plunges to the east. The system is long and so is rich in carbonate,[26] which it is a sizable contributor to the global subducted carbon inventory (Figure 10.2). Allard et al.[12] showed that the Ambrym volcano in the Vanuatu islands pours out slab carbon at a record rate (~5–9% of the global volcanic carbon flux), providing a unique window onto the unusual CO_2 productivity of this young arc system.[27]

Heterogeneous Carbon Subduction in the Contemporary Ocean

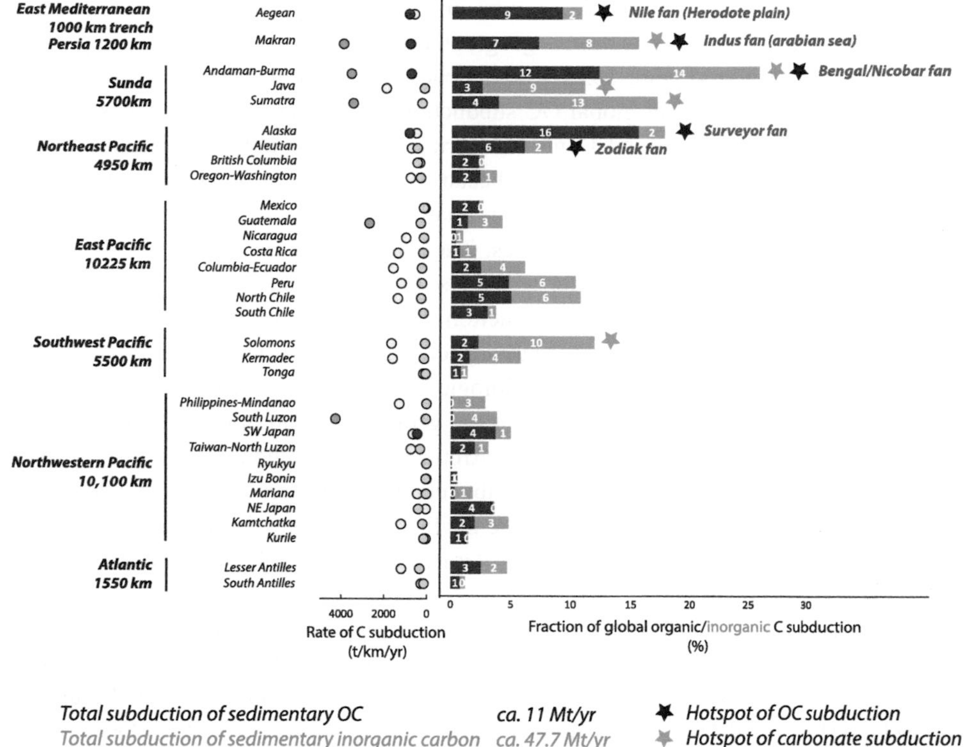

Figure 10.3 Contribution of selected subduction zones to global sedimentary C subduction flux. Fluxes are from Ref. 26, where length of subduction zones and carbon (organic and inorganic) concentrations in trench sediments are compiled (see also Ref. 25). Note that the computation of flux includes corrections for sediment porosity used but not reported in Ref. 26. Rate of carbon subduction are provided in t/km/yr for each subduction zone. Organic carbon is in gray and black, and black circles denote OC hot spots. Inorganic carbon is in pale and dark blue, with dark blue denoting hot spots. The fractional contribution of each subduction to the global sedimentary C flux is indicated as a percentage of total inorganic (blue) or organic (black) carbon subduction, respectively, in the histogram of the right panel (cf. Table 10.1). Note that the flux associated with basalt, gabbro, and ultramafic carbonate is not included. This corresponds to an additional subduction rate of ~850 ± 200 tC/km/yr.

10.2.3 Hot Spot of Organic Carbon Subduction in the Sub-Arctic Pacific Rim (Soft Tissue Pump)

A heterogeneous distribution and diversity characterize the soft tissue pump,[28] as well as the subduction of organic carbon (OC). The burial of subductable OC in pelagic environments represents a flux of up to 10–30 MtC/yr over the last 150 kyr (Figure 10.2),[29] and perhaps during most of the Phanerozoic.[2] Though modest compared to net primary productivity (~100 Gt/yr),[30] this weak but continuous isolation of organic reductants in sediments and rocks is essential for the persistence of atmospheric O_2 levels that are so high and so far from thermodynamic equilibrium[31] over geological timescales.

Just like carbonate, OC subduction is geographically heterogeneous,[26] but its deposition hot spots are elsewhere (Figure 10.3). The trenches of Alaska, the Aleutians, and British Columbia contribute ~23% of global OC subduction, while many other trenches, such as those of the western Pacific and Central America, are comparatively marginal. This is either because the trenches are small or because sediments do not contain much OC (Figure 10.2). This heterogeneity seems primarily controlled by continental erosion, weathering, and marine sedimentation rates. The northern Pacific hot spot is dominated by subduction of thick sequences of terrigenous sediments.[32] Surprisingly, all OC subduction hot spots are associated with the subduction of deep-sea turbiditic sequences, such as Surveyor fan (Alaska) and Indus fan (Makran) (cf. Figure 10.3). The OC fraction derived from terrestrial sources (petrogenic and biologic[33]) can locally reach more than 70% in active margin sediments.[34] This means that the geomorphic and climatic conditions that influence the export of organic-rich tropical, Arctic, and mountainous sediments[35,36] to the marine environment[37] are the same as those that control the pattern of OC subduction, too. Hence further research is needed to elucidate the link between continental configuration in the Wilson cycle, erosion/sedimentation patterns, and OC burial, with special attention paid to active margins.[38]

Unlike carbonates, which conserve their structure across a broad range of pressure and temperature conditions of Earth's interior, aging biomolecules display stunning plasticity in structure and composition[39–41] acquired during transport and transformation through the lithosphere (Figure 10.2). It is often assumed that graphite is the main form of OC in rocks. However, organic macromolecules (e.g. kerogens) may not become proper crystalline graphite before the late stages of the subduction process (i.e. >600°C; Figure 10.1). The transformation of organic materials to their stable forms (e.g. graphite) involves a succession of metastable macromolecular intermediates called kerogens.[42] Increasing temperature and the presence of water accelerates this process and drives the sequential release of CO_2 (i.e. reduction of the kerogen residue) and then hydrocarbons (i.e. oxidation of the kerogens) from the kerogen residue in shallow, unconsolidated sediments[43] and in accretionary wedges (Figure 10.1).[44] This so-called carbonization[42] may decrease the amount of OC effectively subducted by 20–40% or more depending on whether or not hydrocarbons are trapped within the descending slabs. This is still an open question.

$$[CH_2O] \xrightarrow[\textit{reduction (abiotic/biological}^{44})]{CO_2\uparrow,\ H_2O\uparrow \text{ to atmosphere}} \text{kerogen} \xrightarrow[\textit{oxidation}]{CH_4\uparrow,\ H_2O\uparrow \text{ to atmosphere}} \text{graphite/diamond} \tag{10.1}$$

Subsequent collapse of the kerogen structure involves 3D ordering of the carbon-rich aromatic backbone and is favored by pressure and shear.[45] This structural ordering is called graphitization (Figure 10.1). First-principles approaches are now shedding light on the microscopic pathways involved in this process.[46,47] Because the transformations modeled by (10.1) are so slow in nature, it is even conceivable that the coldest slabs (e.g. the Mariana and the Aegean OC subduction hot spots) could reach the diamond stability field before completion of the carbonization and graphitization process. The graphitization process itself is irreversible, and field observations indicate that slabs may not lose much OC beyond 350–400°C.[48,49]

The low reactivity of compact graphitic materials and their intricate weaving within poorly soluble mineral matrices explain why a graphitic fraction dislodged from rocks during erosion and weathering (18–104 MtC/yr[33]) is continuously exported to the ocean (Figure 10.1). How much of this fossil petrogenic material subducts is not yet known, but it may represent 50% of OC in the trenches of South America.[34]

10.2.4 An Ancient Hydrothermal Carbon Sink

A last major contributor to allochthonous carbon incorporation to the oceanic lithosphere is hydrothermal carbonatization (Figure 10.1 and Table 10.1).[50] More details are provided in a companion chapter in this book (Chapter 15). In short, the circulation of cold and carbon-rich seawater through the oceanic crust and mantle dissolves cations from the oceanic lithosphere. Elevated concentrations of cations and rising temperature have been responsible for cycles of carbonate deposition in oceanic hydrothermal systems[51] today and in the past. Occurrences of carbonatized basalts have been reported as early as in the early Archean Craton[52] of eastern Pilbara (Australia) and in Archean greenstone belts.[53] They suggest that: (1) carbonatization of the lithosphere may have been the means by which the atmosphere lost its primordial CO_2[54]; and (2) carbonatization of the lithosphere may have been the only flux of global importance capable of balancing outputs from the deep Earth during the Archean, Proterozoic, and large swaths of the early Phanerozoic (Figure 10.1). Therefore, surface biogeochemical processes involving the water cycle may have controlled the magnitude and longevity of carbon inputs to subduction zones since their initiation, more than 3 Ga.[55] The above shows that the pelagic reservoir is not at steady state over the Cenozoic. What about the subduction zone itself?

10.3 Is the Subduction Zone Carbon Neutral?

To the first order, the parameter that is most likely to affect models of atmosphere–climate–lithosphere evolution over geological time (<100 Myr) is the subduction efficiency, σ (Figure 10.1). This macroscopic parameter is the fraction of subducted carbon that penetrates beyond sub-arc depths in subduction zones (cf. Figure 10.1).[21,56] For a given subduction flux, the lower the parameter σ, the faster the response of the geological carbon cycle to perturbations in flux at the inlet of the subduction system.

Johnston et al.[27] compared inventories of global input to and output from selected subduction zones to determine their efficiency. After correcting for possible mantle and continental contributions,[16,57] Johnston et al. found that 18–70%[21,26] of the input (Table 10.1) might be accounted for by arc carbon emissions derived from the slab, setting σ to ~0.3–0.8. Corrections are needed because the accretionary and subduction cycles intersect when arc magma rises through the continental crust and that magma assimilates a fraction of the carbon that had been accreted to continents during previous collisional events (Figure 10.1).[18] This phenomenon is thought to control the important emissions in Italy (e.g. Etna, Vesuvius), parts of the Andes, and Indonesia[58] today. The estimates of Johnston et al. can be compared with estimates from Table 10.1, we find σ of ~0.4–0.9. This is consistent, and it shows that a sizable fraction of subducted carbon does not take the short path out to the exosphere and continuously accumulates anywhere between the mantle wedge and the transition zone. The mismatch between input and volcanic output of slab carbon suggests that subduction zones, in their present configuration, act as a net sink of carbon over timescales of about 10–50 Myr.

But the meaning of a mismatch itself is unclear. Fundamentally, volcanic arcs are indicators of slab dehydration and partial melting in the mantle,[20,59] and the two processes are only indirectly linked to C loss from slabs. For example, dehydration at sub-arc depths for cool and cold slabs (e.g. Tonga, Mariana) drives increased fluid flux into the mantle wedge, producing higher degrees of partial melting and a greater proportion of carbon transport to the volcanic front (western Pacific).[12] By contrast, hot and warm slabs promote carbon dissolution at shallower depths and may release carbon away from the volcanic fronts (e.g. Cascadia and Central America). Therefore, cooling of Earth is thought to have changed the *pace* of subduction carbon cycling[14] slowly, over a billion-year timescale: a hotter Precambrian Earth would have returned more carbon faster from the carbonatized lithosphere, while cooler slabs that are denser and rich in garnet[60] would promote penetration of carbonates to sub-arc depth and beyond.

This trend is only qualitative, and complementary approaches are required to quantify the fluxes and to assess their controlling factors. We take a broad-level look at three underlying mechanisms that control the partitioning of carbon between rocks and fluids in the most inaccessible parts of the subduction zones – the slab, mantle wedge, and arcs themselves: (1) the problem of carbon solubility in fluids and its underlying mineralogical controls; (2) the problem of fluid production within slabs; and (3) the problem of fluid reactivity along their paths from the subducted lithosphere to the surface.

10.4 How Rocks Influence the Solubility of Carbon

The composition of subducting slabs differs from that entering the subduction system. Open system processes – in which rocks are infiltrated by fluids and melts (also referred to indiscriminately as *liquids*) – control how much and how fast carbon is removed from slabs. The greater the amount of carbon released by dissolution and melting at fore- and sub-arc depths (Figure 10.1), the faster the subduction cycle may respond to any kind of perturbation in flux. Carbonates are the most abundant form of carbon in slabs (Table 10.1), and most box models of the geological carbon cycle[61] represent the fate of carbonate by progress of the so-called reverse *Urey reaction*[62]:

$$CaCO_3 + SiO_2 \rightarrow CaSiO_3 + CO_2 \tag{10.2}$$

However, this reaction is also a model, an aggregation of many processes.[63] It is therefore not a mechanistic explanation of the dissolution process. The last decade has offered valuable insights into the microscopic complexity of Urey-type "decarbonation" reactions[4] and their macroscopic expressions as carbon fluxes.

10.4.1 Dissolution by Rising Pressure and Temperature: Which Silicate Is in Charge?

An essential but implicit meaning of (10.2) is that the release of C from slabs is linked to that of water – the solvent and catalyst for the reaction. Carbon release tracks the main sequence of fore-arc and sub-arc mineral dehydration reactions.[64] The most important of these reactions are the destabilization of lawsonite/epidote during eclogitization of the altered oceanic crust[65] at ~450–700°C and the dehydration of antigorite/chlorite[66] in the serpentinized oceanic mantle. Both take place at around the same temperature, but the second pathway occurs deeper in the subduction zone because deserpentinization and dechloritization reactions occur in the altered ultramafic rocks of the lower and usually colder section of the descending slab (Figures 10.1 and 10.4).

Another insight from (10.2) is that the solubility of carbonates in hydrous fluids is primarily controlled by the relative stability of the various silicates (e.g. wollastonite, garnet[40]) in which they transform; in other words, the rapid rise of C solubility at the onset of garnet formation in most sediment and mafic lithologies[67] reflects the increasing relative stability of calc-silicates with rising temperature:

$$\underset{(Ca,Mg,Mn,Fe)CO_3}{M-\text{carbonate}} + \underset{quartz/lawsonite}{(\text{Al})\text{silicate}} + \underset{H_2O}{\text{solvent}} \rightarrow \underset{CaSiO_3/garnet}{M-(\text{Al})\text{silicate}^{ss}} + \underset{dissolved\,CO_2}{\text{mixed carbonic fluid}} \tag{10.3}$$

This process is an incongruent dissolution,[67] where M is a divalent metal Ca, Mg, Fe, or Mn.

In practice, the congruent or incongruent nature of the dissolution depends on the kinetics of dissolution, the thermodynamic properties of calc–silicates, the kinetics of calc–silicate precipitation, and the rate of fluid transport.[40,67] Recent thermodynamic analysis, however, suggest that congruent pathways may be restricted only to very high pressures and low temperatures[67]; regimes where fluid fluxes tend to be low. In addition, the ubiquitous presence of calc–silicates such as lawsonite down to 350°C in most silicic and aluminous lithologies[68] supports the idea that their formation is not kinetically limited in the subduction zone. Hence, field and thermodynamic evidence suggests that incongruent pathways have been the most important in driving the dissolution of carbon along most subduction geotherms, today and in the geological past).

This general rule may not hold in the case of dolostone and limestone lithologies. These rocks usually lack the alumina and, in the most extreme case, silica that promote vigorous

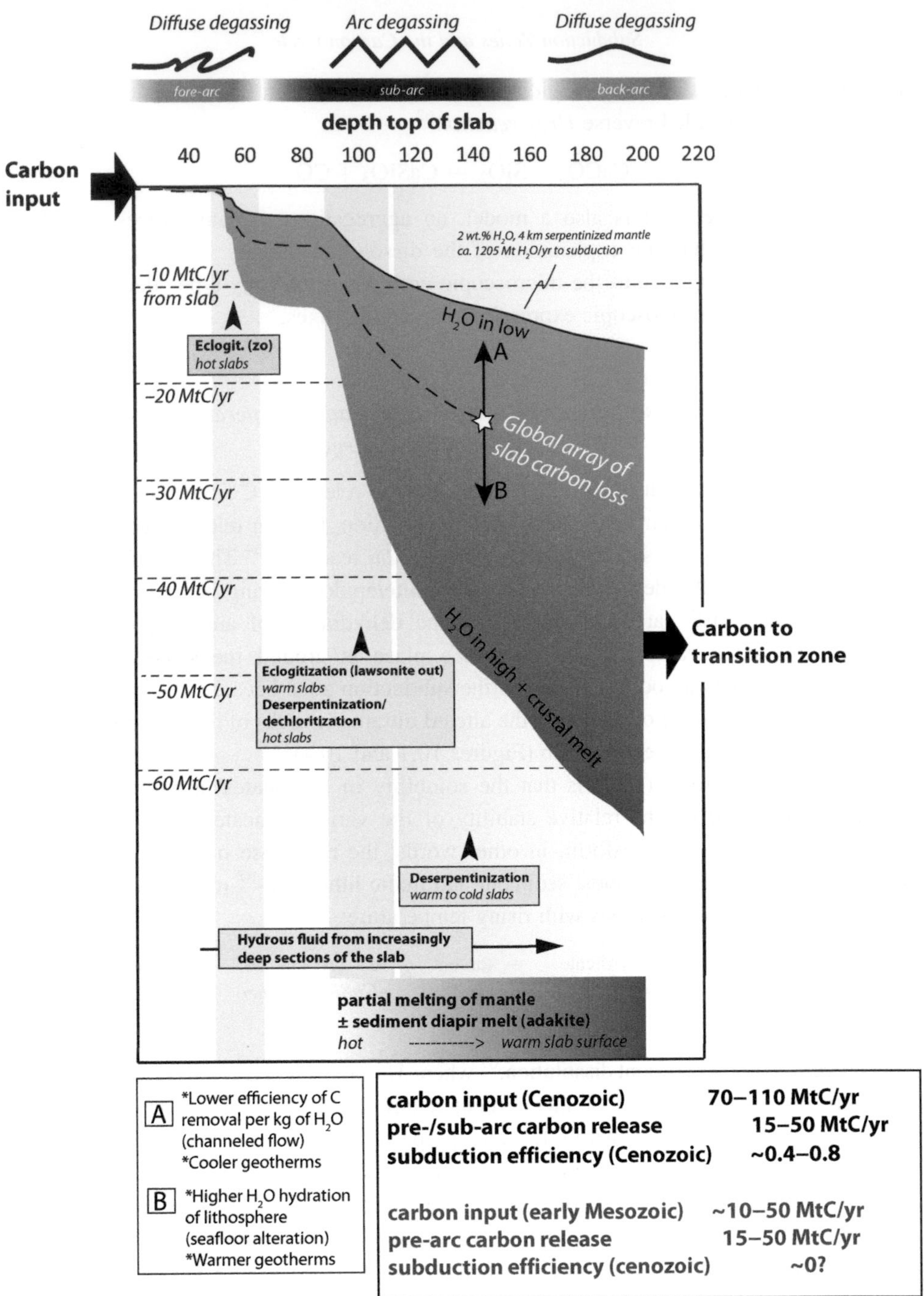

Figure 10.4 Devolatilization pattern (H_2O and C) in a subducting slab and its link to subduction efficiency. The latter is the fraction of subducted carbon released at fore-arc and sub-arcs depths (cf. Ref. 3). The results are only qualitative and are based on ongoing studies investigating the coupling between C, H, Na, K, Si, and Al cycles in open subduction-zone systems.[67] The shallow output flux depends on the hydration structure of the oceanic lithosphere, and conceivably, on the degree of partial melting (red tones). Overall, this flux varies between ~15 and 50 MtC/yr (i.e. in the order of magnitude of arc emissions[21,27]), between 0.1 and 0.6 of the incoming flux, and more likely ~0.4–0.5 in the Cenozoic (correspond to σ of ~0.5–0.6; cf. Figures 10.6 and 10.7). This ratio may have approached 0 in the Paleozoic and Mesozoic (i.e. before the rise of pelagic calcifiers). This should be considered in long-term models of Phanerozoic carbon cycle evolution.

dissolution (10.2 and 10.3). In addition, marble lithologies are also rather dry and also notably impermeable to fluid flow, and therefore not prone to dissolution. Yet dissolution reactions may also be sustained by chemical disequilibrium between contrasted lithologies even when pressure and temperature are invariant. For example, diffusion of volatiles (H_2O[69] or H_2[40]) and nonvolatile elements (Al and Si[70]) into the marbles may modify *a* (CO_2), or the stability of calc–silicates (10.2 and 10.3) and drive reactions (10.2) and (10.3). In fact, field evidence shows that when dissolution of marbles does occur, it tends to be restricted to the vicinity of fractures[70] and tectonic discontinuities,[69,71] precisely where fluids rich in Al and Si can circulate and boost carbon dissolution processes. Therefore, it is likely that incongruent pathways play the dominant role in carbon mobilization from slabs at subsolidus conditions. The growing importance of deep-sea carbonates (carbonate ooze[72]) subduction in the next 100 million years (Table 10.1) makes them obvious targets for future research focusing on their peculiar mechanical properties and open-system behavior at elevated pressures and temperatures.

10.4.2 Carbonate Melts from Hot Slabs and Diapirs?

Partial melting of the slab may start at >900°C in the presence of water. It can affect the top section of hot slabs (e.g. Aleutians/Mexico) when it is infiltrated by water, and it may also occur in unusually hot and buoyant fragments of slab sediments in the mantle wedge.[73] The process forms ionic melts that are rich in carbonate components (i.e. hydrous carbonatites),[74] but the thermodynamic,[75] structural,[76] and rheological properties of those compositionally heterogeneous melts remain largely unknown. Assessing their properties is important because carbonate melts are highly mobile, highly reactive,[77] and important carriers of carbon beyond sub-arc depth anywhere between ~100 km and 400–600 km within the slab and in the mantle wedge. A few implications follow:

(1) In carbonate bearing lithologies, formation of metamorphic garnet (e.g. grossular), the ubiquitous silicate mineral of high-pressure metamorphism, may be the most practical and important indicator of both dehydration[78] and C redistribution (e.g. carbonate dissolution) between rocks and fluids via *Urey-type* processes (Figure 10.4).
(2) Carbon cycling in the slab is a problem of transport[79] where dehydration, decarbonation, desilicification, and dealuminification of rocks are parts of a single overarching problem.[67] Work is underway to link the microscopic properties of geo-fluids (aqueous, carbonate, silicate) and minerals with macroscopic patterns of element transport across shallow lithospheric reservoirs.[48,67,80]
(3) There is a growing consensus, informed by thermodynamic models, that pulses of carbon release from slabs by subsolidus dissolution do occur around 500–800°C, representing an overall flux of between ~10 and 40 MtC/yr between ~80 and 140 km depth below the volcanic arc.[48,81,82] This flux is uncertain, but overall it is consistent with volcanic emissions.[21] The pulsatile character (Figure 10.4) of the flux, however, is primarily due to the timing and volume of H_2O infiltration rather than to

the thermodynamics, or kinetics, of the *Urey-type* reactions. This is important because it means that the subduction efficiency σ, as opposed to total carbon release, is controlled by *where* H_2O is concentrated within the hydrated slab, rather than by the absolute amount of H_2O contained in it. Only the water contained in the uppermost 10 km of slab mediates the fast cycling of carbon in the fore-arc and sub-arc regions.

(4) The parameter σ may not be a mere parameter at all over geological timescales. It is more likely to be a variable that is dependent on the nature of carbon inputs. Indeed, we have shown that limestones and carbonatized lithosphere rocks behave differently, both mechanically and chemically. For example, while σ of ~0.4–0.8 in the Cenozoic (Table 10.1) is characterized by important subduction of limestones (Table 10.1), this value may have vanished during most of the Precambrian, Paleozoic, and early Mesozoic, stranding carbon in shallow surface reservoirs. Interestingly, this idea is qualitatively consistent with recent isotopic and geochemical proxies,[83] which suggest a net growth of the continental and pelagic carbonate reservoir (Table 10.1) over the last 2 Gyr.[2,83]

10.4.3 Where Is the Barrier to Deep Carbon Subduction?

Global budgets are sensitive to thermal models and the permeability/rheology of rocks within the slab. Therefore, slabs that are cold and/or covered by thick and impermeable carbonate layers should retain a significant fraction of their carbon cargo to beyond sub-arc depths. What is the fate of carbon beyond this limit (Figure 10.1)?

If carbon makes it beyond sub-arc depths (~150 km), it may not push much further beyond the transition zone (~410–660 km). The reason for this is the existence of a deep depression in the temperature (1000–1100°C) at which anhydrous carbonated oceanic crust melts at uppermost transition-zone conditions.[84] Thomson et al. showed that the curvature of the solidus of carbonated eclogite is also controlled by the partitioning of nonvolatile elements, chief among them calcium and sodium, between mineral phases[84] in high-pressure eclogites. Most geotherms should intersect this melting barrier.

To the best of our knowledge, there is as yet no experiment that quantifies the effect of water on the topology of the anhydrous solidus of carbonated eclogites in the vicinity of the deep curvature (i.e. at the conditions of the transition zone around ~10–20 GPa). Qualitatively at least, it is now undisputed that increasing activity of water $a(H_2O)$ dampens the solidus of carbonate rocks,[74,85] bringing their temperature of melting closer to 900–1000°C, promoting the formation of carbonate melts. This principle applies to carbonatized basaltic lithologies[86] in particular and to all carbonate-bearing systems in general, typically those rich in iron and manganese[87] that are relevant for the fate of subducted banded iron formation in the Archean Earth. This should limit even more the possibility of carbon subduction through the transition zone (Figure 10.1).

10.4.4 Are Thermal Anomalies the Norm?

There is a growing consensus that the relevant pressure–temperature trajectories of slab materials may be hotter than most canonical models assume.[81] First of all, there is an apparent incompatibility between slab thermal models predicted by theory[88] – usually cold – versus those inferred from phase assemblage relations and mineral chemistry[89] – generally warmer. Second, the mantle wedge may be populated by detached fragments of slab sediments (e.g. diapirs[73]) that are thought to follow hotter pressure–temperature trajectories than the rest of the slab (Figure 10.1). It is important because it implies that the production of hydrous carbonatite melts may be more common than expected, particularly for cold slabs (e.g. Aegean) covered by kilometer-thick buoyant sedimentary piles. Overall, research into the thermal regime of slab materials may eventually redefine what thermal normality is in the subduction environment.

Taken together, the possibility of paths that are hotter than normal for most subducted sediments and the existence of a chemical barrier to deep carbonate subduction (Figure 10.1) means that most subducted carbon must have been stranded above the transition zone through most of Earth's history.[21] However, there remain important uncertainties on the subduction efficiency, in particular, its variability in space and time. This uncertainty hinders quantitative understanding of the response dynamics of the carbon cycle to perturbations.

10.5 Transport and Reactivity of Carbon-Bearing Liquids

Aqueous fluids and carbonatitic melts produced in the slab are predicted to be mobile due to their low viscosity. But crustal liquids are not passive carriers of carbon back to the surface. They re-equilibrate continuously with changing environmental conditions across the crust and mantle. Both carbonate and elemental C (graphite/diamond) may form along the way and delay the return of subducted C to the exogenic cycles. It may take anywhere from thousands to millions of years for the mobilized carbon to return to the exosphere.

Transport of fluids and melts occurs mostly along lithological interfaces and other discontinuities, such as faults and fractures (see Figure 10.4 and Supplementary Materials at the end of chapter) in the heterogeneous interface between slab and mantle,[90] and within the rocky mantle itself.[64] This is observed in the field[71] and via geophysical observations of fluid migration through the mantle wedge.[91] The fate of C-bearing liquids during their ascent through the slab, mantle wedge, and overriding crust is controlled by four critical variables: pressure/temperature, redox state, activity of silica, and activity of water.

10.5.1 Pressure and Temperature

For the same reason that rising pressure and temperature enhances the solubility of carbonates in subducting slabs, downward temperature paths, too, may lead to carbonate precipitation out of fluids ascending along and across the slab. This chromatographic process is predicted thermodynamically[69] and is observed in the field.[92] This illustrates that there is a delay between the time carbon is dissolved from the slab and the time it eventually reaches the surface.

10.5.2 Low- and High-Temperature Redox Processes

The redox state, too, is important for the fate of both carbonates and organic materials, and it is usually measured by the thermodynamic activity of oxygen. Carbon can bond to both oxygen and hydrogen, and the balance between oxygenated species (e.g. CO_2) and hydrogenated carbon species (e.g. CH_4) in a fluid–rock system depends on the nature and abundance of other "redox" elements in the rock, such as Fe (Fe^{2+}, Fe^{3+}), Mn, and multiple S species susceptible to exchanging electrons with carbon.[93] But there is a limit to the amount of C a fluid can contain at typical subduction-zone conditions. This limit is fixed by thermodynamics, and varies with pressure, temperature, and rock composition.[67,94] If graphitic materials are stable – as is usually the case in subducted sediment (Figure 10.4) – then the so-called graphite-saturated COH system imposes a minimum in carbon solubility midway between fluids dominated by CO_2 and fluids dominated by CH_4.[94] The existence of such a minimum is important because it implies that quasi-static changes in pressure, temperature, $a(H_2O)$, or fO_2 of a graphite- or diamond-saturated fluid could involve both carbon dissolution and/or precipitation of the fluid. Therefore, C may be locked in the form of graphite during fluid ascent through the slab[40,95] or through the continental crust,[96,97] creating complex locking and unlocking pathways for deep carbon (Figure 10.5).

Fluctuations in redox conditions may occur anywhere from the shallow to the deep subduction zone. For example, it has been shown that serpentinite assemblages maintain their reducing power in the shallow subduction zone at conditions lower than 10 kbar and

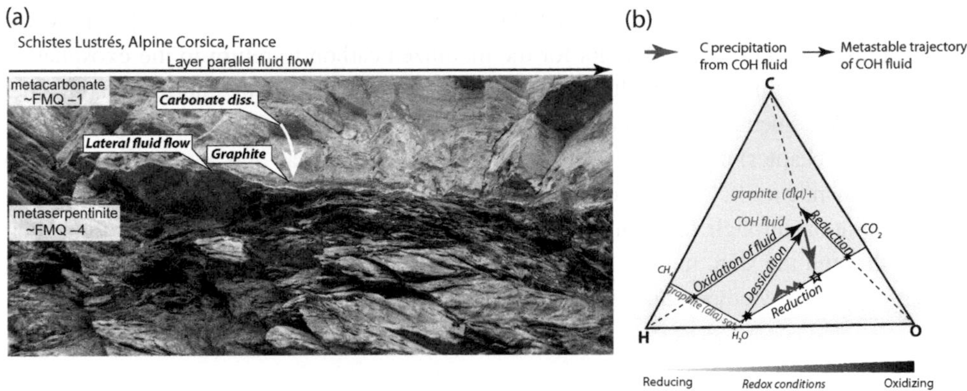

Figure 10.5 Redox pathways in the subduction zone. Graphite precipitation from carbonate at a lithological interface. (a) Field image of the outcrop in Alpine Corsica.[40] (b) Representative COH diagram showing the curvature of the C-saturation surface at the elevated pressures and temperatures typical of subduction zones, illustrating various pathways leading to elemental carbon precipitation.

500°C.[98] Those conditions may cause the spontaneous formation of refractory graphite from carbonate according to:

$$\underset{(carb)}{CO_2} + \underset{(mag/serp)}{4\,FeO} = \underset{(garnet)}{2\,Fe_2O_3} + \underset{(graphite)}{C^0}. \tag{10.4}$$

There is now evidence for such a mechanism occurring at temperatures of no more than 450°C (Figure 10.4).[40,71] A redox mechanism formally analogous to (10.4) is thought to occur below 250 km, too, where carbonatite melts expelled from the slab re-equilibrate with mantle lithologies where metallic iron is stable (Figure 10.1).[99] Mutual re-equilibration between carbonatite melts and mantle rocks involves the spontaneous production of diamond according to:

$$\underset{(melt)}{MgCO_3} + \underset{(metallic\ iron)}{2\,Fe^0} = \underset{(iron\ oxide)}{3\left(Mg_{0.33}, Fe^{2+}{}_{0.67}\right)O} + \underset{(diamond)}{C^0}. \tag{10.5}$$

10.5.3 SiO_2 Activity

Just like $a(SiO_2)$ influences the dissolution of C in hydrous fluids of slabs via *Urey-type* reactions (cf. (10.2) and (10.3)), fluctuations of $a(SiO_2)$ impact the behavior of C in all types of subduction-zone liquid in which it may dissolve. This is particularly true of carbonate and other alkaline melts. Experimental works[77,100] have shown that dolomitic melts formed in the descending slab degas at the contact of peridotitic mineral assemblages characterized by comparatively higher $a(SiO_2)$ (orthopyroxene and clinopyroxene).

$$\underset{\text{base (carbonatitc melt)}}{2CO_3{}^{2-\,\text{melt}}} + \underset{\text{acid (mantle rock)}}{SiO_2{}^{\text{rock}}} = \underset{\text{base (carbonatitic melt)}}{SiO_4{}^{4-,\text{melt}}} + \underset{\text{acid (mantle gas)}}{2CO_2{}^{\text{gas}}}. \tag{10.6}$$

This mechanism may supply large fluxes of diffuse CO_2 outgassing with decompression, as the melt dissolves more silica or any other acidic component. Therefore, cataclysmic eruptions (kimberlites[101]) may ensue if those melts stall for long periods of time[1] in environments such as the base of the continental crust.

10.5.4 Water

Just like surface processes, many processes of the slab–mantle wedge interface involve repeated cycles of fluid desiccation, whereby fast and near-quantitative redistribution of H_2O from fluids toward solid (e.g. mantle wedge serpentinization) or melt phases occurs. Desiccation is a possibly dominant mode of elemental carbon sequestration in the crust and upper mantle, although this mechanism has received little attention so far. Because the solubility of water in silicate melts that are poor in alkalis is greater than that of carbonic species,[102,103] rehydration (below 600°C[104]) and flux melting of the mantle wedge by infiltrating COH fluids may form restitic graphite/diamond (Figure 10.5b) via:

$$\text{reactant}^{\text{rock}} \quad + \quad \text{C(H}_2\text{O)}^{\text{fld}} \quad = \quad \text{residue C}^{\text{gph/dia}} + \text{solid/melt} \cdot \text{H}_2\text{O}. \tag{10.7}$$

This mechanism is supported by theory (Figure 10.5); evidence for it in nature may be found in the Kokshetav massif (north Kazakhstan)[105] or in Lianoning (northeast China),[106] where the association of graphite with jadeites is intriguing.

Overall, a combination of physicochemical processes drive not only carbon loss from slabs, but also its reactivity and fate in the heterogeneous chromatographic columns that separate slab liquids from the surface. Therefore, a carbon flux at the surface of Earth (or of any planet) does not mean an active release is operating.

10.6 Carbon Dynamics at the Subduction/Collision Transition

Surface, subduction, and mantle processes control the fluxes of C. But numerous past instances of climatic and carbon cycle perturbations[107] show that changes in geological fluxes occur continuously on Earth. Subduction zones evolve, may change polarity,[108] and even vanish in complex collision zones. The subduction/collision transition[109] illustrates how quickly the reorganization of Earth's *tectonic building blocks* may cause imbalances in the geological C cycle (Figure 10.5). We refer to these episodic changes as the *pulse* of the carbon cycle.

The collision zone[110] is a dominant pathway of carbon processing,[111] particularly in the Indo-Pacific region today (Ontong Java in the Vanuatu–Solomon arc) and in the former Tethyan orogens.[112,113] Docking of Greater India to Eurasia caused a situation of geodynamic instability marked by the Himalayan orogeny (Figure 10.6) that lasted for at least 50 My[113] and is still going on today at a reduced pace. Are the Himalayas carbon neutral over a geological timescale?[114] This is a question of timescales.

On geological timescales, early workers proposed that the Tethyan orogeny as a whole would operate as a CO_2 sink. The sink is due to enhanced silicate weathering[115] of freshly exhumed rocks and is usually modeled by the forward *Urey-Ebelmen*[4] reaction (cf. (10.2)):

$$CaSiO_3 + CO_2 \rightarrow CaCO_3 + SiO_2. \tag{10.8}$$

Recent studies provide insights into the complexity of this problem. Enhanced erosion, biotic OC burial,[116] and accretion of pelagic carbonates to the growing orogen[112] operate as long-term CO_2 sinks, while the weathering of petrogenic OC[117] and enhanced sulfide weathering[118] counteract CO_2 sources, at least transiently,[118] in much the same way as the interaction of orogenic magma with accreted carbonates[16] or OC[119] in continents do (Figures 10.1 and 10.6). It is not at all clear how the acceleration of phosphorus and other nutrients exported to the marine basins surrounding the Himalayan orogen boosted net primary productivity, at least locally (i.e. another CO_2 sink). The inventory and relevant timescales of these processes – surface or deep – are crucial to establishing a thorough carbon budget for an orogeny over geological timescales.

As part of this inventory effort, we propose that focusing on the orogen itself may still miss an entire – deeper – dimension of the tectonic disequilibrium caused by collision.

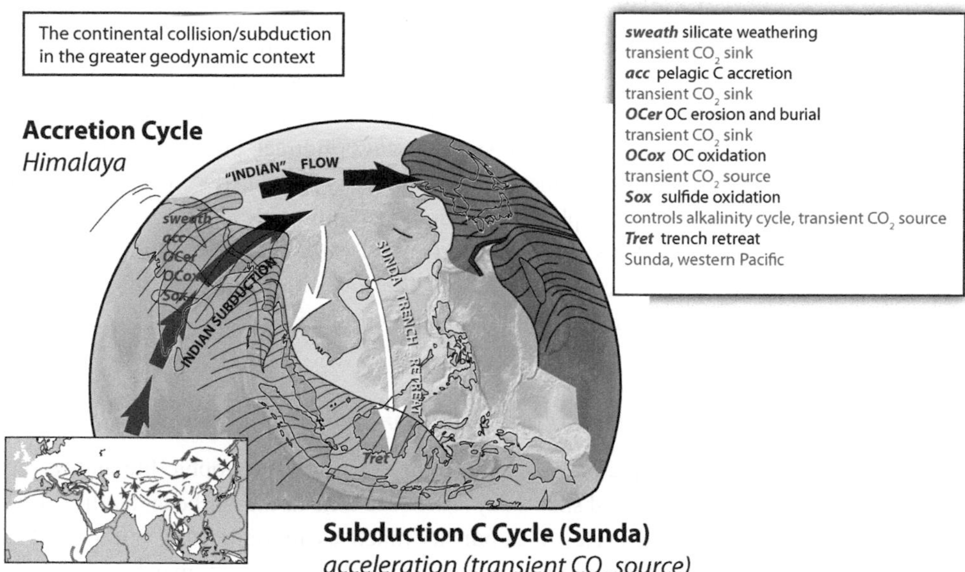

Figure 10.6 Examples of tectonic complexity at subduction zones: the collision factory. Long-range kinematic and dynamic reorganization in the western Pacific and Sundaland triggered by greater India subduction and collision; modified from Ref. 147. The carbon budget of the collision zone[114] may only be resolved at the mesoscale (i.e. one that includes the orogen itself and the evolving subduction zones in its broad periphery).

Figure 10.6 shows how the collision of India with Eurasia since the Eocene involved large-scale coupling between continent–continent collision and slab subduction in the western Pacific and Southeast Asia (Sunda trench). The southeast-ward motion of lithospheric fragments of Eurasia was accommodated by lithospheric and asthenospheric reorganization east of the collision zone.[120] As a consequence, continental subduction and collision to the north have been accompanied by extension tectonics distributed over a 3000-km region across the Sunda plate and western Pacific starting 45 Myr ago.[121] We propose that slab retreat and rollback over this extended area may have boosted the pre-collisional CO_2 source associated with the Sunda and western Pacific subductions, partly compensating for the sinks linked to the growing orogen itself. The mechanisms involved in a subduction acceleration are illustrated in the model of Section 10.7.

The result of tectonic and biological coevolution over the Cenozoic is one of long-term cooling, with a thermal maximum at the onset of the collision. Today, atmospheric CO_2 is at a record low level, and the pace of the pre-anthropic subduction carbon cycle has reached a *kinetic* minimum.[122] Antarctica and Greenland ice sheets grew over the last ~30–40 My. Degassing of the pelagic carbonates that have subducted beyond sub-arc depths over the Cenozoic and enhancements in pelagic sediment subduction in the future may boost the pace of the *non-anthropogenic* carbon cycle sometime in the next 100 Myr.

10.7 A Flavor of Life: A 3 Billion-Year-Old Record

The biosphere and geosphere are not separate entities. They are bound by the electron (redox) and proton (acid–base) transfer reactions of biological metabolism. Therefore, while Earth is the scaffold of biological evolution, the evolving style and longevity of the subduction-zone carbon cycle are in various ways indirect products of biological evolution (Figure 10.7). Four main milestones of this coevolution were: (1) the origin of cell death; (2) the evolution of the photosynthetic water-splitting complex; (3) the rise of multicellular algae and land plants; and (4) the advent of carbonate biomineralization. Life altered the geological cycle of electrons about 2.5 billion years ago and then the geological cycle of alkalinity, in the beginning indirectly, and then more directly.

10.7.1 Biological Evolution Influences Carbon Subduction: First Milestone

Genomic studies[123] and isotopic proxies[19] place the origin of autotrophic life and respiration – the main carbon source and sink today – at some time in the Archean, but life alone did not affect the *geological* cycle of carbon as much as mortality did. Viral lysis of prokaryotes[124,125] and, possibly, genetically encoded death pathways[124] may have played a disproportionate role in driving prokaryote mortality and, therefore, a primordial soft-tissue pump that helped the early biological necromass to spread across all marine habitats, including trenches and subduction zones. The ancestral origin of viruses[126] is at least qualitatively consistent with the idea that a sustained flux of dead organic materials from the sunlit upper ocean to the marine sediments may have been established at some time in the Archean. While it is conceivable that carbonatization of the oceanic crust, and its subduction, had been in place ever since oceans and the oceanic crust coexisted, it is the marine accumulation, isolation, accretion, and subduction[127,128] of organic reductants that sparked an OC cycle that was truly geological in scale.

$$CO_2 + \text{reductant} \rightarrow C_{org} + \text{oxidants}. \tag{10.9}$$

10.7.2 Biological Evolution Influences the Dioxygen Cycle: Second Milestone

About 2.4 Gyr ago, the accumulation of OC in sediments, continents, and the mantle through subduction zones[128] became, for the first time, tied to the *production* and eventual *accumulation* (about 2.33 Ga)[2] of dioxygen gas in the lithosphere (in the form of iron oxides) and then in the exosphere.[129] The evolution of the water-splitting system[130] exploited, for the first time, the abundance of the water molecule as an almost unlimited source of protons and electrons for photosynthesis, releasing O_2 as a highly reactive waste product.[131]

$$H_2O \rightarrow 4e^- + 4H^+ + O_2. \tag{10.10}$$

This redox pathway enhanced the stability of atmospheric CO_2 (causing major glaciation, Figure 10.7), acidified riverine waters,[132] and therefore may have boosted the weathering flux

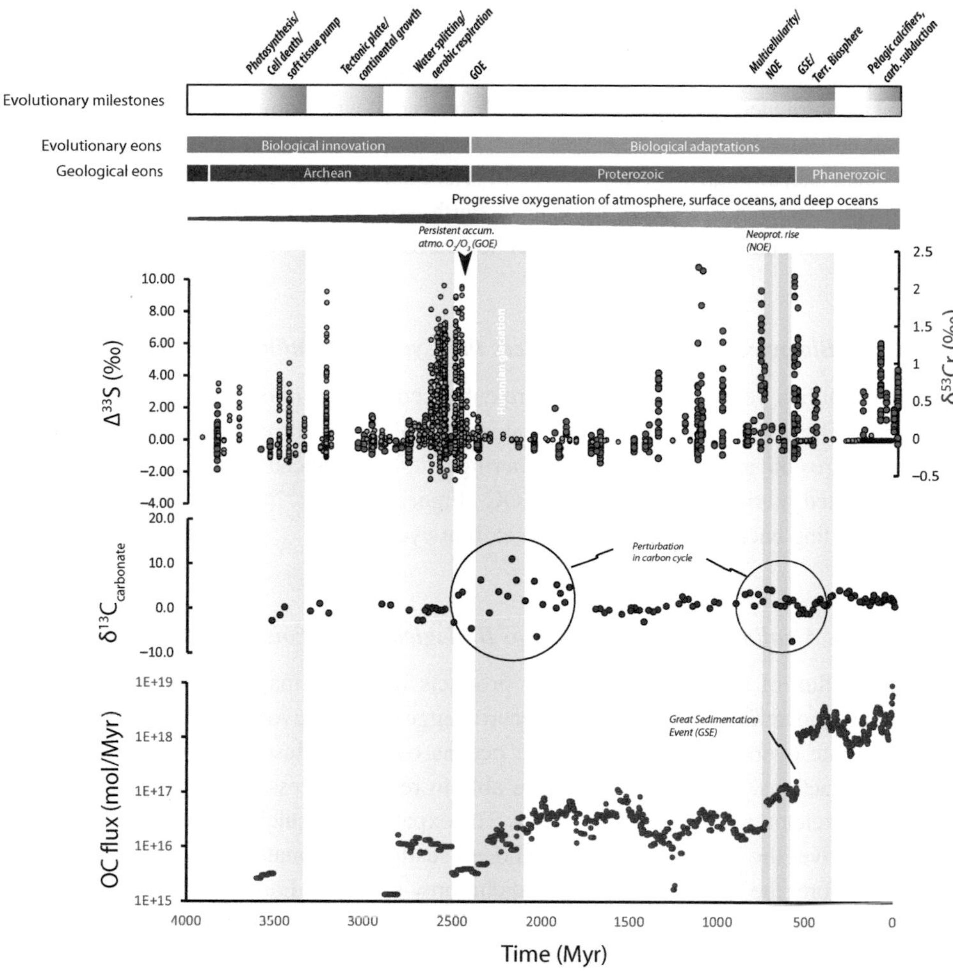

Figure 10.7 Milestones in the coevolution of life, the surface environment, and the tectonic carbon cycle (see also Figure 10.2). The eons of biological innovations and adaptations are from Ref. 131. Green tones denote biological transitions, blue tones denote geological transitions. (Upper Graph) The rise of atmospheric ozone and oxygen ($\Delta^{33}S$ proxy[148]), the meso/neoproterozoic rise of oxygen ($\delta^{56}Cr$ proxy[149,150]), the evolution of bryophytes and vascular plants, and the evolution of pelagic calcifiers are highlighted (adapted from Ref. 151). (Middle Graph) Statistical reconstruction of the carbonate isotope ($\delta^{13}C_{carbonate}$) record through time.[19] (Lower Graph) The statistical reconstruction of the global OC burial flux reconstructed from the North American sedimentary record from Ref. 37 is also appended. Note that the sharp increase in OC burial flux at the “great unconformity”[138] that marks an order of magnitude rise in continental weathering and sedimentation through the early Paleozoic. GOE = Great Oxygenation Event; NOE = Neoproterozoic Oxygenation Event.

of essential nutrients (e.g. trace elements, phosphorus, etc.[133]) from proto-continents to oceans. Life, therefore, exerted its first indirect control of the cycle of alkalinity and caused large-scale perturbations in the carbon cycle (Figure 10.7). A more subtle effect of this gradual redox shift was that it contributed to locking water on Earth, instead of water slowly escaping into space[134] as it did on Venus.[135] This is important because water drives and lubricates the subduction-zone system and, through it, the entire carbon cycle. Therefore, the rise of redox disequilibrium between the fluid and solid compartments of Earth may have laid the ground for the exceptional longevity and vigor of Earth's subduction carbon cycle.

10.7.3 Biological Evolution Influences the Cycle of Alkalinity: Third Milestone

About 1.5 billion years later, the biosphere enhanced its control on the *geological cycle of alkalinity*. The invasion of continents by algae and plants with roots in the early Paleozoic enhanced erosion and continental weathering.[136,137] Enhanced sediment export to the ocean boosted marine sedimentation of OC (Figure 10.7),[37,138,139] yet another potential process feeding back into the subduction carbon cycle.

10.7.4 Alkalinity Feeding Back into Biological Evolution: Fourth Milestone

The rising flux of dissolved weathering products (e.g. Ca) may have promoted further biological innovations. The geological record suggests algae evolved new ways to exploit the carbonate oversaturation state of the oceans over the Mesozoic.[140] By accelerating chemical reactions, these organisms were able to reroute excess oceanic alkalinity to the carbonate skeleton and ultimately to rocks. The expansion of calcifiers to the distal parts of the oceans over the last 150 Myr boosted deep seafloor carbonate deposition and subduction. Therefore, the idea that a clear-cut dichotomy separates Earth's biological processes (dominated by redox processes) and abiotic geochemical reactions (dominated by proton transfer chemistry)[141] does not hold in the context of geological time (Figure 10.2). Fascinatingly, much of the biotic carbonate of the Atlantic and Indian oceans has yet to be subducted; when will the *Atlantic Ring of Fire* form? We do not know.

10.7.5 Response of Climate to the Enhanced Subduction of Pelagic Carbonates

The geological response to pelagic calcifiers is still very much ahead of us. The models of Figures 10.6 and 10.7 illustrate potential scenarios and the timescales involved. They illustrate how subduction zones are components of a larger dynamic system that includes continents, oceans, and biosphere. Figure 10.8 is simplistic, and it builds on much previous work[3,8]; details are provided in the Supplementary Online Materials.

We focus on the geological response of the carbonate cycle and atmospheric temperature to modifications in key geodynamic or biological parameters: β, the partitioning of carbonate between the pelagic and continental (accretion) environment; σ, the fraction of

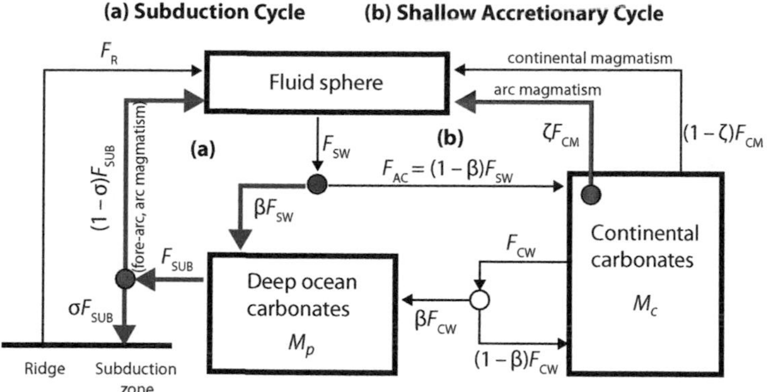

Figure 10.8 Non-steady-state response of the carbonate cycle to geodynamic and biological forcing: model design. The pelagic (M_p) and continental (M_c) carbonate reservoirs evolve by exchange of C (i.e. fluxes F_i) with the atmosphere and mantle (Figure 10.6). Mantle is assumed to be a reservoir of infinite residence time. The silicate weathering flux F_{sw} is linked to atmospheric temperature by a simple polynomial relation from Ref. 61. The C fluxes F_i are linked to M_i via rate constants k_i (e.g. $F_{cm} = k_{cm}M_c$, and $F_{sub} = k_{sub}M_p$). The initial steady state is obtained with starting conditions chosen and updated from Ref. 3 to be consistent with present-day values for C fluxes and reservoir sizes: F_i = 3 Tmol/yr, $F_{sw,i}$ = 10 Tmol/yr (Ref. 152); $F_{cw,i}$ = 15 Tmol/yr (Ref. 152); k_{cm} = 0.00035, k_{sub} = 0.0072, $M_{c,i} = 7.583 \times 10^{21}$ mol,[83] $M_{p,i} = 0.917 \times 10^{21}$ mol,[83] $\sigma = 0.6$ (see above), $\beta = 0.25$,[3] $\zeta = 0$. The system is solved analytically to find its steady state, which is close to present-day condition (i.e. $M_c = 7.909 \times 10^{21}$ mol (residence time $\tau_c \approx 2800$ Myr); $M_p = 0.810 \times 10^{21}$ mol (residence time $\tau_p \approx 140$ Myr); cf. Supplementary Online Materials). This steady state is then used for perturbation analysis (Figure 10.7).

subducted carbon transported beyond sub-arc depths (Figure 10.3); and ζ, a self-amplification factor corresponding to the fraction of continental carbonate degassing that is controlled by arc magmatism. The pelagic (M_p) and continental (M_c) carbonate reservoirs evolve by exchange of C (fluxes F_i) with the atmosphere and mantle (Figure 10.6). The silicate weathering flux F_{sw} is linked to atmospheric temperature by a simple polynomial relation.[61] Starting from a set of initial conditions (steady state), the evolution of the system described in Figure 10.8 obeys a set of coupled differential equations:

$$dM_c/dt = (1-\beta)F_{sw} - \beta F_{cw} - F_{cm}, \tag{10.11}$$

$$dM_p/dt = \beta(F_{sw} + F_{cw}) - F_{sub}. \tag{10.12}$$

For the sake of simplicity, we also impose that the surface reservoir evolves quasi-statically (i.e. F_{sw} responds instantaneously to any change in input flux; e.g. mid-ocean ridge degassing (F_i), the fraction of subducted carbon degassed in the fore-arc and sub-arc ($(1-\sigma)F'_{sub}$), or the flux of continental carbon degassing (F_{cm})), which gives:

$$F_r + F_{cm} + (1-\sigma)F_{sub} - F_{sw} = 0. \tag{10.13}$$

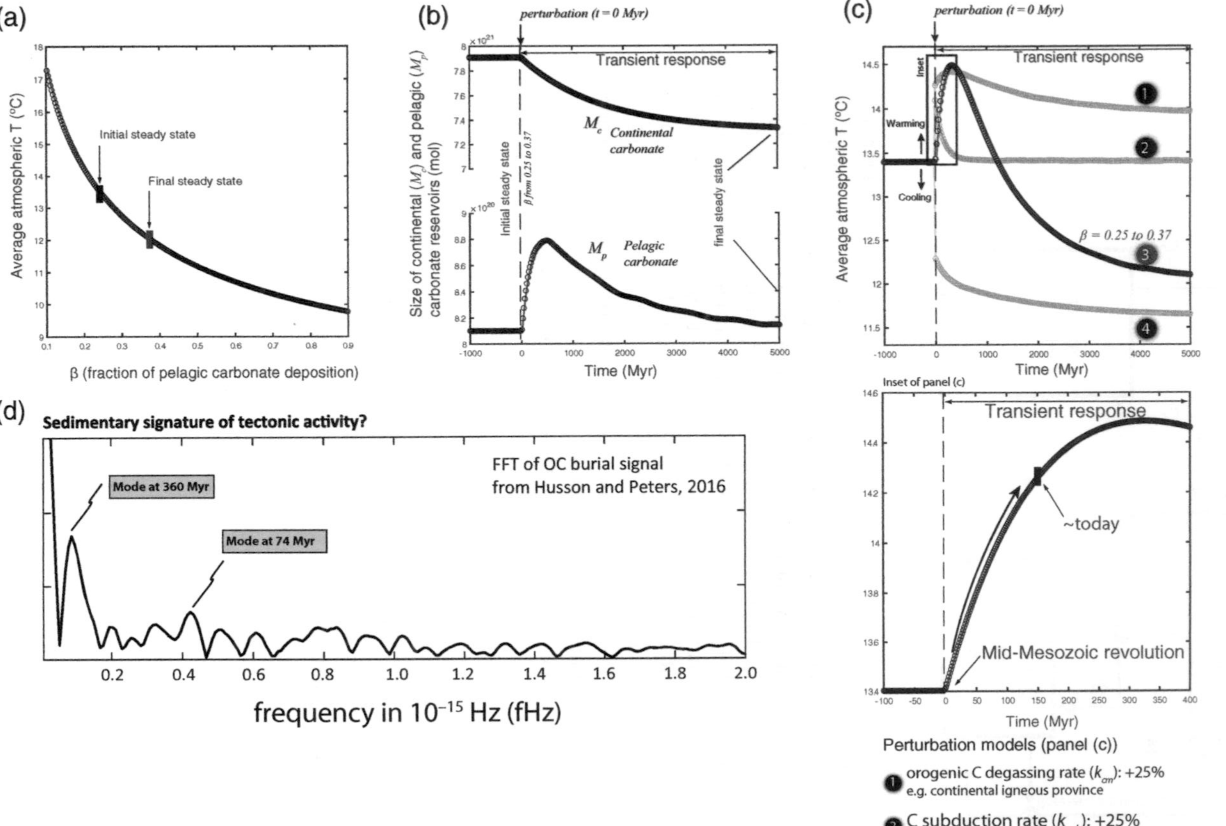

Figure 10.9 Non-steady-state response of the carbonate cycle to geodynamic and biological forcing: perturbation analysis. (a) Average atmospheric temperature obtained by solving the differential system of equations analytically (cf. Supplementary Online Materials) for values of β ranging from 0.1 (continental mode) to 0.9 (pelagic mode). (b) Size of continental (M_c) and pelagic (M_p) reservoirs as a function of time after initial perturbation of the steady state (Figure 10.6). At time 0, β is subjected to a step rise from 0.25 to 0.37 and the system is left to evolve. The transient values of M_c and M_p are tracked for 5 Gyr. (c) (Upper Panel) The shape and response time of the system (average temperature) for perturbations in β, σ, k_{cm}, and k_{sub} are

Perturbation analysis reveals a few important patterns (Figure 10.6). First, the shape and relaxation time varies over two orders of magnitudes. It is shorter when perturbation mostly affects the pelagic reservoir where residence time is short (e.g. σ) and it is longer when it affects both pelagic and continental reservoirs (e.g. β, ζ). Modification of the rate of subduction (k_{sub}) may also induce very long relaxation times because it affects the degassing of continental carbonates by arc magmatism (rate k_{am}) via ζ and thus M_c. Second, the non-steady-state response to augmentation (e.g. in β) involves an initial stage of atmospheric heating (β = 0.25–0.37 leads to ~2°C warming over the next 500 Myr), followed much later by relaxation to an atmospheric temperature that is cooler than the initial condition.

The important point is that paleoclimates most likely reflect a non-steady-state dynamic of the carbon cycle. This behavior prevails over timescales that can reach hundreds of millions of years (Figure 10.9). For example, we find that the signal corresponding to the time evolution of continental OC burial from Husson and Peters[37] contains at least two salient modes at ~0.08 and ~0.4 fHz, equivalent to periods of 360 and 74 Myr, respectively (Figure 10.9d), which are typical of tectonic events; the former is most likely linked to supercontinent cycles. Surprisingly, biological evolution may cause long-term non-steady-state responses from the geological carbon cycle, too (Figure 10.9c). In this case, the timescale of the response is not controlled by the biosphere, but by the large residence time of carbon in lithospheric reservoirs, particularly continents.

10.8 The Way Forward

We have shown that the subduction zone and its processes regulate three major properties of the geological carbon cycle: its steady-state fluxes – its *pace;* its episodic changes in pace – the *pulse;* and its stunning persistence over geological time – its *longevity*. These are the properties of a dynamic system characterized by *self-stabilizing feedback* between exogenic and endogenic carbon on Earth. This feedback is rooted in four levels of complexity – compositional, biological, tectonic, and kinetic – that transcend all scales.

Top-down feedback arises through the heterogeneous *composition* of subducted rocks. The long-term persistence of oceans, which is so important to the functioning of the subduction-zone carbon cycle, was in part linked to *biological evolution* and its role in

Figure 10.9 (*cont.*) compared. It is often the non-steady-state response of the system that matters when studying the long-term climatic (and isotopic[152]) impact of geodynamic transitions and/or biological innovations. (Lower Panel) Magnification of the 400 Myr following the perturbation, with the approximate location of the present time with respect to the mid-Mesozoic revolution.[6] (d) Pulse of the OC cycle on Earth over geological timescales. Fourier transform of the OC burial flux of Ref. 37 showing two dominant main modes. Typical frequencies are in femtohertz. Surprisingly, the signature of the 'supercontinental cycle at around 0.08 fHz seems to be significant, despite the important noise of the signal. FFT = fast Fourier transform.

maintaining an elevated, out-of-equilibrium concentration of O_2 in the atmosphere. Yet carbon deposition itself does not control when, where, and how subduction occurs; there is a *contingency* at the heart of the carbon cycle, contributing to its *kinetic* complexity. The relative slowness of the geological cycle of carbon is important for us, as a society, because it sets the ultimate threshold of carbon fluxes that anthropogenic CO_2 emissions may not exceed without long-lasting consequences for the atmosphere, ocean, and biosphere[142,143]; that is, it is a measure of the *fragility* of the surface geobiosphere.

Subduction zones transform and modify the reactivity of carbon, exerting primary bottom-up feedback control on the *pace* of the carbon cycle and its associated redox processes over geological timescales. We still do not understand well enough the microscopic processes of the liquid and solid carbon carrier phases of carbon that give rise to the fluxes that we observe. Although slow, the subduction carbon cycle proves surprisingly dynamic and subject to periodic accelerations and decelerations. This rhythmicity, or *pulse,* is controlled in part by the collision and subduction dynamics of Earth's *tectonic building blocks*. The dynamism of the southwest Pacific is a natural window onto the pulse of the carbon cycle today and in the past.

Subduction zones are components of a larger planetary system that is dynamic and in constant evolution; this system includes continents, oceans, and the biosphere. Geological hindsight shows that not only the *pace*,[31] but also the *pulse* and *longevity* of an atmospheric composition may record nonequilibrium planetary processes. This may include life, but also active tectonics. On Earth, the complex frequency distribution of the atmospheric compositional signal testifies to the size of the geochemical reservoirs, the vigor and nature of tectonic processes, and the presence of active biological life. Despite large uncertainties, our survey suggests that the heterogeneous subduction system is – today and possibly since the Mesozoic – in disequilibrium. Pelagic deposition exceeds carbonate subduction, which itself exceeds degassing from fore-arc and magmatic arc systems (Table 10.1).

10.9 Limits to Knowledge and Unknowns

There are several remaining questions about the role of subduction in the carbon cycle, including:

- What is the hydration and thermal structure of subducting slabs?
- What was the subduction efficiency *before* the Mid-Mesozoic revolution?
- What is the mechanical behavior and fate of limestones in the subduction zone?
- When and how will the Atlantic Ring of Fire form?
- Is Earth's carbon cycle unique in the universe?

Acknowledgments

We thank Tyler Volk and Jerôme Noir for helpful discussions about box modeling the carbonate cycle, Heather Stoll for discussions of Cenozoic evolution, Peter Ulmer and Olivier

Bachmann for informal reviews, and Mary Edmonds for her comments on an earlier version of the manuscript. Special thanks go to postdocs and graduate students of the "Physical Geobiology Group," the efforts of Xin Zhong to improve the modeling of the carbon cycling, and the insightful comments of Jonathan Viaud Murat, Gabriela Ligeza, and Hichem Ben Lakhdar, who have been precious and constant sources of motivation. Katy Evans reviewed this chapter and provided valuable suggestions for improvements. The authors wish to thank Isabelle Daniel and Beth Orcutt for their tact, patience, and competence in handling the chapter. This project is funded through a Branco Weiss Fellowship (MEG).

Questions for the Classroom

1. How slow is the slow carbon cycle and why does it matter for our society?
2. How does solar energy influence the subduction carbon cycle?
3. Can transitions in the geological carbon cycle influence biological evolution and how?
4. Why is water so critical for biological metabolisms and for the long-term carbon cycle?

References

1. Sleep, N. H. Stagnant lid convection and carbonate metasomatism of the deep continental lithosphere. *Geochemistry, Geophysics, Geosystems* **10**, Q11010 (2009).
2. Hayes, J. M. & Waldbauer, J. R. The carbon cycle and associated redox processes through time. *Philosophical Transactions of the Royal Society of London B: Biological Sciences* **361**, 931–950 (2006).
3. Volk, T. Sensitivity of climate and atmospheric CO_2 to deep-ocean and shallow-ocean carbonate burial. *Nature* **337**, 637–640 (1989).
4. Galvez, M. E. & Gaillardet, J. Historical constraints on the origins of the carbon cycle concept. *Comptes Rendus Geoscience* **344**, 549–567 (2012).
5. Wallmann, K. Controls on the Cretaceous and Cenozoic evolution of seawater composition, atmospheric CO_2 and climate. *Geochimica et Cosmochimica Acta* **65**, 3005–3025 (2001).
6. Ridgwell, A. A Mid Mesozoic Revolution in the regulation of ocean chemistry. *Marine Geology* **217**, 339–357 (2005).
7. Cartapanis, O., Galbraith, E. D., Bianchi, D. & Jaccard, S. Carbon burial in deep-sea sediment and implications for oceanic inventories of carbon and alkalinity over the last glacial cycle. *Climate of the Past* **14**, 1819–1850 (2018).
8. Caldeira, K. Continental–pelagic carbonate partitioning and the global carbonate-silicate cycle. *Geology* **19**, 204–206 (1991).
9. Wilkinson, B. H. & Walker, J. C. Phanerozoic cycling of sedimentary carbonate. *American Journal of Science* **289**, 525–548 (1989).
10. Clift, P. & Vannucchi, P. Controls on tectonic accretion versus erosion in subduction zones: implications for the origin and recycling of the continental crust. *Reviews of Geophysics* **42**, 2003RG000127 (2004).

11. Mottl, M. J., McCollom, T. M., Wheat, C. G. & Fryer, P. Decarbonation of the subducting Pacific plate triggered by the lawsonite-to-epidote transition beneath the Mariana forearc serpentinite mud volcanoes. In: *American Geological Union Fall Meeting Abstracts*, U51C-03 (2008).
12. Allard, P. et al. Prodigious emission rates and magma degassing budget of major, trace and radioactive volatile species from Ambrym basaltic volcano, Vanuatu Island arc. *Journal of Volcanology and Geothermal Research* **322**, 119–143 (2016).
13. Sakai, H. et al. Venting of carbon dioxide-rich fluid and hydrate formation in mid-Okinawa trough backarc basin. *Science* **248**, 1093–1096 (1990).
14. Sleep, N. H. & Zahnle, K. Carbon dioxide cycling and implications for climate on ancient Earth. *Journal of Geophysical Research: Planets (1991–2012)* **106**, 1373–1399 (2001).
15. Lee, C.-T. A. et al. Continental arc–island arc fluctuations, growth of crustal carbonates, and long-term climate change. *Geosphere* **9**, 21–36 (2013).
16. Mason, E., Edmonds, M. & Turchyn, A. V. Remobilization of crustal carbon may dominate volcanic arc emissions. *Science* **357**, 290 (2017).
17. Foley, S. F. & Fischer, T. P. An essential role for continental rifts and lithosphere in the deep carbon cycle. *Nature Geoscience* **10**, 897 (2017).
18. McKenzie, N. R. et al. Continental arc volcanism as the principal driver of icehouse–greenhouse variability. *Science* **352**, 444–447 (2016).
19. Krissansen-Totton, J., Buick, R. & Catling, D. C. A statistical analysis of the carbon isotope record from the Archean to Phanerozoic and implications for the rise of oxygen. *American Journal of Science* **315**, 275–316 (2015).
20. Campbell, I. & Taylor, S. No water, no granites – no oceans, no continents. *Geophysical Research Letters* **10**, 1061–1064 (1983).
21. Kelemen, P. B. & Manning, C. E. Reevaluating carbon fluxes in subduction zones, what goes down, mostly comes up. *Proceedings of the National Academy of Sciences* **112**, E3997–E4006 (2015).
22. Caldeira, K. Enhanced Cenozoic chemical weathering and the subduction of pelagic carbonate. *Nature* **357**, 578–581 (1992).
23. Archer, D. E. An atlas of the distribution of calcium carbonate in sediments of the deep sea. *Global Biogeochemical Cycles* **10**, 159–174 (1996).
24. Zachos, J. C. et al. Rapid acidification of the ocean during the Paleocene–Eocene thermal maximum. *Science*, **308**, 1611–1615 (2005).
25. Plank, T. & Langmuir, C. H. The chemical composition of subducting sediment and its consequences for the crust and mantle. *Chemical Geology* **145**, 325–394 (1998).
26. Clift, P. D. A revised budget for Cenozoic sedimentary carbon subduction. *Reviews of Geophysics* **55**, 97–125 (2017).
27. Johnston, F. K., Turchyn, A. V. & Edmonds, M. Decarbonation efficiency in subduction zones: implications for warm Cretaceous climates. *Earth and Planetary Science Letters* **303**, 143–152 (2011).
28. Arndt, S. et al. Quantifying the degradation of organic matter in marine sediments: a review and synthesis. *Earth-Science Reviews* **123**, 53–86 (2013).
29. Cartapanis, O., Bianchi, D., Jaccard, S. L. & Galbraith, E. D. Global pulses of organic carbon burial in deep-sea sediments during glacial maxima. *Nature Communications* **7**, 10796 (2016).
30. del Giorgio, P. A. & Duarte, C. M. Respiration in the open ocean. *Nature* **420**, 379–384 (2002).

31. Krissansen-Totton, J., Bergsman, D. S. & Catling, D. C. On detecting biospheres from chemical thermodynamic disequilibrium in planetary atmospheres. *Astrobiology* **16**, 39–67 (2016).
32. Cui, X., Bianchi, T. S., Jaeger, J. M. & Smith, R. W. Biospheric and petrogenic organic carbon flux along southeast Alaska. *Earth and Planetary Science Letters* **452**, 238–246 (2016).
33. Galy, V., Peucker-Ehrenbrink, B. & Eglinton, T. Global carbon export from the terrestrial biosphere controlled by erosion. *Nature* **521**, 204 (2015).
34. Blair, N. E. & Aller, R. C. The fate of terrestrial organic carbon in the marine environment. *Annual Review of Marine Science* **4**, 401–423 (2012).
35. Hilton, R. G. et al. Tropical-cyclone-driven erosion of the terrestrial biosphere from mountains. *Nature Geoscience* **1**, 759–762 (2008).
36. Hilton, R. G. et al. Erosion of organic carbon in the Arctic as a geological carbon dioxide sink. *Nature* **524**, 84–87 (2015).
37. Husson, J. M. & Peters, S. E. Atmospheric oxygenation driven by unsteady growth of the continental sedimentary reservoir. *Earth and Planetary Science Letters* **460**, 68–75 (2017).
38. Bao, R. et al. Tectonically-triggered sediment and carbon export to the Hadal zone. *Nature Communications* **9**, 121 (2018).
39. Beyssac, O., Rouzaud, J.-N., Goffé, B., Brunet, F. & Chopin, C. Graphitization in a high-pressure, low-temperature metamorphic gradient: a Raman microspectroscopy and HRTEM study. *Contributions to Mineralogy and Petrology* **143**, 19–31 (2002).
40. Galvez, M. E. et al. Graphite formation by carbonate reduction during subduction. *Nature Geoscience* **6**, 473–477 (2013).
41. Adam, P., Schneckenburger, P., Schaeffer, P. & Albrecht, P. Clues to early diagenetic sulfurization processes from mild chemical cleavage of labile sulfur-rich geomacromolecules. *Geochimica et Cosmochimica Acta* **64**, 3485–3503 (2000).
42. Helgeson, H. C., Richard, L., McKenzie, W. F., Norton, D. L. & Schmitt, A. A chemical and thermodynamic model of oil generation in hydrocarbon source rocks. *Geochimica et Cosmochimica Acta* **73**, 594–695 (2009).
43. Berner, R. A. Burial of organic carbon and pyrite sulfur in the modern ocean: its geochemical and environmental significance. *American Journal of Science* **282**, 451–473 (1982).
44. Petrenko, V. V. et al. Minimal geological methane emissions during the Younger Dryas–Preboreal abrupt warming event. *Nature* **548**, 443–446 (2017).
45. Oohashi, K., Hirose, T. & Shimamoto, T. Shear-induced graphitization of carbonaceous materials during seismic fault motion: experiments and possible implications for fault mechanics. *Journal of Structural Geology* **33**, 1122–1134 (2011).
46. Berthonneau, J. et al. Mesoscale structure, mechanics, and transport properties of source rocks' organic pore networks. *Proceedings of the National Academy of Sciences* **115**, 12365 (2018).
47. Weck, P. F. et al. Model representations of kerogen structures: an insight from density functional theory calculations and spectroscopic measurements. *Scientific Reports* **7**, 7068 (2017).
48. Connolly, J. A. D. & Galvez, M. E. Electrolytic fluid speciation by Gibbs energy minimization and implications for subduction zone mass transfer. *Earth and Planetary Science Letters* **501**, 90–102 (2018).

49. Zhang, S., Ague, J. J. & Brovarone, A. V. Degassing of organic carbon during regional metamorphism of pelites, Wepawaug Schist, Connecticut, USA. *Chemical Geology* **490**, 30–44 (2018).
50. Alt, J. C. et al. Recycling of water, carbon, and sulfur during subduction of serpentinites: a stable isotope study of Cerro del Almirez, Spain. *Earth and Planetary Science Letters* **327–328**, 50–60 (2012).
51. Alt, J. C. et al. The role of serpentinites in cycling of carbon and sulfur: seafloor serpentinization and subduction metamorphism. *Lithos* **178**, 40–54 (2013).
52. Nakamura, K. & Kato, Y. Carbonatization of oceanic crust by the seafloor hydrothermal activity and its significance as a CO_2 sink in the Early Archean. *Geochimica et Cosmochimica Acta* **68**, 4595–4618 (2004).
53. Shibuya, T., Komiya, T., Nakamura, K., Takai, K. & Maruyama, S. Highly alkaline, high-temperature hydrothermal fluids in the early Archean ocean. *Precambrian Research* **182**, 230–238 (2010).
54. Ueda, H., Sawaki, Y. & Maruyama, S. Reactions between olivine and CO_2-rich seawater at 300°C: implications for H_2 generation and CO_2 sequestration on the early Earth. *Geoscience Frontiers* **8**, 387–396 (2017).
55. Dhuime, B., Hawkesworth, C. J., Cawood, P. A. & Storey, C. D. A change in the geodynamics of continental growth 3 billion years ago. *Science* **335**, 1334–1336 (2012).
56. Sieber, M. J., Hermann, J. & Yaxley, G. M. An experimental investigation of C–O–H fluid-driven carbonation of serpentinites under forearc conditions. *Earth and Planetary Science Letters* **496**, 178–188 (2018).
57. Aiuppa, A., Fischer, T. P., Plank, T., Robidoux, P. & Di Napoli, R. Along-arc, inter-arc and arc-to-arc variations in volcanic gas CO_2/ST ratios reveal dual source of carbon in arc volcanism. *Earth-Science Reviews* **168**, 24–47 (2017).
58. Chiodini, G. et al. First $^{13}C/^{12}C$ isotopic characterisation of volcanic plume CO_2. *Bulletin of Volcanology* **73**, 531–542 (2011).
59. Stolper, E. & Newman, S. The role of water in the petrogenesis of Mariana trough magmas. *Earth and Planetary Science Letters* **121**, 293–325 (1994).
60. Bjørnerud, M. G. & Austrheim, H. Inhibited eclogite formation: the key to the rapid growth of strong and buoyant Archean continental crust. *Geology* **32**, 765–768 (2004).
61. Berner, R. A., Lasaga, A. C. & Garrels, R. M. The carbonate–silicate geochemical cycle and its effect on atmospheric carbon dioxide over the past 100 million years. *American Journal of Science* **283**, 641–683 (1983).
62. Goldschmidt, V. M. Die Gesetze der Gesteinsmetamorphose, mit beispielen aus der Geologie des sudlichen Norwegens. *Videnskapsselskapets Skrifter, Mathematiker-naturvissenshaft, KL*, 1–16 (1912).
63. Berner, R. A. The long-term carbon cycle, fossil fuels and the atmospheric CO_2. *Nature* **426**, 323–326 (2003).
64. McGary, R. S., Evans, R. L., Wannamaker, P. E., Elsenbeck, J. & Rondenay, S. Pathway from subducting slab to surface for melt and fluids beneath Mount Rainier. *Nature* **511**, 338 (2014).
65. Poli, S., Franzolin, E., Fumagalli, P. & Crottini, A. The transport of carbon and hydrogen in subducted oceanic crust: an experimental study to 5 GPa. *Earth and Planetary Science Letters* **278**, 350–360 (2009).
66. Ulmer, P. & Trommsdorff, V. Serpentine stability to mantle depths and subduction-related magmatism. *Science* **268**, 858–861 (1995).

67. Galvez, M. E., Connolly, J. A. D. & Manning, C. E. Implications for metal and volatile cycles from the pH of subduction zone fluids. *Nature* **539**, 420–424 (2016).
68. Tsujimori, T. & Ernst, W. Lawsonite blueschists and lawsonite eclogites as proxies for palaeo-subduction zone processes: a review. *Journal of Metamorphic Geology* **32**, 437–454 (2014).
69. Ague, J. J. Release of CO_2 from carbonate rocks during regional metamorphism of lithologically heterogeneous crust. *Geology* **28**, 1123–1126 (2000).
70. Ague, J. J. & Nicolescu, S. Carbon dioxide released from subduction zones by fluid-mediated reactions. *Nature Geoscience* **7**, 355–360 (2014).
71. Galvez, M. E. et al. Metasomatism and graphite formation at a lithological interface in Malaspina (Alpine Corsica, France). *Contributions to Mineralogy and Petrology* **166**, 1687–1708 (2013).
72. Pimm, A. C. Sedimentology and history of the northeastern Indian Ocean from Late Cretaceous to Recent. *Initial Reports of the Deep Sea Drilling Project* **22**, 717–804 (1974).
73. Gerya, T. V. & Yuen, D. A. Rayleigh–Taylor instabilities from hydration and melting propel 'cold plumes' at subduction zones. *Earth and Planetary Science Letters* **212**, 47–62 (2003).
74. Skora, S. et al. Hydrous phase relations and trace element partitioning behaviour in calcareous sediments at subduction-zone conditions. *Journal of Petrology* **56**, 953–980 (2015).
75. Kang, N., Schmidt, M. W., Poli, S., Franzolin, E. & Connolly, J. A. Melting of siderite to 20GPa and thermodynamic properties of $FeCO_3$-melt. *Chemical Geology* **400**, 34–43 (2015).
76. Foustoukos, D. I. & Mysen, B. O. The structure of water-saturated carbonate melts. *American Mineralogist* **100**, 35–46 (2015).
77. Russell, J. K., Porritt, L. A., Lavallée, Y. & Dingwell, D. B. Kimberlite ascent by assimilation-fuelled buoyancy. *Nature* **481**, 352 (2012).
78. Baxter, E. F. & Caddick, M. J. Garnet growth as a proxy for progressive subduction zone dehydration. *Geology* **41**, 643–646 (2013).
79. Frezzotti, M. L., Selverstone, J., Sharp, Z. D. & Compagnoni, R. Carbonate dissolution during subduction revealed by diamond-bearing rocks from the Alps. *Nature Geoscience* **4**, 703–706 (2011).
80. Galvez, M. E., Manning, C. E., Connolly, J. A. & Rumble, D. The solubility of rocks in metamorphic fluids: a model for rock-dominated conditions to upper mantle pressure and temperature. *Earth and Planetary Science Letters* **430**, 486–498 (2015).
81. Gorman, P. J., Kerrick, D. M. & Connolly, J. A. D. Modeling open system metamorphic decarbonation of subducting slabs. *Geochemistry, Geophysics, Geosystems* **7**, Q04007 (2006).
82. Connolly, J. A. D. Computation of phase equilibria by linear programming: a tool for geodynamic modeling and its application to subduction zone decarbonation. *Earth and Planetary Science Letters* **236**, 524–541 (2005).
83. Hirschmann, M. M. Comparative deep Earth volatile cycles: the case for C recycling from exosphere/mantle fractionation of major (H_2O, C, N) volatiles and from H_2O/Ce, CO_2/Ba, and CO_2/Nb exosphere ratios. *Earth and Planetary Science Letters* **502**, 262–273 (2018).
84. Thomson, A. R., Walter, M. J., Kohn, S. C. & Brooker, R. A. Slab melting as a barrier to deep carbon subduction. *Nature* **529**, 76–79 (2016).

85. Wyllie, P. J. & Tuttle, O. F. The System CaO–CO_2–H_2O and the origin of carbonatites. *Journal of Petrology* **1**, 1–46 (1960).
86. Poli, S. Carbon mobilized at shallow depths in subduction zones by carbonatitic liquids. *Nature Geoscience* **8**, 633–636 (2015).
87. Kang, N. & Schmidt, M. W. The melting of subducted banded iron formations. *Earth and Planetary Science Letters* **476**, 165–178 (2017).
88. Syracuse, E. M., van Keken, P. E. & Abers, G. A. The global range of subduction zone thermal models. *Physics of the Earth and Planetary Interiors* **183**, 73–90 (2010).
89. Penniston-Dorland, S. C., Kohn, M. J. & Manning, C. E. The global range of subduction zone thermal structures from exhumed blueschists and eclogites: rocks are hotter than models. *Earth and Planetary Science Letters* **428**, 243–254 (2015).
90. Angiboust, S., Pettke, T., De Hoog, J. C., Caron, B. & Oncken, O. Channelized fluid flow and eclogite-facies metasomatism along the subduction shear zone. *Journal of Petrology* **55**, 883–916 (2014).
91. Worzewski, T., Jegen, M., Kopp, H., Brasse, H. & Castillo, W. T. Magnetotelluric image of the fluid cycle in the Costa Rican subduction zone. *Nature Geoscience* **4**, 108 (2011).
92. Piccoli, F., Brovarone, A. V., Beyssac, O., Martinez, I., Ague, J. J. & Chaduteau, C. Carbonation by fluid–rock interactions at high-pressure conditions: implications for carbon cycling in subduction zones. *Earth and Planetary Science Letters* **445**, 146–159 (2016).
93. Tumiati, S., Godard, G., Martin, S., Malaspina, N. & Poli, S. Ultra-oxidized rocks in subduction mélanges? Decoupling between oxygen fugacity and oxygen availability in a Mn-rich metasomatic environment. *Lithos* **226**, 116–130 (2015).
94. Connolly, J. A. D. & Cesare, B. C–O–H–S fluid composition and oxygen fugacity in graphitic metapelites. *Journal of Metamorphic Geology* **11**, 379–388 (1993).
95. Brovarone, A. V. et al. Massive production of abiotic methane during subduction evidenced in metamorphosed ophicarbonates from the Italian Alps. *Nature Communications* **8**, 14134 (2017).
96. Weis, P. L., Friedman, I. & Gleason, J. P. The origin of epigenetic graphite: evidence from isotopes. *Geochimica et Cosmochimica Acta* **45**, 2325–2332 (1981).
97. Rumble III, D., Duke, E. F. & Hoering, T. L. Hydrothermal graphite in New Hampshire: evidence of carbon mobility during regional metamorphism. *Geology* **14**, 452–455 (1986).
98. Malvoisin, B., Chopin, C., Brunet, F. & Galvez, M. E. Low-temperature wollastonite formed by carbonate reduction: a marker of serpentinite redox conditions. *Journal of Petrology* **53**, 159–176 (2012).
99. Rohrbach, A. & Schmidt, M. W. Redox freezing and melting in the Earth's deep mantle resulting from carbon-iron redox coupling. *Nature* **472**, 209–212 (2011).
100. Dalton, J. A. & Wood, B. J. The compositions of primary carbonate melts and their evolution through wallrock reaction in the mantle. *Earth and Planetary Science Letters* **119**, 511–525 (1993).
101. Jones, A. P., Genge, M. & Carmody, L. Carbonate melts and carbonatites. *Reviews in Mineralogy and Geochemistry* **75**, 289–322 (2013).
102. Stolper, E., Fine, G., Johnson, T. & Newman, S. Solubility of carbon dioxide in albitic melt. *American Mineralogist* **72**, 1071–1085 (1987).

103. Mysen, B. Experimental, in-situ carbon solution mechanisms and isotope fractionation in and between (C–O–H)-saturated silicate melt and silicate-saturated (C–O–H) fluid to upper mantle temperatures and pressures. *Earth and Planetary Science Letters* **459**, 352–361 (2017).
104. Pirard, C. & Hermann, J. Experimentally determined stability of alkali amphibole in metasomatised dunite at sub-arc pressures. *Contributions to Mineralogy and Petrology* **169**, 1–26 (2014).
105. Korsakov, A. V. & Hermann, J. Silicate and carbonate melt inclusions associated with diamonds in deeply subducted carbonate rocks. *Earth and Planetary Science Letters* **241**, 104–118 (2006).
106. Zhang, C., Yu, X.-Y. & Jiang, T.-L. Mineral association and graphite inclusions in nephrite jade from Liaoning, northeast China: implications for metamorphic conditions and ore genesis. *Geoscience Frontiers* **10**, 425–437 (2019).
107. Rothman, D. H., Hayes, J. M. & Summons, R. E. Dynamics of the Neoproterozoic carbon cycle. *Proceedings of the National Academy of Sciences* **100**, 8124–8129 (2003).
108. Cooper, P. A. & Taylor, B. Polarity reversal in the Solomon Islands arc. *Nature* **314**, 428 (1985).
109. Hoareau, G. et al. Did high Neo-Tethys subduction rates contribute to early Cenozoic warming? *Climate of the Past* **11**, 1751–1767 (2015).
110. Pubellier, M. & Meresse, F. Phanerozoic growth of Asia: geodynamic processes and evolution. *Journal of Asian Earth Sciences* **72**, 118–128 (2013).
111. Condie, K. C. & O'Neill, C. The Archean–Proterozoic boundary: 500 My of tectonic transition in Earth history. *American Journal of Science* **310**, 775–790 (2010).
112. Selverstone, J. & Gutzler, D. S. Post-125 Ma carbon storage associated with continent–continent collision. *Geology* **21**, 885–888 (1993).
113. Tapponnier, P. et al. Oblique stepwise rise and growth of the Tibet Plateau. *Science* **294**, 1671–1677 (2001).
114. Gaillardet, J. & Galy, A. Himalaya – carbon sink or source? *Science* **320**, 1727–1728 (2008).
115. Raymo, M. E., Ruddiman, W. F. & Froelich, P. N. Influence of late Cenozoic mountain building on ocean geochemical cycles. *Geology* **16**, 649–653 (1988).
116. Hilton, R. G., Galy, A. & Hovius, N. Riverine particulate organic carbon from an active mountain belt: importance of landslides. *Global Biogeochemical Cycles* **22**, GB1017 (2008).
117. Hilton, R. G., Gaillardet, J., Calmels, D. & Birck, J.-L. Geological respiration of a mountain belt revealed by the trace element rhenium. *Earth and Planetary Science Letters* **403**, 27–36 (2014).
118. Torres, M. A., West, A. J. & Li, G. Sulphide oxidation and carbonate dissolution as a source of CO_2 over geological timescales. *Nature* **507**, 346–349 (2014).
119. Tomkins, A. G., Rebryna, K. C., Weinberg, R. F. & Schaefer, B. F. Magmatic sulfide formation by reduction of oxidized arc basalt. *Journal of Petrology* **53**, 1537–1567 (2012).
120. Sternai, P. et al. On the influence of the asthenospheric flow on the tectonics and topography at a collision–subduction transition zones: comparison with the eastern Tibetan margin. *Journal of Geodynamics* **100**, 184–197 (2016).
121. Royden, L. H., Burchfiel, B. C. & van der Hilst, R. D. The geological evolution of the Tibetan Plateau. *Science* **321**, 1054–1058 (2008).

122. Edmond, J. M. & Huh, Y. Non-steady state carbonate recycling and implications for the evolution of atmospheric PCO_2. *Earth and Planetary Science Letters* **216**, 125–139 (2003).
123. Soo, R. M., Hemp, J., Parks, D. H., Fischer, W. W. & Hugenholtz, P. On the origins of oxygenic photosynthesis and aerobic respiration in Cyanobacteria. *Science* **355**, 1436–1440 (2017).
124. Bidle, K. D. & Falkowski, P. G. Cell death in planktonic, photosynthetic microorganisms. *Nature Reviews Microbiology* **2**, 643–655 (2004).
125. Fuhrman, J. A. Marine viruses and their biogeochemical and ecological effects. *Nature* **399**, 541–548 (1999).
126. Forterre, P. & Prangishvili, D. The great billion-year war between ribosome-and capsid-encoding organisms (cells and viruses) as the major source of evolutionary novelties. *Annals of the New York Academy of Sciences* **1178**, 65–77 (2009).
127. Duncan, M. S. & Dasgupta, R. Rise of Earth's atmospheric oxygen controlled by efficient subduction of organic carbon. *Nature Geoscience* **10**, 387–392 (2017).
128. Godderis, Y. & Veizer, J. Tectonic control of chemical and isotopic composition of ancient oceans; the impact of continental growth. *American Journal of Science* **300**, 434–461 (2000).
129. Knoll, A. H. Biomineralization and evolutionary history. *Reviews in Mineralogy and Geochemistry* **54**, 329–356 (2003).
130. Fischer, W., Hemp, J. & Johnson, J. E. Evolution of oxygenic photosynthesis. *Annual Review of Earth and Planetary Sciences* **44**, 647–683 (2016).
131. Falkowski, P. G. & Godfrey, L. V. Electrons, life and the evolution of Earth's oxygen cycle. *Philosophical Transactions of the Royal Society B: Biological Sciences* **363**, 2705–2716 (2008).
132. Halevy, I. & Bachan, A. The geologic history of seawater pH. *Science* **355**, 1069 (2017).
133. Anbar, A. D. Elements and evolution. *Science* **322**, 1481–1483 (2008).
134. Catling, D. C., Zahnle, K. J. & McKay, C. Biogenic methane, hydrogen escape, and the irreversible oxidation of early Earth. *Science* **293**, 839–843 (2001).
135. Driscoll, P. & Bercovici, D. Divergent evolution of Earth and Venus: influence of degassing, tectonics, and magnetic fields. *Icarus* **226**, 1447–1464 (2013).
136. Schwartzman, D. W. & Volk, T. Biotic enhancement of weathering and the habitability of Earth. *Nature* **340**, 457 (1989).
137. Hemingway, J. D. et al. Microbial oxidation of lithospheric organic carbon in rapidly eroding tropical mountain soils. *Science* **360**, 209–212 (2018).
138. Peters, S. E. & Gaines, R. R. Formation of the Great Unconformity as a trigger for the Cambrian explosion. *Nature* **484**, 363–366 (2012).
139. Berner, R. A. Inclusion of the weathering of volcanic rocks in the geocarbsulf model. *American Journal of Science* **306**, 295–302 (2006).
140. Monteiro, F. M. et al. Why marine phytoplankton calcify. *Science Advances* **2**, e1501822 (2016).
141. Falkowski, P. G., Fenchel, T. & Delong, E. F. The microbial engines that drive Earth's biogeochemical cycles. *Science* **320**, 1034–1039 (2008).
142. DePaolo, D. J. Sustainable carbon emissions: the geologic perspective. *MRS Energy & Sustainability – A Review Journal* **2**, 1–16 (2015).
143. Gruber, N. Warming up, turning sour, losing breath: ocean biogeochemistry under global change. *Philosophical Transactions of the Royal Society of London A: Mathematical, Physical and Engineering Sciences* **369**, 1980–1996 (2011).

144. Canfield, D., Glazer, A. & Falkowski, P. The evolution and future of Earth's nitrogen cycle. *Science* **333**, 192–196 (2010).
145. Smith, R. W., Bianchi, T. S., Allison, M., Savage, C. & Galy, V. High rates of organic carbon burial in fjord sediments globally. *Nature Geoscience* **8**, 450–453 (2015).
146. Burdige, D. J. Preservation of organic matter in marine sediments: controls, mechanisms, and an imbalance in sediment organic carbon budgets? *Chemical Reviews* **107**, 467–485 (2007).
147. Jolivet, L. et al. Mantle flow and deforming continents: from India–Asia convergence to Pacific subduction. *Tectonics* **37**, 2887–2914 (2018).
148. Farquhar, J., Bao, H. & Thiemens, M. Atmospheric influence of Earth's earliest sulfur cycle. *Science* **289**, 756–758 (2000).
149. Cole, D. B. et al. A shale-hosted Cr isotope record of low atmospheric oxygen during the Proterozoic. *Geology* **44**, 555–558 (2016).
150. Canfield, D. E. et al. Highly fractionated chromium isotopes in Mesoproterozoic-aged shales and atmospheric oxygen. *Nature Communications* **9**, 2871 (2018).
151. Reinhard, C. T., Olson, S. L., Schwieterman, E. W. & Lyons, T. W. False negatives for remote life detection on ocean-bearing planets: lessons from the early Earth. *Astrobiology* **17**, 287–297 (2017).
152. Caves, J. K., Jost, A. B., Lau, K. V. & Maher, K. Cenozoic carbon cycle imbalances and a variable weathering feedback. *Earth and Planetary Science Letters* **450**, 152–163 (2016).

Appendix to Chapter 10
How Do Subduction Zones Regulate the Carbon Cycle?

Supplementary Material: Description of the Model

What follows is a description of the model presented in Figure 10.7 (main text).

The main features of the model have been presented in the main text (Figure 10.7). It is simplistic in the sense that it neglects organic carbon and sulfur cycles that also operate as transient sources and sinks of CO_2 over geological timescales.[1,2] That said, the mass of the pelagic (M_p) and continental (M_c) carbonate reservoirs evolve by exchange of C (i.e. fluxes F_i) with the atmosphere and mantle. We distinguish mid-ocean ridge carbon degassing (F_j); carbonate subduction (F_{sub}), where σF_{sub} returns to mantle and $(1 - \sigma)F_{sub}$ returns to exosphere; silicate weathering (F_{sw}), where βF_{sw} precipitates in the pelagic environment and $(1 - \beta)F_{sw}$ is accreted to continents; carbonate weathering (F_{cw}); and continental carbon degassing (F_{cm}), where ζF_{cm} is attributed to continental arc degassing and $(1 - \zeta)$ F_{cm} is attributed to orogenic degassing. The initial steady state is obtained with starting conditions chosen and updated from Volk,[3] to be consistent with present-day values in C flux and reservoir sizes.

Silicate and continental weathering: The silicate weathering flux (F_{sw}) and the continental carbonate weathering flux (F_{cw}) are derived from present-day values ($F_{sw,ref}$ and $F_{cw,ref}$) using a weathering rate constant, f_{wr}. $F_{sw,ref}$ is set at 10 Tmol/yr and $F_{cw,ref}$ at 15 Tmol/yr after Caves et al.[2] M_c affects the carbonate weathering flux as it departs from the present-day size of the continental reservoir, $M_{c,ref}$, estimated at 7583 × 10^{18} mol. The equations are:

$$F_{sw} = F_{sw,ref} f_{wr}, \tag{1}$$

$$F_{cw} = F_{cw,ref} f_{wr} M_c / M_{c,ref}. \tag{2}$$

Link between silicate weathering and atmospheric temperature: The silicate weathering rate constant is linked to atmospheric temperature by a simple polynomial relation from Berner et al.[4]:

$$f_{wr} = 1 + 8.7 \times 10^{-2}(T - T_{ref}) + 1.9 \times 10^{-3}(T - T_{ref})^2, \tag{3}$$

with T_{ref} = 15°C.

Continental carbonate degassing: Orogenic release of CO_2 by accreted carbonates is assumed to be proportional to M_c:

$$F_{cm} = k_{cm} M_c. \tag{4}$$

We attribute ζF_{cm} to continental arc degassing:

$$\zeta F_{cm} = k_{am} M_c, \tag{5}$$

and $(1 - \zeta) F_{cm}$ to orogenic degassing:

$$(1 - \zeta) F_{cm} = k_{om} M_c \tag{6}$$

Therefore, ζ is an *auto-amplification* factor that connects continental C degassing to subduction-zone processes and rates.

Injecting (4) into (6) gives:

$$\zeta = 1 - \frac{k_{om}}{k_{cm}}. \tag{7}$$

We take $k_{cm} = 3.57 \times 10^{-4}$/Myr, which we calibrated to reproduce modern-day flux F_{cm} from Volk.[3] The parameter ζ is free to vary and depends on the proportion of continental and island arcs.[5] For simplicity, the examples depicted in Figure 10.8 consider $\zeta = 0$ (island arcs only).

Carbon subduction: The flux of pelagic carbonates entering the subduction zone is also assumed to be proportional to M_p:

$$F_{sub} = k_{sub} M_p. \tag{8}$$

We take $k_{sub} = 0.0072$ Myr^{-1} to approach a present-day carbon subduction flux of 70 MtC/yr. The parameter σ is the fraction of subducted carbon the subducts to beyond sub-arc depths, and it is taken at 0.6 (cf. main text). The present-day size of the continental reservoir, $M_{p,ref}$, is 0.917×10^{21} mol.[6] We note that this value is about 10 times larger than the value adopted by Volk.[3] While this modifies the response time of the whole system compared to that expected from Volk,[3] it does not change its internal structure (i.e. its fundamental dynamics).

Biological factors: The parameter β is the partitioning of carbonate between pelagic and continental (accretion) environments. It is set at 0.25 for the initial conditions prior to perturbations.

Mid-ocean ridge and other mantle degassing fluxes: The mid-ocean ridge flux is taken to be $F_j = 3$ Tmol/yr.[6]

The whole system (Figure 10.7) evolves dynamically from its initial conditions, but the atmospheric subsystem is assumed to evolve quasi-statically during this process (i.e. its response times to changes in input fluxes are infinitely faster than the characteristic timescales of the whole system). The system is solved analytically to find its steady state, which is close to the present-day condition: $M_c = 7.909 \times 10^{21}$ mol (i.e. residence time

$\tau_c \approx 2800$ Myr); $M_p = 0.810 \times 10^{21}$ mol (i.e. residence time $\tau_p \approx 140$ Myr). This steady state is then used for perturbation analysis (Figure 10.7).

References

1. Torres, M. A., West, A. J. & Li, G. Sulphide oxidation and carbonate dissolution as a source of CO_2 over geological timescales. *Nature* **507**, 346–349 (2014).
2. Caves, J. K., Jost, A. B., Lau, K. V. & Maher, K. Cenozoic carbon cycle imbalances and a variable weathering feedback. *Earth and Planetary Science Letters* **450**, 152–163 (2016).
3. Volk, T. Sensitivity of climate and atmospheric CO_2 to deep-ocean and shallow-ocean carbonate burial. *Nature* **337**, 637–640 (1989).
4. Berner, R. A., Lasaga, A. C. & Garrels, R. M. The carbonate–silicate geochemical cycle and its effect on atmospheric carbon dioxide over the past 100 million years. *American Journal of Science* **283**, 641–683 (1983).
5. Lee, C.-T. A. et al. Continental arc–island arc fluctuations, growth of crustal carbonates, and long-term climate change. *Geosphere* **9**, 21–36 (2013).
6. Hirschmann, M. M. Comparative deep Earth volatile cycles: the case for C recycling from exosphere/mantle fractionation of major (H_2O, C, N) volatiles and from H_2O/Ce, CO_2/Ba, and CO_2/Nb exosphere ratios. *Earth and Planetary Science Letters* **502**, 262–273 (2018).

11

A Framework for Understanding Whole-Earth Carbon Cycling

CIN-TY A. LEE, HEHE JIANG, RAJDEEP DASGUPTA, AND MARK TORRES

11.1 Introduction

Climate, specifically Earth's surface temperature, is controlled by solar insolation, albedo, and the greenhouse gas content in the atmosphere.[1,2] Solar insolation relates to the solar radiative flux, which depends on changes in the sun's luminosity and the planet's distance from the sun. Albedo represents the fraction of the solar radiative flux reflected back into space, which is influenced by clouds, ice sheets, vegetation, the land/ocean ratio, and aerosols. That part of the solar flux not reflected back into space is absorbed by Earth's surface and re-radiated in the form of infrared energy. In the presence of certain gases that absorb infrared radiation, this re-radiated energy heats the atmosphere.

Radiative balance calculations show that, without greenhouse gases, Earth's average surface temperature would be well below the freezing point of water.[1] Geologic evidence for an active hydrologic cycle as far back as the early Archean suggests that Earth's surface temperature has not strayed wildly from the triple point of water, requiring the warming effects of greenhouse gases. Even in the first 2 billion years, when the sun's luminosity was substantially lower than today's, surface temperatures were still conducive to liquid water. What is remarkable, then, is that the amount of greenhouse gases in the atmosphere must vary in a way that buffers Earth's surface temperature, even when other factors like solar luminosity change. If greenhouse gases fall below some threshold, liquid water would freeze over globally, or if a threshold is exceeded, temperatures could rise to a level at which most of the water would be in the form of vapor, not unlike Venus today. This billion-year "stability" of Earth's climate indicates that greenhouse gas contents are regulated by an internal thermostat to maintain a clement climate.[2]

Superimposed on this billion-year climatic stability are shorter-term climatic fluctuations ranging from long-lived (10+ My) greenhouse and icehouse intervals[3,4] to <100-ky oscillations associated with Milankovitch orbital dynamics.[5] Icehouse conditions of the last 30 My have been characterized by large, permanent polar ice sheets (Figure 11.1a) that wax and wane with orbital cycles.[6] In contrast, the late Mesozoic to early Cenozoic greenhouse interval (150–50 Ma) was characterized by average surface temperatures >10°C higher than today and free of permanent ice sheets (Figure 11.1a).[4] Between 600 and 800 Ma, Earth may have temporarily experienced runaway icehouses, known as

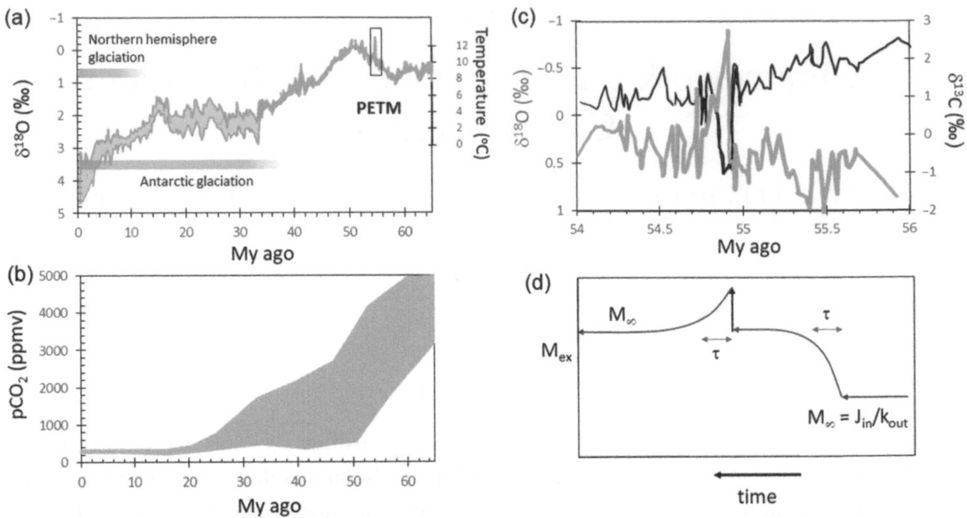

Figure 11.1 (a) Oxygen isotope record of seawater since 65 Ma with corresponding estimates of temperature on the right-hand axis (from Zachos et al.[4]). Onsets of Antarctic and northern hemisphere glaciations are shown. Box denotes the Paleocene–Eocene thermal maximum (PETM), which is expanded in (b). (b) Paleo-pCO_2 estimates for the last 65 My (adapted from Zachos et al.[4]). (c) Carbon and oxygen isotopic composition of seawater across the PETM (adapted from Zachos et al.[4]). (d) Schematic diagram of the mass of C in the exogenic system (M_{ex}) versus time. This diagram shows the concepts of response time (τ) and steady state.

"Snowball" events, where temperatures got cold enough for continental ice sheets to develop at low latitudes.[7,8]

Paleo-proxy data indicate that the partial pressure of atmospheric CO_2, the dominant greenhouse gas for most of Earth's history, is generally high during greenhouse times and low during icehouses (Figure 11.1b).[9,10] Thus, understanding long-term climate variability requires an understanding of what controls the C content of the *exogenic system*, which we define here as the sum of the ocean, atmosphere, biosphere, and the thin veneer of reactive soil or marine sediments that support life (Figures 11.2 and 11.3). These exogenic reservoirs cycle rapidly between each other on timescales of days to thousands of years, allowing for short-term variability of atmospheric CO_2. The exogenic C cycle operates on top of a much longer deep Earth C cycle, wherein the total amount of C in the exogenic system is controlled by fluxes into and out of Earth's interior, which we term the *endogenic system* (Figures 11.2 and 11.3).

The endogenic system includes C in the core, convecting mantle, and lithospheres (crust and lithospheric mantle). Endogenic C enters the exogenic system via volcanic degassing as well as metamorphic degassing and weathering of fossil organic C and carbonate. Carbon from the ocean, atmosphere, and biosphere is removed by carbonate precipitation and organic C burial, the former through an intermediate step of silicate weathering and the latter through photosynthesis. Subduction or deep burial of these carbonates and organic

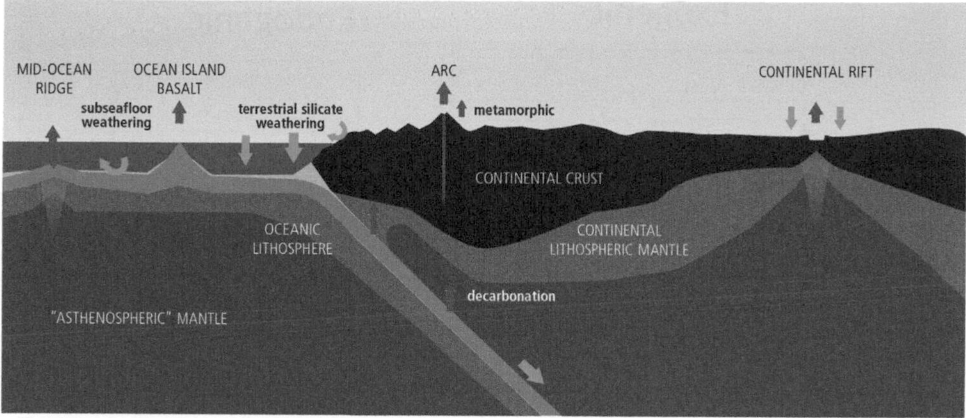

Figure 11.2 Cartoon (not to scale) showing how the whole-Earth C cycle relates to plate tectonics and the dynamics of the mantle. Orange arrows correspond to outgassing from Earth's interior. Blue arrows correspond to carbonate precipitation or organic C burial. Curved blue arrows represent terrestrial silicate weathering and seafloor weathering.

C is the mechanism by which exogenic C is transferred back into the endogenic system. The response time of the exogenic system to perturbations in C inputs or outputs is thought to be ~10–100 ky, which would imply that on My timescales, inputs and outputs are roughly balanced, so that the total exogenic C budget is near steady state.[27,28] The fact that Earth has fluctuated between long intervals of greenhouse or icehouse conditions implies that Earth transitions between different steady states, requiring that inputs from the endogenic system or the efficiency of weathering and organic C burial are not constant over My timescales. This chapter describes how C cycles between the endogenic and exogenic systems.

11.2 Basic Concepts of Elemental Cycling

11.2.1 Steady State and Residence Time

The rate of change of C mass in reservoir *i*, dM_i/dt, is given by

$$\frac{dM_i}{dt} = \sum J_{ji} - \sum J_{ij}, \tag{11.1}$$

where the first term on the right represents the sum of all mass flows of C from reservoirs *j* into *i* and the second term right represents the sum of all mass flows of C out of reservoir *i* to reservoirs *j*. More simply, (11.1) expressed as *total* inputs J_{in} and outputs J_{out} from a reservoir *M*, such as the entire exogenic system M_{ex} (ocean + atmosphere + biosphere), is given by

$$\frac{dM_{ex}}{dt} = J_{in} - J_{out}. \tag{11.2}$$

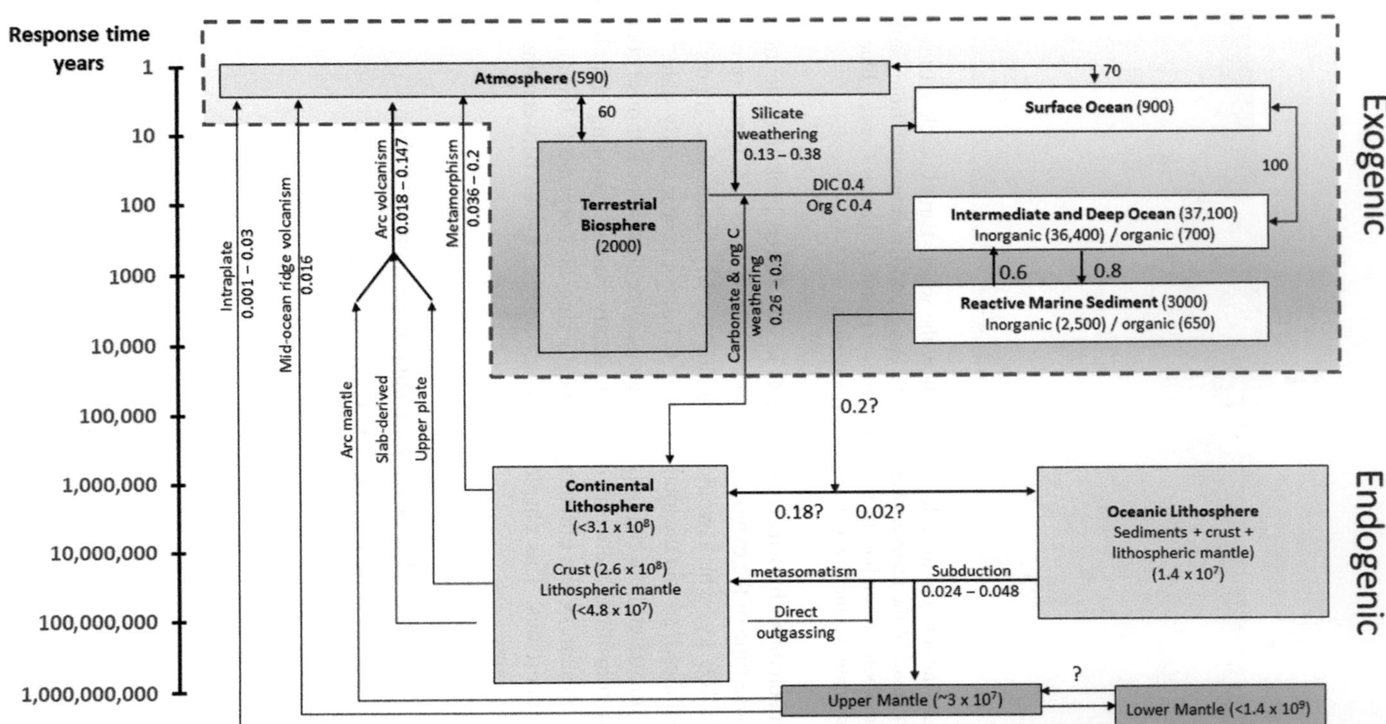

Figure 11.3 Box model describing the whole-Earth C cycle (excluding the core). Boxes within the red outlined region represent reservoirs within the exogenic system: the atmosphere, oceans, and the biosphere. All other boxes represent Earth's interior reservoirs and refer to the endogenic system.

To completely model the CO_2 content of the atmosphere (and not just the total exogenic system), both alkalinity and total C fluxes must be tracked, as the stoichiometry of these fluxes can vary considerably and their relative balance determines the proportion of exogenic C partitioned into the atmosphere. The treatment here, by ignoring this complexity, captures the first-order behavior of the system. J_{in} would then represent all inputs of C (volcanic and metamorphic degassing) and J_{out} the total C drawdown by silicate weathering (followed by carbonate burial) and organic C burial. J_{out} must scale with the amount of C in the system; after all, if there is no C in the exogenic system, then the output must be zero. At the most basic level, we can adopt a linear scaling, such that

$$J_{out} = k_{out} M_{ex}, \tag{11.3}$$

where k_{out} is the sum of all rate constants (s^{-1}) describing the efficiency of different pathways by which C is sequestered from the exogenic system

$$\left(k_{out} = \sum\nolimits_j k_{ij}\right). \tag{11.4}$$

Analogous to radioactive decay, k_{out} represents the probability that an atom of C is removed from the system per unit time. Because the output rate depends on the amount of C in the system, k_{out} represents a *negative feedback*, which essentially counteracts any change in C in the system (see Lasaga[29]). As we will discuss later, k_{out} represents a complex function of many processes. For example, C removal by silicate weathering and carbonate precipitation might be enhanced with greater water precipitation and temperature, both of which increase when atmospheric CO_2 increases.[2] The net effect is a negative feedback (Figure 11.4).

J_{in} must also scale with the mass of C from which the flux derives. If most of J_{in} derives from volcanic degassing, then $J_{in} = k_{in} M_m$, where M_m represents the mass of C in the mantle. If we assume that the mass of C in the mantle is much larger compared to that in the exogenic system, then J_{in} is approximately constant and (11.2) becomes a first-order differential equation:

$$\frac{dM_{ex}}{dt} = J_{in} - k_{out} M_{ex}, \tag{11.5}$$

Figure 11.3 (*cont.*) Arrows represent fluxes of C from one reservoir to another. Numbers by arrows represent fluxes in units of Gton C/y. Numbers in parentheses within each box refer to total reservoir mass of C (Gton C). Reservoirs are placed according to the estimated residence or response time of C in the reservoir (see vertical axis). The vertical extent of each box represents the range of plausible residence/response times. The horizontal size of each box is arbitrary. Image inspired by Sundquist and Visser.[11] C fluxes internal to the exogenic system are taken from Sundquist and Visser.[11–16] Silicate weathering flux is taken from Gaillardet et al.[17] Mantle degassing fluxes are from Dasgupta and Hirschmann,[18] Tucker et al.,[19] and other references discussed in the text. Metamorphic degassing data are from Kerrick and Caldeira.[20] Volcanic arc emissions are from Burton et al.[21] Cretaceous continental arcs are from Lee et al.[22] and Lee and Lackey.[23] Mt. Etna data are from Allard et al.[24] Organic C weathering data are from Petsch,[25] from which carbonate weathering rates are calculated. Anthropogenic emissions are from Friedlingstein et al.[26]

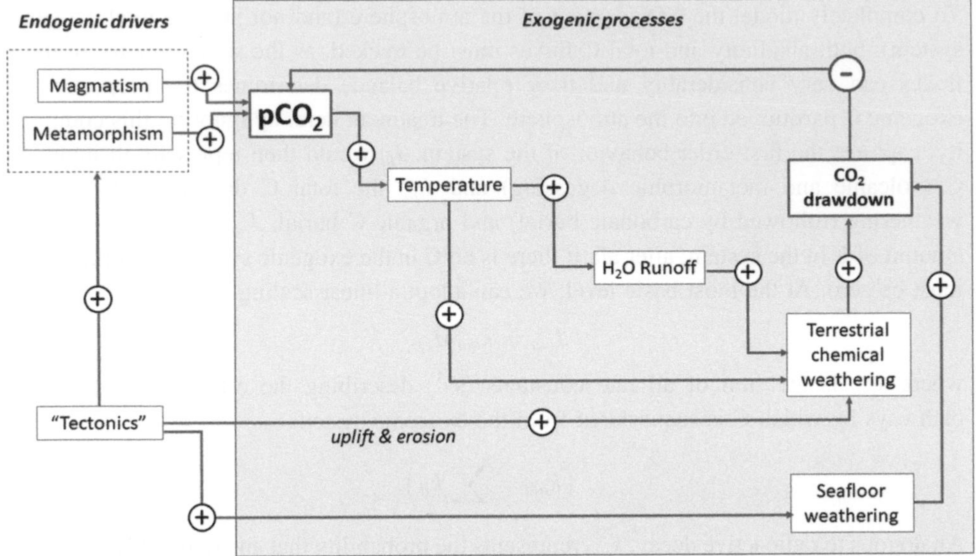

Figure 11.4 Feedback loop describing the silicate weathering feedback acting on the exogenic C system. Arrows correspond to transfer functions, with positive symbols indicating positive feedback and negative symbols representing negative feedback. Increases in pCO_2 increase temperature, which leads to enhanced precipitation and runoff, accelerating chemical weathering rates and drawdown of CO_2. Similarly, an increase in pCO_2 increases seafloor weathering rates, increasing drawdown of CO_2. Tectonics and mantle dynamics drive the entire system by: (1) increasing erosion rates during mountain building, thereby increasing the availability of weatherable substrate and drawdown rates of CO_2; and/or (2) increasing magmatic or metamorphic inputs of CO_2 into the exogenic system. Whether tectonics enhance degassing or drawdown will depend on the nature of the tectonic activity.

with the following solution:

$$M_{ex}(t) = \frac{J_{in}}{k_{out}} + \left(M_{ex}^{o} - \frac{J_{in}}{k_{out}}\right) \exp\left(-k_{out}t\right). \tag{11.6}$$

Here, M_{ex}^{o} represents an initial perturbation to the amount of C in the exogenic system. Equation (11.6) shows that the system will return exponentially to a steady state, given by

$$M_{ex}^{\infty} = J_{in}/k_{out}. \tag{11.7}$$

At steady state, inputs and outputs are balanced, hence the C content of the exogenic system does not evolve (steady state does *not* mean thermodynamic equilibrium). The time for the system to relax back to steady state after a perturbation (*response time*) is roughly the e-fold timescale of the system; that is:

$$\tau_{ex} = 1/k_{out}. \tag{11.8}$$

For a linear feedback, the response time is equivalent to the *residence time* of C in the system (Figure 11.1d). The residence time represents the average time a C atom remains in the exogenic system when the system is at steady state, $dM_{ex}/dt = 0$. At steady state, $J_{in} = J_{out}$, such that the residence time of C in the system is given by

$$\tau^r_{ex} = \frac{M_{ex}}{J_{in}} = \frac{M_{ex}}{J_{out}} = \frac{M_{ex}}{k_{out}M_{ex}} = \frac{1}{k_{out}}. \quad (11.9)$$

When negative feedbacks operate, a system perturbed suddenly, such as by a pulse of CO_2 outgassing, will eventually return to a steady state, defined by the ratio of the external forcings (e.g. the input) to the efficiency of the negative feedback, k_{out} (11.6). On timescales longer than the response time of the system, the C in the system should evolve toward a steady state, after which the only processes that can change the C content in the exogenic system would be temporal changes in the forcings, such as changes in magmatic or metamorphic activity, or changes in the efficiency of carbonate precipitation, which may relate to changes in tectonics and the efficiency of silicate weathering.

The response time for C in the exogenic system has been determined in two ways. One approach is to take the mass of C in the exogenic system and divide it by the measured inputs or outputs to obtain residence times. This approach is only valid if the measured inputs or outputs truly represent steady-state conditions. Another approach is to measure the relaxation time of a perturbed system, with one example being the C isotope excursion observed at the Paleocene–Eocene thermal maximum (PETM).[4] Both approaches suggest that C in the exogenic system has a response time between 10 and 100 ky (Figure 11.1c and d).[4] Thus, on My timescales, the C content of the exogenic system should be close to steady state. However, the fact that atmospheric CO_2 nevertheless changes on My and Gy timescales means that tectonic/magmatic forcings or the global efficiency of the negative feedbacks must change with time as Earth evolves. In other words, Earth's exogenic system is characterized by time-varying steady states or a baseline C content and climate. On short timescales, the exogenic system is not strongly influenced by the deep Earth, but on My timescales, the total budget of the exogenic C is controlled by the deep Earth through degassing and weathering.

11.2.2 Climatic Drivers versus Negative Feedbacks

The most important feature of negative-feedback systems is that steady state, if given enough time, will eventually be attained. Without a negative feedback, where the output does not depend on the amount of C in the exogenic system, a system is unlikely to attain steady state. In such a system, the C content of the exogenic system would rise indefinitely if inputs exceeded outputs or eventually decline to zero if outputs exceeded inputs. Earth's climate has changed through time, but liquid water appears to have been present for almost all of Earth's history, implying that inputs and outputs must be balanced on long timescales, which in turn implies that C in the exogenic system and climate is controlled by a negative-feedback mechanism. Any imbalances in inputs and outputs must be transient and occur on timescales shorter than the response time of exogenic C to perturbations (Figure 11.1d).[27]

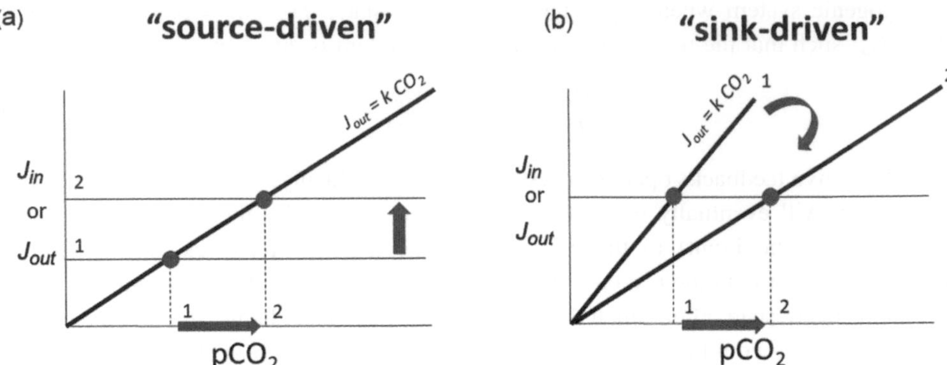

Figure 11.5 Inputs J_{in} and outputs J_{out} as a function of pCO_2. Geologic inputs are assumed not to depend on exogenic C and are thus represented as horizontal lines. Output rate depends on pCO_2, with the slope k representing the sensitivity of the negative weathering feedback. The intersection between J_{in} and J_{out} represents steady state, where $J_{in} = J_{out}$. The pCO_2 at this intersection represents steady-state pCO_2. In (a), we show that steady-state pCO_2 increases by increasing J_{in} while holding the weathering feedback constant (red arrows from states 1 to 2 show an increase in steady-state pCO_2). In (b), we hold inputs constant but decrease the sensitivity of the weathering feedback. This causes steady-state pCO_2 to increase without any change in the input.

Baseline or steady-state C in the exogenic system is driven by the inputs and modulated by the efficiency of the outputs. The greater the kinetic rate constant for C removal from the exogenic system, the more sensitive the negative feedback. In a steady-state system, this concept is illustrated well in a graph of C input or output versus total C in the exogenic system.[30,31] For a linear system, the sensitivity of the negative feedback is the slope of the output as shown in Figure 11.5. Because the input of C from the deep Earth does not depend on the amount of exogenic C, the output function is represented by a horizontal line in Figure 11.5a. The intersection between the two lines defines the steady state. For a given sensitivity, k, the only way to change steady-state exogenic C is to change inputs. If inputs are instead held constant, exogenic C can only change by changing the strength (slope) of the weathering feedback. If exogenic C were to suddenly decrease below the system's steady state, inputs would temporarily exceed outputs, returning exogenic C contents back to steady-state values. Conversely, if exogenic C contents were to suddenly increase, outputs would then exceed inputs, eventually returning the system to steady state. These concepts are shown in Figure 11.5b by keeping track of the intersection of the input and output functions. If the system were to become more efficient at C drawdown (higher k), such as if Earth was characterized by a global increase in mountain building and erosion, steady-state exogenic C would decrease. If drawdown efficiency were to decrease, such as through decreased weatherability of continents, steady-state exogenic C would increase without any change in the input (Figure 11.5b). Most importantly, if the negative-feedback efficiency is strong, then large changes in the inputs are required to change exogenic C, which is to say that the exogenic C content is buffered. This is why Earth's climate, while variable over time, has rarely veered to extreme climatic conditions.

It is often debated whether Earth's climate and exogenic C is input or sink driven. This debate is not about whether inputs or outputs are in balance because, on long timescales, they are in balance.[32] The crux of this debate is whether changes in the C content of the exogenic system are controlled by changes in input or changes in the efficiency of the negative feedback (Figure 11.5). The former involve changes in magmatic, metamorphic, or weathering-related outgassing rates, while the latter involve changing the efficiency of weathering and organic C burial.

11.2.3 When Systems Transition to New Steady States

Despite the overall stability of Earth's climate, there are examples when Earth or other planets have veered toward extreme climatic states. The Snowball climate of the Neoproterozoic, when Earth is thought to have frozen over, is an example of runaway cooling.[8] Mars represents an example of a permanent icehouse, with surface temperatures of –55°C and no active hydrologic cycle.[33,34] Permanent greenhouse characterizes Venus, where surface temperatures are 460°C and the atmosphere is made of 96% CO_2.[33,34] Earth itself may have undergone short-lived excursions to hothouse conditions. These include the PETM and the various hyperthermals around that time.[4,35,36]

In a linear system, runaway processes cannot happen. However, negative-feedback mechanisms need not scale linearly with exogenic C content. For example, if the rate-limiting step for silicate weathering and precipitation of carbonates becomes the rate at which fresh new rock is exposed to the atmosphere, one could envision a "threshold" of atmospheric CO_2 content above which further increases in CO_2 do not result in increased weathering. Similarly, a scenario could be envisioned in which surface temperatures drop to a threshold below which weathering kinetics become negligible, and because surface temperature is in part controlled by the amount of greenhouse gases, this translates to a threshold of atmospheric CO_2 below which weathering rates are insensitive to CO_2. Thus, the functional form for the negative feedback might involve low-sensitivity regimes at low and high atmospheric CO_2 separated by a sensitive, quasi-linear regime at intermediate CO_2 contents (Figure 11.6). For a scenario in which C inputs to the exogenic system do not depend on atmospheric CO_2, one can track how steady-state exogenic C varies as inputs vary. From Figure 11.6, we see that when inputs rise toward an upper threshold, steady-state exogenic C contents can rise uncontrollably, leading to a runaway greenhouse. If inputs decrease below some lower threshold, steady-state exogenic C contents will plummet, resulting in runaway icehouse conditions. Crossing these thresholds may move the planet to a different steady state, where a new type of negative feedback operates (Figure 11.7). Understanding the functional form of global weathering feedbacks is thus critical to understanding where these thresholds lie in terms of exogenic C.

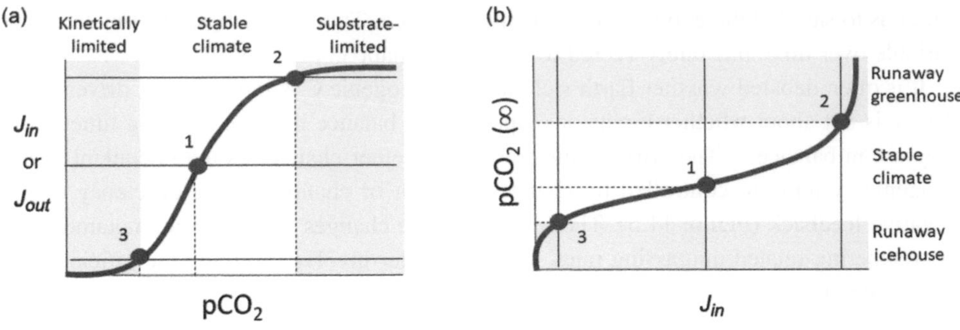

Figure 11.6 (a) Input J_{in} and output J_{out} versus pCO_2. The conceptual model is identical to that of Figure 11.5, but the functional form of the weathering feedback J_{out} is nonlinear, consisting of low-sensitivity regimes at low pCO_2 and high pCO_2. These low-sensitivity regimes are referred to here as kinetically limited and substrate-limited, respectively. At low pCO_2, temperatures are low and kinetics of dissolution are sluggish. At high pCO_2, reaction is limited by the availability of new substrate for weathering. Between these two low-sensitivity regimes lies a quasi-linear high-sensitivity regime, where climate is stable. Intersections of J_{in} (thin horizontal lines) with J_{out} (thick bold line) are denoted by red circles. (b) Steady-state pCO_2 in (a) is plotted versus J_{in}. When inputs rise to some threshold (e.g. from states 1 to 2), further increases in J_{in} will lead to rapid rises in pCO_2. Conversely, when inputs drop below some threshold (states 1 to 3), further decreases in J_{in} will lead to rapid declines of pCO_2. Hothouse and icehouse excursions are controlled by how close the system's baseline climate is to the threshold.

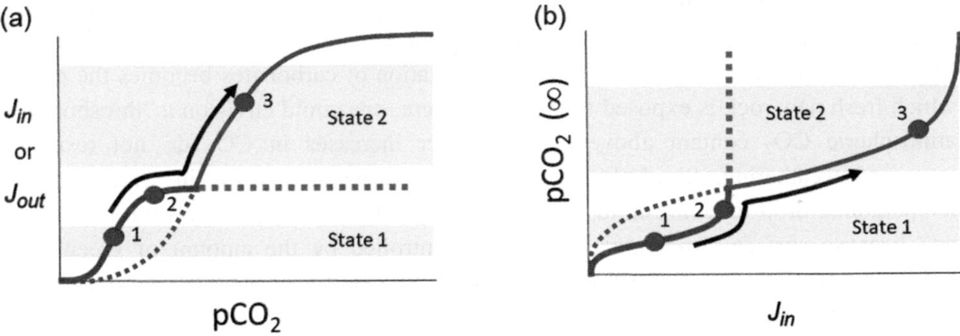

Figure 11.7 (a) and (b) are schematic diagrams that are conceptually identical to those of Figure 11.6, except that two different functional forms of the weathering feedback J_{in} are shown. The two different weathering feedback mechanisms are assumed to operate at different pCO_2 values such that the system can transition into a fundamentally new state after it crosses a pCO_2 threshold (e.g. from states 1 to 2 to 3).

11.3 Carbon Inventories of Earth Reservoirs

11.3.1 Modern and Primitive Mantle Reservoirs

Excluding the core, more than 99.99% of all C on Earth is in the mantle and crust, although exact quantities are highly uncertain.[18] Estimates of the C budget of the uppermost mantle are

Table 11.1 *Reservoirs of carbon*

Reservoirs	Concentration	Gton C	Refs.
Atmosphere	278 ppmv	590	
Oceans		**$(3.7–3.9) \times 10^4$**	[12,13]
Surface layer – inorganic		700–900	[13,14]
Deep layer – inorganic		$(3.6–3.8) \times 10^4$	[12,14]
Total organic		600–700	[15,16]
Reactive marine sediment		**3150**	[12]
Inorganic		2500	[12]
Organic		650	[12]
Terrestrial biosphere and soils		**$(2–2.3) \times 10^3$**	[12,14]
Vegetation		500–600	[12]
Soil		$(1.5–1.7) \times 10^3$	[12]
Continental crust (CC)			
Total CC		**4.2×10^7**	[58]
Total CC		**2.6×10^8**	[59]
Sedimentary (carbonate + organic C)		3.4×10^7	[58]
Igneous and metamorphic rocks		0.8×10^7	[58]
Continental lithospheric mantle	<760 ppm	**$<4.8 \times 10^7$**	This study (= BSE – CC – DM – ROL)
Oceanic crust and lithospheric mantle		**1.4×10^7**	[18]
Mantle (total)			
DM (<670 km)	30 ppm C	3.2×10^7	[19]
Upper mantle (assuming BSE composition)	<350 ppm C	$<3.7 \times 10^8$	This study
Lower mantle (assuming BSE composition)	<350 ppm C	$<1.0 \times 10^9$	This study
BSE	<350 ppm C	$<1.4 \times 10^9$	This study
Fossil fuel reserves		**5×10^3**	[12]

BSE = bulk silicate Earth; DM = depleted mantle; ROL = recycled oceanic lithosphere.

indirectly calculated from the C content of mid-ocean ridge basalts (MORBs), with C itself inferred by bootstrapping to other elements, such as Nb, Ba, H, or noble gases.[37–39] These studies have resulted in estimates of 16–300 ppm C for the MORB source mantle, typically assumed to represent the upper mantle (Figure 11.3). However, reconstructing mantle C contents is faced with the challenge of correcting for degassing of C and other elements. Tucker et al.[19] and Hauri et al. (Chapter 9 of this book) provide the most recent and rigorous reconstructions of the MORB source region and arrive at 30 ppm C (Table 11.1). This estimate

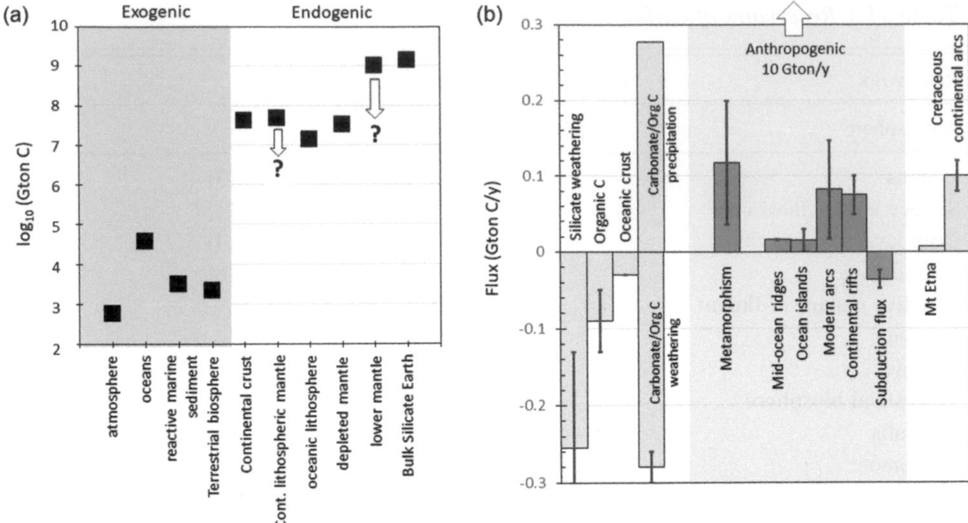

Figure 11.8 (a) Estimates of total C content (Gton C) in various exogenic and endogenic reservoirs. See Table 11.1 and text for details. Estimates of the continental lithospheric mantle and bulk silicate Earth (BSE) are from this study. Downward-pointing arrows denote that these estimates represent maximum bounds. Depleted mantle volume corresponds to mantle above 670 km with C concentration equal to that inferred for the MORB mantle source. Lower mantle corresponds to mantle between 670 km and the core–mantle boundary with a C concentration equivalent to BSE. (b) Graphical representation of C fluxes (Gton C/y) with inputs as positive values and outputs as negative values. Silicate weathering, organic C burial, and seafloor weathering are shown in yellow. Inputs of CO_2 via weathering of organic C and carbonates are combined (negative bars), but such weathering is thought to be balanced by rapid re-precipitation of carbonate and burial of organic carbon (positive bars). Endogenic outgassing is shown in green, while subduction is shown in blue. Degassing through Mt. Etna and Cretaceous continental arcs is shown in order to illustrate the importance of carbonate-intersecting continental arcs. Vertical arrow points to anthropogenic production of CO_2 through fossil fuel burning and cement production (10 Gton C/y).

agrees well with previous estimates of ~15–50 ppm C for the MORB source mantle that used undegassed or minimally degassed MORB CO_2 contents and CO_2/Ba, CO_2/Nb, and H/C ratios (e.g. Refs. 37, 38, 40, and 41). A homogeneous upper mantle (<670 km; mass of 1.06×10^{24} kg[42]) with a MORB source C content would contain ~3.2×10^7 Gton C (Figure 11.8). If the MORB source was representative of the entire mantle down to the core–mantle boundary (4.0×10^{24} kg), the whole mantle would contain 1.2×10^8 Gton C.

Of interest is the C content of the bulk silicate Earth (BSE), which is the primordial mantle prior to extraction of the continents, oceans, and atmosphere (Figure 11.9a). The present mantle represents what remains of the mantle after degassing and silicate differentiation, which we refer to as the Depleted Mantle (DM); the composition of the DM is assumed to be represented by the MORB source mantle.[43,44] Thus, the BSE can be calculated by re-homogenizing the atmosphere, oceans, and continents back into the DM, but this is challenging if the C contents of all of these reservoirs and their sizes are

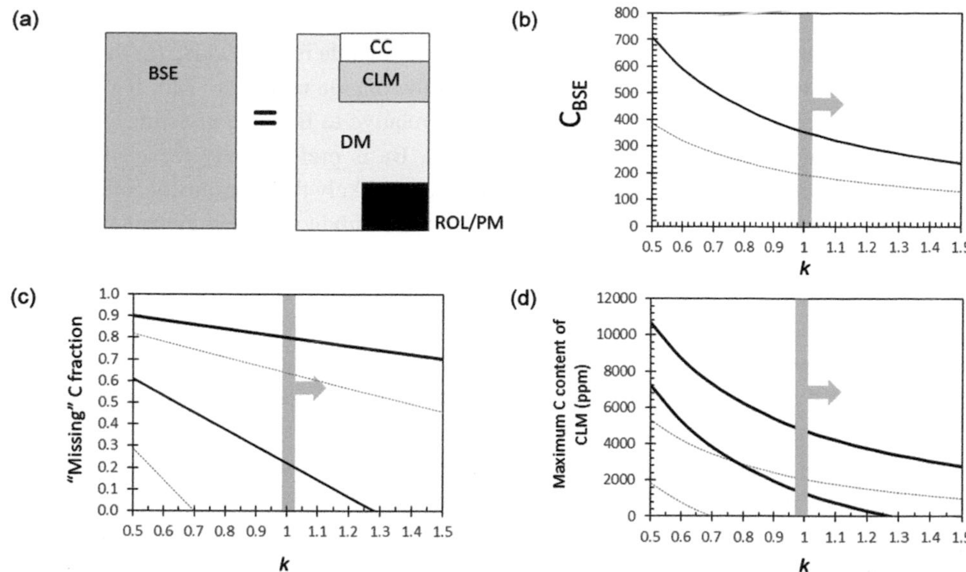

Figure 11.9 (a) Cartoon showing how C concentration in BSE is determined. BSE is the hypothetical primordial mantle composition after combining the DM, continental crust (CC), continental lithospheric mantle (CLM), and all other reservoirs, such as recycled oceanic lithosphere (ROL), as well as oceans, atmosphere, etc. (latter not shown because of small size). (b) C content of BSE as constrained by relative depletion of Ba and U in DM. Horizontal axis represents the relative efficiency k by which C is retained (during melting) or recycled (during subduction) into the mantle compared to Ba or U; k is likely larger than 1. (c) Missing fraction of C from the DM after subtraction of C in continental crust (DM is assumed to be the size of the upper mantle down to 670 km). Thick black lines correspond to estimates based on C/Ba-constrained BSE and thin blues lines are based on C/U-constrained BSE. Two sets of lines for each color are denoted and correspond to calculations using the crustal budgets of C from Wedepohl[58] and Gao et al.[59] This missing C can be stored in the CLM or as recycled components in the mantle. (d) Maximum concentration of C in CLM, assuming all missing C is the CLM.

not known well. A common approach, used for refractory lithophile elements, is to extrapolate magmatic differentiation trends to chondritic ratios, but this cannot be done for C because of its volatility during nebular condensation/planetary accretion and siderophile nature during core formation (e.g. Ref. 45).

Instead, we infer the C of BSE from elements that behave geochemically like C but are refractory in a cosmochemical sense. For example, during mantle melting, C is thought to behave like Ba, a refractory element.[19,41] The depletion of Ba in the DM (0.563 ppm[46]) relative to the BSE (6.6 ppm[47]) – that is, Ba_{DM}/Ba_{BSE} – is ~0.085. If Ba and C also behave similarly during subduction, then the relative depletion of C in the MORB source is equivalent to that of Ba. The BSE concentration of C is then calculated by dividing the C concentration of the DM by the Ba depletion factor ($C_{BSE} = C_{DM}/(Ba_{DM}/Ba_{BSE})$), yielding ~350 ppm C for the BSE (Figure 11.9b). If C does not behave identically to Ba, then a correction factor must be applied, $C_{BSE} = C_{DM}/(k_{C/Ba} \times Ba_{DM}/Ba_{BSE})$, where $k_{C/Ba}$ represents the efficiency by which C is returned to the mantle relative to Ba (or the relative efficiency by which C is retained or recycled into the mantle relative to Ba). If $k_{C/Ba} > 1$, C is preferentially subducted deep into the mantle relative to Ba and our estimates of C in the BSE are hard maximum bounds. If $k_{C/Ba} < 1$, Ba is preferentially recycled and our estimates are minimum bounds (Figure 11.9b). It seems likely that Ba, mostly concentrated in slab sediments, is more efficiently removed from the subducting slab given the extreme enrichments of Ba in arc lavas.[48] While some of the carbonate in subducting sediments might also be removed efficiently,[49] organic C and carbonate in the oceanic crust are likely retained,[50,51] so it seems likely that $k_{C/Ba} > 1$, and thus $C_{BSE} < 350$ ppm. If C during melting behaves more like U, a refractory element with a DM depletion factor of ~0.16 relative to BSE (Figure 11.9b),[46] $C_{BSE} <$ ~192 ppm. In any case, C_{BSE} can be no higher than <350 ppm (Table 11.1), consistent with a recent estimate of C_{BSE} of ~140 ppm.[52] These values of C_{BSE} would imply that the DM reservoir has lost no more than 40–90% of its original C during silicate differentiation; however, the DM reservoir is unlikely to represent the entire convecting mantle, so the fractional loss of C from the whole mantle must be far smaller. We note that the C concentration of the mantle source of ocean island basalts (OIBs) ranges from 33 to 500 ppm C.[18] The source region of OIBs is thought to represent primitive undegassed mantle and/or enriched domains associated with recycled crust.[53,54]

11.3.2 Continental Crust and Continental Lithospheric Mantle

Carbon in the continental crust is represented by ancient sedimentary rocks as well as carbonate and graphite stored in metamorphic and igneous basement rocks. The sedimentary rocks are dominated by carbonates (limestones and dolostones) with organic C-rich sediments making up a smaller fraction.[55,56] The carbonate content of the continental crust is calculated by estimating the volume of carbonate sediments in the continents.[57,58] Due to high variability in the organic C content of sediments, the average organic C content of continents is usually estimated by assuming that organic C makes up ~20% of the total C budget. This approach is based on the assumption that the C isotopic composition of seawater, often assumed to represent the fraction of organic C (f_{org}), has remained relatively constant over most of Earth's history. The C content of the rest of the continental crust is estimated from measurements of basement rocks and extrapolated to the upper, middle, and lower crusts based on a model of the compositional structure of the crust. Because of the high degree of variability in carbonate content of igneous and metamorphic rocks, any estimate of the C content of the nonsedimentary part of the continental crust is fraught with large errors. For example, Wedepohl[58] estimated that the continental basement amounted to one-fifth of the C in sedimentary rocks, yielding a total continental crust budget of 4.2×10^7 Gton C. A more comprehensive study of basement rocks in China found higher concentrations of C in igneous and metamorphic rocks, leading to a much higher estimate of the total continental crust budget of 2.6×10^8 Gton C.[59]

We now turn to the continental lithospheric mantle (CLM), the cold and hence rheologically strong part of the mantle underlying the continental crust, which extends down to depths of ~100 km beneath Phanerozoic terranes and up to ~200–250 km beneath many stable cratons.[60–63] Because of their Gy stability,[64–67] CLM is subjected to numerous metasomatic events, which could lead to hydration or carbonation.[68–75] We can place an upper bound on how much C could be in the CLM. If we assume that the volume of DM is equivalent to the entire upper mantle down to 670 km depth (1.06×10^{24} kg), the total amount of C degassed is $<3.4 \times 10^8$ Gton C as inferred above from C/Ba systematics and $>1.7 \times 10^8$ Gton C from C/U systematics. From these quantities we subtract the amount of C in the continental crust. If we use the higher crustal abundances of Gao et al.[59] and the DM C content inferred from C/Ba, we are left with 20% of unaccounted C that must be housed in the CLM or as recycled components deep in the mantle (Figure 11.9c). If we use the crustal abundances of Wedepohl,[58] 80% is unaccounted for. If we use the lower estimates of C in DM as inferred from C/U systematics, we find that the C budget of the continental crust accounts for or even exceeds the amount of C degassed from the upper mantle, which seems unreasonable; reconciling this discrepancy would require a volume of mantle greater than the upper mantle to have degassed to the extent of the DM reservoir. In any case, if we restrict the DM to the upper mantle and use the C/Ba-constrained DM composition and Gao et al.'s crustal estimate, we find a maximum missing C of 4.8×10^7 Gton C. All of this C in the CLM would translate to a hard upper bound of ~740 ppm C in the CLM, assuming an average CLM thickness of ~150 km (Figure 11.9d and Table 11.1). No doubt, some of the missing C is in the form of recycled crust deep in the mantle, as balancing present-day arc flux requires a small fraction of subducted C to be supplied to sub-arc mantle source regions via fluids or melts (e.g. Ref. 76). Whether the missing C is in the CLM[49] or recycled into the deep mantle[52] is currently debated.

11.3.3 Exogenic Reservoirs

For completeness, we briefly discuss the C budget of the exogenic system (Table 11.1). In total, the exogenic system accounts for <0.01% of Earth's C budget (excluding the core), with reservoir sizes increasing as follows (Figure 11.3): atmosphere < terrestrial biosphere and soil < reactive marine sedimentary carbon < ocean.[11] For a modern preindustrial atmospheric CO_2 content of 280 ppmv, the total amount of pre-anthropogenic C in the atmosphere today is 590 Gton or ~1.4% of the total C in the exogenic system.[11] The oceans account for ~87% of the total exogenic C, with most of the C in the oceans in the form of HCO_3^-. The terrestrial biosphere and soils account for ~4% and the reactive marine sedimentary reservoir accounts for ~8% of the total exogenic C. It is possible that the reactive marine sedimentary reservoir is substantially larger than noted here if the size of gas-hydrate reservoirs has been underestimated.[77]

In summary, although the amount of C in the exogenic system is nearly negligible compared to that of the bulk Earth, all C degassed from the interior of Earth passes through the exogenic system before being buried as carbonates or organic C, and it is through the exogenic system that C influences climate.

11.4 Long-Term Carbon Fluxes

11.4.1 Inputs

11.4.1.1 Volcanic Inputs

On long timescales, the exogenic system is supported by C inputs from the endogenic system through volcanism, metamorphism, and weathering of ancient carbonates and organic C. Outputs from the exogenic system occur through carbonate and organic C burial via the intermediate steps of silicate weathering and photosynthesis. Ultimately, some fraction of carbonate and organic C is removed from the exogenic system by deep burial in the crust or subduction back into the mantle (Figure 11.3).

Fluxing of C into and out of the exosphere can occur through many different species of C (CO_2, CH_4, CO, etc.). We denote all C fluxing, regardless of species, in terms of Gton of elemental C per year. In the case of volcanism and metamorphism, the dominant C species today is CO_2. The most widely used estimate of global volcanic degassing is that of Marty and Tolstikhin.[78] Based on various approaches to estimating CO_2 fluxes through mid-ocean ridges (MORs), arcs, and intraplate magmas (Figure 11.10), they estimated a global volcanic C flux of 0.03–0.13 Gton C/y, with MORs > arcs > intraplate magmatism. These estimates are highly uncertain. In both cases, C fluxes were estimated by measuring $C/^3He$ ratios in gases or hydrothermal plumes and then multiplying by the ^3He flux, which is either measured or calculated. Regardless of these uncertainties, the Marty and Tolstikhin estimates are lower bounds because they did not consider diffuse fluxes, which Burton et al.[21] showed to be significant. Burton et al. estimate that the total subaerial flux, dominated by arc volcanoes, could be as high as 0.15 Gton C/y, significantly greater than the ~0.030 Gton C/y arc flux estimated by Marty and Tolstikhin (Figure 11.8). Diffuse degassing through faults in continental rifts can also be important (Figure 11.10b); diffuse degassing in the East African Rift alone currently emits 0.05–0.10 Gton C/y, comparable to the flux emitted through the entire length of MORs (Figure 11.8).[79] It is also important to consider that some anomalous volcanoes like Mount Etna, whose magmas have interacted with crustal carbonates, have unusually high emissions (~0.006 Gton/y for Mount Etna[24]), with one volcano accounting for a substantial proportion (>10%) of the modern arc flux (Figure 11.8).

Global volcanic inputs of C are thus at least 0.54 Gton C/y if diffuse degassing estimates from Burton et al.[21] are considered. Global volcanic flux of C through geologic time, however, is likely to vary significantly due to variations in plume activity or global oceanic crust production rates.[80,81] It has also been shown that continental arcs, because of magmatic interactions (via assimilation, melt-rock reaction, contact, and regional metamorphism[82–85]) with ancient crustal carbonates stored in the continents emit significantly more CO_2 than island arcs (Figure 11.10c and d); hence, the waxing and waning of

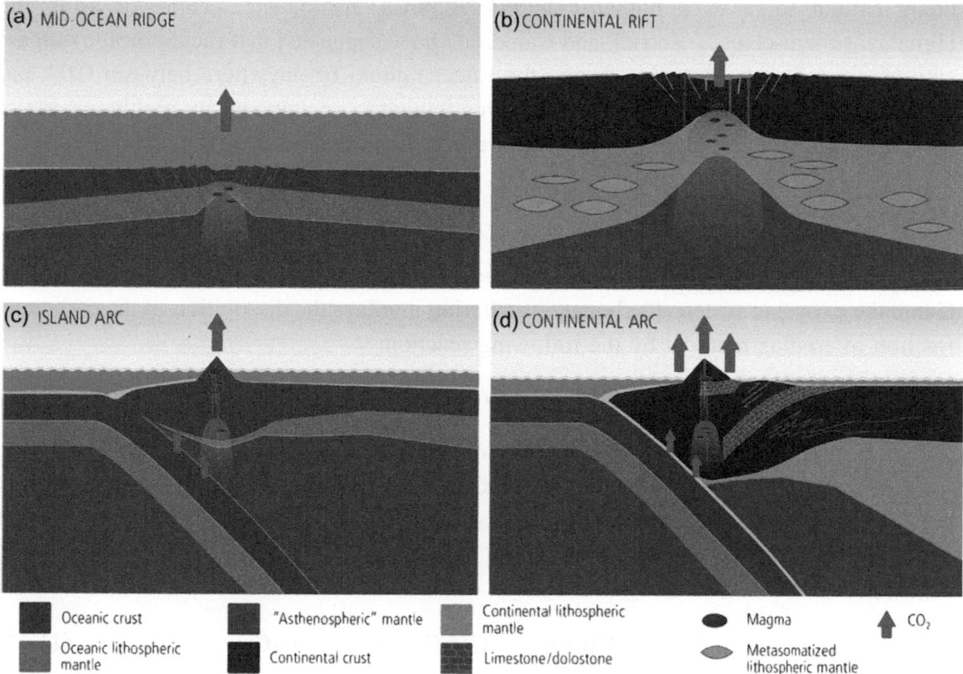

Figure 11.10 The four most important regions of magmatically related degassing (not to scale): (a) MORs, (b) continental rifts, (c) island arcs, and (d) continental arcs. In continental rifts, metasomatized domains within the CLM may contain excess C in the form of carbonate or graphite. Destabilization of such C can enhance magmatic flux during rifting. In continental arcs, magmatic degassing can be enhanced by decarbonation of crustal carbonates.

continental arcs through time will also lead to increases in volcanic CO_2 fluxing.[22,23,86–88] Enhanced continental rifting could also lead to enhanced CO_2 production.[79,89,90]

11.4.1.2 Metamorphic Inputs

Metamorphic inputs of C involve thermal breakdown of crustal carbonates. The dominant decarbonation reactions occur by reaction of carbonate with silica:

$$CaCO_3 + SiO_2 = CaSiO_3 + CO_2, \tag{11.10}$$

to form calc-silicate minerals, such as wollastonite and other pyroxenes.[91,92] These reactions operate at temperatures (400–600°C) that can be achieved over large areas during regional metamorphism. The presence of water, which decreases CO_2 activity, further decreases the temperatures needed for reaction (300°C).[93] Metamorphic decarbonation can thus be pervasive in subduction zones, continental collisions, or continental extension.

Estimates of metamorphic CO_2 degassing rates, particularly on a global scale, are challenging to quantify because metamorphic degassing is diffuse and spatially heterogeneous. It has been shown that the metamorphic CO_2 flux in the Himalayas may be large, being similar to or even higher than that removed by silicate weathering from the Himalayas themselves.[94] Kerrick and Caldeira[20] have suggested that metamorphic outgassing associated with orogenies during the Eocene could be anywhere between 0.04 and 0.20 Gton C/y (Figure 11.8). Importantly, even though these numbers are highly uncertain, their magnitudes are potentially equal to or greater than that of MOR degassing.

11.4.1.3 Carbonate and Organic Carbon Weathering

Weathering of ancient crustal carbonates and organic C represents another large input of C into the exogenic system. Carbonate weathering involves the dissolution of the carbonate fraction in eroding material by the following reaction

$$CaCO_3 + CO_2 + H_2O = Ca^{2+} + 2HCO_3^-. \quad (11.11)$$

For each mole of CO_2 consumed, dissolution releases 2 moles of bicarbonate (HCO_3^-), resulting in a net contribution of 1 mole of CO_2 to the ocean–atmosphere system. However, because carbonate dissolution introduces Ca^{2+} and Mg^{2+} ions as well as bicarbonate ion into the ocean (in silicate weathering, CO_2 is consumed), precipitation of carbonate follows:

$$Ca^{2+} + 2HCO_3^- = CaCO_3 + CO_2 + H_2O. \quad (11.12)$$

On timescales longer than ~10 ky, carbonate dissolution and precipitation balance, such that carbonate weathering is not thought to significantly influence the long-term evolution of exogenic C.[3]

Another source of CO_2 is weathering of organic C, wherein a reduced form of C is oxidized by the atmosphere to form oxidized C; that is, CO_2:

$$O_2 + \text{“}CH_2O\text{”} = CO_2 + H_2O. \quad (11.13)$$

Weathering of “black” shales would be an example of oxidative weathering of organic C. Like carbonate weathering, oxidative weathering of organic C is assumed to be balanced on short timescales by organic C burial:

$$CO_2 + H_2O = O_2 + \text{“}CH_2O\text{”}, \quad (11.14)$$

so that on long timescales, the net effect of organic C weathering on exogenic C is often assumed to be negligible.[3] Such a close balance between organic C fluxes is presumably regulated by atmospheric O_2 concentrations. For example, increases in the oxidative weathering of organic C would lead to lower atmospheric (and seawater) O_2 concentrations, which are thought to enhance organic matter burial. However, there is still uncertainty as to whether the organic C cycle is in balance because many of the redox fluxes are not well constrained. A full treatment of organic C weathering must be coupled with the global oxygen cycle.[3,95]

Although weathering of carbonates and organic C are likely balanced on timescales >100 ky, it is nevertheless important to appreciate their magnitudes (Figure 11.8).

The oxidative weathering rate of organic C is thought to be ~0.05 Gton C/y,[25] and global compilations and river chemistry indicate a modern carbonate weathering flux of 0.148 Gton C/y (these calculations assume stoichiometric weathering).[17] Together, these fluxes are comparable to the total volcanic flux of CO_2 today. To illustrate the magnitude of the carbonate weathering flux, it is now established that the alkalinity in global rivers is dominated by carbonate weathering rather than silicate weathering[17,96,97] due to the fast dissolution kinetics of carbonates compared to silicates. Because these weathering fluxes are large, any imbalances in weathering of carbonates and organic C versus carbonates and organic C burial could lead to large swings in exogenic C on short timescales.[98] The magnitudes and timescales of such swings are ultimately limited by the fact that ocean chemistry adjusts to maintain a relatively constant saturation state with respect to calcium carbonate phases.[99] Estimates of the response timescale of carbonate buffering in the ocean are around 6000 years.[99] Such weathering probably cannot be truly ignored even on >100-ky timescales, especially if the ratio of organic C to carbonate weathering (or deposition) varies with time or if the distribution of carbonate and organic C deposition changes with time (deep sea versus continental shelves). Finally, it has recently been shown that oxidative sulfide weathering in organic-rich sedimentary rocks can lead to a net release of CO_2 if the resulting sulfuric acid reacts with nearby carbonates.[100] Unlike carbonate (by carbonic acid) or organic C weathering, sulfide weathering can serve as a long-term input of CO_2 if there is an imbalance in the S cycle.

11.4.1.4 Carbon Inputs Internal to the Exogenic System

Geologic inputs (volcanic and weathering) of CO_2 into the exogenic system are small compared to fluxes internal to the exogenic system. For context, exchange between the terrestrial biosphere and atmosphere is ~60 Gton C/y and between the surface ocean and atmosphere is ~70 Gton C/y.[11] Present anthropogenic CO_2 emissions from fossil fuels and cement production is ~10 Gton C/y,[26] which exceeds volcanic emissions by more than 50 times. These large fluxes are likely not important on long timescales as they operate on much faster timescales within the exogenic system. Nevertheless, on short timescales, any imbalances in the system associated with these fluxes or sources can lead to large, short-lived excursions in atmospheric CO_2. Intriguingly, terrestrial biomarkers present in river sediments yield apparent radiocarbon "ages" in excess of 1000 years,[101] which implies that imbalances between terrestrial photosynthesis and respiration may play a role in the exogenic C cycle over millennial timescales.

11.4.2 Carbon Outputs

11.4.2.1 Silicate Weathering Chemistry and Carbonate Precipitation

While all silicate minerals undergo dissolution, Ca- and Mg-bearing silicate minerals in the continental crust (e.g. hornblende, plagioclase, pyroxene, and, to a lesser extent, olivine and biotite for Mg only) are often considered the most important in the context of the long-term global C cycle because the dissolved Ca^{2+} and Mg^{2+}, after traveling to the ocean via rivers, eventually precipitate to form calcium and magnesium carbonates in the marine environment. However, alkalinity supplied by the dissolution of Na- and K-bearing silicates dominates the silicate-derived riverine flux and contributes to $CaCO_3$ precipitation via ion-exchange reactions. Additionally, some portion of the Na and K flux is thought to react with Si and Al to form clay minerals in a process termed "reverse" weathering, which, by modifying the seawater alkalinity balance, contributes CO_2 back into the atmosphere.[102]

The dominant Ca-bearing mineral in the average continental crust is plagioclase, but for simplicity, silicate weathering of calc-silicates is often idealized by lumping all calc-silicates into the form of wollastonite, $CaSiO_3$ (however, wollastonite is exceedingly rare in the continental crust!). The sequence of reactions begins with formation of carbonic acid by equilibration with CO_2 in the atmosphere or subsurface pore-space:

$$CO_2 + H_2O = H_2CO_3. \tag{11.15}$$

This is followed by release of H^+ ions:

$$H_2CO_3 = H^+ + HCO_3^-, \tag{11.16}$$

which react with calc-silicates to release Ca^{2+} and Si^{4+} in the form of silica hydroxides into river waters (an identical reaction can be written for $MgSiO_3$):

$$CaSiO_3 + 2H^+ + H_2O = Ca^{2+} + Si(OH)_4. \tag{11.17}$$

Ca^{2+} ions combine with the major carbonate species in seawater, bicarbonate (HCO_3^-), to precipitate calcium carbonate in the marine or lacustrine environments (and dissolved silica precipitates as chert, SiO_2):

$$Ca^{2+} + 2HCO_3^- = CaCO_3 + CO_2 + H_2O. \tag{11.18}$$

If we combine the above equations, we arrive at the commonly cited net reaction for calc-silicate weathering:

$$CaSiO_3 + CO_2 = CaCO_3 + SiO_2. \tag{11.19}$$

The equivalent reaction for plagioclase is:

$$CaAl_2Si_2O_8 + H_2CO_3 + 3H_2O = CaCO_3 + Al_2Si_2O_5(OH)_4. \tag{11.20}$$

Dissolution of calc-silicates followed by precipitation of carbonates results in net consumption of CO_2. Estimates of the terrestrial silicate weathering output of C in the form of carbonate are typically drawn from the dissolved flux of cations in river waters, assuming that most of the Ca and Mg ions and some portion of the Na and K fluxes (via exchange) eventually precipitate as Ca–Mg carbonates. Modern carbonate burial fluxes associated with silicate weathering only are thought to be between 0.13 and 0.38 Gton C/y (Figure 11.8).[11] As discussed above, a fraction of the Ca ion flux in rivers comes from dissolution of carbonate, so the carbonate fraction must be subtracted. According to

Gaillardet et al.,[17] the global terrestrial carbonate-corrected silicate weathering flux may be on the low end of the above range, at ~0.14 Gton C/y. Newer estimates from Moon et al.[103] revise the silicate weathering flux to even lower values (0.09 ± 0.02 Gton C/y), but completely exclude Na and K from these calculations.

Not accounted for in the above estimates of the silicate weathering–carbonate precipitation flux is the role of seafloor basalt alteration. A number of studies have shown that the CO_2 content of altered oceanic crust can be quite high. For example, Mesozoic-aged altered oceanic crust can contain up to 2.5 wt.% CO_2, compared to ~0.5 wt.% CO_2 in late Cenozoic oceanic crust.[104–107] These studies estimate that C sequestration via seafloor alteration could amount to 0.006–0.036 Gton C/y,[104,107,108] similar to modern inputs of C via MORs, but less than the terrestrial silicate weathering sink (Figure 11.8).

11.4.2.2 Photosynthesis and Organic Carbon Burial

Most of the C in the atmosphere–ocean system today is in an oxidized form. A fraction of this C is reduced through autotrophic organisms by oxygenic photosynthesis to form organic C (other types of photosynthesis could have dominated early in Earth's history):

$$CO_2 + H_2O + \text{sun's energy} = \text{“}CH_2O\text{”} + O_2. \tag{11.21}$$

Here, organic C is approximated as "CH_2O" and differs from CO_2 in being composed of reduced C (C^{3+} or less) as compared to oxidized C in CO_2 (C^{4+}). Much of the organic C is respired (oxidized) back to the atmosphere through organic C decomposition or indirectly through consumption by heterotrophic life, reversing the above reaction. In net, photosynthetic production and oxidative destruction of organic C are roughly in balance, even on timescales as short as years or even days in some cases. However, a small fraction of organic C "leaks" out from the exogenic system through sedimentary burial. This burial of organic C, like the burial of carbonate, represents a net long-term sink of C from the exogenic system, but differs from the carbonate cycle in that burial of organic C also generates free O_2, with implications for atmospheric oxygenation. Buried at great enough depths, organic C becomes part of the crustal reservoir in the form of "fossil" C.

Estimates of the organic C burial rate are typically inferred from the C isotopic composition of seawater. Carbon isotopes are fractionated to light values during photosynthesis but are not strongly fractionated during carbonate precipitation (due to a lack of redox change). Assuming the isotopic compositions of long-term C inputs into the exogenic system are mantle-like, the fraction of organic C (commonly referred to as f_{org} in the literature) buried can be estimated from the C isotopic composition of seawater or marine carbonates. Except for short-term excursions, the $^{13}C/^{12}C$ of seawater has remained relatively constant at 4–5 per mil higher than the mantle for most of Earth's history, which translates into an organic C burial fraction of ~20–25% relative to total C burial (carbonate + organic C).[3,9,109]

Whether f_{org} has indeed remained constant and, if so, why, remain important questions.[110] It may be worth exploring whether the isotopic offset between inorganic and organic C remains constant through Earth's history. There is also the possibility that the weathering of carbonates or the metamorphic outgassing efficiencies of crustal carbonates and organic C are different enough that the isotopic composition of the long-term inputs of C into the exogenic system need not balance with mantle-like values at all times.[22,86,111]

11.4.2.3 Subduction Flux

Subduction of the oceanic lithosphere and accompanying marine sediments is the primary mechanism by which C is "permanently" removed from Earth's surface and recycled back into the endogenic system. Subduction itself is not a direct negative feedback in the climate system because it only serves to transport already sequestered C into the mantle. Subduction plays an indirect role in the longer-term exogenic C cycle and climate if it contributes some of its C to arc magmas or it influences the efficiency of C released from the sub-arc mantle wedge and into arc magmas.

Based on estimates of the carbonate and organic C distribution of marine sediments and the amount of C in seafloor altered crust (mostly in the form of carbonate),[104,105,107,112] Dasgupta and Hirschmann[18] estimated a modern global subduction flux of 0.024–0.048 Gton C/y (Figure 11.8). This is comparable to the lower end of estimates of global magmatic degassing. How much of this subducted C is released back into the crust or atmosphere through volcanism and how much continues deep down into the mantle is debated. Because the temperature needed for decarbonation reactions increases with pressure, the subducting slab needs to heat up sufficiently for efficient decarbonation.[113,114] In the absence water, slab decarbonation (either during subsolidus conditions or during partial melting) is not expected to be efficient in most modern subduction zones, except for perhaps the youngest and hottest ones.[51,114] Slab decarbonation would have been more efficient in the Archean if the thermal states of these ancient subduction zones were higher than today.[51] In the presence of water, CO_2 activities decrease, depressing reaction temperatures.[92] Thus, dehydration of serpentine in the slab could give rise to "water-fluxed" decarbonation or carbonate dissolution, enhancing the efficiency of decarbonation (see Refs. 49 and 93). Small amounts of C in the form of graphitized organic C in slabs are still expected to survive fluid-fluxed melting (e.g. Ref. 50). In any case, release from the slab does not guarantee that such C quantitatively returns to the exogenic system through arc volcanoes.

11.5 Efficiency of the Silicate Weathering Feedback

In this section, we discuss the kinetics of silicate weathering (we ignore organic C burial because it is a smaller flux). The kinetics of silicate dissolution are much slower than those of carbonate precipitation in the oceans. The ocean is oversaturated in carbonate and therefore any Ca that enters the ocean tends to precipitate on short timescales. Thus, silicate dissolution is the rate-limiting step. The rate (moles/y) of dissolution should scale as[115,116]:

$$r = kA(a_{H^+})^n, \tag{11.22}$$

where k is a kinetic rate constant (moles/m^2/y) defined relative to a specified surface area A (e.g. geometric or total surface area) and represents a measure of the *efficiency* of dissolution, a_{H^+} is the activity of hydronium ions in solution, and n represents the order of the chemical reaction. Hydronium ion activity a_{H^+} depends on the pCO$_2$ in equilibrium with weathering fluids:

$$K_1 = \frac{p_{CO_2}}{a_{H^+} a_{HCO_3^-}}. \tag{11.23}$$

This relation with pCO$_2$ is one potential source of a negative feedback in the global C cycle. If the pCO$_2$ of weathering fluids increases as the pCO$_2$ of the atmosphere increases (e.g. via changes in primary productivity[117]), silicate dissolution rates increase, which results in more rapid removal of CO_2 from the atmosphere through carbonate precipitation. In the Earth system, the global silicate weathering feedback must also include the effects of atmospheric CO_2 on the hydrologic cycle, vegetation, and so forth, so (11.23) is only meant to conceptually represent the most elementary level of CO_2 feedback.

Increases in pCO$_2$ should also lead to increases in temperature due to the greenhouse effect (~2.4–4.8°C per doubling of pCO$_2$[118]). The negative feedback is enhanced because of the temperature dependence of the kinetics of dissolution[119]; that is:

$$k = k_o \exp\left(-E_A/RT\right), \tag{11.24}$$

where k_o is a pre-exponent constant that depends on the type of mineral, E_A is the activation energy for dissolution, R is the gas constant, and T is temperature. If pCO$_2$ increases, temperature increases, which in turn increases dissolution rates, thereby increasing rates of CO_2 drawdown by silicate weathering (followed by carbonate precipitation).

As we scale up from the mineral to the global scale, additional influences on weathering rate must be considered. For example, if chemical weathering reactions occur close to thermodynamic equilibrium, weathering fluxes will scale with the total water flux through the system.[120] Chemical weathering can also be enhanced by biology, through bioturbation and higher subsurface pCO$_2$ levels. Because hydrology and biology can both accelerate chemical weathering, these processes can act as negative feedbacks if increases in global temperature, perhaps driven by increases in atmospheric CO_2, accelerate the hydrologic cycle.[120,121] Finally, chemical weathering can also be limited by the rate at which fresh rock is supplied by physical weathering and erosion. Erosion can behave as a negative feedback under conditions in which erosion is accelerated by higher hydrologic or biological activity. Under conditions in which erosion is largely controlled by tectonics, which do not depend on climate, the feedback would be poor.

Combining all of these factors results in a complicated functional form of the silicate weathering feedback. Berner[122] proposed the following generalized empirical equation:

$$F_{Si}' = f_w(p_{CO_2}) f_{rf}(t) f_{pl}(t) f_{ro}(t), \tag{11.25}$$

where F_{Si}' represents the silicate weathering flux contributing to carbonate precipitation, f_w is the function describing the sensitivity of mineral dissolution to pCO$_2$, f_{rf} represents mean global relief, f_{pl} represents the effect of plants, and f_{ro} represents average water runoff, with all functions normalized to present conditions. In the Berner model, the temperature effect is subsumed into the effect of temperature on runoff and the effect of pCO_2 on chemical weathering. For the purposes of this chapter, the exact functional relationships are not critical (see Ref. 123). These relationships must be calibrated against field studies to be relevant for a particular region or a given time period in Earth's history.

If we consider f_w, f_{pl}, and f_{ro} to all be related to pCO_2 through a series of transfer functions, we can combine them into one master function f_w^*, which describes the dependency of weathering on pCO_2:

$$F_{Si}' = f_w^*(p_{CO_2}) f_{rf}. \tag{11.26}$$

This leaves the relief term in Berner's model, f_{rf}, as the only function independent of pCO_2. The relief term is fundamental as it relates to erosion rate. Erosion supplies fresh rock to the surface so that it can undergo chemical weathering. At low erosion rates, soil mantles, although chemically depleted due to their long residence times in the weathering regime, no longer contain weatherable minerals. On long timescales, what ultimately drives erosion is topography, which itself is driven by tectonism or magmatism.

We can describe f_{rf} as the weighted average flux of fresh rock supplied by erosion relative to the global mean today:

$$f_{rf} = \frac{\sum E_i A_i}{\sum E_i^o A_i^o}, \tag{11.27}$$

where E_i is the erosion rate (m/y) in a given region i and A_i is the map area of region i, with the quantities in the denominator representing present-day conditions (Figure 11.11). The summation signs denote summation over the entire continental area. What is important is that the area of orogenic activity on Earth is not constant. For example, the hypothesis that mid-Cenozoic cooling was triggered by enhanced erosion associated with the Himalayan orogeny is based on the premise that the mean area of orogenies and mean erosion rates increased.[124,125] A simple way to understand the effect of increased orogenic belts is to consider the case in which the silicate weathering response to changes in pCO_2 is linear. In such a scenario, doubling the length of orogenic belts (i.e. doubling the mass of silicate minerals exposed at the surface per unit time) would increase the sensitivity (e.g. the slope on the silicate weathering-driven burial rate of carbonate versus pCO_2) of the global silicate weathering feedback by a factor of 2 (Figure 11.11). If long-term inputs of CO_2 into the exogenic system do not change, exogenic CO_2 would decrease by a factor of 2, cooling the atmosphere. This simple exercise also implies that in a world devoid of orogenic belts, such as in the extreme case of flat continents riding near sea level, the strength of the negative silicate weathering feedback would be substantially decreased.

Finally, temporal changes in the composition of the bedrock undergoing weathering can be important. It is widely agreed that the average composition of the continental crust is intermediate in composition, equivalent to the composition of a typical andesite.[126] These published averages often obscure the fact that the continental crust is highly heterogeneous,

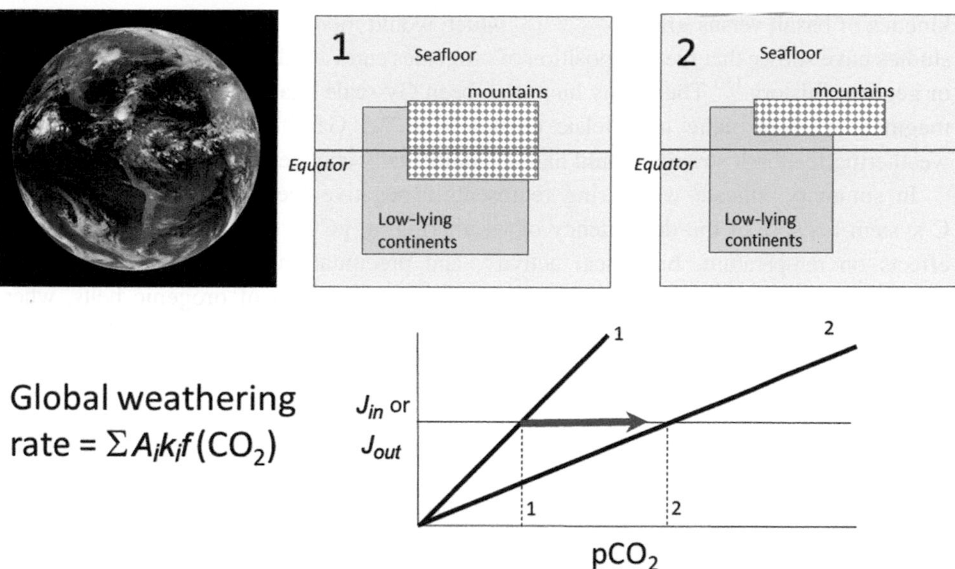

Figure 11.11 The global silicate weathering rate can be described by partitioning Earth into different types of weathering sites, each defined by different weathering kinetics (e.g. seafloor, orogens, and low-relief continents). Changes in the proportions of low-relief continents or orogenic belts will change the global sensitivity of the weathering feedback. A decrease in area of continents and orogens results in a decrease in the weathering feedback strength (states 1 to 2), leading to an increase in pCO_2 (states 1 to 2).

containing rocks of all types, from ultramafic rocks (serpentinites and peridotites) to mafic rocks (basalts) to highly silicic rocks (e.g. rhyolites). Because different lithologies weather at different rates, have different temperature sensitivities, and supply major cations in different proportions, the composition of rocks within an active orogeny matters. For example, the CaO content of a typical basalt/gabbro is between 10 and 15 wt.%, that of andesite/granodiorite is between 6 and 8 wt.%, and that of a rhyolite/granite is less than 3 wt.%. Additionally, available estimates suggest that basalts weather, on average, approximately two times faster than granitic rocks, but this ratio varies as a function of climatic conditions.[127] This compositional effect can be incorporated through the erosional term, f_{rf}:

$$f_{rf} = \frac{\sum E_i A_i \rho_i C_i}{\sum E_i^o A_i^o \rho_i^o C_i^o}, \tag{11.28}$$

where C_i represents the concentration of Ca and Mg (note that not all Mg is sequestered as carbonate) and ρ_i is rock density. Equation (11.28) describes the availability of Ca and Mg due to differences in composition. Depending on the composition of the crust undergoing orogeny, the sensitivity of the regional chemical weathering feedback could vary by a factor of 3–5 due to the availability of Ca and Mg (there is also a small effect of Na and K exchange with Ca in the ocean). Added to this compositional effect will also be the faster dissolution kinetics of basalt versus granite,[17,127–129] which would modify k_i in (11.22). A number of studies have shown that the composition of orogenies can vary depending on orogenic style or geologic history.[130] There may have also been Gy-scale changes in the composition of magmatism, from mafic to a felsic crust around 2.5 Ga,[95,131,132] suggesting that the weathering feedback strength could have fundamentally changed over Earth's history.

In summary, silicate weathering represents a negative feedback in the whole-Earth C system because of the dependency of weathering on pCO_2 through direct and indirect effects on temperature, biological activity, and precipitation/runoff. One of the most important controls of the strength of the feedback is the area of orogenic belts, where chemical weathering is facilitated by high erosion rates. Large fluctuations in the intensity and distribution of orogenic activity could play a key role in modulating the strength of the silicate weathering feedback.

11.5.1 Seafloor Weathering Feedback

In Section 11.4.2, we showed that carbonate sequestration via seafloor alteration was comparable to the CO_2 flux out of MORs, begging the question of how important seafloor alteration is as a negative feedback. Coogan and Dosso[108] inferred that the activation energy for rock dissolution during seafloor alteration is ~92 kJ/mol, similar to that for the dissolution of various silicate minerals under conditions relevant for terrestrial weathering.[115] If direct temperature effects alone were the dominant control on the negative weathering feedback, one would expect the strength of the seafloor alteration feedback to be equivalent to that of the terrestrial feedback, all other variables being equal. However, as discussed above, many other processes operate. In the case of terrestrial weathering, the effects of runoff on erosion will amplify the strength of the negative feedback. In the case of seafloor alteration, there are no erosional mechanisms that replenish fresh rock for dissolution, so the weathering system would be closer to equilibrium and the kinetics of weathering would be slow. These factors could reduce the net sensitivity of seafloor weathering to global temperature. A recent inverse study of the paleoclimate record suggested that seafloor weathering was not as important as continental weathering over the last 100 My.[133] However, during periods of low orogenic activity, the proportional contribution of the seafloor to the global feedback indeed increases (Figure 11.11). For example, silicate weathering could have been dominated by seafloor weathering in the Archean if most of the continents were below sea level.[134] In such a case, the strength of the seafloor weathering feedback becomes critically important to understand.

11.6 Discussion

11.6.1 Framework for Modeling Whole-Earth C Cycling

Earth systems modeling is traditionally rendered in terms of boxes with estimates of reservoir sizes and fluxes (Figure 11.3 and Sections 11.3 and 11.4). However, in such renditions, fluxes and reservoir sizes are snapshots in time and cannot be applied back in time if reservoir sizes change. Instead of using fluxes, it is better to use the kinetics of C transfer between reservoirs, as kinetic rate "constants" do not depend on reservoir size even if the system is not at steady state. Rate constants can of course change if the fundamental physics and chemistry change. If the rate constants are known, (11.1) can be expressed as follows for a linear system:

$$\frac{dM_i}{dt} = \sum k_{ji}M_j - \sum k_{ij}M_i, \tag{11.29}$$

where k_{ij} represents the rate constant describing the transfer of C from reservoir i to j. The first term represents inputs into reservoir j from all reservoirs i. The second term represents outputs from reservoir i into different reservoirs j. Solving (11.29) for a system of interacting reservoirs involves solving a system of equations using a multidimensional $(i \times j)$ matrix k_{ij}, which can be estimated from modern fluxes if the system is near steady state. We can divide the flux associated with any given process by the size of the reservoir from which C is removed, resulting in a fractional rate constant. The sum of all rate constants describing removal of C from a reservoir i is the total rate constant, k_i; that is:

$$k_i = \sum_j k_{ij}. \tag{11.30}$$

The inverse of the total rate constant represents the residence time τ_i of C in reservoir i from which C is extracted[135]:

$$\tau_i = \frac{1}{k_i} = \frac{1}{\sum_j k_{ij}} = \left(\frac{1}{\tau_{ij}}\right)^{-1}. \tag{11.31}$$

In Figure 11.12, we present reservoir residence times (leftmost column) and the rate constant matrix, the latter represented in the form of fractional response times $1/k_{ij}$, which are more intuitive than fractional rate constants. In Figure 11.12, the row denotes the reservoir (i) from which C is being removed and the column denotes the reservoir (j) receiving C from i. Estimates of exogenic rate constants were guided by modern fluxes and reservoir sizes from Table 11.2 and an assumption of near steady state. For endogenic reservoirs, which are larger and may not be at steady state, our estimated rate constants are speculative. The modeler is encouraged to modify the endogenic rate constants to explore the behavior of the system. Empty boxes indicate the absence of flux. Boxes with a question mark indicate that there is a flux, but the magnitude of the flux and the rate constant are poorly known or change considerably with time. For example, degassing of the deep ocean to the atmosphere varies considerably between glacial–interglacial cycles. Transfer of C between the upper mantle and lower mantle is uncertain because it depends on the flux of mantle material between the two reservoirs, which is not known.

11.6.2 Is Earth's Mantle Degassing or "Ingassing"?

A key question is whether mantle degassing rates are balanced by recycling of C back into the mantle through subduction. The only direct way to answer this question is to compare

			Endogenic		ocean			Exogenic			
			Atm	Bio	Oc-S	Oc-D	Oc-rs	Oc Lith	Cont Lith	UM	LM
Endogenic		Atm	~4 y		~10 y	8–10 y					
		Bio	~5–10 ky	~30 y		~3 ky					
	ocean	Oc-S	4–5 y	~10 y			~10 y				
		Oc-D	300–400 y	?		~300 y		40–50 ky			
		Oc-rs	3–4 ky				~5 ky		~100–200 ky ($CaCO_3$)	~15–30 ky ($CaCO_3$)		
Exogenic		Oc Lith	100–300 My	~400/f My arc volc.						~400/f My metasom.	~400/f My mantle wedge	~400/f My deep subduction
		Cont Lith	>1 Gy	0.5–1 Gy? metamorph. degassing								
		UM	~1 Gy	~300 My ridge								?
		LM	>1 Gy	>1 Gy plume degassing							?	

Figure 11.12 Matrix of kinetic rate constants between reservoirs, expressed as fractional response times or the inverse of the kinetic rate constant ($\tau_{ij} = k_{ij}^{-1}$). Each cell corresponds to the transfer of C from reservoir i to j, where i correspond to the row and j corresponds to the column. Empty or gray cells indicate a lack of C transfer from i to j. Cells with a question mark indicate that C transfer occurs, but the magnitude and rate constant are not known or vary significantly in time; these are left for the reader to vary in their own studies. Blue-shaded cells represent exogenic reservoirs. The blue-outlined group of cells represents ocean reservoirs, including reactive marine sediments. The leftmost column shows residence time τ_i of reservoir i. Reservoir symbols as follows: Atm = atmosphere; Bio = terrestrial biosphere; Oc-S = surface ocean; Oc-D = intermediate and deep ocean; Oc-rs = reactive marine sediment; Oc Lith = oceanic lithosphere, which includes oceanic crust and oceanic lithospheric mantle; Cont Lith = continental lithosphere, which includes continental crust (sediments + basement) and CLM; UM = upper mantle (<670 km); LM = lower mantle (>670 km). Earth's core is excluded. "f" corresponds to the fraction of C from the subducting oceanic lithosphere that is lost to a particular reservoir.

mantle degassing fluxes to subduction fluxes, but these quantities are uncertain. A better approach is to explore possibilities through a forward box modeling approach as discussed in Section 11.6.1 and depicted in Figure 11.12. Rate constants can be allowed to vary with time as constrained by the physical evolution of Earth following simple, parameterized convection[136] rather than full-scale coupling of thermodynamic and geodynamic models. For example, it is relatively straightforward to develop petrologically constrained scalings for mantle and crustal degassing as a function of mantle thermal state and convective vigor, with the goal of quantifying degassing at MORs, subducting slabs, arcs, and intraplate environments. Important quantities would be the kinetics of chemical weathering, but also where carbonates are deposited. In the extreme scenario in which all carbonates are deposited on continental shelves, which do not subduct, then there is little recycling of C back into the mantle. As far as we know, there is not yet any whole-Earth box model that

Table 11.2 *Fluxes of carbon*

Fluxes	Gton C/y	Refs.
Exogenic		
Biosphere–atmosphere exchange	60	[12]
Surface ocean–atmosphere exchange	70	[12]
Surface–deep ocean exchange	100	[12]
Carbonate precipitation (silicate weathering)	**0.13–0.38**	[17]
Organic C burial	**0.05–0.13**	This study, calculated from carbonate precipitation with f_{org} = 0.2
Oceanic crust weathering	**0.03**	[107]
Weathering of carbonate and organic C (total)	**0.26–0.30**	
Carbonate weathering	0.22–0.25	This study, calculated from organic C weathering (see below)
Organic C weathering	0.043–0.050	[25]
Metamorphic degassing	**0.036–0.200**	[20]
Mantle degassing (total)	**0.085–0.300**	
MOR	0.0158–0.0163	[19]
Ocean islands	0.001–0.030	[18]
Arcs	0.018–0.147	[21,78]
Modern continental rifts	0.05–0.10	[79]
Subduction flux (sediments + crust + lithospheric mantle)	0.024–0.048	
Miscellaneous		
Mount Etna	0.007	[24]
Cretaceous continental arcs	0.08–0.12	[22]
Anthropogenic	10	[26]

has incorporated the nature of how C is deposited (oceans versus continents). The important controls of where carbonates are deposited will be total area of shallow continental margins, the evolution of life, and the chemistry of the oceans.[137,138] Variations of continental shelf area with time could be one of the most important factors controlling the rate at which C accumulates at the surface, as there are times in Earth's history when continents may have been wholly submerged,[134,139,140] with most C being deposited in epicontinental seas, leaving little to be deposited in the deep sea for subduction.

In any case, we know that sedimentary rocks have accumulated on the continents over Earth's history. We also know that other incompatible elements, such as U, Th, and K, among others, have progressively migrated into the continental crust, leaving behind a depleted upper mantle. If C behaves incompatibly, then the continental crust and CLM C budget have likely grown with time as well. This growing continental reservoir of

C serves as a capacitor, which can enhance outgassing rates by metamorphic or magmatically driven decarbonation of the lithosphere in continental arcs[22] or continental rifts.[79] If direct mantle degassing rates have not changed with time, total outgassing rates from the endogenic system should have increased with time. Alternatively, if mantle degassing rates are thought to have decreased significantly through time, lithospheric amplification could have partly compensated for this decline. Understanding how total outgassing rates from Earth's interior have changed through time will have profound implications for the evolution of other volatiles, whose chemistries may be linked to C. For example, an increasing outgassing flux of C would result in increased photosynthetic production of atmospheric oxygen.[95]

11.6.3 What Drives Greenhouse and Icehouse Conditions?

On 10–100-My timescales, Earth's climate fluctuates between greenhouse and icehouse states, the latter being characterized by permanent polar ice sheets and the former without permanent polar ice. For example, the late Jurassic to the early Cenozoic was characterized by greenhouse conditions, where temperatures were thought to be $>10°C$ higher than today and any ice sheets were thought to be ephemeral. This greenhouse interval was punctuated by a number of short-term (<100-ky) excursions to even warmer conditions, often referred to as "hothouse" events.[141]

Earth began to cool around 50 Ma in the mid-Cenozoic based on the seawater oxygen isotope record, with southern hemisphere glaciation (Antarctica) beginning ~35 Ma, followed by the development of the northern hemisphere ice sheet ~5 Ma.[4,36] Earth in the last several million years has been characterized by bipolar glaciation, and because of the albedo effect of ice, subtle variations in solar insolation driven by orbital dynamics can drive short-term cycles (10–100 ky) of ice sheet growth and retreat. Thus, superimposed on the late Cenozoic icehouse baseline are geologically rapid periodic fluctuations of temperature, ice volume, and eustatic sea level tuned to orbital dynamics. During greenhouse intervals, apparent changes in relative sea level (e.g. sequence boundaries[142,143]) recorded during greenhouse conditions most likely have origins that are different from glacioeustasy, such as external changes in sediment supply. While rapid fluctuations in sea level are a defining characteristic of an icehouse world, short-lived hothouse excursions – a characteristic feature of greenhouse intervals – appear to be absent from icehouse worlds.

Two questions follow. First, what controls baseline climate? Second, what conditions allow for and trigger short-term excursions to extreme warm and cold climates? In general, baseline climate on >10-My timescales is ultimately controlled by the amount of greenhouse gases in the atmosphere, namely CO_2 for most of Earth's history. Orbital dynamics are too rapid, while changes in the sun's luminosity only explain changes over Gy. Understanding long-term baselines comes down to understanding what controls the steady state. The combination of volcanic/metamorphic forcings (input) and the efficiency of the silicate weathering feedback defines a given steady state, and it is the variations in these two quantities that we seek to understand.[30]

11.6.3.1 Greenhouse Intervals

A widely held view is that the late Mesozoic and early Cenozoic greenhouse conditions were driven by a 1.5- to 2-fold increase in MOR spreading rates, leading to an increase in CO_2 output from the mantle.[3,80,81] The contribution from oceanic spreading may have been enhanced by a greater emplacement frequency of large igneous provinces.[80] Mid-Cenozoic cooling toward icehouse conditions is widely thought to have been driven by an increase in the efficiency of the silicate weathering feedback through the collision of India with Eurasia and the uplift of the Himalayas and Tibet.[124,125]

There are, however, reasons to explore other hypotheses. One question is whether enhanced MOR spreading was able to sustain a Cretaceous pCO_2 that was 5–10 times higher than present (Figure 11.1b). If CO_2 concentrations in Mesozoic MORBs were similar to today's, the increase in CO_2 production rates would scale roughly linearly with the increase in spreading rate. Assuming a linear negative feedback, the enhanced spreading rates would lead to only a 1.5- to 2-fold increase in atmospheric CO_2, or the global silicate weathering feedback would have to have been less efficient. The latter seems unlikely because weathering kinetics should increase in the hotter and wetter conditions of greenhouse climates. Furthermore, the late Mesozoic to early Cenozoic was characterized by significant orogenic activity in the form of Andean-style continental arcs. Thus, weathering feedback strength during the late Mesozoic greenhouse should have been high, requiring even higher inputs of CO_2 to sustain the elevated pCO_2.

Recently, it has been suggested that length of continental arcs during the Cretaceous was 1.5–2 times longer than that of the mid-Cenozoic.[22] Due to the large amounts of carbonate stored in the continental crust, continental arcs may degas more CO_2 through metamorphic and magmatically induced decarbonation of crustal carbonate; thus, a global chain of continental arcs might amplify global CO_2 outgassing.[22,86,144] Long-term fluctuations in the length of continental arcs, driven by the assembly and dispersal of continents, have been suggested to drive greenhouse–icehouse oscillations based on studies of the detrital zircon record and arc length through time.[88,145] Lee et al.[22] showed that the long Cretaceous continental arcs may have supported a flux of 0.09–0.13 Gton C/y, more than twice that of global magmatic arcs today (Figure 11.8b). This hypothesis has been contested by Pall et al.,[146] who argue for a weak correlation between carbonate-intersecting continental arcs and high pCO_2. However, this study did not incorporate the Precambrian distribution of carbonates or uncertainties in the paleo-pCO_2 record. In any case, other factors to be considered are the competing effects of enhanced erosion and weathering during continental arc magmatism[147,148] and the slab contribution to CO_2 fluxing through arcs, the latter requiring an understanding of slab composition, slab thermal state, and CO_2 solubility in fluids and melts, among other factors. In any case, faster spreading, faster subduction, more plumes, and a transition to continental arcs over island arcs may operate together,[149] making it difficult to ascribe greenhouse intervals to a single process.

11.6.3.2 Icehouse Drivers

Are long-lived icehouses driven by enhanced silicate weathering feedbacks or decreased outgassing from Earth's interior? The Cretaceous greenhouse and mid-Cenozoic to present icehouse represent an ideal case study. It is widely thought that the Himalayan orogeny played a strong role in Cenozoic cooling beginning ~50 Ma by enhancing the silicate weathering feedback through enhanced erosion rates.[125] However, the rise in $^{87}Sr/^{86}Sr$ often touted as an indicator of terrestrial weathering occurs ~38 Ma, well after cooling initiated.[150,151] Furthermore, radiogenic isotopes do not necessarily correlate with silicate weathering fluxes because source age and composition can be just as important.[152] For example, local weathering of ultramafics or black shales has an outsized influence on seawater $^{187}Os/^{188}Os$.[153,154] Similarly, weathering of carbonates or the very ancient basement influences $^{87}Sr/^{86}Sr$.[152]

These complications in interpreting seawater Sr isotopes raise a number of important questions. Is the Himalayan orogeny, in a regional sense, a net producer or consumer of CO_2?[94,96,97] This, combined with the possibility that regional metamorphism causes metamorphic decarbonation of crustal carbonates, should encourage vigorous discussion on the role of the Himalayan orogeny on global cooling. Recent studies have suggested that the closing of the Tethys ocean, which culminated in a transcontinental orogenic belt, of which the Himalayas are but one segment, involved the obduction of numerous island arc terranes and ophiolites, increasing global silicate weathering efficiency.[155] In an alternative scenario, Kent et al.[156] suggested that the ~65-Ma Deccan flood basalts in India may have increased global weathering efficiency as India migrated north into wet equatorial latitudes in the Cenozoic.

There is also the possibility that cool climates simply represent a decline in magmatic outgassing.[157] Evidence from detrital zircons and arc reconstructions have been used to suggest that long icehouse intervals coincide with periods of reduced continental arc activity, just as greenhouses have been suggested to correlate with enhanced continental arc activity.[22,87,88,145] The rise and fall of continental arcs offer an intriguing process for modulating the exogenic C system. When continental arcs are magmatically active, they are likely to be net producers of CO_2, but after magmatism wanes, the remnant topography transforms continental arcs into regional sinks of CO_2, resulting in a global increase in the efficiency of silicate weathering.[147,148] In such a scenario, magmatic outgassing drives greenhouse intervals, which are followed by icehouse intervals when the same orogens die magmatically but continue to weather (Figure 11.13).

Ice sheets may also influence the long-term climate. Growth of ice sheets increases albedo, which cools the climate and encourages additional ice sheet growth. Could ice sheet expansion have further facilitated Cenozoic cooling by enhancing the silicate weathering associated with glacial erosion? This is a tantalizing thought as the Antarctic ice sheet is situated on the ancient Archean basement, raising the question of whether the onset of Antarctic glaciation at ~35 Ma may have played a role in the rise of seawater Sr isotopes at about the same time. On the other hand, glacial weathering might even result in the net production of CO_2 through the oxidation of sulfides, which in turn drives carbonate

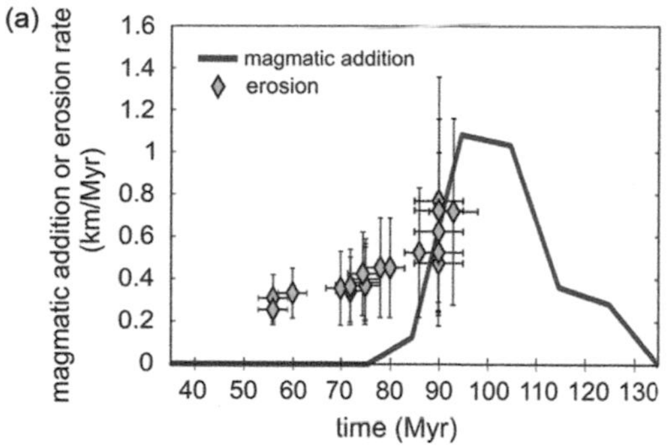

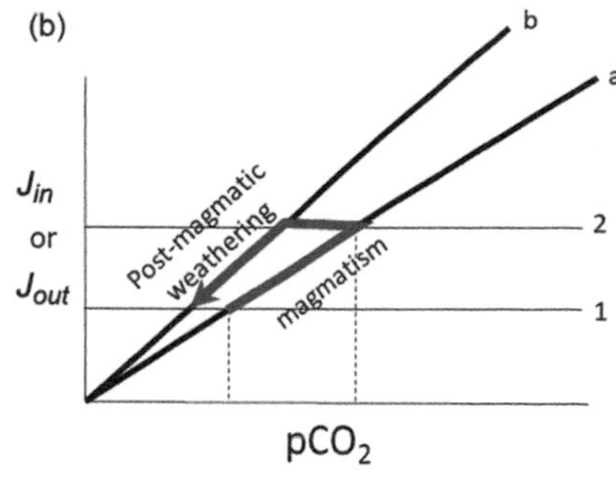

Figure 11.13 (a) Case study of a magmatic orogen: Cretaceous Peninsular Ranges batholith in southern California, USA.[147] Red line shows the observed magmatic production rate and symbols with error bars represent inferred erosion rates. The orogen is characterized by both magmatism (and CO_2 degassing) and erosion, but they are not in phase. After magmatism ends, remnant topography allows for continued erosion and regional drawdown of CO_2 unsupported by magmatic degassing. (b) Global effect of magmatic orogen on steady-state pCO_2. Details of the plot are identical to those of Figure 11.5. Magmatic orogeny increases CO_2 outgassing, so pCO_2 increases from states 1 to 2. Uplift during magmatism enhances erosion, increasing the sensitivity of the global weathering feedback (states a to b), which buffers the rise of pCO_2. After magmatism ends, physical and chemical weathering persist, driving pCO_2 to low levels. Magmatic orogens can potentially drive greenhouses, but are followed by global cooling due to protracted weathering.

dissolution; because the resulting acidity is driven by sulfide oxidation rather than carbonation itself, this results in a net release of CO_2, counteracting the effects of glacial weathering and the ice–albedo positive feedback.[158,159]

11.6.4 Climatic Excursions and Runaways

11.6.4.1 Hothouses

The Mesozoic–early Cenozoic greenhouse was punctuated by short-lived hothouse excursions, including the three oceanic anoxia events in the Cretaceous (OAE-1, -2, and -3) and the late PETM and surrounding hothouse excursions ("hyperthermals").[36,141,160,161] Many of these excursions have in common an increase in temperature and pCO_2, but different triggers. The PETM is characterized by a negative C isotope excursion, leading to suggestion that methane hydrates in the marine system were destabilized and released into the atmosphere.[35] Other mechanisms, such as destabilization of peat, opening of the North Atlantic Ocean, and even cometary impact,[162–164] have also been suggested for the PETM. Emplacement of the Caribbean flood basalt is thought to have driven Cretaceous OAE-2 at the Cenomanian–Turonian boundary,[165,166] with volcanic release of CO_2 leading to global warming, which in turn decreased the solubility of O_2 in the ocean and setting conditions to be favorable for organic C preservation.[166]

Hothouse excursions clearly have different triggers, but occur only during long-lived greenhouse baselines, not during icehouse baselines. Greenhouse baselines may be close to the threshold for hothouse excursions, such that any number of stochastic events could trigger an excursion if the system is close to threshold. The fact that hothouses return rapidly to baseline is further evidence that short-lived episodic events are responsible for triggering hothouses. Flood basalts, methane hydrate destabilization, and cometary impacts, among others, are all singular events that cannot sustain a long-term transition to a new steady state (unless the system is characterized by multiple steady states). These hothouse excursions all return to baseline because of the buffering effects of the silicate weathering feedback.

In some cases, excursions run away to a fundamentally different state. Unlike Earth, Venus is in a permanent hothouse. In part, this is due to continuous accumulation of CO_2 or other greenhouse gases from the Venusian interior, enhanced by the fact that Venus has not had an efficient mechanism (e.g. weathering) for geologically sequestering C. The Venusian crust appears to be continually resurfaced, so the lack of weathering is not due to lack of a continuous supply of fresh material. Rather, surface conditions are so hot that water remains as a vapor, preventing the precipitation of liquid water to serve as a medium to dissolve rocks.[34] An explanation for Venus's perpetual hothouse is that Venus crossed a threshold of atmospheric CO_2, which resulted in enough warming to vaporize significant amounts of water. Water itself is an effective greenhouse gas, which further warms the atmosphere, in turn evaporating more water, until the system runs away to a new hothouse steady state. The greenhouse effect of water vapor on Earth is limited by the fact that water condenses and rains out, preventing runaway. Venus may have been more prone to cross this threshold because it lies closer to the sun.

11.6.4.2 Snowballs

During the Neoproterozoic and the early Paleoproterozoic, our planet appears to have almost completely frozen over ("Snowball") based on the presence of low-latitude glacial sediments. In the Neoproterozoic, there were two or three Snowballs, which may have lasted 10–50 My.[7,8] During Neoproterozoic Snowball times, most of the continents were situated near the equator, which would have resulted in a strong ice–albedo feedback once ice sheets began to grow on these equatorial continents, ultimately leading to runaway glaciation.

It is generally agreed that the long-term baseline pCO_2 during the Neoproterozoic was low, but why this was the case is debated. One hypothesis is that continents were preferentially at low latitudes and would thus have been characterized by higher silicate weathering kinetics. A recent variant of this hypothesis invokes emplacement of the Franklin large igneous province, which increased global weatherability due to the increased surface area of easily weathered basalt.[167] Other hypotheses argue that continents were in a dispersed state during Neoproterozoic times, resulting in increased continental margins, which in turn accelerated biological productivity and enhanced the efficiency of C sequestration through organic C burial.[7,8] Highly positive seawater C isotopes predating the onset of the Snowball intervals are consistent with an unusually high fraction of organic C burial before each Snowball. Others have suggested that there was a paucity of continental arcs during the Neoproterozic, resulting in a decrease in global magmatic outputs.[87,88,145] Importantly, none of these scenarios is unique to the Neoproterozoic, which suggests that without the unique backdrop of low-latitude continents and low magmatic outputs, the conditions needed to cross the threshold would not otherwise have been met.

11.7 Summary and Further Research Directions

This chapter discussed the C cycling and long-term climate from a whole-Earth systems perspective. A conceptual framework for modeling C cycling between each reservoir was presented, along with a summary of the inventories of C in Earth's major reservoirs and the fluxes of C between them. Long-term climate (>1 My) is controlled by the amount of C in the exogenic system (atmosphere, ocean, and biosphere), which itself is controlled by exchange between the endogenic (crust and mantle) and exogenic systems. Total exogenic C is controlled by outgassing from Earth's interior via volcanism/metamorphism and sequestration of carbonate and organic C via silicate weathering and photosynthesis. The response time for silicate weathering is on the order of 100 ky, such that the amount of C in the exogenic system, and by implication climate, is at steady state on timescales longer than 1 My and endogenic inputs and exogenic outputs are balanced. Any long-term climate variability must reflect changing steady states, defined by temporal changes in the magnitude of volcanic and metamorphic inputs or in the efficiency (e.g. strength of weathering kinetics) of silicate weathering feedbacks, which draw down CO_2 by precipitating carbonate. Silicate weathering serves as a negative feedback because CO_2 drawdown increases with increasing pCO_2 through a complex series of transfer functions, in which increasing pCO_2 increases acidity and surface temperature, the latter of which increases precipitation and runoff, all of which enhance dissolution rates. Although tectonics and mantle dynamics drive endogenic outgassing, these endogenic processes can also enhance silicate weathering through tectonically driven erosion.

Long-term greenhouse intervals are likely driven by enhanced MOR spreading or the length of continental rifts and carbonate-intersecting continental arcs. Long-term icehouses have been widely thought to be driven by enhanced global weathering kinetics through an increase in orogenic activity, but decreased endogenic outgassing may be more important. Magmatic orogens may drive greenhouse intervals, but after magmatism ends, erosion and weathering of remnant topography increases global weathering kinetics, causing colder climates to immediately follow. The continental crust and CLM represent large repositories of C in the form of sediments, metamorphic C in the crust, and metasomatized domains in the lithospheric mantle. This continental C can be periodically tapped during rifting and continental arc magmatism, leading to greenhouse states. Below, we list several questions to guide future research.

What controls baseline C in the exogenic system and hence climate? Specifically, what is more effective in changing baseline C: changes in source or in the efficiency of the sink?[168] What are the relative roles of MORs, plumes, continental arcs, continental rifts, and orogens in controlling C inputs? How do global weathering kinetics change with time or with changes in the nature of Earth's surface? What is the functional form of the silicate weathering feedback? Are orogens net regional sources or sinks of C?

How has the lithospheric (crust + lithospheric mantle) reservoir of C changed through time? The answer to this question is key to quantifying how mantle degassing and metamorphic fluxes change with time. Progress may come from characterizing average amounts of carbonate and graphite in igneous and metamorphic basement rocks (see Ref. 169). It will also be important to build a more quantitative budget of carbonate- and organic C-bearing sediments on the continents (see Ref. 170).

How has the spatial and temporal distribution of C deposition changed through time? Specifically, how has C deposition been partitioned between the shallow and deep sea through time? Also of interest is whether organic C and carbonate remain coupled during sedimentary deposition; if not, the bulk redox state of total C will vary regionally, with implications for the redox state of subducted materials.

How did early Earth volatile cycles operate? During much of Earth's history, atmospheric oxygen levels were low. Questions thus arise as to whether CO_2 or some other carbon species, such as methane, was the dominant greenhouse gas during those times. Were Archean atmospheric CO_2 contents much higher than today, as seems to be required by a lower solar luminosity (a faint young sun)?[1] If other greenhouse gases were more important in the Archean, how would weathering feedbacks have behaved?

How have changes in the nature of the C cycle influenced the cycling of other elements on Earth? The dominant form of di-oxygen (O_2) production on Earth is through photosynthetic reduction of CO_2 to organic C. The evolution of atmospheric O_2 is thus controlled by the C cycle. Oxygen availability in turn influences the geochemical behaviors of redox-sensitive elements, such as S, N, and metals, as well as elements like P that complex with or absorb onto metals of a certain oxidation state.

We conclude our review with the caveat that what we have discussed here is strongly biased by our experiences on Earth. Nevertheless, the concepts presented here should be transportable to understanding element cycles and climates on other planets in our solar system or beyond.

Acknowledgments

This work was supported by the US NSF Frontiers of Earth Systems Dynamics grant (OCE-1338842) and a NASA grant 80NSSC18K0828. We thank the organizers Craig Manning and Isabelle Daniel of the 2018 Deep Carbon Science Gordon Research Conference in Rhode Island for providing inspiration. We also thank L. Carter, D. Catling, M. Edmonds, J. Eguchi, D. Grewal, M. Kastner, A. Lenardic, T. Lyons, D. Müller,

T. Plank, K. Siebach, J. Tucker, A. J. West, and L. Yeung for various insights. We thank Jeremy Caves for his insightful review.

References

1. Sagan, C. & Mullen, G. Earth and Mars: evolution of atmospheres and surface temperatures. *Science* **177**, 52–56 (1972).
2. Walker, J. C. G., Hays, P. B. & Kasting, J. F. A negative feedback mechanism for the long-term stabilization of Earth's surface temperature. *J. Geophys. Res.* **86**, 9776–9782 (1981).
3. Berner, R. A. *The Phanerozoic Carbon Cycle: CO_2 and O_2* (Oxford University Press, 2004).
4. Zachos, J. C., Dickens, G. R. & Zeebe, R. E. An early Cenozoic perspective on greenhouse warming and carbon-cycle dynamics. *Nature* **451**, 279–283 (2008).
5. Hays, J. D., Imbrie, J. & Shackleton, N. J. Pacemaker of the ice ages. *Science* **194**, 1121–1132 (1976).
6. Huybers, P. & Wunsch, C. Obliquity pacing of the late Pleistocene glacial terminations. *Nature* **434**, 491–494 (2005).
7. Hoffman, P. F. & Schrag, D. P. The snowball Earth hypothesis: testing the limits of global change. *Terra Nova* **14**, 129–155 (2002).
8. Hoffman, P. F., Kaufman, A. J., Halverson, G. P. & Schrag, D. P. A Neoproterozoic Snowball Earth. *Science* **281**, 1342–1346 (1998).
9. Berner, R. A. A model for atmospheric CO_2 over Phanerozoic time. *Am. J. Sci.* **291**, 339–376 (1991).
10. Royer, D. L., Berner, R. A., Montanez, I. P., Tabor, N. J. & Beerling, D. J. CO_2 as a primary driver of Phanerozoic climate. *Geol. Soc. Am. Today* **14**, 4–10 (2004).
11. Sundquist, E. T. & Visser, K. The geologic history of the carbon cycle. *Treat. Geochem.* **8**, 425–472 (2011).
12. Sundquist, E. T. The global carbon dioxide budget. *Science* **259**, 934–941 (1993).
13. Sundquist, E. T. Geological perspectives on carbon dioxide and the carbon cycle. In: *The Carbon cycle and Atmospheric CO_2: Natural Variations Archean to Present*, Vol. 32 (eds. E. T. Sundquist & W. S. Broecker), 5–60 (American Geophysical Union, 1985).
14. Sarmiento, J. L. & Gruber, N. Sinks for anthropogenic carbon. *Phys. Today* **5**, 30–36 (2002).
15. Doval, M. D. & Hansell, D. A. Organic carbon and apparent oxygen utilization in the western South Pacific and the central Indian Oceans. *Marine Chem.* **68**, 249–264 (2000).
16. Hansell, D. A. & Carlson, C. A. Deep-ocean gradients in the concentration of dissolved organic carbon. *Nature* **395**, 263–266 (1998).
17. Gaillardet, J., Dupre, B., Louvat, P. & Allegre, C. J. Global silicate weathering and CO_2 consumption rates deduced from the chemistry of large rivers. *Chem. Geol.* **159**, 3–30 (1999).
18. Dasgupta, R. & Hirschmann, M. M. The deep carbon cycle and melting in Earth's interior. *Earth Planet. Sci. Lett.* **298**, 1–13 (2010).
19. Tucker, J. M., Mukhopadhyay, S. & Gonnermann, H. M. Reconstructing mantle carbon and noble gas contents from degassed mid-ocean ridge basalts. *Earth Planet. Sci. Lett.* **496**, 108–119 (2018).

20. Kerrick, D. M. & Caldeira, K. Metamorphic CO_2 degassing from orogenic belts. *Chem. Geol.* **145**, 213–232 (1998).
21. Burton, M. R., Sawyer, G. M. & Granieri, D. Deep carbon emissions from volcanoes. *Rev. Mineral. Geochem.* **75**, 323–354 (2013).
22. Lee, C.-T. A. et al. Continental arc-island arc fluctuations, growth of crustal carbonates and long-term climate change. *Geosphere* **9**, 21–36 (2013).
23. Lee, C.-T. A. & Lackey, J. S. Global continental arc flare-ups and their relation to long-term greenhouse conditions. *Elements* **11**, 125–130 (2015).
24. Allard, P. et al. Eruptive and diffuse emissions of CO_2 from Mount Etna. *Nature* **351**, 387–391 (1991).
25. Petsch, S. T. Weathering of organic carbon. *Treat. Geochem.* 12, 217–238 (2014).
26. Friedlingstein, P. et al. Update on CO_2 emissions. *Nat. Geosci.* **3**, 811–812 (2010).
27. Berner, R. A. & Caldeira, K. The need for mass balance and feedback in the geochemical carbon cycle. *Geology* **25**, 955–956 (1997).
28. Broecker, W. S. & Sanyal, A. Does atmospheric CO_2 police the rate of chemical weathering. *Global Biogeochem. Cycles* **12**, 403–408 (1998).
29. Lasaga, A. C. *Kinetic Theory in the Earth Sciences* (Princeton University Press, 1998).
30. Caves, J. K., Jost, A. B., Lau, K. V. & Maher, K. Cenozoic carbon cycle imbalances and a variable silicate weathering feedback. *Earth Planet. Sci. Lett.* **450**, 152–163 (2016).
31. Kump, L. R. & Arthur, M. A. Global chemical erosion during the Cenozoic: weatherability balances the budgets. In: *Tectonic Uplift and Climate Change* (ed. W. F. Ruddiman), 399–426 (Plenum Press, 1997).
32. Zeebe, R. E. & Caldeira, K. Close mass balance of long-term carbon fluxes from ice-core CO_2 and ocean chemistry records. *Nat. Geosci.* **1**, 312–315 (2008).
33. Lodders, K. & Fegley, J. B. *The Planetary Scientist's Companion* (Oxford University Press, 1998).
34. Ingersoll, A. P. *Planetary Climates* (Princeton University Press, 2013).
35. Dickens, G. R., O'Neil, J. R., Rea, D. K. & Owen, R. M. Dissociation of oceanic methane hydrate as a cause of the carbon isotope excursion at the end of the Paleocene. *Paleoceanography* **10**, 965–971 (1995).
36. Zachos, J., Pagani, M., Sloan, L., Thomas, E. & Billups, K. Trends, rhythms, and aberrations in global climate 65 Ma to present. *Science* **292**, 686–693 (2001).
37. Saal, A. E., Hauri, E. H., Langmuir, C. H. & Perfit, M. R. Vapour undersaturation in primitive mid-ocean-ridge basalt and the volatile content of Earth's upper mantle. *Nature* **419**, 451–455 (2002).
38. Hirschmann, M. M. & Dasgupta, R. The H/C ratios of Earth's near-surface and deep reservoirs, and consequences for deep Earth volatile cycles. *Chem. Geol.* **262**, 4–16 (2009).
39. Marty, B. & Jambon, A. $C/^3He$ in volatile fluxes from the solid Earth: implications for carbon geodynamics. *Earth Planet. Sci. Lett.* **83**, 16–26 (1987).
40. Michael, P. J. & Graham, D. W. The behavior and concentration of CO_2 in the suboceanic mantle: inferences from undegassed ocean ridge and ocean island basalts. *Lithos* **236–237**, 338–351 (2015).
41. Rosenthal, A., Hauri, E. H. & Hirschmann, M. M. Experimental determination of C, F, and H partitioning between mantle minerals and carbonated basalt, CO_2/Ba and CO_2/Nb systematics of partial melting, and the CO_2 contents of basaltic source regions. *Earth Planet. Sci. Lett.* **412**, 77–87 (2015).

42. Anderson, D. L. *Theory of the Earth* (Blackwell, 1989).
43. Hofmann, A. W. Chemical differentiation of the Earth: the relationship between mantle, continental crust, and oceanic crust. *Earth Planet. Sci. Lett.* **90**, 297–314 (1988).
44. Zindler, A. & Hart, S. Chemical geodynamics. *Annu. Rev. Earth Planet. Sci.* **14**, 493–571 (1986).
45. Chi, H., Dasgupta, R., Duncan, M. S. & Shimizu, N. Partitioning of carbon between Fe-rich alloy melt and silicate melt in a magma ocean – implications for the abundance and origin of volatiles in Earth, Mars and the Moon. *Geochim. Cosmochim. Acta* **139**, 447–471 (2014).
46. Workman, R. K. & Hart, S. R. Major and trace element composition of the depleted MORB mantle (DMM). *Earth Planet. Sci. Lett.* **231**, 53–72 (2005).
47. McDonough, W. F. & Sun, S.-S. The composition of the Earth. *Chem. Geol.* **120**, 223–253 (1995).
48. Elliott, T., Plank, T., Zindler, A., White, W. & Bourdon, B. Element transport from slab to volcanic front at the Mariana arc. *J. Geophys. Res.* **102**, 14991–15019 (1997).
49. Kelemen, P. B. & Manning, C. E. Reevaluating carbon fluxes in subduction zones, what goes down, mostly comes up. *Proc. Natl Acad. Sci.* **112**, E3997–E4006 (2015).
50. Duncan, M. S. & Dasgupta, R. Rise of Earth's atmospheric oxygen controlled by efficient subduction of organic carbon *Nat. Geosci.* **10**, 387–392 (2017).
51. Dasgupta, R. Ingassing, storage, and outgassing of terrestrial carbon through geologic time. *Rev. Mineral. Geochem.* **75**, 183–229 (2013).
52. Hirschmann, M. M. Comparative deep Earth volatile cycles: the case for C recycling from exosphere/mantle fractionation of major (H_2O, C, N) volatiles and from H_2O/Ce, CO_2/Ba, and CO_2/Nb exosphere ratios. *Earth Planet. Sci. Lett.* **502**, 262–273 (2018).
53. Hofmann, A. W. & White, W. M. Mantle plumes from ancient oceanic crust. *Earth Planet. Sci. Lett.* **57**, 421–436 (1982).
54. Garapić, G., Mallik, A., Dasgupta, R. & Jackson, M. G. Oceanic lavas sampling the high-^{3}He/^{4}He mantle reservoir: primitive, depleted, or re-enriched? *Am. Mineral.* **100**, 2066–2081 (2015).
55. Wilkinson, B. H. & Walker, J. C. G. Phanerozoic cycling of sedimentary carbonate. *Am. J. Sci.* **289**, 525–548 (1989).
56. Wilkinson, B. H. & Algeo, T. J. Sedimentary carbonate record of calcium-magnesium cycling. *Am. J. Sci.* **289**, 1158–1194 (1989).
57. Ronov, A. B. Evolution of rock composition and geochemical processes in the sedimentary shell of the Earth. *Sedimentology* **19**, 157–172 (1972).
58. Wedepohl, K. H. The composition of the continental crust. *Geochim. Cosmochim. Acta* **59**, 1217–1232 (1995).
59. Gao, S. et al. Chemical composition of the continental crust as revealed by studies in East China. *Geochim. Cosmochim. Acta* **62**, 1959–1975 (1998).
60. Jordan, T. H. Composition and development of the continental tectosphere. *Nature* **274**, 544–548 (1978).
61. Lee, C.-T. A., Luffi, P. & Chin, E. J. Building and destroying continental mantle. *Annu. Rev. Earth Planet. Sci.* **39**, 59–90 (2011).
62. Griffin, W. L., O'Reilly, S. Y. & Ryan, C. G. The composition and origin of subcontinental lithospheric mantle. In: *Mantle Petrology: Field Observations and High Pressure Experimentation: A Tribute to R. (Joe) Boyd*, eds. Yingwei Fei, Constance M. Bertka & Bjorn O. Mysen, Vol. 6, 13–45 (Geochemical Society, 1999).

63. Gung, Y., Panning, M. & Romanowicz, B. Global anisotropy and the thickness of continents. *Nature* **422**, 707–711 (2003).
64. Pearson, D. G. et al. Re–Os, Sm–Nd, and Rb–Sr isotope evidence for thick Archaean lithospheric mantle beneath the Siberian craton modified by multistage metasomatism. *Geochim. Cosmochim. Acta* **59**, 959–977 (1995).
65. Pearson, D. G., Carlson, R. W., Shirey, S. B., Boyd, F. R. & Nixon, P. H. Stabilisation of Archaean lithospheric mantle: a Re–Os isotope study of peridotite xenoliths from the Kaapvaal craton. *Earth Planet. Sci. Lett.* **134**, 341–357 (1995).
66. Carlson, R. W., Pearson, D. G. & James, D. E. Physical, chemical, and chronological characteristics of continental mantle. *Rev. Geophys.* **43**, RG1001 (2005).
67. Walker, R. J., Carlson, R. W., Shirey, S. B. & Boyd, F. R. Os, Sr, Nd, and Pb isotope systematics of southern African peridotite xenoliths: implications for the chemical evolution of subcontinental mantle. *Geochim. Cosmochim. Acta* **53**, 1583–1595 (1989).
68. Foley, S. F. Rejuvenation and erosion of the cratonic lithosphere. *Nat. Geosci.* **1**, 503–510 (2008).
69. Li, Z.-X. A., Lee, C.-T. A., Peslier, A. H., Lenardic, A. & Mackwell, S. J. Water contents in mantle xenoliths from the Colorado Plateau and vicinity: implications for the rheology and hydration-induced thinning of continental lithosphere. *J. Geophys. Res.* **113**, B09210 (2008).
70. Ionov, D. A., Doucet, L. S. & Ashchepkov, I. V. Composition of the lithospheric mantle in the Siberian Craton: new constraints from fresh peridotites in the Udachnaya-East kimberlite. *J. Petrol.* **51**, 2177–2210 (2010).
71. Ionov, D. A., Chanefo, I. & Bodinier, J.-L. Origin of Fe-rich lherzolites and wehrlites from Tok, SE Siberia by reactive melt percolation in refractory mantle peridotites. *Contrib. Mineral. Petrol.* **150**, 335–353 (2005).
72. Ionov, D. A., Bodinier, J.-L., Mukasa, S. B. & Zanetti, A. Mechanisms and sources of mantle metasomatism: major and trace element compositions of peridotite xenoliths from Spitsbergen in the context of numerical modelling. *J. Petrol.* **43**, 2219–2259 (2002).
73. Zheng, J. P. et al. Refertilization-driven destabilization of subcontinental mantle and the importance of initial lithospheric thickness for the fate of continents. *Earth Planet. Sci. Lett.* **409**, 225–231 (2015).
74. Saha, S., Dasgupta, R. & Tsuno, K. High pressure phase relations of a depleted peridotite fluxed by CO_2–H_2O-bearing siliceous melts and the origin of mid-lithospheric discontinuity. *Geochem. Geophys. Geosyst.* **19**, 595–620 (2018).
75. Dasgupta, R. Volatile-bearing partial melts beneath oceans and continents – where, how much, and of what compositions? *Am. J. Sci.* **318**, 141–165 (2018).
76. Duncan, M. S. & Dasgupta, R. Pressure and temperature dependence of CO_2 solubility in hydrous rhyolitic melt – implications for carbon transfer to mantle source of volcanic arcs via partial melting of subducting crustal lithologies. *Contrib. Mineral. Petrol.* **169**, 54 (2015).
77. Dickens, G. R. Rethinking the global carbon cycle with a large, dynamic and microbially mediated gas hydrate capacitor. *Earth Planet. Sci. Lett.* **213**, 169–183 (2003).
78. Marty, B. & Tolstikhin, I. N. CO_2 fluxes from mid-ocean ridges, arcs and plumes. *Chem. Geol.* **145**, 233–248 (1998).
79. Lee, H. et al. Massive and prolonged deep carbon emissions associated with continental rifting. *Nat. Geosci.* **9**, 145–149 (2016).

80. Larson, R. L. Latest pulse of Earth: evidence for a mid-Cretaceous superplume. *Geology* **19**, 547–550 (1991).
81. Larson, R. L. Geological consequences of superplumes. *Geology* **19**, 963–966 (1991).
82. Carter, L. B. & Dasgupta, R. Effect of melt composition on crustal carbonate assimilation – implications for the transition from calcite consumption to skarnification and associated CO_2 degassing. *Geochem. Geophys. Geosyst.* **17**, 3893–3916 (2016).
83. Carter, L. B. & Dasgupta, R. Hydrous basalt–limestone interaction at crustal conditions: implications for generation of ultracalcic melts and outflux of CO_2 at volcanic arcs. *Earth Planet. Sci. Lett.* **427**, 202–214 (2015).
84. Deegan, F. M. et al. Magma–carbonate interaction processes and associated CO_2 release at Merapi volcano, Indonesia: insights from experimental petrology. *J. Petrol.* **51**, 1027–1051 (2010).
85. Carter, L. B. & Dasgupta, R. Decarbonation in the Ca–Mg–Fe carbonate system at mid-crustal pressure as a function of temperature and assimilation with arc magmas – implications for long-term climate. *Chem. Geol.* **492**, 30–48 (2018).
86. Mason, E., Edmonds, M. & Turchyn, A. V. Remobilization of crustal carbon may dominate volcanic arc emissions. *Science* **357**, 290–294 (2017).
87. McKenzie, N. R., Hughes, N. C., Gill, B. C. & Myrow, P. M. Plate tectonic influences on Neoproterozoic–early Paleozoic climate and animal evolution. *Geology* **42**, 127–130 (2014).
88. McKenzie, N. R. et al. Continental arc volcanism as the principal driver of icehouse–greenhouse variability. *Science* **352**, 444–447 (2016).
89. Johansson, L., Zahirovic, S. & Muller, R. D. The interplay between the eruption and weathering of large igneous provinces and the deep-time carbon cycle. *Geophys. Res. Lett.* **45**, 5380–5389 (2018).
90. Brune, S., Williams, S. E. & Muller, R. D. Potential links between continental rifting, CO_2 degassing and climate change through time. *Nat. Geosci.* **10**, 941–946 (2017).
91. Ferry, J. M. Dehydration and decarbonation reactions as a record of fluid infiltration. *Rev. Mineral. Geochem.* **26**, 351–391 (1991).
92. Spear, F. S. *Metamorphic Phase Equilibria and Pressure–Temperature–Time Paths.* (Mineralogical Society of America, 1993).
93. Ague, J. J. & Nicolescu, S. Carbon dioxide released from subduction zones by fluid-mediated reactions. *Nat. Geosci.* **7**, 355–360 (2014).
94. Evans, M. J., Derry, L. A. & France-Lanord, C. Degassing of metamorphic carbon dioxide from the Nepal Himalaya. *Geochem. Geophys. Geosyst.* **9**, Q04021 (2008).
95. Lee, C.-T. A. et al. Two-step rise in atmospheric oxygen linked to the growth of continents. *Nat. Geosci.* **9**, 417–424 (2016).
96. Evans, M. J., Derry, L. A. & France-Lanord, C. Geothermal fluxes of alkalinity in the Narayani river system of central Nepal. *Geochem. Geophys. Geosyst.* **5**, Q08011 (2004).
97. Jacobson, A. D., Blum, J. D. & Walter, L. M. Reconciling the elemental and Sr isotope composition of Himalayan weathering fluxes: insights from the carbonate geochemistry of stream waters. *Geochim. Cosmochim. Acta* **66**, 3417–3429 (2002).
98. Walker, J. C. G. & Opdyke, B. C. Influence of variable rates of neritic carbonate deposition on atmospheric carbon dioxide and pelagic sediments. *Paleoceanography* **10**, 415–427 (1995).
99. Zeebe, R. E. & Westbroek, P. A simple model for the $CaCO_3$ saturation state of the ocean: the "Strangelove," the "Neritan," and the "Cretan" Ocean. *Geochem. Geophys. Geosyst.* **4**, 1104 (2003).

100. Torres, M. A., West, A. J. & Li, G. Sulphide oxidation and carbonate dissolution as a source of CO_2 over geological timescales. *Nature* **507**, 346–349 (2014).
101. Galy, V. & Eglinton, T. Protracted storage of biospheric carbon in the Ganges–Brahmaputra basin. *Nat. Geosci.* **4**, 843–847 (2011).
102. Isson, T. T. & Planavsky, N. J. Reverse weathering as a long-term stabilizer of marine pH and planetary climate. *Nature* **560**, 471–475 (2018).
103. Moon, S., Chamberlain, C. P. & Hilley, G. E. New estimates of silicate weathering rates and their uncertainties in global rivers. *Geochim. Cosmochim. Acta* **134**, 257–274 (2014).
104. Coogan, L. A. & Gillis, K. M. Evidence that low-temperature oceanic hydrothermal systems play an important role in the silicate–carbonate weathering cycle and long-term climate regulation. *Geochem. Geophys. Geosyst.* **14**, 1771–1786 (2013).
105. Alt, J. & Teagle, D. A. H. The uptake of carbon during alteration of ocean crust. *Geochim. Cosmochim. Acta* **63**, 1527–1535 (1999).
106. Gillis, K. M. & Coogan, L. A. Secular variation in carbon uptake into the ocean crust. *Earth Planet. Sci. Lett.* **302**, 385–392 (2011).
107. Staudigel, H., Hart, S. R., Schmincke, H.-U. & Smith, B. M. Cretaceous ocean crust at DSDP Sites 417 and 418: carbon uptake from weathering versus loss by magmatic outgassing. *Geochim. Cosmochim. Acta* **53**, 3091–3094 (1989).
108. Coogan, L. A. & Dosso, S. E. Alteration of ocean crust provides a strong temperature dependent feedback on the geological carbon cycle and is a primary dirver of the Sr-isotopic composition of seawater. *Earth Planet. Sci. Lett.* **415**, 38–46 (2015).
109. Schidlowski, M. A 3,800-million-year isotopic record of life from carbon in sedimentary rocks. *Nature* **333**, 313–318 (1988).
110. Krissansen-Totton, J., Buick, R. & Catling, D. C. A statistical analysis of the carbon isotope record from the Archean to Phanerozoic and implications for the rise of oxygen. *Am. J. Sci.* **315**, 275–316 (2015).
111. Shields, G. A. & Mills, B. J. W. Tectonic controls on the long-term carbon isotope mass balance. *Proc. Natl Acad. Sci.* **114**, 4318–4323 (2017).
112. Plank, T. & Langmuir, C. H. The chemical composition of subducting sediment and its consequences for the crust and mantle. *Chem. Geol.* **145**, 325–394 (1998).
113. Kerrick, D. M. & Connolly, J. A. D. Metamorphic devolatilization of subducted oceanic metabasalts: implications for seismicity, arc magmatism and volatile recycling. *Earth Planet. Sci. Lett.* **189**, 19–29 (2001).
114. Kerrick, D. M. & Connolly, J. A. D. Metamorphic devolatilization of subducted marine sediments and transport of volatiles into the Earth's mantle. *Nature* **411**, 293–296 (2001).
115. Kump, L. R., Brantley, S. L. & Arthur, M. A. Chemical weathering, atmospheric CO_2, and climate. *Annu. Rev. Earth Planet. Sci.* **28**, 611–667 (2000).
116. Winnick, M. J. & Maher, K. Relationships between CO_2, thermodynamic limits on silicate weathering, and the strength of the silicate weathering feedback. *Earth Planet. Sci. Lett.* **485**, 111–120 (2018).
117. Pagani, M., Caldeira, K., Berner, R. A. & Beerling, D. J. The role of terrestrial plants in limiting atmospheric CO_2 decline over the past 24 million years. *Nature* **460**, 85–88 (2009).
118. PALAEOSENS. Making sense of palaeoclimate sensitivity. *Nature* **491**, 683–691 (2012).
119. Laidler, K. *Chemical Kinetics*, 3rd edn. (Harper & Row, 1987).

120. Maher, K. & Chamberlain, C. P. Hydrologic regulation of chemical weathering and the geologic carbon cycle. *Science* **343**, 1502–1504 (2014).
121. Held, I. M. & Soden, B. J. Robust responses of the hydrological cycle to global warming. *J. Clim.* **19**, 5686–5699 (2006).
122. Berner, R. A. GEOCARB II: a revised model of atmospheric CO_2 over Phanerozoic time. *Am. J. Sci.* **294**, 56–91 (1994).
123. West, J. Thickness of the chemical weathering zone and implications for erosional and climatic drivers of weathering and for carbon-cycle feedbacks. *Geology* **40**, 811–814 (2012).
124. Raymo, M. E., Ruddiman, W. F. & Froelich, P. N. Influence of late Cenozoic mountain building on ocean geochemical cycles. *Geology* **16**, 649–653 (1988).
125. Raymo, M. E. & Ruddiman, W. F. Tectonic forcing of late Cenozoic climate. *Nature* **359**, 117–122 (1992).
126. Rudnick, R. L. & Fountain, D. M. Nature and composition of the continental crust: a lower crustal perspective. *Rev. Geophys.* **33**, 267–309 (1995).
127. Ibarra, D. E. et al. Differential weathering of basaltic and granitic catchments from concentration-discharge relationships. *Geochim. Cosmochim. Acta* **190**, 265–293 (2016).
128. Dessert, C., Dupre, B., Gaillardet, J., Francois, L. M. & Allegre, C. J. Basalt weathering laws and the impact of basalt weathering on the global carbon cycle. *Chem. Geol.* **202**, 257–273 (2003).
129. Gislason, S. R., Arnorsson, S. & Armannsson, H. Chemical weathering of basalt as deduced from the composition of precipitation, rivers and rocks in SW Iceland. *Am. J. Sci.* **296**, 3–77 (1996).
130. Farner, M. J. & Lee, C.-T. A. Effects of crustal thickness on magmatic differentiation in subduction zone volcanism: a global study. *Earth Planet. Sci. Lett.* **470**, 96–107 (2017).
131. Tang, M., Chen, K. & Rudnick, R. L. Archean upper crust transition from mafic to felsic marks the onset of plate tectonics. *Science* **351**, 372–375 (2016).
132. Keller, C. B. & Schoene, B. Statistical geochemistry reveals disruption in secular lithospheric evolution about 2.5 Gyr ago. *Nature* **485**, 490–493 (2012).
133. Krissansen-Totton, J. & Catling, D. C. Constraining climate sensitivity and continental versus seafloor weathering using an inverse geological carbon cycle model. *Nat. Commun.* **8**, 15423 (2017).
134. Lee, C.-T. A. et al. Deep mantle roots and continental emergence: implications for whole-Earth elemental cycling, long-term climate, and the Cambrian explosion. *Int. Geol. Rev.* **60**, 431–448 (2018).
135. Lee, C.-T. A., Lee, T.-C. & Wu, C.-T. Modeling the compositional evolution of recharging, evacuating, and fractionating (REFC) magma chambers: implications for differentiation of arc magmas. *Geochim. Cosmochim. Acta* **143**, 8–22 (2014).
136. McGovern, P. J. & Schubert, G. Thermal evolution of the earth – effects of volatile exchange between atmosphere and interior. *Earth Planet. Sci. Lett.* **96**, 27–37 (1989).
137. Ridgwell, A. & Zeebe, R. E. The role of the global carbonate cycle in the regulation and evolution of the Earth system. *Earth Planet. Sci. Lett.* **234**, 299–315 (2005).
138. Ridgwell, A. A mid-Mesozoic revolution in the regulation of ocean chemistry. *Marine Geol.* **217**, 339–357 (2005).
139. Korenaga, J., Planavsky, N. J. & Evans, D. A. D. Global water cycle and the coevolution of the Earth's interior and surface environment. *Philos. Trans. A Math. Phys. Eng. Sci.* **375**, 20150393 (2016).

140. Flament, N., Coltice, N. & Rey, P. F. A case for late-Archaean continental emergence from thermal evolution models and hypsometry. *Earth Planet. Sci. Lett.* **275**, 326–336 (2008).
141. Jenkyns, H. C., Forster, A., Schouten, S. & Sinninghe Damste, J. S. High temperatures in the late Cretaceous Arctic ocean. *Nature* **432**, 888–892 (2004).
142. Haq, B. U., Hardenbol, J. & Vail, P. R. Chronology of fluctuating sea levels since the Triassic. *Science* **235**, 1156–1167 (1987).
143. Vail, P. R., Mitchum, R. W. & Thompson, S. Seismic stratigraphy and global changes of sea level, part 4: global cycles of relative changes of sea level. *Am. Assoc. Pet. Geol. Memoirs* **26**, 82–97 (1977).
144. Aiuppa, A., Fischer, T. P., Plank, T., Robidoux, P. & Di Napoli, R. Along-arc, inter-arc and arc-to-arc variations in volcanic gas CO_2/S_T ratios reveal dual source of carbon in arc volcanism. *Earth Sci. Rev.* **168**, 24–47 (2017).
145. Cao, W., Lee, C.-T. A. & Lackey, J. S. Episodic nature of continental arc activity since 750 Ma: a global compilation. *Earth Planet. Sci. Lett.* **461**, 85–95 (2017).
146. Pall, J. et al. The influence of carbonate platform interactions with subduction zone volcanism on palaeo-atmospheric CO_2 since the Devonian. *Clim. Past* **14**, 857–870 (2018).
147. Jiang, H. & Lee, C.-T. A. Coupled magmatism–erosion in continental arcs: reconstructing the history of the Cretaceous Peninsular Ranges batholith, southern California through detrital hornblende barometry in forearc sediments. *Earth Planet. Sci. Lett.* **472**, 69–81 (2017).
148. Lee, C.-T. A., Thurner, S., Paterson, S. R. & Cao, W. The rise and fall of continental arcs: interplays between magmatism, uplift, weathering and climate. *Earth Planet. Sci. Lett.* **425**, 105–119 (2015).
149. Lenardic, A. et al. Continents, super-continents, mantle thermal mixing, and mantle thermal isolation: theory, numerical simulations, and laboratory experiments. *Geochem. Geophys. Geosyst.* **12**, Q10016 (2010).
150. Kent, D. V. & Muttoni, G. Equatorial convergence of India and early Cenozoic climate trends. *Proc. Natl Acad. Sci.* **105**, 16065–16070 (2008).
151. Hodell, D. A. et al. Variations in the strontium isotope composition of seawater during the Paleocene and early Eocene from ODP Leg 208 (Walvis Ridge). *Geochem. Geophys. Geosyst.* **8**, Q09001 (2007).
152. Bataille, C. P., Willis, A., Yang, X. & Liu, X.-M. Continental crust composition: a major control of past global chemical weathering. *Sci. Adv.* **3**, e1602183 (2017).
153. Peucker-Ehrenbrink, B., Ravizza, G. & Hofmann, A. W. The marine 187Os/186Os record of the past 80 million years. *Earth Planet. Sci. Lett.* **130**, 155–167 (1995).
154. Peucker-Ehrenbrink, B. & Ravizza, G. The marine osmium isotope record. *Terra Nova* **12**, 205–219 (2000).
155. Jagoutz, O., Macdonald, F. A. & Royden, L. Low-latitude arc-continent collision as a driver for global cooling. *Proc. Natl Acad. Sci.* **113**, 4935–4940 (2016).
156. Kent, D. V. et al. A case for a comet impact trigger for the Paleocene/Eocene thermal maximum and carbon isotope excursion. *Earth Planet. Sci. Lett.* **211**, 13–26 (2003).
157. Berner, R. A., Lasaga, A. C. & Garrels, R. M. The carbonate–silicate geochemical cycle and its effect on atmospheric carbon dioxide over the past 100 million years. *Am. J. Sci.* **284**, 641–683 (1983).
158. Tranter, M. et al. Geochemical weathering at the bed of Haut Glacier d'Arolla, Switzerland – a new model. *Hydrol. Process* **16**, 959–993 (2002).

159. Torres, M. A., Moosdorf, N., Hartmann, J., Adkins, J. F. & West, A. J. Glacial weathering, sulfide oxidation, and global carbon cycle feedbacks. *Proc. Natl Acad. Sci.* **114**, 8716–8721 (2017).
160. Jenkyns, H. C. Cretaceous anoxic events: from continents to oceans. *J. Geol. Soc. Lond.* **137**, 171–188 (1980).
161. Slotnick, B. S. et al. Early Paleogene variations in the calcite compensation depth: new constraints using old borehole sediments from across Ninetyeast Ridge, central Indian Ocean. *Clim. Past* **11**, 473–493 (2015).
162. Cramer, B. S. & Kent, D. V. Bolide summer: the Paleocene/Eocene thermal maximum as a response to an extraterrestrial trigger. *Palaeogeog. Palaeoclimatol. Palaeoecol.* **224**, 144–166 (2005).
163. Kurtz, A. C., Kump, L. R., Arthur, M. A., Zachos, J. C. & Paytan, A. Early Cenozoic decoupling of the global carbon and sulfur cycles. *Paleoceanography* **18**, 1090 (2003).
164. Storey, M., Duncan, R. A. & Swisher, C. C. Paleocene-Eocene thermal maximum and the opening of the Northeast Atlantic. *Science* **316**, 587–589 (2007).
165. Selby, D., Mutterlose, J. & Condon, D. J. U–Pb and Re–Os geochronology of the Aptian/Albian and Cenomanian/Turonian stage boundaries: implications for time-scale calibration, osmium isotope seawater composition and Re–Os systematics in organic-rich sediments. *Geology* **265**, 394–409 (2009).
166. Leckie, R. M., Bralower, T. J. & Cashman, R. Oceanic anoxic events and plankton evolution: biotic response to tectonic forcing during the mid-Cretaceous. *Paleoceanography* **17**, 13-1–13-29 (2002).
167. Cox, G. M. et al. Continental flood basalt weathering as a trigger for Neoproterozoic Snowball Earth. *Earth Planet. Sci. Lett.* **446**, 89–99 (2016).
168. Godderis, Y. & Francois, L. M. The Cenozoic evolution of the strontium and carbon cycles: relative importance of continental erosion and mantle exchanges. *Chem. Geol.* **126**, 169–190 (1995).
169. White, A. F., Schulz, M. S., Lowenstern, J. B., Vivit, D. V. & Bullen, T. D. The ubiquitous nature of accessory calcite in granitoid rocks: implications for weathering, solute evolution, and petrogenesis. *Geochim. Cosmochim. Acta* **69**, 1455–1471 (2005).
170. Husson, J. M. & Peters, S. E. Atmospheric oxygenation driven by unsteady growth of the continental sedimentary reservoir. *Earth Planet. Sci. Lett.* **460**, 68–75 (2017).

12

The Influence of Nanoporosity on the Behavior of Carbon-Bearing Fluids

DAVID COLE AND ALBERTO STRIOLO

12.1 Introduction

Porosity and permeability are key variables linking the origin, form, movement, and quantity of carbon-bearing fluids that collectively dictate the physical and chemical evolution of fluid–gas–rock systems.[1] The distribution of pores, pore volume, and their connectedness vary widely, depending on the Earth material, its geologic context, and its history. The general tendency is for porosity and permeability to decrease with increasing depth, along with pore size and/or fracture aperture width. Exceptions involve zones of deformation (e.g. fault or shear zones), regions bounding magma emplacement and subduction zones. Pores or fractures display three-dimensional hierarchical structures, exhibiting variable connectivity defining the pore and/or fracture network.[2–5] This network structure and topology control: (1) internal pore volumes, mineral phases, and potentially reactive surfaces accessible to fluids, aqueous solutions, volatiles, inclusions, etc.[6–8]; and (2) diffusive path lengths, tortuosity, and the predominance of advective or diffusive transport.[9–12] For solids dominated by finer networks, transport is dominated by slow advection and/or diffusion.[13–16]

Despite the extensive spatial and temporal scales over which fluid–mineral interactions can occur in geologic systems, interfacial phenomena including fluids at mineral surfaces or contained within buried interfaces such as pores, pore throats, grain boundaries, micro-fractures, and dislocations (Figure 12.1) impact the nature of multiphase flow and reactive transport in geologic systems.[13,17–21] Complexity in fluid–mineral systems takes many forms, including the interaction of dissolved constituents in water, wetting films on mineral surfaces, adsorption of dissolved and volatile species, the initiation of reactions, and transport of mobile species.[22–24] Direct observations and modeling of physical (transport) and chemical properties (reactivity) and associated interactions are challenging when considering the smallest length scales typical of pore and fracture features and their extended three-dimensional network structures.[25,26]

The various void types and their evolution during reaction with fluids are critically important factors controlling the distribution of the fluid-accessible pore volume, flow dynamics, fluid retention, chemical reactivity, and contaminant species transport.[27–32] While fracture-dominated flow can be volumetrically dominant in shallow crustal settings

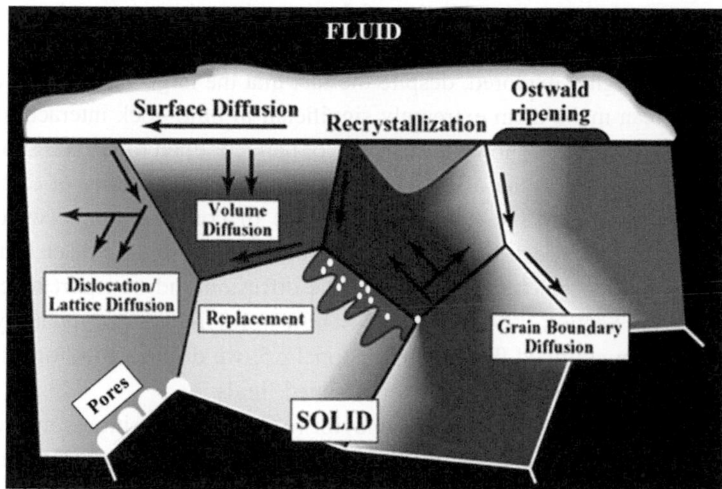

Figure 12.1 Schematic illustration of the various mass transfer and diffusion processes that can control the behavior of fluids and dissolved species in a heterogeneous crystalline matrix.

or in zones of deformation,[33–38] there is a continuum of coupled reaction–transport phenomena at length scales from fractures to the smallest nanopores. Nanoscale pores may interconnect larger volumes, act as pore throats, and play an important role in fracture initiation. Diffusion of chemical constituents from pore fluids along grain boundaries and into solids may drive the textural and chemical evolution of geologic media.[38–41]

There is general agreement that the collective structure and properties of bulk fluids are altered by solid substrates, confinement between two mineral surfaces, or in narrow pores due to the interplay of the intrinsic length scales of the fluid molecules and the length scale of fluid–solid (i.e. confinement) interactions.[22,42–45] The combined effect of confinement and intermolecular forces, which vary among different mineral systems, results in unique (and poorly constrained) perturbations to a wide range of thermodynamic and transport parameters, generally different from those observed in the bulk phase.[46–51] Also impactful is the degree of randomness of the matrix, geometry, surface composition (e.g. silicates, carbonates, organic matter (OM)), fractal nature of the pores, and connectivity of the pore network.[52]

It has also been hypothesized that chemical reactions in pores smaller than 100 nm may be different from those in larger pores because of: (1) confinement-induced changes in fluid properties and associated solutes (e.g. electrolytes, $scCO_2$, CH_4)[53]; (2) shifts in thermodynamic equilibrium composition[54–56]; and (3) overlapping electric double layers and surface curvature, distribution of preferential adsorption sites, surface charge, and counterion behavior.[57] The question as to whether surface chemical processes (e.g. ionization, complexation, adsorption) in nanopores differ from those in macropores ($>$100 nm) remains largely unresolved. Compounding this is the formidable challenge in characterizing the structure and dynamics of fluid species within internal pore or fracture networks.[13] Predictions on the direction and magnitude of changes in chemical reactions have not been thoroughly explored, despite the fact that the large surface area contributed by smaller pores can make them extremely significant in fluid–rock interactions.[6–8,58]

This chapter seeks to show how the structure of nanoconfined fluids is responsible for a variety of effects relevant to the origin, form, movement, and quantity of carbon-bearing fluids. "Form" is related to the molecular structure of confined fluids, which includes segregation of molecular species across the pore. In Section 12.3, we discuss how form is related to "movement," which is reflected by the diffusion and transport mechanisms of confined fluids. In Section 12.4, we discuss how form relates to "quantity" (i.e. the solubility of gases in confined fluids). In Section 12.5, we discuss how form is related to "origin," restricted to the reactivity of nanoconfined fluids.

The richness and complexity of fluid behavior (e.g. phase transitions, molecular orientation and relaxation, diffusion, adsorption, wetting, capillary condensation, etc.) in confined geometries underscores the need to adopt a multidisciplinary approach. In concert with molecular simulations, there is a robust set of experimental approaches used to directly or indirectly probe the behavior of fluids at interfaces or in nanopore confinement. The reader is referred to our previous research[22] and to a number of excellent reviews on molecular dynamics (MD)[50,55,59–64] and experimental methods and outcomes.[65–71]

12.2 Nanopore Earth Materials and Fluids

12.2.1 Nanopore Features

Porosity and permeability are key variables that link the thermal, hydrologic, geomechanical, and geochemical behavior exhibited by most, if not all, fluid–rock systems. In general, there is a strong positive correlation between porosity and permeability in many porous and fractured geologic media (see Nelson[72] for sedimentary rocks and Ingebritsen et al.[73] for crystalline rocks). What is truly remarkable is the extraordinary range in measured permeability (some 16 orders of magnitude), from as high as 10^{-7} m^2 (10^5 Darcy) for well-sorted gravel to as low as 10^{-23} m^2 (sub-nano Darcy) in crystalline rocks, mudstones (shale), fault gouges, and halide deposits.[38,74] An assortment of measurements (e.g. deep borehole, core-scale flooding, geophysical) indicates that both porosity and permeability tend to decrease with increasing depth. For example, permeability measurements of core obtained from the Kola superdeep well (11.4–12.0 km) range from 10^{-17} to 10^{-22} m^2, depending on the effective confining pressure.[75] In a series of seminal papers, Ingebritsen and Manning[76–78] described the permeability of continental crust as a function of depth in the context of specific geologic settings. They demonstrated that permeabilities decreased rapidly through the brittle portion of the crust to values between 10^{-14} and 10^{-16} m^2 at a depth of approximately 10 km. Below this depth through the brittle–ductile transition zone (~10–15 km) extending to depths of roughly 25 km, the permeability becomes more constant with depth, ranging between 10^{-16} and 10^{-18} m^2.

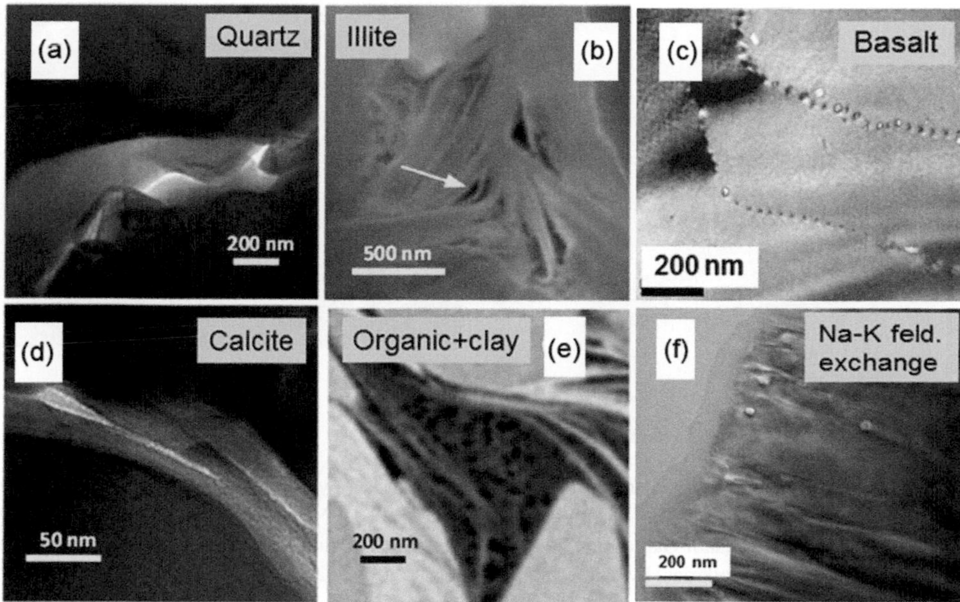

Figure 12.2 Transmission electron microscopy images of different nanoscale pores. (a) St. Peter sandstone, Wisconsin; (b) illite from Utica shale, Ohio; (c) basalt clast, Costa Rica; (d) Hueco limestone, north Mexico; (e) organic embedded in clay, Utica shale, Ohio; (f) experimental hydrothermal replacement of adularia by albite, 600°C, 200 MPa.

Norton and Knapp[39] documented the porosity of numerous crystalline rocks and classified pores (fractures) into three broad categories: flow, diffuse, and residual (isolated). Of these, the first two are relevant to the nature of the permeability regimes, in that the smaller, sometimes dead-ended diffuse pores and pore throats contribute to the chemistry of the larger, throughgoing connected pores and fractures (i.e. flow porosity). The diffuse porosity estimates identified by Norton and Knapp[39] for volcanic and crystalline rocks range from 0.01% to 10%. Bredehoeft and Norton[79] expanded on this by providing estimates of permeability for fractured igneous and metamorphic rocks that range from 10^{-12} to 10^{-16} m^2, and for unfractured equivalents ranging from 10^{-17} to 10^{-20} m^2. Anovitz and Cole[80,81] surveyed the many techniques used to quantify pore features (e.g. advanced electron microscopy, X-ray, and neutron scattering) and indicated that the range in pore size measured in rocks (sedimentary, igneous, and metamorphic) can vary by approximately eight orders of magnitude from the nanometer scale to the millimeter scale.

General nanoscale features observed in many rocks include intergranular or interparticle (Figure 12.2a, c, and d) versus intraparticle porosity (Figure 12.2b, e, and f). The latter pore type is frequently the result of dissolution–precipitation processes and is typically less well connected to the flow porosity compared to intergranular porosity involving grain-to-grain contacts of two different phases known as *interphase boundaries* or *grain boundaries*, defined as the interface of two grains of the same phase. The intergranular pore type is

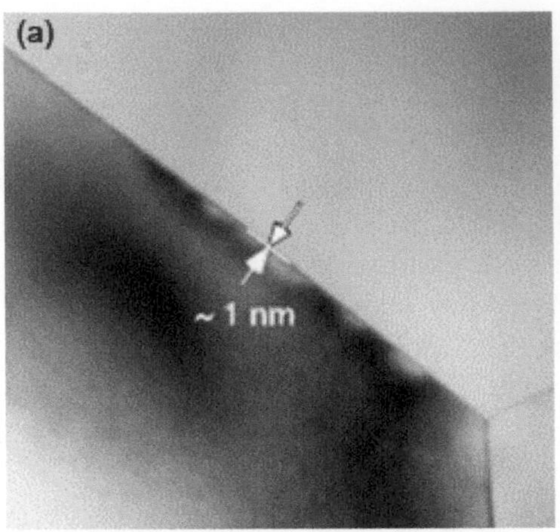

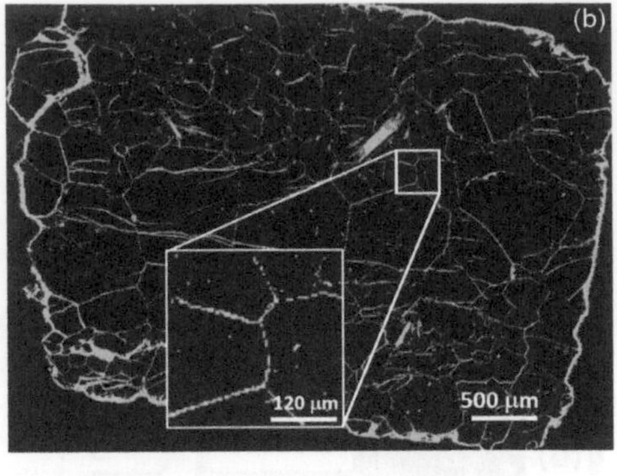

Figure 12.3 (a) High-resolution transmission electron microscopy image of the San Carlos olivine grain boundary and triple junction (from Marquardt and Faul[41]). Note the estimated grain boundary width of ~1 nm. (b) Balsam Gap dunite showing the extent of fluid infiltration and serpentinization along olivine grain boundaries, hydrothermally treated at 300°C, 200 MPa.

Reprinted with permission from Marquardt K, Faul UH. The structure and composition of olivine grain boundaries: 40 years of studies, status and current developments. *Phys Chem Miner* 2018; **45**: 139–172.[41] Copyright 2018 *Physics and Chemistry of Minerals*.

thought to control transport in very tight rocks, especially those encountered deeper in the crust and mantle. As Wark and Watson[82] point out, the premise is that rocks below the brittle–ductile transition (>12 km for continental crust) cannot sustain a fracture network system, and thus fluids – especially at low-volume fractions – must reside in intergranular pores. Intergranular fluids, however, will flow only if the pores (i.e. grain boundaries in this case) are interconnected. Fluid flow may be a misnomer when discussing very deep-seated rocks. The width of space between grains in very deep crustal and mantle rocks (e.g. olivine-rich mantle) has been a source of debate for many years. Marquardt and Faul[41] provided an excellent summary of the structure and composition of olivine grain boundaries and conclude that the width is on the order of 1 nm or less (Figure 12.3a). At this reduced dimension, fluid mobility takes the form of grain boundary diffusion, which can be several orders of magnitude greater than volume (lattice) diffusion through minerals, but many orders of magnitude slower than advective flow.[83–85] Nevertheless, fluids do penetrate these spatially restricted domains, as evidenced by dissolution, grain boundary migration, and phase alteration (e.g. olivine conversion to serpentine) (Figure 12.3b). High-resolution transmission electron microscopy reveals that some olivine-rich rocks like the San Carlos olivine also display amorphous films that are 1–2 nm wide between olivine grains thought to be remnant melt.[86]

The presence of nanoscale pores and fractures seems to be widespread in Earth materials and may be extremely significant for controlling fluid accessibility in deep-seated crustal and upper-mantle rocks depending on the extent of pore network connectivity.

Pores may be either slit-like or cylindrical in shape and can exist along grain boundaries, at the contact of unlike grains (interphase boundaries), or entirely within a given phase. In simplest terms, the pore surface compositions are dominated by silica or silicates, oxides, clay, carbonate, and, where present, OM of varying thermal maturity. As we discuss below, it is at the dimensions of a few nanometers to tens of nanometers that fluid properties deviate significantly from those of the bulk fluids.

12.2.2 Fluid Properties Affected by Nanoconfinement

It has been reported that the collective structure and properties of bulk fluids are altered by confinement between two surfaces or in narrow pores due to the interplay of the intrinsic length scales of the fluid and the length scale due to confinement.[87–89] For example, melting/freezing of water is affected by confinement in geomaterials,[90,91] and confined fluids can be liquid at the thermodynamic conditions (pressure and temperature) at which the correspondent bulk fluids are gaseous. It has also been shown that confinement can promote new phases that simply do not exist in the bulk.[92] This has been an area of particular interest to those who investigate the properties of water. It has been demonstrated that confined water can be prevented from freezing at conditions well inside the "no-man's land."[93] It is possible that similar effects will also be observed for carbon-bearing fluids, although perhaps the high temperature in the subsurface will prevent the formation of exotic phases. In point of fact, almost any fluid property is influenced by nanoconfinement to a varying degree: thermophysical (e.g. density, phase behavior, dielectric constant, compressibility), transport (e.g. diffusion, shear), dynamical motion (translational, rotational, vibrational, and librational), and interfacial interactions (e.g. adsorption, wetting, solubility) are expected to be strongly dependent upon confinement.

12.3 Form and Movement: Transport Mechanisms under Nanoconfinement

The main mechanisms responsible for fluid transport are convective motion and molecular diffusion. When the pore size becomes of the order of magnitude of fluid molecules (i.e. in nanopores), diffusion can become the dominant mechanism. Because in the subsurface many pores are in this size regime, it is important to understand the mechanisms by which molecular diffusion can take place in nanopores. In this section, we discuss how the structure of confined fluids (form) affects their transport (movement). We focus on fluid mixtures. The studies summarized here are selected primarily from the body of work produced by the authors. Future investigations will stress the wide applicability of our extrapolations.

As a first example, we refer to Chiavazzo et al.,[94] who used computer simulation to investigate water transport in "nanoconfined geometries" (i.e. within nanopores, around nanoparticles, carbon nanotubes, and proteins). Almost 60 cases were considered, some being obtained from the literature. The results, reproduced in Figure 12.4, were interpreted using a general phenomenological relationship:

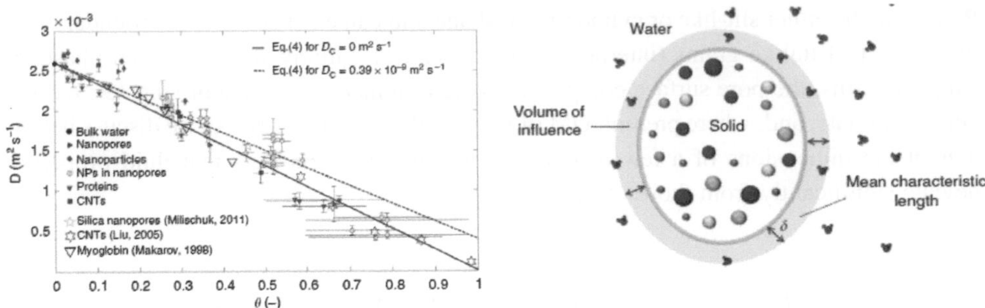

Figure 12.4 (Left) Self-diffusion coefficient for nanoconfined water as a function of the ratio θ between "surface" and total water volumes. The image contains simulation results for many systems. (Right) Schematic illustrating the region of influence of a solid substrate on an interfacial fluid. Figure reproduced from Chiavazzo E, Fasano M, Asinari P, Decuzzi P. Scaling behaviour for the water transport in nanoconfined geometries. *Nat Commun* 2014; **5**: 3565.[94] Permission granted from Creative Commons https://creativecommons.org/licenses/by/3.0.

$$D(\theta) = D_B\left[1 + \left({}^{D_C}/_{D_B} - 1\right)\theta\right]. \tag{12.1}$$

In (12.1), D is the overall self-diffusion coefficient of water calculated in the various systems and D_C and D_B are the self-diffusion coefficients estimated for confined and bulk water, respectively. According to this relationship, D was found to scale linearly with the ratio θ between "surface" and total water volumes. The "surface" water volume is a measure of the interfacial region where water molecules are in contact with the confining material. As the pore width decreases, the amount of "surface" versus total volume increases.

12.3.1 Steric Effects Enhance Surface versus Pore Diffusion

When fluid molecules are confined within nanopores, they can experience a strong attraction to the pore surfaces (e.g. via hydrogen bonds). It is possible that the molecules strongly adsorbed on a surface show very slow diffusion parallel to the surface (i.e. surface diffusion). This is because to move from one adsorption site to an adjacent one, the fluid molecules need to be released from one strong interaction. Hopping mechanisms can be observed, in which one molecule hops from one preferential adsorption site to the next. As the adsorption energy becomes stronger, the frequency of such hops lowers, and diffusion is delayed. If the preferential adsorption sites are so close to each other that hopping is not necessary, it is possible to observe sliding.[95] It is generally assumed that surface diffusion is slower than that near the pore center (i.e. "pore diffusion"). For a pure fluid, as the pore size decreases, the contribution of surface diffusion increases compared to that of surface diffusion, as shown in Figure 12.4 in the case of water.

However, we also know that the opposite situation can also arise. Phan et al.[44] considered a fluid mixture composed of ethanol and water inside a slit-shaped alumina nanopore of width 1 nm. The pore surface exposed a high density of –OH groups. Equilibrium studies on a flat surface[43] showed that water preferentially adsorbs on the alumina substrate and displaces ethanol molecules, should these be present on the surface. When the pore is of 1 nm, the physical state of the system, depending on composition, can present one layer of water on each of the two pore surfaces, with ethanol near the pore center. The pore width is so narrow that steric effects hinder the diffusion of ethanol near the pore center, while water surface diffusion is possible. This effect disappears when the pore width increases. A detailed analysis of the trajectories of the confined water molecules showed that the diffusion of water is characterized by hops. Such hops are from one adsorption site to another one on the surface across the pore volume. This mechanism becomes hindered as the pore width increases because the water molecules would have to transport across a large region filled with ethanol. In this example, increasing pore width delays water surface diffusion while allowing faster ethanol pore diffusion.

Complementary experimental investigations are difficult because it is hard to differentiate between molecules that are adsorbed on the surface from those that accumulate near the pore center. Some experimental results, however, suggest that, in some cases, surface diffusion can be the dominant transport mechanism. For example, Kim et al.[96] synthesized "silylated" mesoporous silica membranes. The chemical functionalization made the pore surfaces more attractive for ethanol than for water. Once exposed to water–ethanol liquid mixtures, ethanol permeated the membranes much faster than water, suggesting that surface diffusion became an important mechanism of transport. Ge et al.[97] used zeolites to enhance polymer-based composite membranes. The zeolites preferentially attracted water, and separation factors of ~10,000 were achieved in water-removal pervaporation processes (i.e. water permeated these membranes much faster than ethanol). For completeness, it should be remembered that the permeability of a membrane is equal to the product of the diffusion coefficient of the fluid through the membrane and the amount adsorbed. Thus, enhanced diffusion might not be the only mechanism responsible for the results just discussed.

12.3.2 Molecular Lubrication Enhances Pore Diffusion

Even in the absence of preferential adsorption sites where nonspecific attractive interactions lead to an accumulation of fluids near pore surfaces, it is generally expected that the diffusion of the fluid molecules in a nano- to meso-pore is slower than that in the bulk. For example, the enhanced fluid density near the pore walls leads to more frequent fluid–fluid and fluid–pore collisions, which reduce the mean free path and the diffusion coefficient. This expectation is demonstrated, for example, by quasielastic neutron scattering (QENS) experiments conducted for ethane in the gas phase. When ethane was confined in two SiO_2 pore-glass materials with pore sizes 11.1 and 41.5 nm, the self-diffusion coefficient was found to be a factor of ~4 slower compared to that measured for bulk ethane at similar temperature–pressure conditions.[98]

If these results are due to the fluid accumulation near the pore interface, then it might be possible to enhance the fluid diffusion by adding small amounts of another fluid that preferentially adsorbs on the solid substrate. For example, Le et al.[99,100] simulated systems containing a hydrocarbon (either butane or octane) and CO_2 in a silica slit-shaped pore of width ~2 nm. CO_2 formed molecularly thin layers on the pore surfaces, while the hydrocarbon filled the pore. At low dosages, CO_2 enhanced hydrocarbon diffusion. These computational results were confirmed by Gautam et al.[101] and Patankar et al.,[98] who used QENS to study the diffusion of propane and ethane in mesoporous SiO_2, respectively, in the presence of CO_2. The ethane experiments, conducted at different temperatures, showed that the activation energy of diffusion decreased in the presence of CO_2 by a factor of ~2 compared to data obtained for pure ethane, which was consistent with the simulations. Others also reported that CO_2 enhanced the diffusion of hydrocarbons through different porous materials.[99,101,102]

12.3.3 Molecular Hurdles Due to Strong Fluid–Fluid Interactions

The enhanced diffusion discussed in Section 12.3.2 is due to the preferential adsorption of CO_2 to the pore surface, but one can ask whether similar observations are possible when a fluid other than CO_2, also strongly attracted to the pore surface, is used. One such fluid could be water. Gautam et al.[103] used QENS to probe whether this was the case. The system used contained propane and D_2O in silica-based, cylindrical pores of diameter ~1.5 nm. The experimental results showed that the propane diffusion was depressed, rather than enhanced, by D_2O. Le et al.[104] conducted MD simulations for a system mimicking the experimental one. The results showed that although water molecules are strongly adsorbed on the pore surface, the pores have such a narrow diameter and the water–water interactions are so strong that water molecules form molecular bridges across the pore volume. These bridges hinder propane diffusion. Simulation snapshots illustrating the phenomenon are shown in Figure 12.5.

The formation of water bridges was also found to delay methane transport across slit-shaped muscovite pores. Ho and Striolo[105] found that methane convective transport within slit-shaped pores carved out of muscovite and partially filled with water is strongly dependent on the relative orientation of the water bridges and that of imposed motion. When the water bridges are perpendicular to the direction of motion, methane transport is regulated by the movement of water, which adheres to the pore surface. When the water bridges are parallel to the direction of motion, fast methane transport is observed.

Phenomena such as those just described are expected to be important when low-solubility or immiscible fluids are simultaneously present within a pore, and they are relevant for the production of shale gas, in which water is often used as the fracturing fluid. Many studies have addressed how methane could escape kerogen. These studies could be relevant for systems in which pores are extremely heterogeneous in geometry and of comparable size to the fluid molecules.

12.3.4 Transport of Guest Molecules in Confined Fluids

It is worth discussing the relation between the structure of fluids in confined water and their transport. Phan et al.[106] simulated the transport of a mixture mimicking natural gas

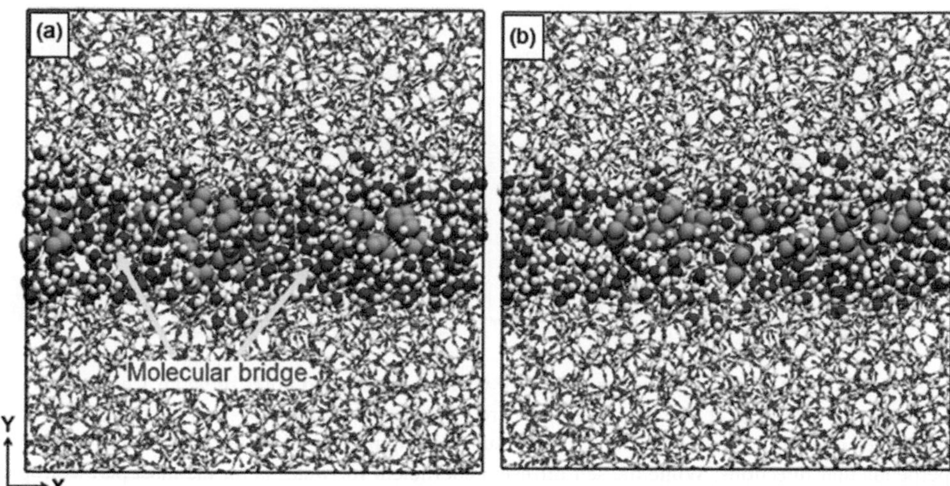

Figure 12.5 Simulation snapshots representing the effect of water molecular bridges (red and white spheres) on the transport of propane (cyan spheres) diffusing through a cylindrical pore carved out of amorphous silica (red and yellow spheres connected by lines). In (a), the water bridges hinder the transport at higher water/gas ratios ($\geq$17), while in (b), the bridges are dissociated and propane diffusion is faster at lower water/gas ratios ($\leq$12).

Reprinted with permission from Le TTB, Striolo A, Gautam S, Cole DR. Propane–water mixtures confined within cylindrical silica nanopores: structural and dynamical properties probed by molecular dynamics. *Langmuir* 2017; **33**: 11310–11320.[104] Copyright 2017 American Chemical Society.

(ethane, methane, and H_2S) through nanopores filled with water. The results show that H_2S permeates the pores much faster than the other compounds and that ethane is too large to permeate the hydrated pores. Bui et al.[107] extended the study to consider different porous materials. In the left-hand panel of Figure 12.6, we report the self-diffusion coefficient estimated for water in various nanopores: confinement reduces the self-diffusion of water, consistent with Figure 12.4. For reference, MD simulations yield a self-diffusion coefficient for bulk liquid water of ~28×10^{-10} m^2/s.[108] The results in Figure 12.6 also show that the chemical properties of the minerals are important: all the pores considered have the same width and the simulation temperature is constant, yet the water self-diffusion coefficient varies by a factor of ~2. In the right-hand panel of Figure 12.6, we report the self-diffusion coefficients estimated for aqueous methane confined in the water-bearing pores considered in the left-hand panel of Figure 12.6. There is a correlation between the self-diffusion coefficient of water and that of aqueous methane, but there are also deviations. Bui et al.[107] highlighted the fact that the diffusion of aqueous methane in calcite nanopores can be anisotropic (the diffusion along the x-axis of the pore is about three times faster than that along the y-axis). For completeness, it is worth pointing out that anisotropic diffusion of hydrocarbons had been previously reported in nanopores carved out of calcite by Franco et al.,[109] suggesting that the surface structure of this solid material indeed affects the transport of the guest molecules.

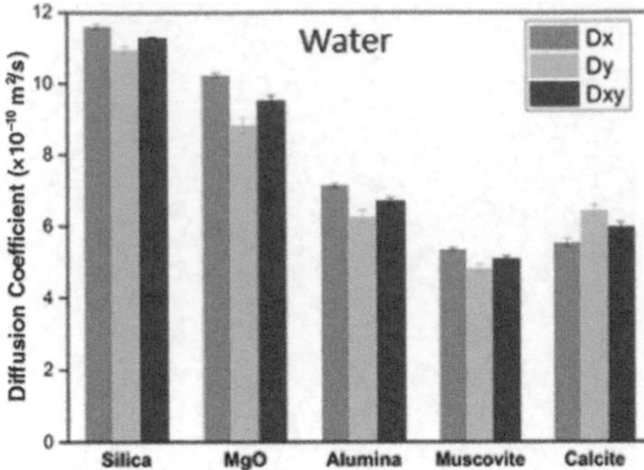

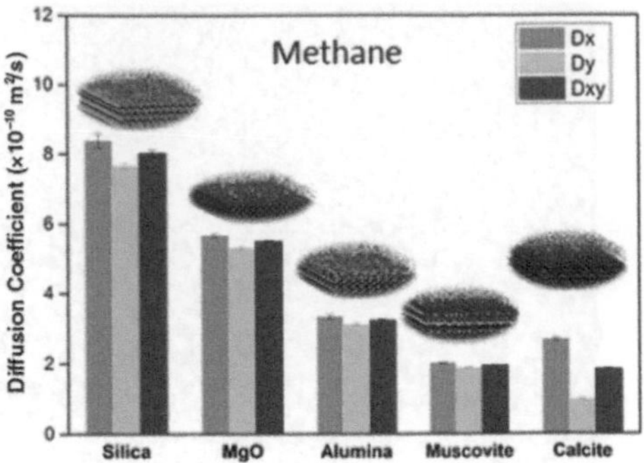

Figure 12.6 (Left) Self-diffusion coefficient for water in slit-shaped pores of width 1 nm carved out of different minerals at the same temperature. (Right) Self-diffusion coefficient for methane molecules dissolved in confined water. The results distinguish the diffusion coefficients along two perpendicular directions. While the self-diffusion coefficient for water is for the most part isotropic, that of aqueous methane in calcite pores is highly anisotropic, with transport along the x-axis being much faster than that along the y-axis.

Reprinted with permission from Bui T, Phan A, Cole DR, Striolo A. Transport mechanism of guest methane in water-filled nanopores. *J Phys Chem C* 2017; **121**: 15675–15686.[107]

12.3.5 Transport of Aqueous Electrolytes in Narrow Pores

Argyris et al.[110] and Ho et al.[111] attempted to quantify the mobility of aqueous electrolytes at low concentration within narrow slit-shaped pores. The goal was to quantify potential differences in the mobility of Cs^+ and Na^+ ions, which have the same valence but significantly different size. The aqueous solutions were electrostatically balanced by Cl^- ions. These MD simulations were conducted at similar conditions (1 M ionic strength, ambient temperature). The cristobalite silica pores were slit-shaped, with widths of ~1.2 nm. The surface chemistry was manipulated to replicate different pH scenarios. The results were quantified in terms of the self-diffusion coefficients of the electrolytes in the direction parallel to the pore surface. A strong correlation was observed between the preferential position of an ion in the direction perpendicular to the pore surface and its mobility. This correlation was interpreted based on prior results[112,113] that revealed that the mobility of interfacial water strongly depends on the distance from the solid substrate, as well as on the protonation state of the surface. The results concerning ion mobility are summarized in Figure 12.7.

It is worth stressing that the results depend strongly on the properties of the pore surface. In fact, MD simulations by Rotenberg et al.[114] have shown how Cs^+ ions tend to accumulate near clay surfaces, presumably because of the difference in surface charge. Because of the importance of preferential adsorption sites (e.g. see the recent contribution by Loganathan and Kalinichev[115]), which include pore edges, and because of the practical importance of quantifying ion migrations in the subsurface, advanced force fields are being developed to enhance the reliability of atomistic simulations,[116–118] and various approaches are being developed to include hydrodynamics and multiscale properties in the calculations.[119–122]

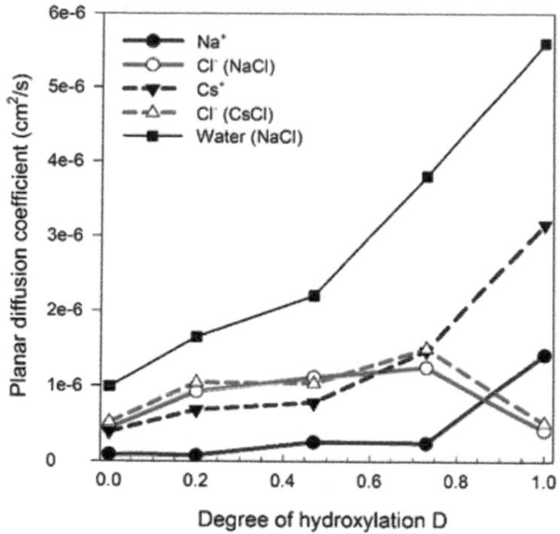

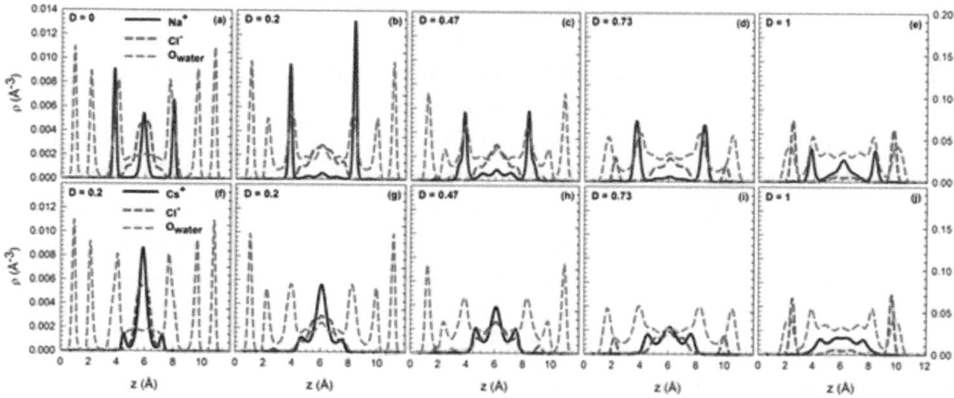

Figure 12.7 (Top) Self-diffusion coefficient estimated for different aqueous electrolytes and water in slit-shaped pores as a function of the degree of protonation D (D=0 in panel (a); D=0.2 in (b); D=0.47 in (c); D=0.73 in (d); and (D=1 in (e)). In all systems, Cs^+ ions are more mobile than Na^+ ions. (Bottom) Density profiles in the direction perpendicular to the solid substrate, with the same degree of protonation as in Top panels (panels f-j). The density profiles discriminate NaCl and CsCl systems. In general, the Cs^+ ions accumulate near the centers of the pores, while Na^+ ions are more closely associated with the pore surfaces. Because the molecular mobility depends on the distance from the surface, slower near the surface, the preferential distribution correlates with the ionic mobility.

Reprinted with permission from Ho TA, Argyris D, Cole DR, Striolo A. Aqueous NaCl and CsCl solutions confined in crystalline slit-shaped silica nanopores of varying degree of protonation. *Langmuir* 2012; **28**: 1256–1266.[111] Copyright 2012 American Chemical Society.

12.4 Form and Quantity: Confinement Effects on Solubility

To estimate the amounts of fluids present in the subsurface, one approach could be first to estimate the volume available (i.e. the accessible pore space) and then to use our understanding of bulk fluids to estimate the amount of fluids in such a volume at the subsurface pressure–temperature conditions. This approach is reasonable as long as the fluid behavior in the bulk is consistent with that in the subsurface, which it is for larger pores (>100 nm). In this section, we discuss the effects nanopores have on the structure of fluids ("form") that deviate markedly from the bulk. One property affected strongly by nanoconfinement is solubility. Another property of importance concerns the behavior of aqueous electrolytes in confinement. Because confinement alters the dynamics of water molecules, could it be that the dielectric constant of water in the deep subsurface differs strongly from that known in the bulk? Should this be the case, what are the implications regarding our still limited understanding of aqueous electrolytes in the subsurface?

12.4.1 Volatile Gas Solubility in Confined Liquids

Several studies reported an enhanced solubility of gases in liquids confined in small pores,[123,124] a phenomenon referred to as "oversolubility." Luzar and Bratko[125] reported 5–10-fold enhancements of N_2 and O_2 solubility in water in 38–43-Å hydrophobic pores. Experiments support these estimates. Pera-Titus et al.[126] studied the solubility of H_2 in $CHCl_3$, CCl_4, n-hexane, ethanol, and water within γ-alumina, silica, and MCM-41 and found that H_2 solubility was enhanced by up to 15 times when the pore size was <15 nm. Rakotovao et al.[127] confirmed these results using ^1H nuclear magnetic resonance (NMR).

According to Ho et al.,[123] oversolubility could be due to: (1) more favorable interactions between solute and solid than between solvent and solid, favoring solute adsorption close to the pore walls; (2) the pore being partially filled, resulting in a gas–solvent interface; and (3) adsorption of the gas being favored in regions of low solvent densities due to layering. Gadikota et al.[128] showed that solute size and the presence of salt in confined water are also important in modulating the dissolution of various gases in confined water. Using both experiments and simulations, these authors demonstrated greater solubility for CO_2 and Ar in Na-montmorillonite compared to bulk water, whereas confinement reduced CH_4. Gadikota et al.[128] analyzed the contributions to the free energy of solvation due to cavity formation and affinity to the water-filled pore. The free energy change due to cavity formation was found to be generally positive (i.e. it introduces a barrier to solubility), while the enthalpic contribution is generally favorable to solubilization. Both effects depend on the size of the gas molecule.

Phan et al.[129] revealed approximately one order of magnitude higher solubility of methane in water confined in a partially filled 1 nm-wide silica pore. Methane molecules accumulated in narrow regions within the pores, where water molecular density was low. It was found that water molecules surrounding the guest methane were organized in hydrogen-bonded structures resembling gas hydrates. However, because of steric constraints, these clathrates were not stable. Phan et al.[130] also considered other 1-nm slit-pore systems (MgO, Al_2O_3). The results demonstrated that methane solubility in confined water strongly depends on the confining material (Figure 12.8) because the molecular structure of confined water differs markedly within the three pore types. A direct correlation was observed between methane diffusion coefficients and water molecular density fluctuations.

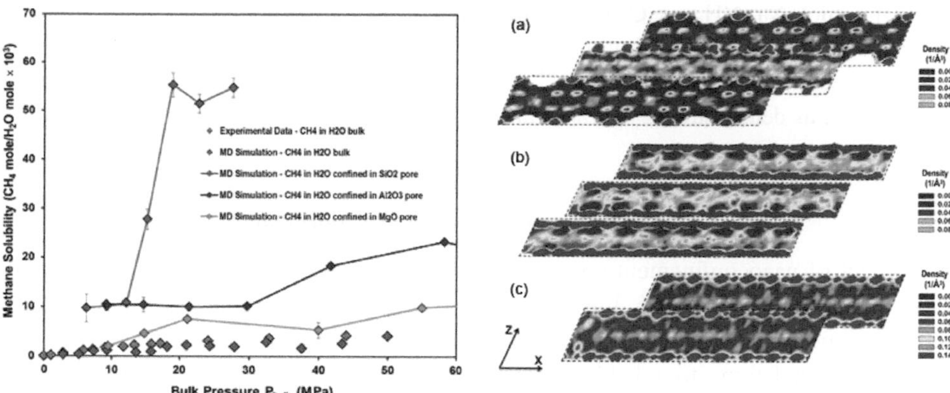

Figure 12.8 (Left) Methane solubility in confined water as a function of bulk pressure in MPa – SiO_2 blue, Al_2O_3 red, MgO green. Simulated bulk solubility of methane in liquid water at 298 K from Sakamaki et al.[131] (Right) Variability for in-plane density distributions of water oxygen in layers parallel to the x–z plane at several locations along the y-axis within (a) SiO_2, (b) Al_2O_3, and (c) MgO. The results are for pores containing only water. Densities are expressed in number of molecules per cubic Å.

Reprinted with permission from Phan A, Cole DR, Striolo A. Factors governing the behaviour of aqueous methane in narrow pores. *Philos Trans A Math Phys Eng Sci* 2015; **374**: 20150019.[130] Copyright and usage: the authors.

These observations are in contrast to those by Badmos et al.,[132] who estimated, using molecular dynamics, the solubility of H_2S in water confined in silica nanopores at 313 K. The simulation results suggested that confinement reduces the H_2S solubility in water and that the solubility increases with increasing pore size. These results are due to perturbations on the coordination of water molecules around H_2S due to confinement. The perturbations are stronger in the narrowest pores considered. These results are qualitatively consistent with those reported for aqueous NaCl reported by Malani et al.[133] This group reported a lower NaCl solubility in water confined within graphene nanopores. However, NaCl is a salt, and therefore it is likely that the mechanisms responsible for the reduced solubility of H_2S and NaCl in confined water are due to different interfacial processes.

The interest in geologic CO_2 sequestration resulted in numerous studies on the form and movement of CO_2 through porous networks, chiefly focusing on montmorillonite. Bowers et al.[134] studied the incorporation of supercritical CO_2 in smectite by *in situ* ^{13}C and ^{23}Na magic angle spinning NMR. They observed line broadening in the $^{13}CO_2$ resonance at 90°C and 50 bar CO_2 pressure for Na-hectorite, consistent with CO_2 pressure in the interlayers. Kirkpatrick et al.[135] reviewed efforts using NMR and molecular modeling to explore mineral surfaces and interlayer filling by H_2O, cations, anions, CO_2, and natural OM in swelling clays. In addition to entering the interlayer, there is evidence that CO_2 dissolves in the confined water. Schaef et al.[136] used NMR, x-ray diffraction and Monte Carlo simulation to explore the intercalation of dry supercritical CO_2 into three swelling clays in which Na, NH_4, or Cs occupied the interlayer. In the absence of water, CO_2 did enter the NH_4 and Cs clays, but was excluded from the Na clay. A weak intercalation of CH_4 has been observed for smectites,[137] but is more pronounced for 4-nm weakly hydrated silica pores, as demonstrated by Ok et al.[138]

12.4.2 Aqueous Electrolytes in Confinement

The effect of nanoconfinement on the transport of aqueous electrolytes has been briefly summarized in Section 12.3.5. It is possible that, because confinement affects the structure and dynamics of confined water, it indirectly affects the solubility of electrolytes. For molecular simulations to reliably predict the properties of confined electrolytes, it is important that adequate force fields describe the interactions between water, electrolytes, and pore constituents, as well as all the cross-interactions between these components.[139,140] The reliability of the force fields used to describe the electrolytes depends on the model chosen to simulate water, and it is essential to tune the ion–ion interactions if one seeks to replicate the experimental salt solubility.[141,142] Svoboda and Lísal[140] recently employed the grand canonical Monte Carlo algorithm to predict the solubility of NaCl in montmorillonite pores. The simulations were conducted at 365 K and 275 bar. At these conditions, the simulation models implemented (SPC/E model for water) predict a saturation NaCl concentration in water of 3.14 mol/kg,[139] which underestimates the experimental value of 6.1 mol/kg.[143] The pore widths considered ranged from 1.0 to 3.2 nm, and they were simulated in equilibrium with saturated bulk aqueous NaCl solutions. The results show that the concentrations of the adsorbed ions in the nanopores are lower than in the bulk aqueous system and decrease as the pore width narrows. Moucka et al.[139] used simulations to compare the aqueous solubility of NaCl in Na-montmorillonite and pyrophyllite pores of the same width. They were able to distinguish the ionic solubility near the center of the pore, where the properties are similar to those found in bulk water, and those near the pore surfaces, where the properties are significantly different from those found in bulk water. As the pore width decreases, the contribution of the region near to the pore surface increases compared to that of the fluids near the pore center, and consequently the NaCl solubility strongly decreases compared to that observed in bulk water. In addition to structural effects, one could attempt to explain these observations based on the changes in the dielectric constant of water due to confinement, discussed below.

12.4.3 Dielectric Constant of Nanoconfined Water

The dielectric properties of water and electrolytes in nanopores are still poorly constrained for most substrates relevant to the subsurface. According to Renou et al.,[144] the static dielectric constant (ε) characterizes the capacity of a material to transmit an electric field and controls the charge migration and dipole reorientation. Compared to other fluids types, bulk water has a high ε due to the strong bonds between hydrogen and oxygen as well as hydrogen bonding between water molecules. This large ε means that ionic substances tend to dissociate in water, yielding ionic solutions. Thus, the water ε is essential to understanding the solvation of ions and their transport. For example, Sverjensky et al.[145] extracted from the research of Pan et al.[146] the theoretical estimates of the water dielectric constant at extreme pressures and temperatures to calculate the solubility of quartz and corundum to 600 MPa and 1200°C. This effort indicates that ε ranges from as low as ~2 at 1200°C and 100 PMa to ~80 for ambient conditions.

Of interest to this chapter is how much of an effect nanoconfinement might have on ε, and further, how this change might impact mineral solubility. In confinement or near interfaces, ε is no longer a simple scalar quantity, but it becomes tensorial. The radial component of water $\varepsilon_\perp$ is fundamentally important to the solvation and transport of ions, in contrast to the axial component ($\varepsilon_{\|}$), which is important for transport behavior along the pore wall. However, there are conflicting results with respect to the magnitude of the ε values. For example, a number of MD studies have demonstrated a local interfacial increase in the axial (tangential) component of the dielectric constant ($\varepsilon_{\|}$) relative to the bulk water value and to values exhibited by the radial $\varepsilon_\perp$.[147–149] An enhanced $\varepsilon_{\|}$ was also predicted in other studies that addressed the role of cylindrical silica and carbon nanopores.[144,149–151] Conversely, values for the radial $\varepsilon_\perp$ tend to be equal to or much less than the bulk ε. This anisotropy is most pronounced in hydrophobic pores.

Unlike the increase of the axial dielectric permittivity near the pore wall noted above, Renou et al.[144] observed a significant decrease in $\varepsilon_{\|}$ relative to the ambient bulk water value with increasing surface charge density using MD. More recently, Fumagalli et al.[152] observed anomalously low ε (approaching 1) of confined water that was two to three molecules thick in dielectric imaging experiments involving graphene and boron nitrite. Similarly, Zhu et al.[153] reported on the anomalous behavior of ε for aqueous solutions of NaCl of varying concentration (0–1.5 M) confined in a 2.4-nm diameter cylindrical silica pore (Figure 12.9). They noted that the dielectric constant profile near the surface correlates with that of the water density. Using classical MD simulations, Zhu et al.[153] demonstrated: (1) a marked anisotropy between $\varepsilon_{\|}$ and $\varepsilon_\perp$; (2) that the values for $\varepsilon_{\|}$ decrease relative to those of pure water, but show enhancement near the pore wall; and (3) that $\varepsilon_\perp$ shows a nonlinear NaCl concentration dependency. Renou et al.[154] addressed the structural and dielectric behavior of confined NaCl, NaI, $MgCl_2$, and Na_2SO_4 (1 M concentration) in nanoporous hydrophilic 1.2-nm silica pores. They also reported a dielectric anisotropy of confined water and an unusual increase in the radial component of permittivity ($\varepsilon_\perp$) of confined solutions.

It is notable that the range of dielectric constant values reported for water confined in nanopores encompasses the range of ε values estimated for bulk water from ambient to extreme conditions. This suggests the distinct possibility that the bulk values for ε currently used to estimate fluid properties in the subsurface might not be appropriate, especially for pores such as those illustrated in Figures 12.2 and 12.3. We lack a comprehensive assessment of ε for nanoconfined environments relevant to subsurface conditions (i.e. at elevated pressure and temperature with widely varying solution compositions) and

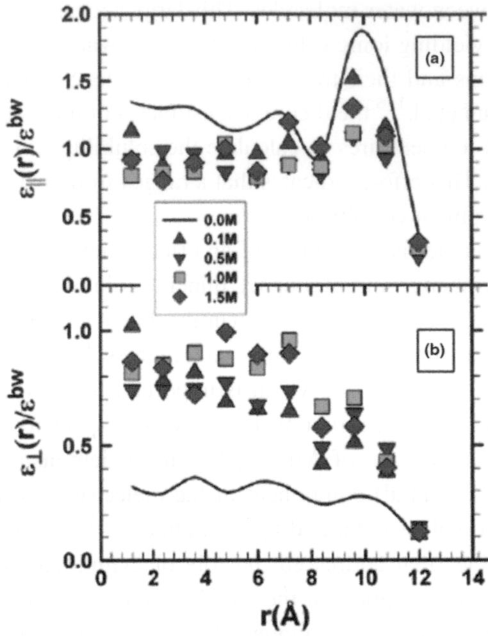

Figure 12.9 Profiles of the ratios of $\varepsilon_{||}$ (a) and $\varepsilon_{\perp}$ (b) for NaCl solutions of varying concentrations confined in 2.4-nm diameter cylindrical SiO_2 pores to the dielectric constant of bulk water (ε^{bw}). Reprinted with permission from Zhu H, Ghoufi A, Szymczyk A, Balannec B, Morineau D. Anomalous dielectric behavior of nanoconfined electrolytic solutions. *Phys Rev Lett* 2012; **109**: 107801.[153] Copyright 2012 American Physical Society.

a quantification of whether these "local" effects yield important consequences on, for example, salt solubility and other important observables.

12.5 Form and Origin: Confinement Effects on Reactivity

12.5.1 General Concept

In this section, we address the reactivity of carbon-bearing fluids and how confinement can affect their behavior. Based on thermodynamic arguments, Shock[155] predicted that, in the subsurface, the most common form of fluid carbon would be that of CO_2, not CH_4. This prediction was based on the analysis of the thermodynamic equilibrium of the reaction:

$$CO_2 + 4H_2 \leftrightarrow CH_4 + 2H_2O. \qquad (12.2)$$

The equilibrium was assessed based on the expected pH as buffered by the rocks. However, nanoconfinement could potentially affect fluid reactivity via three main mechanisms: (1) the closeness of the rock surfaces provides catalytic sites that speed up some reaction pathways, thus affecting the kinetics of the reactions; (2) the confinement is so restrictive that some of the reactants or the products are either not able to react or are not allowed to escape from the porous matrix; and (3) the preferential adsorption of some of the fluid components onto the rock surface can affect the overall equilibrium composition via a molecular-level application of Le Chatelier's principle. The first of these three scenarios concerns kinetic effects. We only note that some olivines contain Ni, Co, and/or Cr, which could act as catalysts to increase the rate of chemical reactions, including the oxidation of CH_4 to CO_2.[156]

12.5.2 Methanation of Carbon Dioxide

Using reactive ensemble Monte Carlo simulation, Le et al.[157] explored whether confinement could alter the bulk expectations regarding reaction (12.2). In particular, they tested whether the preferential adsorption of either CO_2 or H_2O on the pore surface could affect the equilibrium conversion of CO_2 to CH_4. The system simulated represented a nanoporous matrix in contact with a larger microfracture occupied by the bulk fluid. Water, hydrogen, carbon dioxide, and methane were free to exchange between the nanopores and the microfracture and to react within the various environments. The microfracture was large enough to provide a "bulk" system. The conceptual framework is summarized in Figure 12.10.

Le et al.[157] generated slit-shaped pores (1 and 2 nm) carved out of β-cristobalite silica. The results showed that the thermodynamic equilibrium composition shifted toward methane production, suggesting that thermal hydrocarbon synthesis near hydrothermal

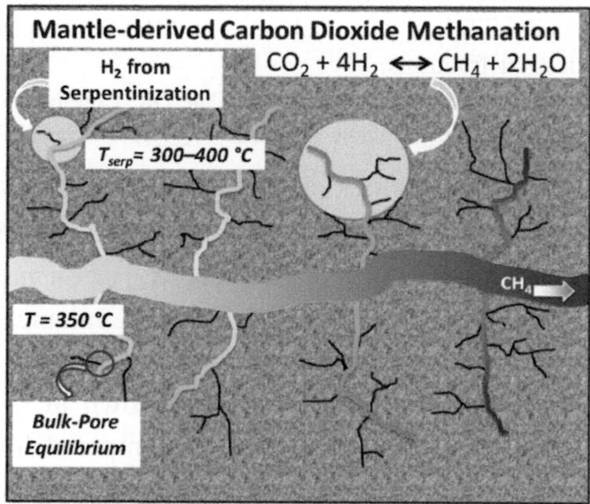

Figure 12.10 Schematic for a possible carbon dioxide methanation process. Within oceanic crust, mantle-derived melt is emplaced at shallow depths and heats the ultramafic ocean crust driving seawater circulation along natural fractures and microfractures where serpentinization reactions take place. The H_2 produced could come into contact with carbon dioxide derived from the mantle.
Reprinted with permission from Le T, Striolo A, Turner CH, Cole DR. Confinement effects on carbon dioxide methanation: a novel mechanism for abiotic methane formation. *Sci Rep* 2017; **7**: 9021.[157]

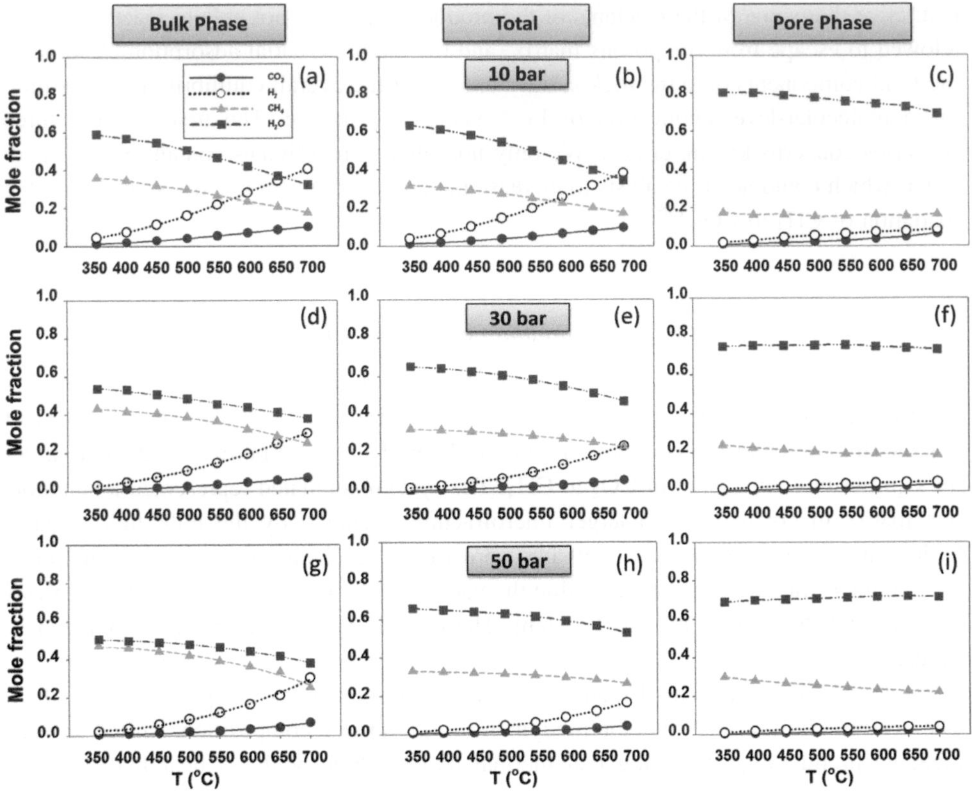

Figure 12.11 Product fraction of CO_2 methanation (reaction (12.1)) for the (a, d, and g) bulk phase and (c, f, and i) pore phase in equilibrium with each other at different temperatures and pressures. (b), (e), and (h) show total molecular fractions.

Reprinted with permission from Le T, Striolo A, Turner CH, Cole DR. Confinement effects on carbon dioxide methanation: a novel mechanism for abiotic methane formation. *Sci Rep* 2017; **7**: 9021.[157]

vents and deeper in the magma-hydrothermal system is possible. The results showed strong dependency of the reaction equilibrium conversions, X_{CO_2}, on nanopore size, nanopore chemistry, and nanopore morphology. All conditions that enhance water adsorption (i.e. increasing pore hydrophilicity or surface roughness) resulted in an enhancement of reaction yield. Representative results for 1-nm pores are shown in Figure 12.11, where the composition in the bulk phase (left-hand panels) is compared to that within the pores (right-hand panels). In situations in which the pore space is dominant compared to the bulk space, it is conceivable that the overall system composition strongly deviates from bulk expectations.

These results could contribute to an important scientific debate focusing on the possibility of the abiotic synthesis of hydrocarbons during oceanic crust–seawater interactions, which could reshape the core hypotheses of the origin of life.[158,159]

12.6 Summary and Opportunities

Exploring the behavior of complex geofluids confined in micro- and meso-porous networks provides the foundation for quantifying geologically relevant systems where mass and energy exchange occur. Nanoconfinement in porous systems gives rise to rich behavior that results from the interplay of geometrical restriction imposed by different pore features (size, geometry, chemical composition, etc.), the interaction between the porous material and the fluid, and the nature of the confined fluid itself. Fluid structure (or "form") affects most if not all fundamental properties of interest. Since nanopores affect structure, they impact movement, reactivity ("origin"), and quantity. While in relatively large mesopores fluid "movement" might occur at different timescales as a function of location in the pore, greater geometric confinement can lead to severely constrained motion, such as the molecular diffusion mechanisms discussed in Section 12.3. Within a given porous matrix, timescales of motion of confined fluids can exhibit interesting dependencies on the molecular size and concentration of the confining fluid, such as the levitation effect of confined hydrocarbons or anomalous pressure dependence on gas dynamics in mesoporous media.

Water and associated dissolved constituents, as a confined fluid, exhibit very rich structural and dynamical behavior due to the effects of hydrogen bonding, compounded by the other effects identified above. All of these effects lead to a wide variety of length and time scales relevant to the study of confined fluids and as such require a variety of experimental and computational tools for a thorough study of the behavior of geologically relevant confined fluids. The techniques that probe the behavior of confined fluids at the molecular level depend heavily on the interpretation of the measured data and as such can be prone to subjective bias. Therefore, if possible, it is advantageous to look at each solid–fluid system from different perspectives, both experimental and computational. Much progress is expected when synergistic approaches are implemented. For example, one obvious next step would be the detailed experimental and computational investigation of the structural properties of fluid mixtures in increasingly realistic systems. This includes complex fluid mixtures, as well as porous materials that are heterogeneous and better representative of the subsurface. Once these analyses are obtained, one should challenge the mechanisms discussed in Sections 12.3 and 12.4. A second target of opportunity pertains to whether the presence of ions and associated hydration affect the transport and accumulation of hydrocarbons in nanoporous matrices. The presence of concentrated salts disrupts water structure, slows down its dynamics, and alters the water free energy landscape, which impacts transport of guest molecules.[106,130] Clearly, how one fluid component invades and displaces and/or competes with another for surface sites will have a profound effect on these types of transport mechanisms. The dielectric properties of water and aqueous systems under nanoconfinement are poorly constrained for Earth-relevant systems. Recent studies have exhibited a wide range of values that are not always in agreement.

Finally, we point to the effects of confinement on the equilibrium composition of reactive systems. The results summarized in Section 12.5 suggest that thermogenic methane production could be enhanced by nanoconfinement, albeit within a specific window of pressure–temperature–composition conditions. To quantify whether those simulation results are relevant across a broader temperature and pressure landscape, a few research questions remain to be addressed, in particular:

1. Will the CO_2 methanation reaction encounter kinetic barriers within the pores or will the pore surfaces provide active sites for catalyzing the reaction?
2. How much nanopore volume in the subsurface is needed for achieving levels of methane production that are relevant for hydrocarbons synthesis?
3. Can these insights be reproduced by experiments?

Future studies could be implemented to address these and related questions and to quantify how these phenomena affect the carbon cycle.

Acknowledgments

AS acknowledges financial support from the European Union via the Marie Curie Career Integration Grant No. 2013-CIG-631435, as well as from the European Union's Horizon 2020 research and innovation program under Grant No. 640979 and Grant No. 764810. DC was supported by the US Department of Energy, Office of Basic Energy Sciences, Division of Chemical Sciences, Geosciences and Biosciences under grant DE-SC0006878. DC and AS both receive partial funding from the Sloan Foundation-funded Deep Carbon Observatory. Special thanks to Ms. Tran Le for her help in assembling the reference list.

Questions for the Classroom

1. Are nanopores observed in all Earth materials? How are these observed? What are their shapes, sizes, and distributions? Is there a correlation between these properties and depth?
2. Discuss quantitatively how confinement affects the thermophysical properties of a fluid, whether pure or a mixture. Discuss the effects due to strong pore–fluid interactions and how changing the pore width affects these outcomes. Discuss in particular such effects with respect to solubility, reactivity, phase equilibria, and melting.
3. Discuss the differences between surface and pore diffusion. What are the mechanisms responsible for each and when is surface diffusion expected to dominate over pore diffusion in determining the overall fluid transport across a porous material? What mechanisms become dominant as the pore size becomes comparable to the size of a fluid molecule?
4. How does the presence of a pore affect the reactivity of a fluid? Distinguish between kinetic and thermodynamic effects.

References

1. Anovitz LM, Cole DR. Characterization and analysis of porosity and pore structures. *Rev Mineral Geochem* 2015; **80**: 61–164.
2. Keller LM, Holzer L, Wepf R, Gasser P. 3D geometry and topology of pore pathways in Opalinus clay: implications for mass transport. *Appl Clay Sci* 2011; **52**: 85–95.
3. Keller LM, Schuetz P, Erni R, Rossell MD, Lucas F, Gasser P et al. Characterization of multi-scale microstructural features in Opalinus clay. *Microporous Mesoporous Matter* 2013; **170**: 83–94.
4. Mehmani Y, Balhoff MT. Mesoscale and hybrid models of fluid flow and solute transport. *Rev Mineral Geochem* 2015; **80**: 433–459.
5. Peth S, Horn R, Beckmann F, Donath T, Fischer J, Smucker AJM. Three-dimensional quantification of intra-aggregate pore-space features using synchrotron-radiation-based microtomography. *Soil Sci Soc Am J* 2008; **72**: 897–907.
6. Landrot G, Ajo-Franklin JB, Yang L, Cabrini S, Steefel CI. Measurement of accessible reactive surface area in a sandstone, with application to CO_2 mineralization. *Chem Geol* 2012; **318**: 113–125.
7. Peters CA. Accessibilities of reactive minerals in consolidated sedimentary rock: an imaging study of three sandstones. *Chem Geol* 2009; **265**: 198–208.
8. Beckingham LE, Mitnick EH, Steefel CI, Zhang S, Voltolini M, Swift AM et al. Evaluation of mineral reactive surface area estimates for prediction of reactivity of a multi-mineral sediment. *Geochim Cosmochim Acta* 2016; **188**: 310–329.
9. Bernabe Y, Li M, Maineult A. Permeability and pore connectivity: a new model based on network simulations. *J Geophys Res* 2010; **115**: B10203.
10. Ewing RP, Liu CX, Hu QH. Modeling intragranular diffusion in low-connectivity granular media. *Water Resour Res* 2012; **48**: W03518.
11. Keller LM, Holzer L, Wepf R, Gasser P, Munch B, Marschall P. On the application of focused ion beam nanotomography in characterizing the 3D pore space geometry of Opalinus clay. *Phys Chem Earth* 2011; **36**: 1539–1544.
12. Lindquist WB, Venkatarangan A, Dunsmuir J, Wong TF. Pore and throat size distributions measured from synchrotron X-ray tomographic images of Fontainebleau sandstones. *J Geophys Res Solid Earth* 2000; **105**: 21509–21527.
13. Hay MB, Stoliker DL, Davis JA, Zachara JM. Characterization of the intragranular water regime within subsurface sediments: pore volume, surface area, and mass transfer limitations. *Water Resour Res* 2011; **47**: W10531.
14. Dultz S, Simonyan AV, Pastrana J, Behrens H, Plotze M, Rath T. Implications of pore space characteristics on diffusive transport in basalts and granites. *Environ Earth Sci* 2013; **69**: 969–985.
15. Navarre-Sitchler A, Steefel CI, Yang L, Tomutsa L, Brantley SL. Evolution of porosity and diffusivity associated with chemical weathering of a basalt clast. *J Geophys Res* 2009; **114**: F02016.
16. Polak A, Elsworth D, Yasuhara H, Grader AS, Halleck PM. Permeability reduction of a natural fracture under net dissolution by hydrothermal fluids. *Geophys Res Lett* 2003; **30**: 2020.
17. Stack AG, Fernandez-Martinez A, Allard LF, Banuelos JL, Rother G, Anovitz LM et al. Pore-size-dependent calcium carbonate precipitation controlled by surface chemistry. *Environ Sci Technol* 2014; **48**: 6177–6183.
18. DePaolo DJ, Cole DR. Geochemistry of geologic carbon sequestration: an overview. *Rev Mineral Geochem* 2013; **77**: 1–14.

19. Steefel CI, DePaolo DJ, Lichtner PC. Reactive transport modeling: an essential tool and a new research approach for the Earth sciences. *Earth Planet Sci Lett* 2005; **240**: 539–558.
20. Steefel CI, Molins S, Trebotich D. Pore scale processes associated with subsurface CO_2 injection and sequestration. *Rev Mineral Geochem* 2013; **77**: 259–303.
21. Tokunaga TK, Wan JM. Capillary pressure and mineral wettability influences on reservoir CO_2 capacity. *Rev Mineral Geochem* 2013; **77**: 481–503.
22. Cole DR, Ok S, Striolo A, Anh P. Hydrocarbon behavior at nanoscale interfaces. *Rev Mineral Geochem* 2013; **75**: 495–545.
23. Stack AG. Precipitation in pores: a geochemical frontier. *Rev Mineral Geochem* 2015; **80**: 165–190.
24. Zachara J, Brantley S, Chorover J, Ewing R, Kerisit S, Liu CX et al. Internal domains of natural porous media revealed: critical locations for transport, storage, and chemical reaction. *Environ Sci Technol* 2016; **50**: 2811–2829.
25. Mehmani A, Mehmani Y, Prodanovic M, Balhoff M. A forward analysis on the applicability of tracer breakthrough profiles in revealing the pore structure of tight gas sandstone and carbonate rocks. *Water Resour Res* 2015; **51**: 4751–4767.
26. Steefel CI, Appelo CAJ, Arora B, Jacques D, Kalbacher T, Kolditz O et al. Reactive transport codes for subsurface environmental simulation. *Comput Geosci* 2015; **19**: 445–478.
27. DiCarlo DA, Aminzadeh B, Roberts M, Chung DH, Bryant SL, Huh C. Mobility control through spontaneous formation of nanoparticle stabilized emulsions. *Geophys Res Lett* 2011; **38**: L24404.
28. DiCarlo D, Aminzadeh B, Chung DH, Roberts M, Huh C, Bryant SL. Effect of nanoparticles on the displacement pattern of CO_2 injection in porous media. Presented at: *Goldschmidt Conference*, Session 275, 2012.
29. Bryant S, Yu H, Murphy M, Zhang T, Worthen, Yoon KY et al. Mechanisms governing nanoparticle transport in porous media. Presented at: *Goldschmidt Conference*, Session 275, 2012.
30. Ramanan B, Holmes WM, Sloan WT, Phoenix VR. Investigation of nanoparticle transport inside coarse-grained geological media using magnetic resonance imaging. *Environ Sci Technol* 2012; **46**: 360–366.
31. Phoenix V, Lakshmanan S, Sloan W, Holmes WM. Opening the black box: imaging nanoparticle transport through rock with MRI. Presented at: *Goldschmidt Conference*, Session 172, 2012.
32. Deutch J, Moniz EJ. *The Future of Coal: Options for a Carbon-Constrained World. An Interdisciplinary MIT Study*. Cambridge, MA: Massachusetts Institute of Technology, 2007.
33. Titley SR, ed. *Evolution and Style of Fracture Permeability in Intrusive-Centered Hydrothermal Systems*. Washington, DC: National Academic Press, 1990, pp. 50–63.
34. Ingebritsen SE, Sanford WE, eds. *Groundwater in Geologic Processes*. Cambridge: Cambridge University Press, 1998.
35. Laubach SE, Ward ME. Diagenesis in porosity evolution of opening-mode fractures, Middle Triassic to Lower Jurassic La Boca Formation, NE Mexico. *Tectonophysics* 2006; **419**: 75–97.
36. Laubach SE, Reed RM, Olson JE, Lander RH, Bonnell LM. Coevolution of crack-seal texture and fracture porosity in sedimentary rocks: cathodoluminescence observations of regional fractures. *J Struct Geol* 2004; **26**: 967–982.

37. Gale JFW, Lander RH, Reed RM, Laubach SE. Modeling fracture porosity evolution in dolostone. *J Struct Geol* 2010; **32**: 1201–1211.
38. Ague JJ, ed. *Fluid Flow in the Deep Crust*. Oxford: Elsevier, 2014, pp. 203–247.
39. Norton D, Knapp R. Transport phenomena in hydrothermal systems: the nature of porosity. *Am J Sci* 1977; **277**: 913–936.
40. Petford N, Koenders MA. Shear-induced pressure changes and seepage phenomena in a deforming porous layer – I. *Geophys J Int* 2003; **155**: 857–869.
41. Marquardt K, Faul UH. The structure and composition of olivine grain boundaries: 40 years of studies, status and current developments. *Phys Chem Miner* 2018; **45**: 139–172.
42. Rother G, Krukowski EG, Wallacher D, Grimm N, Bodnar RJ, Cole DR. Pore size effects on the sorption of supercritical CO_2 in mesoporous CPG-10 Silica. *J Phys Chem C* 2012; **116**: 917–922.
43. Phan A, Cole DR, Striolo A. Liquid ethanol simulated on crystalline alpha alumina. *J Phys Chem B* 2013; **117**: 3829–3840.
44. Phan A, Cole DR, Striolo A. Preferential adsorption from liquid water–ethanol mixtures in alumina pores. *Langmuir* 2014; **30**: 8066–8077.
45. Gubbins KE, Long Y, Sliwinska-Bartkowiak M. Thermodynamics of confined nano-phases. *J Chem Thermodyn* 2014; **74**: 169–183.
46. Firincioglu T, Ozkan E, Ozgen C. Thermodynamics of multiphase flow in unconventional liquids-rich reservoirs. Presented at: *SPE Annual Technical Conference and Exhibition Society of Petroleum Engineers*, 2012.
47. Wu Y-S. *Multiphase Fluid Flow in Porous and Fractured Reservoirs*. Houston, TX: Gulf Professional Publishing, 2015.
48. Teklu TW, Alharthy N, Kazemi H, Yin XL, Graves RM, AlSumaiti AM. Phase behavior and minimum miscibilitypressure in nanopores. *Spec Reserv Eval Eng* 2014; **17**: 396–403.
49. Akkutlu IY, Rahmani DB. *SPE International Symposium on Oilfield Chemistry*. Richardson, TX: Society of Petroleum Engineers, 2013.
50. Striolo A. Understanding interfacial water and its role in practical applications using molecular simulations. *MRS Bull* 2014; **39**: 1062–1068.
51. Clarkson CR, Haghshenas B, Ghanizadeh A, Qanbari F, Williams-Kovacs JD, Riazi N et al. Nanopores to megafractures: current challenges and methods for shale gas reservoir and hydraulic fracture characterization. *J Nat Gas Sci Eng* 2016; **31**: 612–657.
52. Ho TA, Criscenti LJ, Wang YF. Nanostructural control of methane release in kerogen and its implications to wellbore production decline. *Sci Rep* 2016; **6**: 28053.
53. Ferreira DR, Schulthess CP, Giotto MV. An investigation of strong sodium retention mechanisms in nanopore environments using nuclear magnetic resonance spectroscopy. *Environ Sci Technol* 2012; **46**: 300–306.
54. Turner DD, Ferrare RA, Brasseur LAH, Feltz WF. Automated retrievals of water vapor and aerosol profiles from an operational Raman lidar. *J Atmos Ocean Technol* 2002; **19**: 37–50.
55. Turner CH, Brennan JK, Lisal M, Smith WR, Johnson JK, Gubbins KE. Simulation of chemical reaction equilibria by the reaction ensemble Monte Carlo method: a review. *Mol Simulat* 2008; **34**: 119–146.
56. Santiso E, Firoozabadi A. Curvature dependency of surface tension in multicomponent systems. *AIChE J* 2005; **52**: 311–322.

57. Abbas M, Nadeem R, Zafar MN, Arshad M. Biosorption of chromium (III) and chromium (VI) by untreated and pretreated *Cassia fistula* biomass from aqueous solutions. *Water Air Soil Pollut* 2008; **191**: 139–148.
58. Beckingham LE, Steefel CI, Swift AM, Voltolini M, Yang L, Anovitz LM et al. Evaluation of accessible mineral surface areas for improved prediction of mineral reaction rates in porous media. *Geochim Cosmochim Acta* 2017; **205**: 31–49.
59. Allen MP, Tildesley DJ, eds. *Computer Simulation of Liquids*. Oxford: Oxford University Press, 2017.
60. Marx D, Hutter J, eds. *Ab Initio Molecular Dynamics, Basic Theory and Advanced Methods*. Cambridge: Cambridge University Press, 2009.
61. Frenkel D, Smit B, eds. *Understanding Molecular Simulation*. Amsterdam: Elsevier, 2001.
62. Brovchenko I, Oleinikova A. Which properties of a spanning network of hydration water enable biological functions? *Chemphyschem* 2008; **9**: 2695–2702.
63. Striolo A, Michaelides A, Joly L. The carbon–water interface: modeling challenges and opportunities for the water-energy nexus. *Annu Rev Chem Biomol Eng* 2016; **7**: 533–556.
64. Senftle TP, Hong S, Islam MM, Kylasa SB, Zheng YX, Shin YK et al. The ReaxFF reactive force-field: development, applications and future directions. *NPJ Comput Mater* 2016; **2**: 15011.
65. Fenter P, Sturchio NC. Mineral–water interfacial structures revealed by synchrotron X-ray scattering. *Prog Surf Sci* 2004; **77**: 171–258.
66. Cole DR, Herwig K, Mamontov E, Larese LZ. Neutron scattering and diffraction studies of fluids and fluid–solid interactions. In: Wenk H-R, ed. *Neutron Scattering in Earth Science, Reviews in Mineralogy and Geochemistry, Vol. 63*. Chantilly, VA: Mineralogical Society of America, 2006, pp. 313–362.
67. Cole DR, Mamontov E, Rother G. Structure and dynamics of fluids in microporous and mesoporous earth and engineered materials. In: Liang L, Rinaldi R, Schober H, eds. *Neutron Applications in Earth, Energy, and Environmental Sciences*. Berlin: Springer, 2009, pp. 547–570.
68. Karger J, Ruthven DM, Theodorou DN, eds. *Diffusion in Nanoporous Materials*. Hoboken, NJ: Wiley, 2012.
69. Melnichenko Y, ed. *Small-Angle Scattering from Confined and Interfacial Fluids. Applications to Energy Storage and Environmental Fluids*. Berlin: Springer, 2016.
70. Gautam S. Use of quasielastic neutron scattering and molecular dynamics simulation to study molecular dynamics under confinement. In: Reimer A, ed. *Horizons in World Physics Vol. 290*. New York: Nova Science Publishers, 2017, pp. 25–44.
71. Gautam S, Le T, Striolo A, Cole D. Molecular dynamics simulations of propane in slit shaped silica nano-pores: direct comparison with quasielastic neutron scattering experiments. *Phys Chem Chem Phys* 2017; **19**: 32320–32332.
72. Nelson PH. Permeability–porosity relationships in sedimentary rocks. *Log Anal* 1994; **35**: 38–62.
73. Ingebritsen SE, Sanford WE, Neuzil CE, eds. *Groundwater in Geologic Processes*. Cambridge: Cambridge University Press, 2006.
74. Gleeson T, Ingebritsen SE, ed. Introduction. In *Crustal Permeability*. Hoboken, NJ: Wiley, 2017, pp. 1–5.
75. Lockner DA, Byerlee JD, Kuksenko V, Ponomarev A, Sidorin A. Quasi-static fault growth and shear fracture energy in granite. *Nature* 1991; **350**: 39–42.

76. Ingebritsen SE, Manning CE. Geological implications of a permeability–depth curve for the continental crust. *Geology* 1999; **27**: 1107–1110.
77. Ingebritsen SE, Manning CE. Diffuse fluid flux through orogenic belts: implications for the world ocean. *Proc Natl Acad Sci USA* 2002; **99**: 9113–9116.
78. Ingebritsen SE, Manning CE. Permeability of the continental crust: dynamic variations inferred from seismicity and metamorphism. *Geofluids* 2010; **10**: 193–205.
79. Bredehoeft JD, Norton DL. Mass and energy transport in a deforming Earth's crust. In: *The Role of Fluids in Crustal Processes*. Washington, DC: National Academy Press, 1990. pp. 27–41.
80. Anovitz LM, Cole DR, Jackson AJ, Rother G, Littrell KC, Allard LF et al. Effect of quartz overgrowth precipitation on the multiscale porosity of sandstone: a (U)SANS and imaging analysis. *Geochim Cosmochim Acta* 2015; **158**: 199–222.
81. Anovitz LM, Cole DR. Analysis of the pore structures of shale using neutron and X-ray small-angle scattering. In: Vialle S, Ajo-Franklin J, Carey W, eds. *Geological Carbon Storage: Subsurface Seals and Caprock*, Hoboken, NJ: Wiley, 2018, pp. 71–119.
82. Wark DA, Watson EB. Grain-scale permeabilities of texturally equilibrated, monomineralic rocks. *Earth Planet Sci Lett* 1998; **164**: 591–605.
83. Cole DR, Chakraborty S. Rates and mechanisms of isotopic exchange. *Rev Mineral Geochem* 2001; **43**: 83–223.
84. Watson EB, Baxter EF. Diffusion in solid-Earth systems. *Earth Planet Sci Lett* 2007; **253**: 307–327.
85. Farver JR. Oxygen and hydrogen diffusion in minerals. *Rev Mineral Geochem* 2010; **72**: 447–507.
86. Wirth R. Thin amorphous films (1–2 nm) at olivine grain boundaries in mantle xenoliths from San Carlos, Arizona. *Contrib Mineral Petrol* 1996; **124**: 44–54.
87. Gelb LD, Gubbins KE, Radhakrishnan R, Sliwinska-Bartkowiak M. Phase separation in confined systems. *Rep Prog Phys* 1999; **62**: 1573–1659.
88. Radhakrishnan R, Gubbins KE, Sliwinska-Bartkowiak M. Global phase diagrams for freezing in porous media. *J Chem Phys* 2002; **116**: 1147–1155.
89. Alba-Simionesco C, Coasne B, Dosseh G, Dudziak G, Gubbins KE, Radhakrishnan R et al. Effects of confinement on freezing and melting. *J Phys Condens Matter* 2006; **18**: R15–R68.
90. Wang J, Kalinichev AG, Kirkpatrick RJ. Molecular modeling of the 10-Å phase at subduction zone conditions. *Earth Planet Sci Lett* 2004; **222**: 517–527.
91. Wang JW, Kalinichev AG, Kirkpatrick RJ. Structure and decompression melting of a novel, high-pressure nanoconfined 2-D ice. *J Phys Chem B* 2005; **109**: 14308–14313.
92. Radhakrishnan R, Gubbins KE, Sliwinska-Bartkowiak M. Existence of a hexatic phase in porous media. *Phys Rev Lett* 2002; **89**: 076101.
93. Gallo P, Arnann-Winkel K, Angell CA, Anisimov MA, Caupin F, Chakravarty C et al. Water: a tale of two liquids. *Chem Rev* 2016; **116**: 7463–7500.
94. Chiavazzo E, Fasano M, Asinari P, Decuzzi P. Scaling behaviour for the water transport in nanoconfined geometries. *Nat Commun* 2014; **5**: 3565.
95. Ho TA, Papavassiliou DV, Lee LL, Striolo A. Liquid water can slip on a hydrophilic surface. *Proc Natl Acad Sci USA* 2011; **108**: 16170–16175.
96. Kim HJ, Brunelli NA, Brown AJ, Jang KS, Kim WG, Rashidi F et al. Silylated mesoporous silica membranes on polymeric hollow fiber supports: synthesis and permeation properties. *ACS Appl Mater Interfaces* 2014; **6**: 17877–17886.

97. Ge Q, Wang Z, Yan Y. High-performance zeolite NaA membranes on polymer–zeolite composite hollow fiber supports. *J Am Chem Soc* 2009; **131**: 17056–17057.
98. Patankar S, Gautam S, Rother G, Podlesnyak A, Ehlers G, Liu T et al. Role of confinement on adsorption and dynamics of ethane and an ethane–CO_2 mixture in mesoporous CPG silica. *J Phys Chem C* 2016; **120**: 4843–4853.
99. Le T, Striolo A, Cole DR. CO_2–C_4H_{10} mixtures simulated in silica slit pores: relation between structure and dynamics. *J Phys Chem C* 2015; **119**: 15274–15284.
100. Le T, Ogbe S, Striolo A, Cole DR. N-octane diffusivity enhancement via carbon dioxide in silica slit-shaped nanopores – a molecular dynamics simulation. *Mol Simulat* 2016; **42**: 745–752.
101. Gautam S, Liu TT, Rother G, Jalarvo N, Mamontov E, Welch S et al. Dynamics of propane in nanoporous silica aerogel: a quasielastic neutron scattering study. *J Phys Chem C* 2015; **119**: 18188–18195.
102. Chathoth SM, He L, Mamontov E, Melnichenko YB. Effect of carbon dioxide and nitrogen on the diffusivity of methane confined in nano-porous carbon aerogel. *Microporous Mesoporous Matter* 2012; **148**: 101–106.
103. Gautam S, Jalarvo N, Rother G, Liu T, Mamontov E, Dai S et al. Quasieleastic neutron scattering studies on propane in nanoporous silica. Presented at: *German Conference on Neutron Scattering*, 2016.
104. Le TTB, Striolo A, Gautam SS, Cole DR. Propane–water mixtures confined within cylindrical silica nanopores: structural and dynamical properties probed by molecular dynamics. *Langmuir* 2017; **33**: 11310–11320.
105. Ho TA, Striolo A. Water and methane in shale rocks: flow pattern effects on fluid transport and pore structure. *AIChE J* 2015; **61**: 2993–2999.
106. Phan A, Cole DR, Weiss RG, Dzubiella J, Striolo A. Confined water determines transport properties of guest molecules in narrow pores. *ACS Nano* 2016; **10**: 7646–7656.
107. Bui T, Phan A, Cole DR, Striolo A. Transport mechanism of guest methane in water-filled nanopores. *J Phys Chem C* 2017; **121**: 15675–15686.
108. Mark P, Nilsson L. Structure and dynamics of the TIP3P, SPC, and SPC/E water models at 298 K. *J Phys Chem A* 2001; **105**: 9954–9960.
109. Franco LFM, Castier M, Economou IG. Anisotropic parallel self-diffusion coefficients near the calcite surface: a molecular dynamics study. *J Chem Phys* 2016; **145**: 084702.
110. Argyris D, Cole DR, Striolo A. Ion-specific effects under confinement: the role of interfacial water. *ACS Nano* 2010; **4**: 2035–2042.
111. Ho TA, Argyris D, Cole DR, Striolo A. Aqueous NaCl and CsCl solutions confined in crystalline slit-shaped silica nanopores of varying degree of protonation. *Langmuir* 2012; **28**: 1256–1266.
112. Ho TA, Argyris D, Papavassiliou DV, Striolo A, Lee LL, Cole DR. Interfacial water on crystalline silica: a comparative molecular dynamics simulation study. *Mol Simulat* 2011; **37**: 172–195.
113. Argyris D, Cole DR, Striolo A. Dynamic behavior of interfacial water at the silica surface. *J Phys Chem C* 2009; **113**: 19591–19600.
114. Rotenberg B, Marry V, Dufreche JF, Malikova N, Giffaut E, Turq P. Modelling water and ion diffusion in clays: a multiscale approach. *CR Chim.* 2007; **10**: 1108–1116.
115. Loganathan N, Kalinichev AG. Quantifying the mechanisms of site-specific ion exchange at an inhomogeneously charged surface: case of Cs^+/K^+ on hydrated muscovite mica. *J Phys Chem C* 2017; **121**: 7829–7836.

116. Levesque M, Marry V, Rotenberg B, Jeanmairet G, Vuilleumier R, Borgis D. Solvation of complex surfaces via molecular density functional theory. *J Chem Phys* 2012; **137**: 224107.
117. Tesson S, Louisfrema W, Salanne M, Boutin A, Rotenberg B, Marry V. Classical polarizable force field to study dry charged clays and zeolites. *J Phys Chem C* 2017; **121**: 9833–9846.
118. Pouvreau M, Greathouse JA, Cygan RT, Kalinichev AG. Structure of hydrated gibbsite and brucite edge surfaces: DFT results and further development of the ClayFF classical force field with metal–O–H angle bending terms. *J Phys Chem C* 2017; **121**: 14757–14771.
119. Botan A, Marry V, Rotenberg B, Turq P, Noetinger B. How electrostatics influences hydrodynamic boundary conditions: poiseuille and electro-osmostic flows in clay nanopores. *J Phys Chem C* 2013; **117**: 978–985.
120. Carof A, Marry V, Salanne M, Hansen JP, Turq P, Rotenberg B. Coarse graining the dynamics of nano-confined solutes: the case of ions in clays. *Mol Simulat* 2014; **40**: 237–244.
121. Jeanmairet G, Marry V, Levesque M, Rotenberg B, Borgis D. Hydration of clays at the molecular scale: the promising perspective of classical density functional theory. *Mol Phys* 2014; **112**: 1320–1329.
122. Rotenberg B, Marry V, Salanne M, Jardat M, Turq P. Multiscale modelling of transport in clays from the molecular to the sample scale. *CR Geosci* 2014; **346**: 298–306.
123. Ho LN, Clauzier S, Schuurman Y, Farrusseng D, Coasne B. Gas uptake in solvents confined in mesopores: adsorption versus enhanced solubility. *J Phys Chem Lett* 2013; **4**: 2274–2278.
124. Hu YF, Huang LL, Zhao SL, Liu HL, Gubbins KE. Effect of confinement in nanoporous materials on the solubility of a supercritical gas. *Mol Phys* 2016; **114**: 3294–3306.
125. Luzar A, Bratko D. Gas solubility in hydrophobic confinement. *J Phys Chem B* 2005; **109**: 22545–22552.
126. Pera-Titus M, El-Chahal R, Rakotovao V, Daniel C, Miachon S, Dalmon J. Direct volumetric measurement of gas oversolubility in nanoliquids: beyond Henry's Law. *Chemphyschem* 2009; **10**: 2082–2089.
127. Rakotovao V, Ammar R, Miachon S, Pera-Titus M. Influence of the mesoconfining solid on gas oversolubility in nanoliquids. *Chem Phys Lett* 2010; **485**: 299–303.
128. Gadikota G, Dazas B, Rother G, Cheshire MC, Bourg IC. Hydrophobic solvation of gases (CO_2, CH_4, H_2, noble gases) in clay interlayer nanopores. *J Phys Chem C* 2017; **121**: 26539–26550.
129. Phan A, Cole DR, Striolo A. Aqueous methane in slit-shaped silica nanopores: high solubility and traces of hydrates. *J Phys Chem C* 2014; **118**: 4860–4868.
130. Phan A, Cole DR, Striolo A. Factors governing the behaviour of aqueous methane in narrow pores. *Philos Trans A Math Phys Eng Sci* 2015; **374**: 20150019.
131. Sakamaki R, Sum AK, Narumi T, Ohmura R, Yasuoka K. Thermodynamic properties of methane/water interface predicted by molecular dynamics simulations. *J Chem Phys* 2011; **134**: 144702.
132. Badmos SB, Striolo A, Cole DR. Aqueous hydrogen sulfide in slit-shaped silica nanopores: confinement effects on solubility, structural, and dynamical properties. *J Phys Chem C* 2018; **122**: 14744–14755.
133. Malani A, Ayappa KG, Murad S. Effect of confinement on the hydration and solubility of NaCl in water. *Chem Phys Lett* 2006; **431**: 88–93.

134. Bowers GM, Hoyt DW, Burton SD, Ferguson BO, Varga T, Kirkpatrick RJ. In situ ^{13}C and ^{23}Na magic angle spinning NMR investigation of supercritical CO_2 incorporation in smectite–natural organic matter composites. *J Phys Chem C* 2014; **118**: 3564–3573.
135. Kirkpatrick RJ, Kalinichev AG, Bowers GM, Yazaydin AO, Krishnan M, Saharay M et al. NMR and computational molecular modeling studies of mineral surfaces and interlayer galleries: a review. *Am Mineral* 2015; **100**: 1341–1354.
136. Schaef HT, Loganathan N, Bowers GM, Kirkpatrick RJ, Yazaydin AO, Burton SD et al. Tipping point for expansion of layered aluminosilicates in weakly polar solvents: supercritical CO_2. *Appl Matter Interfaces* 2017; **9**: 36783–36791.
137. Bowers GM, Loring JS, Schaef HT, Walter ED, Burton SD, Hoyt DW et al. Interaction of hydrocarbons with clays under reservoir conditions: *in situ* infrared and nuclear magnetic resonance spectroscopy and X-ray diffraction for expandable clays with variably wet supercritical methane. *ACS Earth Space Chem* 2018; **2**: 640–652.
138. Ok S, Hoyt DW, Andersen A, Sheets J, Welch SA, Cole DR et al. Surface interactions and confinement of methane: a high pressure magic angle spinning. *Langmuir* 2017; **33**: 1359–1367.
139. Moucka F, Svoboda M, Lisal M. Modelling aqueous solubility of sodium chloride in clays at thermodynamic conditions of hydraulic fracturing by molecular simulations. *Phys Chem Chem Phys* 2017; **19**: 16586–16599.
140. Svoboda M, Lísal M. Concentrated aqueous sodium chloride solution in clays at thermodynamic conditions of hydraulic fracturing: insight from molecular dynamics simulations. *J Chem Phys* 2018; **148**: 222806.
141. Nezbeda I, Moucka F, Smith WR. Recent progress in molecular simulation of aqueous electrolytes: force fields, chemical potentials and solubility. *Mol Phys* 2016; **114**: 1665–1690.
142. Smith WR, Nezbeda I, Kolafa J, Moucka F. Recent progress in the molecular simulation of thermodynamic properties of aqueous electrolyte solutions. *Fluid Phase Equilib* 2018; **466**: 19–30.
143. Stephen H, Stephen T, eds. *Solubilities of Inorganic and Organic Compounds*. New York: Macmillan, 1963.
144. Renou R, Szymczyk A, Maurin G, Malfreyt P, Ghoufi A. Superpermittivity of nanoconfined water. *J Chem Phys* 2015; **142**: 184706.
145. Sverjensky DA, Harrison B, Azzolini D. Water in the deep Earth: the dielectric constant and the solubilities of quartz and corundum to 60kb and 1200°C. *Geochim Cosmochim Acta* 2014; **129**: 125–145.
146. Pan D, Spanu L, Harrison B, Sverjensky DA, Galli G. Dielectric properties of water under extreme conditions and transport of carbonates in the deep Earth. *Proc Natl Acad Sci USA* 2013; **110**: 6646–6650.
147. Bonthuis DJ, Gekle S, Netz RR. Dielectric profile of interfacial water and its effect on double-layer capacitance. *Phys Rev Lett* 2011; **107**: 166102.
148. Bonthuis DJ, Netz RR. Unraveling the combined effects of dielectric and viscosity profiles on surface capacitance, electro-osmotic mobility, and electric surface conductivity. *Langmuir* 2012; **28**: 16049–16059.
149. Schaaf C, Gekle S. Spatially resolved dielectric constant of confined water and its connection to the non-local nature of bulk water. *J Chem Phys* 2016; **145**: 084901.
150. Ghoufi A, Szymczyk A, Renou R, Ding M. Calculation of local dielectric permittivity of confined liquids from spatial dipolar correlations. *EPL J* 2012; **99**: 37008.

151. Zhu PP, Stebner AP, Brinson LC. A numerical study of the coupling of elastic and transformation fields in pore arrays in shape memory alloy plates to advance porous structure design and optimization. *Smart Mater Struct* 2013; **22**: 094009.
152. Fumagalli L, Esfandiar A, Fabregas R, Hu S, Ares P, Janardanan A et al. Anomalously low dielectric constant of confined water. *Science* 2018; **360**: 1339–1342.
153. Zhu H, Ghoufi A, Szymczyk A, Balannec B, Morineau D. Anomalous dielectric behavior of nanoconfined electrolytic solutions. *Phys Rev Lett* 2012; **109**: 107801.
154. Renou R, Ghoufi A, Szymczyk A, Zhu H, Neyt JC, Malfreyt P. Nanoconfined electrolyte solutions in porous hydrophilic silica membranes. *J Phys Chem C* 2013; **117**: 11017–11027.
155. Shock EL. Chemical environments of submarine hydrothermal systems. *Origins Life Evol B* 1992; **22**: 67–107.
156. Foley SF, Prelevic D, Rehfeldt T, Jacob DE. Minor and trace elements in olivines as probes into early igneous and mantle melting processes. *Earth Planet Sci Lett* 2013; **363**: 181–191.
157. Le T, Striolo A, Turner CH, Cole DR. Confinement effects on carbon dioxide methanation: a novel mechanism for abiotic methane formation. *Sci Rep* 2017; **7**: 9021.
158. Holm NG, ed. *Marine Hydrothermal Systems and the Origin of Life – Report of SCOR Working Group 91*. Dordrecht: Kluwer, 1992.
159. Brack A, ed. *The Molecular Origins of Life: Assembling Pieces of the Puzzle*. Cambridge: Cambridge University Press, 1998.

13

A Two-Dimensional Perspective on CH_4 Isotope Clumping

Distinguishing Process from Source

EDWARD D. YOUNG

13.1 Introduction

Isotope ratios have been used extensively to trace the origins of methane gases (e.g. Schoell 1980). For this purpose, the stable isotope ratios $^{13}C/^{12}C$ and D/H have been paramount. These ratios refer to the atomic abundances of the rare isotopes of carbon and hydrogen relative to the more abundant isotopes, in aggregate, and inclusive of all of the methane isotopic molecular species in a sample of gas. We therefore refer to these ratios as "bulk" isotope ratios. The term "isotopologue" refers to specific isotopic versions of the *molecules*. For example, the "$^{12}CH_3D$ isotopologue" refers to the $^{12}CH_3D$ + $^{12}CH_2DH$ + $^{12}CHDH_2$ + $^{12}CDH_3$ permutations of the D-substituted isotopic species of CH_4 collectively. In the geosciences, the term "clumping" denotes more than one heavy isotope in a single molecule or molecular unit (e.g. $^{13}C^{18}O^{16}O^{16}O$ + $^{13}C^{16}O^{18}O^{16}O$ + $^{13}C^{16}O^{16}O^{18}O$ in the CO_3^{2-} moiety within the $CaCO_3$ crystalline structure). In this chapter, the results of recent studies of the relative abundances of the clumped methane species $^{13}CH_3D$ and $^{12}CH_2D_2$ measured at the University of California, Los Angeles (UCLA) are summarized.

We begin with a description of the goals of this research program. The original excitement about making use of the ^{13}C–^{18}O multiply substituted isotopologue of CO_2, $^{13}C^{18}O^{16}O$ + $^{13}C^{16}O^{18}O$, derived from acid digestion of carbonate was due to the prospect of removing the various logical degeneracies that have historically plagued our interpretations of the significance of $^{18}O/^{16}O$ (usually expressed as $\delta^{18}O$ values, the per mil differences in $^{18}O/^{16}O$ from a standard material) in carbonates (Eiler et al. 2005). Marine carbonate oxygen isotope ratios can vary in response to temperature, ice volume, or secular variations in the $\delta^{18}O$ of the oceans. By using the temperature-dependent propensity of ^{13}C and ^{18}O to form bonds as a homogeneous (as in a single-phase as opposed to heterogeneous fractionation between two separate phases, in this case carbonate and water) thermometer, the relationship between bulk $\delta^{18}O$ in the carbonate and that in the water becomes irrelevant if the goal is to deduce temperature of formation. The goal, therefore, was to develop an isotopic tracer in which the bulk isotope ratios are normalized out.

The CH_4 clumping project at UCLA, in collaboration with the Carnegie Institution of Science, began in 2008 with the prospects for funding by the Deep Carbon Observatory and an eye toward replicating the powerful aspects of carbonate clumping for methane.

Figure 13.1 Mass spectrometer used to separate the two rare mass-18 isotopologues of CH_4 gas molecules. The instrument – the Panorama – is housed at UCLA and is the first of its kind (Young et al. 2016).

In particular, the group originally sought to distinguish biotic from abiotic methane based largely on temperature of formation, but also based on reaction pathways. Even before delivery of our unique mass spectrometer built for this purpose (Figure 13.1), methane experts (in particular Barbara Sherwood Lollar, University of Toronto) warned of the naivety of assuming that methane molecules would have simple, single-stage histories, and emphasized in particular the important role of mixing, the breadth of abiotic organic reactions over the range of temperatures that might be involved, and the complexity of microbial methanogenesis pathways and associated kinetic fractionation effects. In response, we developed predictions for these various phenomena in preparation for the new data to come.

The work described herein is the realization of those original goals to use two mass-18 isotopologues of CH_4 to disambiguate the multiple possible sources of methane gases using a process-oriented approach. Just as carbonate isotope clumping can break the degeneracy between seawater $\delta^{18}O$ and temperature of formation, the use of two methane mass-18 isotopologues should remove the ambiguities arising from variations in the $^{13}C/^{12}C$ and D/H ratios of the carbon and hydrogen sources of methane. We seek to eliminate, or at least mitigate, the uncertainties surrounding the use of $\delta^{13}C$ and δD as the primary arbiters for the origin of methane gas. Even when combined with other important tracers of gas provenance like C_1/C_2+ (the ratio of CH_4 to $C_2H_6 + C_3H_8$, etc.), ambiguities regarding the provenance of methane often remain when relying on bulk isotope ratios. Rules of thumb have been constructed based on years of observations. One example is that a decrease in $^{13}C/^{12}C$ ($\delta^{13}C$) with increasing carbon number for the alkanes can be evidence for abiotic formation. But these rules are, for the most part, empiricisms and fraught with exceptions. The goal of the studies described here is to make use of the relative concentrations of $^{13}CH_3D$ and $^{12}CH_2D_2$ together in methane gas to remove the uncertainties plaguing other tracers while avoiding adding just another layer of complexity.

This means using what these isotopologues are telling us in order to gain a fresh vantage point for our views on the origin of methane and to find out if we might be led astray by some of the earlier criteria for provenance.

Perhaps the most succinct way to summarize this perspective is that the isotopic bond ordering, or clumping, traces process, while bulk isotope ratios trace both process and source isotopic compositions. By following the isotopic bond ordering, we are isolating process from variations in source material.

13.2 Temperature

The concept of using "clumping" of heavy isotopes in a molecule as a thermometer is understood by referencing the null condition of the purely stochastic distribution of isotopes among the molecules of interest. In the case of methane, one is concerned with the fraction of carbon that is composed of the heavy isotope, $X(^{13}C) = {}^{13}C/({}^{13}C + {}^{12}C)$, and the fraction of the hydrogen isotopes that is composed of deuterium, $X(D) = D/(D + H)$. In the case of purely random distributions of isotopes among molecules, the products of fractional isotope abundances yield the joint probabilities, or predicted concentrations, for the isotopologues. For example, $X(^{12}CH_3D) = X(^{12}C)(X(H))^3X(D)$, where the exponent accounts for the simultaneous occurrence of H at three positions. Isotopologue abundances are not generally random, however. The temperature-dependent exchange of isotopes between isotopologues occurs by reactions such as

$$^{13}CH_4 + {}^{12}CH_3D \rightleftarrows {}^{12}CH_4 + {}^{13}CH_3D. \tag{13.1}$$

The equilibrium constant for this reaction is

$$k_{Eq,\,^{13}CD} = \frac{\left[^{13}CH_3D\right]\left[^{12}CH_4\right]}{\left[^{13}CH_4\right]\left[^{12}CH_3D\right]}, \tag{13.2}$$

where the square brackets denote concentrations that can be equated with the fractional abundances of the isotopologues by number. At high temperatures ($\geq$1000 K), the distributions of isotopologues are effectively random (stochastic) and substitution of the fractional abundances into the equilibrium constant in (13.2) yields (e.g. Richet et al. 1977):

$$k_{Eq,\,^{13}CD} = \frac{\left[^{13}CH_3D\right]\left[^{12}CH_4\right]}{\left[^{13}CH_4\right]\left[^{12}CH_3D\right]} = \frac{4X\left(^{13}C\right)\ (X(H))^3X(D)\ X\left(^{12}C\right)\ (X(H))^4}{4X\left(^{12}C\right)\ (X(H))^3X(D)\ X\left(^{13}C\right)\ (X(H))^4} = 1. \tag{13.3}$$

Equation (13.3) shows that an equilibrium constant of unity in this case corresponds to a purely random distribution of isotopes among the molecules. At lower temperatures, the stabilizing effects of two heavy isotopes bonded together take hold and the equilibrium constant in (13.2) gets progressively larger as temperature decreases. This enhancement in rare multiply substituted isotopologues, or clumping, can be expressed in delta notation and in per mil units as

$$\Delta_{^{13}CH_3D} = 10^3\left(\frac{X(^{13}CH_3D)}{X(^{13}CH_3D)_{Stochastic}} - 1\right). \tag{13.4}$$

Similarly, internal isotope exchange leading to doubly deuterated methane can be described by the reaction

$$2^{12}CH_3D \rightleftarrows {}^{12}CH_2D_2 + {}^{12}CH_4, \tag{13.5}$$

with the equilibrium constant

$$k_{Eq,CH_2D_2} = \frac{[^{12}CH_2D_2][^{12}CH_4]}{[^{12}CH_3D]^2}. \tag{13.6}$$

The stochastic value for the equilibrium constant for reaction (13.6) is

$$k_{Eq,^{12}CH_2D_2} = \frac{[^{12}CH_2D_2][^{12}CH_4]}{[^{12}CH_3D]^2} = \frac{6X(^{12}C)\ (X(H))^2X(D)^2\ X(^{12}C)\ (X(H))^4}{\left[4X(^{12}C)(X(H))^3X(D)\right]^2} = \frac{3}{8}, \tag{13.7}$$

such that a stochastic distribution of isotopes leads to an equilibrium constant of 3/8. Per mil departures from this stochastic ratio are quantified using

$$\Delta_{^{12}CH_2D_2} = 10^3\left(\frac{X(^{12}CH_2D_2)}{X(^{12}CH_2D_2)_{Stochastic}} - 1\right). \tag{13.8}$$

The two parameters $\Delta^{12}CH_2D_2$ and $\Delta^{13}CH_3D$ are independent intramolecular thermometers where thermodynamic equilibrium applies. The relationships between temperature and both $\Delta^{12}CH_2D_2$ and $\Delta^{13}CH_3D$ are calculable (e.g. Ma et al. 2008; Webb and Miller 2014; Liu and Liu 2016) and loci of thermodynamic equilibrium in plots of $\Delta^{12}CH_2D_2$ versus $\Delta^{13}CH_3D$ serve as useful references in the plots to follow (Figure 13.2).

13.3 Criteria for Intra-CH_4 Thermodynamic Equilibrium

Criteria for establishing whether or not methane molecules are in isotopologue equilibrium with one another is a first-order requisite for making use of these new data. A datum for a sample of methane gas that plots on the curve in Figure 13.2 would be regarded as representing thermodynamic equilibrium at the indicated temperature by virtue of the concordant temperatures recorded by both clumped species. However, in the absence of data for both $\Delta^{13}CH_3D$ and $\Delta^{12}CH_2D_2$, Figure 13.2 cannot be used to establish equilibrium or nonequilibrium.

While the large-geometry mass spectrometer that permits use of Figure 13.2 was being built for UCLA, two other groups published several papers based on application of the single mass-18 isotopologue $^{13}CH_3D$, or unresolved $^{13}CH_3D$ and $^{12}CH_2D_2$, where the two

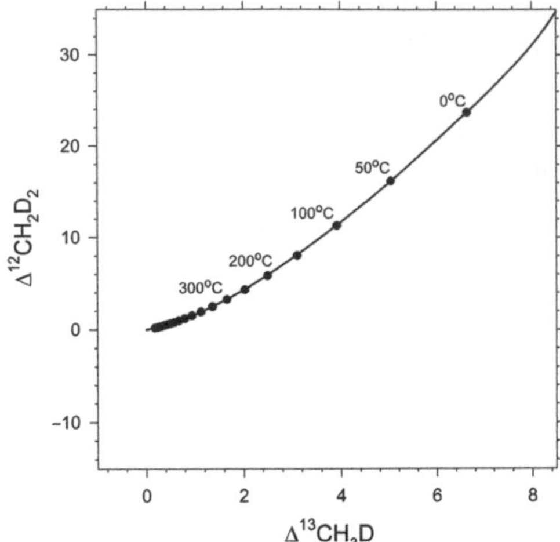

Figure 13.2 Plot of thermodynamic equilibrium among methane isotopologues as a function of temperature in $\Delta^{12}CH_2D_2$ versus $\Delta^{13}CH_3D$ space (after Young et al. 2017).

species are lumped together in one measurement (Stolper et al. 2014, 2015; Wang et al. 2016). Their conclusions were reviewed by Douglas et al. (2017), who describe the "reversibility of methanogensis" hypothesis that was used in several of these papers to explain why temperatures derived using the relative abundances of $^{13}CH_3D$ are not always reasonable (Wang et al. 2015). However, our studies of $\Delta^{13}CH_3D$ and $\Delta^{12}CH_2D_2$ suggest that the rate of methanogenesis is not the primary factor that controls isotope clumping in microbialgenic methane.

The reversibility of the methanogenesis concept is based on earlier work relating D/H and $^{13}C/^{12}C$ in microbial methanogenesis to the chemical potential gradient between reactants H_2, CO_2, or methyl groups and the product CH_4 (Valentine et al. 2004; Penning et al. 2005). This hypothesis posits that the degree of isotopic equilibrium among CH_4 molecules will depend on the rate of methanogenesis (see below). In these studies, the primary tool for establishing whether or not $\Delta^{13}CH_3D$ (or the similar parameter Δ_{18}, which is used where the two mass-18 species are unresolved, since normally $^{13}CH_3D \gg {}^{12}CH_2D_2$ in abundance) represents equilibrium is whether or not the observed D/H partitioning between water and methane, $\alpha_{D/H}$ (H_2O–CH_4) (the fractionation factor $\alpha_{D/H}$ between water and methane is defined as D/H of water divided by D/H of methane), is consistent with the temperature deduced from the single mass-18 clumping temperature. In this scenario, if $\Delta^{13}CH_3D$ is out of equilibrium but progresses toward equilibrium, there will be a correlation between the $\Delta^{13}CH_3D$ (Δ_{18}) and δD of the methane (unless, by coincidence, the methane is already in equilibrium with water). Of course, the obvious disadvantage of this approach as the primary criterion for intra-methane equilibrium is that one has to have the δD of water coexisting with the methane of interest, and there are circumstances where water samples are either impractical to obtain or nonexistent.

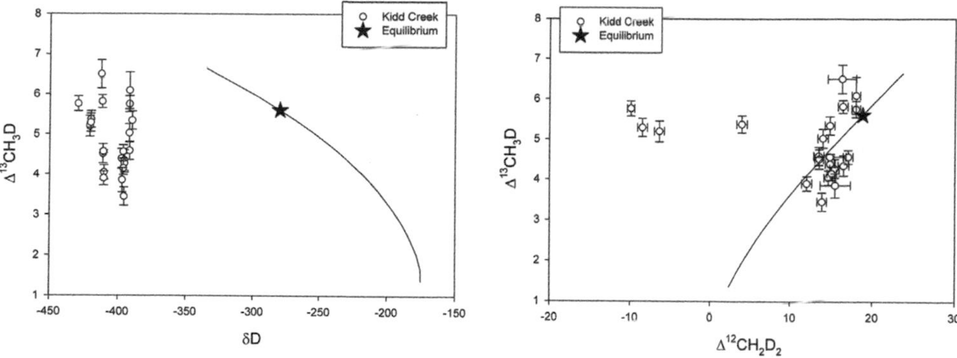

Figure 13.3 Kidd Creek Mine CH_4 data showing the $\Delta^{13}CH_3D$ versus two different potential measures of the degree of intra-CH_4 isotopic equilibrium, δD, and $\Delta^{12}CH_2D_2$. Curves show equilibrium (at fixed δD water = –32‰) from 0°C to 350°C. The stars represent equilibrium at 30°C. The data show a time evolution toward equilibrium in $\Delta^{12}CH_2D_2$ (Young et al. 2017).

In addition to being inconvenient, the water δD criterion for methane isotopologue equilibrium is not universal, as illustrated using $\Delta^{13}CH_3D$ and $\Delta^{12}CH_2D_2$ data from the Kidd Creek Mine, Timmins, Ontario, Canada (Young et al. 2017). Figure 13.3 shows $\Delta^{13}CH_3D$ versus δD and versus $\Delta^{12}CH_2D_2$ for CH_4 from Kidd Creek. Because the δD values of the waters in the mine are generally within a few per mil of –30‰ standard mean ocean water (SMOW), the δD of CH_4 is a measure of $\alpha_{D/H}$ (H_2O–CH_4). One sees that while $\Delta^{12}CH_2D_2$ shows a clear progression toward equilibrium, δD does not. At this location, δD is not especially useful as an indicator of equilibrium, but $\Delta^{12}CH_2D_2$ is. This confirms that heterogeneous (water versus gaseous CH_4) disequilibrium in D/H can persist even as homogeneous equilibrium (exchange among gaseous CH_4 molecules) is approached.

13.4 Kinetics

An understanding of the effects of kinetics on methane isotope clumping is also crucial for interpreting these new data. The idea that the *rate* of methanogensis controls the degree of departure from thermodynamic equilibrium among methane isotopologues has been put forward several times in the literature, perhaps in most complete form in the supplement to Wang et al. (2015) in which the authors invoke “Michaelis–Menten” kinetics for the H addition steps to build methane. This kinetic formalism is commonly used in the biochemistry of enzymes. The equations represent simple transition state-like kinetics where the transition state is replaced by the enzyme–substrate complex:

$$E + S \underset{k_r}{\overset{k_f}{\rightleftarrows}} ES \xrightarrow{k_{cat}} E + P, \tag{13.9}$$

where E is the enzyme, S is the substrate, ES is the substrate bound to the catalyzing enzyme (the enzyme–substrate complex), and P is the product. The forward and reverse rate constants for binding are shown in (13.9), and the subscript "cat" signifies the catalyzed rate constant. The binding of the substrate to the enzyme by a reversible (equilibrium) process is the foundation of Michaelis–Menten kinetics, just as it is foundational for transition state theory. For this equilibrium, the forward and reverse rates are by definition equal:

$$k_f[E][S] = k_r[ES]. \tag{13.10}$$

Mass balance requires $[E] = [E]_0 - [ES]$, where the "0" subscript signifies the initial concentration. After some manipulation, (13.10) leads to the simple first-order rate equation for product P:

$$\frac{d[P]}{dt} = k_{cat}\frac{[E]_o[S]}{\left(\frac{k_r}{k_f}\right) + [S]}. \tag{13.11}$$

Comparing the above to the often-cited Michaelis–Menten equation for an enzyme-catalyzed reaction:

$$\frac{d[P]}{dt} = \frac{V_{max}[S]}{k_m + [S]}, \tag{13.12}$$

where k_m is the "affinity" for the enzyme plus substrate, one sees by inspection that (13.11) is equivalent to (13.12), where the "affinity" for the bound substrate is here equivalent to k_r/k_f and the maximum rate used in the Michaelis–Menten formulation, V_{max}, is $k_{cat}[E]_0$. Notice that when the substrate is plentiful such that $[S] \gg k_r/k_f$, the rate of product production is maximized since $d[P]/dt \approx V_m = k_{cat}[E]_0$. Conversely, where the concentration of the substrate is very small, leading to starvation and thus $[S] \ll k_r/k_f$, we have $d[P]/dt \approx k_{cat}[E]_0[S]/(k_r/k_f)$, which is the maximum rate multiplied by $[S]k_f/k_r$, and so smaller than the maximum rate of reaction. What is more, in this case, both the forward and reverse reactions to form the ES complex are involved, leading to the suggestion that reversibility, or equilibrium, controls the attachment of the substrate to the enzyme.

While (13.11) and (13.12) are handy for describing how the rate of product formation depends on three postulated rate constants, they offer little or no fundamental insights into the effects on isotope clumping. For example, we do not know *a priori* which reaction in the sequence of steps leading to CH_4 formation is properly described by (13.9). Importantly, the kinetic isotope effects (KIEs) and/or equilibrium isotope effects (EIEs) for each of the steps in (13.9) can be adjusted to fit the data, but this relegates (13.11) and (13.12) to little more than a framework for applying fit parameters. Measurements of both $\Delta^{13}CH_3D$ and $\Delta^{12}CH_2D_2$ suggest that the "reversibility of methanogenesis" hypothesis is not the best explanation for the clumping of isotopes into methane of microbial origin.

Here, we consider an alternative approach to explaining the effects of microbial methanogenesis on methane mass-18 isotopologue abundances based on the large degree of disequilibrium exhibited by the low $\Delta^{12}CH_2D_2$ values characteristic of microbialgenic gas in combination with highly variable $\Delta^{13}CH_3D$ values. For this purpose, we take the reaction scheme used by Young et al. (2017) as an analogue for hydrogenotrophic methanogenesis and the steps leading to the net reaction $CO_2 + 4H_2 \rightarrow CH_4 + 2H_2O$. Young et al. (2017) used the reaction scheme for the Fischer–Tropsch synthesis of methane on a cobalt (Co) catalyst suggested by Qi et al. (2014). Methanation of CO_2 can occur by conversion first to CO, with the subsequent steps being the same as those for direct methanation of CO (Wang et al. 2011). The kinetic model presented here is therefore relevant to methane production from CO_2 as well. We added isotope exchange between methane molecules on the surface, desorption and adsorption of CH_4 gas, and attack of CH_4 by OH to the elementary steps leading to methane formation. The set of reactions can be represented by these basic reactions and their isotopically substituted equivalents:

$$\begin{aligned}
&CO_g \rightarrow CO^* \\
&H_{2\,g} \rightarrow 2H^* \\
&CO^* + H^* \rightarrow HCO^* \\
&HCO^* + H^* \rightarrow HCOH^* \\
&HCOH^* \rightarrow CH^* + OH^* \\
&CH^* + H^* \rightarrow CH_2{}^* \\
&CH_2{}^* + H^* \rightarrow CH_3{}^* \\
&CH_3{}^* + H^* \rightarrow CH_4{}^* \\
&OH^* + H^* \rightarrow H_2O_g \\
&CH_4{}^* + OH^* \rightarrow H_2O^* + CH_3{}^* \\
&CH_4{}^* \rightleftarrows CH_4{}^* \quad \text{(isotope exchange)} \\
&CH_4{}^* \rightarrow CH_{4g} \\
&CH_{4g} \rightarrow CH_4{}^*,
\end{aligned} \tag{13.13}$$

where an asterisk signifies a surface-adsorbed species and a subscript "g" refers to a gas species. With all isotopologues and isotopomers, the model consists of 124 species and 796 reactions. The rate constants for the reactions are of the form

$$k_f = Q_{\text{Tun}} \frac{k_b T}{h} \frac{q^+}{\prod_r q_r} \left(\frac{-Ea}{k_b T} \right), \tag{13.14}$$

where k_b is the Boltzmann constant, h is the Planck constant, Ea is the activation energy, q_r are the partition functions for reactant species r, q^+ is the partition function for the transition state, and Q_{Tun} is a correction for quantum tunneling. The 124 ordinary differential equations comprising the model are solved numerically using the Lawrence Livermore ordinary differential equation solver (*DLSODE*). Activation energies for the methane formation reactions were taken from Qi et al. (2014). Values for Ea for the reaction CH_4 + OH were taken from Haghnegahdar et al. (2015). The tunneling correction is from Bell (1959, 1980).

Young et al. (2017) used forward rate constants throughout the reaction network to characterize the effects of the kinetics alone on product methane isotopologues. We concluded in that work that the very negative $\Delta^{12}CH_2D_2$ values were due either to the influence of quantum tunneling or to the combinatorial effect of accessing multiple sources of hydrogen produced by very different fractionation factors involving D/H during the assembly of methane. We favored tunneling in particular for our measurements of products from the Sabatier reaction in the laboratory ($CO_2 + 4H_2 \rightarrow CH_4 + 2H_2O$) mainly because invoking a slight difference in tunneling for ^{13}C–H bonds versus ^{12}C–H bonds could account for the combination of very positive $\Delta^{13}CH_3D$ values along with the negative $\Delta^{12}CH_2D_2$ values (note: an error in that paper has the ratio of tunneling length scales $a_D/a_H = 1.005$, when in fact it is the ratio of $a_{13C\text{–}H,D}/a_{12C\text{–}H,D}$ that was fit to 1.005 to explain the high $\Delta^{13}CH_3D$ values in the Sabatier reaction; see below). We noted that a combinatorial effect (e.g. Yeung et al. 2015; Rockmann et al. 2016; Yeung 2016) may be important for microbial methanogenesis, but we also expressed concern that the required differences in D/H between pools of hydrogen would need to be very large.

However, Cao et al. (2019) have argued recently that adding certain equilibrium steps into our reaction network can indeed produce the requisite large differences in D/H among pools of hydrogen that can reproduce the very low $\Delta^{12}CH_2D_2$ values observed in microbialgenic methane gases. Here, we take their suggestion and investigate the effects of reversibility on the clumping in the reaction network. For these calculations, the KIE and EIE fractionation factors for each reaction step provided by Qi et al. (2014) are used.

Why would a metal-catalyzed reaction scheme have anything to do with the microbial production of CH_4? What advantage is there in trying to use this reaction scheme to address the microbialgenesis of methane? The basic conceit is that the metal surface simulates the enzymes that afford catalyzed transfer of electrons and protons back and forth during the reaction scheme, and that it is the nature of the various C_1 C–H moieties and their bond stiffness that actually control the isotope effects, regardless of the molecule to which they are bound. This may be useful insofar as at least we are using well-characterized KIE and EIE fractionation factors rather than making them up to explain the data. The logic is that despite the potential failings of the analogy, there is benefit in using a real physicochemical model to try to explain what we are seeing. This model is used in lieu of a detailed understanding of the isotope effects attending each step of the enzymatically catalyzed reactions.

With this in mind, one can observe that the steps of building CH_4 molecules from $CO_2 + H_2$ by enzymes and from $CO + H_2$ on metals are broadly similar. The reaction pathway for hydrogenotrophic addition of hydrogen to form methane as given by Lieber et al. (2014) for *Methanococcus maripaludis* is

$$\begin{array}{l} CO_2 \xrightarrow{(1)Fwd} \text{COH-MF} \xrightarrow{(2)Ftr} \text{COH-H}_4\text{MPT} \xrightarrow{(3)Mch} \text{CH} \equiv \text{H}_4\text{MPT} \xrightarrow{(4)Hmd} \\ \text{CH}_2 = \text{H}_4\text{MPT} \xrightarrow{(5)Mer} \text{CH}_3 - \text{H}_4\text{MPT} \xrightarrow{(6)Mtr} \text{CH}_3 - \text{CoM} \xrightarrow{(7)Mcr} \text{CH}_4 \end{array}, \quad (13.15)$$

where the facilitating enzymes are shown above the arrows (e.g. *Ftr* = formyltransferase; numbers are for reference only) and the C_1 carriers are shown as the molecule acronyms (e.g. H_4MPT = tetrahydromethanopterin). As described by Shima et al. (2002), these steps comprise a series of C_1 molecules that act as terminal electron acceptors (oxidants that themselves are reduced) where carbon is progressively reduced, with each transfer of C_1 facilitated by the indicated enzyme. For example, COH is transferred from the methanofuran molecule to the next C_1 carrier H_4MPT by the enzyme *Ftr*. The subsequent conveyance of COH to CH (step (3)) is facilitated by methenyl-H_4MPT cyclohydrolase (*Mch*). The last step in which the CH_3 in methyl-coenzyme M (CH_3–CoM) is reduced to methane is facilitated by the methyl-coenzyme M reductase (*Mcr*); *Mcr* is the focus of a lot of attention because it is common to methanogens and anaerobic methanotrophs and is involved in that last step that is common to both. Now compare this with the reaction network in (13.13) written in similar fashion:

$$\begin{array}{l} CO \xrightarrow{(1)} CO* \xrightarrow{(2)} HCO* \xrightarrow{(3)} HCOH* \xrightarrow{(4)} \\ CH* \xrightarrow{(5)} CH_2* \xrightarrow{(6)} CH_3* \xrightarrow{(7)} CH_4* \end{array} . \tag{13.16}$$

The reactions in (13.15) and in (13.16) both utilize H_2 as the main source of electrons for reduction. In both reaction sequences, C_1 passes from oxidized carbon through an alcohol/formyl molecule to a formaldehyde-like molecule followed by progressive reduction by proton addition.

The effects of kinetics versus equilibrium for each of the steps that produce methane are investigated by using the reactions in (13.16) as analogues for the reactions in (13.15) and examining the consequences of imposing various combinations of equilibrium versus kinetics for each step. Borrowing terminology suggested by Cao and Bao, we can refer to this using the vector [0, 0, 0, 0, 0, 0, 0] to signify where all seven steps are kinetic, or [1, 1, 1, 1, 1, 1, 1] to indicate where all seven steps are completely reversible. Intermediate circumstances are represented accordingly; for example, where the first three steps are equilibrium, we have [1, 1, 1, 0, 0, 0, 0]. For comparison, note that Stolper et al. (2015) assume that the rate-limiting step, the only kinetic step, is step (7) in (13.15), which is mediated by *Mcr*, corresponding to [1, 1, 1, 1, 1, 1, 0]. This seems to be a common assumption.

The results of this exercise are summarized in Figure 13.4. The results for different assumptions about reversibility based on this reaction scheme show clearly that the very low $\Delta^{12}CH_2D_2$ values and moderately low $\Delta^{13}CH_3D$ values of the gases produced by microbial methanogenesis are well explained by scenarios in which equilibrium for the first two or three steps in the reaction sequence in (13.16) obtains, with subsequent steps being kinetic. Indeed, Qi et al. (2014) suggested that step (3) in (13.16) is the likely rate-limiting step, with steps (1) and (2) being reversible. We note that the EIEs (defined as the reaction rate constant for hydrogen only divided by the rate constant for deuterium only) have typical values of order 0.6 compared with the KIEs defined similarly for the same reactions, which usually have values >1. The result is the production of extreme D/H ratios for the various pools of hydrogen atoms along the reaction sequence. For example, in the successful model [1, 1, 0, 0, 0, 0, 0], δD relative to the reactant hydrogen gas (0‰ in this case) is +125.6‰ for CH, 0.4‰ for CH_2, and −104.9‰ for CH_3 in this reaction scheme. Quantum tunneling of protium (H) and deuterium (D) may also play a role in the enzymatically catalyzed reactions, but we find that we can use a more conservative tunneling distance to fit the data than was used previously for the Sabatier reaction (see the caption to Figure 13.4). Notice that the results with all irreversible steps cannot explain the microbial methanogenesis data (Figure 13.4). Also, the [1, 1, 1, 1, 1, 1, 0] scheme in which the final step is rate limiting gives nonsensical results, with $\Delta^{12}CH_2D_2$ and

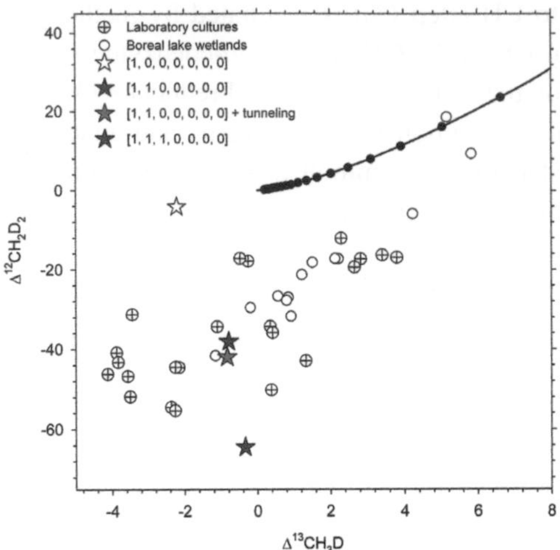

Figure 13.4 Summary of modeling reactions in (13.13) and (13.16). The stars represent model results for different combinations of equilibrium (signified by a 1 in the square brackets) or kinetic (signified by a 0 in the square brackets) steps during methane formation for the seven steps numbered in (13.16), as described in the text. Black dots denote equilibrium in +50°C intervals starting with 0°C at the upper right. Circles show natural (open) and laboratory culture (crossed) data for microbial methanogenesis. The H tunneling distance used in the model that includes tunneling is 3×10^{-10} m with no carbon isotope effect. For the [1, 1, 0, 0, 0, 0] calculation, only steps (1) and (2) include isotopic equilibrium.

$\Delta^{13}CH_3D$ values of +138.4 and +11.3, showing that this scheme, analogous to that used by Stolper et al. (2015), is unlikely to be correct.

The reaction schemes resembling the microbial data in Figure 13.4 also give realistic CH_4 δD and $\delta^{13}C$ bulk values for the product methane. For example, for the [1, 1, 0, 0, 0, 0, 0] scheme, we obtain CH_4 $\delta D = -161.2$‰ and $\delta^{13}C = -44.1$‰ relative to the hydrogen and carbon substrates.

Gruen et al. (2018) confirm that all four H atoms comprising CH_4 from hydrogenotrophic methanogenesis are from water, and yet we see profound disequilibrium in ΔCH_2D_2. While this simple model explains the first-order $\Delta^{12}CH_2D_2$ and $\Delta^{13}CH_3D$ characteristics of microbial methanogenesis, there is more to the kinetic story with respect to methane formation. Also, it should be pointed out that the highest $\Delta^{12}CH_2D_2$ and $\Delta^{13}CH_3D$ values near the equilibrium curve in Figure 13.4 are boreal lake analyses that are evidently the result of methanotrophy (see below) rather than methanogenesis.

13.5 The Microbial Array

The positively sloping array in $\Delta^{12}CH_2D_2$ versus $\Delta^{13}CH_3D$ space defined by the *in vitro* and *in vivo* microbial methanogenesis data (Figure 13.4) cannot be explained easily by simply varying the degree of reversibility for the reaction steps. Quantum tunneling

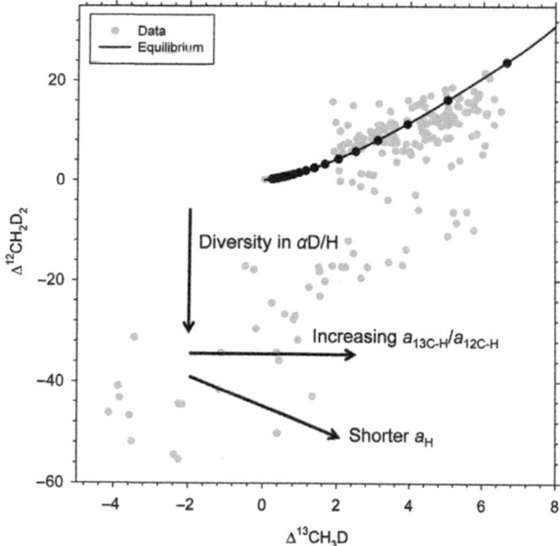

Figure 13.5 Vectors showing the effects of the key kinetic parameters controlling clumping during CH_4 formation.

involving both D/H and $^{13}C/^{12}C$ seems to be required. The involvement of carbon isotopes may at first seem unlikely. However, Miller et al. (1981) point out that it is incorrect to think of hydrogen tunneling in isolation. It is the reacting system as a whole that experiences tunneling, and if carbon atoms are involved in the reaction coordinate, the isotopic composition of the carbon can have an effect on tunneling. Using the same model reaction sequence described above, I have explored the interaction between various reversibility scenarios and the effects of tunneling. Generalized vectors showing the effects of the various parameters based on this exploration of parameter space are shown in Figure 13.5. Disparities in reversibility, and thus large differences in fractionation factors at each step, cause decreases in $\Delta^{12}CH_2D_2$ at constant $\Delta^{13}CH_3D$. This occurs with only moderately low δD for the product methane compared with the more extreme values caused by tunneling. Shorter hydrogen/deuterium tunneling distances increase the effects of quantum tunneling. Where the tunneling distance a_H is greater for reactions involving ^{13}C (as expected) than for ^{12}C, the slope of the enhancement in tunneling in $\Delta^{12}CH_2D_2$ versus $\Delta^{13}CH_3D$ space is negative (Figure 13.5).

The effect of differences in tunneling distances involving ^{13}C versus ^{12}C can be visualized using a Marcus theory representation of tunneling shown in Figure 13.6; the lower energy of vibration for the ^{13}C–H bond relative to the C–H bond, for example, will necessitate a greater tunneling distance. The effect of $a_{13C\text{–}H}/a_{12C\text{–}H} > 1$ on $\Delta^{13}CH_3D$ can be understood as follows: the longer tunneling distance for the ^{13}C case gives less advantage to H relative to D in bonding with ^{13}C relative to the ^{12}C case, all else equal, and so $^{13}CH_3D$ is preferred relative to $^{13}CH_4$. Thus, $\Delta^{13}CH_3D$ goes up.

The principles described in Figures 13.5 and 13.6 can be used to describe the "kinetic array" that lies well below the equilibrium curve in Figure 13.4. There is a hint that these

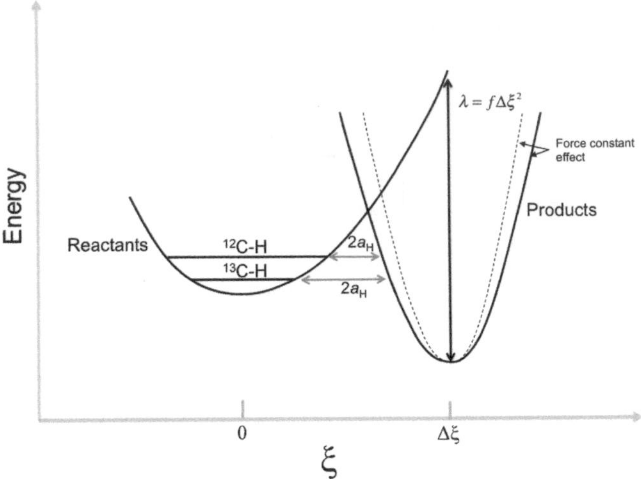

Figure 13.6 Schematic illustrating the relationship between carbon isotope substitution and tunneling. The ordinate is potential energy and the abscissa is the reaction coordinate ζ representing the interatomic motions leading to the reaction. The parabolas represent the potential energy of reactants and products in terms of interatomic distances along the reaction coordinate. The gray arrows shows the "tunneling" of hydrogen through the energy barrier associated with the transition from reactants to products. $2a_H$ is the tunneling distance for a hydrogen bonded to either ^{13}C or ^{12}C (see text). λ refers to the "reorganization" energy as defined in Marcus theory that depends on the force constant f and the separation between the two energy minima Δξ.

principles may have applications beyond just microbial methanogenesis. The methane gases from the deeper levels of the Kidd Creek Mine study fall on the high $\Delta^{12}CH_2D_2$ and $\Delta^{13}CH_3D$ end of the microbial array (Young et al. 2017). These gases are confidently regarded as abiotic with no connection to microbial methanogenesis, as far as we know (Sherwood Lollar et al. 2007). Therefore, their continuity with the microbialgenic gases implies a fundamental kinetic link.

The high end of the kinetic array occupied by some of the microbial data and the Kidd Creek data can be explained by a larger role for both kinetics and tunneling. Removing the reversibility in all but step (1) and adjusting a_H and a_{13C-H}/a_{12C-H} to 5.2×10^{-11} m and 1.016, respectively, yields a fit to the Kidd Creek data and the high end of the microbial kinetic array (black star in Figure 13.7). The $^{13}C/^{12}C$ tunneling effect of 1.016 is squarely in the range found by Miller et al. (1981). The δD obtained in this model is −470.7‰ relative to the source of hydrogen, similar to the low values for the Kidd Creek methane.

Taken together, the two results for the two ends of the "kinetic array" based on the analogue model for methane formation suggest that the positively sloping array in $\Delta^{12}CH_2D_2$ versus $\Delta^{13}CH_3D$ space can be explained by a gradation from reversibility in some early key steps and a more limited role for tunneling on the lower end toward less reversibility and a larger role for tunneling on the higher end, including more participation of carbon in the reaction coordinate. In the case of the Kidd Creek data, it may make sense

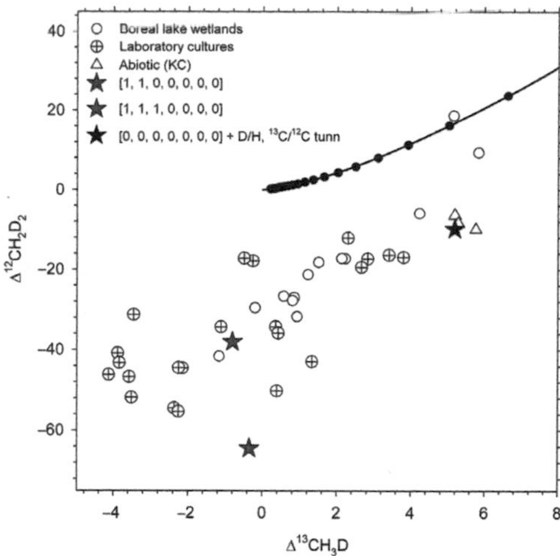

Figure 13.7 Comparison between the "kinetic array" in $\Delta^{12}CH_2D_2$ versus $\Delta^{13}CH_3D$ space and several relevant reaction rate models based on the reaction scheme shown in (13.13) and (13.16). The array may be defined by a "mixing" of the processes, similar to those depicted by the blue and black stars. See text for discussion. KC = Kidd Creek.

that catalysis of methane formation by metal surfaces would permit more of a role for carbon motions than in the case of the C_1 host molecules facilitating microbial methanogenesis.

13.6 Is This New Information Helping?

Is this new information about processes orthogonal to differences in isotopic reservoirs actually helping us determine the provenance of methane gases? We examine this question with the aid of Figure 13.8. Here, we plot ~170 natural CH_4 samples analyzed and vetted in our laboratory at UCLA to date as well as the culture experiments and several high-temperature Fischer–Tropsch experiments in $\Delta^{12}CH_2D_2$ versus $\Delta^{13}CH_3D$ space on the left and in the traditional Schoell plot (δD versus $\delta^{13}C$) on the right. Based on the clumping diagram, three groups of samples are selected as examples of "process end members" as a sort of test of the efficacy of considering process as a means of deducing provenance. Two of the three groups are easily defined and substantiated by laboratory experiments. The first of these are the natural samples that represent purely microbial methanogenesis based on their similarity to the combination of $\Delta^{12}CH_2D_2$ and $\Delta^{13}CH_3D$ values obtained in our *in vitro* culture experiments (in most cases these are from boreal wetlands). The second group comprises gases produced at high temperatures (meaning >100°C) that lie at or near to isotopologue thermodynamic equilibrium. While many of these gases are thermogenic, we are purposefully avoiding the traditional classification terminology at this stage because these terms can connote both process and the sources of carbon and hydrogen (e.g.

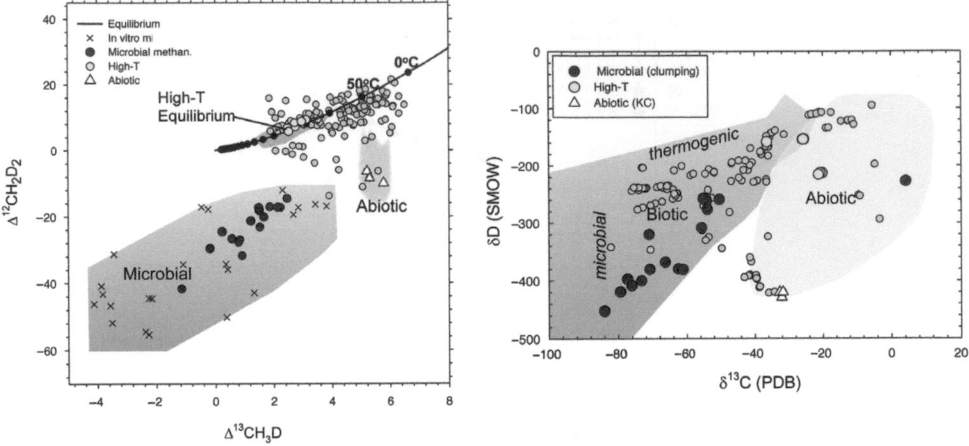

Figure 13.8 Comparison of data categorized by process in $\Delta^{12}CH_2D_2$ versus $\Delta^{13}CH_3D$ space (left panel) to their positions in the bulk isotope "Schoell" plot (right panel). Details are discussed in the text. Fields on the left are from the mass-18 isotopologue work, while the fields in bulk isotope space are from Etiope and Schoell (2014) and Etiope (2017). Gray symbols are data not categorized explicitly in this plot. The data labeled "microbial methane" are dominantly, although not entirely, from boreal lakes. PDB = Pee Dee Belemnite standard.

Sherwood Lollar et al. 2006; Etiope and Sherwood Lollar 2013). The third group represents "abiotic" methane. Here, we have immediately deviated from the strategy outlined in the previous sentence and introduced a measure of circularity into the classification by convolving it with source material. This is justified because of the overwhelming geological and geochemical evidence suggesting that the Kidd Creek Mine gases owe their origins to abiotic processes deep in the crust.

The three end-member processes are well separated on the left-hand panel of Figure 13.8. Those same data points, with the same color coding, are plotted on the Schoell plot on the right in Figure 13.8. I have added two major methane source fields, biotic and abiotic, as inspired by the fields shown in Etiope and Schoell (2014) and Etiope (2017). The microbial and thermogenic process subfields of the biotic field are also shown. Overall, the microbial gases as defined by the isotopic bond ordering plot in the microbial subfield for biotic gases in the Schoell plot. The high-temperature equilibrium gases tend to plot in the thermogenic subfield of the biotic field in the Schoell plot, and the Kidd Creek abiotic gases are at the lower edges of the abiotic field in the Schoell plot. Have we learned anything?

The answer appears to be yes because the isotopologue fields separate process from source material unambiguously. For example, the vertical (δD) positions of boreal wetland microbial methanogenesis data in the Schoell plot are lower than those for many "typical" microbial gases because the water δD values for these arctic environs are approximately −200‰ (e.g. Douglas et al. 2016 and references therein), rather lower than for waters from lower latitudes. The [1, 1, 0 ... 0] kinetic model (blue star in Figure 13.7) predicts a downward shift in δD of ~160‰ relative to the source hydrogen, yielding bulk δD values of about −360‰, consistent with the boreal wetland data in Figure 13.8. The breadth of the microbial field in the Schoell plot is therefore dictated in part by the range of source hydrogen and carbon samples independent of process. In isotopic bond ordering space, this effect is normalized out.

The importance of normalizing away reactant bulk isotopic compositions is best exemplified by the fact that there are instances where data falling squarely in the microbial field in $\Delta^{13}CH_3D$–$\Delta^{12}CH_2D_2$ space are not in the biotic field on the Schoell plot. The reasons for this can be traced to unexpected source materials (as enumerated in publications in preparation). Similarly, gases equilibrated at relatively high temperatures as evidenced in the $\Delta^{13}CH_3D$–$\Delta^{12}CH_2D_2$ plot span the biotic and abiotic fields in the Schoell plot.

Most of the other data points that appear in Figure 13.8 (undifferentiated data are shown in gray) that do not fall within one of the fields in $\Delta^{13}CH_3D$–$\Delta^{12}CH_2D_2$ space are the result of mixing, fractionation by molecular mass (e.g. Giunta et al. 2018), or possibly processes that are still under study (see Section 13.7).

As an illustrative example of both the high information content and the complexity afforded by CH_4 isotopologue data, we consider here a methane sample from a seafloor vent from the North Atlantic. It is generally thought that aqueous alteration of ultramafic rocks can produce abiotic methane as a by-product of serpentinization that releases hydrogen gas by a reaction resembling $3Fe_2SiO_4 + 2H_2O \rightarrow 2Fe_3O_4 + 3SiO_2 + 2H_2(g)$. However, the process is less clear than once thought due to severe kinetic limitations (McCollom 2016). The temperatures or kinetic pathways of methane formation in serpentinizing systems would be valuable arbiters for competing hypotheses for CH_4 formation in these settings. One well-studied site for methane production associated with active serpentinization is the Atlantis Massif that lies east of the intersection of the Mid-Atlantic Ridge and the Atlantis transform fault on the North Atlantic Ocean seafloor. The Lost City hydrothermal field from the southern Atlantis Massif has been of particular interest as the archetypical example of an off-axis seafloor hydrothermal alteration zone (Kelley et al. 2005). Fluids are venting at temperatures of ~30–90°C and both archaeal methanogens and methanotrophs are evidenced in the system (Kelley et al. 2005). Earlier studies indicated that serpentinization at the site was seawater dominated at temperatures of ~150–250°C (Allen and Seyfried Jr. 2004; Boschi et al. 2008). Proskurowski et al. (2006) used apparent D/H partitioning between CH_4, H_2, and H_2O to suggest that methane formation linked to serpentinization occurred at temperatures of ~110–150°C. Wang et al. (2018) used $\Delta^{13}CH_3D$ to estimate methane formation temperatures for several seafloor vent systems, including Lost City. Their $\Delta^{13}CH_3D$ value for the Beehive vent at Lost City is 1.84 ± 0.60‰ (95% confidence), from which they derive a methane formation temperature of 270 +104/−68°C. These authors discuss the evidence supporting this temperature, which is higher than many previous estimates.

At UCLA, we measured a methane sample from the Beehive vent at Lost City (collected in 2005 and provided by Marvin Lilly, University of Washington) and obtained $\Delta^{13}CH_3D = 1.95‰ \pm 0.40\ (2\sigma)$ and $\Delta^{12}CH_2D_2 = 11.6‰ \pm 1.2\ (2\sigma)$ (Figure 13.9). This

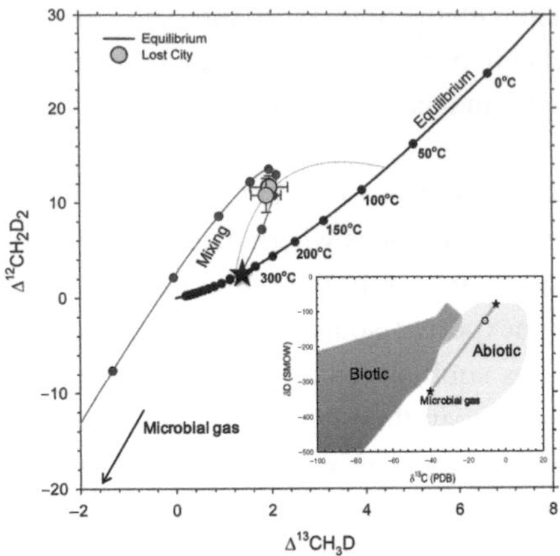

Figure 13.9 The mass-18 isotopologue data and bulk $\delta^{13}C$ and δD values for a sample from the Beehive vent of the Lost City hydrothermal field and two possible mixing scenarios to explain the data discussed in the text. Stars show end-member compositions for the missing scenario involving microbial gas described in the text. Solid dots on the mixing curve mark proportions in 10% increments. The Lost City data are the gray circles in both plots. Error bars are 2σ. PDB = Pee Dee Belemnite standard.

$\Delta^{13}CH_3D$ value is indistinguishable from the Wang et al. (2018) value, and thus at face value it is consistent with the high formation temperature (260 +58/–45°C in our case) obtained in the Wang et al. study. However, the $\Delta^{12}CH_2D_2$ value is 7.5‰ higher than the equilibrium value at that temperature. Since the datum is displaced from the equilibrium curve (Figure 13.9), the $\Delta^{13}CH_3D$ value of 1.95 cannot be taken *a priori* as a direct measure of methane bond formation temperature.

The position of the Lost City sample above the equilibrium curve in Figure 13.9 suggests that the gas is a mixture. One mixture that fits the datum within error is composed of 20% microbial gas and 80% high-temperature (~350°C) abiotic gas (blue curve in Figure 13.9). The bulk isotopic compositions of these two end members and the associated mixing curve in bulk isotope ratio space are shown in the inset in Figure 13.9. The end-member $\delta^{13}C$ and δD values required by the fit to the $\Delta^{13}CH_3D$ and $\Delta^{12}CH_2D_2$ data are reasonable, if imperfect, representatives of microbial methanogenesis and abiotic gas, respectively. In this case, the methane isotopologue data may be indicative of a microbial methanogenesis component to the Lost City gas, whereas any temperature information for the mixture end members is model dependent. While the mixing scenario described above is plausible, the Lost City datum by itself does not lead to an entirely unique interpretation. For example, a fit to the isotopologue data can also be obtained by mixing 375°C and 75°C gases with abiotic-like $\delta^{13}C$ and δD values of −20 and −60‰ and −3.5 and −183‰, respectively (green curve in Figure 13.9).

Possible explanations for the Lost City data other than mixing are: (1) the bulk isotopic composition of the gas was altered by a process that fractionates isotopologues by molecular mass, moving the gas along a slope-1 line in Figure 13.9 and thus suggesting a low formation temperature of ~75°C by extrapolation back to the equilibrium curve; or (2) combinations of mixing and fractionation.

The Lost City example shows at once the potential power of the methane isotopologue data, but also the complexity in the interpretations that can arise, especially where the data are sparse. Of course, suites of data representing ranges in composition are more diagnostic than individual data points. Among the data shown in Figure 13.8 are examples of linear trends in $\Delta^{12}CH_2D_2$ versus $\Delta^{13}CH_3D$ space with slopes of unity that are telltale signs of fractionation by molecular mass, for example.

The ability to use rare isotopologue abundances rather than bulk isotope ratios as means of tracing the origins of methane gases could have important applications elsewhere in the solar system. Measuring $\Delta^{12}CH_2D_2$ and $\Delta^{13}CH_3D$ in methane from the atmospheres of Mars or Saturn's moon Titan, for example, would provide powerful evidence for the formation mechanisms of Martian and Titanian methane even where the meaning of bulk $^{13}C/^{12}C$ and D/H in the methane is hampered by unknown chemical cycles on those bodies.

13.7 The Effects of Oxidation on $\Delta^{12}CH_2D_2$ and $\Delta^{13}CH_3D$

It appears that the combination of values for $\Delta^{12}CH_2D_2$ and $\Delta^{13}CH_3D$ provides a signature of microbialgenic methane (the microbial array) that is independent of bulk isotope ratios, and therefore independent of sources of carbon and hydrogen. The uniqueness of the low $\Delta^{13}CH_3D$ and extremely low $\Delta^{12}CH_2D_2$ values that in combination signify microbial methanogenesis depends on whether other processes might mimic this effect. We have *in vitro* evidence that anaerobic oxidation of methane (AOM) tends to drive isotopologue compositions of methane toward equilibrium and away from the microbial array. However, at time of writing, the effects of bacterial aerobic oxidation of methane remain unknown.

The data clustered near the equilibrium curve in the upper right of the $\Delta^{12}CH_2D_2$ versus $\Delta^{13}CH_3D$ diagram (e.g. left panel of Figure 13.8) present a vexing problem. At face value, these gases are in near isotopologue equilibrium at <50°C. Two prominent examples are a few of the boreal wetland samples and some of the deep mine samples. However, it seems unlikely that CH_4 gas would equilibrate at such low temperatures absent a reaction that breaks and reforms methane molecules (as opposed to isotope exchange with water, for example, which is slow; e.g. Sessions et al. 2004). What is this process?

Young et al. (2017) postulated that the evolution of CH_4 effusing from the Kidd Creek Mine fluids over several years from the disequilibrium abiotic field toward the equilibrium curve is the result of cycling between microbial methanogenesis and methanotrophy. The shifts over time are seen mainly in $\Delta^{12}CH_2D_2$, which is initially in gross disequilibrium, but not in $\Delta^{13}CH_3D$, where the latter is in apparent equilibrium from the start (Figure 13.3).

This cycling may or may not be distinct from the "cryptic methane cycle" described recently by Maltby et al. (2018) and Xiao et al. (2017); the cryptic methane cycle refers to high SO_4 noncompetitive environs, while experiments suggest that methane reemitted by AOM is driven by low concentrations of the electron acceptor (e.g. SO_4; see below). In addition, based on samples from Baltic Sea sediments, Ash et al. (2019) find evidence that isotopic bond-order equilibrium in these sediments is the result of AOM. Stolper et al. (2015) had hypothesized that AOM might yield thermodynamic equilibrium among methane isotopologues, suggesting that isotopic equilibrium between CH_4, H_2, and CO_2 would ensue. The Baltic Sea results suggest that CH_4 equilibration occurs without equilibrium with water. The concept that AOM has the capacity to yield methane in isotopic bond-order equilibrium builds on the earlier work of Holler et al. (2011) and Yoshinaga et al. (2014) showing that AOM can lead to $^{13}C/^{12}C$ equilibrium. AOM shares enzymatic machinery with microbial methanogenesis, leading to the notion that AOM is a reversal of methanogenesis (Scheller et al. 2010; Timmers et al. 2017). Indeed, Yan et al. (2018) report Fe-based AOM by the methanogen *Methanosarcina acetivorans* in which oxidation of methane occurs by the reversal of the biochemical pathway for acetoclastic and CO_2 reduction to methane (see (13.17)). The influence of AOM on the isotopologue composition of CH_4 in nature seems plausible; Yoshinaga et al. (2014) point out that methanogens represent a small contingent of the microbial communities in AOM-active sediments and that $^{14}CO_2$ is converted to $^{14}CH_4$ in sulfate–methane transition zones (Orcutt et al. 2005).

Confounding the identification of the CH_4 isotopologue signature of AOM in natural settings is the contradictory evidence for its bulk isotopic effects. While Holler et al. (2009) found increases in bulk $^{13}C/^{12}C$ and D/H in residual methane left behind by AOM in the laboratory, Yoshinaga et al. (2014) emphasized that methane from sulfate-limited AOM horizons has a relatively low $\delta^{13}C$ value, not high as expected from simple classical kinetics. They suggest that this is due to equilibration of $^{13}C/^{12}C$ in the back reaction to convert dissolved inorganic carbon (CO_2 dissolved in water) to CH_4.

Douglas et al. (2017) refer to "differential reversibility of methanogenesis" in the context of explaining various degrees of apparent equilibration of $\Delta^{13}CH_3D$ (or Δ_{18}) values for methane. They suggest that hydrogenotrophic, H_2-limited methanogenesis (Valentine 2011) causes reversibility. They attribute the equilibrium to rapid H/D exchange with water. This interpretation is in the context of the "reversibility of methanogensis" hypothesis rather than reversibility in AOM. These earlier suggestions notwithstanding, the $\Delta^{12}CH_2D_2$ versus $\Delta^{13}CH_3D$ data suggest that interspecies (e.g. CH_4 versus H_2O) equilibration is not a requisite for intraspecies (CH_4 alone) equilibration.

Because of the difficulty in equilibrating isotopologues at low (near room) temperatures and because of the circumstantial evidence for AOM in several of the sites where low-temperature equilibration in methane gas is evidenced in the isotopologue data, experiments are underway attempting to characterize the $\Delta^{12}CH_2D_2$ versus $\Delta^{13}CH_3D$ effects of AOM. These include experiments by J. Gregory Ferry's group at Penn State University that comprise exchange of the methyl moiety in methyl-coenzyme M (CH_3-SCOM) and CH_4 by the reaction couple

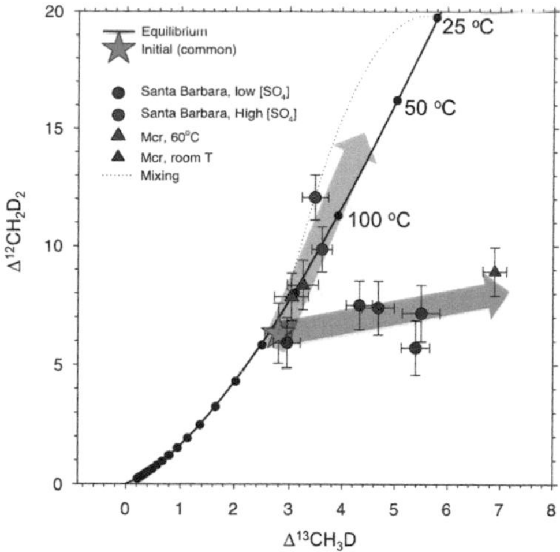

Figure 13.10 Summary of preliminary AOM *in vitro* experiments in $\Delta^{12}CH_2D_2$ versus $\Delta^{13}CH_3D$ space. Symbol shapes distinguish the two sets of experiments. The star shows the initial gas composition of the Santa Barbara slurry experiments. The *Mcr* experiments are migrated so that they are relative to the Santa Barbara initial gas composition. The arrows shows the two different trends that are evident in the data as described in the text. Equilibrium temperatures strictly apply only to the Santa Barbara experiments due to the migration of the *Mcr* data.

$$\begin{array}{c} CH_4 + \text{CoB-S-S-CoM} \rightarrow \text{HSCoB} + CH_3\text{-SCoM} \\ CH_3\text{-SCoM} + \text{HSCoB} \rightarrow \text{CoB-S-S-CoM} + CH_4, \end{array} \tag{13.17}$$

where HSCoB is coenzyme B that provides the proton and CoB-S-S-CoM is the hetero-disulfide of coenzymes M and B that serves as the oxidant in the reaction. Reaction (13.17) is catalyzed by the *Mcr* enzyme and is a reversible instance of the final step in (13.15). This experiment is particularly informative as it isolates the step that is often cited as the source of equilibration. The hypothesis is that the reversibility of this step might capture the mechanism for CH_4 isotopologue equilibration by AOM. Accordingly, the expectation was that we would see reactant methane in the headspace for this cocktail of coenzymes progress toward isotopologue equilibration.

Some preliminary results from the Penn State experiments are shown in Figure 13.10. The data define two trends. At temperatures below those optimal for the *Mcr* used in the experiments (near room temperature), there is a trend toward equilibrium values for $\Delta^{13}CH_3D$ but little change in $\Delta^{12}CH_2D_2$. At 60°C, the optimal temperature for the thermophile-derived enzyme, there are modest but discernible increases in both $\Delta^{13}CH_3D$ and $\Delta^{12}CH_2D_2$ values (Figure 13.10). The higher-temperature experiments presumably involve greater turnover of methane by enzymatic activity and thus more exchange of hydrogen. Based on the modeling summarized in Figure 13.4, the positive shift in $\Delta^{13}CH_3D$ at relatively constant $\Delta^{12}CH_2D_2$ observed in the low-temperature experiments suggests that carbon played a larger role than hydrogen in affecting the isotopologue abundances resulting from exchange.

In a set of parallel experiments by Tina Treude's laboratory at UCLA, methane head-space gas was consumed by SO_4-limited AOM in *in vitro* marine sediment slurries using samples collected from active methane seeps in the Santa Barbara channel off the coast of Santa Barbara, California. Methane incubations lasted for various time intervals in the order of days to weeks with concentrations of SO_4 ranging from 38 mM to <500 μM. The low-SO_4 incubations resemble the low-temperature *Mcr* experiments in which the $\Delta^{13}CH_3D$ values of residual headspace methane increased toward equilibrium while $\Delta^{12}CH_2D_2$ remained relatively constant. However, the higher concentration of SO_4 results resemble the higher-temperature *Mcr* experiments, with both $\Delta^{12}CH_2D_2$ and $\Delta^{13}CH_3D$ increasing with methane consumption (Figure 13.10). One datum from the Santa Barbara sediment slurry experiments lies above the equilibrium curve and might be explained by mixing between initial gas and equilibrated gas, as shown by the dotted curve in Figure 13.10.

These experiments are ongoing, and more detailed reports are in preparation, but our preliminary conclusion is that AOM can push residual methane toward $\Delta^{12}CH_2D_2$ and $\Delta^{13}CH_3D$ equilibrium on trajectories in $\Delta^{12}CH_2D_2$ and $\Delta^{13}CH_3D$ space that depend on the conditions. Where the enzymatically facilitated exchange is active, both $\Delta^{12}CH_2D_2$ and $\Delta^{13}CH_3D$ are affected, while under more restricted exchange activity, only $\Delta^{13}CH_3D$ is affected. Under no conditions do the effects of AOM resemble those for microbial methanogenesis in $\Delta^{12}CH_2D_2$ and $\Delta^{13}CH_3D$ space.

Aerobic microbial oxidation effects on $\Delta^{13}CH_3D$ were studied by Wang et al. (2016), in which $\Delta^{13}CH_3D$ values were observed to decrease by several per mil with methane consumption. The effects on $\Delta^{12}CH_2D_2$ are as yet not known from experiments. Oxidation of methane in the atmosphere by Cl or OH radicals was investigated experimentally by Whitehill et al. (2017) for $^{12}CH_4$, $^{13}CH_4$, $^{12}CH_3D$, and $^{13}CH_3D$ (i.e. $\Delta^{13}CH_3D$). The effects of Cl and OH oxidation of methane on $^{12}CH_2D_2$ (i.e. $\Delta^{12}CH_2D_2$) as well as on the other species was modeled by Haghnegahdar et al. (2017). The predictions for $\Delta^{12}CH_2D_2$ effects of Cl and OH oxidation have been verified by measurements of the experimental products produced by Whitehill et al. (2017) at UCLA. All of these results on the Cl and OH oxidation of methane indicate large shifts in clumping down to extremely low $\Delta^{12}CH_2D_2$ values (negative tens of per mil) and low $\Delta^{13}CH_3D$ values well below zero. What is more, in each of the studies, the isotopic KIE values (kinetic isotope fractionation factors) associated with the simple reactions $CH_4 + OH \rightarrow CH_3 + H_2O$ and $CH_4 + Cl \rightarrow CH_3 + HCl$ closely approximate the "rule of the geometric mean" (RGM) (Bigeleisen 1955) in which the KIE for $^{13}CH_3D/CH^4$, for example, is the product of those for $^{13}CH_4/CH_4$ and $^{12}CH_3D/CH_4$, etc. Where the RGM applies, the relative abundances of the multiply substituted isotopologues – $^{13}CH_3D$ and $^{12}CH_2D_2$ in this application – are controlled entirely by the classical KIEs and not by the equilibrium zero-point energy effects specific to each of the isotopologues. The results of the classical kinetics in which the lighter isotopic species of methane react with Cl and OH more rapidly than the heavier isotopologues are increases in bulk $^{13}C/^{12}C$ and D/H in the residual methane left behind and concomitant decreases in $\Delta^{12}CH_2D_2$ and $\Delta^{13}CH_3D$ (see Question 1 in the "Questions for the Classroom" section for the proof of this behavior). One expects the RGM to apply for single-step kinetic reactions such as those that oxidize CH_4 in the atmosphere. The RGM does not apply to the methane isotopologue effects of AOM. Whether the RGM applies to aerobic bacterial oxidation is at present not known, but experiments are planned.

13.8 Conclusions

Isotopic bond ordering in methane gas molecules traces process, while bulk isotope ratios trace both process and source isotopic compositions. By measuring the isotopic bond ordering in CH_4 gas, it is possible to isolate processes from sources. The $^{13}C/^{12}C$ ratios of microbialgenic methane, for example, are controlled by the substrate carbon as well as by the processes of methane formation. The D/H ratios of methane are controlled by the source of hydrogen (often water), as well as the reaction path to formation. The combination of $\Delta^{13}CH_3D$ and $\Delta^{12}CH_2D_2$ yields a measure of the process of formation irrespective of the $^{13}C/^{12}C$ or the D/H ratios of the source carbon and hydrogen. While more experimental work is required to investigate all possible reaction pathways, thus far it appears that the position of a methane datum in $\Delta^{13}CH_3D$ versus $\Delta^{12}CH_2D_2$ space can be used to identify processes of formation independent of the uncertainties in the source material. One can predict that the ability to trace the origins of methane independent of bulk carbon and hydrogen isotope ratios could prove invaluable for assessing the origins of CH_4 gas on other solar system bodies where the meaning of bulk isotope ratios would not be accurately known.

13.9 Limits to Knowledge and Unknowns

A known unknown is the effect of "cracking" of organics in $\Delta^{13}CH_3D$ versus $\Delta^{12}CH_2D_2$ space. Thus far, it seems that low – even negative – values for $\Delta^{12}CH_2D_2$ result from the multiple steps of building a CH_4 molecule as in microbial methanogenesis. For comparison, Shuai et al. (2018) found that while nonhydrous pyrolysis of coal can generate methane with equilibrium $\Delta^{13}CH_3D$ (Δ_{18}) values at temperatures ranging from ~400°C to ~500°C and from ~600°C to 700°C, disequilibrium in Δ_{18} by about 2‰ results from ~500°C to ~600°C. The degree of disequilibrium also depends on the rate of heating. Based on these results, an important question is whether such processes could ever lead to confusion between microbial methanogenesis and cracking in $\Delta^{13}CH_3D$ versus $\Delta^{12}CH_2D_2$ space. Hydrous pyrolysis of shale evidently leads to equilibrium isotopologue distributions in the product methane (Shuai et al. 2018). It is perhaps relevant in this context that the temperature window leading to disequilibrium in methane isotopologues due to pyrolysis seems to be rather narrow, suggesting that this may not be a generally important process. More experiments are required.

Acknowledgments

This chapter is based on a compilation of work by a long list of collaborators, many of whom been funded by the Deep Energy community of the Deep Carbon Observatory (Sloan Foundation). This chapter is not meant to replace the primary literature produced by these workers, but rather is an update on the status of results obtained with some novel new data. The author acknowledges the critical input and efforts (in no particular order) by Barbara Sherwood Lollar (University of Toronto), Gius Etiope (Istituto Nazionale di Geofisica e Vulcanologia), Thomas Giunta (University of Toronto), Jeanine Ash (UCLA, Rice University), Jabrane Labidi (UCLA), Issaku Kohl (Thermo Fisher Scientific), Douglas Rumble III (Geophysical Laboratory, Carnegie Institution for Science), Tina Treude (UCLA), Sebastian Krause (UCLA), Rachel Harris (Princeton University), James Gregory Ferry (Penn State University), Divya Prakash (Penn State University), Alexis Templeton (University of Colorado), Daniel Blake Nothaft (University of Colorado), L. Taras Bryndzia (Shell International Exploration and Production, Inc.), Marvin Lilly (University of Washington), Mojhgan Haghnegahdar (UCLA), and Edwin Schauble (UCLA), among many others. The conclusions put forward here are those of the author and may not be shared by all of the people engaged in the various studies.

Questions for the Classroom

1. Why would partial consumption of isotopically light methane molecules leave the residual methane with lower $\Delta^{13}CH_3D$ and $\Delta^{12}CH_2D_2$ values relative to the initial isotopologue composition of the gas in the absence of isotopic exchange among the methane molecules?
2. Why do lower temperatures favor greater $\Delta^{13}CH_3D$ and $\Delta^{12}CH_2D_2$ values relative to the stochastic values of zero?
3. Why are $\Delta^{13}CH_3D$ and $\Delta^{12}CH_2D_2$ values independent of bulk isotope ratios?

References

Allen D. E. and Seyfried Jr. W. E. (2004) Serpentinization and heat generation: constraints from Lost City and Rainbow hydrothermal systems. *Geochemica et Cosmochimica Acta* **68**(6): 1347–1354.

Ash J. L., Egger M., Treude T., Kohl I., Cragg B., Parkes R., Slomp C., Sherwood Lollar B. and Young E. D. (2019) Exchange catalysis during anaerobic methanotrophy revealed by $^{12}CH_2D_2$ and $^{13}CH_3D$ in methane. *Geochemical Perspectives Letters* **10**: 26–30.

Bell R. P. (1959) The tunnel effect correction for parabolic potential barriers. *Transactions of the Faraday Society* **55**: 1–4.

Bell R. P. (1980) *The Tunnel Effect in Chemistry*. London, Chapman & Hall.

Bigeleisen J. (1955) Statistical mechanics of isotopic systems with small quantum corrections. I. General considerations and the rule of the geometric mean. *Journal of Chemical Physics* **23**(12): 2264–2267.

Boschi C., Dini A., Fruh-Green G. L. and Kelley D. S. (2008) Isotopic and element exchange during serpentinization and metasomatism at the Atlantis Massif (MAR 30°N): insights from B and Sr isotope data. *Geochemica et Cosmochimica Acta* **72**: 1801–1823.

Cao X., Bao H. and Peng Y. (2019) A kinetic model for isotopologue signatures of methane generated by biotic and abiotic CO_2 methanation. *Geochimica et Cosmochimica Acta* **249**: 59–75.

Douglas P. M., Stolper D. A., Eiler J. M., Sessions A. L., Lawson M., Shuai Y., Bishop A., Podlaha O. G., Ferreira A. A., Neto E. V. S., Niemann M., Steen A. S., Huang L., Chimiak L., Valentine D. L., Fiebig J., Luhmann A. J., Seyfried W. E. J., Etiope G., Schoell M., Inskeep W. P., Moran J. J. and Kitchen N. (2017) Methane clumped isotopes: progress and potential for a new isotopic tracer. *Organic Geochemistry* **113**: 262–282.

Douglas P. M. J., Stolper D. A., Smith D. A., Anthony K. M. W., Paull C. K., Dallimore S., Wik M., Crill P. M., Winterdahl M., Eiler J. M. and Sessions A. L. (2016) Diverse origins of Arctic and subarctic methane point source emissions identified with multiply-substituted isotopologues. *Geochemica et Cosmochimica Acta* **188**: 163–188.

Eiler J. M., Ghosh P., Affek H., Schauble E., Adkins J., Schrag D. and Hoffman P. (2005). Carbonate plaeothermometry based on abundances of 13C–18O bonds. Presented at: *Goldschmidt Conference*, Moscow, Idaho, A127l.

Etiope G. (2017) Methane origin in the Samail ophiolite: comment on "Modern water/rock reactions in Oman hyperalkaline peridotite aquifers and implications for microbial habitability" [*Geochem. Cosmochim. Acta* 179 (2016) 217–24]. *Geochemica et Cosmochimica Acta* **197**: 467–470.

Etiope G. and Schoell M. (2014) Abiotic gas: atypical, but not rare. *Elements* **10**: 291–296.

Etiope G. and Sherwood Lollar B. (2013) Abiotic methane on Earth. *Reviews of Geophysics* **51**: 276–299.

Giunta T., Young E. D., Warr O., Kohl I., Ash J. L., Martini A. M., Mundle S. A. C., Rumble D., Perez-Rodriquez I., Wasley M., LaRowe D. E., Gilbert A. and Sherwood Lollar B. (2018) Methane sources and sinks in continental sedimentary basins: new insights from paired clumped isotopologues $^{13}CH_3D$ and $^{12}CH_2D_2$. *Geochemica et Cosmochimica Acta* **245**: 327–351 (2019).

Gruen D. S., Wang D. T., Konneke M., Topcuoglu B. D., Stewart L. C., Goldhammer T., Holden J. F., Hinrichs K.-U. and Ono S. (2018) Experimental investigation on the controls of clumped isotopologue and hydrogen isotope ratios in microbial methane. *Geochemica et Cosmochimica Acta* **237**: 339–356.

Haghnegahdar M., Schauble E. A. and Young E. D. (2015) Constructing an atmospheric methane budget using $^{13}CH_3D$ and CH_2D_2 in sources and sinks. Presented at: *American Geophysical Union*, 82743l.

Haghnegahdar M. A., Schauble E. and Young E. D. (2017) A model for $^{12}CH_2D_2$ and $^{13}CH_3D$ as complementary tracers for the budget of atmospheric CH_4. *Global Biogeochemical Cycles* **31**: 1387–1407.

Holler T., Wegener G., Knittel K., Boetius A., Brunner B., Kuypers M. M. and Widdel F. (2009) Substantial $^{13}C/^{12}C$ and D/H fractionation during anaerobic oxidation of methane by marine consortia enriched *in vitro*. *Environmental Microbiology Reports* **1**: 370–376.

Holler T., Wegener G., Niemann H., Deusner C., Ferdelman T. G., Boetius A., Brunner B. and Widdel F. (2011) Carbon and sulfur back flux during anaerobic microbial oxidation of methane and coupled sulfate reduction. *Proceedings of the National Academy of Sciences* **108**: E1484–E1490.

Kelley D. S., Karson J. A., Fruh-Green G. L., Yoerger D. R., Shank T. M., Butterfield D. A., Hayes J. M., Shrenk M. O., Olson E. J., Proskurowski G., Jakuba M., Bradley A., Larson B., Ludwig K., Glickson D., Buckman K., Bradley A. S., Brazelton W. J., Roe K., Elend M. J., Delacour A., Bernasconi S. M., Lilley M. D., Baross J. A., Summons R. E. and Sylva S. P. (2005) A serpentinite-hosted ecosystem: the Lost City hydrothermal field. *Science* **307**: 1428–1434.

Lieber D. J., Catlett J., Madayiputhiya N., Nandakumar R., Lopez M. M., Metcalf W. M. and Buan N. R. (2014) A multienzyme complex channels substrates and electrons through acetyl-CoA and methane biosynthesis pathways in Methanosarcina. *PLoS ONE* **9**(9): e107563.

Liu Q. and Liu Y. (2016) Clumped-isotope signatures at equilibrium of CH_4, NH_3, H_2O, H_2S and SO_2. *Geochemica et Cosmochimica Acta* **175**: 252–270.

Ma Q., Wu S. and Tang Y. (2008) Formation and abundance of doubly-substituted methane isotopologues ($^{13}CH_3D$) in natural gas systems. *Geochemica et Cosmochimica Acta* **72**(22): 5446–5456.

Maltby J., Steinle L., Loscher C. R., Bange H. W., Fischer M. A., Schmidt M. and Treude T. (2018) Microbial methanogensis in the sulfate-reducing zone of sediments in the Eckernforde Bay, W Baltic Sea. *Biogeosciences* **15**: 137–157.

McCollom T. M. (2016) Abiotic methane formation during experimental serpentinization of olivine. *Proceedings of the National Academy of Sciences* **113**: 13965–13970.

Miller D. J., Subramanian R. and Saunders W. H. J. (1981) Mechanisms of elimination reactions. 33. Carbon isotope effects in E2 reactions of (2-phenylethyl-2-^{14}C)trimehylammonium ion. The role of tunneling. *Journal of the American Chemical Society* **103**: 3519–3522.

Orcutt B., Boetius A., Elbert M., Samarkin V. and Joye S. B. (2005) Molecular biogeochemistry of sulfate reduction, methanogenesis and anaerobic oxidation of methane at Gulf of Mexico cold seeps. *Geochemica et Cosmochimica Acta* **69**: 4267–4281.

Penning H., Plugge C. M., Galand P. E. and Conrad R. (2005) Variation of carbon isotope fractionation in hydrogenotrophic methanogenic microbial cultures and environmental samples at different energy status. *Global Change Biology* **11**(12): 2103–2113.

Proskurowski G., Lilley M. D., Kelley D. S. and Olson E. J. (2006) Low temperature volatile production at the Lost City hydrothermal field, evidence from a hydrogen stable isotope geothermometer. *Chemical Geology* **229**: 331–343.

Qi Y., Yang J., Duan X., Zhu Y.-A., Chen D. and Holmen A. (2014) Discrimination of the mechanism of CH_4 formation in Fischer–Tropsch synthesis on Co catalysts: a combined approach of DFT, kinetic isotope effects and kinetic analysis. *Catalysis Science & Technology* **4**: 3534–3542.

Richet P., Bottinga Y. and Javoy M. (1977) A review of hydrogen, carbon, nitrogen, oxygen, sulphur, and chlorine stable isotope fractionation among gaseous molecules. *Annual Reviews in Earth and Planetary Science* **5**: 65–110.

Rockmann T., Popa M. E., Krol M. C. and Hofmann M. E. G. (2016) Statistical clumped isotope signatures. *Scientific Reports* **6**: 31947.

Scheller S., Goenrich M., Boecher R., Thauer R. K. and Juan B. (2010) The key nickel enzyme of methanogenesis catalyses the anaerobic oxidation of methane. *Nature* **465**: 606–609.

Schoell M. (1980) The hydrogen and carbon isotopic composition of methane from natural gases of various origins. *Geochemica et Cosmochimica Acta* **44**: 649–661.

Sessions A. L., Sylva S. P., Summons R. E. and Hayes J. M. (2004) Isotopic exchange of carbon-bound hydrogen over geologic timescales. *Geochimica et Cosmochimica Acta* **68**: 1545–1559.

Sherwood Lollar B., Lacrampe-Couloume G., Slater G. F., Ward J., Moser D. P., Gihring T. M., Lin L.-H. and Onstott T. C. (2006) Unravelling abiogenic and biogenic sources of methane in the Earth's deep subsurface. *Chemical Geology* **226**: 328–339.

Sherwood Lollar B., Voglesonger K., Lin L.-H., Lacrampe-Couloume G., Telling J., Abrajano T. A., Onstott T. C. and Pratt L. M. (2007) Hydrogeologic controls on episodic H_2 release from Precambrian fractured rocks – energy for deep subsurface life on Earth and Mars. *Astrobiology* **7**(6): 971–986.

Shima S., Warkentin E., Thauer R. K. and Ermler U. (2002) Structure and function of enzymes involved in the methanogenic pathway utilizing carbon dioxide and molecular hydrogen. *Journal of Bioscience and Bioengineering* **93**: 519–530.

Shuai Y., Douglas P. M. J., Zhang S., Stolper D. A., Ellis G. S., Lawson M., Lewan M., Formolo M., Jingkui M., He K., Hu G. and Eile J. M. (2018) Equilibrium and non-equilibrium controls on the abundances of clumped isotopologues of methane during thermogenic formation in laboratory experiments: implications for the chemistry of pyrolysis and the origins of natural gases. *Geochimica et Cosmochimica Acta* **223**: 159–174.

Stolper D. A., Lawson M., Davis C. L., Ferreira A. A., Santos Neto E. V., Ellis G. S., Lewan M. D., Martini A. M., Tang Y., Schoell M., Sessions A. L. and Eiler J. M. (2014) Formation temperatures of thermogenic and biogenic methane. *Science* **344**: 1500–1503.

Stolper D. A., Martini A. M., Clog M., Douglas P. M., Shusta S. S., Valentine D. L., Sessions A. L. and Eiler J. M. (2015) Distinguishing and understanding thermogenic and biogenic sources of methane using multiply substituted isotopologues. *Geochemica et Cosmochimica Acta* **161**: 219–247.

Timmers P. H. A., Welte C. U., Koehorst J. J., Plugge C. M., Jetten M. S. M. and Stams A. J. M. (2017) Reverse methanogenesis and respiration in methanotrophic archaea. *Archaea* **2017**: 1654237.

Valentine D. L. (2011) Emerging topics in marine methane biogeochemistry. *Annual Review of Marine Science* **3**: 147–171.

Valentine D. L., Chidthaisong A., Rice A., Reeburgh W. S. and Tyler S. C. (2004) Carbon and hydrogen isotope fractionation by moderately thermophilic methanogens. *Geochemica et Cosmochimica Acta* **68**: 1571–1590.

Wang D. T., Gruen D. S., Sherwood Lollar B., Hinrichs K.-U., Stewart L. C., Holden J. F., Hristov A. N., Pohlman J. W., Morrill P. L., Könneke M., Delwiche K. B., Reeves E. P., Sutcliffe C. N., Ritter D. J., Seewald J. S., McIntosh J. C., Hemond H. F., Kubo M. D., Cardace D., Hoehler T. M. and Ono S. (2015) Nonequilibrium clumped isotope signals in microbial methane. *Science* **348**: 428–431.

Wang D. T., Reeves E. P., McDermott J. M., Seewald J. S. and Ono S. (2018) Clumped isotopologue constraints on the origin of methane at seafloor hotsprings. *Geochemica et Cosmochimica Acta* **223**: 141–158.

Wang D. T., Welander P. V. and Ono S. (2016) Fractionation of the methane isotopologues $^{13}CH_4$, $^{12}CH_3D$, and $^{13}CH_3D$ during aerobic oxidation of methane by Methylococcus capsulatus (Bath). *Geochemica et Cosmochimica Acta* **192**: 186–202.

Wang W., Wang S., Ma X. and Gong J. (2011) Recent advances in catalytic hydrogenation of carbon dioxide. *Chemical Society Reviews* **40**: 3703–3727.

Webb M. A. and Miller T. F., III (2014) Position-specific and clumped stable isotope studies: comparison of the Urey and path-integral approaches for carbon dioxide, methane, and propane. *Journal of Physical Chemistry* **118**: 467–474.

Whitehill A. R., Joelsson L. M. T., Schmidt J. A., Wang D. T., Johnson M. W. and Ono S. (2017) Clumped isotope effects during OH and Cl oxidation of methane. *Geochemica et Cosmochimica Acta* **196**: 307–325.

Xiao K.-Q., Beulig F., Kjeldsen K. U., Jorgensen B. B. and Risgaard-Petersen N. (2017) Concurrent methane production and oxidation in surface sediment from Aarhus Bay, Denmark. *Frontiers in Microbiology* **8**: 1198.

Yan Z., Joshi P., Gorski C. A. and Ferry J. G. (2018) A biochemical framework for anaerobic oxidation of methane driven by Fe(III)-dependent respiration. *Nature Communications* **9**: 1642.

Yeung L. Y. (2016) Combinatorial effects on clumped isotopes and their significance in biogeochemistry. *Geochemica et Cosmochimica Acta* **172**: 22–38.

Yeung L. Y., Ash J. L. and Young E. D. (2015) Biological signatures in clumped isotopes of O_2. *Science* **348**(6233): 431–434.

Yoshinaga M. Y., Holler T., Goldhammer T., Wegener G., Pohlman J. W., Brunner B., Kuypers M. M. M., Hinrichs K.-U. and Elvert M. (2014) Carbon isotope equilibration during sulphate-limited anaerobic oxidation of methane. *Nature Geoscience* **7**: 190–194.

Young E. D., Kohl I. D., Sherwood Lollar B., Etiope G., Rumble III D., Li S., Haghnegahdar M., Schauble E. A., McCain K. A., Foustoukos D. I., Sutcliffe C. N., Warr O., Ballentine C. J., Onstott T. C., Hosgormez H., Neubeck A., Marques J. M., Perez-Rodriquez I., Rowe A. R., LaRowe D. E., Magnabosco C., Yeung L. Y., Ash J. L. and Bryndzia L. T. (2017) The relative abundances of resolved $^{12}CH_2D_2$ and $^{13}CH_3D$ and mechanisms controlling isotopic bond ordering in abiotic and biotic methane gases. *Geochemica et Cosmochimica Acta* **203**: 235–264.

Young E. D., Rumble III D., Freedman P. and Mills M. (2016) A large-radius high-mass-resolution multiple-collector isotope ratio mass spectrometer for analysis of rare isotopologues of O_2, N_2, CH_4 and other gases. *International Journal of Mass Spectrometry* **401**: 1–10.

14

Earth as Organic Chemist

EVERETT SHOCK, CHRISTIANA BOCKISCH, CHARLENE ESTRADA, KRISTOPHER FECTEAU, IAN R. GOULD, HILAIRY HARTNETT, KRISTIN JOHNSON, KIRTLAND ROBINSON, JESSIE SHIPP, AND LYNDA WILLIAMS

14.1 Introduction: The Disconnect between Earth and the Lab

Earth is a powerful organic chemist, transforming vast quantities of carbon through complex processes leading to diverse suites of products that include the fossil fuels upon which modern societies depend. When exploring how Earth operates as an organic chemist, it is tempting to turn to how organic reactions are traditionally studied in chemistry labs. While highly informative, especially for insights gained into reaction mechanisms, doing so can also be a source of frustration, as many of the reactants and conditions employed in chemistry labs have few or no parallels to geologic processes. It is difficult, for example, to find natural conditions where laboratory reagents such as concentrated sulfuric acid are available, or where extreme oxidants such as chromate and permanganate or reductants such as lithium aluminum hydride are in abundance. Likewise, organic solvents other than the complex mixtures in petroleum and high-pressure natural gases are impossible to find. Instead, the most Earth-abundant fluid that could serve as a reaction medium is water, which is often excluded from organic chemistry procedures and labs. Nevertheless, Earth uses water at high temperatures and pressures as a reactant, catalyst, and solvent for organic transformations on a massive scale.

A common approach to understanding traditional organic reactions is to analyze them in terms of the strengths of the bonds that are broken versus the bonds that are made. When weaker bonds in the reactants are transformed into stronger bonds in the products, the energy of the electrons decreases and the reaction is considered likely to proceed. This is a common approach because *it usually works*. Those reactions that form stronger bonds are the reactions that are observed to occur, and they are those that are included in traditional organic chemistry textbooks. This is a purely enthalpic view of chemical reactions; the role of entropy is not included. With the exception of some fragmentation reactions that form more product molecules than there were reactant molecules, enthalpic effects tend to dominate the majority of traditional organic chemistry at ambient laboratory conditions. Organic reactions under hydrothermal conditions occur at higher temperatures and pressures than ambient, by definition, and the entropic contribution to the reaction free energy is thus larger than at ambient, to the extent that reactions can start to be controlled by entropic rather than enthalpic effects. Several examples of this contrast in thermodynamic influence are employed in this chapter as starting points for exploring reaction mechanisms through experiments.

A strength of traditional organic chemistry is a rich attention to mechanistic detail and descriptions of reactive intermediates and transition states. Another common concept for understanding organic reactions at ambient laboratory conditions is that the favorable reaction will be the one that proceeds via the lowest-energy (most stable) intermediate or transition state (i.e. the fastest reaction *wins*). That this approach usually works is consistent with the majority of traditional organic chemical reactions occurring under kinetic control. In contrast, reactions at higher temperatures are more likely to be reversible, and indeed, reversibility is a common feature of hydrothermal organic reactions. Under these conditions, reactions are more likely to occur under thermodynamic rather than kinetic control. Therefore, much of the difference between organic reactivity at hydrothermal and ambient conditions can be understood if reactions at ambient are controlled mainly by enthalpy and kinetics and hydrothermal reactions are controlled more by entropy and thermodynamics. The primary goals of this chapter are to provide examples of predicting thermodynamic influences and to use the predictions to design experiments that reveal the mechanisms of how reactions occur at the elevated temperatures and pressures encountered in Earth. This work is ongoing, and we hope this chapter can inspire numerous and diverse experimental and theoretical advances in hydrothermal organic geochemistry.

14.2 The Setting for Organic Transformations in the Deep Carbon Cycle

Earth's organic carbon cycle has a small surface component with short residence times (years to centuries) coupled to an enormous deeper component with extraordinarily long residence times (hundreds of millions of years). As shown in Figure 14.1, the vast majority (>99.5%) of organic carbon on Earth exists in the deep pool as dispersed organic matter in shales and sediments (15,000,000 petagrams (Pg) C; Hedges, 1992). The global fossil fuel reserves (4500 Pg C) are a tiny fraction of the organic carbon in rocks. The current anthropogenic fossil fuel use (~7.2 Pg y^{-1}), however, dwarfs the physical weathering of dispersed organic matter in continental rocks (~0.7 Pg y^{-1}), which illustrates the profound effects of human activities on the short-term cycle and potentially the long-term deep carbon cycle. The massive deep organic reservoir is supplied by a very small trickle of organic matter (0.1 Pg C y^{-1}) derived from continental and marine primary production (Hedges, 1992; Bianchi, 2011) and consisting largely of lipids, lignin, and complex carbohydrate derivatives such as cellulose. Using these starting materials, Earth generates suites of new organic products.

Temperature and pressure changes with depth enable the transformation of organic compounds within Earth, which are driven by mismatches among oxidation states set by dominant mineral reactions and those of carbon in organic and inorganic compounds. Iron and sulfur are the most abundant elements that have multiple oxidation states in major rock-forming minerals. As a consequence, assemblages of iron- and

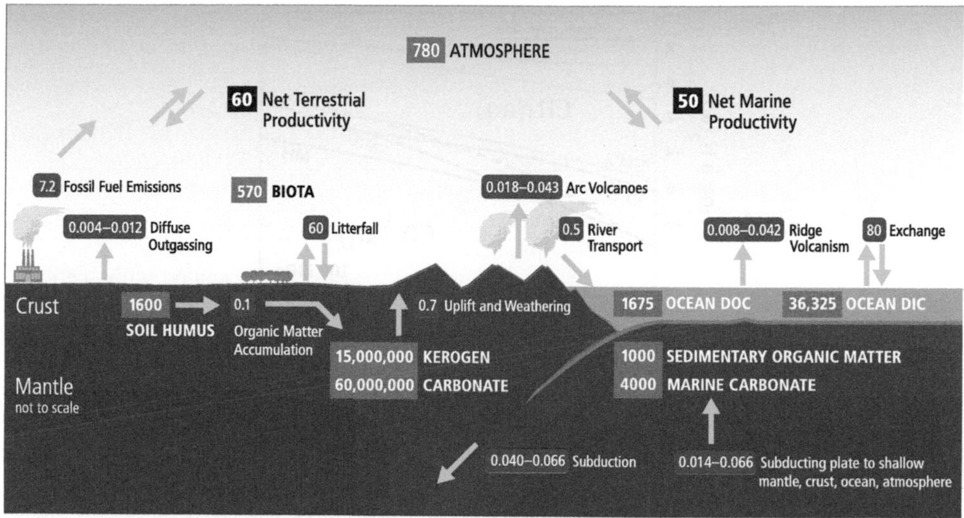

Figure 14.1 Surface and deep carbon cycles, which are linked at subduction zones. Numbers in orange boxes represent reservoirs of carbon in Pg (petagrams, 10^{15} g) and arrows with red boxes show fluxes in Pg y^{-1}. Data summarized from Hedges (1992), Bianchi (2011), and Kelemen and Manning (2015). DIC = dissolved inorganic carbon; DOC = dissolved organic carbon.

sulfur-bearing minerals commonly control the oxidation states in the subsurface. Occasionally, as in petroleum reservoirs, black shales, and coal seams, organic compounds are abundant enough to control subsurface oxidation states (Helgeson et al., 1993; 2009; Shock, 1994; Lammers et al., 2015). But in most geologic environments, the rocks call the shots.

Reactions among minerals that can be balanced by H_2, or H_2 together with H_2O, serve as reference points for estimating relative oxidation states during geochemical processes and can be compared with reactions among organic compounds to uncover how mineral–water–organic reactions may re-equilibrate. An example used widely as a reference frame for the oxidation state of the upper-mantle source regions of basaltic melts is the mineral assemblage fayalite–magnetite–quartz (FMQ; Mattioli & Wood, 1986; O'Neill & Wall, 1987; Cottrell & Kelley, 2011), where fayalite represents the ferrous-iron end member of olivine solid-solution phases and for which a reaction can be written that includes H_2 and H_2O as:

$$3Fe_2SiO_4(\text{fayalite}) + 2H_2O \Leftrightarrow 2Fe_3O_4(\text{magnetite}) + 3SiO_2(\text{quartz}) + 2H_2(aq). \quad (14.1)$$

Oxidation–reduction equilibrium among the constituents of aqueous fluids and basalts can be approximated by FMQ, which will set a specific value of the activity of $H_2(aq)$, $aH_2(aq)$, at each combination of temperature and pressure. Calculated values of log $aH_2(aq)$ set by reaction of the FMQ assemblage and water at 100 MPa and 0–500°C are

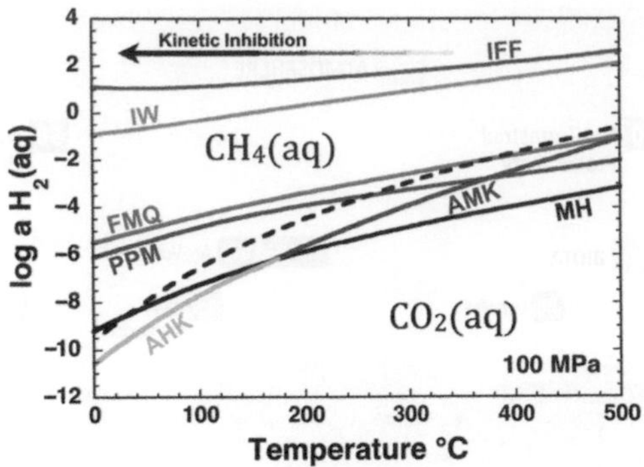

Figure 14.2 Oxidation states of rocks and carbon depicted as the activity of H_2(aq) versus temperature at 100 MPa. Solid curves represent values of log aH_2(aq) set by equilibrium among mineral assemblages and H_2O; reactions given in the text. Dashed curve shows values of log aH_2(aq) corresponding to equal activities of aqueous carbon dioxide and aqueous methane. Note that IFF and IW fall on the methane-dominated side of the diagram, MH, AMK, and AHK fall on the carbon dioxide side of the diagram and that the curves for FMQ and PPM cross from CO_2(aq) to CH_4(aq) dominance regions with decreasing temperature. The arrow depicts schematically the kinetic inhibition that drives departures from stable equilibrium in the C–O–H chemical system with decreasing temperature. IFF = iron–ferrosilite–fayalite; IW = iron–wüstite; FMQ = fayalite–magnetite–quartz; PPM = pyrrhotite–pyrite–magnetite; MH = magnetite–hematite; AMK = annite–magnetite–K-feldspar; AHK = annite–hematite–K-feldspar.

shown as one of the curves falling roughly in the middle of Figure 14.2. Higher values of log aH_2(aq), corresponding to more reduced conditions, are set by the iron–wüstite (IW) and iron–ferrosilite–fayalite (IFF) assemblages given by:

$$\text{Fe (iron)} + H_2O \Leftrightarrow \text{FeO (wüstite)} + H_2(aq), \tag{14.2}$$

and

$$\text{Fe (iron)} + FeSiO_3(\text{ferrosilite}) + H_2O \Leftrightarrow Fe_2SiO_4(\text{fayalite}) + H_2(aq), \tag{14.3}$$

where ferrosilite corresponds to the ferrous-iron end member of orthopyroxene solid-solution phases. These assemblages may reflect oxidation states that prevail during the alteration of ultramafic rocks. At more oxidized conditions, lower values of log aH_2(aq) are reflected by the pyrrhotite–pyrite–magnetite (PPM) and magnetite–hematite (MH) assemblages consistent with

$$3/2FeS\ (\text{pyrrhotite}) + H_2O \Leftrightarrow 3/4FeS_2(\text{pyrite}) + 1/4Fe_3O_4(\text{magnetite}) + H_2(aq), \tag{14.4}$$

and

$$2Fe_3O_4(\text{magnetite}) + H_2O \Leftrightarrow 3Fe_2O_3(\text{hematite}) + H_2(aq). \tag{14.5}$$

Values of log $aH_2(aq)$ set by the annite–magnetite–K-feldspar (AMK) and annite–hematite–K-feldspar (AHK) assemblages form curves with steeper trajectories across Figure 14.2, and they correspond to:

$$KFe_3AlSi_3O_{10}(OH)_2(\text{annite}) + 1/2H_2O \Leftrightarrow 3/2Fe_2O_3(\text{hematite}) + KAlSi_3O_8(\text{K-feldspar}) + H_2(aq), \quad (14.6)$$

and

$$KFe_3AlSi_3O_{10}(OH)_2(\text{annite}) \Leftrightarrow Fe_3O_4(\text{magnetite}) + KAlSi_3O_8(\text{K-feldspar}) + H_2(aq). \quad (14.7)$$

Only the stable portions of the AMK and AHK curves are shown in Figure 14.2. Annite is the ferrous-iron end member of biotite solid-solution phases, which can be found with K-feldspar and iron oxides in granites and rhyolites. Analogous diagrams at other pressures look nearly identical (Shock, 1992; Manning et al., 2013).

Also shown in Figure 14.2 is a dashed curve that represents equal activities of the aqueous forms of CO_2 and CH_4 at stable equilibrium with respect to the reaction

$$CH_4(aq) + 2H_2O \Leftrightarrow CO_2(aq) + 4H_2(aq). \quad (14.8)$$

Note that the AMK and AHK curves, and most of the MH curve, fall on the $CO_2(aq)$ side of the dashed curve. This means that carbon in fluids at stable equilibrium with these mineral assemblages will be predominantly in the form of $CO_2(aq)$ and $CH_4(aq)$ will be far less abundant. In contrast, the IFF and IW curves fall well on the side of the dashed curve where $CH_4(aq)$ is dominant. The other two mineral assemblage curves, PPM and FMQ, show the remarkable behavior of being crossed by the dashed curve. This means that the stable equilibrium speciation of carbon consistent with these mineral assemblages shifts from $CO_2(aq)$ predominance at high temperatures to $CH_4(aq)$ predominance as temperature decreases. Oxidation states at or near FMQ are commonly encountered for basalts, and those set by the PPM assemblage are common in hydrothermal systems hosted in basalt, such as those that are commonly found at mid-ocean ridges. The PPM assemblage also closely approximates conditions in many sulfide deposits found in sedimentary rocks. The interplay between carbon chemistry and rock-buffered oxidation states helps explain why natural gas and other organic-rich accumulations are found in lower-temperature sedimentary basins and why high-temperature volcanic gases are dominated by CO_2.

As indicated by Figure 14.2, there are thermodynamic drives to convert $CO_2(aq)$ to $CH_4(aq)$ as temperature decreases at the oxidation states set by common rock compositions. This conversion may happen at high temperatures (Shock, 1990; 1992; McDermott et al., 2015; 2018; Wang et al., 2018), but becomes less and less likely as temperature decreases, as indicated by the arrow labeled "Kinetic Inhibition." As a result, $CO_2(aq)$ and, by extension, carbonate minerals are metastable at FMQ and PPM oxidation states as temperatures decrease. Likewise, it is on the lower-temperature side of the dashed curve in

Figure 14.2 that organic compounds are metastably preserved against conversion to the stable $CH_4(aq)$. As temperatures decrease, reaction mechanisms among metastable compounds determine how Earth operates as an organic chemist.

The diversity and complexity of organic molecules found in Earth systems, from hydrocarbons, to functionalized compounds containing O, N, S, and other elements, to large geopolymeric materials, hint at the wide range of organic reactions conducted within Earth. Wide-ranging oxidation states of carbon, from –4 to +4, are found among methane, hydrocarbons, organic matter, and carbonate, implying diverse organic oxidation–reduction, hydration–dehydration, addition–elimination, substitution, and disproportionation reactions. How such reactions happen is the focus of this chapter.

Modern organic chemistry, both in the classroom and in industry, focuses on rates and specificities of reaction. While Earth organic chemistry is not always fast or specific, it is continuous and effective. In an era when chemistry questions ranging from the emergence of life to habitable planets, biofuel production, and green chemistry are at the forefront of scientific inquiry, it is time to grasp how Earth promotes its broad range of complex organic reactions. One path forward is to follow organic compounds into hydrothermal conditions, predict what can happen based on thermodynamic drives, and conduct experiments to determine which of those possibilities are granted mechanistic permission.

14.3 Hydration/Dehydration as Examples of Elimination Reactions

A striking example of how conventional organic chemistry fails to translate well to geologic conditions is provided by reactions in which organic compounds dehydrate in water. Predictions made from enthalpies of reaction and considerations of bond energies, coupled with misapplication of simple concepts like Le Chatelier's principle and uncertainty about changes with temperature, could lead to the misconception that dehydration is unlikely in aqueous solution. In contrast, geologists are comfortable with the idea of hydrous minerals dehydrating as temperature and pressure increase, and, given that H_2O is the inescapable product, they raise no concerns about mineral dehydration occurring in the presence of an aqueous fluid (Shock, 1993). Do minerals behave differently from organic compounds? Or is dehydration an inevitable consequence of increases in temperature and pressure, even in aqueous solution?

An approach to answering these questions is provided by combining thermodynamic analysis with hydrothermal experiments that yield kinetic data. As shown in Figure 14.3, the equilibrium constant for dehydration of an alcohol – 1-butanol in this case – favors the persistence of the alcohol in solution at ambient conditions. There is only the faintest hint of a thermodynamic drive for the alkene to form at low temperatures, which is consistent with geologists' blissful lack of concern for the dehydration of ethanol while drinking a cold beer. The explanation for this is that, at ambient conditions, equilibrium favors the alcohol, referred to as the alkene/water addition product in organic chemistry textbooks. Earth, which is not limited to ambient conditions in chemistry labs, did not learn organic

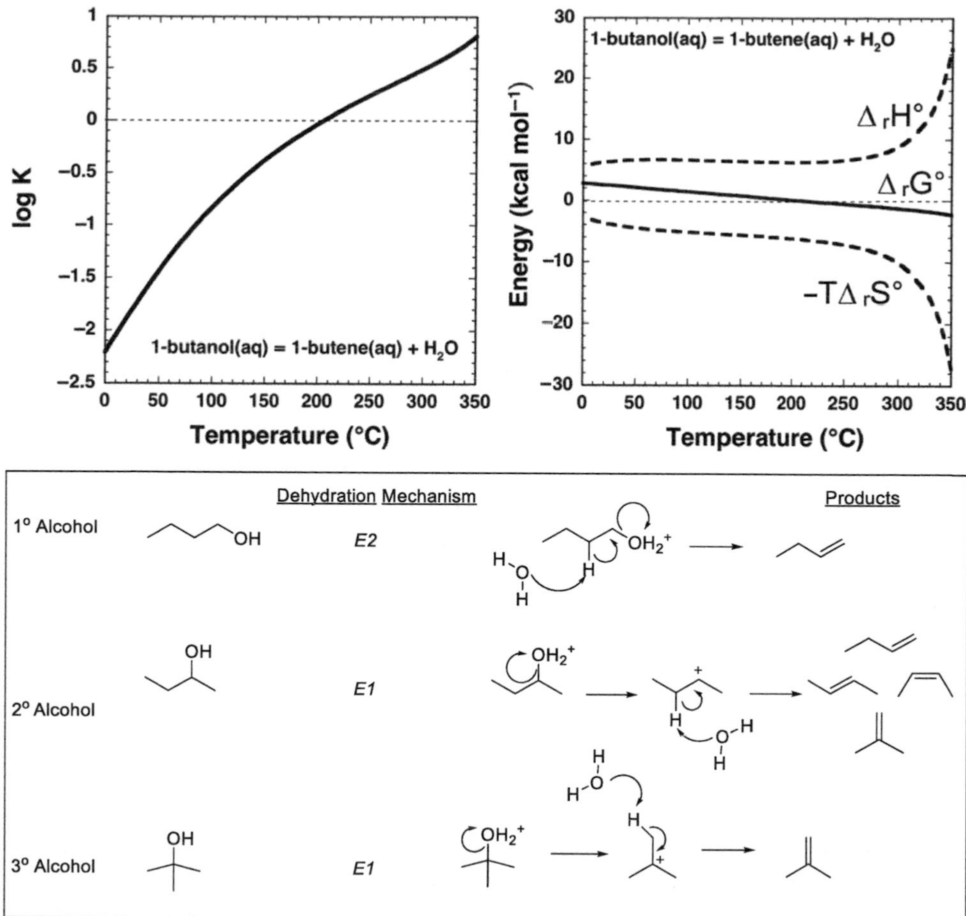

Figure 14.3 Thermodynamics and mechanisms of aqueous alcohol dehydration reactions. The equilibrium constant for reaction (14.9) as a function of temperature at P_{sat}, calculated with equations and parameters from Shock and Helgeson (1990) and Shock et al. (1992), is shown in the upper-left plot. Note that the sign of log K changes at about 200°C. This change in sign of log K corresponds to a change in sign of $\Delta_rG°$ for reaction (14.9) shown in the upper-right plot, which is caused by the increasingly negative contribution of the $-T\Delta_rS°$ term surpassing the decreasingly positive contribution of the $\Delta_rH°$ term as temperature increases. As a consequence, 1-butanol can dehydrate to 1-butene in aqueous solutions at elevated temperature and pressure. When reaction (14.9) occurs in hydrothermal solution, it most likely follows a bimolecular elimination (E2) mechanism involving the alcohol and a molecule of H_2O as shown in the lower panel, which is common for primary alcohols. In contrast, secondary and tertiary alcohols are likely to dehydrate in aqueous solution by unimolecular elimination (E1) mechanisms triggered by protonation of the hydroxyl group in the alcohol (Bockisch et al., 2018). Note that alkene products of dehydration differ depending on the structure of the parent alcohol.

chemistry from textbooks and conducts organic reactions over wide ranges of temperature and pressure. As examination of Figure 14.3 shows, with an increase in temperature at P_{sat} (the saturation pressures corresponding to vapor–liquid equilibrium for H_2O at these temperatures), the sign of the logarithm of the equilibrium constant (log K) for the reaction

$$C_4H_9OH(aq) \Leftrightarrow C_4H_8(aq) + H_2O \tag{14.9}$$

changes, and the thermodynamic drive above 200°C is in the direction of generating considerable 1-butene from 1-butanol by dehydration *in aqueous solution*. The same reaction now exhibits properties exactly the reverse of how it is portrayed in organic chemistry textbooks.

The origins of this reversal in the sign of log K for reaction (14.9) are revealed by comparing the contributions of standard enthalpy ($\Delta_r H°$) and entropy ($\Delta_r S°$) changes to the standard Gibbs energy ($\Delta_r G°$) change of the reaction in Figure 14.3, together with the relations $\Delta_r G° = \Delta_r H° - T\Delta_r S°$ and $\Delta_r G° = -2.303RT \log K$, where the subscript "r" stands for reaction. At low temperatures, both the standard enthalpy and standard Gibbs energy changes are positive for reaction (14.9). With increasing temperature, the growing magnitude of the $-T\Delta_r S°$ term overwhelms the increasingly positive contribution from $\Delta_r H°$, leading to a decrease in the magnitude of $\Delta_r G°$ and ultimately a change in sign. At higher temperatures, the thermodynamic drive for the reaction favors dehydration in aqueous solution.

Having a thermodynamic drive for a reaction is one thing; determining whether a mechanism exists so that the reaction can occur at hydrothermal conditions typically requires experiments. Dehydration of an alcohol to form an alkene is possible at ambient laboratory conditions by using concentrated acids accompanied by physical separation of reactants and products through distillation (Friesen and Schretzman, 2011). Even the conversion of an alkene and water to an alcohol at ambient requires Brønsted or Lewis acid catalysis, even though the reaction is understood as being favorable in textbook fashion owing to formation of a stronger σ-bond at the expense of a weaker π-bond (Anslyn and Dougherty, 2005). Therefore, it can come as a surprise that alcohol dehydration is rapid and effective in hydrothermal experiments (Kuhlman et al., 1994; Antal et al., 1998; Katritzky et al., 2001; Hietala and Savage, 2015; Bockisch et al., 2018).

Mechanistically, hydrothermal dehydration of primary (1°) alcohols, in which there is one carbon bonded to the carbon that is bonded to the hydroxyl (–OH) group, likely involves protonation of the –OH followed by simultaneous loss of H_2O and deprotonation to form the alkene, as shown in the lower panel of Figure 14.3. This makes it a bimolecular elimination, termed an E2 reaction (Xu et al., 1991). At high temperatures and pressures, an E2 mechanism is possible because protonation is an option owing to the enhanced dissociation of H_2O, which raises the proton concentration in neutral hydrothermal solution. In contrast, tertiary (3°) alcohols, in which the carbon bonded to the hydroxyl group is also bonded to three other carbon atoms, are likely to react by a unimolecular elimination (E1) mechanism involving protonation of the hydroxyl followed by elimination of H_2O to form a tertiary carbocation that subsequently deprotonates to form the alkene (Xu et al.,

1997). Recent experimental results (Bockisch et al., 2018) show that for many secondary (2°) alcohols, in which the carbon bonded to the hydroxyl group is bonded to two other carbons, the E1 mechanism dominates over the E2 mechanism except in cases where the molecular structure of the alcohol is selected to only allow the E2 mechanism to occur. A specific example is a cyclohexane ring in which the bonds to the hydrogen and the hydroxyl group that are simultaneously eliminated are constrained to be anti- and coplanar. Bockisch et al. (2018) conclude that the E1 mechanism is associated with a positive entropy of activation, whereas the E2 mechanism is associated with a negative entropy of activation, suggesting that E2 elimination becomes less favorable compared to other possible competing reactions, including E1 at higher temperatures. Because the E1 mechanism leads to faster reaction rates than the E2 mechanism at higher temperatures, a prediction from the existing experimental data is that 1° alcohols should persist in hydrothermal fluids longer than other types of alcohols.

14.4 Dehydrogenation/Hydrogenation Reactions

Removal of one or more moles of H_2, which is required to convert an alkane to an alkene, a linear alkane to a cyclic one, or a cyclic hydrocarbon to a fully aromatic product, is generally considered to be unfavorable under ambient conditions. Even the common name for alkanes as a class, paraffins (from the Latin, *parum affinis*, meaning “insufficient affinity”), emphasizes the apparently low reactivity of these molecules. Indeed, a review of organic chemistry textbooks reveals that reactions of alkanes require heat, catalysis, or *very* strong acids. In particular, accomplishing the dehydrogenation of an alkane at ambient laboratory conditions is difficult – industrial processes typically require temperatures over 500°C (Panizza et al., 2003). One recent and promising method lowers the reaction temperature to 200°C using molecular oxygen as the oxidant, but it requires palladium and titanium dioxide as the catalysts (Dummer et al., 2010). The reverse reaction – hydrogenation of an alkene to form an alkane – is easier to do at ambient laboratory conditions, but still requires over-pressures of H_2 and noble metal catalysts (Rylander, 1973).

As illustrated in Figure 14.4, the oxidation of cyclohexane to benzene in the lab can be accomplished at high temperatures – above 300°C in the gas phase – with a platinum catalyst (House, 1972). Intermediate reductions and dehydrogenations involving cyclohexene and cyclohexadiene can proceed at somewhat lower temperatures and hydrogen pressures, but still require noble metal catalysts (Rylander, 1973). Meanwhile, the hydration of cyclohexene to cyclohexanol can be accomplished in water with a strong Brønsted acid catalyst, but a Lewis acid catalyst such as mercuric acetate ($Hg(OAc)_2$) is normally used (Zweifel and Nantz, 2006), while the dehydration of cyclohexanol requires concentrated sulfuric acid and elevated temperatures (Streitwieser et al., 1995). Oxidation of cyclohexanol to cyclohexanone is usually performed using chromium (VI) reagents (Wiberg, 1965) and the reverse reduction reaction with sodium borohydride (Brown and

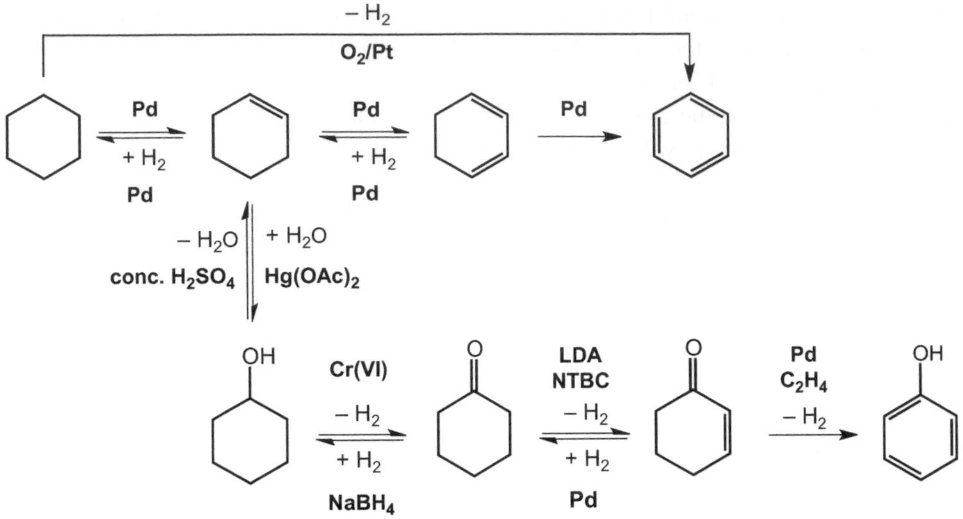

Figure 14.4 Pathways for converting cyclohexane to other organic compounds, including benzene and phenol, as implemented in organic chemistry laboratories. The use of noble metal catalysts, concentrated acids, and brutal oxidants and reductants such as chromate and borohydride, together with other compounds without natural sources, differentiates how organic chemistry is conducted in laboratories and industry from how it occurs in Earth.

Ramachandran, 1996). Further oxidation of cyclohexanone to cyclohexenone has been reported using the rather specialized oxidant N-*tert*-butylbenzenesulfinimidoyl chloride (NTBC) with lithium diisopropylamide (LDA) in tetrahydrofuran at –78°C (Matsuo and Aizawa, 2005), and the reverse reduction requires hydrogen and a palladium catalyst (Rylander, 1973). The final dehydrogenation step to form phenol can be accomplished by coupling it to the reduction of hydrogen acceptors such as ethene with a Pd catalyst (El-Deeb et al., 2017). None of these steps has a plausible geochemical version.

Nevertheless, the presence of alkanes and corresponding alkenes and aromatics in fumarolic systems (Tassi et al., 2015; Venturi et al., 2017) indicates that such transformation reactions do in fact occur in high-temperature geologic systems, suggesting that alkane–alkene reactions are reversible. As summarized in Figure 14.5, a thermodynamic assessment of the transformation of aqueous cyclohexane (C_6H_{12}) to aqueous benzene (C_6H_6) via

$$C_6H_{12}(aq) \Leftrightarrow C_6H_6(aq) + 3\ H_2(aq) \tag{14.10}$$

reveals that log K for reaction (14.10) becomes less negative with increasing temperature, suggesting a shift toward the production of benzene from cyclohexane. However, unlike the hydration/dehydration reactions discussed above, analysis of the fate of this reaction from standard-state thermodynamic properties is complicated by its stoichiometry. The shift in $\Delta_r G°$ to less positive values, which is explained by the powerful influence of the

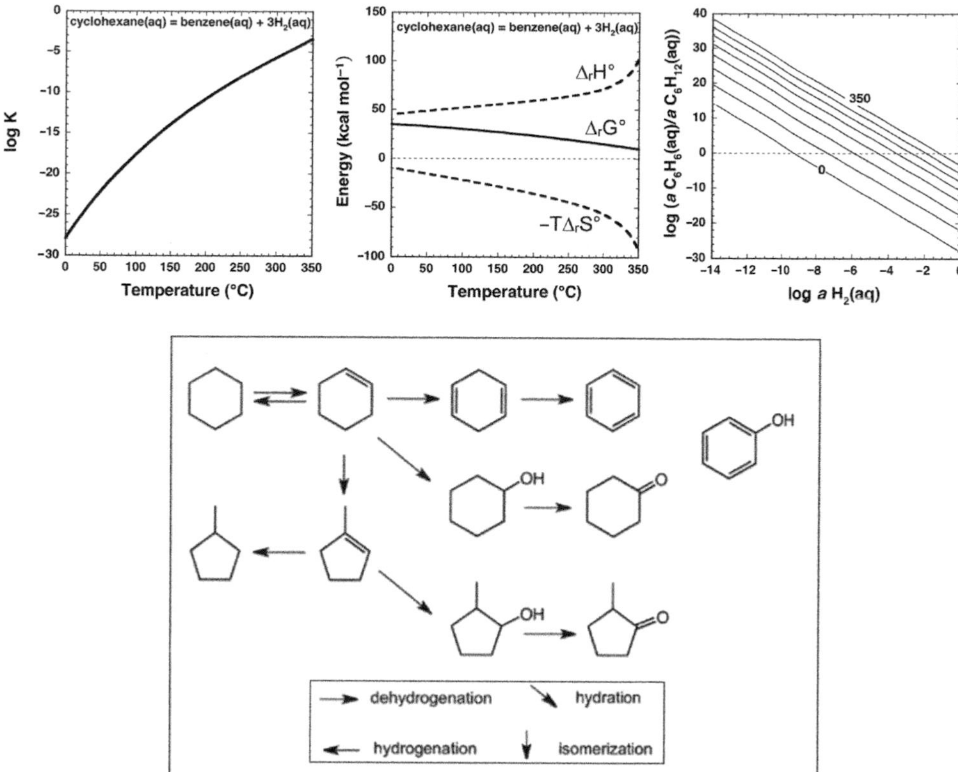

Figure 14.5 Thermodynamics of hydrothermal conversion of cyclohexane to benzene, which requires none of the extreme measures outlined in Figure 14.4, together with reaction paths from hydrothermal experiments. Note that log K for reaction (14.10) in the upper-left plot becomes increasingly positive with increasing temperature, as calculated with equations and parameters from Shock et al. (1989; 1992) and Plyasunov and Shock (2001). The positive increase in log K is reflected in the increasingly less positive values of $\Delta_r G^\circ$ shown in the upper-middle plot, which also shows how the increasing negative contribution of the $-T\Delta_r S^\circ$ term surpasses the decreasingly positive contribution of the $\Delta_r H^\circ$ term as temperature increases. Contours of the equilibrium ratio of benzene to cyclohexane as functions of log aH_2(aq) and temperature from 0°C to 350°C (every 50°C) are shown in the upper-right plot. Following an isotherm, the ratio of benzene to cyclohexane increases with decreasing log aH_2(aq), and at constant log aH_2(aq) the ratio increases with increasing temperature. The bottom panel shows reaction paths determined by Venturi et al. (2017) in hydrothermal experiments at 300°C and 8.5 MPa. Arrows indicate the type of reaction leading to the suite of products from cyclohexane. The lack of arrows to phenol, which was a product in the experiments, indicates that the reaction pathways were not resolved in the experiments.

entropy contribution compared to the enthalpy contribution (Figure 14.5), is also consistent with an increasing thermodynamic drive to produce benzene, but the magnitude of that shift is difficult to resolve from values of $\Delta_r G^\circ$ owing to the involvement of 3 moles of H_2 in the production of each mole of benzene. This is where the third plot in Figure 14.5

becomes particularly useful. Each equilibrium constant combined with a value of the activity of H_2(aq) sets the equilibrium ratio of benzene to cyclohexane. The temperature-dependent contours of those equilibrium ratios are shown as functions of log aH_2(aq), revealing that, at lower values of aH_2(aq), benzene can be considerably more abundant than cyclohexane. This aligns with the notion that low values of aH_2(aq) imply more oxidized conditions, and the average oxidation state of carbon in benzene (–1) is considerably more oxidized than the corresponding value for cyclohexane (–2). Similar conclusions were reached by Venturi et al. (2017), who made comparisons of equilibrium ratios with oxidation states set by mineral equilibria.

In this case, thermodynamic analysis helps to explain the observations of natural systems via a drive toward a metastable equilibrium state involving benzene, cyclohexane, and hydrogen in aqueous solution, but can, of course, provide no evidence of how the reaction happens. Hydrothermal experiments with cyclohexane and common sulfide and oxide minerals (Venturi et al., 2017) demonstrate that conversion of cyclohexane to cyclohexene (C_6H_{10}) and benzene is possible at 300°C and P_{sat}. A summary of these results is provided in the lower panel of Figure 14.5. The oxidation–reduction reaction between cyclohexane and cyclohexene was shown to be reversible. Additional oxidation reactions from cyclohexene leading to cyclohexadiene and benzene occurred, as did isomerization reactions yielding methylcyclopentene isomers. Hydration of both cycloalkenes led to alcohols and ketones. Phenol production may have come from benzene or cyclohexanol; the reaction pathway awaits further experimental investigation.

Conversion of cycloalkanes into their aromatic equivalents is of commercial interest as a component of petroleum reformation. Dehydrogenation of cyclohexane to benzene and cyclohexanol to phenol requires heterogeneous catalysis and high-temperature gas-phase reactions. Cyclohexane dehydrogenation generally requires metallic palladium as a catalyst and temperatures from 350°C to 550°C (e.g. see Heinemann et al., 1953). Cyclohexanol can be dehydrogenated under somewhat milder conditions (i.e. 200–300°C), but still requires metallic catalysts such as copper or zinc oxide (Romero et al., 2011). The major conclusions from the experiments of Venturi et al. (2017) are that the presence of minerals yields products that are unexpected based on the reactions of the organic compounds in water alone, and that reactions that are traditionally thought to require metal catalysts can be catalyzed by simple minerals under these conditions.

The influence of minerals on the reactions of alkanes may be far more widespread than is currently appreciated. As an example, Shipp et al. (2014) used sphalerite (ZnS) at 300°C and 100 MPa to enhance alkane reaction rates and improve reaction specificity. This work revealed that the influence of the sphalerite surface on the reaction of 1,2-dimethylcyclohexane, which was complex and reversible in water alone, was to generate only the corresponding *cis-* or *trans*-stereoisomer. Experiments with sphalerite and 1,2-dimethylcyclohexane in D_2O further demonstrated a single deuterium incorporation into the product. This indicated that the C–H bonds were activated individually rather than in a concerted mechanism. Mechanistic clues were also gathered by Shipp et al. (2013), who showed that both monomethyl and dimethyl cyclohexane oxidation reactions to produce the corresponding alkenes are reversible (albeit often slowly) at 300°C and 100 MPa and that a complex suite of products, including both hydrogenation/dehydrogenation and hydration/dehydration reactions, are possible.

Reversibility of hydrogenation/dehydrogenation reactions was also demonstrated at 300°C and 100 MPa for ketones and alcohols by Shipp et al. (2013) for 2-methylcyclohexanone and 2-methylcyclohexanol, as well as several dimethylcyclohexanone and dimethylcyclohexanol isomers, and by Yang et al. (2012) for dibenzylketone and the corresponding alcohol at 300°C and 70 MPa. As mentioned above, production of cyclohexanol and cyclohexanone in experiments starting with cyclohexene by Venturi et al. (2017) also suggests that hydrogenation/dehydrogenation reactions of ketones and alcohols are reversible. Taken together, reversible hydrogenation/dehydrogenation reactions among pairs of alkanes and alkenes, as well as alcohols and ketones, may be driven by changes in the activity of H_2(aq), which is itself often controlled by mineral–fluid equilibria in reactions involving Fe-bearing oxides, sulfides, and silicates, as discussed above. Experimental evidence for the influence of mineral assemblages on the fate of organic compounds is summarized by Seewald (2001; 2003; 2017), McCollom and Seewald (2007), and Yang et al. (2018). There may also be conditions in organic-rich sedimentary rocks, petroleum reservoirs, and coal beds where organic reactions determine the oxidation state of Fe in sedimentary rocks (Helgeson et al., 1993; 2009; Shock, 1994).

14.5 Organic Oxidations

The hydrogenation/dehydrogenation reactions summarized above represent manifestations of oxidation–reduction reactions involving organic compounds in geochemical processes. Studies of ore deposits have repeatedly reached the conclusion that such reactions involving organic compounds and metals in minerals or in solution are essential for certain types of ore deposition (see below). Exploring these organic–inorganic connections in hydrothermal experiments provides insights into novel reaction coupling that may be unexpected based on organic oxidation experiments at ambient conditions.

An example of the unexpected is the series of organic oxidation reactions involving dissolved $CuCl_2$ at 250°C and 4.0 MPa studied by Yang et al. (2015). Based on laboratory experience in organic chemistry at ambient conditions, divalent copper is thought to be an ineffective oxidant for organic reactions. This stems from the relatively low oxidation potential of the Cu(II)–Cu(I) pair at ambient, which is not nearly as attractive as oxidation by Cr(VI) or Mn(VII). Of course, the notion of the *strength* of an inorganic oxidant depends on how its oxidation potential relates to the oxidation potential of the organic compound being oxidized. As oxidation potentials are strongly and diversely temperature dependent (Amend and Shock, 2001), conditions can change dramatically at elevated temperatures and pressures.

Dramatic change with temperature characterizes the standard-state thermodynamic properties of organic compound oxidation coupled to Cu(II) reduction in aqueous solution

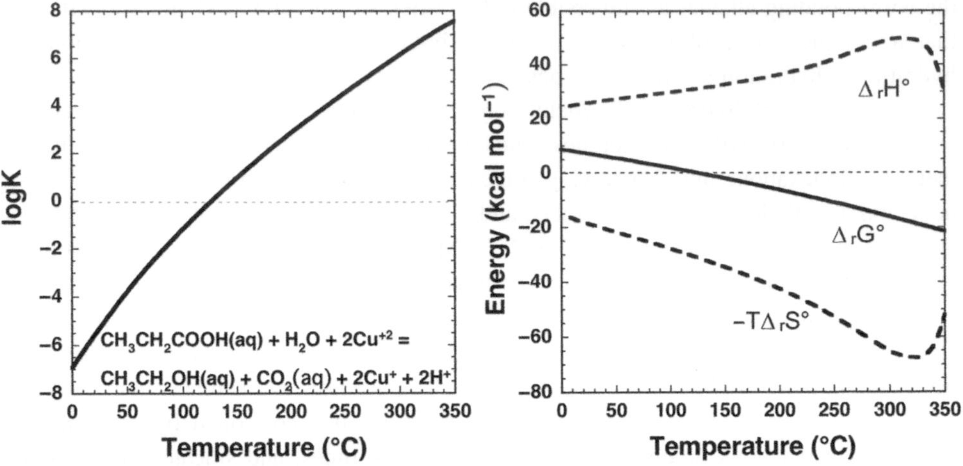

Figure 14.6 Thermodynamics of the aqueous oxidation of propanoic acid to ethanol and CO_2 coupled to the reduction of Cu^{+2} to Cu^+ given by reaction (14.11). Equilibrium constants are shown in the left panel and standard-state thermodynamic properties in the right panel. Calculations were conducted with equations and parameters from Shock et al. (1989; 1992; 1997), Shock and Helgeson (1990), and Shock (1995). At ambient laboratory conditions, log K for this reaction is negative, consistent with organic chemistry experience that reduction of Cu^{+2} cannot be coupled to the oxidation of organic compounds. However, as temperature increases, log K for reaction (14.11) becomes positive and $\Delta_r G°$ becomes negative owing to the dominant influence of the $-T\Delta_r S°$ term as temperature increases. At these conditions, Cu^{+2} can be used to oxidize organic solutes as demonstrated in experiments on phenylacetic acid by Yang et al. (2015) (see text).

as shown in Figure 14.6. In this example, propanoic acid (CH_3CH_2COOH) is oxidized to ethanol (CH_3CH_2OH) and CO_2 by reduction of Cu^{+2} to Cu^+, as in

$$CH_3CH_2COOH(aq) + H_2O + 2Cu^{+2} \rightarrow CH_3CH_2OH(aq) + CO_2(aq) + 2Cu^+ + 2H^+. \quad (14.11)$$

The equilibrium constant for reaction (14.11) at and near ambient conditions indicates that it is highly unfavorable for the reaction to proceed, consistent with organic laboratory experience. At ambient conditions, where log K is strongly negative, Cu^{+2} would not be chosen as an oxidant for this organic oxidation. But, as temperature increases, log K and $\Delta_r G°$ change sign at about 125°C. Note that at all temperatures $\Delta_r H°$ has a large positive contribution to $\Delta_r G°$, and that $-T\Delta_r S°$ has a large negative contribution. The effect of temperature is to drive a milder positive shift in $\Delta_r H°$ and a more robust negative shift in $-T\Delta_r S°$, such that the influence of the latter term causes $\Delta_r G°$ to become negative. At the higher temperatures encountered in hydrothermal fluids, Cu(II) apparently can become an effective oxidant for organic reactions.

The potential for Cu(II) to become an effective organic oxidant in water at elevated temperatures and pressures was documented by Yang et al. (2015) for a suite of reactions leading from phenylacetic acid ($C_6H_5CH_2COOH$) all the way to benzoic acid (C_6H_5COOH) using $CuCl_2$. In summary, the stepwise oxidation reactions can be written as

$$C_6H_5CH_2COOH + Cu^{+2} \rightarrow C_6H_5CH_2 + CO_2 + H^+ + Cu^+, \quad (14.12)$$

$$C_6H_5CH_2 + Cu^{+2} \rightarrow C_6H_5CH_2{}^+ + Cu^+, \quad (14.13)$$

followed by hydration to form benzyl alcohol ($C_6H_5CH_2OH$)

$$C_6H_5CH_2{}^+ + H_2O \rightarrow C_6H_5CH_2OH + H^+, \quad (14.14)$$

$$C_6H_5CH_2OH + Cu^{+2} \rightarrow C_6H_5CHOH + H^+ + Cu^+, \quad (14.15)$$

$$C_6H_5CHOH + Cu^{+2} \rightarrow C_6H_5CHO + H^+ + Cu^+, \quad (14.16)$$

$$C_6H_5CHO + Cu^{+2} + H_2O \rightarrow C_6H_5C(OH)_2 + H^+ + Cu^+, \quad (14.17)$$

and

$$C_6H_5C(OH)_2 + Cu^{+2} \rightarrow C_6H_5COOH + H^+ + Cu^+, \quad (14.18)$$

keeping in mind that chloride complexes will dominate the speciation of Cu^{+2} and Cu^+ at the experimental conditions. Note that reduction of 6 moles of Cu(II) is required to oxidize 1 mole of phenylacetic acid to benzoic acid.

Mechanistic details deduced for these reactions include the likely involvement of aqueous complexes between Cu^{+2} and phenylacetate that trigger reaction (14.12), leading quickly (via reactions (14.13) and (14.14)) to benzyl alcohol, which is considerably more stable in aqueous solution than its radical ($C_6H_5CH_2{}^{\bullet}$) and carbocation ($C_6H_5CH_2{}^+$) precursors. As documented by Yang et al. (2015), the lack of rate effects by adding electron-withdrawing or electron-donating functional groups to the aromatic ring supports the argument that the mechanism is likely to involve the formation of a Cu (II)-phenylacetate complex in solution. In contrast, use of ring substituents in a series of benzyl alcohol experiments strongly supports the formation of a positive charge on the aromatic ring in the rate-determining step to form benzaldehyde (C_6H_5CHO), and the mechanisms for reactions (14.15) and (14.16) are proposed to involve one-electron oxidations via benzylic radicals. The final step in the chain of oxidations produces benzoic acid from benzaldehyde (reactions (14.17) and (14.18)), which requires the contribution of an oxygen atom from H_2O. The prevalence of organic dehydration reactions in hydrothermal solutions discussed above implies that conventional oxidation driven by initial formation of a hydrate from the aldehyde, followed by its oxidation, is unlikely in these experiments. Yang et al. (2015) concluded that radical cation formation followed by H_2O addition is a more likely mechanism, supported by mild ring-substituent effects indicating development of a positive charge on the benzene ring. These results show how coupling of metal reductions and organic oxidations that are unexpected based on ambient laboratory conditions may be common when Earth acts as an organic chemist.

14.6 Amination/Deamination as Examples of Substitution Reactions

Organic nitrogen compounds are essential for all life on Earth; nucleobases and amino acids serve as building blocks for DNA and proteins, respectively. The availability of different forms of nitrogen throughout Earth's history has therefore strongly influenced the emergence and evolution of the biosphere (Mancinelli and McKay, 1988). At the surface of Earth, biology is predominantly responsible for the relatively rapid cycling of nitrogen between its organic and inorganic forms (Galloway et al., 2004). Over longer timescales, however, the fluxes of nitrogen to and from the biosphere must be dependent on global geologic processes, such as burial of sediments, plate tectonics, and volcanism (Berner, 2006; Boudou et al., 2008). Such geochemical settings may be outside the conditions that permit biochemistry, and also where the reactions of organic nitrogen compounds are less understood owing to wide ranges of temperature, pressure, and compositional variables that have yet to be explored.

In aqueous environments, the main class of abiotic reactions that biologically relevant organic nitrogen compounds undergo is deamination, the loss of an amino group ($-NH_2$ or $-NH_3^+$). Certain nucleobases are known to readily deaminate via hydrolysis (reaction initiated by H_2O) under near-ambient conditions (Garrett and Tsau, 1972; Wang and Hu, 2016), while amino acids are more recalcitrant, requiring higher temperatures before deamination is observable on laboratory timescales (Cox and Seward, 2007). These experimental observations seem to account for the fact that, compared to nucleobases, amino acids are better preserved in the geologic record (Bada et al., 1999) and much more abundant in meteorites (Pizzarello and Shock, 2010). Indeed, the metastability of aliphatic amines, including amino acids and alkyl amines, is further supported by synthetic organic chemistry techniques aimed at room-temperature deamination, which requires harsh conditions such as the addition of nitrous acid (Monera et al., 1989), nitrosyl chloride (White and Scherrer, 1961), or high-energy radiation (Dale and Davies, 1950; Shadyro et al., 2003).

Underlying thermodynamic reasons for such extreme laboratory requirements at ambient conditions are revealed by the calculations shown in Figure 14.7 for the aqueous deamination of alanine ($C_2H_5NH_2COOH$) to lactic acid ($C_2H_5OHCOOH$) via the organic substitution reaction

$$C_2H_5NH_2COOH(aq) + H_2O \Leftrightarrow C_2H_5OHCOOH(aq) + NH_3(aq). \tag{14.19}$$

Note that log K is negative for this reaction at all temperatures at P_{sat}, and only approaches 0 at the highest temperatures. Once again, the contribution of $-T\Delta_rS°$ causes $\Delta_rG°$ to become less positive with increasing temperature, but cannot overcome the positive contribution of $\Delta_rH°$ except at the highest temperatures (>350°C). As in the case of cyclohexane oxidation to benzene shown in Figure 14.5, the stoichiometry of this reaction can inhibit an intuitive understanding of the influence of ammonia abundance. The right-hand panel of Figure 14.7 shows how the equilibrium activity ratio of lactic acid to alanine depends on the activity of aqueous ammonia, revealing

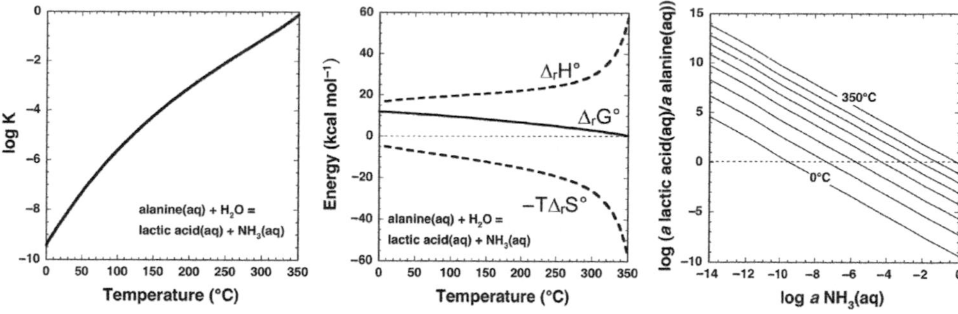

Figure 14.7 Thermodynamics of the deamination of alanine to lactic acid consistent with reaction (14.19) calculated with equations and parameters from Shock et al. (1989; 1992), Shock (1995), and Dick et al. (2006). Note that the values of log K at P_{sat} (left) are negative at low temperatures, but become less negative as temperature increases. Likewise, positive values of $\Delta_rG°$ (middle) become increasingly less positive with increasing temperature. These changes suggest an increase in the thermodynamic drive for deamination as temperature increases, which can be assessed with the plot on the right. Contours of the equilibrium ratio of lactic acid to alanine every 50°C from 0°C to 350°C are shown as functions of log $a\mathrm{NH_3(aq)}$, indicating that the ratio increases with increasing temperature at constant log $a\mathrm{NH_3(aq)}$. Alternately, the plot can reveal temperatures at which equilibrium requires significant deamination of alanine; at log $a\mathrm{NH_3(aq)} = -6$, the ratio exceeds unity at just below 100°C, and at log $a\mathrm{NH_3(aq)} = -3$, the ratio exceeds unity above 200°C.

that deamination in the presence of micromolal concentrations of NH_3(aq) favors lactic acid above about 75°C, while at millimolal concentrations of NH_3(aq), temperatures above about 200°C will favor deamination of alanine to lactic acid (assuming molality equals activity, which is a reasonable approximation for a neutral solute in a dilute solution). Although laboratory deamination conditions are generally irrelevant to natural settings, various experimental tools developed by physical organic chemists show promise for making predictions about amino-group reactivity across a diversity of geochemical environments.

Using some of these tools, Robinson et al. (2019) investigated deamination rates and mechanisms for model amines under acidic, hydrothermal conditions (250°C, 4.0 MPa, pH 3.3), relevant to amino acid deamination in granitic or rhyolitic hydrothermal systems. Distinguishing reaction mechanisms enables improved prediction of reactivity in diverse natural systems because individual mechanisms depend differently on geochemical conditions. Using a variety of benzylamine derivatives, Robinson et al. (2019) elucidated two separate deamination substitution mechanisms, with unimolecular (S_N1) and bimolecular (S_N2) rate-limiting steps, as shown at the top of Figure 14.8. Both mechanisms result in the same hydration product, benzyl alcohol, and were resolved by taking advantage of a traditional organic chemistry technique using Hammett relationships (Hammett, 1935). These are linear free energy kinetic

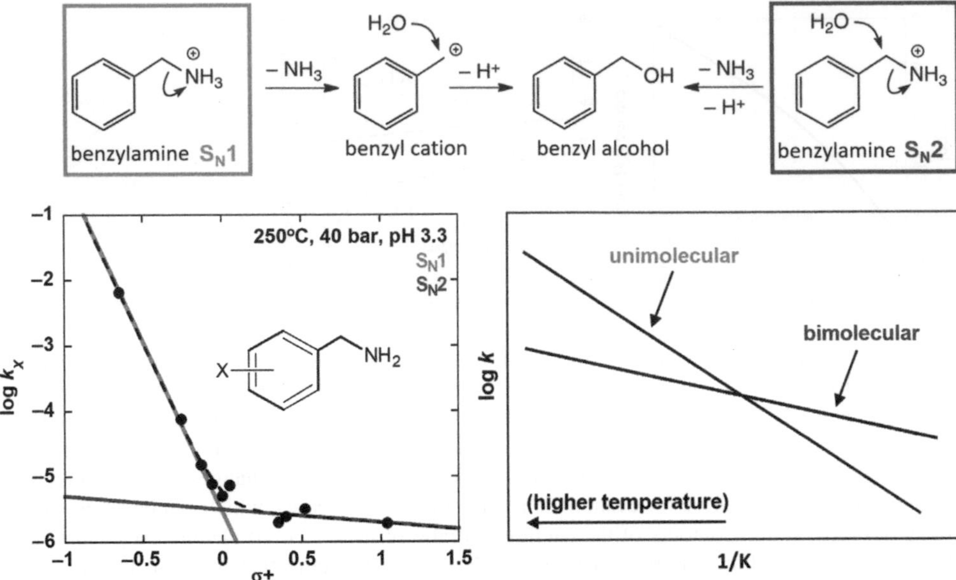

Figure 14.8 Reaction scheme (top) showing unimolecular substitution (S_N1) and bimolecular substitution (S_N2) deamination mechanisms for benzylamine (after Robinson et al., 2019). A Hammett plot (bottom left) shows deamination rate constants for benzylamine derivatives that possess different ring substituents (X) versus their associated σ^+ values. The data set is fit with a curved dotted line, which was produced with two additive linear functions that indicate the presence of both an S_N1 (orange) and an S_N2 (purple) mechanism. A conceptual Arrhenius plot (bottom right) illustrates the expected temperature dependence for unimolecular and bimolecular reaction rates relative to one another.

relationships that provide insight into the sign and the extent of charge buildup in the transition states of various mechanisms. The kinetics of deamination were compared for different ring-substituted benzylamines, which were assigned empirically derived Hammett σ^+ values according to the electron-withdrawing or electron-donating character of their substituents (Gordon and Ford, 1972), as shown in the lower left of Figure 14.8. The nonlinear Hammett plot revealed the presence of both S_N1 and S_N2 mechanisms. By comparison, a secondary amine (α-methylbenzylamine) exhibited only an S_N1 mechanism.

The strength of predictive reaction rate models is greatly enhanced by this type of effort to distinguish mechanisms. In this case, the reaction rates for the S_N1 and S_N2 mechanisms would be expected to have different temperature dependencies owing to the entropic favorability of a unimolecular rate-limiting step over a bimolecular one. This expectation is illustrated in the lower-right plot in Figure 14.8, which shows a conceptual Arrhenius plot of competing linear relationships between the log of the rate constant (k) for single-reaction mechanisms versus the inverse temperature in Kelvin (1/K). The results from

Robinson et al. (2019) demonstrate that hydrothermal deamination of benzylamine has two mechanisms and will not follow a single linear Arrhenius trend. Therefore, extrapolation of observed deamination rates across temperature with a single trend would be unreliable. However, the identification of S_N1 and S_N2 mechanisms allows the rate for each to be characterized across temperature independently, which will permit much more accurate extrapolation and prediction.

14.7 When Organic Molecules Combine: Disproportionation Reactions and Electrophilic Aromatic Substitutions

The preceding examples illustrate the hydrothermal transformation of one organic compound into another through elimination and substitution reactions. Organic compounds can also react with one another to form new products at hydrothermal conditions. In disproportionation reactions two molecules react to generate two new molecules, and in electrophilic aromatic substitution (EAS) reactions two or more molecules can combine to generate greater molecular complexity. Examples of both of these processes are found in the reactions of aldehydes and ketones.

Disproportionation is the primary reaction pathway available to aldehydes that lack α-hydrogens (i.e. hydrogens bonded to a carbon that is bonded to the carbonyl carbon), such as benzaldehyde (C_6H_5COH). In the process, two molecules of the aldehyde react to yield equal amounts of the corresponding alcohol and carboxylic acid, as in

$$C_6H_5COH + C_6H_5COH + H_2O \rightarrow C_6H_5CH_2OH + C_6H_5COOH. \tag{14.20}$$

Thermodynamic data equivalent to those presented above for other reactions are lacking in this case; however, insight into this reaction and its mechanism was obtained from detailed kinetic studies (Fecteau et al., 2019). The mechanism of this reaction involves bimolecular reaction of the aldehydes since the reaction is characterized by clean second-order kinetics, and the reaction rate increases with increasing benzaldehyde concentration as expected for a second-order process. Eyring analysis of temperature-dependent kinetic data yields an entropy of activation of -161 J mol^{-1} K^{-1}, which is also consistent with a bimolecular transition state at the rate-limiting step.

In the laboratory, aldehyde disproportionation is catalyzed by strongly basic conditions such as those obtained by adding sodium hydroxide, the classic example being the Cannizzaro reaction of benzaldehyde (Cannizzaro, 1853; Swain et al., 1979). Typically, an aqueous sodium hydroxide solution is mixed with a protic organic solvent such as methanol to enhance the solubility of the aldehyde (Swain et al., 1979). Under hydrothermal conditions, solubility is dramatically enhanced, and the reaction occurs without the addition of base (Katritzky et al., 1990a; Tsao et al., 1992; Ikushima et al., 2001; Nagai et al., 2004; Fecteau et al., 2019). Experiments as a function of pH show that the hydrothermal reaction can also be catalyzed by hydroxide at higher pH, and at lower pH water serves as the nucleophile for hydrate formation, which then donates hydride to another benzaldehyde molecule in the rate-limiting step.

Another disproportionation found by Fecteau et al. (2019) involves benzaldehyde and benzyl alcohol given by

$$C_6H_5COH + C_6H_5CH_2OH \rightarrow C_6H_5COOH + C_6H_5CH_3. \tag{14.21}$$

Kinetic studies suggest a similar mechanism to the hydrothermal benzaldehyde disproportionation. The products of this disproportionation reaction – a hydrocarbon and a carboxylic acid – represent members of the most abundant organic compounds found in sedimentary basin fluids (petroleum and aqueous), suggesting that reaction (14.21) may represent a class of common geochemical reactions.

Aldehydes are common reactants in the synthesis of larger organic compounds through reactions that create carbon–carbon bonds, such as the aldol condensation and a closely related reaction, the Claisen–Schmidt condensation. These reactions require aldehydes with α-hydrogens, and at ambient they are promoted by both acidic and basic conditions, leading to enol or enolate ion intermediates, respectively. Aldol reactions have been observed under hydrothermal conditions (Katritzky et al., 1990b) and compete with the disproportionation of the aldehyde.

EAS is a reaction where a hydrogen atom of an aromatic ring is substituted by an organic electrophile acting as a Lewis acid toward the π-electrons of the aromatic ring. The Lewis acid site is centered on a carbon atom, which allows new carbon–carbon bonds to form larger, more complex organic structures. In ambient laboratory applications, the Friedel–Crafts reaction, which relies upon a Lewis acid catalyst such as aluminum trichloride to generate the organic electrophile, is perhaps the most prevalent example of EAS. A common industrial example is the production of ethylbenzene, an intermediate in the production of styrene and the plastics produced from it. Under hydrothermal conditions (250°C, P_{sat}), EAS has been observed to readily occur with certain organic compounds that form stable carbocations that can act as electrophiles, such as benzyl alcohol (Fecteau et al., 2019; Robinson et al., 2019). Benzyl alcohol forms such a carbocation upon loss of water (i.e. dehydration), a step that is promoted by the increased concentrations of hydronium ions afforded by the enhanced dissociation of water at hydrothermal conditions compared to ambient. The benzyl cation is a powerful carbon-centered Lewis acid, and formation of the cation from benzyl alcohol is promoted by simple Brønsted acids – even hydronium ions derived from H_2O – and does not require the addition of catalytic Lewis acids. A striking example of EAS at these conditions is the production of the polycyclic aromatic hydrocarbon anthracene from two benzyl alcohol molecules via two consecutive EAS reactions, the second of which is intramolecular. Assignment of the mechanism of each step to EAS is supported by detailed kinetic studies of the reaction. These EAS reactions yield 9,10-dihydroanthracene, which can dehydrogenate to form the more thermodynamically stable anthracene (Fecteau et al., 2019).

14.8 Summary of Hydrothermal Organic Transformations

As described above, numerous organic transformations happen readily in water at elevated temperatures and pressures, and the same transformations require concentrated acids,

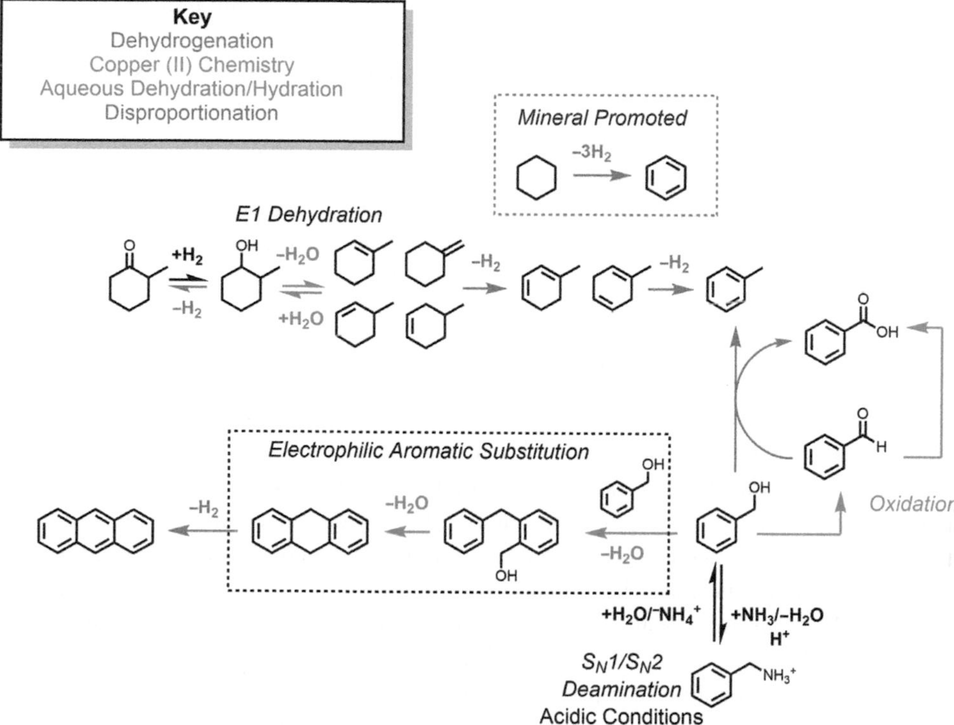

Figure 14.9 Examples of the hydrothermal organic reactions summarized in this chapter. Color coding groups reactions by type; mechanisms are in italics.

extreme oxidants or reductants, or other unnatural conditions in the laboratory at ambient conditions. The dramatic shift in the dielectric constant of H_2O with increasing temperature makes its solvent properties for organic compounds similar to those of organic solvents at ambient conditions (Shock, 1992). Additionally, the pH and oxidation state of hydrothermal environments are buffered by reactions among fluids and rock-forming minerals and can control the organic reactions taking place.

The potential connections among several hydrothermal organic reactions are summarized in Figure 14.9. An alcohol, represented by the methylcyclohexanol in the upper part of the figure, can dehydrogenate reversibly to form the corresponding ketone via an elimination mechanism, or dehydrate reversibly to form alkenes. Further hydrogenation/dehydrogenation reactions link various hydrocarbons. In Figure 14.9, this is represented by methylcycloalkenes dehydrogenating to form methylcyclodienes and ultimately toluene, as well as the mineral-assisted dehydrogenation of cyclohexane to benzene. Another hydrothermal path toward toluene starts from benzyl alcohol in the lower-right corner, which can react with benzaldehyde to produce benzoic acid and toluene as an example of a disproportionation reaction. Hydrothermal oxidation of benzyl alcohol to benzaldehyde, as well as benzaldehyde oxidation to benzoic acid, can each be coupled to reduction of Cu(II), which is utterly unfamiliar at ambient laboratory conditions. This is an example of how an oxidation state imposed by minerals can influence or determine organic reaction paths and products. In low-pH environments, which can occur in high-temperature, organic-rich sediments and in hydrothermal systems hosted in silicic igneous rocks, benzylamine can reversibly deaminate to produce benzyl alcohol through simultaneous unimolecular and bimolecular substitution mechanisms. Benzyl alcohol can also react via an EAS mechanism with itself, undergoing two steps of hydrothermal dehydration to generate 9,10-dihydroanthracene and a dehydrogenation to produce anthracene.

By generating results along time series, the same experiments that reveal mechanisms of hydrothermal reactions can also supply kinetic rate data as summarized in Figure 14.10. Experimental results are summarized as the first half-lives of the indicated reactions at either 250°C (top) or 300°C (bottom). Dehydration of tertiary alcohols is by far the most rapid reaction, and decarbonylation of aldehydes is the slowest in the hydrothermal experiments summarized here. Currently available information reveals that certain types of reactions exhibit wide ranges in rates that are likely tied to variations in mechanism. As an example, tertiary alcohols dehydrate faster than secondary alcohols because the primary intermediate is a carbon-centered cation, and tertiary cations are more stable than secondary cations and are therefore formed more quickly (see Bockisch et al., 2018). In addition, deamination of a secondary amine can be faster than deamination of a primary amine owing to a change in mechanism that is somewhat more subtle. In the case of the secondary amine, only the unimolecular substitution (S_N1) reaction is involved, whereas the primary amine reacts through unimolecular (S_N1) and more sluggish bimolecular (S_N2) mechanisms simultaneously (see Robinson et al., 2019).

While Figure 14.10 serves to summarize our current state of knowledge about hydrothermal organic transformation rates, which have underlying mechanistic explanations, it also serves as a starting point for mapping what we do not yet know. In cases where we have sufficient data, we can tell that some reactions (e.g. dehydration, decarboxylation, deamination) can exhibit variable rates, and mechanistic explanations for those variations are possible. In other cases, data are insufficient to conclude that wide-ranging comprehension is possible. We know what rates of oxidation with Cu(II) are like, but that does not mean that we can predict what oxidation rates will be like with Au(I), Hg(II), Fe(III), Mn(III), V(V), or U(VI). Each of these elements is found to be enriched in organic-rich geologic materials, suggesting that coupled metal–organic oxidation–reduction reactions could be common geochemical processes (Gize, 1999; Greenwood et al., 2013; Shock et al., 2013; Hu et al., 2015; 2017; Cumberland et al., 2016), but mechanistic revelations about reactions that prevail under a given set of conditions are just beginning to be reached. Likewise, known rates of hydrothermal EASs are unlikely to be sufficient for predicting rates of transitions from lignin to coal to anthracite that are accompanied by increasing temperatures and pressures in sedimentary basins. Nevertheless, the results currently available and the experimental techniques that allow

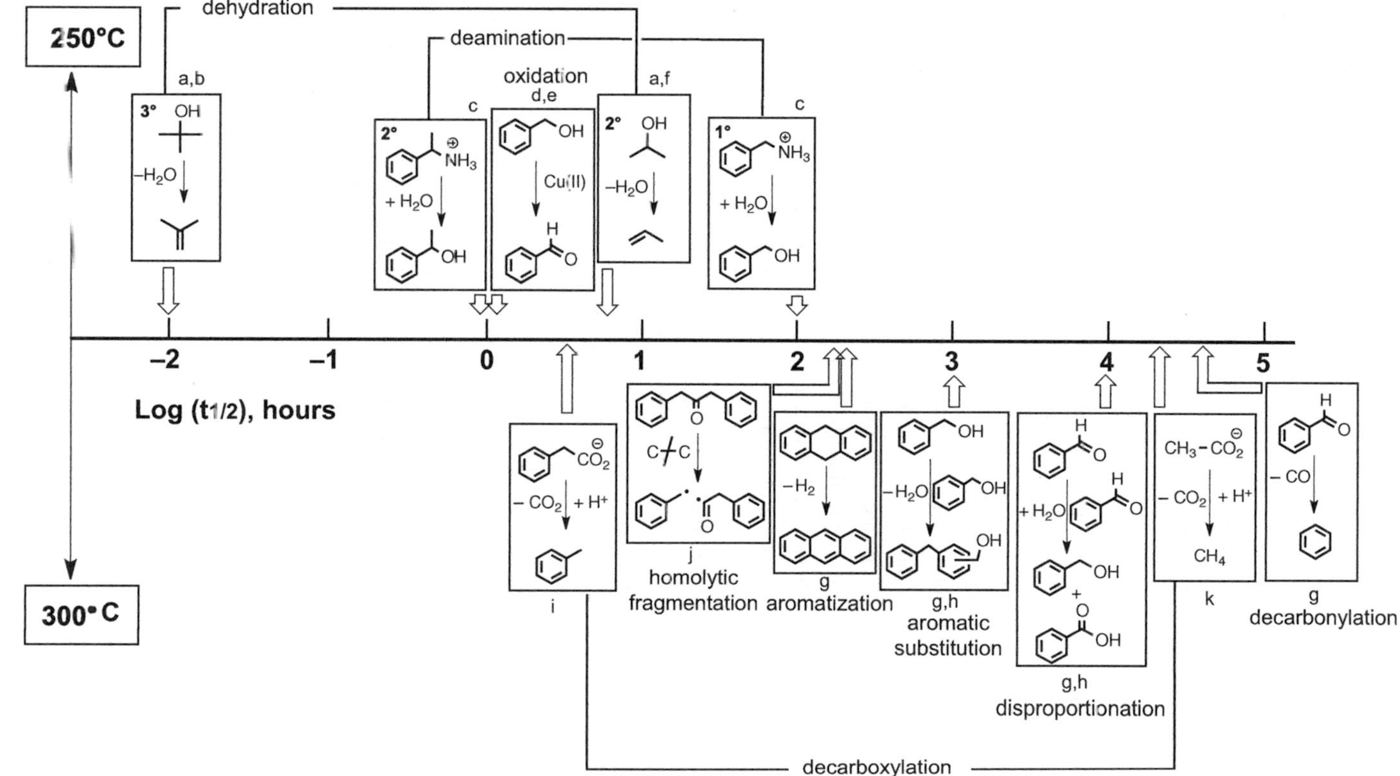

Figure 14.10 Approximate relative rates of hydrothermal reactions given as the logarithm of the reactions half-life ($t_{1/2}$, the time to reach half of the starting reactant concentration). The reactions above the half-life timeline were performed at 250°C, while those below were performed at 300°C. a = reaction proceeds via the protonated oxygen; b = Xu and Antal (1994); c = Robinson et al. (2019); d = for reaction with 0.05 m copper(II) chloride; e = Yang et al. (2015); f = Bockisch et al. (2018); g = Fecteau et al. (2019); h = for reaction at 0.1 m; i = Glein et al. (2019); j = Yang et al. (2012); k = Palmer and Drummond (1986).

mechanistic insight set the stage for a rapid expansion of knowledge about how Earth operates as an organic chemist.

14.9 Organic Reactions in the Deep Carbon Cycle

We may ask: How relevant are the results summarized above, and the chemistry of organic compounds in general, to the large-scale deep carbon cycle? The experiments described here were performed at pressures that are typical of shallow crustal pressures, yet conditions at higher pressures, such as those encountered in subduction zones, may drive similar reaction pathways. In support of this suggestion, Huang et al. (2017) report the results of experiments at 300°C and much higher pressures (from 2.4 to 3.5 GPa) using a diamond anvil cell in which the organic compound isobutane was formed from aqueous sodium acetate. Solid Na_2CO_3 was another product of this reaction, suggesting an overall disproportionation reaction occurs, related to those described in Section 14.8, in which the carbon that is initially present in acetate is both reduced to isobutane and oxidized to carbonate. Huang et al. (2017) also found that the newly generated isobutane formed an immiscible hydrocarbon liquid phase that coexisted with minerals at their experimental conditions, and they go on to suggest that hydrocarbon fluids may be part of subduction-zone assemblages.

Evidence that organic compounds, including hydrocarbons, exist and survive at temperatures and pressures that are much more extreme than those of the experiments above is documented by the extraction of organic compounds from metamorphic and intrusive rocks (Salvi and Williams-Jones, 1997; 2006; Price and DeWitt, 2001; Taran and Giggenbach, 2003; Krumrei et al., 2007; Sawada et al., 2008; Nivin, 2011; Potter et al., 2013) and by indications of organic contributions to carbon budgets in rocks associated with subduction (Alt et al., 2012a; 2012b; 2013; Clift, 2017). Nevertheless, characterization of the types of organic compounds, specifically which functional groups dominate, is extremely limited at present. This kind of information will first be required in order to take advantage of the mechanistic information more easily obtained at lower temperatures and pressures. Mechanistic information will be essential to understanding, for example, which compositions of functional groups and which reactions are sluggish enough to preserve organic compounds under these conditions.

Existing high-pressure/temperature techniques, including but not limited to diamond anvil cell experiments, make it possible to test mechanistic hypotheses and will permit quantitative extrapolations of reaction rates to extreme conditions. Specifically, focusing on reactions through which biomarker compounds can be transformed would lead to an enriched interpretation of the geologic history of geochemical transformations following the incorporation of biomolecules into geologic materials. It is possible to imagine a time in the future when knowing which functional groups dominate compositional mixtures of organic material in geologic samples and what their dominant reactions, mechanisms, and kinetics are could be used to deduce the physical conditions and durations of geologic processes in ways that are complementary to better developed methods involving radiogenic isotopes. We hope that this review points the way to further experiments that will enable such possibilities.

14.10 Geomimicry as a New Paradigm for Green Chemistry

A major goal of chemistry has always been the controlled transformation of materials. Attaining that goal is a primary focus of organic chemistry, which has both blessed and cursed the world with new products that can save lives or cause pollution and disease. Meanwhile, Earth functions as a consummate organic chemist conducting an enormous range of chemical reactions involving organic compounds in aqueous solution without using toxic reagents, expensive noble metal catalysts, or extreme oxidants, reductants, or concentrated acids. Benchtop organic chemists have spent decades developing the techniques and tools to make these reactions faster or more specific, but ultimately, nearly every reaction imaginable in the lab also occurs at natural geochemical conditions.

As summarized above, hydrothermal organic transformations occur at moderately elevated temperatures and pressures. The properties of hydrothermal water enable organic chemistry. Its dielectric constant is substantially lower, mimicking organic solvents, and its extent of dissociation into hydronium and hydroxide is greater, promoting both acid- and base-catalyzed reactions. Thus, hydrothermal water can act as a solvent, a reagent, and a catalyst. In addition, virtually all hydrothermal chemistry takes place in the presence of solid-phase minerals, whose surfaces can also act as reagents and heterogeneous catalysts.

The emerging topic of *geomimicry* seeks answers to the practical questions: *How does Earth perform chemistry? And how can we learn from it?* By capitalizing on reaction chemistries and mechanisms informed by geologic processes and enabled under hydrothermal conditions, reactions can be promoted that use less toxic and more abundant natural materials. Hydrothermal chemistry offers new reaction pathways toward sustainable fuel production to reduce our dependence on fossil carbon. As we learn to control hydrothermal reactions, new capabilities in reaction selectivity could revitalize fine chemical production. Most importantly, geomimicry offers a paradigm shift for green chemistry toward altogether new processes that will not pollute because they are already Earth's ways of doing organic chemistry.

Questions for the Classroom

1 Discuss the preservation potential of biomarker molecules in geologic materials if they are surrounded by silica versus if they are surrounded by hematite. Propose a set of relevant reactions using your choice of biomarkers.
2 Rank hydrothermal organic reactions in terms of their potential to be affected by a stainless steel reaction vessel used in experiments.

3 What would be the effect of conducting the same reactions at the same temperature and pressure in the presence of a PPM assemblage?
4 The maturation of buried organic matter such as kerogen over geologic timescales involves decreases in the abundances of hydrogen, nitrogen, and oxygen relative to carbon in the kerogen structure. Write some mechanistically plausible reactions that could lead to kerogen maturation.
5 Use Figure 14.9 to argue how organic reactions may differ during burial of pelagic sediment, continental-derived sediment, altered mid-ocean ridge basalt, and altered peridotite.
6 Rate data in Figure 14.10 are for either 250°C or 300°C, but not both. Propose additional experiments that would expand our understanding of the temperature dependence of these rates, making the case for your choice of reactions using geologic conditions and processes on Earth and other ocean worlds of the solar system.
7 Using the results summarized in Figures 14.9 and 14.10, predict the fate of compounds in organic-rich sedimentary rocks during episodes of tectonic uplift.
8 Use the *SUPCRT* or *CHNOSZ* codes to evaluate equilibrium constants for the dehydration of an alcohol to the corresponding alkene and show how the dehydration reaction is affected by changes in temperature and pressure. Now couple the dehydration of the alcohol to a mineral hydration reaction of your choice and assess the combined fate of organic and mineral transformations.
9 Changes in pH can have dramatic effects on the mechanisms of organic reactions in Earth. How and why would changes in pH influence the transformation of a long-chain alcohol, an amino acid, and a membrane lipid?
10 What model organic compounds would you predict to be present under CO_2 ice on Mars and what products would you expect during heating from a meteorite impact?

References

Alt, J. C., Garrido, C. J., Shanks, W. C. III, Turchyn, A., Padrón-Navarta, J. A., Sánchez-Vizcaíno, V. L., Goméz Pugnaire, M. T. and Marchesi, C. (2012a) Recycling of water, carbon, and sulfur during subduction of serpentinites: a stable isotope study of Cerro del Almirez, Spain. *Earth Planet. Sci. Lett.* **327–328**, 50–60.

Alt, J. C., Shanks, W. C. III, Crispini, L., Gaggero, L., Schwarzenbach, E. M., Früh-Green, G. L. and Bernasconi, S. M. (2012b) Uptake of carbon and sulfur during seafloor serpentinization and the effects of subduction metamorphism in Ligurian peridotites. *Chem. Geol.* **322–323**, 268–277.

Alt, J. C., Schwarzenbach, E. M., Früh-Green, G. L., Shanks, W. C., Bernasconi, S. M., Garrido, C. J., Crispini, L., Gaggero, L., Padrón-Navarta, J. A. and Marchesi, C. (2013) The role of serpentinites in cycling of carbon and sulfur: seafloor serpentinization and subduction metamorphism. *Lithos* **178**, 40–54.

Amend, J. P. and Shock, E. L. (2001) Energetics of overall metabolic reactions of thermophilic and hyperthemophilic Archaea and Bacteria. *FEMS Microbiol. Rev.* **25**, 175–243.

Anslyn, E. and Dougherty, D. (2005) *Modern Physical Organic Chemistry*. Sausalito, CA: University Science Books.

Antal, M. J., Carlsson, M., Xu, X. and Anderson, D. G. M. (1998) Mechanism and kinetics of the acid-catalyzed dehydration of 1- and 2-propanol in hot compressed liquid water Ind. *Eng. Chem. Res.* **37**, 3820–3829.

Bada, J. L., Wang, X. S. and Hamilton, H. (1999) Preservation of key biomolecules in the fossil record: current knowledge and future challenges. *Philos. Trans. R. Soc. Lond. B Biol. Sci.* **354**, 77–87.

Berner, R. A. (2006) Geological nitrogen cycle and atmospheric N_2 over Phanerozoic time. *Geology* **34**, 413–415.

Bianchi, T. S. (2011) The role of terrestrially derived organic carbon in the coastal ocean: a changing paradigm and the priming effect. *Proc. Natl Acad. Sci. USA* **108**, 19473–19481.

Bockisch, C., Lorance, E. D., Hartnett, H. E., Shock, E. L. and Gould I. R. (2018) Kinetics and mechanisms of dehydration of secondary alcohols under hydrothermal conditions. *ACS Earth Space Chem.* **2**, 821–832.

Boudou, J. P., Schimmelmann, A., Ader, M., Mastalerz, M., Sebilo, M. and Gengembre, L. (2008) Organic nitrogen chemistry during low-grade metamorphism. *Geochim. Cosmochim. Acta* **72**, 1199–1221.

Brown, H. C. and Ramachandran, P. V. (1996) Sixty years of hydride reductions. In Abdel-Magid, A. F. (ed.). *Reductions in Organic Synthesis: Recent Advances and Practical Applications. ACS Symposium Series*, Vol. **641**, pp. 1–30, Washington, DC: American Chemical Society.

Cannizzaro S. (1853) Über den der Benzoësaüre entsprechenden Alkohol. *Justus Liebigs Ann. Chem.* **88**, 129–130.

Clift, P. D. (2017) A revised budget for Cenozoic sedimentary carbon subduction. *Rev. Geophys.* **55**, 97–125.

Cottrell, E. and Kelley, K. A. (2011) The oxidation state of Fe in MORB glasses and the oxygen fugacity of the upper mantle. *Earth Planet. Sci. Lett.* **305**, 270–282.

Cox, J. S. and Seward, T. M. (2007) The hydrothermal reaction kinetics of aspartic acid. *Geochim. Cosmochim. Acta* **71**, 797–820.

Cumberland, S. A., Douglas, G., Grice, K., and Moreau, J. W. (2016) Uranium mobility in organic matter-rich sediments: a review of geological and geochemical processes. *Earth Sci. Rev.* **159**, 160–185.

Dale, W. M. and Davies, J. V. (1950) Deamination of aqueous solutions of L-serine by X-radiation. *Nature* **166**, 1121.

Dick, J. M., LaRowe, D. E. and Helgeson, H. C. (2006) Temperature, pressure, and electrochemical constraints on protein speciation: group additivity calculation of the standard molal thermodynamic properties of ionized unfolded proteins, *Biogeosciences* **3**, 311–336.

Dummer, N. F., Bawaked, S., Hayward, J., Jenkins, R. and Hutchings, G. J. (2010) Oxidative dehydrogenation of cyclohexane and cyclohexene over supported gold, palladium and gold–palladium catalysts. *Catal. Today* **154**, 2–6.

El-Deeb, I. Y., Tian, M., Funakoshi, T., Matsubara, R. and Hayashi, M. (2017) Conversion of cyclohexanones to alkyl aryl ethers by using a Pd/C-ethylene system. *Eur. J. Org. Chem.* **2017**, 409–413.

Fecteau, K. M., Gould, I. R., Glein, C. R., Williams, L. B., Hartnett, H. E. and Shock, E. L. (2019) Production of carboxylic acids from aldehydes under hydrothermal conditions: a kinetic study of benzaldehyde. *ACS Earth Space Chem.* **3**, 170–191.

Friesen, J. B. and Schretzman, R. (2011) Dehydration of 2-methyl-1-cyclohexanol: new findings from a popular undergraduate laboratory experiment. *J. Chem. Educ.* **88**, 1141–1147.

Galloway, J. N., Dentener, F. J., Capone, D. G., Boyer, E. W., Howarth, R. W., Seitzinger, S. P., Asner, G. P., Cleveland, C. C., Green, P. A., Holland, E. A. Karl, D. M., Michaels, A. F., Porter, J. H., Townsend, A. R. and Vörösmarty C. J. (2004) Nitrogen cycles: past, present, and future. *Biogeochemistry* **70**, 153–226.

Garrett, E. R. and Tsau, J. (1972) Solvolyses of cytosine and cytidine. *J. Pharm. Sci.* **61**, 1052–1061.

Gize, A. P. (1999) Organic alteration in hydrothermal sulfide ore deposits. *Econ. Geol.* **94**, 967–979.

Glein, C., Gould, I., Lorance, E., Hartnett, H. and Shock, E. (2019) Mechanisms of decarboxylation of phenylacetic acids and their sodium salts in water at high temperature and pressure. *Geochim. Cosmochim. Acta* (submitted).

Gordon, A. J. and Ford, R. A. (1972) *Chemist's Companion: A Handbook of Practical Data, Techniques, and References.* New York: Wiley.

Greenwood, P. F., Brocks, J. J., Grice, K., Schwark, L., Jaraula, C. M. B., Dick, J. M. and Evans, K. A. (2013) Organic geochemistry and mineralogy. I. Characterisation of organic matter associated with metal deposits. *Ore Geol. Rev.* **50**, 1–27.

Hammett, L. P. (1935) Some relations between reaction rates and equilibrium constants. *Chem. Rev.* **17**, 125–136.

Hedges, J. I. (1992) Global biogeochemical cycles: progress and problems. *Mar. Chem.* **39**, 67–93.

Heinemann, H., Mills, G. A., Hattman, J. B. and Kirsch, F. W. (1953) Houdriforming reactions: studies with pure hydrocarbons. *Ind. Eng. Chem.* **45**, 130–134.

Helgeson, H. C., Knox, A. M., Owens, C. E. and Shock, E. L. (1993) Petroleum, oil field waters and authigenic mineral assemblages: are they in metastable equilibrium in hydrocarbon reservoirs? *Geochim. Cosmochim. Acta* **57**, 3295–3339.

Helgeson, H. C., Richard, L., McKenzie, W. F., Norton, D. L. and Schmitt, A. (2009) A chemical and thermodynamic model of oil generation in hydrocarbon source rocks. *Geochim. Cosmochim. Acta* **73**, 594–695.

Hietala, D. C. and Savage, P. E. (2015) Reaction pathways and kinetics of cholesterol in high-temperature water. *Chem. Eng. J.* **265**, 129–137.

House, H. O. (1972) *Modern Synthetic Reactions*, 2nd edn. New York: W. A. Benjamin.

Hu, S., Evans, K., Craw, D., Rempel, K., Bourdet, J., Dick, J. and Grice, K. (2015) Raman characterization of carbonaceous material in the Macraes orogenic gold deposit and metasedimentary host rocks, New Zealand. *Ore Geol. Rev.* **70**, 80–95.

Hu, S.-Y., Evans, K., Craw D., Rempel, K. and Grice, K. (2017) Resolving the role of carbonaceous material in gold precipitation in metasediment-hosted orogenic gold deposits. *Geology* **45**, 167–170.

Huang, F., Daniel, I., Cardon, H., Montagnac, G. and Sverjensky, D. A. (2017) Immiscible hydrocarbon fluids in the deep carbon cycle. *Nat. Commun.* **8**, 15798.

Ikushima, Y., Hatakeda, K., Sato, O., Yokoyama, T. and Arai, M. (2001) Structure and base catalysis of supercritical water in the noncatalytic benzaldehyde disproportionation using water at high temperatures and pressures. *Angew. Chem., Int. Ed.* **40**, 210–213.

Katritzky, A. R., Balasubramanian, M. and Siskin, M. (1990a) Aqueous high-temperature chemistry of carbo- and heterocycles. 2. Monosubstituted benzenes: benzyl alcohol, benzaldehyde and benzoic acid. *Energy Fuels* **4**, 499–505.

Katritzky, A. R., Luxem, F. J. and Siskin, M. (1990b) Aqueous high-temperature chemistry of carbo- and heterocycles. 5. Monosubstituted benzenes with a two carbon atom side chain oxygenated at the β-position. *Energy Fuels* **4**, 514–517.

Katritzky, A. R., Nichols, D. A., Siskin, M., Murugan, R. and Balasubramanian, M. (2001) Reactions in high-temperature aqueous media. *Chem. Rev.* **101**, 837–892.

Kelemen, P. B. and Manning, C. E. (2015) Reevaluating carbon flux in subduction zones, what goes down, mostly comes up. *Proc. Natl Acad. Sci. USA* **112**, 3997–4006.

Krumrei, T. V., Pernicka, E., Kaliwoda, M. and Markl, G. (2007) Volatiles in a peralkaline system: abiogenic hydrocarbons and F–Cl–Br systematics in the naujaite of the Ilímaussaq intrusion, South Greenland. *Lithos* **95**, 298–314.

Kuhlmann, B., Arnett, E. M. and Siskin, M. (1994) Classical organic reactions in pure superheated water. *J. Org. Chem.* **59**, 3098–3101.

Lammers, L. N., Brown, Jr., G. E., Bird, D. K., Thomas, R. B., Johnson, N. C., Rosenbauer, R. J. and Maher, K. (2015) Sedimentary reservoir oxidation during geologic CO_2 sequestration. *Geochim. Cosmochim. Acta* **155**, 30–46.

Mancinelli, R. L. and McKay, C. P. (1988) The evolution of nitrogen cycling. *Orig. Life Evol. Biosph.* **18**, 311–325.

Manning, C. E., Shock, E. L. and Sverjensky, D. A. (2013) The chemistry of carbon in aqueous fluids at crustal and upper-mantle conditions: experimental and theoretical constraints. *Rev. Mineral. Geochem.* **75**, 109–148.

Mattioli, G. S. and Wood, B. J. (1986) Upper mantle oxygen fugacity recorded by spinel lherzolites. *Nature* **322**, 626–628.

Matsuo, J. and Aizawa, Y. (2005) One-pot dehydrogenation of carboxylic acid derivatives to α,β-unsaturated carbonyl compounds under mild conditions. *Tetrahedron Lett.* **46**, 407–410.

McCollom, T. M. and Seewald, J. S. (2007) Abiotic synthesis of organic compounds in deep-sea hydrothermal environments. *Chem. Rev.* **107**, 382–401.

McDermott, J. M., Seewald, J. S., German, C. R. and Sylva, S. P. (2015) Pathways for abiotic organic synthesis at submarine hydrothermal fields. *Proc. Natl Acad. Sci. USA* **112**, 7668–7672.

McDermott, J. M., Sylva, S. P., Ono, S., German, C. R. and Seewald, J. S. (2018) Geochemistry of fluids from Earth's deepest ridge-crest hot-springs: Piccard hydrothermal field, Mid-Cayman Rise. *Geochim. Cosmochim. Acta* **228**, 95–118.

Monera, O. D., Chang, M. K. and Means, G. E. (1989) Deamination of *n*-octylamine in aqueous solution: the substitution/elimination ratio is not altered by a change of 10^8 in hydroxide ion concentration. *J. Org. Chem.* **54**, 5424–5426.

Nagai, Y., Matubayasi, N. and Nakahara, M. (2004) Noncatalytic disproportionation and decarbonylation reactions of benzaldehyde in supercritical water. *Chem. Lett.* **33**, 622–623.

Nivin, V. A. (2011) Variations in the composition and origin of hydrocarbon gases from inclusions in minerals of the Khibiny and Lovozero Plutons, Kola Peninsula, Russia. *Geol. Ore Deposits* **53**, 699–707.

O'Neill, H. St. C. and Wall, V. J. (1987) The olivine–orthopyroxene–spinel oxygen geobarometer, the nickel precipitation curve, and the oxygen fugacity of the Earth's upper mantle. *J. Petrol.* **26**, 1169–1191.

Palmer, D. A. and Drummond, S. E. (1986) Thermal decarboxylation of acetate. Part I. The kinetics and mechanism of reaction in aqueous solution. *Geochim. Cosmochim. Acta* **50**, 813–823.

Panizza, M., Resini, C., Busca, G., Fernández López, E. and Sánchez Escribano, V. (2003) A study of the oxidative dehydrogenation of cyclohexane over oxide catalysts. *Catal. Lett.* **89**, 199–205.

Pizzarello S. and Shock, E. (2010) The organic composition of carbonaceous meteorites: the evolutionary story ahead of biochemistry. In Deamer, D. and Shostak, J. (eds.). *The Origins of Life*, pp. 89–107, Cold Spring Harbor, NY: Cold Spring Harbor Press.

Plyasunov, A. V. and Shock, E. L. (2001) Correlation strategy for determining the parameters of the revised Helgeson–Kirkham–Flowers model for aqueous nonelectrolytes. *Geochim. Cosmochim. Acta* **65**, 3879–3900.

Potter, J., Salvi, S. and Longstaffe, F. J. (2013) Abiogenic hydrocarbon isotopic signatures in granitic rocks: identifying pathways of formation. *Lithos* **182–183**, 114–124.

Price, L. C. and DeWitt, E. (2001) Evidence and characteristics of hydrolytic disproportionation of organic matter during metasomatic processes. *Geochim. Cosmochim. Acta* **65**, 3791–3826.

Robinson, K. J., Gould, I. R., Fecteau, K. M., Hartnett, H. E., Williams, L. B. and Shock, E. L. (2019) Deamination reaction mechanisms of protonated amines under hydrothermal conditions. *Geochim. Cosmochim. Acta* **244**, 113–128.

Romero, A., Santos, A., Escrig, D. and Simon, E. (2011) Comparative dehydrogenation of cyclohexanol and cyclohexanone with commercial copper catalysts: catalytic activity and impurities formed. *Appl. Catal. A* **392**, 19–27.

Rylander, P. N. (1973) *Organic Synthesis with Noble Metal Catalysts*. New York: Academic Press.

Salvi, S. and Williams-Jones, A. E. (1997) Fischer–Tropsch synthesis of hydrocarbons during sub-solidus alteration of the Strange Lake peralkaline granite, Quebec/Labrador, Canada. *Geochim. Cosmochim. Acta* **61**, 83–99.

Salvi, S. and Williams-Jones, A. E. (2006) Alteration, HFSE mineralization and hydrocarbon formation in peralkaline igneous systems: insights from the Strange Lake Pluton, Canada. *Lithos* **91**, 19–34.

Sawada, Y., Sampei Y., Hada, O. and Taguchi, S. (2008) Thermal degradation and polymerization of carbonaceous materials in a metapelite–granitoid magma system in the Ryoke metamorphic belt, SW Japan. *J. Asian Earth Sci.* **33**, 91–105.

Seewald, J. S. (2001) Aqueous geochemistry of low molecular weight hydrocarbons at elevated temperatures and pressures: constraints from mineral buffered laboratory experiments. *Geochim. Cosmochim. Acta* **65**, 1641–1664.

Seewald, J. S. (2003) Organic–inorganic interactions in petroleum-producing sedimentary basins. *Nature* **426**, 327–333.

Seewald, J. S. (2017) Laboratory simulations of organic geochemical processes at elevated temperatures. In White, W. M. (ed.). *Encyclopedia of Geochemistry: Encyclopedia of Earth Science Series*, pp. 789–792, New York: Springer International Publishing.

Shadyro, O. I., Sosnovskaya, A. A. and Vrublevskaya, O. N. (2003) C–N bond cleavage reactions on the radiolysis of amino-containing organic compounds and their derivatives in aqueous solutions. *Int. J. Radiat. Biol.* **79**, 269–279.

Shipp, J., Gould, I. R., Herckes, P., Shock, E. L., Williams, L. B. and Hartnett, H. E. (2013) Organic functional group transformations in water at elevated temperature and pressure: reversibility, reactivity, and mechanisms. *Geochim. Cosmochim. Acta* **104**, 194–209.

Shipp, J. A., Gould, I. R., Shock, E. L., Williams, L. B. and Hartnett, H. E. (2014) Sphalerite is a geochemical catalyst for carbon–hydrogen bond activation. *Proc. Natl Acad. Sci. USA* **111**, 11642–11645.

Shock, E. L. (1990) Geochemical constraints on the origin of organic compounds in hydrothermal systems. *Orig. Evol. Life Biosph.* **20**, 331–367.

Shock, E. L. (1992) Chemical environments of submarine hydrothermal systems. *Orig. Evol. Life Biosph.* **22**, 67–107.

Shock, E. L. (1993) Hydrothermal dehydration of aqueous organic compounds. *Geochim. Cosmochim. Acta* **57**, 3341–3349.

Shock, E. L. (1994) Application of thermodynamic calculations to geochemical processes involving organic acids. In Pittman, E. D. and Lewan M. D. (eds.). *Organic Acids in Geological Processes*, pp. 270–318, Berlin: Springer.

Shock, E. L. (1995) Organic acids in hydrothermal solutions: standard molal thermodynamic properties of carboxylic acids and estimates of dissociation constants at high temperatures and pressures. *Am. J. Sci.* **295**, 496–580.

Shock, E. L. and Helgeson, H. C. (1990) Calculation of the thermodynamic and transport properties of aqueous species at high pressures and temperatures: standard partial molal properties of organic species. *Geochim. Cosmochim. Acta* **54**, 915–945.

Shock, E. L., Helgeson, H. C. and Sverjensky, D. A. (1989) Calculation of the thermodynamic and transport properties of aqueous species at high pressures and temperatures: standard partial molal properties of inorganic neutral species. *Geochim. Cosmochim. Acta* **53**, 2157–2183.

Shock, E. L., Oelkers, E. H., Johnson, J. W., Sverjensky, D. A. and Helgeson, H. C. (1992) Calculation of the thermodynamic properties of aqueous species at high pressures and temperatures: effective electrostatic radii, dissociation constants, and standard partial molal properties to 1000°C and 5 kb. *J. Chem. Soc. Faraday Trans.* **88**, 803–826.

Shock, E. L., Sassani, D. C., Willis, M. and Sverjensky, D. A. (1997) Inorganic species in geologic fluids: correlations among standard molal thermodynamic properties of aqueous ions and hydroxide complexes. *Geochim. Cosmochim. Acta* **61**, 907–950.

Shock, E. L., Canovas, P., Yang, Z., Boyer, G., Johnson, K., Robinson, K., Fecteau, K., Windman, T. and Cox, A. (2013) Thermodynamics of organic transformations in hydrothermal fluids. *Rev. Mineral. Geochem.* **76**, 311–350.

Streitwieser, A., Heathcock, C. H. and Kosower, E. M. (1995) *Introduction to Organic Chemistry*, 4th edn. Upper Saddle River, NJ: Prentice-Hall.

Swain, C. G., Powell, A. L., Sheppard, W. A. and Morgan, C. R. (1979) Mechanism of the Cannizzaro reaction. *J. Am. Chem. Soc.* **101**, 3576–3583.

Taran, Y. A. and Giggenbach, W. F. (2003) Geochemistry of light hydrocarbons in subduction-related volcanic and hydrothermal fluids. In Simmons, S. F. and Graham, I. (eds.). *Volcanic, Geothermal, and Ore-Forming Fluids; Rulers and Witnesses of Processes within the Earth*, pp. 61–74, Littleton, CO: Society of Economic Geologists.

Tassi, F., Venturi, S., Cabassi, J., Capecchiacci, F., Nisi, B. and Vaselli, O. (2015) Volatile organic compounds (VOCs) in soil gases from Solfatara crater (Campi Flegrei, southern Italy): geogenic source(s) vs. biogeochemical processes. *Appl. Geochem.* **56**, 37–49.

Tsao, C. C., Zhou, Y., Liu, X. and Houser, T. J. (1992) Reactions of supercritical water with benzaldehyde, benzylidenebenzylamine, benzyl alcohol, and benzoic acid. *J. Supercrit. Fluids* **5**, 107–113.

Venturi, S., Tassi, F., Gould, I. R., Shock, E. L., Hartnett, H. E., Lorance, E. D., Bockisch, C., Fecteau, K. M., Capecchiacci, F. and Vaselli, O. (2017) Mineral-assisted production of benzene under hydrothermal conditions: insights from experimental studies on C_6 cyclic hydrocarbons. *J. Volcanol. Geothermal Res.* **346**, 21–27.

Wang, D. T., Reeves, E. P., McDermott, J. M., Seewald, J. S. and Ono, S. (2018) Clumped isotopologue constraints on the origin of methane at seafloor hot springs. *Geochim. Cosmochim. Acta* **223**, 141–158.

Wang, S. and Hu, A. (2016) Comparative study of spontaneous deamination of adenine and cytosine in unbuffered aqueous solution at room temperature. *Chem. Phys. Lett.* **653**, 207–211.

White, E. H. and Scherrer, H. (1961) The triazene method for the deamination of aliphatic amines. *Tetrahedron Lett.* **2**, 758–762.

Wilberg, K. B. (1965) Oxidation by chromic acid and chromyl compounds. In Wilberg, K. B. (ed.). *Oxidation in Organic Chemistry*, pp. 69–184, New York: Academic Press.

Xu, X. and Antal, M. J. (1994) Kinetics and mechanism of isobutene formation from *t*-butanol in hot liquid water. *AIChE J.* **40**, 1524–1534.

Xu, X., Antal, M. J. and Anderson D. G. M. (1997) Mechanism and temperature-dependent kinetics of the dehydration of tert-butyl alcohol in hot compressed liquid water. *Ind. Eng. Chem. Res.* **36**, 23–41.

Xu, X., De Almeida, C. P. and Antal, M. J., Jr. (1991) Mechanism and kinetics of the acid-catalyzed formation of ethene and diethyl ether from ethanol in supercritical water. *Ind. Eng. Chem. Res.* **30**, 1478–1485.

Yang, Z., Gould, I. R., Williams, L. B., Hartnett, H. E. and Shock, E. L. (2012) The central role of ketones in reversible and irreversible hydrothermal organic functional group transformations. *Geochim. Cosmochim. Acta* **98**, 48–65.

Yang, Z., Hartnett, H. E., Shock, E. L. and Gould, I. R. (2015) Organic oxidations using geomimicry. *J. Org. Chem.* **80**, 12159–12165.

Yang, Z., Williams, L. B., Hartnett, H. E., Gould, I. R. and Shock, E. L. (2018) Effects of iron-containing minerals on hydrothermal reactions of ketones. *Geochim. Cosmochim. Acta* **223**, 107–126.

Zweifel, G. S. and Nantz, M. H. (2006) *Modern Organic Synthesis: An Introduction*. New York: W. H. Freeman.

15

New Perspectives on Abiotic Organic Synthesis and Processing during Hydrothermal Alteration of the Oceanic Lithosphere

MURIEL ANDREANI AND BÉNÉDICTE MÉNEZ

15.1 Introduction

The main known organic compounds on Earth are biologically derived, whether they are direct products of biological activity or the result of thermal degradation of bio-derived material. While the synthesis of organic compounds from inorganic reactants is a common process in the chemical industry, it remains an unverified component of the deep carbon cycle on Earth and possibly on other planetary bodies. Abiotic organic synthesis is central to life emergence and sustainability, and possibly to "geo-inspired" resources. Intensive efforts are still needed to unravel the possible forms, sources, quantities, and formation mechanisms of abiotic carbon compounds under geologically relevant conditions. An improved knowledge of their processing within the lithosphere is also mandatory to better quantify their impact on biogeochemical cycles and their contribution to C fluxes between Earth's external and internal envelops. Their presence in fluids and rocks may also affect the kinetics of fluid–silicate reactions and the fates of other elements, particularly the redox-sensitive ones (e.g. transition metals, S).

Abiotic organic compounds in the lithosphere can have two main origins: either rising from a deep volcanic source in the mantle or formed *in situ* in the upper lithosphere during hydrothermal processes from mantellic or seawater inorganic carbon compounds (see Refs. 1 and 2 for reviews). In the present chapter, we focus on lithospheric hydrothermal processes that include late magmatic stages and fluid–rock reactions.

In fluids, natural occurrences of recognized abiotic organic volatiles of hydrothermal origin include methane (CH_4), short-chain alkanes, and small organic acids. They have mainly been observed in geothermal systems or continental seepages within ophiolites and Precambrian shields[3] and at hydrothermal vents near mid-ocean ridges[4] and subduction forearcs.[5] Hence, they are not necessarily associated with deep active volcanism, and they attest to the contribution of lithospheric hydrothermal processes to abiotic organic synthesis. In most cases, these occurrences of organic volatiles are associated with high concentrations of H_2 reached by reduction of water during aqueous alteration of ferrous iron-bearing minerals. The latter are particularly abundant in mantle and olivine-rich lithologies whose alteration process is known as serpentinization.

CH_4 and short-chain hydrocarbons have attracted scientific attention over recent decades, and different reactions have been proposed to explain their possible abiotic formation.[1,3] The ones occurring in the oceanic lithosphere include re-speciation and high-temperature (T) reactions (>400–500°C) of magmatic C–O–H fluids during cooling, carbonate decomposition to CH_4 (<800°C), and inorganic carbon reduction by H_2 at lower T (<400–500°C). The latter has been the most investigated pathway and may occur through Fischer–Tropsch-type (FTT) reactions under gaseous or aqueous conditions. However, the exact mechanism and the nature of potential metal catalysts are still debated (see reviews in Refs. 1 and 6). Up to now, experimental results have converged toward kinetic inhibition of CH_4 formation at low T (<300–400°C) according to the low methane yield achieved in most experiments (see Section 15.3.1). This is in agreement with recent work putting forward a dominant deep source for CH_4 venting at mid-ocean ridges.[7,8] This deep CH_4 may result from the entrapment and re-speciation at high T (>400°C) of mantle-derived CO_2 within fluid inclusions or vesicles in magmatic rocks,[9,10] the fluids being later released during hydrothermal alteration of oceanic rocks. Conversely, no consensus has yet emerged to explain the origin of CH_4 seepages observed in continental settings where fluid–rock reactions occur at even lower T (<100°C). Such settings may offer favorable environments for gas-phase reactions that are more efficiently catalyzed than aqueous ones.[11] However, most of these assumptions are based on CH_4 carbon isotopic signatures that can be highly similar to those produced by microbial methanogenesis under alkaline conditions at high levels of dissolved inorganic carbon.[12,13]

Among organic acids, formate is the only one so far recognized as being of possible abiotic origin in nature, resulting from the equilibration in aqueous solution of CO_2 or carbonate ions and H_2 under neutral to alkaline conditions.[14–16] A variety of other (e.g. methanethiol, amino acids) and heavier (aliphatic and aromatic hydrocarbons, organic acids) organic compounds were also observed in hydrothermal vent fluids and chimneys, but they are often considered as biologically derived.[16–24] Should abiotic synthesis of such compounds occur, they are likely to be too diluted in hydrothermal fluids to be distinguished from background biological contributions.[24]

Bulk analyses of hard rocks from the oceanic lithosphere describe up to 1500 ppm of total organic carbon (TOC) in serpentinized abyssal peridotites.[25–28] If some are present as volatile organic compounds adsorbed on minerals, a fraction is possibly present as solid carbonaceous matter (CM). The nature, diversity, and origin of such organic phases remain poorly constrained, and one cannot preclude a biological contribution to their formation[29–31] and a thermogenic origin as for kerogens.[30,31] Part of this controversy is due to the fact that occurrences were found in potentially colonized environments and that spectroscopic, chemical, and isotopic evidence lacks unequivocal criteria to discriminate between abiotic and biotic origins. In addition, most of the conditions and reactions leading abiotically to carbonaceous compounds are mostly unknown, including their relationship with methane. The experimental difficulty to produce sufficient CH_4 during hydrothermal synthesis compared to what is observed in nature suggests alternative organic products being preferentially formed and leaves the door open to the existence of metastable compounds in the shallow lithosphere.[32–35]

All of these data force one to consider new paradigms for abiotic organic synthesis in hydrothermal settings, not being CH_4 centered, but rather including the potential formation of organic carbon compounds with intermediate oxidation states, possibly mediated by rock-forming minerals. In this chapter, we discuss natural occurrences of carbonaceous compounds within the hard rocks of the oceanic lithosphere or analogs to complement the numerous studies and reviews focused on organic compounds in fluids discharged at oceanic hydrothermal vents or continental seeps (e.g. Refs. 3 and 20). Data on natural rocks affected by low-T (<400°C) and high-T (>400°C) hydrothermal reactions are compared to the most pertinent experimental and theoretical data currently available. Some resulting assertions remain hypothetical but provide new schemes for considering and investigating organic synthesis in lithospheric hydrothermal environments.

15.2 Carbonaceous Matter in Hydrothermally Altered, Mantle-Derived Rocks

Accumulations of CM can be found from the Proterozoic to the present time in volcanic and igneous rocks from a diversity of geodynamic contexts, including spreading zones, oceanic hot spots, island arcs, and continental rifts.[36,37] We focus here on the organic carbon suspected to be nonbiological based on investigations of the most favorable rocky environments for abiotic organic synthesis (i.e. during hydrothermalism affecting mantle-derived rocks).

15.2.1 Bulk Rock Investigations

The first mention of the presence of CM in mafic and ultramafic rocks came from the vast Russian literature published since the 1950s. At that time, the so-called Russian–Ukrainian School had spent significant effort in exploring hydrocarbons in mafic and ultramafic rocks to support a deep mantle origin hypothesis for hydrocarbons deposits. We refer to Sephton and Hazen[2] for a historical perspective and an inventory of achievements in this field. Although a large part of these studies remains untranslated, a review of the occurrences of putatively abiotic condensed naphtides showed a large diversity of compounds associated with volcanic and igneous rocks in a wide range of geodynamic contexts.[37] These compounds include aromatic and aliphatic hydrocarbons and their N-, O-, and S-bearing derivatives. Notably, they mainly occur in mafic and ultramafic rocks. The first study targeting oceanic rocks has focused on the Rainbow (36°14′N) and Logatchev (14°45′N) hydrothermal fields (Mid-Atlantic Ridge (MAR)).[38] In addition to the presence of low-molecular-weight alkanes and isoalkanes, this study reported the presence of CM in serpentinized rocks, in metalliferous sediments, and in secondary Fe–Ni sulfides. The Fe–Ni sulfides presented the highest concentrations and the broadest diversity of viscous and solid organic compounds depicted as resinous bitumen. Notably, a compositional

Table 15.1 *Organic minerals related to hydrothermal alteration in mafic and ultramafic rocks.*[38,47] *With the exception of evenkite, which is classified as an alkane mineral, all are PAH minerals (adapted from Echigo and Kimata*[39]*).*

Mineral	Chemical formula	Chemical nomenclature	Refs.
Karpatite	$C_{24}H_{12}$	Coronene	48
Idrialite	$C_{22}H_{14}$	Picene	49
Kratochvilite	$(C_6H_4)CH_2(C_6H_4)$	Fluorene	50
Simonellite	$C_{19}H_{24}$	1,1-Dimethyl-7-isopropyl-1,2,3,4-tetrahydro-phenanthrene	51
Evenkite	$(CH_3)(CH_2)_{22}(CH_3)$	*n*-Tetracosane	52

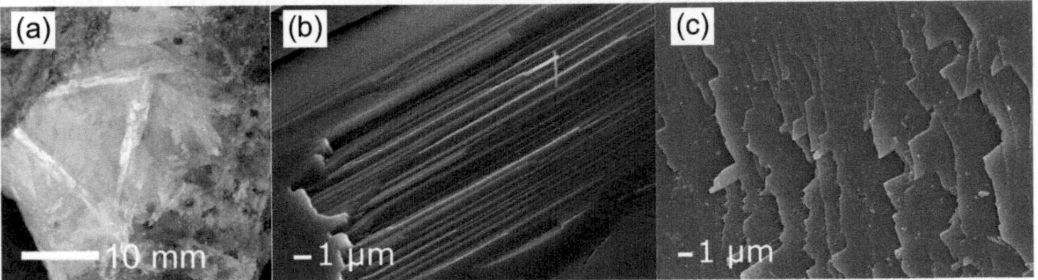

Figure 15.1 (a) A vein of karpatite (yellow crystals) surrounded by quartz (white crystals) and cinnabar (red spots). Scanning electron micrographs of (b) the broken surface of native karpatite and (c) its layering at the end of the layered structure.

relationship was established between these resinous bitumen and hydrothermal crystalline hydrocarbons identified as karpatite ($C_{24}H_{12}$) and idrialite ($C_{22}H_{14}$) found in the same rock samples and thought to derive from serpentinization reactions.[38]

Karpatite (Figure 15.1) and idrialite correspond to polycyclic aromatic hydrocarbon (PAH) minerals. PAH minerals are the most typical molecular organic minerals with well-defined chemical compositions and crystallographic properties.[39] These molecular crystals can form thanks to the high stability of PAHs in hydrothermal fluids,[40] allowing transport without thermal degradation and subsequent concentration and crystallization at the hydrothermal discharge zone.[41,42] Coronene and phenanthrene, the precursors of karpatite and of a number of natural molecular organic minerals (Table 15.1), were shown to derive from the hydrothermal alteration of organic matter present in sediments,[43] but can also be produced during basalt cooling (Section 15.3.4).[44,45] Although the mechanisms of PAH concentration prior to crystallization is still unknown, occurrences of molecular organic minerals may attest to localized high concentrations in (poly)aromatic compounds in the oceanic lithosphere.

More recently, three studies characterized by bulk molecular approaches (1) serpentinized peridotites and gabbroic rocks recovered at the Atlantis Massif (30°N MAR) at both the Lost City hydrothermal field and Hole U1309D (Expeditions 304/305, Integrated Ocean Drilling Program (IODP)),[26] (2) peridotites from the Ashadze (12°58′N) and Logatchev hydrothermal sites along the MAR,[53] and (3) the fossil ocean–continent transition recorded in the Swiss Alps.[54] All of these studies used gas chromatography and/or gas chromatography–mass spectrometry. Based on the presence of biomarkers (e.g. pristane, phytane, squalane, hopanes, and steranes) or higher relative abundances of n-C_{16} to n-C_{20} alkanes,[26,54] amino acids, and long-chain n-alkanes[53] identified in the solvent-extracted fraction, they all concluded that the organic carbon was of biological origin, as supported by TOC isotopic analysis of the oceanic rocks.[25,28] However, biological contamination may have overprinted any possible abiotic geochemical signatures,[26] since these oceanic basement rocks have been affected by long-lived hydrothermal alteration and present-day microbial ecosystems.[29–31]

15.2.2 *In Situ* Investigations at the Microscale

Development of *in situ* techniques currently allows imaging of rock-hosted CM along with their relationship with mineral parageneses. The scarce data for oceanic rocks report that the carbon phases exhibit a structural organization varying from amorphous to well-organized graphite-like material (Figure 15.2a and b). Organic compounds have also been detected as thin films or gel-like materials embedding minerals and filling microfractures (Figure 15.2c). As is detailed below, some have an origin that is still debated, depending on the geological setting and textural criteria.

The Hyblean basaltic diatreme (Sicily, southern Italy), recognized as a paleo-oceanic serpentinite-hosted hydrothermal system comparable to those found along slow-spreading ridge segments,[57] is one of the most studied places where the presence of large amounts of putatively abiotic carbonaceous compounds has been related to hydrothermal activity. A series of papers reported the occurrence of heavy hydrocarbons in deep-seated xenoliths including metasomatic gabbroic xenoliths[47] and highly serpentinized peridotite xenoliths.[58] In the gabbroic xenoliths, the presence of saturated aliphatic and aliphatic–aromatic hydrocarbons was highlighted by bulk rock analysis using electron impact-direct pyrolysis mass spectra. These observations were supported by Fourier-transform infrared (FTIR) spectroscopy showing carbon-cored clayey vesicles and dull-black, tiny pellets associated with the clayey mat pervasively intruding into fractures of the host rocks.[47] Organic crystalline phases were also detected using X-ray diffraction (Table 15.1), and thermal decrepitation allowed the analysis of hydrocarbons trapped in fluid inclusions. Similarly, in the extensively serpentinized and carbonated ultramafic xenoliths, microscopic accumulations of sulfur-bearing organic matter commonly occur between hydrothermal minerals

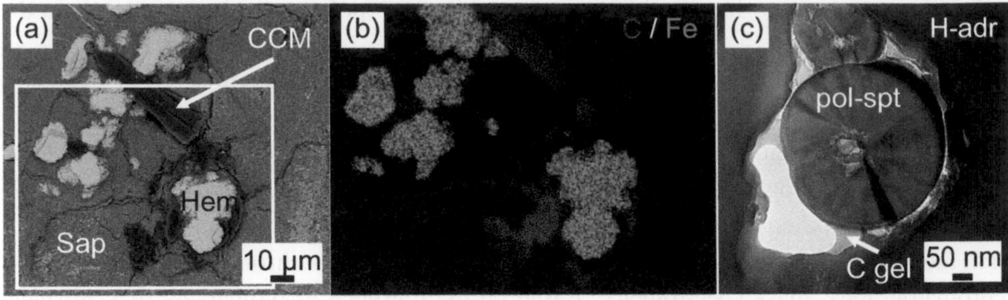

Figure 15.2 Examples of occurrences of organic carbon in serpentinized oceanic rocks. (a) Scanning electron micrograph of O-bearing condensed carbonaceous matter (CCM) abiotically formed jointly with hematite (Hem) and saponite (Sap) during the low-T alteration (T < 150°C) of oceanic serpentinites of the Ligurian Tethyan ophiolites. (b) Associated elemental distributions of carbon (red) and iron (green) within the square area in (a). Reproduced with permission of Springer Nature, from Sforna et al. (2018), *Nat Commun*, **9**, 5049, figure 2c and d.[55] (c) Transmission electron micrograph showing polyhedral serpentine (pol-spt) sections wet by a jelly film of organic carbon interfacing between the pol-spt and an andraditic hydrogarnet (H-adr) in serpentinites from the MAR (4-6°N). In these rocks, organic carbon was shown to mediate the nucleation and growth of polyhedral and polygonal serpentine from the hydrogarnet.
Reproduced with permission of Elsevier, from Ménez et al. (2018), *Lithos*, **323**, 262–276.[56]

(i.e. secondary calcites and fibrous phyllosilicates), forming occasionally coarse bituminous patches.[58] The organic matter was estimated to represent 3–6% of the whole rock. Micro-FTIR spectroscopy highlighted the presence of condensed aromatic rings with aliphatic tails consisting of a few C atoms, which is suggestive of asphaltene-like structures. Such structures were confirmed by solubility tests in toluene and *n*-hexane, thermogravimetric analyses and differential thermal analysis. X-ray photoelectron spectroscopy (XPS) also indicated minor S and O functional groups.

In these studies, it has been proposed that asphaltenes and high-molecular-weight hydrocarbons, respectively, derived from *in situ* aromatization or progressive polymerization and polycondensation reactions of light aliphatic hydrocarbons formed by FTT reactions.[47,58] The hydrocarbon-bearing xenoliths possibly represent any level of the hydrothermal system, including its deepest and hottest parts at ~400°C.[47] Nevertheless, both biotic and abiotic origins could have been possible. It is challenging to infer the possible processes and formation conditions giving rise to heavy hydrocarbon accumulations in such a complex geologic setting. Mantle xenoliths experienced transport and decompression during the formation of the diatreme, with possible consequences for CM that are unrelated to hydrothermal circulation.

Indeed, CM commonly occurs in basalts and mantle xenoliths, which were thoroughly investigated during the 1980s and 1990s.[59–62] At that time, the aim was to assess the mantle carbon content by using a suite of techniques including electron microscopy, chemical imaging, XPS, and carbon isotopes,[59,61–63] as well as thermal desorption surface analysis by laser ionization and low-energy electron diffraction.[62] In these rocks, CM was observed as discrete platy lumps of up to 20–200 μm in size or as thin, amorphous films of a few nanometers located on quench-produced crack surfaces, grain boundaries, and the walls of fluid inclusions. They consisted dominantly of graphite-intercalated compounds along with ill-defined complex mixtures of graphite-like compounds and organic material composed of C, H, and possibly N. All of these studies concluded that CM had an abiotic origin based on: (1) its preferential concentration on sulfide spherules attached to vesicle walls,[60] where the sulfides may have played a catalytic role in organic compound formation and concentration as also observed for hydrothermal sulfides;[38] or (2) the close association of carbon with Si, Al, alkalis, halogens, and/or transition metals, which are elements that were likely present in the volcanic gas at the origin of these carbon accumulations.[62] Again, to account for the production of CM in basalts and mantle xenoliths, these studies invoked: (1) FTT reactions involving volcanic gases degassed of host lava and reacting with fresh and chemically active crack surfaces formed by thermal stresses during eruption, decompression, and cooling; and (2) subsequent evolution of the condensate during cooling. Although heterogeneous catalysis at the mineral surface was the favored hypothesis, organics may have been alternatively assimilated into the volcanic gases prior to eruption and deposited on cracks formed during eruption and cooling.[62]

The possible role of mineral surfaces in the abiotic formation of CM in the oceanic lithosphere was pointed out in a series of recent studies. Various associations between minerals and condensed CM were documented using scanning electron microscopy (SEM) and Raman spectroscopy within magma-impregnated, mantle-derived serpentinites of the Ligurian Tethyan ophiolites, in a context of common occurrences in the lower oceanic crust (Figure 15.2a and b).[55] Three distinct types of CM in paragenetic equilibrium with low-T mineralogical assemblages have been sequentially formed at decreasing T during the hydrothermal alteration of the rock assemblage. The first type corresponds to thin films of aliphatic chains, coating hydroandraditic garnets in bastitized pyroxenes. The second type forms micrometric aggregates associated with the alteration rims of spinel and plagioclase. The third and most massive type appears as large aggregates (up to 200 μm in size) bearing highly aromatic carbon and short aliphatic chains associated with hematite and Fe-saponite assemblages replacing the pseudomorphoses after plagioclase (Figure 15.2a and b). The systematic association of a given type of CM with a specific mineral paragenesis indicates that condensed CM precipitated simultaneously to the growth of the host mineralogical assemblage, overall in favor of an abiotic endogenesis. The mineral formation would have been accompanied by the production of H_2 able to reduce carbon species, as is the case when ferric hydrogarnets and ferric serpentines form[64–66] or when spinel oxidizes to Cr-magnetite forming ferritchromite rims.[67] This possible abiotic pathway was not considered in former studies performed on hydrogarnet-hosted carbonaceous compounds that suggested, based on their biological Raman and FTIR spectroscopy signatures, that this disordered CM may have resulted from the hydrothermal alteration of cryptoendolithic microbial ecosystems.[30,31] Overall, while the involved mechanisms and reactants are still unknown, this recent study emphasizes a key role of local parageneses forming unique microenvironments prone to the synthesis of

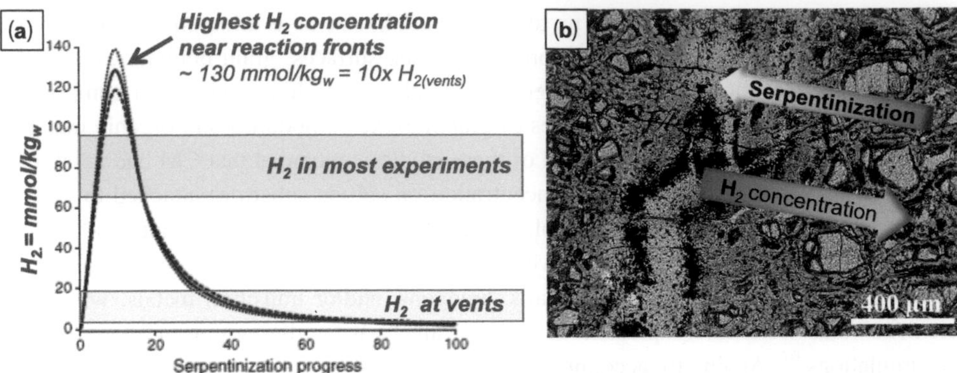

Figure 15.3 (a) Evolution of H_2 concentration in hydrothermal fluid as a function of serpentinization degree estimated from mass balance calculation on Fe^{2+} and Fe^{3+} in abyssal peridotite. (b) Differences in H_2 concentrations, generated by the spatial heterogeneity of reaction, create a redox gradient between partly and fully serpentinized areas down to the micrometric scale.
Modified with permission of Elsevier, from Andreani et al. (2013), *Lithos*, **178**, 70–83, figure 9b.[64]

abiotic hydrocarbon.[55] The productivity could have been also controlled by the local presence of catalysts such as Cr^{3+} in hydrogarnets or Fe-saponite in addition to specific redox conditions among the large redox potential gradients existing in serpentinizing systems down to the microscale (Figure 15.3).[64]

The catalytic role of phyllosilicates in abiotic organic synthesis has been suggested by the syngenetic link between phyllosilicates and organic compounds in several hydrothermal systems. The abundance of long-chain aliphatic and aromatic hydrocarbons (identified using FTIR spectroscopy) has been reported in a diapir of saponite-dominated clays intruded in a diatremic tuff–breccia deposit.[68] The clays formed during the high-T (350–400°C) hydrothermal alteration of mafic and ultramafic lithologies in the Hyblean crustal basement. Based on the close association of hydrocarbons with hydrothermal clays, the lack of fossils, and the resemblance with the previously described xenolith-hosted organic compounds,[47,58] a possible abiogenic origin was suggested for the hydrocarbons via FTT reactions mediated by the organoclay. A close association of organic compounds with clays minerals such as saponite has been previously described in interplanetary dust[69] and in carbonaceous meteorites that have undergone significant aqueous alteration processes.[70–72] A similar association was recently highlighted at a micrometric scale using FTIR spectroscopy in oceanic serpentinites collected at ~170 m depth below seafloor during the IODP Expeditions 304/305 targeting the Atlantic Massif.[73]

Overall, all of these studies highlight that hydrothermally derived organic carbon trapped within the upper lithosphere as heavy and aromatic compounds can be chemically and structurally diverse, although formation mechanisms are not yet well-known. Whether aromatization occurs *in situ* from CO_2/CO and H_2 during serpentinization or derives from progressive polymerization and polycondensation reactions of light aliphatic ± aromatic hydrocarbons inherited from higher-T FTT reactions still needs to be addressed. This latter scenario implies that reactants for the low-T (<400°C) hydrothermal synthesis of abiotic organic compounds that may occur during serpentinization might be more diverse than H_2 and CO_2/CO and account for aliphatic and aromatic compounds, as supported by the presence of molecular organic minerals in serpentinized rocks (Sections 15.2.1 and 15.2.2).[38,47]

15.2.3 Carbon in Fluid Inclusions Trapped in the Oceanic Lithosphere

Fluid inclusions are used to document the composition of high-T fluids (either of magmatic or seawater origin) circulating in the lower crustal component of hydrothermal systems. Hence, they can inform on the intermediate abiotic processes that may happen at higher T and then feed with diverse reactants the low-T hydrothermal reactions, including organic synthesis. Fluid inclusions are made of liquid and/or gas with or without tiny daughter minerals that are trapped within a crystal structure during either primary crystallization or secondary healing of fluid-filled cracks. They occur throughout the gabbroic and peridotitic plutonic crust[74] and have been vastly studied by Raman spectroscopy and microthermometry. In ophiolites, fluid inclusions formed in sub-seafloor rocks may be preserved during tectonic uplift and maintain their original chemical signatures. This has been shown for pure methane fluid inclusions occurring in olivine from partially serpentinized harzburgites and dunites from the Nidar ophiolite complex (eastern Ladakh, India)[75]. In oceanic rocks, all studies have pointed to the presence of $CH_4 \pm H_2 \pm CO_2$ within fluid inclusions,[9,10,74,76,77] although propane and ethane were also sometimes reported in the gaseous phases.[78] Analyses of fluid inclusions in primitive olivine gabbros, oxide gabbros, and evolved granitic rocks recovered from the slow-spreading Southwest Indian Ridge at Ocean Drilling Program Hole 735B (Atlantis II fracture zone) recorded CH_4 concentrations 15–40 times those of hydrothermal vent fluids from sediment-poor environments and of basalt-derived volcanic gases.[10,74] These studies also frequently reported the presence of disordered graphite, carbonaceous compounds, and putative graphite coating inclusion walls.[9,74,76]

While several processes were proposed to explain the presence of organic compounds in high-T hydrothermal fluids,[79,80] fluid inclusions are generally believed to represent evolved magmatic fluids dominated by CO_2 or CO.[9,10,74] Indeed, subsequent to devolatilization and re-equilibration, graphite precipitation is promoted by re-speciation of magmatic CO_2, and the formation of CH_4-enriched fluids is promoted by cooling from 800°C to 500°C of these H_2-rich fluids. Alternatively, re-speciation and reduction of entrapped CO_2-bearing magmatic fluids by diffusion of external H_2 into the inclusions during cooling was also considered.[9,74] Once trapped within crystals as fluid inclusions, the fluids can still evolve following reactions between the fluid and the host mineral. Direct evidence for *in situ* H_2 production and strongly reducing conditions within serpentinizing olivine fluid inclusions was provided for the Mineoka ophiolite complex considered as an analog to serpentinite-hosted hydrothermal vent systems.[81] In the latter study, mineral-phase equilibria indicated that CH_4–H_2-bearing fluids were trapped under equilibrium conditions at T below 300°C, and the absence of CO_2 was suggestive of extensive reduction of CO_2 to CH_4 within the inclusions.[81] Whether CH_4 was also produced at higher T remains an open question. The presence of H_2 and CH_4 together with secondary mineral microinclusions was similarly reported in olivine and orthopyroxene crystals from a harzburgite from the northern Oman ophiolite considered to be similar to abyssal peridotite.[82] In olivine, the mineral inclusions mainly consisted of lizardite and brucite with small amounts of magnetite, while in orthopyroxene they were made of talc and chromian spinel. The differential *in situ* production of reduced gas and secondary phases was related to the presence of either magnetite or a magnetite component in chromium spinels.[82] Alternatively, both H_2 and CH_4 inherited from high-T magmatic processes could have hydrothermally evolved differentially depending on the reactivity of their mineral host (e.g. its capacity to produce *in situ* H_2 or to form mineral byproducts able to catalyze organic synthesis, as discussed in Section 15.2.2).

The presence of reduced carbon species in fluid inclusions within oceanic gabbros and mantle peridotites could represent a potentially important source of organic compounds in hydrothermal fluids.[7,8,10] In particular, CH_4-rich aqueous fluids trapped in the oceanic lithosphere within fluid inclusions recently redrew attention as they could constitute the main source of methane venting at unsedimented mid-ocean ridge hydrothermal fields.[9,10,74,77,83] Similar processes were invoked to explain elevated concentrations of both CH_4 and H_2 in fluids from the Menez Gwen (37°50′N, MAR), Lucky Strike (37°17′N, MAR), and Piccard (Mid-Cayman Ridge) vent fields usually referred to as basalt-hosted vents supposedly less rich in reduced gases compared to ultramafic rock-hosted vents,[20] despite the possible production of non-negligible amounts of H_2 by diking-eruptive events.[84]

Overall, in agreement with Section 15.2.2, studies on fluid inclusions emphasize the need to consider a larger range of reactants in addition to CO_2/CO and H_2 for low-T hydrothermal synthesis of abiotic organic compounds. This includes methane and light hydrocarbons, but also potentially reactive graphitic phases. Due to their tiny size and their entrapment in minerals, the latter are extremely challenging to analyze, and most of the previous studies seldom went further in characterizing the chemical diversity and degree of structural order of this CM. The precipitation of highly crystalline hydrothermal graphite from aqueous fluids containing CO_2 and CH_4 was reported at T as low as 500°C during the propylitic hydrothermal alteration of volcanic host rocks.[85] Other studies, targeting different geodynamical contexts, have shown that precipitation of CM in fluid inclusions, notably on their walls, can lead to graphitic material with varying degrees of crystallinity and disorder.[86–89] Disordering in graphite is principally caused by in-plane defects and/or heteroatoms (e.g. O, N, S).[90] While crystalline graphite was shown to be highly refractory and chemically inert, the chemical reactivity of graphitic carbon increases with structural disorder and the abundance of heteroatoms and unsaturations.[91] Hence, this possibly impacts the diversification of abiotic organic compounds observed in the altering oceanic lithosphere.

15.3 Comparison with Experiments and Thermodynamic Predictions

The heterogeneity of natural systems and the multistage character of hydrothermal alteration seem crucial for organic synthesis. This considerably complicates the identification of organic products, their origin, and the abiotic reaction sequences. A complementary experimental and theoretical approach is therefore mandatory to unraveling reaction paths and rates and determining the most favorable conditions that may lead to the abiotic formation of CM. The literature reviewed in the next sections highlights the gap between natural observations and experiments (Sections 15.3.1–15.3.3) and opens a wide field for future investigation, supported by thermodynamic predictions (Section 15.3.4).

15.3.1 Experimental Approach

Most of the experimental work dedicated to hydrothermal organic synthesis has focused on fluids, especially on the production of volatile organic compounds, using either monophasic (pure liquid) or biphasic (liquid + gas) systems. Solid mineral phases have not been systematically introduced in the experimental devices, and potential carbonaceous phases were rarely characterized after experiments. When present, minerals were either used as reactants, such as olivine to produce H_2, or as redox buffers such as the hematite–magnetite (HM), hematite–magnetite–pyrite (HMP), pyrite–pyrrhotite–magnetite (PPM), or quartz–fayalite–magnetite (QFM) assemblages. Minerals have also been introduced as potential catalysts of organic reactions.[92–94] Unfortunately, they were never characterized after experiments, precluding the finding of CM occurrences.

Hence, the role of minerals on the formation of carbonaceous compounds under hydrothermal conditions, as emphasized from natural observations, has remained largely unexplored. This is partly explained by the analytical difficulty of detecting such a small organic fraction within the solid products (concentration close to or below detection limits) and of locating and characterizing it with high-resolution methods. In addition, contamination issues (from the experimental setup or deriving from organic compounds preexisting in the minerals used; e.g. trapped in fluid inclusions) remain central when looking for low levels of organic products whose nature is unknown and difficult to address.

The role of the experimental container (vessel or capsules) on CO_2 reduction or H_2 production reactions under hydrothermal conditions is potentially non-negligible while not clearly established. Stainless steel reactors have been shown to accelerate FTT reactions[95,96] compared to quartz, glass, or Au and TiO_2 reactors. However, stainless steel reactor walls may be passivated after several hours of experimental runs.[95] That is also the case for titanium, whose oxidation can be a source of H_2 if a preoxidation is not done prior to experiments. Gold, in the form of nanorods or particles, is well known to catalyze aqueous CO_2 reduction to CO during electrochemical experiments,[97,98] but it is not usually considered for hydrothermal experiments, probably because CO was either not measured or not abundant in products, and also because the gold liner is seen as a smooth surface. Hence, vessels or capsules made of gold or oxidized titanium are usually thought to be the most inert materials and are often preferred at high pressure (P) and T to stainless steel, platinum, alumina, or hastelloy (Ni–Fe-rich alloy). The effect of hastelloy may depend on the reaction of interest. It was infrequently used for FTT reactions, probably because of the potential catalytic properties of the Ni-rich metal alloy, but it was used for H_2 production.[99] Its catalytic effect has not yet been demonstrated for such reactions. The use of Teflon™ especially may be a direct source of carbon contamination with increasing T. Concerning reactor permeability to gas, H_2 has a high diffusivity in most metals, but H_2 loss is expected to be relatively low at T < 400°C.[100]

Hence, there is no homogeneity in the type of reactor used in the literature, which often precludes comparison between results, especially when the mineral effect has to be unraveled. In addition, the possible precipitation of a carbonaceous phase on the reactor wall has never been investigated.

15.3.2 Carbon-Bearing Reactants in Experiments

Oceanic hydrothermal systems developing near ridge axes have two main sources of inorganic carbon: (1) mantle-derived carbon delivered by magmatic activity in which CO_2 dominates in the C–O–H system, with very minor CO and CH_4;[1] and (2) dissolved inorganic species that result from the equilibration of atmospheric CO_2 and seawater. Hence, aqueous CO_2 ($CO_{2(aq)}$) and bicarbonate and carbonate ions (ΣCO_2) are usually the most abundant species in solutions and are the preferential sources of carbon used in experiments and models designed to test abiotic organic synthesis under hydrothermal conditions mimicking natural systems.

Nevertheless, ΣCO_2 are not the only single-carbon compounds possibly available at equilibrium in hydrothermal fluids, especially if H_2 is available. Experiments have shown that the speciation of aqueous single-carbon compounds in the C–O–H system for T = 150–300°C and P = 35 MPa is controlled by reactions between ΣCO_2, CO, ΣHCOOH (HCOOH, formic acid + $HCOO^-$, formate), CH_2O (formaldehyde), CH_3OH (methanol), and CH_4 (Figure 15.4).[101] Indeed, the water–gas shift reaction (15.1) under aqueous hydrothermal conditions leads to the formation of ΣHCOOH as an intermediate product (Figure 15.4):

$$CO + H_2O = CO_2 + H_2. \tag{15.1}$$

ΣHCOOH reaches a redox-dependent equilibrium with methanol, formaldehyde, and CH_4 within few days at T > 150°C, but may need years at T < 100°C.[101] Formaldehyde and CH_4 were close to below detection limits in the experiments, suggesting kinetic inhibition at least for the P–T range tested.[101] In the absence of CH_4, the relative concentrations of single-carbon compounds in fluids strongly depended on T, H_2 fugacity (fH_2), and pH. Under neutral and acidic conditions, CO_2 largely dominated, with minor amounts of CO and ΣHCOOH (several orders of magnitude < CO_2) at 350°C. At 150°C and similar

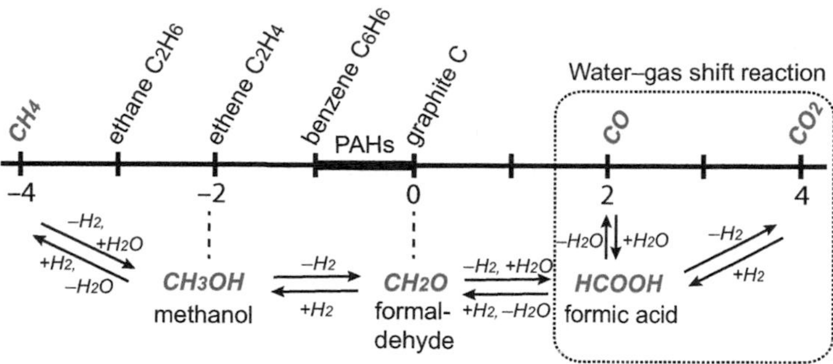

Figure 15.4 Oxidation states of carbon in some single-carbon organic compounds. The water–gas shift reaction (15.1) is represented along with the successive reversible redox reactions that control the speciation of single-carbon compounds under hydrothermal conditions.[44,101]

pH, methanol is predicted to be only one order of magnitude below CO_2 with minor CO and ΣHCOOH, but it could exceed CO_2 if the H_2 concentration increased to values similar to those measured at ultramafic rock-hosted hydrothermal vents (i.e. ~10 mM[102]). Under alkaline conditions, the formate concentration can be equivalent to bicarbonate with decreasing T or increasing H_2 concentrations.

Direct formation of methanol (~1 mM max) from a H_2–CO_2-rich vapor phase can also occur, with a limited conversion rate (10^{-4}%) at 300–350°C (18 MPa), provided magnetite surfaces are available.[95] This could be realistic at mid-ocean ridges (e.g. following dike emplacement with volatile exsolution and migration in adjacent oxide gabbro or in serpentinized peridotites where magnetite is abundant). Magnetite shows decreasing surface reactivity with time and can be regenerated with increasing T, suggesting that it may serve as a catalyst during successive diking events, for instance. The authors do not preclude possible intermediates such as CO and ΣHCOOH in the redox reaction, but these compounds were not detected or measured, respectively, and no graphitic phase or alkanes were observed.[95]

These works show that a large variety of single-carbon species can be available in fluids in addition to ΣCO_2 (mainly methanol and formate), especially in natural serpentinizing systems with highly variable pH and local H_2 levels.[64] This is particularly true in low-T serpentinizing environments that may be dominated by methanol or formate depending on pH, provided equilibrium is reached. This diversity reinforces the conclusions of Section 15.2 about the need to consider a wider range of carbon-bearing reactants for organic synthesis in natural systems and consequently in experiments. Natural systems also pointed to the possible availability of PAHs, CM, and deep magmatic CH_4 for low-T hydrothermal reactions. Except for formic acid, whose decomposition in aqueous fluid (Figure 15.4) is frequently exploited to experimentally produce H_2 and CO_2, these compounds have rarely (e.g. methanol[103]) or never been used as reactants in geologically relevant experiments.

CH_4, which is abundant in hydrothermal fluids or cold seepages associated with the serpentinization of mantle rocks,[3,102] was usually considered as the reduction product of inorganic carbon sources rather than as a possible reactant. An important experimental effort was deployed to reproduce methane synthesis by FTT reactions under moderate to low-T conditions ($\leq$500°C), but systematically failed to produce abundant CH_4. The natural H_2/CH_4 ratio measured in serpentinization-related oceanic hydrothermal environments (<30)[102] remains much lower than experimental ones (>500; except values at 42[104] and 17;[105] e.g. see McCollom[106] for a review). This difference was proposed as an indication of the abiotic versus biotic origin of CH_4 in natural systems, where intense H_2 consumption and associated CH_4 production can be attributed to biological activity.[107] This simplistic criterion was refuted because it cannot account for the complex processes occurring in the environment,[108] and it also disregarded some other parameters. The H_2/CH_4 ratio may vary with T, decreasing as T increases from 200°C to 500°C at 300 MPa.[109] The effect of pressure has been seldom investigated, but a few studies have shown that increased pressure can account for a significant increase in the CH_4 yield between 100 and 350 MPa.[110] CH_4 formation can also compete with CO_2 carbonation depending on fH_2, CO_2 partial pressure, and T. At 200°C, CH_4 formation by CO_2 reduction is limited by H_2 production during olivine alteration if the system is supersaturated with respect to carbonate because the kinetics of carbonate precipitation are faster. Finally, the micromolar levels of CH_4 often found in experiments might not even be produced by *in situ* reactions, but instead may represent contamination. This is often difficult to assess since blank experiments are rarely provided and ^{13}C-labeled experiments are scarce. Contamination issues are critical in very-low-T experiments (<100°C) where product levels are much lower and display contrasting values of H_2 and CH_4 despite similar protocols.[13,67,111–115] The most complete investigation conducted so far at T $\leq$ 100°C led to the formation of low-molecular-weight organic acids (mainly formate and acetate); no CH_4 was observed during H_2 production by serpentinization.[13] This suggests that at such low-T and moderately alkaline conditions (pH ~8), the inorganic carbon (CO_2, CO, or bicarbonate) tends to equilibrate with organic acids instead of CH_4 in a metastable assemblage as previously described for higher T (150–350°C).[14,15,101]

Hence, the high H_2/CH_4 ratio observed in most experiments (e.g. run at ~300°C, 30–50 MPa) highlights the kinetic inhibition of CH_4 formation from the reduction of dissolved CO_2 species or formate under aqueous hydrothermal conditions.[14,15] For this reason, the role of realistic catalysts (magnetite, hematite, chromite, Fe–Ni alloys, or sulfides) has been tested to circumvent this limitation. Most of them allowed very limited conversion of inorganic carbon to CH_4 (<1% max) and even less for longer-chain alkanes ($\ll$0.1% max) under laboratory timescales (several months), with a final CO_2/CH_4 ratio ranging from 100[116] to 1000.[14,15,117–119] Lower values of CO_2/CH_4, as small as ~0.1–1.0 and 10, were obtained in studies using Fe–Ni alloys (porous awaruite)[104] or Ni-sulfides[119] and Co-bearing magnetite,[105] respectively, pointing to some very specific catalysts possibly available in serpentinized peridotites. Conversely to CO_2, hematite, magnetite, and Fe–Ni alloys did not affect the stability of formate and formic acid, at least for T of 170–260°C.[15]

The presence of a gaseous phase in the system, which creates conditions closer to the industrial Fisher–Tropsch process, also seems to favor a higher yield of CH_4 and the formation of more complex hydrocarbons,[15,96,106] even in the absence of a catalyst[110] (if olivine and gold reactor walls are excluded). At very-low-T conditions,[120] Ru-bearing chromitite exhibits efficient catalytic properties for CH_4 production under gaseous conditions. The slow kinetics of the first step (CO_2 to CO reduction) in a gas phase is already well known in the industry, and chemists have strived to improve catalyst efficiency, notably by increasing the local concentration of CO_2 on the catalyst surface by altering the catalyst's nature, shape, and size.[98,121]

Overall, these experiments seldom report on the potential presence of heavier organic compounds or on the organic content of the liquid[122] and solid phases, if any are present (see Section 15.3.1), leaving the possibility for other type of products and reaction mechanisms to account for low CH_4 contents. Mineral reactants also need to be systematically introduced in future works for a more realistic approach that fits better with the conditions of natural systems (Section 15.2).

15.3.3 Experimental Occurrences of Carbonaceous Material

Very few experiments have investigated CM phases and their relationship with minerals using bulk or *in situ* methods such as SEM or transmission electron miscroscopy (TEM) with or without XPS, Raman, and FTIR spectroscopy. As for the characterization of natural occurrences, these methods are seldom used together, limiting comparisons, and the exact nature of CM is usually unidentified.

As described in Section 15.2.2, in the oceanic lithosphere, CM can be first deposited at high T (>400°C) on mineral surfaces and cracks during magmatic degassing at depth and subsequent migration and cooling of mantle-derived fluids. Experiments designed to reproduce this process simulated the sudden cooling of a C–O–H magmatic gas over freshly cracked olivines.[123] Evidence for direct carbon precipitation on newly formed cracks has been reported despite thermodynamically unfavorable experimental conditions for graphite precipitation.[123] In this study, it was proposed that freshly cracked surfaces of olivine, and possibly of other silicates, offer chemically active areas facilitating the heterogeneous nucleation of CM during interaction with C–O–H gases, at least for initial T between 400°C and 800°C. The deposited carbonaceous film displayed a more complex structure than graphite. Bonding was dominated by C–C and C–H species under the more oxidizing conditions tested, whereas the more reducing ones showed similar amounts of C–C, C–H, C–O, and metal–C species such as carbides (SiC or MgC).[123] Carbides, mainly observed at 400°C, could have acted as reaction intermediates for FTT reactions, but this has not been proven since the analyses of companion gases could not be performed in this study.

The thermal decomposition of siderite under water vapor conditions (300°C) and saturated vapor pressure (P_{sat}) also resulted in the formation of CM.[124] Siderite alone provides both H_2 from water reduction by Fe^{2+} oxidation and inorganic carbon transformed into a reduced carbon phase (15.2):

$$6FeCO_3 \rightarrow 2Fe_3O_4 + 5CO_2 + C. \tag{15.2}$$

Although siderite may not be abundant in the oceanic lithosphere, other carbonates are present and may locally decompose (e.g. during magmatic injections) and react if H_2 is available. After solvent extraction, the organic products were identified as dominantly alkylated and hydroxylated aromatic compounds,[124] which considerably differs from the FTT products (aliphatic chains) that would have been catalyzed by the abundantly formed magnetite. In addition, discrepancies between the relative H/C ratios of reactants and products suggested the presence of an unidentified product with a low H/C ratio, which could be an insoluble C-rich phase remaining in the solid phase.[124]

Under aqueous conditions more representative of low-T hydrothermal processes, two types of experiments report the formation of a poorly crystallized CM: (1) alteration of ferromagnesian silicates by a CO_2-enriched fluid; and (2) carbonate dissolution experiments.

Experiments involving the carbonation of olivine (400–500°C, 100 MPa, static capsules[125]) and of a sandstone made of Fe^{2+}-rich volcanic clasts (100°C, 10 MPa, flow-through reactor[126]) reported the precipitation of a poorly crystallized graphitic phase (Figure 15.5a and b). In both experiments, the graphitic phase accounts for a non-negligible part of the reaction products in addition to phyllosilicates and carbonates. As in natural systems (Section 15.2.2), it is embedded in phyllosilicates, corresponding to serpentine (Figure 15.5a) or Fe-rich chlorite (chamosite; Figure 15.5b) here, closely associated with magnetite when present.[126] CM formation was attributed to the reduction of CO_2 (or CO at high P–T[125]) by H_2 initially present in solution (Eqs. (15.3) and (15.4))[125] or by Fe^{2+}-bearing minerals.[126]

$$CO_2 + 2H_2 = C + 2H_2O. \tag{15.3}$$

$$CO + H_2 = C + H_2O. \tag{15.4}$$

The two protocols described above resulted in different oxygen fugacities (fO_2). It was estimated to be close to the CCO buffer (C as graphite–CO) in the high-P–T runs,[125] which prevented the formation of magnetite, in opposition to the low-P–T runs.[126] This indicates that magnetite is not mandatory for the formation of graphitic material, whose nature remains poorly constrained, however. Organic volatiles have not been investigated in these studies.

Siderite dissolution experiments at 200°C and 300°C and 50 MPa described two types of carbonaceous products (Figure 15.5c and d)[127]. A poorly structured hydrated or hydrogenated CM occurred without spatial relationship with minerals in all runs (Figure 15.5c). At 200°C, a more ordered graphitic carbon formed on the surface of the neo-formed magnetite grains or near iron oxides at the siderite surface (Figure 15.5d). CO_2 was the only gas detected at 200°C, while additional small amounts of H_2 and CH_4 were detected at 300°C. A blank experiment showed that trace amounts (<mM) of dissolved organic compounds were present as contaminants.[127] Although they were much less

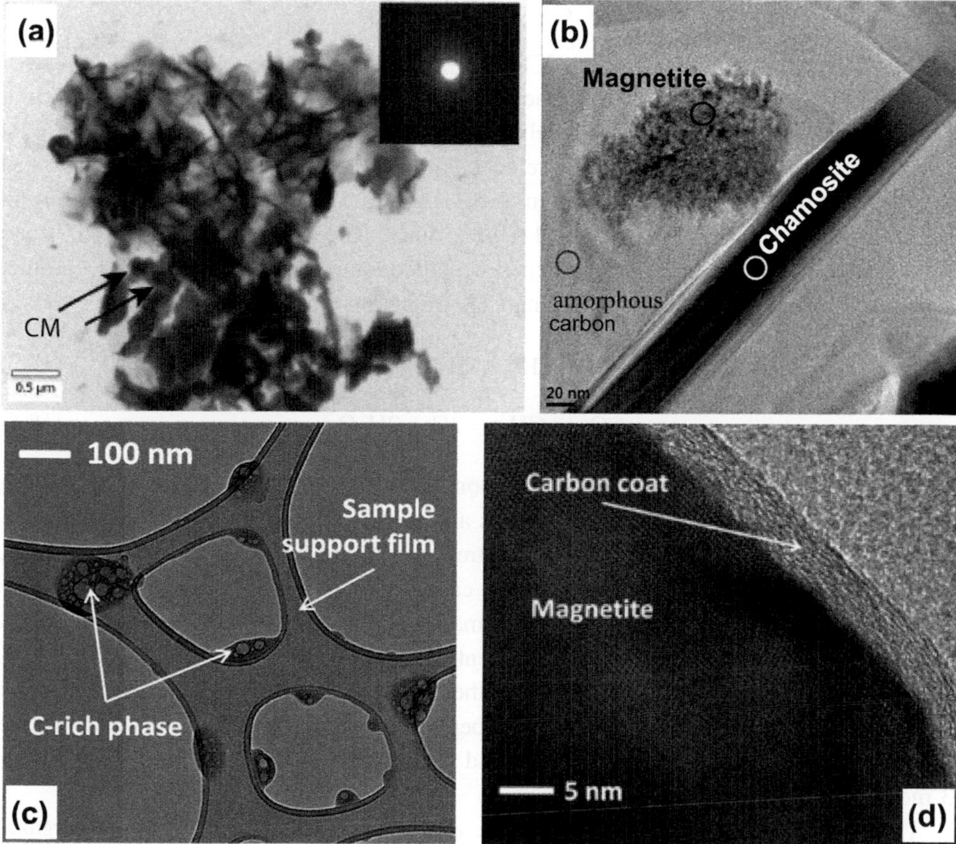

Figure 15.5 Examples of CM that precipitated in hydrothermal experiments. (a) Poorly crystallized graphitic phase (round particles) formed during high-T (400–500°C) carbonation of olivine. (b) Amorphous carbon precipitation during low-T (100°C) carbonation of a sandstone made of Fe^{2+}-rich volcanic clasts. (c, d) Two different types of CM precipitated during low-T (200–300°C) siderite dissolution: a poorly structured hydrated or hydrogenated carbonaceous phase (c) and a more ordered graphitic phase (d).

Reproduced with permission of Elsevier, from (a) Dufaud et al. (2009), *Chem Geol*, **265**, 79–87, figure 6,[125] (b) Luquot et al. (2012), *Chem Geol*, **294–295**, 75–88, figure 10,[126] and (c, d) Milesi et al. (2015), *Geochim Cosmochim Acta*, **154**, 201–211, figures 4 and 5a.[127]

abundant than CO_2, there remain open questions regarding whether they contributed to the formation of low levels of CH_4 and the presence of carbonaceous phases. Whatever the case may be, the results of this work fit well with kinetic inhibition of CH_4 formation from CO_2 and H_2 at low T (<400°C; Section 15.3.2) and suggest a preferential precipitation of CM with decreasing T through reactions (15.3) and (15.5).

$$3FeCO_3 + H_2O_{(aq)} = Fe_3O_4 + 3CO_{2(aq)} + H_{2(aq)}. \qquad (15.5)$$

$$C + 2H_2 = CH_4. \quad (15.6)$$

The inefficiency of magnetite as a FTT reaction catalyst is also pointed out here (see Section 15.3.2). Poisoning of the magnetite surface by carbon coatings (Figure 15.5d) at low T has been questioned; alternatively, the carbon coating may only form when CH_4 cannot (i.e. at low T). A syngenetic relationship between CM and CH_4 has also been proposed but not demonstrated. Equations (15.3) and (15.6) on the one hand and (15.7) and (15.8) on the other show two reversal genetic links. The second one (Eqs. (15.7) and (15.8)) appears less likely under the experimental conditions tested ($\leq$300°C) according, again, to the kinetic inhibition hampering CH_4 formation (15.7).

$$CO_2 + 4H_2 = CH_4 + 2H_2O. \quad (15.7)$$

$$CO_2 + CH_4 = 2C + 2H_2O. \quad (15.8)$$

Equation (15.6), which suggests that carbon coating may serve as an intermediate step to CH_4 formation at least at 300°C where its absence would result from its full consumption, may be a more relevant hypothesis, as already mentioned in Section 15.2.3. Similarly, it was previously suggested that background carbon, possibly more reduced than graphite on the magnetite surface, facilitates CH_4 formation.[118]

At lower T compatible with life, experimental data are lacking on the formation of CM, while this is the condition under which it is the most difficult to discriminate the origin of CM in natural systems (Section 15.2). In serpentinization experiments run at $\leq$100°C (see Section 15.3.2), no CH_4 was produced, and the strong decrease of CO_2 and H_2 after 3 months was not fully explained.[13]

The limited data available to describe the association of or interplay between organic volatiles and CM limit the conclusions so far on their genetic relationship.

If all occurrences of CM in experimental solid products are due to *in situ* abiotic reactions from an inorganic source (not contamination), they suggest that at least small amounts of CM easily form under a wide range of hydrothermal conditions during short runs of a few days to a few months (100–500°C, 10–110 MPa, fO_2 between CCO and QFM buffers). The rare available images report different aspects (spherules, film-like surficial layers, or porous accumulations) similar to some of those observed in natural systems (Figure 15.2). They seem to condense from gas or precipitate from fluid preferentially on the mineral surfaces where they possibly evolve. Minerals used in experiments (mainly magnetite) were not as diverse as in nature (Section 15.2), and other mineral surfaces such as hematite, sulfides, hydrogarnets, phyllosilicates, epidote, and quartz, not yet introduced in experiments, may allow the formation of carbonaceous compounds, possibly when the reducing conditions are not optimal.[38,55,60,85,123]

In agreement with natural observations (Section 15.2.2), one experimental study has demonstrated that smectite clays, and particularly montmorillonites, can promote and preserve organic compounds formed under seafloor hydrothermal conditions (300°C, 100 MPa).[103] While $^{13}CH_4$ was the dominant organic product in the gas-phase, aromatic compounds, including PAHs (up to C_{20}), were extracted from the solid products that reacted for 6 weeks with labeled ^{13}C-methanol (10.4 M). The absence of organic material within the illite sample compared to the smectite one showed that the clay interlayer properties exert a control on organic synthesis, notably the negatively charged layer of smectite. The exact reaction path was not constrained; the role of alkanes as intermediate reactants or products and the presence of methanol as a major reactant are worth noting. This study also highlighted the possible protection of organic compounds by clays from thermal alteration at 300°C, making them available for further reactions under lower-T conditions.

15.3.4 Thermodynamic Predictions

The nature of hydrothermal CM observed in natural and experimental works related to the alteration of mantle-derived rocks is not clarified yet. It ranges from graphite to very different forms in terms of structural order, nature and quantity of heteroatoms, unsaturations, and hence reactivity. There has been a considerable amount of work to quantify the thermodynamics of C–C bond formation in graphite and other phases.[32] We selected here the main results related to CM formation under hydrothermal conditions for comparison with previous sections.

The formation of graphite is predicted from C–O–H fluids under a wide range of hydrothermal conditions, provided there is no kinetic inhibition. Graphite can readily precipitate from the fluid; this is favored by a decrease in temperature or an increase in pressure along with a decrease in water content.[36,45,128,129] Both T decrease (deep fluid ascent) and water decrease (e.g. during hydration reactions) correspond well to natural oceanic scenarios in the deeper part of the lithosphere. Graphite formation results from redox reactions unless the reactant carbon is at the same oxidation state (i.e. zero). This does not necessarily require very strong reducing conditions, and graphite can be produced through either oxidation or reduction reactions (Figure 15.4).[44,130] In addition to graphite, CO_2 or CH_4 should be the dominant carbon species in hydrothermal fluids depending on temperature and oxygen fugacity. For a given oxygen fugacity, CO_2 should predominate at high-T conditions, while CH_4 is favored at the lower T[35,131] according to the equilibrium described in (15.7).

When methane and graphite formation is inhibited (i.e. when the kinetic barriers hamper access to a stable equilibrium,[35] as in most natural and experimental systems; Sections 15.2, 15.3.2, and 15.3.3), there is considerable potential for the hydrothermal synthesis of a wide range of metastable organic molecules provided that fH_2 is high enough.[33,44,45] The reversible character of several organic reactions in hydrothermal fluids contributes to this variety (e.g. hydration/dehydration or oxidation/reduction reactions).[132] Calculations have shown that organic synthesis of metastable phases is even possible in environments where the amount of H_2 is usually below those observed during the alteration of mantle-dominated environments;[44,45] as an example, the rapid cooling of volcanic gases containing CO, CO_2, H_2O, and H_2 below ~250°C provides a thermodynamic drive for the abiotic synthesis of metastable hydrocarbons such as condensed *n*-alkane and PAHs, depending on the H/C ratio of the gas. Such products may relate to the complex CM and molecular organic minerals observed in natural samples and in CH_4-poor experiments. In more reducing environments, the energetic drive of metastable compound formation is increased and reaction temperatures can be shifted to higher values.[44,45] Decreasing temperature allows for metastable mixtures of different condensed phases, in agreement with the variety of compounds found in natural rock textures (Section 15.2.2).

The potential abiotic formation of CM during serpentinization that is central to this chapter has been investigated by comparing experimental fluid compositions with natural hydrothermal compositions across a wide range of CO_2 concentrations.[33] Predicted mineral assemblages consisted of serpentine and brucite with or without magnetite and carbonates depending on the relative activities of H_2 and CO_2. $CO_{2(aq)}$ was considered at equilibrium with graphitic compounds that share the thermodynamic properties of graphite and a hydrogenated aromatic carbon compound (i.e. anthracene, a PAH made of three benzene rings) in order to mimic the experimental products (Figure 15.5c and d).[33,127] Results show that serpentinization fluids can equilibrate with both the graphitic and the hydrogenated carbon when the formation of alkanes is prohibited (kinetic inhibition). As the precipitation of a hydrogenated carbon compound requires a higher hydrogen activity, its formation predicted after the serpentinization reaction significantly progresses (i.e. >70% complete), similarly to the classical serpentinization degree attained in oceanic peridotites (e.g. Andreani et al.[133]).

15.4 Summary

According to natural, experimental, and theoretical results, the accumulation of CM within mantle-derived rocks of the oceanic lithosphere seems a very likely process, yet it is poorly considered during hydrothermalism. Even if CM formation under such conditions has been largely unexplored experimentally compared to volatile products, this outcome fits well with recent models that restrict CH_4 formation to high-T processes (>400°C) at depth, leaving open the possibility for various metastable carbon phases to form at shallower levels. Although the total amount of abiotic CM is not quantified yet, and might represent a small fraction of the TOC contained in the present-day oceanic lithosphere, a better understanding of its formation mechanisms, nature, and reactivity is of prime importance.

The large range of aspects, structural orders, and compositions of the CM trapped within rock textures, ranging from stable graphite usually at high T (>~500°C) to variably ordered, hydrogenated, O-, N- and S-bearing materials and PAHs, attest to the lithosphere's organic wealth available for evolved chemical reactions and ecosystem development. The structure and functions of ecosystems inhabiting the shallow levels of the oceanic lithosphere indeed suggest that despite the general thought that CH_4 or inorganic carbon species represent the main sustainable feedstocks of carbon in those settings, they may not be the main sources of carbon for microorganisms that rather use formate[134] or more diverse abiotic organic compounds, including PAHs.[135–137] The variety of organic compounds reflects the characteristics of natural systems, which are dynamic and highly heterogeneous chemically, structurally, and hydrodynamically, therefore creating chemical and redox gradients that are able to drive metastable reactions over a wide range of physicochemical conditions. They also provide various mineral substrates that have been shown to facilitate CM accumulations and likely further transformation. Whether minerals can affect the nature and crystallinity of the CM remains to be addressed. According to the very different crystallo-chemistry and surface properties of mineral families, different types of organic compounds may even be expected within different mineral substrates, rendering them as micro-factories with high specificity as observed in natural serpentinites.[55]

Observations force us to consider natural rocks as being built up of several far-from-equilibrium microenvironments that are expected to evolve with time and system fluctuations (fH_2, P, T, pH, and fluid transport and chemistry), each offering at a given stage a specific mineral assemblage and chemical conditions that are propitious to a given series of reactions. They considerably deviate from static, buffered, bulk mineral assemblages undergoing interactions with fluids near equilibrium, as is often considered. Such microenvironments may be the locations of organic reactions simply succeeding each other or competing together, but also inorganic reactions such as carbonate precipitation and biological activity, depending on the alteration history and fluid paths. Indeed, the oceanic crust is subjected to a progressive and multistage evolution of its organic pool, including both abiotic reactions and multiple recolonizations, all along the way from ridge to subduction, hence modifying the initial signature of a compound of interest.

We attempted to summarize the main stages of abiotic carbon processing in a heterogeneous oceanic lithosphere (Figure 15.6a) from deep to shallow levels in the sketch of Figure 15.6. We recommend going beyond the formation of CH_4 and the FTT reaction by considering the heterogeneity and multistage character of the system at all scales. First, CM (or graphite) can directly precipitate on fresh mineral surfaces and in vesicles during cooling and re-speciation of CO_2-rich magmatic fluids at depth (>400°C and up to 800°C[123]). Fluids can also get trapped within fluid inclusions that display C_1–C_3 *n*-alkanes initially in the fluid or formed *in situ* and CM (or graphite) formed *in situ* (see Section 15.2.3), possibly helped by subsequent retrograde reactions (stage 1, Figure 15.6b). Then, provided relatively low fO_2 is available, CM can continue to form at $\leq$400°C (stages 2 and 3, Figure 15.6b) among the new mineral assemblage from aqueous C-bearing hydrothermal fluids (magmatic or seawater derived) thought to carry a large variety of organic and inorganic carbon species formed at deeper lithospheric levels (see Sections 15.2.2 and 15.3.2). The strongest thermodynamic drive for the formation of most metastable organic molecules (see Section 15.3.4), including *n*-alkanes, organic acids, and PAHs, is below 200°C (stage 3, Figure 15.6b) for typical redox conditions occurring in the mafic component of the lithosphere (near the PPM buffer). In the dominant ultramafic component, conditions should be far more favorable to the abiotic formation of metastable products, and extending to higher temperatures (at least 300°C at the QFM buffer). Indeed,

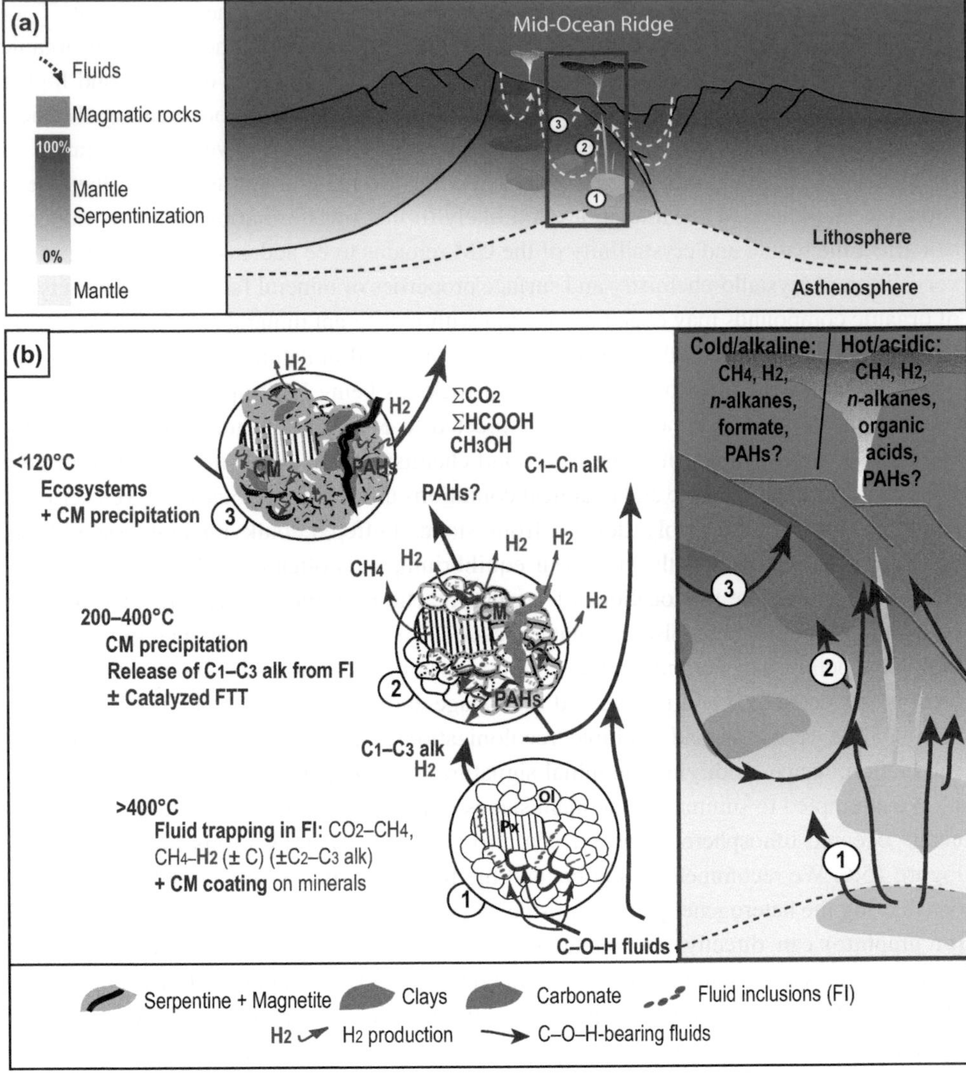

Figure 15.6 Sketch depicting the main stages of carbon processing in the oceanic lithosphere during the multistage fluid–rock reactions[133] recorded in such dynamic systems. (a) Geologic setting of a slow-spreading ridge where mantle is tectonically exhumed, simultaneously with localized and ephemeral magmatic injections. These environments appear to be the most favorable ones for abiotic organic synthesis according to the available natural observations, experiments, and thermodynamic calculations. (b) Summary of the three main stages of carbon processing in a column of the oceanic lithosphere shown in (a), illustrated through a magnification of the mantle rock textures along with the hypothetical nature of the percolating fluids. See the text for details. alk = alkanes.

serpentinization provides high levels of H_2 over a wide range of T and pH conditions (100°C, pH 9 to 350°C, pH 3)[13,102] that favors reduction reactions while providing variable single-carbon reactants in the fluid, especially at low T < ~200°C (ΣCO_2, $\Sigma HCOOH$, methanol).[101] A body of experimental work has shown that the most efficient production of H_2 from serpentinization occurs at ~300°C, notably with modified seawater after interaction with magmatic lithologies.[138,139] Alteration of ultramafic rocks also releases specific transition metals such as Fe and Ni and is known to change sulfur speciation, leading to the formation of potential metallic–sulfide catalysts.[140,141] The fH_2 of those systems can be highly variable down to the micrometer scale, ranging from QFM buffer conditions in magnetite-dominated domains to fH_2 values of several orders of magnitude higher to form Fe–Ni alloys such as awaruite near serpentinization fronts.[141] Serpentinization is also characterized by the abundant formation of phyllosilicates, oxides, and hydroxides that are thought to facilitate CM precipitation, condensation, and transformation (stages 2 and 3, Figure 15.6b). The ephemeral magmatic activity associated with mantle-rock exhumation in a slow-spreading environment (Figure 15.6) may also drive fluctuating T or pH conditions and C–O–H fluid inputs. This can shift the local thermal regime from a cold-alkaline to a hot-acidic environment, and vice versa, as observed at the Rainbow massif.[142] The tectono-magmatic activity near the ridge axis also affects transport processes that control the fluid residence time in the system, a determinant of C–O–H fluid speciation, especially at low T.[101]

15.5 Limits to Knowledge and Unknowns

To achieve a better understanding of the mechanisms and processes leading to the abiotic formation of CM in the oceanic lithosphere and to carbon cycling under those environments in general, one of the major challenges future studies will have to face is to improve criteria for establishing the origin of organic carbon accumulations found within the habitable oceanic rocks. As highlighted in this chapter, graphitic compounds occur from deep levels where solely abiotic reactions can account for their generation, up to colonizable depths, where H_2 and abiotic organic compounds are valuable sources of energy and carbon for microbial communities[27,143] below the T limit of life currently established at 122°C.[144] This deep life indeed has the ability to capitalize on the steady stream of both inorganic and organic serpentinization by-products, converting C into biomass and competing against abiotic reactions of C reduction.[23] These rock-hosted microbial ecosystems, along with the photosynthesis-derived organic compounds contained in seawater and injected into the hydrothermal conduits, experience hydrothermal degradation and can be transformed, similarly to sedimentary organic matter, into a kerogen-like or bitumen-like macromolecular material[18,19,31] with structural and chemical similarities with the hydrothermally formed CM. Up to now, more attention has been paid to CM in meteorites[145] than in hard rock samples from Earth's lithosphere. Meteorites testify to the capacity of hydrothermal activity to drive complex abiotic organic synthesis.

Identifying CM origins will have to go through more systematic CM characterization in rocks from various geological settings. This may also contribute to the identification of new organic minerals (see mineralchallenge.net).

The role (in terms of kinetics and selectivity) of mineral phases in organic synthesis, particularly in CM formation, evolution, and preservation, should be a prime direction for future research, too. New experimental work is required in which minerals should be introduced and carefully characterized to possibly identify CM, which has been neglected so far, along with its reactivity and formation mechanisms. We encourage a more systematic investigation of the liquids, gases, and solids altogether in future hydrothermal experiments in order to unravel possible genetic relationships and to determine their respective roles in the organic synthesis factory.

Tracking reactions *in situ* in both natural and experimental solids requires maintaining the ongoing effort to develop microscale to nanoscale investigation methods and to identify biotic/abiotic criteria for use both in natural samples and in experiments that will have to face contamination issues. As detailed in Section 15.2, most of the evidence for CM abiogenicity was derived from the mineral assemblages and textural relationships between CM and mineral phases. Compared to bulk analysis, microscale characterization using high-resolution imaging techniques will allow a comprehensive description of the textural context in which organic carbon occurs and will provide constraints regarding whether or not the carbon compounds are indigeneous.[145] Microscale techniques can also capture the co-occurrence and relationships of CM with metals and minerals, and are thus particularly well suited to tracking organic matter production, evolution, and transport within the oceanic lithosphere.

Acknowledgments

The authors want to thank the Deep Carbon Observatory, the French CNRS (INSU, PIE, EPOV, and MITI Défi Origines programs), and the French National Research Agency (ANR-14-CE01-0008) for funding. This chapter is also the result of numerous fruitful discussions with several partners, notably the participants of the deepOASES project (PI B. Ménez).

Questions for the Classroom

1 Why is it important to better constrain the abiotic organic reactions on Earth?
2 Which impact (if any) could abiotic organic synthesis have on Earth today and have had in the past?
3 What is the main message of this chapter and what do you think of it?
4 Why is the oceanic lithosphere a propitious place to investigate abiotic organic synthesis?
5 Which other settings on Earth would be of interest, as well as elsewhere in the solar system? What would be the implications?

6 Which organic compounds are thermodynamically stable? How and why can metastable organic compounds be formed and subsist through time in rocks?
7 What are the main abiotic organic compounds formed through the hydrothermal alteration of the oceanic lithosphere? How do they vary as a function of temperature? What are the other important parameters?
8 What are the strengths of micro-imaging and *in situ* micro-spectroscopic approaches for unraveling the reaction paths leading to abiotic organic synthesis in rocks?
9 Can you think of potential markers for discriminating between the abiotic and biologic origins of carbonaceous compounds in rocks? Can they be addressed with available techniques of analyses or do they require technological development?

References

1. McCollom, T. M. Laboratory simulations of abiotic hydrocarbon formation in Earth's deep subsurface. In: *Carbon in Earth. Reviews in Mineralogy & Geochemistry*, Vol. **75** (eds. R. M. Hazen, A. P. Jones & J. A. Baross), 467–494 (Mineralogical Society of America, 2013).
2. Sephton, M. A. & Hazen, R. M. On the origins of deep hydrocarbons. In: *Carbon in Earth. Reviews in Mineralogy & Geochemistry*, Vol. **75** (eds. R. M. Hazen, A. P. Jones & J. A. Baross), 449–465 (Mineralogical Society of America, 2013).
3. Etiope, G. & Sherwood Lollar, B. Abiotic methane on Earth. *Rev Geophys* **51**, 276–299 (2013).
4. Proskurowski, G. et al. Abiogenic hydrocarbon production at Lost City hydrothermal field. *Science* **319**, 604–607 (2008).
5. Haggerty, J. A. & Fisher, J. B. Short-chain organic acids in interstitial waters from Mariana and Bonin forearc serpentines: Leg 125. In: *Proceedings of the Ocean Drilling Program*, Vol. **125** (eds. P. Fryer, J. A. Pearce & L. B. Stokking), 387–395 (Ocean Drilling Program, 1992).
6. McCollom, T. M. & Seewald, J. S. Abiotic synthesis of organic compounds in deep-sea hydrothermal environments. *Chem Rev* **107**, 382–401 (2007).
7. Wang, D. T., Reeves, E. P., McDermott, J. M., Seewald, J. S. & Ono, S. Clumped isotopologue constraints on the origin of methane at seafloor hot springs. *Geochim Cosmochim Acta* **223**, 141–158 (2018).
8. McDermott, J. M., Seewald, J. S., German, C. R. & Sylva, S. P. Pathways for abiotic organic synthesis at submarine hydrothermal fields. *Proc Natl Acad Sci USA* **112**, 7668–7672 (2015).
9. Kelley, D. S. Methane-rich fluids in the oceanic crust. *J Geophys Res* **101**, 2943–2962 (1996).
10. Kelley, D. S. & Früh-Green, G. L. Abiogenic methane in deep-seated mid-ocean ridge environments: insights from stable isotope analyses. *J Geophys Res* **104**, 10439–10460 (1999).
11. Etiope, G. et al. Widespread abiotic methane in chromitites. *Sci Rep* **8**, 8728 (2018).
12. Etiope, G. Methane origin in the Samail ophiolite: comment on "Modern water/rock reactions in Oman hyperalkaline peridotite aquifers and implications for microbial habitability". *Geochim Cosmochim Acta* **197**, 467–470 (2017).

13. Miller, H. M. et al. Modern water/rock reactions in Oman hyperalkaline peridotite aquifers and implications for microbial habitability. *Geochim Cosmochim Acta* **179**, 217–241 (2016).
14. McCollom, T. M. & Seewald, J. S. A reassessment of the potential for reduction of dissolved CO_2 to hydrocarbons during serpentinization of olivine. *Geochim Cosmochim Acta* **65**, 3769–3778 (2001).
15. McCollom, T. M. & Seewald, J. S. Experimental constraints on the hydrothermal reactivity of organic acids and acid anions: I. Formic acid and formate. *Geochim Cosmochim Acta* **67**, 3625–3644 (2003).
16. Lang, S. Q., Butterfield, D. A., Schulte, M., Kelley, D. S. & Lilley, M. D. Elevated concentrations of formate, acetate and dissolved organic carbon found at the Lost City hydrothermal field. *Geochim Cosmochim Acta* **74**, 941–952 (2010).
17. Holm, N. G. & Charlou, J. L. Initial indications of abiotic formation of hydrocarbons in the Rainbow ultramafic hydrothermal system, Mid-Atlantic Ridge. *Earth Planet Sci Lett* **191**, 1–8 (2001).
18. Simoneit, B. R. T., Lein, A. Y., Peresypkin, V. I. & Osipov, G. A. Composition and origin of hydrothermal petroleum and associated lipids in the sulfide deposits of the Rainbow field (Mid-Atlantic Ridge at 36°N). *Geochim Cosmochim Acta* **68**, 2275–2294 (2004).
19. Konn, C. et al. Hydrocarbons and oxidized organic compounds in hydrothermal fluids from Rainbow and Lost City ultramafic-hosted vents. *Chem Geol* **258**, 299–314 (2009).
20. Konn, C., Charlou, J. L., Holm, N. G. & Mousis, O. The production of methane, hydrogen, and organic compounds in ultramafic-hosted hydrothermal vents of the Mid-Atlantic Ridge. *Astrobiology* **15**, 381–399 (2015).
21. Reeves, E. P., McDermott, J. M. & Seewald, J. S. The origin of methanethiol in mid-ocean ridge hydrothermal fluids. *Proc Natl Acad Sci USA* **111**, 5474–5479 (2014).
22. McCollom, T. M., Seewald, J. S. & German, C. R. Investigation of extractable organic compounds in deep-sea hydrothermal vent fluids along the Mid-Atlantic Ridge. *Geochim Cosmochim Acta* **156**, 122–144 (2015).
23. Lang, S. Q. et al. Microbial utilization of abiogenic carbon and hydrogen in a serpentinite-hosted system. *Geochim Cosmochim Acta* **92**, 82–99 (2012).
24. Lang, S. Q., Frûh-Green, G. L., Bernasconi, S. M. & Butterfield, D. A. Sources of organic nitrogen at the serpentinite-hosted Lost City hydrothermal field. *Geobiology* **11**, 154–169 (2013).
25. Alt, J. C. et al. Uptake of carbon and sulfur during seafloor serpentinization and the effects of subduction metamorphism in Ligurian peridotites. *Chem Geol* **322–323**, 268–277 (2012).
26. Delacour, A., Früh-Green, G. L., Bernasconi, S. M., Schaeffer, P. & Kelley, D. S. Carbon geochemistry of serpentinites in the Lost City hydrothermal system (30°N, MAR). *Geochim Cosmochim Acta* **72**, 3681–3702 (2008).
27. Früh-Green, G. L., Connolly, J. A. D., Plas, A., Kelley, D. S. & Grobety, B. Serpentinization of oceanic peridotites: implications for geochemical cycles and biological activity. In: *The Subseafloor Biosphere at Mid-Ocean Ridges*, Vol. **144**, *Geophysical Monograph Series* (eds. W. S. D. Wilcock et al.), 119–136 (American Geophysical Union, 2004).
28. Schwarzenbach, E. M., Früh-Green, G. L., Bernasconi, S. M., Alt, J. C. & Plas, A. Serpentinization and carbon sequestration: a study of two ancient peridotite-hosted hydrothermal systems. *Chem Geol* **351**, 115–133 (2013).

29. Klein, F. et al. Fluid mixing and the deep biosphere of a fossil Lost City-type hydrothermal system at the Iberia Margin. *Proc Natl Acad Sci USA* **112**, 12036–12041 (2015).
30. Ménez, B., Pasini, V. & Brunelli, D. Life in the hydrated suboceanic mantle. *Nat Geosci* **5**, 133–137 (2012).
31. Pasini, V. et al. Low temperature hydrothermal oil and associated biological precursors in serpentinites from Mid-Ocean Ridge. *Lithos* **178**, 84–95 (2013).
32. Amend, J. P., LaRowe, D. E., McCollom, T. M. & Shock, E. L. The energetics of organic synthesis inside and outside the cell. *Philos Trans R Soc Lond B Biol Sci* **368**, 20120255 (2013).
33. Milesi, V., McCollom, T. M. & Guyot, F. Thermodynamic constraints on the formation of condensed carbon from serpentinization fluids. *Geochim Cosmochim Acta* **189**, 391–403 (2016).
34. Shipp, J. et al. Organic functional group transformations in water at elevated temperature and pressure: reversibility, reactivity, and mechanisms. *Geochim Cosmochim Acta* **104**, 194–209 (2013).
35. Shock, E. L. Geochemical constraints on the origin of organic compounds in hydrothermal systems. *Origins Life Evol Biosphere* **20**, 331–367 (1990).
36. Rumble, D. Hydrothermal graphitic carbon. *Elements* **10**, 427–433 (2014).
37. Zubkov, V. S. Tendencies in the distribution and hypotheses of the genesis of condensed naphthides in magmatic rocks from various geodynamic environments. *Geochem Int* **47**, 741–757 (2009).
38. Pikovskii, Y. I., Chernova, T. G., Alekseeva, T. A. & Verkhovskaya, Z. I. Composition and nature of hydrocarbons in modern serpentinization areas in the ocean. *Geochem Int* **42**, 971–976 (2004).
39. Echigo, T. & Kimata, M. Crystal chemistry and genesis of organic minerals: a review of oxalate and polycyclic aromatic hydrocarbon minerals. *Can Mineral* **48**, 1329–1357 (2010).
40. Stein, S. On the high temperature chemical equilibria of polycyclic aromatic hydrocarbons. *J Phys Chem* **82**, 566–571 (1978).
41. Murdoch, J. Pendletonite, a new hydrocarbon mineral from California. *Am Mineral* **52**, 611–616 (1967).
42. Echigo, T., Kimata, M. & Maruoka, T. Crystal-chemical and carbon-isotopic characteristics of karpatite ($C_{24}H_{12}$) from the Picacho Peak Area, San Benito County, California: evidences for the hydrothermal formation. *Am Mineral* **92**, 1262–1269 (2007).
43. Simoneit, B. R. T. & Lonsdale, P. F. Hydrothermal petroleum in mineralized mounds at the seabed of Guaymas Basin. *Nature* **295**, 198 (1982).
44. Zolotov, M. & Shock, E. Abiotic synthesis of polycyclic aromatic hydrocarbons on Mars. *J Geophys Res* **104**, 14033–14049 (1999).
45. Zolotov, M. Y. & Shock, E. L. A thermodynamic assessment of the potential synthesis of condensed hydrocarbons during cooling and dilution of volcanic gases. *J Geophys Res* **105**, 539–559 (2000).
46. Potticary, J., Jensen, T. T. & Hall, S. R. Nanostructural origin of blue fluorescence in the mineral karpatite. *Sci Rep* **7**, 9867 (2017).
47. Ciliberto, E. et al. Aliphatic hydrocarbons in metasomatized gabbroic xenoliths from Hyblean diatremes (Sicily): genesis in a serpentinite hydrothermal system. *Chem Geol* **258**, 258–268 (2009).

48. Piotrovskii, G. L. Karpatite (carpathite) – a new organic mineral from Transcarpathia. *Lvov Geol Obs Miner Sb* **9**, 120–127 (1955).
49. Strunz, H. & Contag, B. Evenkite, flagstaffite, idrialite, and refikite. *Neu Jb Mineral, Mh* **1**, 19–25 (1965).
50. Rost, R. The minerals in the burning shafts at Kladno. *Ceská Ákad Rozpravky* **11**, 1–19 (1937).
51. Foresti, E. & Riva di Sanseverino, L. X-ray crystal and molecular structure of an organic mineral: simonellite, $C_{19}H_{24}$. *Atti Accad Naz Lincei* **47**, 41–54 (1969).
52. Skropyshev, A. V. A paraffin in a polymetallic vein. *Dokl Akad Nauk SSSR* **88**, 717–719 (1953).
53. Bassez, M.-P., Takano, Y. & Ohkouchi, N. Organic analysis of peridotite rocks from the Ashadze and Logatchev hydrothermal sites. *Int J Mol Sci* **10**, 2986–2998 (2009).
54. Mateeva, T. et al. Preserved organic matter in a fossil ocean continent transition in the Alps: the example of Totalp, SE Switzerland. *Swiss J Geosci* **110**, 457–478 (2017).
55. Sforna, M. C. et al. Abiotic formation of condensed carbonaceous matter in the hydrating oceanic crust. *Nat Commun* **9**, 5049 (2018).
56. Ménez, B. et al. Mineralizations and transition metal mobility driven by organic carbon during low-temperature serpentinization. *Lithos* **323**, 262–276 (2018).
57. Scribano, V., Sapienza, G., Braga, R. & Morten, L. Gabbroic xenoliths in tuff–breccia pipes from the Hyblean Plateau: insights into the nature and composition of the lower crust underneath South-eastern Sicily, Italy. *Miner Petrol* **86**, 63–88 (2006).
58. Scirè, S. et al. Asphaltene-bearing mantle xenoliths from Hyblean diatremes, Sicily. *Lithos* **125**, 956–968 (2011).
59. Mathez, E. A. Carbonaceous matter in mantle xenoliths: composition and relevance to the isotopes. *Geochim Cosmochim Acta* **51**, 2339–2347 (1987).
60. Mathez, E. A. & Delaney, J. R. The nature and distribution of carbon in submarine basalts and peridotite nodules. *Earth Planet Sci Lett* **56**, 217–232 (1981).
61. Mathez, E. A., Dietrich, V. J. & Irving, A. J. The geochemistry of carbon in mantle peridotites. *Geochim Cosmochim Acta* **48**, 1849–1859 (1984).
62. Tingle, T. N., Mathez, E. A. & Michael, F. H. Carbonaceous matter in peridotites and basalts studied by XPS, SALI, and LEED. *Geochim Cosmochim Acta* **55**, 1345–1352 (1991).
63. Muenow, D. W. High temperature mass spectrometric gas-release studies of Hawaiian volcanic glass: Pele's tears. *Geochim Cosmochim Acta* **37**, 1551–1561 (1973).
64. Andreani, M., Munoz, M., Marcaillou, C. & Delacour, A. μXANES study of iron redox state in serpentine during oceanic serpentinization. *Lithos* **178**, 70–83 (2013).
65. Klein, F. et al. Magnetite in seafloor serpentinite – some like it hot. *Geology* **42**, 135–138 (2014).
66. Plümper, O., Beinlich, A., Bach, W., Janots, E. & Austrheim, H. Garnets within geode-like serpentinite veins: implications for element transport, hydrogen production and life-supporting environment formation. *Geochim Cosmochim Acta* **141**, 454–471 (2014).
67. Mayhew, L. E., Ellison, E. T., McCollom, T. M., Trainor, T. P. & Templeton, A. S. Hydrogen generation from low-temperature water-rock reactions. *Nat Geosci* **6**, 478–484 (2013).

68. Manuella, F. C., Carbone, S. & Barreca, G. Origin of saponite-rich clays in a fossil serpentinite-hosted hydrothermal system in the crustal basement of the Hyblean Plateau (Sicily, Italy). *Clays Clay Miner* **60**, 18–31 (2012).
69. Pizzarello, S., Cooper, G. W. & Flynn, G. J. The nature and distribution of the organic material in carbonaceous chondrites and interplanetary dust particles. In: *Meteorites and the Early Solar System II* (eds. D. S. Lauretta & H. Y. McSween Jr.), 625–651 (University of Arizona Press, 2006).
70. Le Guillou, C. & Brearley, A. Relationships between organics, water and early stages of aqueous alteration in the pristine CR3.0 chondrite MET 00426. *Geochim Cosmochim Acta* **131**, 344–367 (2014).
71. Pearson, V. K. et al. Clay mineral–organic matter relationships in the early solar system. *Meteorit Planet Sci* **37**, 1829–1833 (2002).
72. Zega, T. J. et al. Mineral associations and character of isotopically anomalous organic material in the Tagish Lake carbonaceous chondrite. *Geochim Cosmochim Acta* **74**, 5966–5983 (2010).
73. Pisapia, C., Jamme, F., Duponchel, L. & Ménez, B. Tracking hidden organic carbon in rocks using chemometrics and hyperspectral imaging. *Sci Rep* **8**, 2396 (2018).
74. Kelley, D. S. & Früh-Green, G. L. Volatile lines of descent in submarine plutonic environments: insights from stable isotope and fluid inclusion analyses. *Geochim Cosmochim Acta* **65**, 3325–3346 (2001).
75. Sachan, H. K., Mukherjee, B. K. & Bodnar, R. J. Preservation of methane generated during serpentinization of upper mantle rocks: evidence from fluid inclusions in the Nidar ophiolite, Indus Suture Zone, Ladakh (India). *Earth Planet Sci Lett* **257**, 47–59 (2007).
76. Vanko, D. A. & Stakes, D. S. Fluids in oceanic layer 3: evidence from veined rocks, Hole 735B, Southwest Indian Ridge. In: *Proceeding of the Ocean Drilling Program*, Vol. **118** (eds. R. P. V. Herzen, R. P. Fox, A. Palmer & P. T. Robinson), 181–215 (Ocean Drilling Program, 1997).
77. Kelley, D. S. Fluid evolution in slow-spreading environments. In: *Proceeding of the Ocean Drilling Program*, Vol. **153** (eds. J. A. Karson, M. Cannat, D. J. Miller & D. Elthon), 399–415 (Ocean Drilling Program, 1997).
78. Bortnikov, N. S. et al. The Rainbow serpentinite-related hydrothermal field, Mid-Atlantic Ridge, 36°14'N: mineralogical and geochemical features. In: *Mineral Deposits at the Beginning of the 21st Century* (eds. A. Piestrzyhski et al.), 265–268 (Swets & Zeitlinger Publishers Lisse, 2001).
79. Elthon, D. Petrology of gabbroic rocks from the Mid-Cayman Rise Spreading Center. *J Geophys Res* **92**, 658–682 (1987).
80. Nakamura, K. et al. Serpentinized troctolites exposed near the Kairei hydrothermal field, Central Indian Ridge: insights into the origin of the Kairei hydrothermal fluid supporting a unique microbial ecosystem. *Earth Planet Sci Lett* **280**, 128–136 (2009).
81. Katayama, I., Kurosaki, I. & Hirauchi, K.-I. Low silica activity for hydrogen generation during serpentinization: an example of natural serpentinites in the Mineoka ophiolite complex, central Japan. *Earth Planet Sci Lett* **298**, 199–204 (2010).
82. Miura, M., Arai, S. & Mizukami, T. Raman spectroscopy of hydrous inclusions in olivine and orthopyroxene in ophiolitic harzburgite: implications for elementary processes in serpentinization. *J Miner Petrol Sci* **106**, 91–96 (2011).

83. Kelley, D. S., Gillis, K. M. & Thompson, G. Fluid evolution in submarine magma–hydrothermal systems at the Mid-Atlantic Ridge. *J Geophys Res* **98**, 19579–19596 (1993).
84. Holloway, J. R. & O'Day, P. A. Production of CO_2 and H_2 by diking-eruptive events at mid-ocean ridges: implications for abiotic organic synthesis and global geochemical cycling. *Int Geol Rev* **42**, 673–683 (2000).
85. Luque, F. J. et al. Deposition of highly crystalline graphite from moderate-temperature fluids. *Geology* **37**, 275–278 (2009).
86. Pasteris, J. D. Occurrence of graphite in serpentinized olivines in kimberlite. *Geology* **9**, 356–359 (1981).
87. Pasteris, J. D. & Chou, I. M. Fluid-deposited graphitic inclusions in quartz: comparison between KTB (German Continental Deep-Drilling) core samples and artificially reequilibrated natural inclusions,. *Geochim Cosmochim Acta* **62**, 109–122 (1998).
88. Satish-Kumar, M. Graphite-bearing CO_2–fluid inclusions in granulites: insights on graphite precipitation and carbon isotope evolution. *Geochim Cosmochim Acta* **69**, 3841–3856 (2005).
89. Wopenka, B. & Pasteris, J. D. Structural characterization of kerogens to granulite-facies graphite: applicability of Raman microprobe spectroscopy. *Am Mineral* **78**, 533–557 (1993).
90. Beny-Bassez, C. & Rouzaud, J. N. Characterization of carbonaceous materials by correlated electron and optical microscopy and Raman microscopy. In: *Scanning Electron Microscopy*, 119–132 (SEM Inc., 1985).
91. Beyssac, O. & Rumble, D. Graphitic carbon: a ubiquitous, diverse, and useful geomaterial. *Elements* **10**, 415–420 (2014).
92. Shipp, J., Gould, I. R., Shock, E. L., Williams, L. B. & Hartnett, H. E. Sphalerite is a geochemical catalyst for carbon–hydrogen bond activation. *Proc Natl Acad Sci USA* **111**, 11642–11645 (2014).
93. Venturi, S. et al. Mineral-assisted production of benzene under hydrothermal conditions: insights from experimental studies on C_6 cyclic hydrocarbons. *J Volcanol Geothermal Res* **346**, 21–27 (2017).
94. Yang, Z., Gould, I. R., Williams, L. B., Hartnett, H. E. & Shock, E. L. Effects of iron-containing minerals on hydrothermal reactions of ketones. *Geochim Cosmochim Acta* **223**, 107–126 (2018).
95. Voglesonger, K. M., Holloway, J. R., Dunn, E. E., Dalla-Betta, P. J. & O'Day, P. A. Experimental abiotic synthesis of methanol in seafloor hydrothermal systems during diking events. *Chem Geol* **180**, 129–139 (2001).
96. McCollom, T. M. & Seewald, J. S. Abiotic formation of hydrocarbons and oxygenated compounds during thermal decomposition of iron oxalate. *Origins Life Evol Biosphere* **29**, 167–186 (1999).
97. Chen, Y., Li, C. W. & Kanan, M. W. Aqueous CO_2 reduction at very low overpotential on oxide-derived Au nanoparticles. *J Am Chem Soc* **134**, 19969–19972 (2012).
98. Liu, M. et al. Enhanced electrocatalytic CO_2 reduction via field-induced reagent concentration. *Nature* **537**, 382–386 (2016).
99. Marcaillou, C., Munoz, M., Vidal, O., Parra, T. & Harfouche, M. Mineralogical evidence for H_2 degassing during serpentinization at 300°C/300 bar. *Earth Planet Sci Lett* **303**, 281–290 (2011).
100. Chou, I. M. Permeability of precious metals to hydrogen at 2kb total pressure and elevated temperatures. *Am J Sci* **286**, 638–658 (1986).

101. Seewald, J. S., Zolotov, M. Y. & McCollom, T. M. Experimental investigation of single carbon compounds under hydrothermal conditions. *Geochim Cosmochim Acta* **70**, 446–460 (2006).
102. Fouquet, Y. et al. Geodiversity of hydrothermal processes along the Mid-Atlantic ridge and ultramafic-hosted mineralization: a new type of oceanic Cu–Zn–Co–Au volcanogenic massive sulfide deposit. In: *Diversity of Hydrothermal Systems on Slow Spreading Ocean Ridges*, Vol. **188**, *Geophysical Monograph Series* (eds. P. Rona, C. Devey, J. Dyment & B. Murton), 321–367 (American Geophysical Union, 2010).
103. Williams, L. B., Canfield, B., Voglesonger, K. M. & Holloway, J. R. Organic molecules formed in a "primordial womb". *Geology* **33**, 913–916 (2005).
104. Horita, J. & Berndt, M. E. Abiogenic methane formation and isotopic fractionation under hydrothermal conditions. *Science* **285**, 1055–1057 (1999).
105. Ji, F., Zhou, H. & Yang, Q. The abiotic formation of hydrocarbons from dissolved CO_2 under hydrothermal conditions with cobalt-bearing magnetite. *Origins Life Evol Biosphere* **38**, 117–125 (2008).
106. McCollom, T. M. Abiotic methane formation during experimental serpentinization of olivine. *Proc Natl Acad Sci USA* **113**, 13965–13970 (2016).
107. Oze, C., Jones, L. C., Goldsmith, J. I. & Rosenbauer, R. J. Differentiating biotic from abiotic methane genesis in hydrothermally active planetary surfaces. *Proc Natl Acad Sci USA* **109**, 9750–9754 (2012).
108. Lang, S. Q. et al. H_2/CH_4 ratios cannot reliably distinguish abiotic vs. biotic methane in natural hydrothermal systems. *Proc Natl Acad Sci USA* **109**, E3210–E3210 (2012).
109. Huang, R. et al. The H_2/CH_4 ratio during serpentinization cannot reliably identify biological signatures. *Sci Rep* **6**, 33821 (2016).
110. Lazar, C., Cody, G. D. & Davis, J. M. A kinetic pressure effect on the experimental abiotic reduction of aqueous CO_2 to methane from 1 to 3.5 kbar at 300°C. *Geochim Cosmochim Acta* **151**, 34–48 (2015).
111. Hellevang, H., Huang, S. & Thorseth, I. H. The potential for low-temperature abiotic hydrogen generation and a hydrogen-driven deep biosphere. *Astrobiology* **11**, 711–724 (2011).
112. Neubeck, A., Duc, N. T., Bastviken, D., Crill, P. & Holm, N. G. Formation of H_2 and CH_4 by weathering of olivine at temperatures between 30 and 70°C. *Geochem Trans* **12**, 6 (2011).
113. Neubeck, A. et al. Olivine alteration and H_2 production in carbonate-rich, low temperature aqueous environments. *Planet Space Sci* **96**, 51–61 (2014).
114. Neubeck, A., Nguyen, D. T. & Etiope, G. Low-temperature dunite hydration: evaluating CH_4 and H_2 production from H_2O and CO_2. *Geofluids* **16**, 408–420 (2016).
115. McCollom, T. M. & Donaldson, C. Generation of hydrogen and methane during experimental low-temperature reaction of ultramafic rocks with water. *Astrobiology* **16**, 389–406 (2016).
116. Berndt, M. E., Allen, D. E. & Seyfried, W. E., Jr. Reduction of CO_2 during serpentinization of olivine at 300°C and 500 bar. *Geology* **24**, 351–354 (1996).
117. Foustoukos, D. I. & Seyfried, W. E. Hydrocarbons in hydrothermal vent fluids: the role of chromium-bearing catalysts. *Science* **304**, 1002–1005 (2004).
118. Fu, Q., Sherwood Lollar, B., Horita, J., Lacrampe-Couloume, G. & Seyfried, W. E. Abiotic formation of hydrocarbons under hydrothermal conditions: constraints from chemical and isotope data. *Geochim Cosmochim Acta* **71**, 1982–1998 (2007).

119. Lazar, C., McCollom, T. M. & Manning, C. E. Abiogenic methanogenesis during experimental komatiite serpentinization: implications for the evolution of the early Precambrian atmosphere. *Chem Geol* **326–327**, 102–112 (2012).
120. Etiope, G. & Ionescu, A. Low-temperature catalytic CO_2 hydrogenation with geological quantities of ruthenium: a possible abiotic CH_4 source in chromitite-rich serpentinized rocks. *Geofluids* **15**, 438–452 (2015).
121. Lu, Q. et al. A selective and efficient electrocatalyst for carbon dioxide reduction. *Nat Commun* **5**, 3242 (2014).
122. Rushdi, A. I. & Simoneit, B. R. T. Lipid Formation by aqueous Fischer–Tropsch-type synthesis over a temperature range of 100C to 400C. *Origins Life Evol Biosphere* **31**, 103–118 (2001).
123. Tingle, T. N. & Hochella, M. F. Formation of reduced carbonaceous matter in basalts and xenoliths: reaction of C–O–H gases on olivine crack surfaces. *Geochim Cosmochim Acta* **57**, 3245–3249 (1993).
124. McCollom, T. M. Formation of meteorite hydrocarbons from thermal decomposition of siderite ($FeCO_3$). *Geochim Cosmochim Acta* **67**, 311–317 (2003).
125. Dufaud, F., Martinez, I. & Shilobreeva, S. Experimental study of Mg-rich silicates carbonation at 400 and 500 °C and 1 kbar. *Chem Geol* **265**, 79–87 (2009).
126. Luquot, L., Andreani, M., Gouze, P. & Camps, P. CO_2 percolation experiment through chlorite/zeolite-rich sandstone (Pretty Hill Formation – Otway Basin – Australia). *Chem Geol* **294–295**, 75–88 (2012).
127. Milesi, V. et al. Formation of CO_2, H_2 and condensed carbon from siderite dissolution in the 200–300°C range and at 50 MPa. *Geochim Cosmochim Acta* **154**, 201–211 (2015).
128. Ferry, J. M. & Baumgartner, L. Thermodynamic models of molecular fluids at the elevated pressures and temperatures of crustal metamorphism. In: *Thermodynamic Modeling of Geological Materials: Minerals, Fluids and Melts. Reviews in Mineralogy and Geochemistry*, Vol. **17** (eds. I. S. E. Carmichael & H. P. Eugster), 323–365 (Mineralogical Society of America, 1987).
129. Holloway, J. R. Graphite–CH_4–H_2O–CO_2 equilibria at low-grade metamorphic conditions. *Geology* **12**, 455–458 (1984).
130. Frost, B. R. Mineral equilibria involving mixed-volatiles in a C–O–H fluid phase; the stabilities of graphite and siderite. *Am J Sci* **279**, 1033–1059 (1979).
131. McCollom, T. M. The influence of minerals on decomposition of the *n*-alkyl-α-amino acid norvaline under hydrothermal conditions. *Geochim Cosmochim Acta* **104**, 330–357 (2013).
132. Shock, E. L. et al. Thermodynamics of organic transformations in hydrothermal fluids. *Rev Mineral Geochem* **76**, 311–350 (2013).
133. Andreani, M., Mével, C., Boullier, A. M. & Escartín, J. Dynamic control on serpentine crystallization in veins: constraints on hydration processes in oceanic peridotites. *Geochem Geophys Geosyst* **8**, Q02012 (2007).
134. Lang, S. Q. et al. Deeply-sourced formate fuels sulfate reducers but not methanogens at Lost City hydrothermal field. *Sci Rep* **8**, 755 (2018).
135. Mason, O. U. et al. First investigation of the microbiology of the deepest layer of ocean crust. *PLoS One* **5**, e15399 (2010).
136. Pisapia, C. et al. Mineralizing filamentous bacteria from the Prony bay hydrothermal field give new insights into the functioning of serpentinization-based subseafloor ecosystems. *Front Microbiol* **8**, 57 (2017).

137. Trias, R. et al. High reactivity of deep biota under anthropogenic CO_2 injection into basalt. *Nat Commun* **8**, 1063 (2017).
138. Malvoisin, B., Brunet, F., Carlut, J., Rouméjon, S. & Cannat, M. Serpentinization of oceanic peridotites: 2. Kinetics and processes of San Carlos olivine hydrothermal alteration. *J Geophys Res* **117**, B04102 (2012).
139. Pens, M., Andreani, M., Daniel, I., Perrillat, J.-P. & Cardon, H. Contrasted effect of aluminum on the serpentinization rate of olivine and orthopyroxene under hydrothermal conditions. *Chem Geol* **441**, 256–264 (2016).
140. Debret, B., Andreani, M., Delacour, A., Rouméjon, S. & Trcera, N. Assessing sulfur redox state and distribution in abyssal serpentinites using XANES spectroscopy. *Earth Planet Sci Lett* **466**, 1–11 (2017).
141. Klein, F. & Bach, W. Fe–Ni–Co–O–S phase relations in peridotite–seawater interactions. *J Petrol* **50**, 37–59 (2009).
142. Andreani, M. et al. Tectonic structure, lithology, and hydrothermal signature of the Rainbow massif (Mid-Atlantic Ridge 36°14'N). *Geochem Geophys* **15**, 3543–3571 (2014).
143. McCollom, T. M. & Bach, W. Thermodynamic constraints on hydrogen generation during serpentinization of ultramafic rocks. *Geochim Cosmochim Acta* **73**, 856–875 (2009).
144. Takai, K. et al. Cell proliferation at 122°C and isotopically heavy CH_4 production by a hyperthermophilic methanogen under high-pressure cultivation. *Proc Natl Acad Sci USA* **105**, 10949–10954 (2008).
145. Steele, A., McCubbin, F. M. & Fries, M. D. The provenance, formation, and implications of reduced carbon phases in Martian meteorites. *Meteorit Planet Sci* **51**, 2203–2225 (2016).

16

Carbon in the Deep Biosphere

Forms, Fates, and Biogeochemical Cycling

SUSAN Q. LANG, MAGDALENA R. OSBURN, AND ANDREW D. STEEN

16.1 Introduction

The form, fate, and biogeochemical cycling of carbon in subsurface environments impacts and reflects microbial activity and has important implications for global elemental fluxes. Photosynthetically derived organic matter (OM) is transported to a depth where it can continue to fuel life far from solar inputs. Alternative energy-yielding reactions such as the oxidation of minerals and reduced gases can fuel life in the rocky subsurface of both the ocean and continents, altering the distribution and characteristics of carbon compounds. Nonbiological reactions such as the precipitation of calcium carbonate influence the availability of dissolved inorganic carbon for lithoautotrophs and, simultaneously, the carbon cycle over geologic time. The abundances, characteristics, and distributions of carbon in the subsurface can therefore provide an integrated history of biotic and abiotic processes and a template for interpreting similar patterns from other planetary bodies.

The goal of this chapter is to compile insights from disparate environments in order to build a mechanistic understanding of the controls on carbon abundance and distribution in the subsurface. The sections below summarize what is known from the oceanic and continental subsurface, realms that are often studied separately. We synthesize commonalities across these environments, highlight what remains unknown, and propose ideas for future directions.

One challenge with working across the marine–continental divide is that the terminology used to describe organic carbon varies between the two. We will use the following terms and abbreviations: particulate organic carbon (POC), dissolved organic carbon (DOC), and dissolved inorganic carbon (DIC). Another discrepancy between communities is in the use of units, with ppm or mg/L dominating the continental literature and μM or mM in the marine literature. We will use molar units throughout for comparison's sake. Finally, while the soil community has moved away from the terms "refractory" and "recalcitrant" OM, they are still common in the marine community. Here, these terms refer to OM that has escaped remineralization due to its inherent molecular structure, physical associations with minerals, energetically unfavorable conditions, or the lack of a specific microbial community adapted to carry out the necessary degradative processes.

16.2 Oceanic Sedimentary Subsurface

Approximately 1.68×10^{14} g of organic carbon per year are buried in marine and estuarine sediments (1). Burial of organic carbon in sediments represents a transfer of reducing equivalents from Earth's surface to the subsurface, thereby allowing persistence of oxidized compounds such as O_2 at the surface (2). The rate of burial of organic carbon in marine sediments therefore has an important influence on the redox state, and thus habitability, of Earth's surface.

Broadly, marine sediments can be divided into river and estuarine delta systems, continental shelves and slopes, and abyssal plains (Figure 16.1). Sediments may be more finely divided into provinces based on microbial community composition, grain size, OM content, and benthic communities, among other variables (3).

The oxidation of organic carbon in sediments is carried out by a series of heterotrophic organisms. Macrofauna have their greatest influence on the surface sediments of continental shelfs, while the role of meiofauna and microorganisms increases with depth and where oxygen is limited (4). Remineralization within anoxic sediments is dominated by microorganisms and is most prevalent at temperatures below ~80°C, constituting ~75% of Earth's total sediment volume of 3.01×10^8 km^3 (5). The composition, abundance, and activity of heterotrophic microorganisms in marine sediments therefore has a strong influence on the burial rate and chemical nature of organic carbon. While these reactions are catalyzed by enzymes, they are ultimately controlled by thermodynamics. This section will briefly review the chemical and biological factors that regulate organic carbon oxidation and burial rates, as well as some of the models that can be constructed to describe and predict those rates.

The burial rate of organic carbon in marine sediments is controlled by a range of biological and geological processes, including sedimentation rate, primary productivity, biological activity, sediment organic carbon content, chemical and physical form of organic molecules, and concentrations of oxidants (electron acceptors), as described below and in several reviews and syntheses (6–13). These factors are interrelated: rapid sedimentation rates influence the quality of OM delivered to the sediment surface, which in turns affects oxidation rates, oxygen exposure time (OET), quantity and composition of heterotrophic microbial communities, and concentrations of potential electron acceptors.

16.2.1 Chemical Composition

OM is delivered to marine sediments from marine sources such as sinking plankton and consumers and from terrigenous sources such as plant litter and soil OM. The chemical composition of fresh biomass is relatively well constrained and consists predominately of carbohydrates, proteins, and lipids. The composition of terrestrial material transferred by fluvial or aeolian processes ranges from fresh biomass to highly degraded and altered material. Lignin phenols synthesized solely by vascular plants have long been used to track

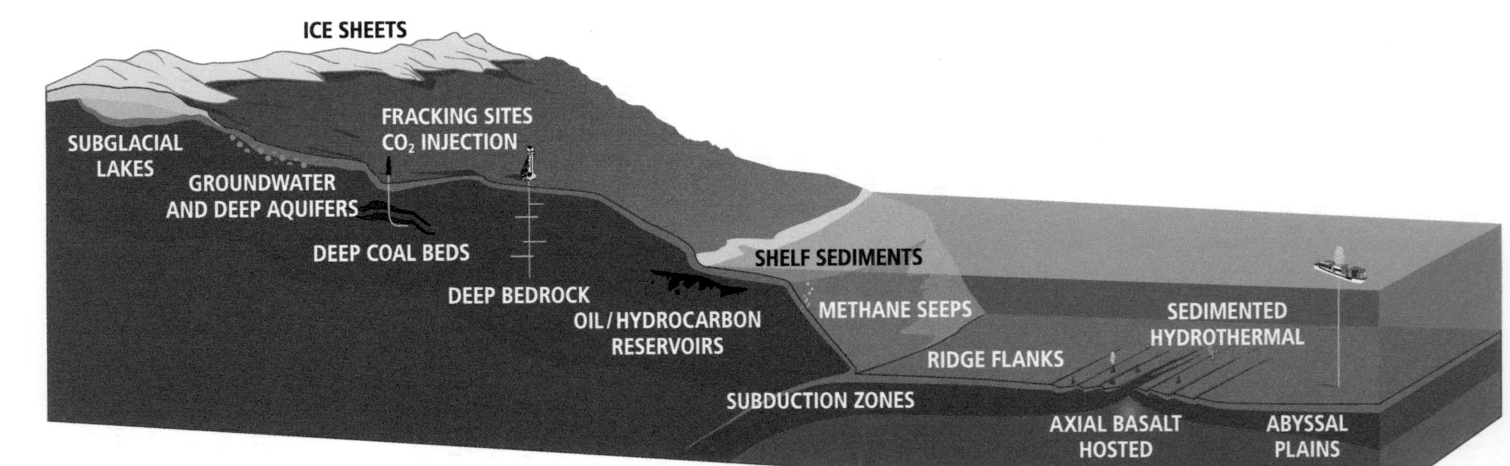

Figure 16.1 Deep biosphere locations on the continents and in the ocean.

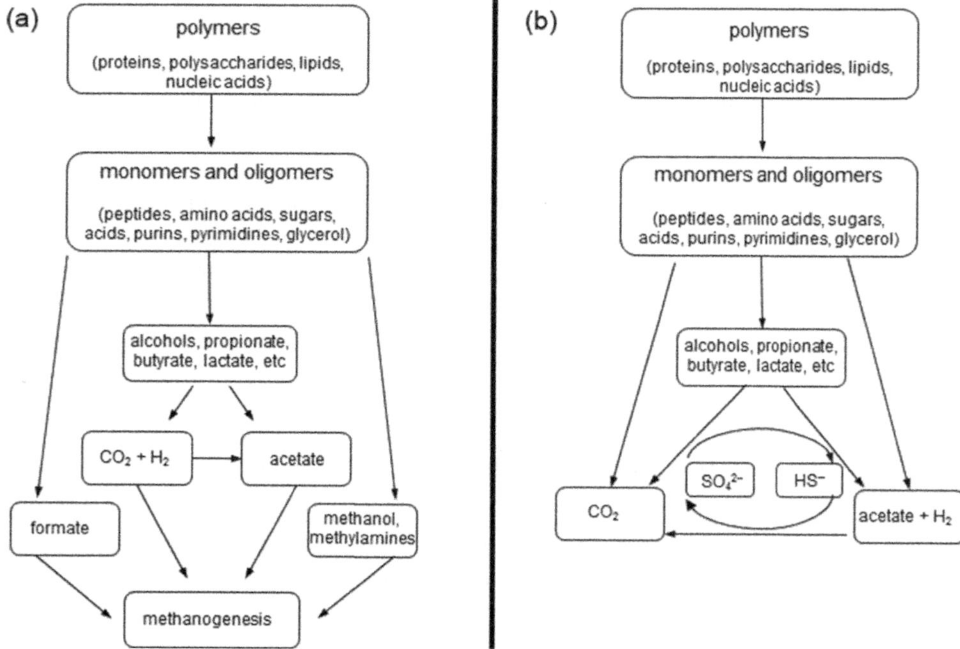

Figure 16.2 Anaerobic breakdown of OM by microorganisms via (a) methanogenesis and (b) sulfidogenesis.
Adapted from (25).

terrestrial inputs into the ocean (14). Ancient and recycled petrogenic carbon can also be remobilized from the weathering of sedimentary rocks (15). This suite of compounds is subject to biotic and abiotic alteration en route to marine deposition, which further diversifies the range of organic compounds present. Physical processes within the catchment are a major control on the composition and reactivity of OM delivered to the ocean by rivers (16), with larger inputs of both recently synthesized and ancient petrogenic organic carbon delivered in regions of higher erosional rates such as small mountainous streams (10,17–19) and some Arctic rivers (20–23).

Within the sediments, heterotrophic organisms and abiotic processes such as condensation reactions or sulfurization can alter the chemical structures of OM. In general terms, the heterotrophic remineralizaiton of larger organic molecules under anoxic conditions proceeds by the breakdown of polymers into monomers and oligomers, followed by smaller alcohols and organic acids, and finally methane and CO_2 (Figure 16.2) (24,25). As a result, small organic molecules such as acetate, ethane, propane, and methane build up in the porewaters of anaerobic sediments, with additional contributions from acetogenesis and hydrogenotrophic methanogenesis (24,26–29).

Ultimately, the vast majority of OM produced in the upper water column is respired, with only 1% of gross primary production escaping remineralization to be buried in the deep sediments (6–8,30). Some molecules survive due to chemical structures that are inherently recalcitrant, a process called "selective preservation." The role of this pathway is disputed. While some compounds such as highly cross-linked macromolecules are inherently less bioavailable than others, microorganisms are capable of metabolizing even ancient and highly altered OM relatively quickly under favorable conditions. Molecules may become less bioavailable due to nonbiological alteration. Abiotic sulfurization of organic molecules, deamination of peptides, and condensation of nitrogen-containing heterocyclic molecules all appear to promote the preservation of organics in sediments (7,31). Random recombination of molecules or the production of altered metabolites by heterotrophic microorganisms can also rapidly convert labile organic carbon into far less reactive material (32,33). Temperature can promote some of these transformations, as discussed in detail in Section 16.4.

16.2.2 Bulk Controls on OM Preservation

In locations with relatively rapid sediment accumulation rates such as river deltas and continental shelf sediments, greater OM preservation is most closely associated with higher mineral surface areas and shorter OETs (7,34). These factors have proven broadly predictive of organic carbon distributions, although they do little to reveal the underlying mechanisms of preservation, nor do they allow for predictions of future responses to changing environmental conditions.

A prevailing paradigm is that microorganisms access POC only after it has been solubilized into DOC (35). Organic molecules enter cells via general uptake porins, which can only accommodate molecules in the size range of 600–1000 Da (36). Organic molecules in seawater, sediment porewater, and soils that are larger than 1000 Da are, however, more bioavailable than small molecules on average (37–40), apparently because smaller molecules tend to be more extensively modified than larger molecules (41). Therefore, microbial extracellular enzymes appear necessary for the uptake and utilization of the most bioavailable organic carbon in sediments. Consistent with this paradigm, extracellular enzyme activity has been observed in deep, old sediments, including 217,000-year-old Mediterranean sapropels (42,43) and Baltic Sea sediments that are up to 10,000 years old (44).

Several findings complicate the view that extracellular enzymes catalyze the rate-limiting step in biological organic carbon oxidation. Extracellular enzyme activities can outstrip the ability of sediment microbes to take up hydrolysate on timescales of days to years, leading to accumulations of apparently bioavailable low-molecular-weight DOM (45). Further, cells do not exclusively take up organic compounds via general uptake porins. Active transporters, for instance, use energy gradients to pass specific molecules through the cell membrane. These can be extremely large: for instance, certain TonB-dependent transporters can import intact proteins up to 69 kDa (46). Additionally, in seawater and in cow rumen, some cells are able to take up larger oligosaccharides into their periplasm, store them over extended periods, and then metabolize them when conditions are right (47,48). The extent to which these mechanisms are important in sediments is not known, but temporal decoupling between macromolecule hydrolysis and metabolism could have implications for the dynamics of sediment OM oxidation.

16.2.3 Sorption

It has been observed for nearly 40 years that the volume-specific quantity of mineral surface area in sediments is correlated with organic carbon content (49). The mechanism underlying this relationship is not precisely understood. Sediments tend to accumulate quantities of organic carbon that are roughly equivalent to the amount that would be required to cover minerals in an organic monolayer (50,51). Sedimentary organic carbon, however, exists in discrete "blebs" (Figure 16.3), so the fact that the average quantity of OM per unit of mineral surface area is roughly monolayer equivalent appears to be essentially coincidental (52).

Several mechanisms appear responsible for the protection of OM by mineral surfaces. First, OM may be occluded between mineral grains, within minerals themselves, or even within a matrix of more recalcitrant sorbed organic compounds (53,54). Encased OM represents a sterile microenvironment in which biological oxidation is impossible. Second, even when sorbed OM is physically accessible to microorganisms, sorption slows or halts the diffusion of organic compounds to cell membranes (55). Finally, sorption distorts the physical structure of extracellular enzymes, preventing them from functioning normally, while simultaneously protecting enzymes from degradation and thereby substantially extending their active lifetimes (56,57). Associations with iron oxides, which include

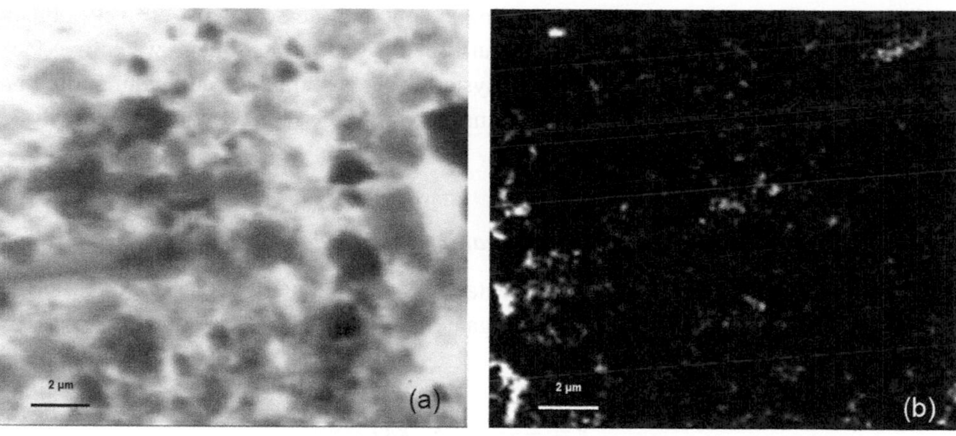

Figure 16.3 (a) Scanning transmission X-ray microscope image and (b) optical density map of the organic carbon distribution of sediments from 1.75 m below seafloor at Integrated Ocean Drilling Program Site 1231 Hole B, Peru Basin. The optical density map was generated by subtracting a pre-edge X-ray image from a post-edge X-ray image; brighter pixels correspond to higher concentrations of organic carbon. OM associated with particles is not distributed evenly over the surface.
Image courtesy of Dr. E. Estes, University of Delaware.

chelation, coprecipitation, and noncovalent bonding to oxide surfaces, accounts for an average of 20% of organic carbon in sediments (58,59). A full understanding of the mechanisms of microbial OM oxidation in sediments requires consideration of both the interactions between organic carbon and sediment minerals and the effects of mineral surfaces on the metabolisms of microorganisms.

16.2.4 Oxygen Exposure Time

Typical marine sediments underlying oxygenated seawater contain oxic porewater near the sediment–water interface, which becomes anoxic with increasing depth due to heterotrophic organic carbon oxidation. The depth of the oxic layer can vary dramatically, from millimeters or less in rapidly accumulating, organic-rich sediments to meters in ocean gyres. The presence of oxygen enhances the remineralization of organic molecules (60–62), and the term "oxygen exposure time" was coined to quantify the average time that sedimentary OM is exposed to "oxic" conditions, which can range from days to thousands of years (8,34,62,63). Organic carbon oxidation is substantially faster in oxic sediments than anoxic sediments because the greater free energy of reaction of organic carbon with oxygen allows for a denser microbial community capable of catalyzing faster oxidation and because specific reactions (e.g. the oxidation of lignin via an oxygen radical intermediate) are not possible, or are vastly slower, in the absence of molecular oxygen (64). Thus, shorter exposure times are associated with higher organic carbon burial efficiencies and the preservation of less degraded materials (7,8,13,34,63,65,66). This correlation is not absolute, however. Large provinces of ocean sediment underlying gyres are oxic to the basement, representing as much as 86 million years of OET (67). In such sediments, sedimentation rates are exceedingly slow and the sediments are very organic poor, and most oxidation apparently occurs directly at the sediment–water interface. Organic carbon oxidation in "rich" anoxic systems such as rapidly accumulating estuarine sediments can exceed 100 μM C day^{-1}, primarily via sulfate reduction, compared with ~3 $\times 10^{-6}$ μM C day^{-1} in oxic gyre sediments.

16.2.5 Models of Organic Carbon Diagenesis

Due to the chemical complexity of sedimentary OM, sedimentary diagenetic models have focused on the transformation of bulk organic carbon to CO_2. One common class of models assumes the following form:

$$r = -\frac{dG}{dt} = \sum_{i=1}^{n} k_i G_i, \qquad (16.1)$$

where r is the bulk rate of CO_2 production, equivalent to the rate disappearance of bulk organic carbon (G), which in turn is the sum of the oxidation rates of different carbon pools (G_i), each of which is oxidized according to a different, characteristic rate constant (k_i) (68). Frequently, these "multi-G" models only include two or three reactivity classes of

OM: usually a fast-reacting "labile" pool, an unreactive "recalcitrant" pool, and sometimes an intermediate "semi-labile" pool. Related models include the reactivity-continuum model, which assumes an infinite number of reactivity pools (69,70), and that of Middelburg (71), in which a single time-dependent reactivity rate constant is assumed. These models are mathematically straightforward, but they are somewhat mechanistically disconnected from the reality of sediment OM, which is tremendously chemically complex (7,72).

Recently, models that include a broader set of parameters, such as microbial biomass, enzyme substrate specificity, and temperature–rate relationships, have been successfully employed in soils and sediments (73,74). By including a wider range of processes, these models have the capacity to both quantitatively fit bulk organic carbon concentration data and make reasonable predictions about systems' likely responses to changing environments.

16.3 Oceanic Rocky Subsurface

Below ocean sediments, the igneous ocean crust hosts ~2% of the total volume of the ocean, making it the largest aquifer system on Earth (75). Seawater actively circulates through this aquifer and drives the transfer of heat and elements between fluids and rocks with ramifications for ocean chemistry (76–79) and for the thermal, physical, and geochemical structure of the crust and mantle (80,81). Microbial life is widespread in the rocky oceanic subsurface and both exploits and influences these exchanges (82–84), altering the abundance and form of carbon. Fluid flow through the rocky subsurface is ultimately driven by a source of heat such as cooling magma or hot rocks (80,85). Heated fluids rise buoyantly and ultimately exit the sub-seafloor, drawing cool seawater into the crust to replace it.

The carbon characteristics of the fluids and rocks in hydrothermal systems and the igneous basement differ greatly depending on the type of host rock, the temperature of the system, and the presence or absence of sediments. Some systems are further influenced by factors such as phase separation, magma injections, seismic activity, extent of subduction, and even tides (86–89). Below, carbon transformations are described in some of the primary types of hydrothermal circulation systems (Figure 16.4).

16.3.1 Characteristics of Recharge Water

The chemical composition of the seawater that enters into the rocky subsurface has a strong influence on subsequent water–rock and microbial reactions. Deep seawater carries DIC in concentrations of 2.1–2.3 mM (90) and DOC in concentrations of ~34–48 μM (2,91). DOC is composed of a complex set of molecules, some of which turnover rapidly on timescales of hours to years. The majority of DOC, however, is slow to remineralize and has the potential to be stored for millennia in the ocean's interior (see (92) for review). Refractory

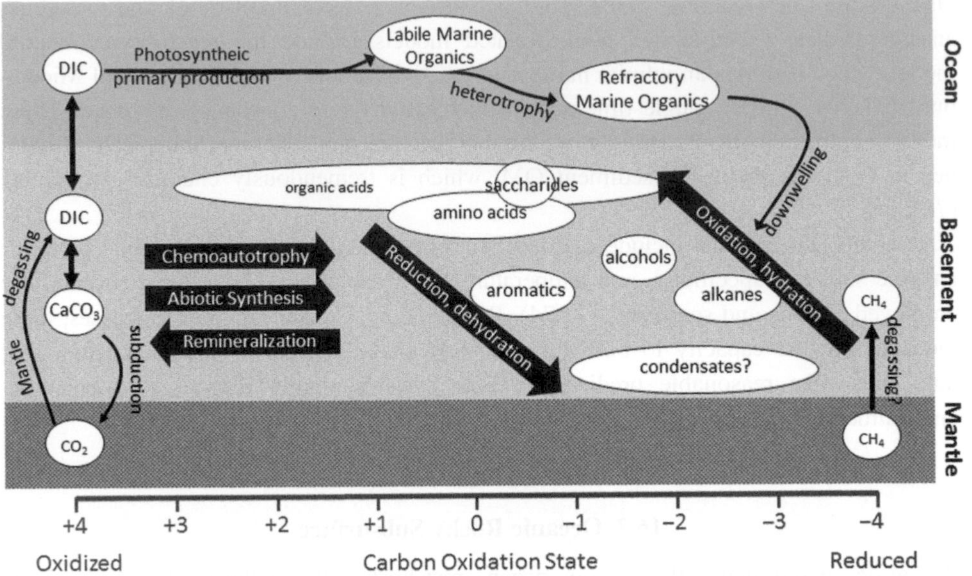

Figure 16.4 The abundance and composition of organic molecules in hydrothermal fluids will reflect a complex reaction history. While chemoautotrophy and abiotic synthesis involve the reduction of inorganic carbon into organic molecules, remineralization will do the reverse. Oxidation and dehydration reactions produce smaller, more polar compounds that are generally more labile and more easily consumed by heterotrophic microorganisms. Reduction and dehydration reactions may produce larger and more apolar material that is more resistant to microbial degradation and may be sequestered in the subsurface or persist for long periods of time in the deep ocean.

DOC is highly degraded, has few recognizable biomarkers, and has a ^{14}C age of 4000–6000 years, substantially longer than the mixing time of the ocean (93). DOC isolated from seawater and subjected to nuclear magnetic resonance and Fourier-transform ion cyclotron resonance mass spectrometry is composed primarily of carboxyl-rich aliphatic matter (94), acylated polysaccharides (95), and carotenoid degradation products (96) (see (97) for review).

16.3.2 Axial High Temperature, Basalt Hosted

The most widely recognized hydrothermal systems are close to axial spreading centers, where new injections of magma maintain high temperatures (Figure 16.1). The host rock is mafic and fluids exit through chimney structures at temperatures that can reach >400°C (98). Exiting fluids are rich in dissolved metals that, upon mixing with cold seawater, precipitate the sulfide minerals that give them the name "black smokers." In the water column, the hot fluids mix further with seawater, cool, reach neutral buoyancy, and spread away from the vent field. The chemical signatures from these plumes of water can be detected thousands of kilometers away from the field (99,100).

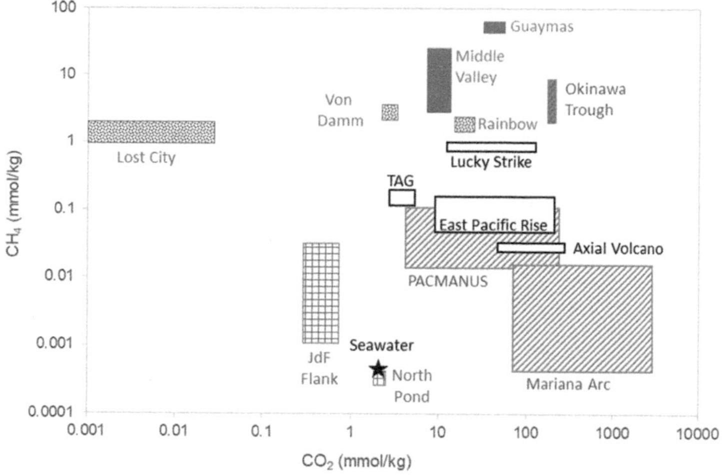

Figure 16.5 Range of methane and CO_2 concentrations in basalt-hosted high-temperature (black outline; Axial Volcano, Trans-Atlantic Geotraverse (TAG), 9°N East Pacific Rise, Lucky Strike), ultramafic-hosted (green diamonds; Lost City, East Summit of Von Damm, Rainbow), ridge flank (blue checkers; Juan de Fuca ridge flank, North Pond), back-arc basins (orange diagonal; PACMANUS, Mariana Arc, Okinawa Trough), and sedimented systems (gray boxes; Guaymas, Middle Valley, Okinawa Trough). Seawater composition is included for comparison. Methane concentrations at North Pond are plotted at the reported detection limit of the analysis (0.5 μM). References are given in Table 16.1.

The majority of high-temperature vent fluids have DIC concentrations equal to or greater than deep seawater due to inputs of magmatic CO_2 (Figure 16.5 and Table 16.1) (101). DIC concentrations are generally 3–30 mM, or they can be higher when fluids are impacted by phase separation, fresh inputs of magma, or sedimentary degradation (see (101) for review). Additions of magmatic CO_2 are identified by $\delta^{13}C$ isotopic signatures (–9‰ to –4‰) (102,103) that are markedly different from deep seawater DIC (–0.5‰ to 1.0‰) (104). The lack of ^{14}C in CO_2 from some hydrothermal fluids demonstrates that the DIC carried with recharge water can be fully removed during sub-seafloor circulation in some cases (105). Calcium carbonate veins in basalts and gabbros have isotope values consistent with precipitation of marine DIC at the relatively low temperatures of seawater recharge (106,107).

Methane concentrations in sediment-free, high-temperature axial fluids (~7–200 μM) are higher than those of seawater (0.0003 μM), but generally low when compared to sedimented or ultramafic-influenced systems (Figure 16.5 and Table 16.1). For example, vent fluid CH_4 concentrations range from 7 to 213 μM from high-temperature vents from along the East Pacific Rise (111,123,147–150), while those from along the Mid Atlantic Ridge (MAR) range from 8 to 147 μM (98,117–119,151–153). Concentrations can spike as a result of volcanic eruptions and due to outgassing after a dike injection (88,89,113).

The majority of the DOC carried with deep seawater is destroyed during circulation through mafic hydrothermal systems. The first evidence for this removal came from a study

Table 16.1 *Characteristics and carbon contents of representative oceanic sub-seafloor fluids.*

System type	Seawater	Sedimentary porewater		Basalt hosted, unsedimented, high temperature			Basalt hosted, unsedimented, diffuse/mixed fluids	
Example system	Below 1500 m[a]	Cascadia Margin[b] 0–65 mbsf	Cascadia Margin[b] 65–189 mbsf	East Pacific Rise (9°50'N)[c]	Axial Volcano[d]	TAG[e]	East Pacific Rise diffuse (9°50'N)[f]	Axial Volcano[g]
Temperatures (°C)	2–5	≤9	~8–12	275–371	217–328	290–321	23–55	3–78
pH (at 23°C)	7.8	7.7–8.0	7.7–8.4	3.5–4.2	3.5–4.4	3.1–3.5	5.8–6.4	4.6–5.8
H_2 (mM)	0.0003	–	–	0.27–8.4	<0.1–0.8	0.10–0.37	BDL–0.0058	<0.5
Dissolved iron (μM)	<0.001	–	–	8–5150	12–1065	1640–5170	<2–277	35–400
ΣCO_2 (mM)	2.1–2.3	7–24	18–29	9.4–219	50–285	2.9–5.0	3.0–11.8	–
$\delta^{13}C_{CO2}$ (‰)	–0.6 to 0.4	–24.4 to 25.6	27.9–33.6	–4.2 to –3.7	–	–13.0 to –6.9	–4.2 to –2.2	–
$F^{14}C_{CO2}$	0.7511–0.9677	–	–	–	–	–	–	–
CH_4 (mM)	0.0003	BDL–69	71–236	0.05–0.16	0.025	0.12–0.16	0.003–0.500	<0.6
$\delta^{13}C_{CH4}$ (‰)	–	–59.7 ± 7.1	–46.7 to –41.7	–34.6 to –16.8	–	–9.5 to –8.0	–	–
$F^{14}C_{CH4}$	–	–	–	–	–	–	–	–
$\Sigma(C_2H_6–C_4H_8)$ (μM)	BDL	0–35 ppmv[q]	0.5–7318 ppmv[q]	–	–	–	–	–
$\delta^{13}C_{C2H6,C4H8}$ (‰)	–	–	–	–	–	–	–	–
CO (μM)	BDL	–	–	BDL–2.0	–	BDL	5.7–17.8	–
CH_3SH (nM)	BDL	–	–	2.4–4.9	–	12	–	–
DOC (μM)	35–45	400–3200	1700–5100	–	8–24	–	–	34–71
$\delta^{13}C_{DOC}$ (‰)	–20 to –22	–23.6 to –22.1	–20.2 ± 0.4	–	–	–	–	–18.6
$F^{14}C_{DOC}$	0.444–0.767	–	–	–	–	–	–	0.481[r]
Formate (μM)	BDL	–	–	–	–	–	–	–
Acetate (μM)	BDL	5–57	14–89	–	–	–	–	–
Hydrolizable amino acids (μM)	80–160	–	–	–	–	–	–	–

"–" is used where no reports available in the literature.

BDL = below detection limit; mbsf = neters below seafloor; TAG = Trans-Atlantic Geotraverse.

[a] (93, 104, 108, 109).

[b] (29, 110).

[c] (109, 111–113).

[d] (114–116).

[e] (98, 109, 117–119).

[f] (109, 111, 120).

[g] (86, 114, 116, 121).

[h] (109, 122–126).

[i] (127).

[j] (116, 128–133).

[k] (134, 135).

[l] (109, 136).

[m] (109, 119, 137–139).

[n] (109, 112, 140–143).

[o] (144).

[p] (145, 146).

[q] Headspace gas concentrations in equilibrium with sediments.

[r] $F^{14}C$ of ultrafiltrated DOC (>1000 Da).

Basalt hosted, sedimented, high temperature		Ridge Flank ("warm")	Ridge Flank ("cool")	Ultramafic influenced, high temperature		Ultramafic dominated	Silicic back-arc	
Guaymas Basin[h]	Middle Valley[i]	Juan de Fuca Ridge Flank[j]	North Pond Basement[k]	Von Damm (East Summit)[l]	Rainbow[m]	Lost City[n]	PACMANUS[o]	Okinawa Trough (sedimented)[p]
100–315	40–281	64	3.1–3.8	226	350–367	30–91	152–358	>220–320
4.5–6.1	–	7.5	7.4–7.6	5.6	2.8–3.4	9–11	2.3–4.7	4.7–5.4
0.52–3.30	1.9–8.2	0.3–0.7	–	18.2–19.2	12.3–16.5	1–14	0.0084–0.306	0.05
17–180	–	0.6–1.1	–	–	23,700–24,050	<3.5 μM	76–14,600	–
35–54	8.2–13	0.2–0.6	2.0–2.4	2.80	16.0–24.6	0.0001–0.026	4.4–274	198–200
–9.4	–34.6 to –20.7	–9.7 to –1.3	–0.16 to 0.67	0.8–0.9	–3.15 to –2.5	~–9	–5.7 to –2.3	–5.0 to –4.7
0.056	–	0.083–0.233	0.595–0.865	0.0251–0.0373	–	–	–	–
44.2–58.8	3.0–22.6	0.001–0.030	<0.0005	2.81	1.6–2.5	0.9–2.0	0.014–0.085	2.4–7.1
–43.8	–55.5 to –50.8	–58 to –23	–	–15.6 to –15.3	–17.7 to –15.8	–13.6 to –9.3	–20.8 to –7.4	–41.2 to –36.1
0.077		–	–	0.0056–0.0064	–	0.0017–0.0062	–	–
–	14–310	–	–	695	0.84	1.0–2.0	–	–
–	–25.3 to –18.7	–	–	–12.9 to –9.8	–	–16.0 to –13.0	–	–
27–92.4	–	–	–	n.d.	5.0–7.4	BDL	0.006–0.17	–
11–10,000	–	–	–	22	7.4–10.3	1.4–1.9	–	–
111–2112	–	11–18	18–33	–	–	68–106	–	–
–	–	–34.5 to –24.8	–26.6 to –23.9	–	–	–21.0 to –10.5	–	–
–	–	0.166–0.230[q] 0.186–0.204	0.352–0.472	–	–	–	–	–
<40	–	–	–	88.2	–	36–158	–	–
BDL–295	–	–	–	–	–	1–35	–	–
5.2	–	0.043–0.089	–	–	–	0.7–2.3	–	–

of amino acids in the sediment-covered Guaymas Basin (26,122). Concentrations of dissolved free amino acids in high-temperature fluids (>150°C) were below detection limits and below deep ocean concentrations, with the losses attributed to the instability of organic compounds at high temperatures (122,154). The DOC content of black smoker vents on the unsedimented portions of the Juan de Fuca spreading center is less than half that of deep seawater (<17 versus 36 μM) (116). Concentrations of DOC that can be isolated onto solid-phase extraction (SPE-DOC) phases are ~92% lower in unsedimented black smokers from Juan de Fuca and the MAR than in deep seawater (155). It is possible to experimentally reproduce losses of OM by heating (125,155–157), though this does not conclusively rule out alternative removal mechanisms such as sorption onto mineral surfaces or heterotrophy.

16.3.3 Axial Diffuse Vents, Basalt Hosted

Adjacent to axial, high-temperature systems, local seawater enters the crust, creating "diffuse vents." The mixing of oxygenated seawater and reduced hydrothermal fluids results in chemical disequilibria that microorganisms can exploit for metabolic energy (158). Due to mixing and conductive cooling of fluids, temperatures are often well below the upper temperature limits of life (122°C) (159). As a result, these zones are thriving sub-seafloor microbial habitats (3,82,84,160,161). Microbial activity can alter fluid chemistry, resulting in losses of H_2S and H_2 and gains of CH_4 relative to high-temperature fluids (111,112,115,162).

In diffuse vents on the Juan de Fuca Ridge, DOC is elevated over local deep seawater (~47 versus 36 μM), attributed in part to sub-seafloor autotrophic production (116). This DOC has a lower ^{14}C content and a more positive $\delta^{13}C$ value than local seawater, consistent with a contribution of chemolithoautotrophs incorporating a pre-aged carbon source such as mantle CO_2 (121).

16.3.4 Ridge Flanks

Fluid continues to flow through the rocky subsurface far from the ridge axis, as rocks cool in the absence of new magma injections (Figure 16.1). The extent of advective flow through these "ridge flank" systems can be determined from discrepancies between modeled conductive heat loss and heat flow measurements that indicate the convective flow of water in crust that is 0–65 Ma (75,80). Sediment cover precludes fluid transport into and out of the crust; bare-rock seamounts are therefore the primary locations of advective transport (128). Even in regions with thick sedimentary layers, however, exchange of water, carbon, elements, and nutrients continues between deep sedimentary porewater and basement fluids.

Based on magnesium budgets, fluid fluxes through "cool" ridge flank systems (<45°C) are substantially larger than those through warmer systems (77,163). Cool basement fluids

(<20°C) have been accessed by Integrated Ocean Drilling Program drilling in the North Pond sedimented basin on the MAR (161). Dorado Outcrop on the Cocos Plate has also been confirmed to vigorously vent large quantities of water at temperatures of 10–20°C (164). The "warm" ridge flank system on the Juan de Fuca ridge has been intensely studied for decades, including via series of Ocean Drilling Program boreholes that have been drilled perpendicular to the ridge to allow direct access to the basement (165).

DIC is substantially lower in Juan de Fuca ridge flank fluids than in seawater (0.1–0.9 versus 2.6 mmol/kg; Table 16.1), likely due to precipitation of calcium carbonate in the subsurface (130,166,167). In contrast, fluids from the lower-temperature Dorado and North Pond systems have DIC concentrations that are similar to seawater (134,164). In many cases, the $\delta^{13}C$ values of DIC are lower than that of seawater, suggesting an input from remineralization of organic carbon or CO_2 trapped in basaltic vesicles (130,133,135). The apparent ^{14}C age of DIC is often used as a measure of fluid residence time, although this must be treated with caution, as mixing with older water masses, remobilization of calcium carbonate, input of basalt vesicle CO_2, and remineralization of ^{14}C-depleted OM can influence these signatures (133,168–170).

Methane concentrations are low but detectable in Juan de Fuca ridge flank fluids (1–32 μmol/kg) (131,132). The isotopic signatures of methane (–58.0‰ to –22.5‰) indicate a mixture of processes, including biogenic production and oxidation (132). Methane concentrations at North Pond were below detection (134).

DOC concentrations are lower than seawater in ridge flank fluids on the Juan de Fuca ridge and at North Pond (116,131,133–135). In both cases, this DOC has a lower ^{14}C content and $\delta^{13}C$ signatures that are more negative than those of starting seawater (121,133,135). This pattern was initially attributed to a complete removal of seawater DOC, followed by an input of chemosynthetically derived organic material (121). New data suggest that the isotopic signatures could instead be attributed to the selective oxidation and removal of portions of the seawater DOC pool (133,135). Diffusion of porewater from the sediments covering the ridge flank may also contribute some organic compounds to the fluids (124), as this exchange impacts the inorganic chemistry (129,131,165).

16.3.5 Ultramafic Influenced

Systems hosted on ultramafic rocks undergo water–rock reactions that are distinct from those of mafic environments. Ultramafic systems can be located on spreading centers and influenced by magmatic injections, but they can also be far from the spreading center or along ultra-slow-spreading centers with little to no magmatic influence. The compositional differences between ultramafic rocks derived predominantly from Earth's mantle and mafic rocks such as basalt and gabbro give rise to fluids with distinct chemical signatures. Fluids that have reacted peridotites are strongly enriched in H_2 and CH_4 and, in some cases, have drastically lower metal contents (Figure 16.5 and Table 16.1).

The earliest recognitions of an ultramafic hydrothermal signature in the ocean came from high ratios of CH_4 to Mn and suspended particulate matter in the water column on the MAR (171–173). Subsequently, the Logatchev, Rainbow, Menez Gwen, Ashadze, and Nibelungen hydrothermal fields were identified along the MAR, with fluid chemistries that exhibit a mixture of magmatic influences such as high temperatures (200–372°C), acidic pHs (2–4 at 25°C), and high metal contents (e.g. millimolar concentrations of Fe and hundreds of micromolar concentrations of Mn), but also ultramafic influences such as millimolar concentrations of CH_4 and H_2 (Table 16.1; for reviews, see 119,174). Peridotite-influenced systems have since been identified on the Mid-Cayman Rise (136,175) and Marianas Forearc (163,176). The ultramafic-dominated system in the Lost City Hydrothermal field has minimal interaction with magmatic processes, resulting in lower fluid temperatures (40–91°C), alkaline pHs (9–11 at 23°C), and low metal contents (<100 nM of Fe and <50 nM of Mn) (140,143,177). A magmatic influence is still evident, however, in elevated the ^3He content of fluids (112). Ultramafic-dominated, low-temperature, alkaline systems are also present in the shallow waters of Prony Bay in New Caledonia, fed by meteoric water (178), and on the Southern Mariana Forearc at the Shinkai Seep Field (179).

The inorganic carbon concentration in ultramafic-influenced systems is highly dependent on pH and magmatic inputs. In low-pH ultramafic systems, concentrations of ΣCO_2 can reach as high as those observed in magmatic systems, at ~4–20 mM (Table 16.1; see (119) for a review). The $\delta^{13}C$ values of this CO_2 display "typical" mid-ocean ridge values of –4‰ to –2‰ in some cases such as the Rainbow vent field (119). In other locations such as the Logatchev field, it is unusually positive, up to +9.5‰, even in fluids with ΣCO_2 concentrations higher than seawater (119,180). In alkaline ultramafic systems such as Lost City, the high pHs lead to the rapid precipitation of calcium carbonate and therefore vanishingly low concentrations of ΣCO_2 in end-member fluids (112,140). This removal likely occurs throughout the fluid circulation pathway. Carbonate mineralization is common in ultramafic rocks (181), and isotope signatures indicate precipitation occurs both at cold seawater temperatures and at warmer (65–95°C) temperatures, where $\delta^{13}C$ values indicate that the source ΣCO_2 has a substantial mantle component (107).

Methane concentrations in ultramafic systems are frequently an order of magnitude higher than those in unsedimented, basalt-hosted systems (Figure 16.5 and Table 16.1), and substantial methane anomalies along the MAR have been attributed to exports from these systems (119,171–173). Estimates from mantle ^3He exports suggest serpentinization of ultramafic rocks could account globally for about 75% of the methane flux from mid-ocean ridge systems (182). Isotopic signatures point to a nonbiological source for this methane (112,119,183), although in most systems more CH_4 is present than would be expected in thermodynamic equilibrium with CO_2 (for reviews, see (101,183)). One possibility is that the methane was formed long ago, at higher temperatures than the present day, and is subsequently stripped from vesicles in the rocks (136,184), which contain high CH_4 and CO_2 contents (185,186). Biologically derived methane from methanogenesis may also contribute (187), albeit at relatively low levels when compared to the dominant nonbiological signature.

Short-chain hydrocarbons such as ethane, propane, and butane have been found in the low-micromolar concentrations in a wide range of ultramafic systems (Table 16.1) (112,136,180). Isotopic values that decrease with increasing chain length have been used to demonstrate that these species are not derived from the decomposition of sediments and could have a nonbiological origin such as Fischer–Tropsch-type reactions (112). At the Lost City and Von Damm hydrothermal fields, the concentrations of these compounds increase in conjunction with methane concentrations (112,136), indicating similar processes may lead to their formation and/or cycling.

Formate and acetate have been reported in elevated concentrations in multiple ultramafic systems including Lost City (formate: 36–158 μM; acetate: 1–35 μM) (141), Von Damm (formate: below detection to 669 μM) (136), and Prony Bay (formate: ~4 μM; acetate: ~70 μM) (188). At Lost City, the isotopic composition of formate indicates it is synthesized by two pathways: abiotic synthesis in the subsurface that results in ^{14}C-free formate with a $\delta^{13}C$ signature (–13.0‰ to –8.9‰) similar to methane and short-chain hydrocarbons; and near-surface biological synthesis that incorporates modern DIC, resulting in formate with substantial ^{14}C and a more positive $\delta^{13}C$ signature (–9.1‰ to –4.3‰) (189). At the Von Damm vent field, higher concentrations of formate are found in hot mixed fluids than in pure end-member hydrothermal fluids, demonstrating that this species forms abiotically on timescales of hours to days (136). At Lost City, the $\delta^{13}C$ of acetate (–27‰ to –17‰) could be attributed to a mixture of anaerobic fermentation and acetogenesis (141,189). Given the high abundances of microorganisms in the chimneys, the acetate could also be due to a thermocatalytic breakdown of complex organics in the biomass (141,189).

Hydrolyzable amino acids are present in high abundances in Lost City fluids and chimneys (142). In the fluids, the highest concentrations were observed in locations where concentrations of H_2 had been drawn down by sulfate reducers living in the sub-seafloor or chimney. The ^{13}C of amino acids isolated from the chimneys had fractionation patterns consistent with synthesis by a chemolithoautotrophic source (142). In high-temperature fluids (>300°C) from Rainbow and Ashadze, dissolved free amino acids were detected in the picomolar concentration range, with tryptophan, phenylalanine, and leucine detected in the fluids but not in deep seawater (180). Tryptophan and phenylalanine contain aromatic rings that may assist in molecular stability at high temperatures (190).

16.3.6 Fluxes between the Ocean and Crust

Hydrothermal circulation is the primary means of transferring materials between the crust and ocean (78,191). The net flux of constituents includes both input and removal processes, though these may be geographically and temporally distinct. The impact of hydrothermal circulation on the carbon budget of the ocean remains unconstrained in many ways.

Inorganic carbon is transferred to the deep ocean via magma degassing and removed by carbonate precipitation in the sub-seafloor at roughly similar rates (Figures 16.4 and 16.5). Degassing of mantle volatiles through high-temperature venting is estimated to input ~1×10^{12} mol C yr^{-1} of ΣCO_2 into the ocean (191,192). Carbonate precipitation is estimated to remove 1–3 $\times$ 10^{12} mol C yr^{-1} in ridge flanks (167,193,194). Seawater passing through peridotites results in a loss of 0.4–2.0 $\times$ 10^{11} mol C yr^{-1}, although stable isotope signatures indicate that approximately half of the carbon sequestered into the rock is in the form of organic carbon (195,196).

Export and removal fluxes of DOC can be estimated by combining changes in concentrations with water fluxes through different types of hydrothermal systems (75,80,163). If high-temperature vents remove an average of ~20 μmol of seawater DOC per liter, approximately 0.7–1.4 $\times$ 10^{10} g C yr^{-1} would be lost globally (116). A similar scale loss of $1.4 \pm 0.7 \times 10^{10}$ g C yr^{-1} has been estimated based on changes of SPE-DOC concentrations (155). Ridge flank regions where crustal temperatures are "warm" (>45°C) have more substantial chemical changes in circulating fluids but smaller fluid fluxes than regions where crustal temperatures are "cool" (77,163). If concentrations from the "warm" Juan de Fuca ridge flank system are typical of such systems, 2–13 $\times$ 10^{10} g C yr^{-1} would be removed (116). Due to the larger water fluxes, if DOC concentrations through the "cool" crust at North Pond are globally representative, losses would be an order of magnitude higher at ~9–14 $\times$ 10^{11} g C yr^{-1} or ~5% of the total annual deep oceanic DOC loss (135).

16.4 Sedimented Hydrothermal Systems

Where spreading centers occur under thick sediment packages, hot water rapidly alters the OM fueling heterotrophic communities (e.g. (125,156,190)), releasing inorganic carbon (89), influencing local physiochemical conditions, or forming complex oil-like materials (e.g. (197,198)). The form and fate of carbon in heated sediments depend on its origin (terrigenous versus marine versus chemoautotrophic), temperature, and flow rate. Upon heating, a series of reactions similar to those that give rise to petroleum proceeds, with important differences due to the more water-rich conditions. The production of petroleum is generally considered to begin at ~50–70°C (199). Weak bonds that sorb organic molecules onto surfaces break most easily, followed by bonds involving oxygen, sulfur, or nitrogen. Carbon–carbon bonds require the most energy – and therefore greater temperature or time – to break (199).

Small polar compounds can be mobilized through enhanced desorption and the destruction of noncovalent bonds. The most labile material is removed from the solid phase due to microbial activity, pyrolysis, and/or desorption (156). Over time, the amount of OM transferred into the aqueous phase decreases as the material is physically transported out of the system or biodegraded by microorganisms (125,156,157).

Unlike the dry "cracking" reactions that dominate petroleum reservoirs, breaking carbon–carbon bonds in the presence of water results in more oxidized products. Cracking reactions proceed at temperatures above ~100°C and result in CH_4 and low-molecular-weight hydrocarbons (199–201). In contrast, in the presence of water and minerals, *n*-alkanes will instead degrade to oxygenated products such as alcohols, ketones, carboxylic acids, and, ultimately, CO_2 and CH_4 (89,202). Sediments heated in aqueous environments produce copious amounts of acetate in particular. The reaction temperature impacts the products, with higher temperatures favoring more oxidized products such as CO_2 over CH_4 and propanol over propane (203).

Reduction, condensation, and dehydration reactions proceed at higher temperatures to form macromolecules and aromatics, causing compounds to revert to their most stable states (Figure 16.4) (204,205). Polycyclic aromatic hydrocarbons and cyclic polysulfides, major components of some hydrothermal oils, form only under very high heat (>~300°C) and are signatures of elevated temperatures (190,197,198). Polypeptides form through dehydration and reduction, while lipids crack and recombine (206).

Water washing will selectively transport more soluble components from the subsurface to the surface and leave behind larger condensates (125,190,199,207). Smaller alkanes ($<C_{10}$), aromatic volatiles, compounds containing C–N–S bonds, oligosaccharides, and oligopeptides are often missing in sediments subjected to "water washing," while fluids and plumes contain higher concentrations of these compounds (190,204,207).

These released compounds are highly biodegradable and fuel heterotrophic organisms. The labile amino acids released from sterilized sediments, for example, are utilized and reworked by microorganisms in parallel, nonsterilized experiments (125). In general, the low-molecular-weight organic acids that are primary breakdown products of heating sediments in the presence of water, particularly acetate, are important substrates for anaerobic microorganisms (208).

The residual OM that is not removed with water washing is enriched in less soluble material, leading to "hydrothermal petroleum." Cooling near the sediment–water interface can help trap less soluble compounds through differential condensation and solidification (190,207,209). The distribution of compounds and the maturity of these oils are highly variable.

16.5 Continental Subsurface

Geological heterogeneity produced through plate tectonics diversifies and segments the continental deep subsurface and its constituent biospheres differently from in the marine realm. Mountain and basin formation juxtaposes reactive rocks and minerals and creates new hydrological flow paths. Rock and water ages on the continents range from modern to billions of years (210,211). Terrestrial vegetation supplies vast quantities of organic carbon, although this influence is attenuated with increasing depth. The water age, hydrological connectivity, and major element chemistry of continental subsurface sites dramatically impact carbon cycling and the nature of *in situ* biospheres.

The continental deep subsurface extends downward from the base of the critical zone (212,213), although specific depths and thresholds have yet to be defined, particularly on the upper boundary. The penetration of life into the continental crust appears to be limited not strictly by depth, but rather by temperature, permeability, and perhaps aridity, with clear life detection in even the deepest boreholes and mines. Sites lacking identifiable life are few and far between and appear to be limited by temperature (e.g. German continental deep drilling program (KTB) cores in the Black Forest (214)) or aridity (213).

Estimates of the size of the continental deep biosphere are large (ranging from 2.3×10^{15} to 10^{17} g C), mirroring similar estimates of the marine deep biosphere (4.1×10^{15} g C) and rivaling terrestrial soils (2.6×10^{16} g C) (215–217). The uncertainty in these calculations spans orders of magnitude and has not changed significantly since the original estimates by Whitman (215; Chapter 17, this volume), although the trend is downward (Chapter 17, this volume). However, increasing levels of inquiry applied globally using advanced methodologies have identified abundant, taxonomically diverse communities within the continental deep subsurface, giving credence to vast amounts of carbon contained and being cycled by these ecosystems (218–222).

16.5.1 Types of Continental Deep Subsurface Environments

Continental deep subsurface environments can be broadly divided between sedimentary and crystalline host rocks, but even within this framework they range significantly in carbon content, isolation from the surface, and dominant carbon cycling processes (Figure 16.1). The best-studied sites are found in shallow sedimentary and igneous aquifers owing largely to their relevance to human water supplies (223–231). Hydrocarbon reservoirs contain vast quantities of organic carbon and have distinct microbiology associated with their formation waters (232–234). A recent emphasis on deep coal beds and their constituent carbon cycling has come to the scientific forefront due to their importance in gas extraction via deep fracking technologies (235). Deep crystalline bedrock sites feature the oldest, deepest, and most isolated deep biosphere environments (211,236–238). Caves, in contrast, sit at the interface between the surface and the deep and are covered more completely in other reviews (239–241). This section will describe the forms, cycling, and fate of organic carbon in each environment.

While the deep subsurface biosphere is pervasive, it is difficult to access reliably. Common access points are wells, boreholes, mines, and caves. Each approach has the potential to impact *in situ* processes and must be considered when evaluating data sets. Natural springs are often considered as “portals” or “windows” into the deep biosphere, often showing a mix of surface and subsurface communities (242,243). The last 10 years has seen the establishment of a number of deep subsurface observatories including into permafrost (Permafrost Tunnel Research Facility, AL, USA), deep crystalline bedrock (Deep Mine Microbial Observatory (DeMMO), SD, USA; Coast Range Ophiolite Microbial Observatory (CrOMO), CA, USA; Äspö Hard Rock Laboratory, Sweden; and many others), and sedimentary aquifers (Deep Biosphere in Terrestrial Systems (DEBITS), New Zealand; Savannah River Site, SC, USA).

16.5.2 Continental Carbon Cycling

Organic carbon in the continental deep biosphere may derive from surficial inputs, *in situ* autotrophic carbon fixation, water–rock reactions, or ancient sedimentary sources. The relative balance of these sources depends sharply on geology, both by surface connectivity and host lithology. The following sections describe this balance in sedimentary and igneous aquifers, hydrocarbon reservoirs, deep coal beds, and deep crystalline bedrock.

Key processes in subsurface carbon cycling depend on the relative recalcitrance of ancient OM, supply of labile organic carbon, input of metabolic oxidants and reductants, and aquifer porosity and permeability. Microbial carbon fixation produces labile organic carbon and methane, whereas heterotrophic microbial processes consume both labile and recalcitrant subsurface organic carbon. The relative importance of these two end members broadly suggests autotrophic processes dominate in crystalline and deep rock aquifers whereas heterotrophic processes are more abundant within sedimentary systems, although numerous counterexamples exist and both processes (e.g. (251)) must be active for a functioning ecosystem (225). Organic acids and short-chain hydrocarbons are key microbial products and substrates within most continental subsurface settings with typical concentrations in the 10–100 μM range (Table 16.2). Due to rock dissolution and other processes, DIC can be very high, and may get much higher as aquifers are targeted for anthropogenic carbon sequestration (222).

16.5.3 Sedimentary and Igneous Aquifers

Both sedimentary (e.g. Atlantic coastal plain) and igneous aquifers (e.g. Columbia River basalt aquifer) have been shown to contain vibrant microbial communities and have been the subject of intensive study due to their economic and social importance as sources of drinking and industrial water, as well as their vulnerability to anthropogenic contamination (225,226,231). Recharge timescales of aquifers vary over many orders of magnitude (months to millions of years), controlling the relative supply of exogenous DOC and electron acceptors. In many systems, significant supply of young sedimentary carbon produces relatively high DOC, methane, and organic acid concentrations. The composition of this DOC can be complex, including significant amounts of nitrogen- and sulfur-bearing organic molecules (252). While oligotrophic compared to surface environments, aquifers are relatively carbon rich for the subsurface and can support correspondingly high cell densities (e.g. 10^5 cells/mL), even in oligotrophic crystalline aquifers (246).

Primary productivity within aquifers varies tremendously based on exogenous and sedimentary organic carbon supply, but is significant in some settings. Hydrogen production can be large and may support autotrophic populations, particularly in igneous and ultrabasic host environments, fueling the so-called subsurface lithoautotrophic microbial ecosystems (223,251,253). Utilization of iron oxide minerals as terminal electron acceptors for both autotrophic and heterotrophic metabolisms is common, producing high concentrations of dissolved ferrous iron in many groundwaters

Table 16.2 *Summary of characteristics and carbon contents of different types of continental subsurface systems.*

System type	Shallow sedimentary aquifers	Shallow igneous aquifers	Hydrocarbon reservoir formation waters	Coal beds	Deep bedrock
Example systems	Lower Saxony, Germany[a]	Columbia River flood basalts[b]	Palo Duro Basin[c]	German lignite deposits[d]	South African gold mines[e]
Defining characteristics	Shallow confined and unconfined aquifers with abundant sedimentary OM, fresh waters	Thick basalt deposits with confined aquifers, interbedded sediments, fracture-based porosity, and relatively fresh NaCl-dominated fluids	Sedimentary hydrocarbons interfacing with aqueous brines, organic alkalinity may exceed bicarbonate alkalinity	Extremely organic-rich sediments at varying stages of thermal maturity, limited porosity and permeability	Deep (1.0–3.3 km) fracture-based fluids with thousand- to million-year recharge times
Temperatures	Low to moderate	Low to moderate	Moderate to high	Low to moderate	Low to high
Recharge timescale	Rapid to moderate	Rapid to moderate	Moderate to long	Moderate to long	Moderate to very long
pH	Circumneutral	7.5–8.5 shallow, 8.0–10.5 deep	5–8	6.8–7.2	7.4–9.4
H_2 (mM)	–	Up to 0.06	–	–	7.4
SO_4 (mM)	–	Generally <0.5, but up to 2	Up to 25	Up to 0.148	0.623
Total DIC (mM)	–	0.125–2.800	–	19.8–43.6	0.09–2.40
$\delta^{13}C_{CO2}$ (‰)	–20	–30 to 20; mostly –10	–	–14 to 20	–43 to –5
CH_4 (mM)	0.00089–2.68000	Up to 160	Very high	0.010–0.100	0.026–8.800
$\delta^{13}C_{CH4}$	–110 to 20 (mean –70)	–	Variable	–81 to –71	–58 to –37
Short-chain hydrocarbons $\Sigma(C_2H_6–C_4H_8)$ (μM)	~3 (median)	–	High	–	<0.1–201

DOC (mM)	0.17–0.30	0.16–0.39	0.05–14.75	0.19–0.95	n.d. to 0.410
Formate	–	–	–	2.22–31.1 μmol/g sed	0.44–34.00 μM
Acetate	–	–	6.08 mM	1.7–8.5 μmol/g sed[f]	0.07–28.00 μM
Amino acids	–	–	–	–	0.0133–2.6700 μM
Typical cell density (cells/mL)	10^3–10^8	10^3–10^5	High	10^7	10–10^5

Ranges are reported for the example systems, where there are data available. Ranges and qualitative measures are given where there is significant variability reported or differences between reports. "–" is used to indicate no available reports.

sed = sediment.

[a] (229, 244, 245).

[b] (223, 246).

[c] (228, 232, 233).

[d] (228, 247)

[e] (248–250).

[f] Aqueous extracts.

(254,255). Iron and sulfur oxidative metabolisms are also found where microaerophilic conditions or sufficient nitrate concentrations exist (254,256). Sulfate is a less dominant anion in continental settings relative to its ubiquity in the marine realm, but where present, it can fuel significant populations of autotrophic and heterotrophic sulfate-reducing bacteria (254,255).

Heterotrophic processes rely on the input of DOC from the mineralization of sedimentary carbon, aquifer recharge, or *in situ* microbial activity. In 85% of US aquifers, DOC concentrations were <175 μM (median 42 and 58 μM). These ranges were not significantly different between sedimentary and crystalline aquifers (ranging from 8 to 275 μM, median 42 μM) (257). A more recent analysis of DOC concentration across UK aquifers showed a range of 15–1550 μM (257 μM average) (226), although this sampling includes evidence for significant contamination from agriculture and concomitant microbial respiration that introduced OM. Locally, high concentrations of organic acids (up to 60 μM formate) can be produced by microbial degradation of complex sedimentary OM, particular in shale horizons, which may then diffuse to more porous sediments, driving respiration (258). For shallow aquifer systems, periodic environmental changes related to seasonal shifts, water table fluctuations, or land use may transport both DOC and oxidants to depth, driving increases in heterotrophic respiration (244,259).

Methane is a ubiquitous reservoir of organic carbon in sedimentary aquifers. Methane concentrations are extremely variable but sometimes can reach extremely high values (e.g. 0.9 nM to 2.7 mM in the Lower Saxony region of Germany (229) and 3.1 nM to 293 μM across Great Britain (230)). Sources of methane vary and include abiotic and biotic sources, including microbial methanogenesis (including hydrogenotrophic, acetoclastic, and methyl fermentation) as well as thermogenic cracking of buried OM (224,229,230). The isotopic composition of methane and co-occurring short-chain hydrocarbons can be used to assess methane sources and suggest active microbial CO_2 reduction as the primary source in both German and British aquifers (229,230). High concentrations tend to correlate to organic-rich, low-SO_4 geological formations (229).

16.5.4 Hydrocarbon Reservoirs

Hydrocarbon reservoirs were among the earliest studied continental deep biospheres, with experiments beginning in the 1920s by Colwell and D'Hondt (213). These systems are characterized by large accumulations of liquid and gaseous hydrocarbons, providing abundant sources of carbon and electron donors, but they tend to be correspondingly depleted in oxidants and nutrients. Extremely high concentrations of volatile organic acids (particularly acetate) comprise the majority of DOC in the water phases of hydrocarbon-bearing basins, reaching concentrations of up to hundreds of mM (232,260,261).

The most significant metabolisms in hydrocarbon reservoirs are sulfate reduction, methanogenesis, acetogenesis, iron reduction, and fermentation (260–262), the balance of which is determined by electron acceptor supply. Spatially, biodegradation of oil is concentrated at the oil–water interface and is limited by reservoir temperature, with limited activity being observed above 80°C (234,263). Anthropogenic influence through drilling, water introduction, casing, and fracturing of reservoirs and the introduction of exogenous microbes can significantly change *in situ* carbon cycling, most notoriously causing reservoir souring by stimulating sulfate-reducing populations in previously methanogenic reservoirs. For more a complete description of carbon cycling and biodegradation in hydrocarbon reservoirs, see the reviews by Larter et al. (263), Means and Hubbard (232), and Head et al. (260).

16.5.5 Deep Coal Beds

Coal is formed through the burial and diagenesis of large accumulations of terrestrial plant matter and therefore contains an extremely high organic carbon content. The bioavailablity of this OM to deep subsurface microbes depends on thermal maturity and burial history, which control the form and speciation of OM as well as the sterilization history of resident microbial populations (235). Low-maturity (rank) coals are the most bioavailable and actively accumulate biogenic methane. Aqueous extracts of low-maturity coals and lignites produce extremely high concentrations of organic acids, including acetate, formate, and oxalate in the range of 0.37–2.5.0 mg/g sediment (247,264). Yields of labile OM decrease significantly with increasing thermal maturity (247,264).

Biogenic methane appears to universally accumulate in coals at $<80°C$ (235). Microbial processing of coal to methane is a multistep process that requires and supports an ecosystem of microbes. First, organic polymers are fragmented into hydrocarbon intermediates, followed by a secondary fermentation to methanogenic substrates like CO_2, H_2, organic acids, and alcohols. These substrates then fuel acetoclastic, methylotrophic, and hydrogenotrophic methanogens (235). The rate and efficiency of these processes in different coal deposits and the accessibility of this methane for extraction are of considerable economic importance. For a more complete review of coal bed biogeochemistry, see Strapoc et al. (235).

16.5.6 Deep Bedrock

Deep crystalline bedrock-hosted biospheres stand in contrast to the aforementioned settings in their constituent reservoirs and fluxes of carbon and energy. Here, inputs from the surface are limited, with water residence times reaching millions to billions of years (e.g. (211,265)) and sedimentary carbon (where present) is recalcitrant to graphitic carbon (266). The largest pool of organic carbon is often as methane, although considerable variability is present with depth and lithology (248,266). Porosity and permeability is fracture based, adding a stochastic temporal dynamic to fluxes and mixing (265,267).

Hydrogen, methane, sulfate, and iron cycling drive primary production in deep crystalline bedrock settings. The relative importance of these processes is variable with depth, host lithology, and fluid chemistry (221,248). Precambrian rocks, which constitute the best-studied deep crystalline biospheres, are prolific producers of hydrogen (up to mM concentrations) (238,267), which can serve as the terminal electron donor for either sulfate reduction or CO_2 reduction-based metabolisms (249,265,268–272).

Analysis of subsurface genomes shows that enzymes of hydrogen metabolism are overrepresented, emphasizing the potential dominance of this metabolic strategy (273). Metagenomic surveys suggest that carbon fixation is performed primarily using the reductive acetyl CoA pathway (256,271,272). Extreme metabolic flexibility has been observed in the cosmopolitan subsurface dweller *Candidatus Desulforudis audaxviator*, which can grow in near monoculture in isolated fracture systems (220) and has been found globally (221,274).

Heterotrophic microbes and metabolisms have been found to dominate in some deep crystalline settings, despite the apparent limited availability of exogenous carbon. Abiogenic sources of methane in addition to limited populations of microbial methanogens supply a significant flux of methane to fuel methanotrophic communities reaching tens of mM concentrations (248,266,275,276). Methane cycling has been observed to be most active at moderate depths (0.5–1.5 km) rather than in the deepest, most isolated settings (248,266,277). Other sources of carbon for heterotrophic communities include biofilm-based small organic compounds (265,278), free organic acids formed through fermentation or abiogenesis, and ancient organic carbon (279).

Mineral and biofilm-based metabolisms may be particularly important in deep crystalline settings. Increasing evidence of extensive adaptation to life in biofilms is emerging from these environments in the form of physical adaptations like grappling appendages observed in putative *Candidatus "Altiarchaeum"* (256), as well extracellular electron transport in subsurface isolates (280). In high-pH settings, autotrophic populations may depend on solid carbonate minerals due to carbon speciation in ultrabasic environments (269). Differences between the attached and planktonic communities have long been observed in crystalline aquifer settings (221,277,281), often with orders of magnitude higher cell densities present within the biofilms (278). The net suggestion of these observations is that mineral and biofilm-based lifestyles are the norm for the deep continental subsurface, but are as yet undersampled. Efforts to cultivate and characterize the metabolic capacities of these attached communities are underway.

16.6 Conclusion

16.6.1 Broad Similarities across Systems

The deep biosphere spans an incredible range of physical and chemical conditions. Despite their heterogeneity, some broad similarities are present across systems. Organic carbon concentrations reach their highest levels in regions that have large inputs from primary producers, either presently (continental margins, shallow sedimentary aquifers, diffuse hydrothermal vents) or in the past (hydrocarbon reservoirs, coal beds). In contrast, concentrations are lower in rocky areas with little sedimentary input and low amounts of chemolithoautotrophy in both continental (shallow igneous aquifers, deep bedrock) and marine (ridge flanks, high-temperature hydrothermal vents) systems. Elevated concentrations of methane are related to the anaerobic breakdown of OM and associated methanogenesis (sedimentary porewaters, sedimented hydrothermal systems, hydrocarbon reservoirs, coal beds), but also due to hydrogenotrophic methanogenesis, with the hydrogen supplied by water–rock reactions (basalts) and from mantle inputs (hydrothermal systems). Somewhat surprising is the persistence of some forms of organic carbon that are generally thought to be readily accessible to microorganisms, such as acetate, in many systems (sedimentary porewaters, shallow igneous aquifers, deep bedrock).

16.6.2 Limits to Knowledge and Unknowns

(1) Exchange/transformation of carbon between aqueous and solid phases. A characteristic of subsurface environments is the ubiquitous presence of a solid phase, be it from surface-derived particles or crystalline rocks. The exchange and transformation of carbon between the aqueous and solid phases is therefore a major mechanism for controlling the form and fate of carbon in the subsurface. Major questions remain as to what controls these exchanges and the degree to which they are catalyzed by minerals and microorganisms.

(2) Bioavailability of organic carbon. Microbial respiration has been invoked to account for the oxidation of OM that is millions of years old in sediments (67) and the removal of oceanic dissolved OM that is thousands of years old in the basaltic basement (135). Even 365 million-year-old shale carbon can be incorporated into cellular biomass given the right conditions (282). These studies raise the intriguing question of whether all OM is, ultimately, bioavailable give enough time and a favorable setting, or whether there is some pool that will resist remineralization to CO_2 under all circumstances. This question ties directly into point (3) below.

(3) Controls on reaction rates of biogeochemical processes. The rate at which carbon is transformed or remineralized is fundamentally important to understanding the short- and long-term controls on the global carbon cycle and to identifying the distribution of subsurface life. The processes occurring in the marine and continental subsurface are inherently difficult to accurately mimic in laboratory experiments. While short-term experiments can address the more reactive portions of the organic pool, our understanding of the transformations that occur over century or millennium timescales, particularly when uncultured microorganisms mediate the reactions, is more challenging but no less important.

(4) Predictive ability. This review describes what types of carbon are present in distinct geological, geochemical, and biological environments. Ultimately, however, the reverse is a major goal: the ability to have such a fundamental grasp of the mechanistic controls on carbon cycling that it is possible to accurately predict what types and abundances of carbon will be present in a given system.

(5) Characterization of OM. Despite decades of effort and major progress on several fronts, the molecular structure of the vast majority of OM in the subsurface remains uncharacterized. This gap in our knowledge will continue to inhibit our understanding of carbon biogeochemical cycling in the subsurface.

Acknowledgments

Funding support for SQL is from C-DEBI (NSF grant OIA-0939564) and National Science Foundation grant OCE-1536702. Funding support for ADS is from C-DEBI and NSF OCE-1357242. Funding support for MRO is from NASA Exobiology (NNH14ZDA001N) and NAI Life Underground (NNH12ZDA002C). This is C-DEBI contribution 475.

Questions for the Classroom

1 Continental and oceanic subsurface crystalline aquifers are similar in many ways. How do the characteristics of carbon in, for example, the oceanic North Pond and the continental South African gold mine systems compare and differ? Why?
2 Are the sites studied thus far representative of globally relevant locations where carbon is processed in the subsurface? What locations or geological systems are missing? Why have these not yet been studied? What are the prospects for studying these locations?
3 What effect are humans having on carbon in the deep biosphere?
4 What are the next steps to improve our ability to computationally model different forms of carbon in the subsurface and how they change in the subsurface?
5 Imagine a hypothetical microorganism that is capable of remineralizing any type of nonbioavailable OM back to CO_2. If this microorganism proliferated in subsurface environments, what would the effect be?
6 Ultraviolet radiation can create radical species (compounds that are highly reactive due to the presence of an unpaired electron) that can oxidize organic molecules via random reactions. What is the likely effect of a change in atmospheric ozone concentrations, and therefore ultraviolet flux to Earth's surface, on organic carbon burial rates?

References

1. Martiny JBH, Bohannan BJM, Brown JH, Colwell RK, Fuhrman JA, Green JL, et al. Microbial biogeography: putting microorganisms on the map. *Nature Reviews Microbiology*. 2006;4(2):102–112.
2. Hedges JI. Global biogeochemical cycles – progress and problems. *Marine Chemistry*. 1992;39(1–3):67–93.

3. Schrenk M, Huber J, Edwards K. Microbial provinces in the subseafloor. *Annual Review of Marine Science*. 2010;2:279–304.
4. Rex MA, Etter RJ, Morris JS, Crouse J, McClain CR, Johnson NA, et al. Global bathymetric patterns of standing stock and body size in the deep-sea benthos. *Marine Ecology Progress Series*. 2006;317:1–8.
5. LaRowe DE, Burwicz E, Arndt S, Dale AW, Amend JP. Temperature and volume of global marine sediments. *Geology*. 2017;45(3):275–278.
6. Burdige DJ. Burial of terrestrial organic matter in marine sediments: a re-assessment. *Global Biogeochemical Cycles*. 2005;19(4):GB4011.
7. Burdige DJ. Preservation of organic matter in marine sediments: controls, mechanisms, and an imbalance in sediment organic carbon budgets? *Chemical Reviews*. 2007;107(2):467–485.
8. Hedges JI, Keil RG. Sedimentary organic matter preservation – an assessment and speculative synthesis. *Marine Chemistry*. 1995;49(2–3):81–115.
9. Zonneveld KAF, Versteegh GJM, Kasten S, Eglinton TI, Emeis KC, Huguet C, et al. Selective preservation of organic matter in marine environments; processes and impact on the sedimentary record. *Biogeosciences*. 2010;7(2):483–511.
10. Blair NE, Aller RC. The fate of terrestrial organic carbon in the marine environment. *Annual Review of Marine Science*. 2012;4:401–423.
11. Keil RG, Mayer LM. Mineral matrices and organic matter. In: Holland HD, Turekian KK, eds. *Treatise on Geochemistry*, 2nd edn. Oxford: Elsevier, 2014, pp. 337–359.
12. Keil R, Annual R. Anthropogenic forcing of carbonate and organic carbon preservation in marine sediments. *Annual Review of Marine Sciences*, 2017;9:151–172.
13. Middelburg JJ. Reviews and syntheses: to the bottom of carbon processing at the seafloor. *Biogeosciences*. 2018;15(2):413–427.
14. Hedges JI, Keil RG, Benner R. What happens to terrestrial organic matter in the ocean? *Organic Geochemistry*. 1997;27(5–6):195–212.
15. Blair NE, Leithold EL, Ford ST, Peeler KA, Holmes JC, Perkey DW. The persistence of memory: the fate of ancient sedimentary organic carbon in a modern sedimentary system. *Geochimica et Cosmochimica Acta*. 2003;67(1):63–73.
16. Galy V, Peucker-Ehrenbrink B, Eglinton T. Global carbon export from the terrestrial biosphere controlled by erosion. *Nature*. 2015;521(7551):204–207.
17. Hilton RG, Galy A, Hovius N, Horng MJ, Chen H. Efficient transport of fossil organic carbon to the ocean by steep mountain rivers: an orogenic carbon sequestration mechanism. *Geology*. 2011;39(1):71–74.
18. Goni MA, Hatten JA, Wheatcroft RA, Borgeld JC. Particulate organic matter export by two contrasting small mountainous rivers from the Pacific Northwest, USA. *Journal of Geophysical Research – Biogeosciences*. 2013;118(1):112–134.
19. Bao HY, Lee TY, Huang JC, Feng XJ, Dai MH, Kao SJ. Importance of Oceanian small mountainous rivers (SMRs) in global land-to-ocean output of lignin and modern biospheric carbon. *Scientific Reports*. 2015;5:16217.
20. Opsahl S, Benner R, Amon RMW. Major flux of terrigenous dissolved organic matter through the Arctic Ocean. *Limnology and Oceanography*. 1999;44(8): 2017–2023.
21. Dittmar T, Kattner G. The biogeochemistry of the river and shelf ecosystem of the Arctic Ocean: a review. *Marine Chemistry*. 2003;83(3–4):103–120.
22. Raymond PA, McClelland JW, Holmes RM, Zhulidov AV, Mull K, Peterson BJ, et al. Flux and age of dissolved organic carbon exported to the Arctic Ocean: a carbon

isotopic study of the five largest arctic rivers. *Global Biogeochemical Cycles*. 2007;21(4):GB4011.

23. Feng XJ, Vonk JE, van Dongen BE, Gustafsson O, Semiletov IP, Dudarev OV, et al. Differential mobilization of terrestrial carbon pools in Eurasian Arctic river basins. *Proceedings of the National Academy of Sciences of the United States of America*. 2013;110(35):14168–14173.
24. Blair NE, Carter WD, Boehme SE. Diagenetic isotope effects in an anoxic marine sediment. *Abstracts of Papers of the American Chemical Society*. 1991;201:34-GEOC.
25. Schmitz RA, Daniel R, Deppenmeier U, Gottschalk G. *The Anaerobic Way of Life. Prokaryotes: A Handbook on the Biology of Bacteria*, Vol. 2, 3rd edn: *Ecophysiology and Biochemistry*. Washington, DC: American Chemical Society, 2006, pp. 86–101.
26. Blair NE, Martens CS, Desmarais DJ. Natural abundances of carbon isotopes in acetate from a coastal marine sediment. *Science*. 1987;236(4797):66–68.
27. Gelwicks JT, Risatti JB, Hayes JM. Carbon isotope effects associated with aceticlastic methanogenesis. *Applied and Environmental Microbiology*. 1994;60(2):467–472.
28. Hinrichs KU, Hayes JM, Bach W, Spivack AJ, Hmelo LR, Holm NG, et al. Biological formation of ethane and propane in the deep marine subsurface. *Proceedings of the National Academy of Sciences of the United States of America*. 2006;103 (40):14684–14689.
29. Heuer VB, Pohlman JW, Torres ME, Elvert M, Hinrichs KU. The stable carbon isotope biogeochemistry of acetate and other dissolved carbon species in deep subseafloor sediments at the northern Cascadia Margin. *Geochimica et Cosmochimica Acta*. 2009;73(11):3323–3336.
30. Suess E. Particulate organic–carbon flux in the oceans – surface productivity and oxygen utilization. *Nature*. 1980;288(5788):260–263.
31. Abdulla H, Burdige D, Komada T. Accumulation of deaminated peptides in anoxic sediments of Santa Barbara Basin. *Geochimica et Cosmochimica Acta*. 2018;223: 245–258.
32. Jiao N, Herndl G, Hansell D, Benner R, Kattner G, Wilhelm S, et al. Microbial production of recalcitrant dissolved organic matter: long-term carbon storage in the global ocean. *Nature Reviews Microbiology*. 2010;8(8):593–599.
33. Lechtenfeld OJ, Kattner G, Flerus R, McCallister SL, Schmitt-Kopplin P, Koch BP. Molecular transformation and degradation of refractory dissolved organic matter in the Atlantic and Southern Ocean. *Geochimica et Cosmochimica Acta*. 2014;126: 321–337.
34. Hartnett HE, Keil RG, Hedges JI, Devol AH. Influence of oxygen exposure time on organic carbon preservation in continental margin sediments. *Nature*. 1998;391 (6667):572–574.
35. Hee CA, Pease TK, Alperin MJ, Martens CS. Dissolved organic carbon production and consumption in anoxic marine sediments: a pulsed-tracer experiment. *Limnology and Oceanography*. 2001;46(8):1908–1920.
36. Benz R, Bauer K. Permeation of hydrophilic molecules through the outer membrane of Gram-negative bacteria. *European Journal of Biochemistry*. 1988;176:1–19.
37. Burdige DJ, Gardner KG. Molecular weight distribution of dissolved organic carbon in marine sediment pore waters. *Marine Chemistry*. 1998;62(1–2):45–64.
38. Simpson AJ, Kingery WL, Hayes MHB, Spraul M, Humpfer E, Dvortsak P, et al. Molecular structures and associations of humic substances in the terrestrial environment. *Naturwissenschaften*. 2002;89(2):84–88.

39. Benner R, Amon RMW. The size–reactivity continuum of major bioelements in the ocean. *Annual Review of Marine Science*. 2015;7:185–205.
40. Walker BD, Beaupre SR, Guilderson TP, McCarthy MD, Druffel ERM. Pacific carbon cycling constrained by organic matter size, age and composition relationships. *Nature Geoscience*. 2016;9(12):888–891.
41. Kelleher BP, Simpson AJ. Humic substances in soils: are they really chemically distinct? *Environmental Science & Technology*. 2006;40(15):4605–4611.
42. Coolen MJL, Overmann J. Functional exoenzymes as indicators of metabolically active bacteria in 124,000-year-old sapropel layers of the eastern Mediterranean Sea. *Applied and Environmental Microbiology*. 2000;66(6):2589–2598.
43. Coolen MJL, Cypionka H, Sass AM, Sass H, Overmann J. Ongoing modification of Mediterranean Pleistocene sapropels mediated by prokaryotes. *Science*. 2002;296 (5577):2407–2410.
44. Schmidt JM. *Microbial Extracellular Enzymes in Marine Sediments: Methods Development and Potential Activities in the Baltic Sea Deep Biosphere*. Masters thesis, University of Tennessee, 2016.
45. Robador A, Bruchert V, Steen AD, Arnosti C. Temperature induced decoupling of enzymatic hydrolysis and carbon remineralization in long-term incubations of Arctic and temperate sediments. *Geochimica et Cosmochimica Acta*. 2010;74(8):2316–2326.
46. Noinaj N, Guillier M, Barnard TJ, Buchanan SK. TonB-dependent transporters: regulation, structure, and function. *Annual Review of Microbiology*. 2010;64:43–60.
47. Arnosti C, Repeta DJ. Extracellular enzyme activity in anaerobic bacterial cultures – evidence of pullulanase activity among mesophilic marine bacteria. *Applied and Environmental Microbiology*. 1994;60(3):840–846.
48. Cuskin F, Lowe EC, Temple MJ, Zhu YP, Cameron EA, Pudlo NA, et al. Human gut Bacteroidetes can utilize yeast mannan through a selfish mechanism. *Nature*. 2015;517(7533):165–186.
49. Tanoue E, Handa N. Distribution of particulate organic carbon and nitrogen in the Bering Sea and Northern North Pacific Ocean. *Journal of the Oceanographical Society of Japan*. 1979;35:47–62.
50. Mayer LM. Relationships between mineral surfaces and organic carbon concentrations in soils and sediments. *Chemical Geology*. 1994;114(3–4):347–363.
51. Mayer LM. Surface area control of organic carbon accumulation in continental shelf sediments. *Geochimica et Cosmochimica Acta*. 1994;58(4):1271–1284.
52. Mayer LM. Extent of coverage of mineral surfaces by organic matter in marine sediments. *Geochimica et Cosmochimica Acta*. 1999;63(2):207–215.
53. Knicker H, Hatcher PG. Survival of protein in an organic-rich sediment: possible protection by encapsulation in organic matter. *Naturwissenschaften*. 1997;84(6): 231–234.
54. Moore EK, Nunn BL, Goodlett DR, Harvey HR. Identifying and tracking proteins through the marine water column: insights into the inputs and preservation mechanisms of protein in sediments. *Geochimica et Cosmochimica Acta*. 2012;83: 324–359.
55. Wu SC, Gschwend PM. Sorption kinetics of hydrophobic organic compounds to natural sediments and soils. *Environmental Science & Technology*. 1986;20(7): 717–725.
56. Espeland E, Wetzel R. Complexation, stabilization, and UV photolysis of extracellular and surface-bound glucosidase and alkaline phosphatase: implications for biofilm microbiota. *Microbial Ecology*. 2001;42(4):572–585.

57. Tietjen T, Wetzel R. Extracellular enzyme–clay mineral complexes: enzyme adsorption, alteration of enzyme activity, and protection from photodegradation. *Aquatic Ecology*. 2003;37(4):331–339.
58. Lalonde K, Mucci A, Ouellet A, Gelinas Y. Preservation of organic matter in sediments promoted by iron. *Nature*. 2012;483(7388):198–200.
59. Barber A, Brandes J, Leri A, Lalonde K, Balind K, Wirick S, et al. Preservation of organic matter in marine sediments by inner-sphere interactions with reactive iron. *Scientific Reports*. 2017;7:366.
60. Cowie GL, Hedges JI. The role of anoxia in organic matter preservation in coastal sediments – relative stabilities of the major biochemicals under oxic and anoxic depositional conditions. *Organic Geochemistry*. 1992;19(1–3):229–234.
61. Canfield DE. Factors influencing organic carbon preservation in marine sediments. *Chemical Geology*. 1994;114(3–4):315–329.
62. Keil RG, Hu FS, Tsamakis EC, Hedges JI. Pollen in marine sediments as an indicator of oxidation of organic matter. *Nature*. 1994;369(6482):639–641.
63. Hedges JI, Hu FS, Devol AH, Hartnett HE, Tsamakis E, Keil RG. Sedimentary organic matter preservation: a test for selective degradation under oxic conditions. *American Journal of Science*. 1999;299(7–9):529–555.
64. Kirk TK, Farrell RL. Enzymatic combustion – the microbial-degradation of lignin. *Annual Review of Microbiology*. 1987;41:465–505.
65. Cowie GL, Hedges JI, Prahl FG, Delange GJ. Elemental and major biochemical changes across an oxidation front in an relict turbidite – an oxygen effect. *Geochimica et Cosmochimica Acta*. 1995;59(1):33–46.
66. Cowie GL, Calvert SE, Pedersen TF, Schulz H, von Rad U. Organic content and preservational controls in surficial shelf and slope sediments from the Arabian Sea (Pakistan margin). *Marine Geology*. 1999;161(1):23–38.
67. Røy H, et al. Aerobic microbial respiration in 86-million-year-old deep sea red clay. *Science*. 2012;336:922–925.
68. Westrich JT, Berner RA. The role of sedimentary organic matter in bacterial sulfate reduction – the G model tested. *Limnology and Oceanography*. 1984;29(2): 236–249.
69. Boudreau BP, Ruddick BR. On a reactive continuum representation of organic-matter diagenesis. *American Journal of Science*. 1991;291(5):507–538.
70. Tarutis WJ. On the equivalence of the power and reactive continuum models of organic-matter diagenesis. *Geochimica et Cosmochimica Acta*. 1993;57(6):1349–1350.
71. Middelburg JJ. A simple rate model for organic matter decomposition in marine sediments. *Geochimica et Cosmochimica Acta*. 1989;53(7):1577–1581.
72. Hedges JI, Eglinton G, Hatcher PG, Kirchman DL, Arnosti C, Derenne S, et al. The molecularly-uncharacterized component of nonliving organic matter in natural environments. *Organic Geochemistry*. 2000;31(10):945–958.
73. Wieder WR, Bonan GB, Allison SD. Global soil carbon projections are improved by modelling microbial processes. *Nature Climate Change*. 2013;3(10):909–912.
74. Bradley JA, Amend JP, LaRowe DE. Bioenergetic controls on microbial ecophysiology in marine sediments. *Frontiers in Microbiology*. 2018;9:180.
75. Johnson HP, Pruis MJ. Fluxes of fluid and heat from the oceanic crustal reservoir. *Earth and Planetary Science Letters*. 2003;216(4):565–574.
76. Edmond JM, Measures C, McDuff RE, Chan LH, Collier R, Grant B, et al. Ridge crest hydrothermal activity and the balances of the major and minor elements in the ocean – Galapagos data. *Earth and Planetary Science Letters*. 1979;46(1):1–18.

77. Mottl MJ, Wheat CG. Hydrothermal circulation through midocean ridge flanks – fluxes of heat and magnesium. *Geochimica et Cosmochimica Acta*. 1994;58 (10):2225–2237.
78. Elderfield H, Schultz A. Mid-ocean ridge hydrothermal fluxes and the chemical composition of the ocean. *Annual Review of Earth and Planetary Sciences*. 1996;24:191–224.
79. German CR, Casciotti KA, Dutay JC, Heimburger LE, Jenkins WJ, Measures CI, et al. Hydrothermal impacts on trace element and isotope ocean biogeochemistry. *Philosophical Transactions of the Royal Society A –Mathematical Physical and Engineering Sciences*. 2016;374(2081):20160035.
80. Stein CA, Stein S. Constraints on hydrothermal heat-flux through the oceanic lithosphere from global heat flow. *Journal of Geophysical Research – Solid Earth*. 1994;99(B2):3081–3095.
81. Alt JC. Sulfur isotopic profile through the oceanic-crust – sulfur mobility and seawater–crustal sulfur exchange during hydrothermal alteration. *Geology*. 1995;23 (7):585–588.
82. Summit M, Baross JA. A novel microbial habitat in the mid-ocean ridge subseafloor. *Proceedings of the National Academy of Sciences of the United States of America*. 2001;98(5):2158–2163.
83. Santelli CM, Orcutt BN, Banning E, Bach W, Moyer CL, Sogin ML, et al. Abundance and diversity of microbial life in ocean crust. *Nature*. 2008;453(7195): 653–657.
84. Orcutt BN, Sylvan JB, Knab NJ, Edwards KJ. Microbial ecology of the dark ocean above, at, and below the seafloor. *Microbiology and Molecular Biology Reviews*. 2011;75(2):361–422.
85. Sclater JG, Jaupart C, Galson D. The heat flow through oceanic and continental crust and the heat loss of the Earth. *Reviews of Geophysics*. 1980;18(1):269–311.
86. Butterfield DA, McDuff RE, Mottl MJ, Lilley MD, Lupton JE, Massoth GJ. Gradients in the composition of hydrothermal fluids from the Endeavor Segment vent field – phase-separation and brine loss. *Journal of Geophysical Research – Solid Earth*. 1994;99(B5):9561–9583.
87. Tivey MK, Bradley AM, Joyce TM, Kadko D. Insights into tide-related variability at seafloor hydrothermal vents from time-series temperature measurements. *Earth and Planetary Science Letters*. 2002;202(3–4):693–707.
88. Lilley MD, Butterfield DA, Lupton JE, Olson EJ. Magmatic events can produce rapid changes in hydrothermal vent chemistry. *Nature*. 2003;422(6934):878–881.
89. Seewald J, Cruse A, Saccocia P. Aqueous volatiles in hydrothermal fluids from the Main Endeavour Field, northern Juan de Fuca Ridge: temporal variability following earthquake activity. *Earth and Planetary Science Letters*. 2003;216 (4):575–590.
90. Sarmiento JL, Gruber N. *Ocean Biogeochemical Dynamics*. Princeton, NJ: Princeton University Press, 2006.
91. Hansell DA, Carlson CA, Repeta DJ, Schlitzer R. Dissolved organic matter in the ocean a controversy stimulates new insights. *Oceanography*. 2009;22(4):202–211.
92. Hansell DA. Recalcitrant dissolved organic carbon fractions. *Annual Review of Marine Science*. 2013;5:421–445.
93. Druffel ERM, Williams PM, Bauer JE, Ertel Jr. Cycling of dissolved and particulate organic-matter in the open ocean. *Journal of Geophysical Research – Oceans*. 1992;97(C10):15639–15659.

94. Hertkorn N, Benner R, Frommberger M, Schmitt-Kopplin P, Witt M, Kaiser K, et al. Characterization of a major refractory component of marine dissolved organic matter. *Geochimica et Cosmochimica Acta*. 2006;70(12):2990–3010.
95. Aluwihare LI, Repeta DJ, Chen RF. A major biopolymeric component to dissolved organic carbon in surface sea water. *Nature*. 1997;387(6629):166–169.
96. Arakawa N, Aluwihare LI, Simpson AJ, Soong R, Stephens BM, Lane-Coplen D. Carotenoids are the likely precursor of a significant fraction of marine dissolved organic matter. *Science Advances*. 2017;3(9):e1602976.
97. Repeta DJ. Chemical characterization and cycling of dissolved organic matter. In: Hansell D, ed. *Biogeochemistry of Marine Dissolved Organic Matter*. Washington, DC: American Association for the Advancement of Science, 2015, pp. 21–63.
98. Campbell AC, Palmer MR, Klinkhammer GP, Bowers TS, Edmond JM, Lawrence JR, et al. Chemistry of hot springs on the Mid-Atlantic Ridge. *Nature*. 1988;335 (6190):514–519.
99. Conway TM, John SG. Quantification of dissolved iron sources to the North Atlantic Ocean. *Nature*. 2014;511(7508):212–215.
100. Resing JA, Sedwick PN, German CR, Jenkins WJ, Moffett JW, Sohst BM, et al. Basin-scale transport of hydrothermal dissolved metals across the South Pacific Ocean. *Nature*. 2015;523(7559):200–203.
101. McCollom TM. Observational, experimental, and theoretical constraints on carbon cycling in mid-ocean ridge hydrothermal systems. In: Lowell RP, Seewald JS, Metaxas A, Perfit MR, eds. *Magma to Microbe: Modeling Hydrothermal Processes at Ocean Spreading Centers*. Washington, DC: American Geophysical Union, 2008. pp. 193–213.
102. Pineau F, Javoy M. Carbon isotopes and concentrations in mid-ocean ridge basalts. *Earth and Planetary Science Letters*. 1983;62(2):239–257.
103. DesMarais DJ, Moore JG. Carbon and its isotopes in mid-ocean basaltic glasses. *Earth and Planetary Science Letters*. 1984;69(1):43–57.
104. Kroopnick PM. The distribution of C-13 of sigma-CO_2 in the world ocean. *Deep-Sea Research Part A – Oceanographic Research Papers*. 1985;32(1):57–84.
105. Proskurowski G, Lilley MD, Brown TA. Isotopic evidence of magmatism and seawater bicarbonate removal at the endeavour hydrothermal system. *Earth and Planetary Science Letters*. 2004;225(1–2):53–61.
106. Alt JC, Teagle DAH. Hydrothermal alteration of upper oceanic crust formed at a fast-spreading ridge: mineral, chemical, and isotopic evidence from ODP Site 801. *Chemical Geology*. 2003;201(3–4):191–211.
107. Eickmann B, Bach W, Rosner M, Peckmann J. Geochemical constraints on the modes of carbonate precipitation in peridotites from the Logatchev Hydrothermal Vent Field and Gakkel Ridge. *Chemical Geology*. 2009;268(1–2):97–106.
108. Beaupre SR, Druffel ERM. Constraining the propagation of bomb-radiocarbon through the dissolved organic carbon (DOC) pool in the northeast Pacific Ocean. *Deep-Sea Research Part I – Oceanographic Research Papers*. 2009;56(10):1717–1726.
109. Reeves EP, McDermott JM, Seewald JS. The origin of methanethiol in midocean ridge hydrothermal fluids. *Proceedings of the National Academy of Sciences of the United States of America*. 2014;111(15):5474–5479.
110. Riedel M, Collett TS, Malone M, Scientists E. Expedition 311 synthesis: scientific findings. *Proceedings of the Integrated Ocean Drilling Program*. 2006;311:2.
111. Von Damm KL, Lilley MD. Diffuse flow hydrothermal fluids from 9-degrees 50' N East Pacific Rise: origin, evolution and biogeochemical controls. In: William

SDW, Edward FD, Deborah SK, John AB, Craig SC, eds. *The Subseafloor Biosphere at Mid-Ocean Ridges. Geophysical Monograph Series 144*. Washington, DC: American Geophysical Union, 2004, pp. 245–268.

112. Proskurowski G, Lilley MD, Seewald JS, Früh-Green GL, Olson EJ, Lupton JE, et al. Abiogenic hydrocarbon production at Lost City hydrothermal field. *Science*. 2008;319(5863):604–607.
113. Yucel M, Luther GW. Temporal trends in vent fluid iron and sulfide chemistry following the 2005/2006 eruption at East Pacific Rise, 9 degrees 50' N. *Geochemistry, Geophysics, Geosystems*. 2013;14(4):759–765.
114. Butterfield DA, Massoth GJ, McDuff RE, Lupton JE, Lilley MD. Geochemistry of hydrothermal fluids from Axial Seamount hydrothermal emissions study vent field, Juan de Fuca Ridge – subseafloor boiling and subsequent fluid-rock interaction. *Journal of Geophysical Research – Solid Earth and Planets*. 1990;95(B8): 12895–12921.
115. Butterfield DA, Roe KK, Lilley MD, Huber JA, Baross JA, Embley RW, et al. Mixing, reaction and microbial activity in the sub-seafloor revealed by temporal and spatial variation in diffuse flow vents at Axial Volcano. *Subseafloor Biosphere at Mid-Ocean Ranges*. 2004;144:269–289.
116. Lang SQ, Butterfield DA, Lilley MD, Johnson HP, Hedges JI. Dissolved organic carbon in ridge-axis and ridge-flank hydrothermal systems. *Geochimica et Cosmochimica Acta*. 2006;70(15):3830–3842.
117. Charlou JL, Donval JP. Hydrothermal methane venting between 12-degrees-N and 26-degrees-N along the Mid-Atlantic Ridge. *Journal of Geophysical Research – Solid Earth*. 1993;98(B6):9625–9642.
118. Charlou JL, Donval JP, Jean-Baptiste P, Dapoigny A, Rona PA. Gases and helium isotopes in high temperature solutions sampled before and after ODP Leg 158 drilling at TAG hydrothermal field (26N MAR). *Geophysical Research Letters*. 1996;23:3491–3494.
119. Charlou JL, Donval JP, Konn C, Ondreas H, Fouquet Y, Jean-Baptiste P, et al. High production and fluxes of H_2 and CH_4 and evidence of abiotic hydrocarbon synthesis by serpentinization in ultramafic-hosted hydrothermal systems on the Mid-Atlantic Ridge. In: Rona PA, ed. *Diversity of Hydrothermal Systems on Slow Spreading Ocean Ridges. Geophysical Monograph Series 188*. Washington, DC: American Geophysical Union, 2010, pp. 265–296.
120. Proskurowski G, Lilley MD, Olson EJ. Stable isotopic evidence in support of active microbial methane cycling in low-temperature diffuse flow vents at 9 degrees 50' N East Pacific Rise. *Geochimica et Cosmochimica Acta*. 2008;72(8):2005–2023.
121. McCarthy M, Beaupre S, Walker B, Voparil I, Guilderson T, Druffel E. Chemosynthetic origin of C-14-depleted dissolved organic matter in a ridge-flank hydrothermal system. *Nature Geoscience*. 2011;4(1):32–36.
122. Haberstroh PR, Karl DM. Dissolved free amino-acids in hydrothermal vent habitats of the Guaymas Basin. *Geochimica et Cosmochimica Acta*. 1989;53(11):2937–2945.
123. Von Damm KL. Seafloor hydrothermal activity: black smoker chemistry and chimneys. *Annual Review of Earth and Planetary Sciences*. 1990;18:173–204.
124. Lin HT, Amend JP, LaRowe DE, Bingham JP, Cowen JP. Dissolved amino acids in oceanic basaltic basement fluids. *Geochimica et Cosmochimica Acta*. 2015;164: 175–190.
125. Lin Y, Koch B, Feseker T, Ziervogel K, Goldhammer T, Schmidt F, et al. Near-surface heating of young rift sediment causes mass production and discharge of reactive dissolved organic matter. *Scientific Reports*. 2017;7:44864.

126. McKay L, Klokman VW, Mendlovitz HP, LaRowe DE, Hoer DR, Albert D, et al. Thermal and geochemical influences on microbial biogeography in the hydrothermal sediments of Guaymas Basin, Gulf of California. *Environmental Microbiology Reports*. 2016;8(1):150–161.
127. Cruse AM, Seewald JS. Geochemistry of low-molecular weight hydrocarbons in hydrothermal fluids from Middle Valley, northern Juan de Fuca Ridge. *Geochimica et Cosmochimica Acta*. 2006;70(8):2073–2092.
128. Wheat C, Mottl M, Fisher A, Kadko D, Davis E, Baker E. Heat flow through a basaltic outcrop on a sedimented young ridge flank. *Geochemistry, Geophysics, Geosystems*. 2004;5:Q12006.
129. Wheat CG, Hulme SM, Fisher AT, Orcutt BN, Becker K. Seawater recharge into oceanic crust: IODP Exp 327 Site U1363 Grizzly Bare outcrop. *Geochemistry, Geophysics, Geosystems*. 2013;14(6):1957–1972.
130. Walker BD, McCarthy MD, Fisher AT, Guilderson TP. Dissolved inorganic carbon isotopic composition of low-temperature axial and ridge-flank hydrothermal fluids of the Juan de Fuca Ridge. *Marine Chemistry*. 2008;108(1–2):123–136.
131. Lin HT, Cowen JP, Olson EJ, Amend JP, Lilley MD. Inorganic chemistry, gas compositions and dissolved organic carbon in fluids from sedimented young basaltic crust on the Juan de Fuca Ridge flanks. *Geochimica et Cosmochimica Acta*. 2012;85:213–227.
132. Lin HT, Cowen JP, Olson EJ, Lilley MD, Jungbluth SP, Wilson ST, et al. Dissolved hydrogen and methane in the oceanic basaltic biosphere. *Earth and Planetary Science Letters*. 2014;405:62–73.
133. Lin H-T, Repeta DJ, Xu L, Rappe MS. Selective removal of isotopically enriched dissolved organic carbon from basalt-hosted deep subseafloor fluids of the Juan de Fuca Ridge. *Earth and Planetary Science Letters*. 2019;513:156–165.
134. Meyer JL, Jaekel U, Tully BJ, Glazer BT, Wheat CG, Lin HT, et al. A distinct and active bacterial community in cold oxygenated fluids circulating beneath the western flank of the Mid-Atlantic ridge. *Scientific Reports*. 2016;6:22541.
135. Walter SRS, Jaekel U, Osterholz H, Fisher AT, Huber JA, Pearson A, et al. Microbial decomposition of marine dissolved organic matter in cool oceanic crust. *Nature Geoscience*. 2018;11(5):334–339.
136. McDermott JM, Seewald JS, German CR, Sylva SP. Pathways for abiotic organic synthesis at submarine hydrothermal fields. *Proceedings of the National Academy of Sciences of the United States of America*. 2015;112(25):7668–7672.
137. Douville E, Charlou JL, Oelkers EH, Bienvenu P, Colon CFJ, Donval JP, et al. The rainbow vent fluids (36 degrees 14’ N, MAR): the influence of ultramafic rocks and phase separation on trace metal content in Mid-Atlantic Ridge hydrothermal fluids. *Chemical Geology*. 2002;184(1–2):37–48.
138. Charlou JL, Donval JP, Fouquet Y, Jean-Baptiste P, Holm N. Geochemistry of high H_2 and CH_4 vent fluids issuing from ultramafic rocks at the Rainbow hydrothermal field (36 degrees 14’ N, MAR). *Chemical Geology*. 2002;191(4):345–359.
139. Jean-Baptiste P, Fourre E, Charlou JL, German CR, Radford-Knoery J. Helium isotopes at the Rainbow hydrothermal site (Mid-Atlantic Ridge, 36 degrees 14’ N). *Earth and Planetary Science Letters*. 2004;221(1–4):325–335.
140. Kelley DS, Karson JA, Früh-Green GL, Yoerger DR, Shank TM, Butterfield DA, et al. A serpentinite-hosted ecosystem: the Lost City hydrothermal field. *Science*. 2005;307(5714):1428–1434.

141. Lang SQ, Butterfield DA, Schulte M, Kelley DS, Lilley MD. Elevated concentrations of formate, acetate and dissolved organic carbon found at the Lost City hydrothermal field. *Geochimica et Cosmochimica Acta*. 2010;74(3):941–952.
142. Lang SQ, Früh-Green GL, Bernasconi SM, Butterfield DA. Sources of organic nitrogen at the serpentinite-hosted Lost City hydrothermal field. *Geobiology*. 2013;11(2):154–169.
143. Seyfried WE, Pester NJ, Tutolo BM, Ding K. The Lost City hydrothermal system: constraints imposed by vent fluid chemistry and reaction path models on subseafloor heat and mass transfer processes. *Geochimica et Cosmochimica Acta*. 2015;163:59–79.
144. Reeves EP, Seewald JS, Saccocia P, Bach W, Craddock PR, Shanks WC, et al. Geochemistry of hydrothermal fluids from the PACMANUS, Northeast Pual and Vienna Woods hydrothermal fields, Manus Basin, Papua New Guinea. *Geochimica et Cosmochimica Acta*. 2011;75(4):1088–1123.
145. Sakai H, Gamo T, Kim ES, Tsutsumi M, Tanaka T, Ishibashi J, et al. Venting of carbon-dioxide rich fluid and hydrate formation in Mid-Okinawa Trough Backarc Basin. *Science*. 1990;248(4959):1093–1096.
146. Ishibashi J, Sano Y, Wakita H, Gamo T, Tsutsumi M, Sakai H. Helium and carbon geochemistry of hydrothermal fluids from the Mid-Okinawa Trough Back-arc Basin, southwest of Japan. *Chemical Geology*. 1995;123(1–4):1–15.
147. Welhan JA, Craig H. Methane, hydrogen and helium in hydrothermal fluids at 21N on the East Pacific Rise. In: Rona PA, ed. *Hydrothermal Processes at Seafloor Spreading Centers*. New York: Plenum, 1983, pp. 391–409.
148. Lilley MD, Baross JA, Gordon LI. Reduced gases and bacteria in hydrothermal fluids: the Galapagos Spreading Center and 21 deg N East Pacific Rise. In: Rona PA, Bostrom K, Leubier L, Smith Jr. KL, eds. *Hydrothermal Processes at Seafloor Spreading Centers*. New York: Plenum, 1983, pp. 411–449.
149. Evans WC, White LD, Rapp JB. Geochemistry of some gases in hydrothermal fluids from the southern Juan-de-Fuca-Ridge. *Journal of Geophysical Research – Solid Earth and Planets*. 1988;93(B12):15305–15313.
150. Merlivat L, Pineau F, Javoy M. Hydrothermal vent waters at 13-degrees-N on the East Pacific Rise – isotopic composition and gas concentration. *Earth and Planetary Science Letters*. 1987;84(1):100–108.
151. Jeanbaptiste P, Charlou JL, Stievenard M, Donval JP, Bougault H, Mevel C. Helium and methane measurements in hydrothermal fluids from the Mid-Atlantic Ridge – the Snake Pit Site at 23-degrees-N. *Earth and Planetary Science Letters*. 1991;106 (1–4):17–28.
152. James RH, Elderfield H, Palmer MR. The chemistry of hydrothermal fluids from the Broken Spur Site, 29-degrees-N Mid-Atlantic Ridge. *Geochimica et Cosmochimica Acta*. 1995;59(4):651–659.
153. Lein AY, Grichuk DV, Gurvich EG, Bogdanov YA. A new type of hydrogen- and methane-rich hydrothermal solutions in the rift zone of the Mid-Atlantic Ridge. *Doklady Earth Sciences*. 2000;375(9):1391–1394.
154. Martens CS. Generation of short chain organic-acid anions in hydrothermally altered sediments of the Guaymas Basin, Gulf of California. *Applied Geochemistry*. 1990;5:71–76.
155. Hawkes JA, Rossel PE, Stubbins A, Butterfield D, Connelly DP, Achterberg EP, et al. Efficient removal of recalcitrant deep-ocean dissolved organic matter during hydrothermal circulation. *Nature Geoscience*. 2015;8(11):856–860.

156. Wellsbury P, Goodman K, Barth T, Cragg B, Barnes S, Parkes R. Deep marine biosphere fuelled by increasing organic matter availability during burial and heating. *Nature*. 1997;388(6642):573–576.
157. Seewald JS, Seyfried WE, Thornton EC. Organic-rich sediment alteration – an experimental and theoretical study at elevated temperatures and pressures. *Applied Geochemistry*. 1990;5:193–209.
158. McCollom TM, Shock EL. Geochemical constraints on chemolithoautotrophic metabolism by microorganisms in seafloor hydrothermal systems. *Geochimica et Cosmochimica Acta*. 1997;61(20):4375–4391.
159. Takai K, Nakamura K, Toki T, Tsunogai U, Miyazaki M, Miyazaki J, et al. Cell proliferation at 122 degrees C and isotopically heavy CH_4 production by a hyperthermophilic methanogen under high-pressure cultivation. *Proceedings of the National Academy of Sciences of the United States of America*. 2008;105 (31):10949–10954.
160. Karl DM, Wirsen CO, Jannasch HW. Deep-sea primary production at the Galapagos hydrothermal vents. *Science*. 1980;207(4437):1345–1347.
161. Edwards KJ, Becker K, Colwell F. The deep, dark energy biosphere: intraterrestrial life on Earth. *Annual Review of Earth and Planetary Sciences*. 2012;40:551–568.
162. Wankel SD, Germanovich LN, Lilley MD, Genc G, DiPerna CJ, Bradley AS, et al. Influence of subsurface biosphere on geochemical fluxes from diffuse hydrothermal fluids. *Nature Geoscience*. 2011;4(7):461–468.
163. Mottl MJ. Partitioning of energy and mass fluxes between mid-ocean ridge axes and flanks at high and low temperature. In: Halbach PE, Tunnicliffe V, Hein JR, eds. *Energy and Mass Transfer in Marine Hydrothermal Systems*. Berlin: Dahlem University Press, 2003, pp. 271–286.
164. Wheat CG, Fisher AT, McManus J, Hulme SM, Orcutt BN. Cool seafloor hydrothermal springs reveal global geochemical fluxes. *Earth and Planetary Science Letters*. 2017;476:179–188.
165. Elderfield H, Wheat CG, Mottl MJ, Monnin C, Spiro B. Fluid and geochemical transport through oceanic crust: a transect across the eastern flank of the Juan de Fuca Ridge. *Earth and Planetary Science Letters*. 1999;172(1–2):151–165.
166. Mottl MJ, Wheat G, Baker E, Becker N, Davis E, Feely R, et al. Warm springs discovered on 3.5 Ma oceanic crust, eastern flank of the Juan de Fuca Ridge. *Geology*. 1998;26(1):51–54.
167. Sansone FJ, Mottl MJ, Olson EJ, Wheat CG, Lilley MD. CO_2-depleted fluids from mid-ocean ridge-flank hydrothermal springs. *Geochimica et Cosmochimica Acta*. 1998;62(13):2247–2252.
168. Maloszewski P, Zuber A. Influence of matrix diffusion and exchange-reactions on radiocarbon ages in fissured carbonate aquifers. *Water Resources Research*. 1991;27 (8):1937–1945.
169. Bethke CM, Johnson TM. Paradox of groundwater age. *Geology*. 2002;30 (2):107–110.
170. Bethke CM, Johnson TM. Groundwater age and groundwater age dating. *Annual Review of Earth and Planetary Sciences*. 2008;36:121–152.
171. Rona PA, Widenfalk L, Bostrom K. Serpentinized ultramafics and hydrothermal activity at the Mid-Atlantic Ridge crest near 15-degrees-N. *Journal of Geophysical Research – Solid Earth and Planets*. 1987;92(B2):1417–1427.
172. Charlou JL, Bougault H, Appriou P, Jeanbaptiste P, Etoubleau J, Birolleau A. Water column anomalies associated with hydrothermal activity between 11-degrees-40' and

13-degrees-N on the east pacific rise – discrepancies between tracers. *Deep-Sea Research Part A – Oceanographic Research Papers*. 1991;38(5):569–596.

173. Rona PA, Bougault H, Charlou JL, Appriou P, Nelsen TA, Trefry JH, et al. Hydrothermal circulation, serpentinization, and degassing at a rift-valley fracture-zone intersection – Mid-Atlantic Ridge near 15-degrees-N, 45-degrees-W. *Geology*. 1992;20(9):783–786.
174. Schrenk MO, Brazelton WJ, Lang SQ. Serpentinization, carbon, and deep life. In: Hazen RM, Jones AP, Baross JA, eds. *Carbon in Earth. Reviews in Mineralogy & Geochemistry. 75*. Chantilly: Mineralogical Society of America, 2013, pp. 575–606.
175. German CR, Bowen A, Coleman ML, Honig DL, Huber JA, Jakuba MV, et al. Diverse styles of submarine venting on the ultraslow spreading Mid-Cayman Rise. *Proceedings of the National Academy of Sciences of the United States of America*. 2010;107(32):14020–14025.
176. Haggerty JA. Evidence from fluid seeps atop serpentine seamounts in the mariana fore-arc – clues for emplacement of the seamounts and their relationship to fore-arc tectonics. *Marine Geology*. 1991;102(1–4):293–309.
177. Kelley DS, Karson JA, Blackman DK, Früh-Green GL, Butterfield DA, Lilley MD, et al. An off-axis hydrothermal vent field near the Mid-Atlantic Ridge at 30 degrees N. *Nature*. 2001;412(6843):145–149.
178. Monnin C, Chavagnac V, Boulart C, Menez B, Gerard M, Gerard E, et al. Fluid chemistry of the low temperature hyperalkaline hydrothermal system of Prony Bay (New Caledonia). *Biogeosciences*. 2014;11(20):5687–5706.
179. Ohara Y, Reagan MK, Fujikura K, Watanabe H, Michibayashi K, Ishii T, et al. A serpentinite-hosted ecosystem in the Southern Mariana Forearc. *Proceedings of the National Academy of Sciences of the United States of America*. 2012;109(8):2831–2835.
180. Konn C, Charlou JL, Holm NG, Mousis O. The production of methane, hydrogen, and organic compounds in ultramafic-hosted hydrothermal vents of the Mid-Atlantic Ridge. *Astrobiology*. 2015;15(5):381–399.
181. Bonatti E, Lawrence JR, Hamlyn PR, Breger D. Aragonite from deep-sea ultramafic rocks. *Geochimica et Cosmochimica Acta*. 1980;44(8):1207–1214.
182. Keir RS. A note on the fluxes of abiogenic methane and hydrogen from mid-ocean ridges. *Geophysical Research Letters*. 2010;37:L24609.
183. McCollom TM, Seewald JS. Abiotic synthesis of organic compounds in deep-sea hydrothermal environments. *Chemical Reviews*. 2007;107(2):382–401.
184. Wang DT, Reeves EP, McDermott JM, Seewald JS, Ono S. Clumped isotopologue constraints on the origin of methane at seafloor hot springs. *Geochimica et Cosmochimica Acta*. 2018;223:141–158.
185. Kelley DS, Früh-Green GL. Abiogenic methane in deep-seated mid-ocean ridge environments: insights from stable isotope analyses. *Journal of Geophysical Research – Solid Earth*. 1999;104(B5):10439–10460.
186. Kelley DS, Früh-Green GL. Volatile lines of descent in submarine plutonic environments: insights from stable isotope and fluid inclusion analyses. *Geochimica et Cosmochimica Acta*. 2001;65(19):3325–3346.
187. Bradley AS, Summons RE. Multiple origins of methane at the Lost City hydrothermal field. *Earth and Planetary Science Letters*. 2010;297(1–2):34–41.
188. Pisapia C, Gerard E, Gerard M, Lecourt L, Lang SQ, Pelletier B, et al. Mineralizing filamentous bacteria from the Prony Bay hydrothermal field give new insights into the functioning of serpentinization-based subseafloor ecosystems. *Frontiers in Microbiology*. 2017;8:18.

189. Lang SQ, Früh-Green GL, Bernasconi SM, Brazelton WJ, Schrenk MO, McGonigle JM. Deeply-sourced formate fuels sulfate reducers but not methanogens at Lost City hydrothermal field. *Scientific Reports.* 2018;8:755.
190. Kawka OE, Simoneit BRT. Survey of hydrothermally-generated petroleums from the Guaymas Basin spreading center. *Organic Geochemistry.* 1987;11(4): 311–328.
191. Kadko D, Baross J, Alt J. The magnitude and global implications of hydrothermal flux. In: Humphris SE, Zierenberg RA, Mullineaux LS, Thomson RE, eds. *Seafloor Hydrothermal Systems: Physical, Chemical, Biological, and Geological Interactions. Geophysical Monograph Series.* Washington, DC: American Geophysical Union, 1995, pp. 446–466.
192. Saal AE, Hauri EH, Langmuir CH, Perfit MR. Vapour undersaturation in primitive mid-ocean-ridge basalt and the volatile content of Earth's upper mantle. *Nature.* 2002;419(6906):451–455.
193. Staudigel H, Hart SR, Schmincke HU, Smith BM. Cretaceous ocean crust at DSDP Site-417 and Site-418 – carbon uptake from weathering versus loss by magmatic outgassing. *Geochimica et Cosmochimica Acta.* 1989;53(11):3091–3094.
194. Alt JC, Teagle DAH. The uptake of carbon during alteration of ocean crust. *Geochimica et Cosmochimica Acta.* 1999;63(10):1527–1535.
195. Schwarzenbach EM, Lang SQ, Frah-Green GL, Lilley MD, Bemasconi SM, Mehay S. Sources and cycling of carbon in continental, serpentinite-hosted alkaline springs in the Voltri Massif, Italy. *Lithos.* 2013;177:226–244.
196. Alt JC, Schwarzenbach EM, Frueh-Green GL, Shanks WC, Bernasconi SM, Garrido CJ, et al. The role of serpentinites in cycling of carbon and sulfur: seafloor serpentinization and subduction metamorphism. *Lithos.* 2013;178:40–54.
197. Simoneit BRT, Lonsdale PF. Hydrothermal petroleum in mineralized mounds at the seabed of Guaymas Basin. *Nature.* 1982;295(5846):198–202.
198. Kvenvolden KA, Simoneit BRT. Hydrothermally derived petroleum – examples from Guaymas Basin, Gulf of California, and Escanaba Trough, northeast Pacific-Ocean. *AAPG Bulletin – American Association of Petroleum Geologists.* 1990;74(3): 223–237.
199. Tissot BP, Welte DH. *Petroleum Formation and Occurrence.* Berlin: Springer, 1984.
200. Tissot B, Espitalie J. Thermal evolution of organic-matter in sediments – application of a mathematical simulation – petroleum potential of sedimentary basins and reconstructing thermal history of sediments. *Revue De L Institut Francais Du Petrole.* 1975;30(5):743–777.
201. Lewan MD, Winters JC, McDonald JH. Generation of oil-like pyrolyzates from organic-rich shales. *Science.* 1979;203(4383):897–899.
202. Seewald JS. Aqueous geochemistry of low molecular weight hydrocarbons at elevated temperatures and pressures: constraints from mineral buffered laboratory experiments. *Geochimica et Cosmochimica Acta.* 2001;65(10):1641–1664.
203. Shock EL, Canovas P, Yang ZM, Boyer G, Johnson K, Robinson K, et al. Thermodynamics of organic transformations in hydrothermal fluids. *Thermodynamics of Geothermal Fluids.* 2013;76:311–350.
204. Simoneit BRT. Hydrothermal petroleum – genesis, migration, and deposition in Guaymas Basin, Gulf of California. *Canadian Journal of Earth Sciences.* 1985; 22(12):1919–1929.
205. Didyk BM, Simoneit BRT. Hydrothermal oil of Guaymas Basin and implications for petroleum formation mechanisms. *Nature.* 1989;342(6245):65–69.

206. Rushdi AI, Simoneit BRT. Condensation reactions and formation of amides, esters, and nitriles under hydrothermal conditions. *Astrobiology*. 2004;4(2):211–224.
207. Simoneit B, Lein A, Peresypkin V, Osipov G. Composition and origin of hydrothermal petroleum and associated lipids in the sulfide deposits of the Rainbow Field (Mid-Atlantic Ridge at 36 degrees N). *Geochimica et Cosmochimica Acta*. 2004;68(10):2275–2294.
208. Parkes RJ, Taylor J, Jorckramberg D. Demonstration, using *Desulfobacter* sp., of 2 pools of acetate with different biological availabilities in marine pore water. *Marine Biology*. 1984;83(3):271–276.
209. Simoneit BRT, Kawka OE, Brault M. Origin of gases and condensates in the Guaymas Basin hydrothermal system (Gulf of California). *Chemical Geology*. 1988;71(1–3):169–182.
210. Murdoch LC, Germanovich LN, Wang H, Onstott TC, Elsworth D, Stetler L, et al. Hydrogeology of the vicinity of Homestake mine, South Dakota, USA. *Hydrogeology Journal*. 2012;20(1):27–43.
211. Holland G, Lollar BS, Li L, Lacrampe-Couloume G, Slater GF, Ballentine CJ. Deep fracture fluids isolated in the crust since the Precambrian era. *Nature*. 2013;497 (7449):357–360.
212. Pedersen K. Exploration of deep intraterrestrial microbial life: current perspectives. *FEMS Microbiology Letters*. 2000;185(1):9–16.
213. Colwell FS, D'Hondt S. Nature and extent of the deep biosphere. *Carbon in Earth*. 2013;75:547–574.
214. Huber H, Huber R, Lüdemann H-D, Stetter KO. Search for hyperhermophilic microorganisms in fluids obtained from the KTB pump test. *Scientific Drilling*. 1994;4:127–129.
215. Whitman WB, Coleman DC, Wiebe WJ. Prokaryotes: the unseen majority. *Proceedings of the National Academy of Sciences of the United States of America*. 1998; 95(12):6578–6583.
216. Kallmeyer J, Pockalny R, Adhikari RR, Smith DC, D'Hondt S. Global distribution of microbial abundance and biomass in subseafloor sediment. *Proceedings of the National Academy of Sciences of the United States of America*. 2012;109 (40):16213–16216.
217. McMahon S, Parnell J. Weighing the deep continental biosphere. *FEMS Microbiology Ecology*. 2014;87(1):113–120.
218. Fredrickson JK, Onstott TC, eds. *Biogeochemical and Geological Significance of Subsurface Microbiology*. New York: Wiley-Liss, 2001.
219. Sahl JW, Schmidt RH, Swanner ED, Mandernack KW, Templeton AS, Kieft TL, et al. Subsurface microbial diversity in deep-granitic-fracture water in Colorado. *Applied and Environmental Microbiology*. 2008;74(1):143–152.
220. Chivian D, Brodie EL, Alm EJ, Culley DE, Dehal PS, DeSantis TZ, et al. Environmental genomics reveals a single-species ecosystem deep within earth. *Science*. 2008;322(5899):275–278.
221. Osburn MR, LaRowe DE, Momper LM, Amend JP. Chemolithotrophy in the continental deep subsurface: Sanford Underground Research Facility (SURF), USA. *Frontiers in Microbiology*. 2014;5:610.
222. Probst AJ, Castelle CJ, Singh A, Brown CT, Anantharaman K, Sharon I, Hug LA, Burstein D, Emerson JB, Thomas BC, Banfield JF. Genomic resolution of a cold subsurface aquifer community provides metabolic insights for novel microbes adapted to high CO_2 concentrations. *Environmental Microbiology*. 2016;19:459–474.

223. Stevens TO, McKinley JP. Lithoautotrophic microbial ecosystems in deep basalt aquifers. *Science*. 1995;270(5235):450–454.
224. Veto I, Futo I, Horvath I, Szanto Z. Late and deep fermentative methanogenesis as reflected in the H–C–O–S isotopy of the methane-water system in deep aquifers of the Pannonian Basin (SE Hungary). *Organic Geochemistry*. 2004;35(6):713–723.
225. Fredrickson JK, Balkwill DL. Geomicrobial processes and biodiversity in the deep terrestrial subsurface. *Geomicrobiology Journal*. 2006;23(6):345–356.
226. Lapworth DJ, Gooddy DC, Butcher AS, Morris BL. Tracing groundwater flow and sources of organic carbon in sandstone aquifers using fluorescence properties of dissolved organic matter (DOM). *Applied Geochemistry*. 2008;23(12):3384–3390.
227. Vetter A, Mangelsdorf K, Wolfgramm M, Rauppach K, Schettler G, Vieth-Hillebrand A. Variations in fluid chemistry and membrane phospholipid fatty acid composition of the bacterial community in a cold storage groundwater system during clogging events. *Applied Geochemistry*. 2012;27(6):1278–1290.
228. Grundger F, Jimenez N, Thielemann T, Straaten N, Luders W, Richnow HH, et al. Microbial methane formation in deep aquifers of a coal-bearing sedimentary basin, Germany. *Frontiers in Microbiology*. 2015;6:200.
229. Schloemer S, Elbracht J, Blumenberg M, Illing CJ. Distribution and origin of dissolved methane, ethane and propane in shallow groundwater of Lower Saxony, Germany. *Applied Geochemistry*. 2016;67:118–132.
230. Bell RA, Darling WG, Ward RS, Basava-Reddi L, Halwa L, Manamsa K, et al. A baseline survey of dissolved methane in aquifers of Great Britain. *Science of the Total Environment*. 2017;601:1803–1813.
231. Murphy EM, Schramke JA, Fredrickson JK, Bledsoe HW, Francis AJ, Sklarew DS, et al. The influence of microbial activity and sedimentary organic-carbon on the isotope geochemistry of the middendorf aquifer. *Water Resources Research*. 1992;28(3):723–740.
232. Means JL, Hubbard N. Short-chain aliphatic acid anions in deep subsurface brines – a review of their origin, occurrence, properties, and importance of new data on their distribution and geochemical implications in the paleo-duro-basin, Texas. *Organic Geochemistry*. 1987;11(3):177–191.
233. Lundegard PD, Kharaka YK, eds. *Distribution and Occurrence of Organic Acids in Subsurface Waters*. Berlin: Springer, 1994.
234. Aitken CM, Jones DM, Larter SR. Anaerobic hydrocarbon biodegradation in deep subsurface oil reservoirs. *Nature*. 2004;431(7006):291–294.
235. Strapoc D, Mastalerz M, Dawson K, Macalady J, Callaghan AV, Wawrik B, et al. Biogeochemistry of microbial coal-bed methane. *Annual Review of Earth and Planetary Sciences*. 2011;39:617–656.
236. Pedersen K. Diversity and activity of microorganisms in deep igneous rock aquifers of the Fennoscandian Shield. In: Fredrickson JK, Fletcher M, eds. *Subsurface Microbiology and Biogeochemistry*, New York: Wiley-Liss, 2001, pp. 97–139.
237. Amend JP, Teske A. Expanding frontiers in deep subsurface microbiology. *Palaeogeography, Palaeoclimatology, Palaeoecology*. 2005;219(1–2):131–155.
238. Lollar BS, Onstott TC, Lacrampe-Couloume G, Ballentine CJ. The contribution of the Precambrian continental lithosphere to global H_2 production. *Nature*. 2014;516(7531):379–382.
239. Boston PJ, Spilde MN, Northup DE, Melim LA, Soroka DS, Kleina LG, et al. Cave biosignature suites: microbes, minerals, and Mars. *Astrobiology*. 2001;1(1):25–55.

240. Northup DE, Lavoie KH. Geomicrobiology of caves: a review. *Geomicrobiology Journal*. 2001;18(3):199–222.
241. Barton HA, Northup DE. Geomicrobiology in cave environments: past, current and future perspectives. *Journal of Cave and Karst Studies*. 2007;69(1):163–178.
242. Suzuki S, Ishii S, Wu A, Cheung A, Tenney A, Wanger G, et al. Microbial diversity in The Cedars, an ultrabasic, ultrareducing, and low salinity serpentinizing ecosystem. *Proceedings of the National Academy of Sciences of the United States of America*. 2013;110(38):15336–15341.
243. Magnabosco C, Tekere M, Lau MCY, Linage B, Kuloyo O, Erasmus M, et al. Comparisons of the composition and biogeographic distribution of the bacterial communities occupying South African thermal springs with those inhabiting deep subsurface fracture water. *Frontiers in Microbiology*. 2014;5:679.
244. Baker MA, Valett HM, Dahm CN. Organic carbon supply and metabolism in a shallow groundwater ecosystem. *Ecology*. 2000;81(11):3133–3148.
245. Kotelnikova S. Microbial production and oxidation of methane in deep subsurface. *Earth-Science Reviews*. 2002;58(3–4):367–395.
246. Fry NK, Fredrickson JK, Fishbain S, Wagner M, Stahl DA. Population structure of microbial communities associated with two deep, anaerobic, alkaline aquifers. *Applied and Environmental Microbiology*. 1997;63(4):1498–1504.
247. Vieth A, Mangelsdorf K, Sykes R, Horsfield B. Water extraction of coals – potential for estimating low molecular weight organic acids as carbon feedstock for the deep terrestrial biosphere. *Organic Geochemistry*. 2008;39(8):985–991.
248. Simkus DN, Slater GF, Lollar BS, Wilkie K, Kieft TL, Magnabosco C, et al. Variations in microbial carbon sources and cycling in the deep continental subsurface. *Geochimica et Cosmochimica Acta*. 2016;173:264–283.
249. Lau MCY, Kieft TL, Kuloyo O, Linage-Alvarez B, Van Heerden E, Lindsay MR, et al. An oligotrophic deep-subsurface community dependent on syntrophy is dominated by sulfur-driven autotrophic denitrifiers. *Proceedings of the National Academy of Sciences of the United States of America*. 2016;113(49):E7927–E7936.
250. Kieft TL, Walters CC, Higgins MB, Mennito AS, Clewett CFM, Heuer V, et al. Dissolved organic matter compositions in 0.6–3.4 km deep fracture waters, Kaapvaal Craton, South Africa. *Organic Geochemistry*. 2018;118:116–131.
251. Stevens T. Lithoautotrophy in the subsurface. *FEMS Microbiology Reviews*. 1997;20 (3–4):327–337.
252. Longnecker K, Kujawinski EB. Composition of dissolved organic matter in groundwater. *Geochimica et Cosmochimica Acta*. 2011;75(10):2752–2761.
253. Chapelle FH, O'Neill K, Bradley PM, Methe BA, Ciufo SA, Knobel LL, et al. A hydrogen-based subsurface microbial community dominated by methanogens. *Nature*. 2002;415(6869):312–315.
254. Griebler C, Lueders T. Microbial biodiversity in groundwater ecosystems. *Freshwater Biology*. 2009;54(4):649–677.
255. Flynn TM, Sanford RA, Ryu H, Bethke CM, Levine AD, Ashbolt NJ, et al. Functional microbial diversity explains groundwater chemistry in a pristine aquifer. *BMC Microbiology*. 2013;13:146.
256. Probst AJ, Ladd B, Jarett JK, Geller-McGrath DE, Sieber CMK, Emerson JB, et al. Differential depth distribution of microbial function and putative symbionts through sediment-hosted aquifers in the deep terrestrial subsurface. *Nature Microbiology*. 2018;3(3):328–336.

257. Leenheer JA, Malcolm RL, McKinley PW. Occurrence of dissolved organic carbon in selected ground-water samples in the United States. *Journal of Research of the US Geological Survey*. 1974;2:361–369.
258. McMahon PB, Chapelle FH. Microbial production of organic acids in aquitard sediments and its role in aquifer geochemistry. *Nature*. 1991;349(6306):233–235.
259. Kusel K, Totsche KU, Trumbore SE, Lehmann R, Steinhauser C, Herrmann M. How deep can surface signals be traced in the critical zone? Merging biodiversity with biogeochemistry research in a central German Muschelkalk landscape. *Frontiers in Earth Science*. 2016;4:32.
260. Head IM, Jones DM, Larter SR. Biological activity in the deep subsurface and the origin of heavy oil. *Nature*. 2003;426(6964):344–352.
261. Nazina TN, Shestakova NM, Ivoilov VS, Kostrukov NK, Belyaev SS, Ivanov MV. Radiotracer assay of microbial processes in petroleum reservoirs. *Advances in Biotechnology & Microbiology*. 2017;2(4):1–9.
262. Magot M, Ollivier B, Patel BKC. Microbiology of petroleum reservoirs. *Antonie Van Leeuwenhoek International Journal of General and Molecular Microbiology*. 2000;77(2):103–116.
263. Larter S, Huan H, Adams J, Bennett B, Jokanola O, Oldenburg T, et al. The controls on the composition of biodegraded oils in the deep subsurface: part II – geological controls on subsurface biodegradation fluxes and constraints on reservoir-fluid property prediction. *AAPG Bulletin*. 2006;90(6):921–938.
264. Zhu YL, Vieth-Hillebrand A, Wilke FDH, Horsfield B. Characterization of water-soluble organic compounds released from black shales and coals. *International Journal of Coal Geology*. 2015;150:265–275.
265. Onstott TC, Lin LH, Davidson M, Mislowack B, Borcsik M, Hall J, et al. The origin and age of biogeochemical trends in deep fracture water of the Witwatersrand Basin, South Africa. *Geomicrobiology Journal*. 2006;23(6):369–414.
266. Kietavainen R, Ahonen L, Niinikoski P, Nykanen H, Kukkonen IT. Abiotic and biotic controls on methane formation down to 2.5 km depth within the Precambrian Fennoscandian Shield. *Geochimica et Cosmochimica Acta*. 2017;202:124–145.
267. Lollar BS, Voglesonger K, Lin LH, Lacrampe-Couloume G, Telling J, Abrajano TA, et al. Hydrogeologic controls on episodic H_2 release from Precambrian fractured rocks – energy for deep subsurface life on Earth and Mars. *Astrobiology*. 2007;7(6):971–986.
268. Suzuki Y, Konno U, Fukuda A, Komatsu DD, Hirota A, Watanabe K, et al. Biogeochemical signals from deep microbial life in terrestrial crust. *PLoS One*. 2014;9(12):e113063.
269. Suzuki S, Kuenen JG, Schipper K, van der Velde S, Ishii S, Wu A, et al. Physiological and genomic features of highly alkaliphilic hydrogen-utilizing Betaproteobacteria from a continental serpentinizing site. *Nature Communications*. 2014;5:3900.
270. Li L, Wing BA, Bui TH, McDermott JM, Slater GF, Wei S, et al. Sulfur mass-independent fractionation in subsurface fracture waters indicates a long-standing sulfur cycle in Precambrian rocks. *Nature Communications*. 2016;7:13252.
271. Momper L, Jungbluth SP, Lee MD, Amend JP. Energy and carbon metabolisms in a deep terrestrial subsurface fluid microbial community. *ISME Journal*. 2017;11(10):2319–2333.
272. Magnabosco C, Ryan K, Lau MCY, Kuloyo O, Lollar BS, Kieft TL, et al. A metagenomic window into carbon metabolism at 3 km depth in Precambrian continental crust. *ISME Journal*. 2016;10(3):730–741.

273. Colman DR, Poudel S, Stamps BW, Boyd ES, Spear JR. The deep, hot biosphere: twenty-five years of retrospection. *Proceedings of the National Academy of Sciences of the United States of America*. 2017;114(27):6895–6903.
274. Jungbluth SP, del Rio TG, Tringe SG, Stepanauskas R, Rappe MS. Genomic comparisons of a bacterial lineage that inhabits both marine and terrestrial deep subsurface systems. *PeerJ*. 2017;5:e3134.
275. Lollar BS, Lacrampe-Couloume G, Slater GF, Ward J, Moser DP, Gihring TM, et al. Unravelling abiogenic and biogenic sources of methane in the Earth's deep subsurface. *Chemical Geology*. 2006;226(3–4):328–339.
276. Kadnikov VV, Frank YA, Mardanov AV, Beletskii AV, Ivasenko DA, Pimenov NV, et al. Uncultured bacteria and methanogenic archaea predominate in the microbial community of western Siberian deep subsurface aquifer. *Microbiology*. 2017;86 (3):412–415.
277. Momper L, Reese BK, Zinke L, Wanger G, Osburn MR, Moser D, et al. Major phylum-level differences between porefluid and host rock bacterial communities in the terrestrial deep subsurface. *Environmental Microbiology Reports*. 2017;9 (5):501–511.
278. Wanger G, Southam G, Onstott TC. Structural and chemical characterization of a natural fracture surface from 2.8 kilometers below land surface: biofilms in the deep subsurface. *Geomicrobiology Journal*. 2006;23(6):443–452.
279. Purkamo L, Bomberg M, Nyyssonen M, Kukkonen I, Ahonen L, Itavaara M. Heterotrophic communities supplied by ancient organic carbon predominate in deep fennoscandian bedrock fluids. *Microbial Ecology*. 2015;69(2):319–332.
280. Jangir Y, French S, Momper LM, Moser DP, Amend JP, El-Naggar MY. Isolation and characterization of electrochemically active subsurface *Delftia* and *Azonexus* species. *Frontiers in Microbiology*. 2016;7:756.
281. Lehman RM, Colwell FS, Bala GA. Attached and unattached microbial communities in a simulated basalt aquifer under fracture- and porous-flow conditions. *Applied and Environmental Microbiology*. 2001;67(6):2799–2809.
282. Petsch ST, Eglinton TI, Edwards KJ. ^{14}C-dead living biomass: evidence for microbial assimilation of ancient organic carbon during share weathering. *Science*. 2001;292(5519):1127–1131.

17

Biogeography, Ecology, and Evolution of Deep Life

CARA MAGNABOSCO, JENNIFER F. BIDDLE, CHARLES S. COCKELL, SEAN P. JUNGBLUTH, AND KATRINA I. TWING

17.1 Subsurface Biomes and Their Inhabitants

When we ponder the existence of life extending deep into Earth, a phrase from the movie *Jurassic Park* is often used: that "life finds a way." Numerous investigations into the continental and marine subsurface have shown that life indeed finds a way to exist deep into the subsurface, provided that physical influences, particularly heat, allow for the existence of biomolecules. In this chapter, we will review what is known about the biogeography, ecology, and evolution of deep life, acknowledging along the way that this field is rapidly developing with every new set of experiments and continued exploration.

The subsurface biosphere is loosely defined as the habitable region beneath the soil and sediments where the limits of habitability are typically defined by some physical process (also see Chapter 19, this volume). Current estimates of the habitable volume of the subsurface range from ~2.0 to 2.3 $\times$ 10^9 km^3, or roughly twice the volume of our oceans (Table 17.1). This large biosphere is estimated to hold ~70% of all bacterial and archaeal cells (Figures 17.1 and 17.2) and potentially over 80% all bacterial and archaeal species (for a review, see 1). A variety of habitats and sampling techniques to study the subsurface biosphere have been explored by scientists for nearly a century and are further described throughout this chapter (Sections 17.1.1–17.1.5; also see Figure 16.1 in Chapter 16, this volume).

17.1.1 Continental Subsurface

The earliest investigations into the deep subsurface biosphere were performed in oil fields and coal beds within the continental subsurface in the mid-1920s (13–15; for a review on the history of continental subsurface research, see 16). Since then, many different deep continental biomes have been explored, including (but not limited to) groundwater and deep aquifers, oil and gas reservoirs, deep bedrock, evaporite deposits, and subglacial ecosystems (Figure 17.1). Early reviews on the microorganisms inhabiting these subsurface biomes focused on cell numbers (17,18), cultivation-based measurements of activity (17,18), and/or large collections of clone libraries from amplified regions of the 16S rRNA gene (19,20).

Table 17.1 *Recent estimates of subsurface habitable zones.*

Biome	Volume (km^3)	Definition	Refs.
Continental subsurface	7×10^8	≤85°C isotherm	(1)
Continental subsurface	1×10^9	≤122°C isotherm	(1)
Groundwater	2×10^7	≤2 km depth	(2)
Sub-seafloor sediments	3×10^8	All sediments (3–5)	(6)
Marine sediment porewater	8×10^7	All sediments (3–5)	(6)
Marine crust	~10^9	≤120°C isotherm	(7)
All subsurface	*~2.0–2.3* $\times 10^9$		
Oceans	1×10^9	*All oceans*	(7)

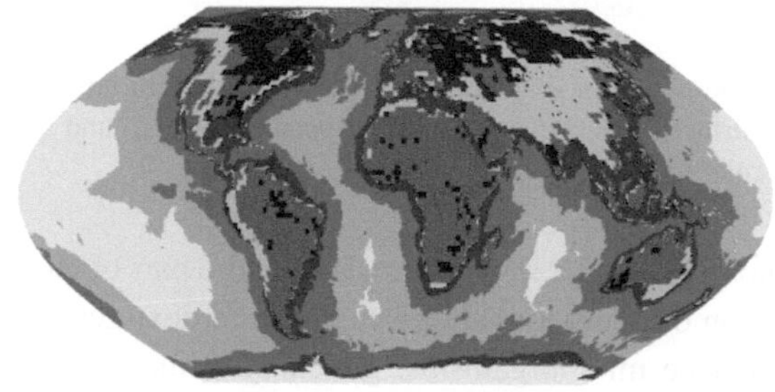

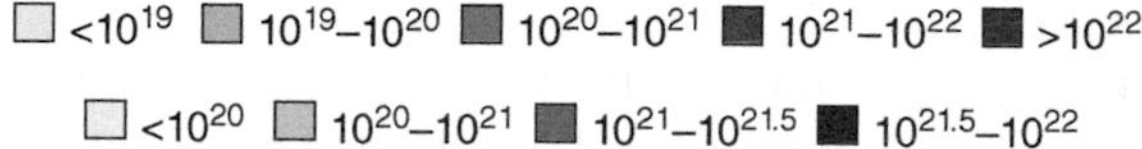

Figure 17.1 Map of sub-seafloor sediment and continental subsurface cell numbers. The distributions of bacterial and archaeal cells in sub-seafloor sediments (blues; adapted from Kallmeyer et al. (8)) and the continental subsurface (browns; adapted from Magnabosco et al. (1)) are shown. Uncertainties in cellular estimates of the marine crust (1,7,9) prevent mapping the distribution of cells throughout the marine crust at this time.

Over the past decade, the applications of metagenomics and next-generation sequencing have allowed researchers to further examine the identities and lifestyles of organisms residing in the continental subsurface. The first deep subsurface metagenome was generated from DNA extracted from the fracture fluids of a 2.8-km deep borehole in South Africa (21) and revealed a "single-species ecosystem" containing a population of chemoautotrophic Firmicutes, "*Candidatus* Desulforudis audaxviator," capable of performing sulfate reduction and fixing nitrogen and carbon (further discussion in Chapter 18, this volume). Although the number of species observed in this early investigation and other subsurface clone libraries (19,20) was low, next-generation sequencing has revealed that single-species continental subsurface ecosystems are an exception to an otherwise highly

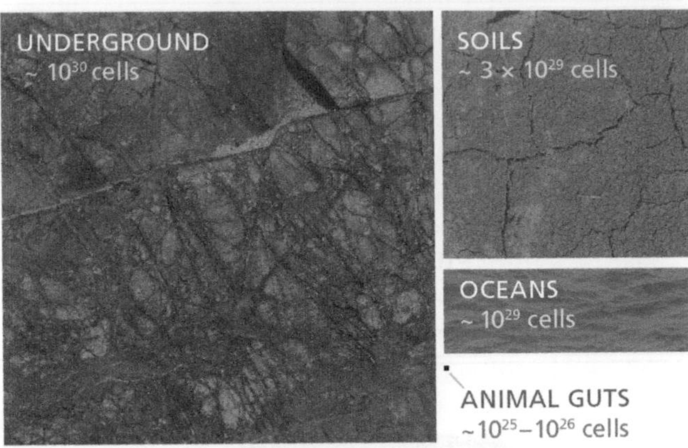

Figure 17.2 Estimated numbers of bacteria and archaea throughout various biomes. Cellular estimates for the subsurface (1,8–10), soils (11), oceans (9,11), and animal guts (9,12) are illustrated to show the relative sizes of each biome.

diverse subsurface biosphere (for a review, see 1). A large part of the increased α-diversity is due to the observation of many low-abundance taxa in next-generation sequencing data sets. These low-abundance microorganisms, generally termed the "rare biosphere" (22), are now accepted as a common feature of environmental microbial communities and may persist in order to preserve a diverse collection of metabolic strategies for survival in changing environments.

Within the continental subsurface, natural fluctuations in fluid chemistries are just beginning to be understood (23); however, large disruptions to subsurface environments through human activities have been reported to dramatically alter continental subsurface communities (24–29). Events that have been shown to increase the overall salinity of deep fluids such as hydraulic fracturing (fracking) (30–36) and drilling (37) are often associated with decreases in α-diversity and increases in halo-tolerant bacteria. More recently, an in-depth analysis of the metabolic potential for 31 unique metagenome-assembled genomes (MAGs) showed that the persisting halo-tolerant bacteria and archaea were capable of fermenting chemical additives brought in during the injection process (26). A similar trend has been reported in the oil wells of the Enermark Field (Alberta, Canada). There, native oil phases support low-diversity communities of methanogens and acetogens (38), while emergent aqueous phases (oil + water) exhibit bursts in populations of sulfate-reducing Deltaproteobacteria that change the overall subsurface community composition (39,40). These "blooms" of sulfate-reducing bacteria (SRB) in water-flooded reservoirs can dramatically raise concentrations of hydrogen sulfide (H_2S) due to their consumption of hydrocarbons coupled to the reduction of sulfate. Commonly referred to as reservoir souring, this microbial by-product can lead to corrosion of pipelines, plugging of oil-bearing rock, and contamination of the extracted oil (41), but it can be controlled via the addition of higher-energy electron acceptors such as nitrate (42). On the other hand, the increased SRB activity and growth observed after H_2 injection in the Opalinus clay (Switzerland) has been proposed as a way to control H_2 generated in underground nuclear waste repositories (25).

In many of the aforementioned examples, it is difficult to determine whether or not the changes in community composition are driven by the introduction of foreign organisms or the expansion of a native member. Long-term monitoring of subsurface fluids in "underground labs" is beginning to reveal how these ecosystems change over time (23) and the variations between the attached and planktonic members of the microbial community (43). After nearly a century of research into life beneath the continents, it is now apparent that the continental subsurface is home to a wide variety of fluid chemistries and lithologies (44–46). A comparison of 326 bacterial and archaeal 16S rRNA gene surveys from a variety of continental subsurface environments revealed a correlation between community composition and sample lithology (1), yet the variety of microbial metabolisms encountered beneath the continents (e.g. 44,47–51), how these metabolisms arose, and the interactions between these organisms (52), viruses (53,54), and the environment are just beginning to be understood.

17.1.2 Sub-seafloor Sediments

Early work on sub-seafloor sediments mainly focused on continental margins and quickly recognized a wide distribution of bacteria and archaea and an importance of heterotrophic metabolisms (55; as described in Chapter 16, this volume). In deeply buried sediments sampled through the scientific drilling program – currently the International Ocean Discovery Program (IODP) – microbes are responsible for large-scale geochemical shifts, including consumption of methane and sulfate (55–57). Further investigations have shown that below oceanic gyres, the deep biosphere may be an oxic environment and interact with deep hydrothermal recharging (58,59). With the noted impacts of deep life on processes relating to global biogeochemistry, continued work is focusing on constraining the rates and limits of these processes (for further discussion on this topic, see Chapter 19, this volume).

Marine sediments are a heterogeneous environment and record Earth history along with the modern life that may be living *in situ*. Tremendous advances were made through the first drilling expedition to focus solely on the marine deep biosphere, Ocean Drilling Program Leg 201, which visited the Peru Margin in 2002. This expedition showed that cells increased in areas of potential geochemical energy (55), that archaeal cells were active and heterotrophic throughout the sediment column (59; Chapter 16, this volume), and that the metagenomic signature of sediments was distinct from any other sampled environment up to that time (60). This expedition also yielded the first metatranscriptome of marine sediments, which showed dominant transcripts for fermentation (61) and that SRB may invest in different cellular strategies based on energy availability (62). Investigations on the Gulf of Mexico (63), Nankai Trough (64,65), Guaymas Basin (66,67), Baltic Sea (68), and

Shimokita Peninsula (69), among others, have shown that sub-seafloor microbial life is found wherever it can exist. The interplay of bacteria and archaea is still being investigated, as areas of the Andaman Sea contain no detectable archaea (70).

The establishment and propagation of the sediment-hosted deep biosphere is still under investigation, but likely includes a combination of selection from the surface environment (71) and persistence of cells with depth (72). What stimulates the deep biosphere, in addition to chemical interfaces, may include the continued influence of depositional conditions (68), tectonic activity (73), geological shifts under pressure (74), and the internal heating of Earth (75). It is still unknown exactly how the community in deep sediment responds to the stress of sedimentation. It is theorized that necromass (76) or radiolysis (77; Chapter 19, this volume) may help support deep communities. Evidence of subsurface acclimation to changing surface conditions exists, showing that the marine deep biosphere is responsive and may be capable of more activity than expected (77). Life in the deep marine biosphere has been reviewed extensively (78–81), and with new investigations underway, discoveries are still abundant.

17.1.3 Oceanic Crust

Despite the larger volume of the basaltic crust environment relative to marine sediments (Table 17.1), relatively little is known about the abundance and extent of microbial biomass in the deep oceanic basement because it is heterogeneous and largely inaccessible (81). Early investigations into microbial activity in the oceanic crust focused on the presence of microchannels and alteration patterns associated with DNA (e.g. 71,72), as the colocation of nucleic acids and unique microtextures (82) are taken together to represent biogenic alteration in basaltic rock. Overall, most biogenic alteration is restricted to the upper 250 m of the oceanic crust, corresponding to a predicted temperature range of 15–80°C (7). Sulfur and iron isotopic information suggests that the oxidation of basaltic crust mainly occurs in the relatively young ridge flank (<20 million years old) (83). Combined, these observations have directed more recent and current IODP drilling expeditions to focus on searching for microbial life in relatively young oceanic crust.

17.1.3.1 Warm Anoxic Basement

The first investigation of the sediment-buried seafloor biosphere was conducted in 1998 by collecting fluids from an undersea Circulation Obviation Retrofit Kit (CORK) observatory in the 3.5-million-year-old basaltic crust of the Northwest Pacific Ocean's Juan de Fuca Ridge (JdFR) flank (84). In this pilot investigation, warm (65°C) fluids originating from hundreds of meters below the seafloor were collected for gene cloning and sequencing and revealed a low-diversity environment with bacteria and archaea (no Eukarya) that was dominated by a Firmicute lineage later shown to be a close genomic relative of the terrestrial subsurface lineage "*Ca.* Desulforudis audaxviator" (21,84). Eight years after the initial CORK installation in the JdFR, scientists sampled a black rust scraping exposed to reducing fluids from the CORK observatory (85) and fluids emanating from an exposed rocky outcrop near the CORK observatory (84) and discovered an abundance of thermophilic lineages, indicating that the deep crustal biosphere is, at least in part, adapted to life at relatively high temperatures.

A new generation of borehole observatories equipped with microbiologically friendly sampling materials (86) was installed near the original JdFR CORK observatory in 2004. A 3-year sampling campaign at the JdFR identified low microbial cellular abundances (~10^4 cells mL^{-1}) and revealed a microbial community whose major lineages changed each year (87). This dynamic subsurface community contrasts with deep marine sediment communities, which are stable on longer timescales and are more similar to communities observed at deep continental sites. In parallel to (87), the first successful retrieval of basement rocks for molecular microbiology analysis was recovered from JdFR (88) and identified methane- and sulfur-cycling bacteria and archaea that were most related to uncultivated marine sediment lineages. Incubations of rocks surrounding the CORK observatories (89,90) revealed colonization of previously sterilized minerals by lineages that were more similar to the microorganisms identified in the crustal fluids than in collected rocks and consistent with a capacity for iron, nitrogen, and sulfur cycling. An incubation of mineral chips at the seafloor was less successful in reproducing the thermophilic and anoxic conditions found at depth, highlighting the difficulty of working in an extreme deep-sea sub-seafloor reducing environment (86).

In 2010, scientists sampled two boreholes separated by ~67 km to investigate the temporal and spatial dynamics of microorganisms residing in JdFR basement fluids. Crustal fluids were retrieved from the location of the original CORK observatory and, for the first time, from younger (1.2 million years) and cooler (39°C) ocean crust. Gene cloning and sequencing revealed that the original CORK observatory was compromised in its ability to produce clean samples (91); this is perhaps expected given the 10+ years of corrosion that has occurred on the CORK parts exposed to both a reducing and oxic environment since installation (92). In contrast, samples collected from the cooler basement location contained evidence for Deltaproteobacteria involved in sulfur cycling and *Clostridia* related to "*Ca.* Desulforudis." The identification of "*Ca.* Desulforudis" in the cooler oceanic basement fluids suggests that relatives of "*Ca.* Desulforudis audaxiator" inhabit a wider range of habitats than one might expect given its prevalence in the South African subsurface (Section 17.1.1). Combined with sulfate reduction rate measurements (93), these results indicate that anaerobic processes may play a major role in the degradation of organic matter in the upper oceanic crust.

Third-generation borehole observatories went online in JdFR in 2011, featuring the most microbiology-friendly components yet adapted. 16S rRNA amplicon sequencing of samples recovered from these observatories revealed an abundance of new lineages within Archaeoglobi, Aminicenantes, and Acetothermia that had not previously been identified in the oceanic crust (93,94) and were later confirmed by metagenome sequencing and genome binning (95). From a functional perspective, microbial communities in the deep subseafloor appear similar to those found in terrestrial hot springs. Together, this work has helped to identify novel microbial diversity and stable members in the deep, warm, anoxic basement biosphere.

Although the JdFR is a popular location to study the warm, anoxic basement crust, a CORK observatory on the Costa Rica margin has been sampled for deep basement microbiology. Warm fluids (58°C) collected from the CORK observatory revealed novel lineages of sulfur-oxidizing bacteria; however, little overlap was found between the Costa Rica site and JdFR (96), and this may, in part, be due to differences in fluid flow and organic matter delivery rates. Further experimentation and sampling from additional locations will be required to constrain the biogeographic patterns and to elucidate the ecology of microorganisms living in the ocean crust.

17.1.3.2 Cold, Oxic Basement

Exploration of the warm, anoxic deep basement biosphere has led the way in understanding the oceanic crust; however, the cold (<20°C) and oxic basement rock is the more abundant of the sub-seafloor igneous biomes, and therefore of critical global biogeochemical importance. To this end, several CORK observatories were installed at North Pond, a sediment pond on the western flank of the Mid-Atlantic Ridge that overlies a hydrologically active ocean crust of ~8 Ma (97). When multiple depth horizons were sampled within the igneous basement, distinctive heterotrophic and autotrophic microbial communities largely composed of Proteobacteria were identified (98). Metagenomic sequencing and subsequent genome binning were used to identify members of the microbial community that were capable of exploiting hypoxic or anoxic conditions (99). Interestingly, enrichment cultures concocted with additions of nitrate and ammonia stimulated the microbial community, which together provide evidence for a heterogeneous functional population in the cold, oxic basement (100,101). Additional investigations surveying a range of seafloor environment types (e.g. crustal ages, mineralogy, redox states, etc.) are needed to further refine global estimates of primary production in the oceanic crust.

17.1.4 Ultra-basic Sites

Serpentinization is a subsurface geochemical process that results in ultra-basic fluids (pH $>$ 10) and abiotically produces methane and small-chain hydrocarbons through Fischer–Tropsch-type synthesis (102–105). The hydrogen, methane, and abiotically generated organic molecules produced by serpentinization can serve as energy for microbial metabolisms in the otherwise energy-limited deep biosphere. One caveat, however, is that serpentinite fluids are depleted in electron acceptors (relative to electron donors) (106), meaning that this potential energy source may not be bioavailable in some environmental settings.

In the marine setting, the limitation of electron acceptors is alleviated as end-member serpentinite fluids emanating from hydrothermal vents mix with surrounding seawater. At the Lost City hydrothermal field (LCHF), vent fluids exhibit moderate temperature, high pH, and millimolar concentrations of hydrogen and methane (104,107–109). Actively venting carbonate chimneys are dominated by a single taxon of methane-cycling archaea in the anoxic chimney interiors (110) and by methanotrophic and sulfur-oxidizing bacteria in the chimney exteriors (111). The older, inactive chimneys are much more diverse, hosting many bacterial taxa as well as anaerobic methane-oxidizing archaea (ANME) (111,112). The microbial communities at LCHF are likely stimulated by, if not entirely dependent on, the H_2, methane, and other carbon sources produced by serpentinization (113). Another site of marine serpentinization, Prony is a shallow marine hydrothermal vent located near New Caledonia that vents fluids elevated in hydrogen and methane (114). Both the pH 11 fluids and chimneys from Prony exhibit similarly low archaeal diversity; however, they have a greater diversity of bacteria than Lost City, including the taxa Chloroflexi, Deinococcus-Thermus, Firmicutes, and Proteobacteria (115,116).

In the continental setting, serpentinization can take place underground in ophiolite complexes. Sampling of these subsurface processes often takes place at springs or pools, where the subsurface fluids come naturally to the surface. Various studies have used chemical proxies to differentiate subsurface, surface, and mixing-zone fluids (117,118) and found consistent trends in the microbial communities of these different zones. The oxic/anoxic mixing zone of continental serpentinite sites is often dominated by a single Betaproteobacterium (106). Early diversity studies identified this organism as *Hydrogenophaga* (119–122). This bacterium was recently isolated from the Cedars, an ultra-basic site in northern California, resulting in a proposed name change to "*Ca.* Serpentinomonas" (123). Multiple studies have shown that the *Hydrogenophaga*/"*Ca.* Serpentinomonas" organisms from serpentinite sites have 99–100% sequence identity (123–125). The *Hydrogenophaga*/ "*Ca.* Serpentinomonas" strains that dominate serpentine systems are alkaliphilic (optimum pH of 11) and autotrophic with growth on hydrogen, oxygen, and calcium carbonate (124). The more end-member serpentinite fluids tend to be host slightly higher diversity and contain anaerobes from the phyla Firmicutes (106,120–122,124,126) and Chloroflexi (122) and the candidate division Parcubacteria (122,126). These strong similarities in the community composition of these geographically distinct sites suggest that these organisms may be specially adapted to withstand the challenging conditions of the serpentinite environment.

Given the importance of methane-cycling microorganisms to marine serpentinite ecosystems (111,113,127), scientists have wondered whether methane is of similar importance in continental settings. Previous diversity studies based on 16S rRNA gene amplicon data have suggested the presence of putative methane-cycling organisms at continental serpentinite sites (121,122,126,128). More recently, metagenomic methods have been employed to look more deeply into this question (129,130). Both metagenomic and experimental evidence for methanogenesis by members of the Methanobacteriaceae and methanotrophy by members of the Methylococcaceae were found in extremely low-biomass sample from the Ligurian ophiolite in Italy (129). At the Santa Elena ophiolite in Costa Rica, all of the genes for diverse methanogenesis pathways were present (e.g. acetoclastic, hydrogenotrophic, and from formate) in metagenomes from a pH 11 serpentinite spring (130).

The aforementioned studies were conducted by sampling the surface expression of a subsurface process through natural springs or pools. These features grant access to an otherwise inaccessible environment, but they represent opportunistic sampling at locations where the subsurface environment interacts with the surface. The Coast Range Ophiolite Microbial Observatory (CROMO) was established in northern California by drilling wells into the actively serpentinizing subsurface environment to access end-member fluids directly (131). A comparison of microbial communities from a wide range of geochemical gradients at CROMO found that the dominant taxa in the system were strongly correlated with pH and the concentrations of methane and carbon monoxide (124).

Studies have recently demonstrated that subsurface fluids and subsurface rocks from the same site exhibit differences in microbial community structure (43). To date, little research on serpentinite rocks has been published, save a study at the Leka ophiolite in Norway, which found that serpentinite groundwaters were dominated by the hydrogen-oxidizing Betaproteobacterium *Hydrogenophaga* (a close relative of the aforementioned "*Ca.* Serpentimonas"), while mineral-associated communities contained microbes involved in nitrite, iron, manganese, and ammonium oxidation (132). At the Ligurian ophiolite in Italy, surface-exposed travertine deposits at serpentinite springs were investigated and archaeal species putatively involved in methane cycling and diverse bacterial species putatively involved in hydrogen oxidation were found, suggesting that these surface organisms could be fueled by deep serpentinization below (128). Additional studies are currently underway to investigate the microbial ecology of serpentinite rocks at continental (131,133) and marine (134) sites of serpentinization. These studies, combined with those investigating the microbial ecology of serpentinite fluids, will give us a more complete understanding of life in the ultra-basic subsurface environment.

17.1.5 Other Subsurface Environments

Submarine volcanoes like the Suiyo (135) and Axial (136) Seamounts emit fluids that are at a lower pH than their surrounding seawater. These localities contain high amounts of H_2S and often harbor large populations of sulfur-oxidizing Epsilonproteobacteria. On the other hand, submarine mud volcanoes and cold methane seeps release large amounts of methane that fuel sizeable populations of ANME (125–127). On the continents, the Rio Tinto in Spain (137) is an acidic environment for which the subsurface microbial communities were investigated. The Mars Astrobiology Research and Technology Experiment (MARTE) project identified three zones within the Rio Tinto's subsurface: (1) a near-surface to ~30 m below sea level (mbsl) zone that supports fungal populations and is primarily driven by heterotrophy and aerobic respiration with seasonal rainfall; (2) a 30–43-mbls zone in which iron and sulfur oxidation of sulfide minerals occurs under aerobic conditions by aerobic iron and sulfur oxidizers; and (3) an anaerobic deep zone (>43 mbsl) that contains organisms inferred to be carrying out anaerobic iron and sulfur oxidation, with SRB potentially producing H_2S and thence pyrite by reaction with host rocks.

While a diverse array of subsurface environments exists on our planet, it is important to consider the implications of the adaptations and lifestyles of subsurface organisms for the habitability of extraterrestrial subsurface environments. The surface of Mars is inhospitable today on account of desiccation, ultraviolet and ionizing radiation, and oxidants (138), yet the subsurface may have been more habitable throughout Martian history (139). In the outer solar system, the detection of hydrogen and silica in fluids ejected from the south polar of Saturn's icy moon Enceladus suggests fluid–rock interaction within that moon, showing that deep subsurface rock–water interactions on Earth may provide analogous insights into the limits of habitability of these environments (140). While the constraints on the fluid compositions of these extraterrestrial environments are poor, the lack of a connection to a surface photosynthetic biosphere means that these environments are likely to be carbon poor and that the primary available redox couples are chemolithotrophic.

All extraterrestrial environments receive an infall of meteoritic material, such as cometary or carbonaceous chrondrite material as well as endogenous reduced organic material. A crude calculation estimates the infall of unaltered carbon on Mars to be ~16 g of unaltered carbon km^{-2} $year^{-1}$ (141), although much of this is in recalcitrant polycyclic aromatic hydrocarbon material. Compared to the estimated net primary productivity on Earth (~2×10^8 g km^{-2}; 142), about seven orders of magnitude more carbon is available to ultimately make its way into the deep subsurface of Earth than of Mars. However, as in Earth's deep subsurface, radiolysis, serpentinization, and reduced volcanic gases may provide H_2 as an electron donor largely independent of carbon availability (143). In the case of icy moons, the quantity of exogenous carbon that is recirculated into the subsurface oceans is unknown (e.g. 144), although detection of low- and high-molecular-weight carbon compounds in the fluids of Enceladus (145) could suggest an endogenous source of carbon, such as from a core with a chondritic composition. Although there is much to be learned about extraterrestrial deep subsurface environments, it is clear that a growing knowledge of the physicochemical conditions in the terrestrial deep subsurface and how they restrict life (Section 17.3.1), as well as the role of the carbon cycle in constraining energy availability (Chapter 19, this volume), provides a foundation not only for understanding extraterrestrial environments, but also for prioritizing the measurements required to better constrain their habitability.

17.2 Global Trends in Subsurface Microbiology

While the previous section focused on the distribution of life throughout the subsurface, this section provides a more comprehensive overview of the organisms residing within the deep subsurface and their interactions between one another.

17.2.1 Archaea and Bacteria

The deep ocean is typically enriched in archaeal cells, but nearly all other marine environments show a dominance of bacterial lineages (146). This was initially challenged by observations on the Peru Margin, which showed active archaeal cells in dominant abundances (147). Further investigation of archaeal intact polar lipids showed that archaeal lipid dominance tracked with organic carbon content in sediments (148); however, it was later shown that these lipid profiles may be problematic due to long-lived phospholipids (149) and that initial measurements should be revised. Subsequent work and methodological clarifications (150) have suggested that archaea and bacteria may have equal abundances in some sediments, yet a range of conditions persist, particularly in locations where no archaea have been found (70). Recent evidence suggests that initial bioturbation may be one of the most significant impacts on archaeal versus bacterial dominance in sediments (151).

Under the continents, the majority of sample sites show a dominance of chemotrophic bacterial lineages (1). Despite small numbers, hydrogenotrophic methanogens are frequently detected, and there is an important link between the available dissolved inorganic C pool and the larger bacterial community (152,153). In the Olkiluoto underground laboratory (Finland), populations of methane-oxidizing archaea become the dominant members of the microbial community within a sulfate–methane mixing zone located around 250–350 mbsl (45). Although subsurface methanogens and ANME have been relatively well characterized, other members of the continental subsurface archaeal communities are less understood. Targeted analysis on archaeal bins and single-cell genomes recovered from the continental subsurface are providing a new, in-depth look at archaea that were previously recognized only by their 16S rRNA gene sequences (for more details, see Chapter 18, this volume).

A notable example is the recent effort to characterize members of the South African Gold Mine Miscellaneous Euryarchaeal Group (SAGMEG) that are frequently reported in both marine and continental subsurface 16S rRNA gene surveys (154). The genomes of four SAGMEGs were recovered from diverse environmental samples and compared, revealing that these organisms most likely derive energy through the oxidation of carbon monoxide coupled to water or nitrite reduction. The comparative analysis also improved the phylogenetic placement of SAGMEG, resulting in a proposed reclassification of SAGMEG to a new group, "Hadesarchaea." The prevalence of "Hadesarchaea" and other candidate phyla (155) within subsurface environments is rapidly expanding our understanding of microbial diversity and changing the way we view the tree of life (156).

17.2.2 Subsurface Isolates and Interactions

While the number of novel taxa identified through DNA sequencing continues to grow at a rapid rate, the isolation and study of subsurface taxa lag behind. An important by-product of the isolation of subsurface microorganisms is the ability to test bacterial and archaeal responses to extreme temperature, pressure, pH, and salinity (see further discussion in Section 17.3), and also sense their contributions to carbon cycling in the deep subsurface (Chapter 18, this volume). A notable example is the isolation and study of the thermoacidophilic archaeon, *Aciduliprofundum boonei*, which was isolated from the hydrothermal fluids of the East Pacific Rise and Eastern Lau spreading center (157). *A. boonei* is capable of growing in fluids as acidic as pH 3.3 and temperatures as high as 75°C. Due to its prevalence in hydrothermal vent 16S rRNA gene surveys, the isolation and study of *A. boonei* has been particularly informative to the study of thermoacidophilic archaea and sulfur and iron reduction in the subsurface.

While isolates provide tremendous insights into the physiology of microorganisms, the syntrophic and symbiotic relationships between microorganisms are ignored. Syntrophic interactions and interconnected metabolisms, however, are commonly cited as important components of subsurface ecology (47,52). One of the most well-studied examples of subsurface syntrophy is the partnership between ANME and SRB (52,158,159) (for more details, see Chapter 18, this volume). Other examples include methogens and ANME (52), methanogens and H_2 producers like *Thermococcus paralvinellae* (160), and sulfur oxidizers and SRB (52). In the oceanic crust, a spatial–temporal comparison of MAGs assembled from the fluids of North Pond on the Mid-Atlantic Ridge revealed a high degree of functional redundancy despite changes in community membership, suggesting that a consistent and stable set of metabolic interactions is necessary for life to succeed in North Pond (99). On the other hand, subsurface symbionts are less understood; however, symbiotic relationships between nanoarchaea and autrophic "*Ca.* Altiarchaeaum sp." in the subsurface fluids of Crystal Geyser, Utah, were recently proposed due to co-occurrence patterns observed from metagenomic data (161).

17.2.3 Subsurface Eukaryotes

The discovery of the nematode, *Halicephalobus mephisto*, within the South African subsurface revealed that complex, multicellular organisms are able to withstand the pressure, isolation, and temperature of the subsurface over 1 km underground (162). A recent effort to sequence the genome of *H. mephisto* revealed that an expanded repertoire of 70-kDa heat-shock protein (Hsp70) may be an important attribute of the *H. mephisto* genome, aiding in the nematode's tolerance to elevated temperatures at depth (163). Following the discovery of subsurface Nematoda, other multicellular eukaryotes including Platyhelminthes, Rotifers, Annelids, and Arthropoda and unicellular Protozoa and Fungi have been identified at depth in the South African subsurface (164). Recent efforts to identify the source and transport of these eukaryotes underground point to freshwater sources and seismic activity (165).

While only a handful of subsurface sites have identified viable multicellular life, several continental (164,166–170) and marine subsurface (171–175) localities have identified diverse populations of fungi through sequencing and isolation. While the means of survival for subsurface fungi are still unknown, many believe subsurface fungi play a role in the degradation and recycling of nutrients in the subsurface via fermentation. It has been proposed that hydrogenosome-containing anaerobic fungi may produce H_2 during carbohydrate degradation and subsequently form syntrophic interactions with methanogens and/or SRB (176,177).

17.2.4 Subsurface Viruses

Viruses in the open ocean are now recognized as important players in marine biogeochemical cycles, initiating an estimated 10^{23} viral infections a second (178) and delivering up to 150 Gt of C to the photic zone each year through cell lysis (179). The role of viruses and their interactions with subsurface life, however, are less understood (for reviews dedicated to the topic of subsurface viruses, see 180–182). Many of the earliest investigations into subsurface viruses focused on enumeration-based methods such as virus particle-to-cell ratios (54,183–185). In the marine environment, viruses are ~2–25 times more abundant than marine bacteria (186), but in subsurface settings, this ratio can be as large as 225:1 (185).

Different hypotheses have been proposed to explain the elevated virus-to-cell ratios in the subsurface, and many have suggested that viral predation is an active process in the subsurface (180,182). Evidence of viable lytic viruses (53), active infections via single-cell sequencing (187), and acquired viral immunity via CRISPR have been reported (21,26,188). On the other hand, long-term preservation of viral particles and low to no rates of viral infections in diffusion-limited marine sediments like the South Pacific Gyre have been proposed to explain elevated virus-to-cell ratios in the subsurface (185).

Although there has been some experimental work to characterize the host range and infection frequency of subsurface viruses (53,188), the majority of these viruses are still uncharacterized (189). Advances in sampling procedures and sequencing technologies are improving the genetic characterization of viruses through (meta)genomics (190) and metatranscriptomics (189). As we continue to learn more about subsurface viruses, an important question will be how viruses influence the evolution of life in the subsurface.

17.3 Subsurface Ecology and Evolution

The abundance and diversity of life in surface environments are constrained by physical and chemical extremes such as temperature, pressure, salinity, and pH (Figure 17.3).

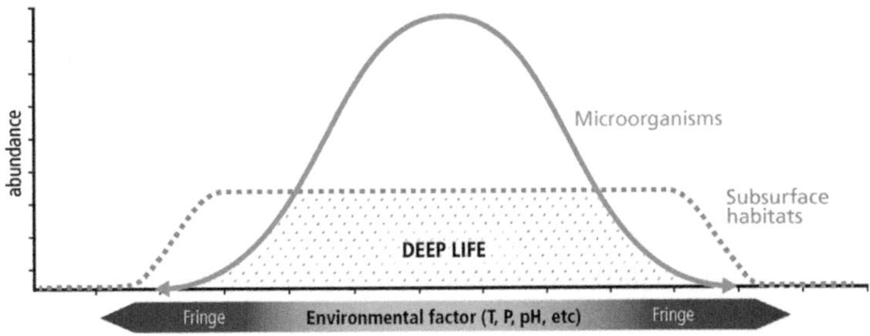

Figure 17.3 Schematic of the distribution of microbial life throughout the subsurface. An illustration of the abundance of microorganisms (blue) and subsurface habitats across a gradient of environmental factors (x-axis; green = habitable, red = uninhabitable) is shown. The overlap of the two curves represents the abundance and distribution of deep life.

As only a few subsurface sites have been sampled near these extremes, documented examples of correlations between these extremes and subsurface life are limited. However, numerous experiments and models have provided additional insights into the physical pressures of life underground.

17.3.1 Physical Extremes in the Deep Subsurface

17.3.1.1 Diffusivity

As a first-order problem, some subsurface environments may be physically restrictive to organisms. Low porosity has been hypothesized to be a limit to life in some environments, an example being the low-porosity Clay Mesa shales that exhibit lower cellular abundances than the adjoining sandstones (191). At sufficient depth, high pressures and small particle sizes (~1 μm or less) are likely to limit physical space and fluid movement (192). In cases where fracturing of deep subsurface substrates has occurred, these restrictions may be overcome. Cockell et al. (193) report an increase in cell abundance in the deep subsurface of the Chesapeake Bay meteorite impact structure and suggest that increases in porosity and fracturing of the rock in the impact-derived suevite layer increase both pore space and potentially fluid flow for microbial growth (193). When sufficient porosity is available, many subsurface environments are found to be energy restricted (for an in-depth discussion of the energy limitation problem, see Chapter 19, this volume).

Among the early successes in understanding how diffusivity influences microbial diversity are the microbial transport experiments performed at the US Department of Energy's field site in Oyster, Virginia (for a review on this topic, see 194). Experiments at the South Oyster Site highlighted the variability in attachment rate of microorganisms due to differences in surface charge and the importance of hydraulic conductivity to transport. More recent work has explored how endospores can be released from the subsurface and transported throughout the marine environment (195,196), with current efforts focused on understanding how these processes influence biogeographic patterns in the subsurface.

17.3.1.2 pH

In environments where space and energy are permissive for growth, other physical and chemical extremes may exert an influence on life, although many of these extremes are poorly studied. As with surface environments, there is no known subsurface environment where pH is known to limit life as a sole stressor. In hydrothermal vent environments, pH values as low as 1.6 are reported (197), although the movements of fluids though these systems make it difficult to determine long-term *in situ* pH values. The lowest growth of a subsurface isolate (*A. boonei* at pH 3.3, 157) may not be the lowest pH limit of subsurface life. Alkaline environments are generated in the subsurface (see Section 17.1.4 for more details), such as during serpentinization of ultramafic rocks, which generates pH values greater than 12 (198). Isolates from the continental subsurface can grow at these extremes.

Although cell numbers in the hyperalkaline Chamorro Seamount were found to decline in regions with a pH greater than 11 (199), there is no definitive evidence that pH alone limits growth.

17.3.1.3 Salinity

The isolation of halophiles from deep-sea environments (200) shows that salt-tolerant organisms can be found in the deep subsurface. Sodium chloride solutions with a water activity >0.75 are unlikely to act as limits to life in the deep subsurface. However, solutions of other salts, such as $MgCl_2$, which can have water activities below the limit for life and potentially exert chaotropic effects, can potentially act as limits to life in deep subsurface $MgCl_2$-rich brines (201). In the continental deep subsurface, evaporites at 1 km depth in the Boulby Mine (Zechstein sequence) have been observed to produce solutions with a water activity of 0.56, below the limit for life and at which no life was detected (221). Deep subsurface evaporites may therefore contain habitats where salinity extremes limit life. These findings were obtained due to Deep Carbon Observatory support for metagenome sequencing of DNA purified and extracted from the evaporitic brines. The metagenomes have also yielded insights into carbon cycling in deep subsurface evaporites. For example, complete pathways for autotrophy were absent, but many complete pathways for metabolism of carbohydrates, amino acids, and other organic molecules were found, suggesting that the communities in the Zechstein sequence use either carbon derived from surface photosynthetic carbon entrained in surface-derived fluids or subsurface hydrocarbons (221).

17.3.1.4 Temperature

Because of the geothermal gradient, temperature is a ubiquitous extreme in the deep subsurface. Geothermal gradients can be extreme, such as in geothermally active regions (e.g. Yellowstone National Park), where the temperature may exceed the upper temperature limit for life (currently 121°C) within centimeters. On the other hand, geothermal gradients at subduction zones may be ~7°C km^{-1}, resulting in a theoretical 121°C isotherm of ~17 km (202). The temperature at which natural communities of organisms are limited is not known. Investigations of deep subsurface marine sediments suggest that cell abundances drop to below detectable levels in sections corresponding to hyperthermophilic growth at less than the 121°C isotherm (203). A study of the Middle Valley sediments over an inactive hydrothermal sulfidic vent (Ocean Drilling Program (ODP) site 1035) showed that cell abundances dropped to below detection when temperatures were in the range 80–110°C at 70–170 m below seafloor (mbsf) (204). A temperature limitation-induced limit to life was similarly suggested for an active hydrothermal system (ODP legs 139 and 169), where cell abundances were found to increase laterally away from the vent, but to decrease substantially with depth and falling below detection at 20–25 mbsf (204).

In addition to temperature extremes, the limits of life may additionally be restricted by energy limitations or the imposition of other stressors (e.g. salinity or extreme pH), making the limit lower than laboratory-determined investigations. However, one should also note that stressors do not always work in synergy to limit life. High pressure was found to shift the upper temperature limit of *Desulfovibrio indonesiensis* from 48°C to 50°C in the laboratory (192). Nevertheless, the lower cell abundances reported in numerous subsurface environments at temperatures that exceed ~70°C compared to more mesophilic temperatures generally support the hypothesis that temperatures limit the deep biosphere in a region consistent with the upper temperature limits recorded for laboratory-grown strains.

17.3.2 Adaptations for Survival at the Extremes

The biochemical and physiological adaptations of organisms in the deep subsurface to physical and chemical extremes such as pH have not been well studied. In principle, there seems no reason to expect that adaptations to extremes such as pH would not be similar to the same adaptations observed in surface-dwelling organisms (e.g. transmembrane proton pumping as a means to regulate pH; 205,206).

However, of the range of physical parameters that can potentially influence deep subsurface life, high pressure is a ubiquitous condition (207,208). High pressure will occur in combination with other stressors encountered in specific environments, so that understanding the limits of life under high pressure in combination with other extremes is paramount to knowing the role of pressure in defining the limits to life in the subsurface (Figure 17.3).

The adaptations to high pressure and combinations of high pressure with other extremes have received some attention. A study of *D. indonesiensis* revealed adaptations in the cellular lipids (192). High temperature was found to reduce the proportion of unsaturated lipids, presumably to enhance membrane packing. Although high pressures might be expected to act antagonistically to reduce membrane fluidity, chain length was found to increase and branching was found to decrease membrane fluidity, suggesting that alterations of saturation, fatty acid length, and branching are employed to allow adaptation to both high temperatures and pressures.

17.3.3 Evolution of Deep Life

Mutation, natural selection, gene flow, and genetic drift are generally considered to be the main drivers of evolution on our planet. For deep life, the origins and evolution of subsurface life remains an open question. While phylogenetic methods are essential for inferring the evolutionary history of genes and species, metagenomic and single-cell studies (26,99,209–212) are beginning to provide new insights into the evolution of deep life.

In the subsurface, *mutations* – changes in the nucleotide sequence of a genome – are typically identified as mismatches observed during the alignment of metagenomic reads to a reference or consensus genome. Recently, scientists investigating the sediments of

Aarhus Bay (210) and the hydrothermal fluids of Von Damm and Piccard in the Mid-Cayman Rise (211) used sequence alignment methods to identify mutations within populations of microorganisms derived from various depths and/or habitats. To evaluate whether or not these populations were under positive or purifying selection, both studies compared the ratio of nonsynonymous and synonymous mutations (K_a/K_s) in protein-coding sequences. With the exception of Piccard, these subsurface sites did not exhibit an elevated K_a/K_s, indicating that the majority of observed genetic changes were silent mutations and that the dominant evolutionary process was *purifying selection* (the removal of deleterious mutations over time) rather than *positive selection* (when beneficial mutations arise and sweep the population). These observations are consistent with more general studies that indicate that purifying rather than positive selection is responsible for the majority of diversity patterns we see within populations of microorganisms (213–215).

In addition to single-nucleotide changes, entire genes and cassettes of genes can be transferred from one population to another via horizontal gene transfer (HGT), an example of *gene flow*. HGTs are best identified through genomic and phylogenetic comparisons and play an important role in bacterial and archaeal evolution. A biofilm metagenome from the LCHF identified increased abundances of transposases relative to surface metagenomes, suggesting that HGT is an important evolutionary process in this system (216). In the continental subsurface, it has been shown that populations of subsurface bacteria have acquired horizontally transferred metabolic genes that distinguish them from that their surface- and sub-seafloor sediment-based relatives (21,43,84). A comparison of 11 *Thermotoga maritima*-like genomes from surface marine sites in the Kuril Islands, Italy, and Açores and subsurface oil reservoirs in Japan and the North Sea revealed that these closely related organisms exhibit high amounts of gene flow between populations, suggesting that these *T. maritima*-like organisms are readily migrating to and from oil reservoirs around our planet (209).

While *T. maritima*-like organisms and the endospores described in Section 17.3.1.1 may readily move between surface and subsurface habitats, other subsurface populations may not migrate as easily. For sub-seafloor sediments, low diffusivity likely inhibits movement between the deepest sediment layers and the surface (210). Low replication rates, small population sizes, and restricted mobility may indicate that *genetic drift*, or the changes in genotype frequencies due to random events, may be a stronger driver of evolution than natural selection. The *founder effect* is a particularly interesting case of genetic drift in the context of sub-seafloor evolution. Here, a heterogeneous population gets divided into small populations of migrants due to stochastic processes. By chance, some of the migrant populations will exhibit low amounts of genetic variability and, over time, may look very different from the parent population.

While the interplay between these evolutionary processes continues to be an active field of subsurface research, additional advances have been made to understand the molecular mechanisms that lead to the diversity we observe. In addition to replication error, DNA damage and erroneous repair, and HGT, diversity-generating retroelements (DGRs) have been found to play an important role in the diversification of tail fiber ligand-binding domains in subsurface archaea (217) and a variety of protein targets in subsurface candidate phyla (218). DGRs create site-specific protein hypervariability through a process known as "mutagenic retrohoming" and are especially interesting in the context on subsurface diversification as they create diversity without cell replication or division. As evolutionary research continues in the subsurface, an important step will be to consider the roles that these processes play in the context of the energy, spatial, and population-level limitations each environment provides.

17.4 Conclusion

While much is still unknown, global initiatives like the Census of Deep Life (CoDL; https://vamps2.mbl.edu/portals/CODL) and IODP Expedition 370 (www.jamstec.go.jp/chikyu/e/exp370) are currently generating large data sets to better characterize the biodiversity and limits to life in the subsurface. Already, the CoDL has coordinated the sequencing of over 600 subsurface samples from a wide variety of subsurface habitats, and the scientists of IODP Expedition 370 have collected 112 cores (equivalent to 577.85 m) across the sediment–basalt interface. Although precautions and methodological improvements have been utilized in these important programs, major challenges such as changes in sequencing technology and contamination from the surface still exist. Similar to the efforts of the US Department of Energy, US Geological Survey, and US Environmental Protection Agency to identify indigenous subsurface microbial communities in the mid-1980s (219), a major challenge will be the characterization of authentic and contaminating organisms. However, while Phelps and colleagues focused on the infiltration of drilling fluids into native samples (219), the current challenge is to remove contaminating DNA sequences from the surface and confounding factors from statistical analyses. Early work by Sheik et al. provides an interpretable outline of best practices for identifying and removing contaminating sequences from subsurface 16S rRNA gene data sets (220). The remarkable progress since the first deep subsurface metagenome was sequenced a decade ago (21) highlights the extraordinary potential to advance our understanding of the subsurface biogeography, ecology, and evolution, with countless new discoveries awaiting on the made.

Acknowledgments

We thank Josh Wood for illustrating Figures 17.2 and 17.3 and Rob Pockalny for assisting in the preparation of the map in Figure 17.1. We also thank R Anderson, the Center for Dark Energy Biosphere Investigations (C-DEBI), the Flatiron Institute Center for Computational Biology (CCB), and participants of the C-DEBI/CCB Evolution Workshop funded by NSF award OIA-0939564 and the Simons Foundation for insightful discussions surrounding the topic of microbial evolution in the deep subsurface.

Questions for the Classroom

1 What are the subsurface biomes and how do they vary?
2 What limits cellular abundances in the continental subsurface and marine subsurface?
3 What are some common themes between the continental and marine subsurface biospheres?
4 Where would heterotrophic lifestyles, autotrophic lifestyles, and a combination (i.e. metabolic versatility) be predicted in the subsurface?
5 What methods are available to study subsurface life and how do they vary?
6 What are the physical limits to life on Earth?
7 What might deep life tell us about life on other planets?
8 What are the mechanisms of evolutionary change? How would one identify them?

List of Online Resources

Maps of subsurface studies and cell numbers: https://caramagnabosco.shinyapps.io/SubsurfaceBiologicalStudies.

C-DEBI BCO-DMO data portal: www.bco-dmo.org/program/554979.

References

1. Magnabosco C, Lin L-H, Dong H, Bomberg M, Ghiorse W, Stan-Lotter H, et al. The biomass and biodiversity of the continental subsurface. *Nat Geosci.* 2018;11(10):707.
2. Gleeson T, Befus KM, Jasechko S, Luijendijk E, Cardenas MB. The global volume and distribution of modern groundwater. *Nat Geosci.* 2016;9(2):161–167.
3. Divins DL. Total Sediment Thickness of the World's Oceans and Marginal Seas. NOAA National Geophysical Data Center. www.ngdc.noaa.gov/mgg/sedthick/sedthick.html, 2002.
4. Laske G, Masters G. A global digital map of sediment thickness. *Eos Trans AGU.* 1997;78:F483.
5. Whittaker JM, Goncharov A, Williams SE, Müller RD, Leitchenkov G. Global sediment thickness data set updated for the Australian-Antarctic Southern Ocean. *Geochem Geophys Geosystems.* 2013;14(8):3297–3305.
6. LaRowe DE, Burwicz E, Arndt S, Dale AW, Amend JP. Temperature and volume of global marine sediments. *Geology.* 2017;45(3):275–278.
7. Heberling C, Lowell RP, Liu L, Fisk MR. Extent of the microbial biosphere in the oceanic crust. *Geochem Geophys Geosystems.* 2010;11(8):Q08003.
8. Kallmeyer J, Pockalny R, Adhikari RR, Smith DC, D'Hondt S. Global distribution of microbial abundance and biomass in subseafloor sediment. *Proc Natl Acad Sci.* 2012;109(40):16213–16216.
9. Bar-On YM, Phillips R, Milo R. The biomass distribution on Earth. *Proc Natl Acad Sci.* 2018;115(25):6506–6511.
10. Parkes RJ, Cragg B, Roussel E, Webster G, Weightman A, Sass H. A review of prokaryotic populations and processes in sub-seafloor sediments, including biosphere: geosphere interactions. *Mar Geol.* 2014;352:409–425.

11. Whitman WB, Coleman DC, Wiebe WJ. Prokaryotes: the unseen majority. *Proc Natl Acad Sci.* 1998;95:6578–6583.
12. Kieft TL, Simmons KA. Allometry of animal–microbe interactions and global census of animal-associated microbes. *Proc Biol Sci.* 2015;282(1810):20150702.
13. Bastin ES, Greer FE, Merritt CA, Moulton G. The presence of sulphate reducing bacteria in oil field waters. *Science.* 1926;63(1618):21–24.
14. Ginsburg-Karagitscheva TL. Microbiological research in the sulphurous and salty waters of Apsheron. *Azerb Neft Khozjajstvo.* 1926;6–7.
15. Lipman CB. The discovery of living microrganisms in ancient rocks. *Science.* 1928;68(1760):272–273.
16. Onstott TC. *Deep Life: The Hunt for the Hidden Biology of Earth, Mars, and Beyond.* Princeton University Press, 2016.
17. Onstott TC, Phelps TJ, Kieft T, Colwell FS, Balkwill DL, Fredrickson JK, et al. *Enigmatic Microorganisms and Life in Extreme Environments.* Kluwer Academic, 1999.
18. Pedersen K, Ekendahl S. Distribution and activity of bacteria in deep granitic groundwaters of southeastern Sweden. *Microb Ecol.* 1990;20(1):37–52.
19. Fredrickson JK, Balkwill DL. Geomicrobial processes and biodiversity in the deep terrestrial subsurface. *Geomicrobiol J.* 2006;23(6):345–356.
20. Gihring TM, Moser DP, Lin L-H, Davidson M, Onstott TC, Morgan L, et al. The distribution of microbial taxa in the subsurface water of the Kalahari Shield, South Africa. *Geomicrobiol J.* 2006;23:415–430.
21. Chivian D, Brodie EL, Alm EJ, Culley DE, Dehal PS, DeSantis TZ, et al. Environmental genomics reveals a single-species ecosystem deep within Earth. *Science.* 2008;322:275–278.
22. Sogin ML, Morrison HG, Huber J A, Mark Welch D, Huse SM, Neal PR, et al. Microbial diversity in the deep sea and the underexplored "rare biosphere". *Proc Natl Acad Sci.* 2006;103(32):12115–12120.
23. Magnabosco C, Timmers PHA, Lau MCY, Borgonie G, Linage-Alvarez B, Kuloyo O, et al. Fluctuations in populations of subsurface methane oxidizers in coordination with changes in electron acceptor availability. *FEMS Microbiol Ecol.* 2018;94: fiy089.
24. O'Mullan G, Dueker ME, Clauson K, Yang Q, Umemoto K, Zakharova N, et al. Microbial stimulation and succession following a test well injection simulating CO_2 leakage into a shallow Newark Basin aquifer. *PLoS One.* 2015;10 (1):e0117812.
25. Bagnoud A, Chourey K, Hettich RL, De Bruijn I, Andersson AF, Leupin OX, et al. Reconstructing a hydrogen-driven microbial metabolic network in Opalinus Clay rock. *Nat Commun.* 2016;7:12770.
26. Daly RA, Borton MA, Wilkins MJ, Hoyt DW, Kountz DJ, Wolfe RA, et al. Microbial metabolisms in a 2.5-km-deep ecosystem created by hydraulic fracturing in shales. *Nat Microbiol.* 2016;1:16146.
27. Wang L-Y, Ke W-J, Sun X-B, Liu J-F, Gu J-D, Mu B-Z. Comparison of bacterial community in aqueous and oil phases of water-flooded petroleum reservoirs using pyrosequencing and clone library approaches. *Appl Microbiol Biotechnol.* 2014;98(9):4209–4221.
28. Mu A, Boreham C, Leong HX, Haese RR, Moreau JW. Changes in the deep subsurface microbial biosphere resulting from a field-scale CO_2 geosequestration experiment. *Front Microbiol.* 2014;5:209.

29. Rajala P, Carpén L, Vepsäläinen M, Raulio M, Sohlberg E, Bomberg M. Microbially induced corrosion of carbon steel in deep groundwater environment. *Front Microbiol*. 2015;6:647.
30. Waldron PJ, Petsch ST, Martini AM, Nüsslein K. Salinity constraints on subsurface archaeal diversity and methanogenesis in sedimentary rock rich in organic matter. *Appl Environ Microbiol*. 2007;73(13):4171–4179.
31. Davis JP, Struchtemeyer CG, Elshahed MS. Bacterial communities associated with production facilities of two newly drilled thermogenic natural gas wells in the Barnett Shale (Texas, USA). *Microb Ecol*. 2012;64(4):942–954.
32. Murali Mohan A, Hartsock A, Hammack RW, Vidic RD, Gregory KB. Microbial communities in flowback water impoundments from hydraulic fracturing for recovery of shale gas. *FEMS Microbiol Ecol*. 2013;86(3):567–580.
33. Wuchter C, Banning E, Mincer TJ, Drenzek NJ, Coolen MJL. Microbial diversity and methanogenic activity of Antrim Shale formation waters from recently fractured wells. *Front Microbiol*. 2013;4:367.
34. Cluff MA, Hartsock A, MacRae JD, Carter K, Mouser PJ. Temporal changes in microbial ecology and geochemistry in produced water from hydraulically fractured Marcellus Shale gas wells. *Environ Sci Technol*. 2014;48(11):6508–6517.
35. Mohan AM, Bibby KJ, Lipus D, Hammack RW, Gregory KB. The functional potential of microbial communities in hydraulic fracturing source water and produced water from natural gas extraction characterized by metagenomic sequencing. *PLoS One*. 2014;9(10):e107682.
36. Akob DM, Cozzarelli IM, Dunlap DS, Rowan EL, Lorah MM. Organic and inorganic composition and microbiology of produced waters from Pennsylvania shale gas wells. *Appl Geochem*. 2015;60:116–125.
37. Dong Y, Kumar CG, Chia N, Kim P-J, Miller PA, Price ND, et al. Halomonas sulfidaeris-dominated microbial community inhabits a 1.8 km-deep subsurface Cambrian Sandstone reservoir. *Environ Microbiol*. 2014;16(6):1695–1708.
38. Magot M, Ollivier B, Patel BKC. Microbiology of petroleum reservoirs. *Antonie Van Leeuwenhoek*. 2000;77(2):103–116.
39. Kryachko Y, Dong X, Sensen CW, Voordouw G. Compositions of microbial communities associated with oil and water in a mesothermic oil field. *Antonie Van Leeuwenhoek*. 2012;101(3):493–506.
40. Zhang F, She Y-H, Chai L-J, Banat IM, Zhang X-T, Shu F-C, et al. Microbial diversity in long-term water-flooded oil reservoirs with different *in situ* temperatures in China. *Sci Rep*. 2012;2:760.
41. Singh A, Van Hamme JD, Kuhad RC, Parmar N, Ward OP. Subsurface petroleum microbiology. In: Parmar N, Singh A eds. *Geomicrobiology and Biogeochemistry*. Springer, 2014, p. 153–173.
42. Voordouw G. Production-related petroleum microbiology: progress and prospects. *Curr Opin Biotechnol*. 2011;22(3):401–405.
43. Momper L, Reese BK, Zinke L, Wanger G, Osburn MR, Moser D, et al. Major phylum-level differences between porefluid and host rock bacterial communities in the terrestrial deep subsurface. *Environ Microbiol Rep*. 2017;9(5):501–511.
44. Magnabosco C, Ryan K, Lau MCY, Kuloyo O, Lollar BS, Kieft TL, et al. A metagenomic window into carbon metabolism at 3 km depth in Precambrian continental crust. *ISME J*. 2016;10(3):730–741.
45. Bomberg M, Nyyssönen M, Pitkänen P, Lehtinen A, Itävaara M. Active microbial communities inhabit sulphate–methane interphase in deep bedrock fracture fluids in Olkiluoto, Finland. *Biomed Res Int*. 2015;2015:979530.

46. Osburn MR, LaRowe DE, Momper LM, Amend JP. Chemolithotrophy in the continental deep subsurface: Sanford Underground Research Facility (SURF), USA. *Front Microbiol.* 2014;5:610.
47. Anantharaman K, Brown CT, Hug LA, Sharon I, Castelle CJ, Probst AJ, et al. Thousands of microbial genomes shed light on interconnected biogeochemical processes in an aquifer system. *Nat Commun.* 2016;7:13219.
48. Purkamo L, Bomberg M, Nyyssönen M, Kukkonen I, Ahonen L, Itävaara M. Heterotrophic communities supplied by ancient organic carbon predominate in deep Fennoscandian bedrock fluids. *Microb Ecol.* 2015;69(2):319–332.
49. Nyyssönen M, Bomberg M, Kapanen A, Nousiainen A, Pitkänen P, Itävaara M. Methanogenic and sulphate-reducing microbial communities in deep groundwater of crystalline rock fractures in Olkiluoto, Finland. *Geomicrobiol J.* 2012;29 (10):863–878.
50. Momper L, Jungbluth SP, Lee MD, Amend JP. Energy and carbon metabolisms in a deep terrestrial subsurface fluid microbial community. *ISME J.* 2017;11 (10):2319–2333.
51. Lin L-H, Wang P-L, Rumble D, Lippmann-Pipke J, Boice E, Pratt LM, et al. Long-term sustainability of a high-energy, low-diversity crustal biome. *Science.* 2006;314:479–482.
52. Lau MCY, Kieft TL, Kuloyo O, Linage-Alvarez B, Van Heerden E, Lindsay MR, et al. An oligotrophic deep-subsurface community dependent on syntrophy is dominated by sulfur-driven autotrophic denitrifiers. *Proc Natl Acad Sci.* 2016;113(49): E7927–E7936.
53. Eydal HSC, Jägevall S, Hermansson M, Pedersen K. Bacteriophage lytic to Desulfovibrio aespoeensis isolated from deep groundwater. *ISME J.* 2009;3 (10):1139–1147.
54. Kyle JE, Eydal HSC, Ferris FG, Pedersen K. Viruses in granitic groundwater from 69 to 450 m depth of the Äspö hard rock laboratory, Sweden. *ISME J.* 2008;2 (5):571–574.
55. D'Hondt S, Jørgensen BB, Miller DJ, Batzke A, Blake R, Cragg BA, et al. Distributions of microbial activities in deep subseafloor sediments. *Science.* 2004;306 (5705):2216–2221.
56. Parkes RJ, Cragg BA, Bale SJ, Getlifff JM, Goodman K, Rochelle PA, et al. Deep bacterial biosphere in Pacific Ocean sediments. *Nature.* 1994;371(6496):410–413.
57. Parkes JR, Webster G, Cragg BA, Weightman AJ, Newberry CJ, Ferdelman TG, et al. Deep sub-seafloor prokaryotes stimulated at interfaces over geological time. *Nature.* 2005;436(7049):390–394.
58. Yanagawa K, Breuker A, Schippers A, Nishizawa M, Ijiri A, Hirai M, et al. Microbial community stratification controlled by the subseafloor fluid flow and geothermal gradient at the Iheya North hydrothermal field in the Mid-Okinawa Trough (Integrated Ocean Drilling Program Expedition 331). *Appl Environ Microbiol.* 2014;80 (19):6126–6135.
59. Orcutt BN, Wheat CG, Rouxel O, Hulme S, Edwards KJ, Bach W. Oxygen consumption rates in subseafloor basaltic crust derived from a reaction transport model. *Nat Commun.* 2013;4:2539.
60. Biddle JF, Fitz-Gibbon S, Schuster SC, Brenchley JE, House CH. Metagenomic signatures of the Peru Margin subseafloor biosphere show a genetically distinct environment. *Proc Natl Acad Sci.* 2008;105(30):10583–10588.
61. Orsi WD, Edgcomb VP, Christman GD, Biddle JF. Gene expression in the deep biosphere. *Nature.* 2013;499(7457):205–208.

62. Orsi WD, Barker Jørgensen B, Biddle JF. Transcriptional analysis of sulfate reducing and chemolithoautotrophic sulfur oxidizing bacteria in the deep subseafloor. *Environ Microbiol Rep*. 2016;8(4):452–460.
63. Biddle JF, White JR, Teske AP, House CH. Metagenomics of the subsurface Brazos-Trinity Basin (IODP site 1320): comparison with other sediment and pyrosequenced metagenomes. *ISME J*. June 6, 2011;5(6):1038–1047.
64. Newberry CJ, Webster G, Cragg BA, Parkes RJ, Weightman AJ, Fry JC. Diversity of prokaryotes and methanogenesis in deep subsurface sediments from the Nankai Trough, Ocean Drilling Program Leg 190. *Environ Microbiol*. 2004;6(3):274–287.
65. Inagaki F, Nunoura T, Nakagawa S, Teske A, Lever M, Lauer A, et al. Biogeographical distribution and diversity of microbes in methane hydrate-bearing deep marine sediments on the Pacific Ocean Margin. *Proc Natl Acad Sci*. 2006;103(8):2815–2820.
66. Biddle JF, Cardman Z, Mendlovitz H, Albert DB, Lloyd KG, Boetius A, et al. Anaerobic oxidation of methane at different temperature regimes in Guaymas Basin hydrothermal sediments. *ISME J*. 2012;6(5):1018–1031.
67. Teske A, Callaghan AV, LaRowe DE. Biosphere frontiers of subsurface life in the sedimented hydrothermal system of Guaymas Basin. *Front Microbiol*. 2014;5:1–11.
68. Marshall IPG, Karst SM, Nielsen PH, Jørgensen BB. Metagenomes from deep Baltic Sea sediments reveal how past and present environmental conditions determine microbial community composition. *Mar Genomics*. 2017;37:58–68.
69. Inagaki F, Hinrichs K-U, Kubo Y, Bowles MW, Heuer VB, Hong W-L, et al. Exploring deep microbial life in coal-bearing sediment down to ~2.5 km below the ocean floor. *Science*. 2015;349(6246):420–424.
70. Briggs BR, Inagaki F, Morono Y, Futagami T, Huguet C, Rosell-Mele A, et al. Bacterial dominance in subseafloor sediments characterized by methane hydrates. *FEMS Microbiol Ecol*. 2012;81(1):88–98.
71. Walsh EA, Kirkpatrick JB, Rutherford SD, Smith DC, Sogin M, D'Hondt S. Bacterial diversity and community composition from seasurface to subseafloor. *ISME J*. 2016;10(4):979–989.
72. Starnawski P, Bataillon T, Ettema TJG, Jochum LM, Schreiber L, Chen X, et al. Microbial community assembly and evolution in subseafloor sediment. *Proc Natl Acad Sci*. 2017;114(11):2940–2945.
73. Riedinger N, Strasser M, Harris RN, Klockgether G, Lyons TW, Screaton EJ. Deep subsurface carbon cycling in the Nankai Trough (Japan) – evidence of tectonically induced stimulation of a deep microbial biosphere. *Geochem Geophys Geosyst*. 2015;16(9):3257–3270.
74. Arndt S, Brumsack H-J, Wirtz KW. Cretaceous black shales as active bioreactors: a biogeochemical model for the deep biosphere encountered during ODP Leg 207 (Demerara Rise). *Geochim Cosmochim Acta*. 2006;70(2):408–425.
75. Parkes RJ, Wellsbury P, Mather ID, Cobb SJ, Cragg BA, Hornibrook ERC, et al. Temperature activation of organic matter and minerals during burial has the potential to sustain the deep biosphere over geological timescales. *Org Geochem*. 2007;38 (6):845–852.
76. Lomstein BA, Langerhuus AT, D'Hondt S, Jørgensen BB, Spivack AJ. Endospore abundance, microbial growth and necromass turnover in deep sub-seafloor sediment. *Nature*. 2012;484(7392):101–104.
77. Contreras S, Meister P, Liu B, Prieto-Mollar X, Hinrichs K-U, Khalili A, et al. Cyclic 100-ka (glacial–interglacial) migration of subseafloor redox zonation on the Peruvian shelf. *Proc Natl Acad Sci*. 2013;110(45):18098–18103.

78. Orcutt BN, Sylvan JB, Knab NJ, Edwards KJ. Microbial ecology of the dark ocean above, at, and below the seafloor. *Microbiol Mol Biol Rev*. 2011;75 (2):361–422.
79. Schrenk MO, Huber JA, Edwards KJ. Microbial provinces in the subseafloor. *Ann Rev Mar Sci.* 2009;2:85–110.
80. Colwell FS, D'Hondt S. Nature and extent of the deep biosphere. *Rev Mineral Geochem*. 2013;75(1):547–574.
81. Baross JA, Wilcock WSD, Kelley DS, DeLong EF, Craig Cary S. The subsurface biosphere at mid-ocean ridges: issues and challenges. In: *The Subseafloor Biosphere at Mid-Ocean Ridges*. American Geophysical Union, 2004, pp. 1–11.
82. Fisk M, McLoughlin N. Atlas of alteration textures in volcanic glass from the ocean basins. *Geosphere*. 2013;9(2):317–341.
83. Bach W, Edwards KJ. Iron and sulfide oxidation within the basaltic ocean crust: implications for chemolithoautotrophic microbial biomass production. *Geochim Cosmochim Acta*. 2003;67(20):3871–3887.
84. Jungbluth SP, Glavina Del Rio T, Tringe SG, Stepanauskas R, Rappé MS. Genomic comparisons of a bacterial lineage that inhabits both marine and terrestrial deep subsurface systems. *PeerJ.* 2017;5:e3134.
85. Nakagawa S, Inagaki F, Suzuki Y, Steinsbu BO, Lever MA, Takai K, et al. Microbial community in black rust exposed to hot ridge flank crustal fluids. *Appl Environ Microbiol*. 2006;72(10):6789–6799.
86. Baquiran J-PM, Ramírez GA, Haddad AG, Toner BM, Hulme S, Wheat CG, et al. Temperature and redox effect on mineral colonization in Juan de Fuca Ridge Flank subsurface crustal fluids. *Front Microbiol.* 2016;7:396.
87. Jungbluth SP, Grote J, Lin HT, Cowen JP, Rappé MS. Microbial diversity within basement fluids of the sediment-buried Juan de Fuca Ridge flank. *ISME J.* 2013;7(1):161–172.
88. Lever MA, Rouxel O, Alt JC, Shimizu N, Ono S, Coggon RM, et al. Evidence for microbial carbon and sulfur cycling in deeply buried ridge flank basalt. *Science.* 2013;339(6125):1305–1308.
89. Smith A, Popa R, Fisk M, Nielsen M, Wheat CG, Jannasch HW, et al. *In situ* enrichment of ocean crust microbes on igneous minerals and glasses using an osmotic flow-through device. *Geochem Geophys Geosyst*. 2011;12(6):Q06007.
90. Orcutt BN, Bach W, Becker K, Fisher AT, Hentscher M, Toner BM, et al. Colonization of subsurface microbial observatories deployed in young ocean crust. *ISME J.* 2011;5:692–703.
91. Jungbluth SP, Lin HT, Cowen JP, Glazer BT, Rappé MS. Phylogenetic diversity of microorganisms in subseafloor crustal fluids from Holes 1025C and 1026B along the Juan de Fuca Ridge flank. *Front Microbiol*. 2014;5:119.
92. Cowen JP, Giovannoni SJ, Kenig F, Johnson HP, Butterfield D, Rappé MS, et al. Fluids from aging ocean crust that support microbial life. *Science*. 2003;299 (5603):120–123.
93. Robador A, Jungbluth SP, LaRowe DE, Bowers RM, Rappé MS, Amend JP, Cowen JP. Activity and phylogenetic diversity of sulfate-reducing microorganisms in low-temperature subsurface fluids within the upper oceanic crust. *Front Microbiol.* 2015;5:748.
94. Jungbluth SP, Bowers RM, Lin HT, Cowen JP, Rappé MS. Novel microbial assemblages inhabiting crustal fluids within mid-ocean ridge flank subsurface basalt. *ISMEJ*. 2016;10(8):2033–2047.

95. Jungbluth SP, Amend JP, Rappé MS. Metagenome sequencing and 98 microbial genomes from Juan de Fuca Ridge flank subsurface fluids. *Sci Data.* 2017;4:170037.
96. Nigro L, Harris K, Orcutt B, Hyde A, Clayton-Luce S, Becker K, et al. Microbial communities at the borehole observatory on the Costa Rica Rift flank (Ocean Drilling Program Hole 896A). *Front Microbiol.* 2012;3:232.
97. Edwards KJ, Bach W, Kalus A. Expedition 336 Scientists. *Expedition 336 Summary.* College Station, TX: Ocean Drilling Program, 2012.
98. Meyer JL, Jaekel U, Tully BJ, Glazer BT, Wheat CG, Lin H-T, et al. A distinct and active bacterial community in cold oxygenated fluids circulating beneath the western flank of the Mid-Atlantic ridge. *Sci Rep.* 2016;6:22541.
99. Tully BJ, Wheat CG, Glazer BT, Huber JA. A dynamic microbial community with high functional redundancy inhabits the cold, oxic subseafloor aquifer. *ISME J.* 2017;12(1):1.
100. Zhang X, Fang J, Bach W, Edwards KJ, Orcutt BN, Wang F. Nitrogen stimulates the growth of subsurface basalt-associated microorganisms at the western flank of the Mid-Atlantic Ridge. *Front Microbiol.* 2016;7:633.
101. Zhang X, Feng X, Wang F. Diversity and metabolic potentials of subsurface crustal microorganisms from the western flank of the Mid-Atlantic Ridge. *Front Microbiol.* 2016;7:363.
102. Charlou JL, Donval JP, Fouquet Y, Jean-Baptiste P, Holm N. Geochemistry of high H_2 and CH_4 vent fluids issuing from ultramafic rocks at the Rainbow hydrothermal field (36 14' N, MAR). *Chem Geol.* 2002;191(4):345–359.
103. McCollom TM, Seewald JS. Abiotic synthesis of organic compounds in deep-sea hydrothermal environments. *Chem Rev.* 2007;107(2):382–401.
104. Proskurowski G, Lilley MD, Seewald JS, Früh-Green GL, Olson EJ, Lupton JE, et al. Abiogenic hydrocarbon production at Lost City hydrothermal field. *Science.* 2008;319(5863):604–607.
105. McCollom TM. Laboratory simulations of abiotic hydrocarbon formation in Earth's deep subsurface. *Rev Mineral Geochem.* 2013;75(1):467–494.
106. Schrenk MO, Brazelton WJ, Lang SQ. Serpentinization, carbon, and deep life. *Rev Mineral Geochem.* 2013;75(1):575–606.
107. Kelley DS, Karson JA, Blackman DK, Früh-Green GL, Butterfield DA, Lilley MD, et al. An off-axis hydrothermal vent field near the Mid-Atlantic Ridge at 30 N. *Nature.* 2001;412(6843):145–149.
108. Kelley DS, Karson JA, Früh-Green GL, Yoerger DR, Shank TM, Butterfield DA, et al. A serpentinite-hosted ecosystem: the Lost City hydrothermal field. *Science.* 2005;307(5714):1428–1434.
109. Lang SQ, Butterfield DA, Schulte M, Kelley DS, Lilley MD. Elevated concentrations of formate, acetate and dissolved organic carbon found at the Lost City hydrothermal field. *Geochim Cosmochim Acta.* 2010;74(3):941–952.
110. Schrenk MO, Kelley DS, Bolton SA, Baross JA. Low archaeal diversity linked to subseafloor geochemical processes at the Lost City hydrothermal field, Mid-Atlantic Ridge. *Environ Microbiol.* 2004;6(10):1086–1095.
111. Brazelton WJ, Schrenk MO, Kelley DS, Baross JA. Methane- and sulfur-metabolizing microbial communities dominate the Lost City hydrothermal field ecosystem. *Appl Environ Microbiol.* 2006;72(9):6257–6270.
112. Brazelton WJ, Ludwig KA, Sogin ML, Andreishcheva EN, Kelley DS, Shen C-C, et al. Archaea and bacteria with surprising microdiversity show shifts in dominance

over 1,000-year time scales in hydrothermal chimneys. *Proc Natl Acad Sci.* 2010;107 (4):1612–1617.
113. Brazelton WJ, Mehta MP, Kelley DS, Baross JA. Physiological differentiation within a single-species biofilm fueled by serpentinization. *MBio.* 2011;2(4):e00127-11.
114. Quesnel B, Gautier P, Boulvais P, Cathelineau M, Maurizot P, Cluzel D, et al. Syntectonic, meteoric water-derived carbonation of the New Caledonia peridotite nappe. *Geology.* 2013;41(10):1063–1066.
115. Quéméneur M, Bes M, Postec A, Mei N, Hamelin J, Monnin C, et al. Spatial distribution of microbial communities in the shallow submarine alkaline hydrothermal field of the Prony Bay, New Caledonia. *Environ Microbiol Rep.* 2014;6 (6):665–674.
116. Postec A, Quéméneur M, Bes M, Mei N, Benaïssa F, Payri C, et al. Microbial diversity in a submarine carbonate edifice from the serpentinizing hydrothermal system of the Prony Bay (New Caledonia) over a 6-year period. *Front Microbiol.* 2015;6:857.
117. Morrill PL, Kuenen JG, Johnson OJ, Suzuki S, Rietze A, Sessions AL, et al. Geochemistry and geobiology of a present-day serpentinization site in California: The Cedars. *Geochim Cosmochim Acta.* 2013;109:222–240.
118. Szponar N, Brazelton WJ, Schrenk MO, Bower DM, Steele A, Morrill PL. Geochemistry of a continental site of serpentinization, the Tablelands Ophiolite, Gros Morne National Park: a Mars analogue. *Icarus.* 2013;224:286–296.
119. Brazelton WJ, Nelson B, Schrenk MO. Metagenomic evidence for H_2 oxidation and H_2 production by serpentinite-hosted subsurface microbial communities. *Front Microbiol.* 2012;2:268.
120. Brazelton WJ, Morrill PL, Szponar N, Schrenk MO. Bacterial communities associated with subsurface geochemical processes in continental serpentinite springs. *Appl Environ Microbiol.* 2013;79:3906–3916.
121. Tiago I, Veríssimo A. Microbial and functional diversity of a subterrestrial high pH groundwater associated to serpentinization. *Environ Microbiol.* 2013;15: 1687–1706.
122. Suzuki S, Ishii S, Wu A, Cheung A, Tenney A, Wanger G, et al. Microbial diversity in The Cedars, an ultrabasic, ultrareducing, and low salinity serpentinizing ecosystem. *Proc Natl Acad Sci.* 2013;110(38):15336–15341.
123. Suzuki S, Kuenen JG, Schipper K, Van Der Velde S, Ishii S, Wu A, et al. Physiological and genomic features of highly alkaliphilic hydrogen-utilizing Betaproteobacteria from a continental serpentinizing site. *Nat Commun.* 2014;5:3900.
124. Twing KI, Brazelton WJ, Kubo MDY, Hyer AJ, Cardace D, Hoehler TM, et al. Serpentinization-influenced groundwater harbors extremely low diversity microbial communities adapted to high pH. *Front Microbiol.* 2017;8:308.
125. Opatkiewicz AD, Butterfield DA, Baross JA. Individual hydrothermal vents at Axial Seamount harbor distinct subseafloor microbial communities. *FEMS Microbiol Ecol.* 2009;70(3):413–424.
126. Niemann H, Lösekann T, De Beer D, Elvert M, Nadalig T, Knittel K. Novel microbial communities of the Haakon Mosby mud volcano and their role as a methane sink. *Nature.* 2006;443(7113):854.
127. Ruff SE, Biddle JF, Teske AP, Knittel K, Boetius A, Ramette A. Global dispersion and local diversification of the methane seep microbiome. *Proc Natl Acad Sci.* 2015;112(13):4015–4020.

128. Quéméneur M, Palvadeau A, Postec A, Monnin C, Chavagnac V, Ollivier B, et al. Endolithic microbial communities in carbonate precipitates from serpentinite-hosted hyperalkaline springs of the Voltri Massif (Ligurian Alps, Northern Italy). *Environ Sci Pollut Res.* 2015;22(18):13613–13624.
129. Brazelton WJ, Thornton CN, Hyer A, Twing KI, Longino AA, Lang SQ, et al. Metagenomic identification of active methanogens and methanotrophs in serpentinite springs of the Voltri Massif, Italy. *PeerJ.* 2017;5:e2945.
130. Crespo-Medina M, Twing KI, Sánchez-Murillo R, Brazelton WJ, McCollom TM, Schrenk MO. Methane dynamics in a tropical serpentinizing environment: the Santa Elena Ophiolite, Costa Rica. *Front Microbiol.* 2017;8:916.
131. Cardace D, Hoehler T, McCollom T, Schrenk M, Carnevale D, Kubo M, et al. Establishment of the Coast Range ophiolite microbial observatory (CROMO): drilling objectives and preliminary outcomes. *Sci Drill.* 2013;16:45–55.
132. Daae FL, Økland I, Dahle H, Jørgensen SL, Thorseth IH, Pedersen RB. Microbial life associated with low-temperature alteration of ultramafic rocks in the Leka ophiolite complex. *Geobiology.* 2013;11(4):318–339.
133. NERC. Oman Drilling Project. www.omandrilling.ac.uk, 2017.
134. Früh-Green GL, Orcutt BN, Green SL, Cotterill C, Morgan S, Akizawa N, et al. Expedition 357 summary. *Proc Int Ocean Discov Progr.* 2017;357.
135. Higashi Y, Sunamura M, Kitamura K, Nakamura K, Kurusu Y, Ishibashi J, et al. Microbial diversity in hydrothermal surface to subsurface environments of Suiyo Seamount, Izu-Bonin Arc, using a catheter-type in situ growth chamber. *FEMS Microbiol Ecol.* 2004;47(3):327–336.
136. Opatkiewicz AD, Butterfield DA, Baross JA. Individual hydrothermal vents at Axial Seamount harbor distinct subseafloor microbial communities. *FEMS Microbiol Ecol.* 2009;70(3):413–424.
137. Fernández-Remolar DC, Prieto-Ballesteros O, Rodríguez N, Gómez F, Amils R, Gómez-Elvira J, et al. Underground habitats in the Rio Tinto basin: a model for subsurface life habitats on Mars. *Astrobiology.* 2008;8(5):1023–1047.
138. Carrier BL. Next steps forward in understanding Martian surface and subsurface chemistry. *J Geophys Res Planets.* 2017;122(9):1951–1953.
139. Michalski JR, Cuadros J, Niles PB, Parnell J, Rogers AD, Wright SP. Groundwater activity on Mars and implications for a deep biosphere. *Nat Geosci.* 2013; 6(2):133.
140. Waite JH, Glein CR, Perryman RS, Teolis BD, Magee BA, Miller G, et al. Cassini finds molecular hydrogen in the Enceladus plume: evidence for hydrothermal processes. *Science.* 2017;356(6334):155–159.
141. Flynn GJ. The delivery of organic matter from asteroids and comets to the early surface of Mars. In: Rickman H, Valtonen MJ, ed. *Worlds in Interaction: Small Bodies and Planets of the Solar System.* Springer, 1996, pp. 469–474.
142. Field CB, Behrenfeld MJ, Randerson JT, Falkowski P. Primary production of the biosphere: integrating terrestrial and oceanic components. *Science.* 1998;281(5374): 237–240.
143. Sherwood Lollar B, Voglesonger K, Lin L-H, Lacrampe-Couloume G, Telling J, Abrajano TA, et al. Hydrogeologic controls on episodic H2 release from Precambrian fractured rocks – energy for deep subsurface life on Earth and Mars. *Astrobiology.* 2007;7(6):971–986.
144. Kattenhorn SA, Prockter LM. Evidence for subduction in the ice shell of Europa. *Nat Geosci.* 2014;7(10):762.

145. Waite JH, Combi MR, Ip W-H, Cravens TE, McNutt RL, Kasprzak W, et al. Cassini ion and neutral mass spectrometer: Enceladus plume composition and structure. *Science*. 2006;311(5766):1419–1422.
146. Karner MB, DeLong EF, Karl DM. Archaeal dominance in the mesopelagic zone of the Pacific Ocean. *Nature*. 2001;409(6819):507.
147. Biddle JF, Lipp JS, Lever MA, Lloyd KG, Sørensen KB, Anderson R, et al. Heterotrophic archaea dominate sedimentary subsurface ecosystems off Peru. *Proc Natl Acad Sci*. 2006;103(10):3846–3851.
148. Lipp JS, Morono Y, Inagaki F, Hinrichs K-U. Significant contribution of Archaea to extant biomass in marine subsurface sediments. *Nature*. 2008;454(7207):991–994.
149. Schouten S, Middelburg JJ, Hopmans EC, Sinninghe Damsté JS. Fossilization and degradation of intact polar lipids in deep subsurface sediments: a theoretical approach. *Geochim Cosmochim Acta*. 2010;74(13):3806–3814.
150. Lloyd KG, May MK, Kevorkian RT, Steen AD. Meta-analysis of quantification methods shows that archaea and bacteria have similar abundances in the subseafloor. *Appl Environ Microbiol*. 2013;79(24):7790–7799.
151. Chen X, Andersen TJ, Morono Y, Inagaki F, Jørgensen BB, Lever MA. Bioturbation as a key driver behind the dominance of bacteria over archaea in near-surface sediment. *Sci Rep*. 2017;7(1):2400.
152. Katayama T, Yoshioka H, Muramoto Y, Usami J, Fujiwara K, Yoshida S, et al. Physicochemical impacts associated with natural gas development on methanogenesis in deep sand aquifers. *ISME J*. 2015;9(2):436–446.
153. Simkus DN, Slater GF, Lollar BS, Wilkie K, Kieft TL, Magnabosco C, et al. Variations in microbial carbon sources and cycling in the deep continental subsurface. *Geochim Cosmochim Acta*. 2016;173:264–283.
154. Baker BJ, Saw JH, Lind AE, Lazar CS, Hinrichs K-U, Teske AP, et al. Genomic inference of the metabolism of cosmopolitan subsurface Archaea, Hadesarchaea. *Nat Microbiol*. 2016;1:16002.
155. Brown CT, Hug LA, Thomas BC, Sharon I, Castelle CJ, Singh A, et al. Unusual biology across a group comprising more than 15% of domain Bacteria. *Nature*. 2015;523(7559):208.
156. Hug LA, Baker BJ, Anantharaman K, Brown CT, Probst AJ, Castelle CJ, et al. A new view of the tree of life. *Nat Microbiol*. 2016;1:16048.
157. Reysenbach A-L, Liu Y, Banta AB, Beveridge TJ, Kirshtein JD, Schouten S, et al. A ubiquitous thermoacidophilic archaeon from deep-sea hydrothermal vents. *Nature*. 2006;442(7101):444.
158. Knittel K, Boetius A. Anaerobic oxidation of methane: progress with an unknown process. *Annu Rev Microbiol*. 2009;63:311–334.
159. Skennerton CT, Chourey K, Iyer R, Hettich RL, Tyson GW, Orphan VJ. Methane-fueled syntrophy through extracellular electron transfer: uncovering the genomic traits conserved within diverse bacterial partners of anaerobic methanotrophic archaea. *MBio*. 2017;8(4):e00530-17.
160. Topçuoğlu BD, Stewart LC, Morrison HG, Butterfield DA, Huber JA, Holden JF. Hydrogen limitation and syntrophic growth among natural assemblages of thermophilic methanogens at deep-sea hydrothermal vents. *Front Microbiol*. 2016;7:1240.
161. Probst AJ, Ladd B, Jarett JK, Geller-McGrath DE, Sieber CMK, Emerson JB, et al. Differential depth distribution of microbial function and putative symbionts through sediment-hosted aquifers in the deep terrestrial subsurface. *Nat Microbiol*. 2018;3(3):328.

162. Borgonie G, García-Moyano A, Litthauer D, Bert W, Bester A, van Heerden E, et al. Nematoda from the terrestrial deep subsurface of South Africa. *Nature.* 2011;474:79–82.
163. Allen SE. *Horizontal Gene Transfer as a Mechanism of Adaptation to an Extreme Subterranean Environment by the Nematode Halicephalobus mephisto.* American University, 2017.
164. Borgonie G, Linage-Alvarez B, Ojo AO, Mundle SOC, Freese LB, Van Rooyen C, et al. Eukaryotic opportunists dominate the deep-subsurface biosphere in South Africa. *Nat Commun.* 2015;6:8952.
165. Borgonie G, Magnabosco C, Garcia-Moyano A, Linage-Alvarez B, Ojo AO, Freese LB, et al. New ecosystems in the deep subsurface follow the flow of water driven by geological activity. *Sci Rep.* 2019;9:3310.
166. Pedersen K. *Preliminary Investigations of Deep Ground Water Microbiology in Swedish Granitic Rock.* Svensk Kärnbränslehantering, 1988.
167. Pedersen K, Arlinger J, Ekendahl S, Hallbeck L. 16S rRNA gene diversity of attached and unattached bacteria in boreholes along the access tunnel to the Äspö hard rock laboratory, *Sweden.* 1996;19(4):249–262.
168. Palumbo AV, Zhang C, Liu S, Scarborough SP, Pfiffner SM, Phelps TJ. Influence of media on measurement of bacterial populations in the subsurface. *Appl Biochem Biotechnol.* 1996;57(1):905.
169. Ekendahl S, O'neill AH, Thomsson E, Pedersen K. Characterisation of yeasts isolated from deep igneous rock aquifers of the Fennoscandian Shield. *Microb Ecol.* 2003;46(4):416–428.
170. Sohlberg E, Bomberg M, Miettinen H, Nyyssönen M, Salavirta H, Vikman M, et al. Revealing the unexplored fungal communities in deep groundwater of crystalline bedrock fracture zones in Olkiluoto, Finland. *Front Microbiol.* 2015;6:573.
171. Sinclair JL, Ghiorse WC. Distribution of aerobic bacteria, protozoa, algae, and fungi in deep subsurface sediments. *Geomicrobiol J.* 1989;7(1–2):15–31.
172. Raghukumar C, Raghukumar S, Sheelu G, Gupta SM, Nagender Nath B, Rao BR. Buried in time: culturable fungi in a deep-sea sediment core from the Chagos Trench, Indian Ocean. *Deep Sea Res Part I Oceanogr Res Pap.* 2004;51(11):1759–1768.
173. Edgcomb VP, Beaudoin D, Gast R, Biddle JF, Teske A. Marine subsurface eukaryotes: the fungal majority. *Environ Microbiol.* 2011;13(1):172–183.
174. Orsi W, Biddle JF, Edgcomb V. Deep sequencing of subseafloor eukaryotic rRNA reveals active fungi across marine subsurface provinces. *PLoS One.* 2013;8(2): e56335.
175. Ivarsson M, Bengtson S, Skogby H, Lazor P, Broman C, Belivanova V, Marone F. A fungal–prokaryotic consortium at the basalt–zeolite interface in subseafloor igneous crust. *PLoS One.* 2015;10(10):e0140106.
176. Ivarsson M, Schnürer A, Bengtson S, Neubeck A. Anaerobic fungi: a potential source of biological H_2 in the oceanic crust. *Front Microbiol.* 2016;7:674.
177. Drake H, Ivarsson M, Bengtson S, Heim C, Siljeström S, Whitehouse MJ, et al. Anaerobic consortia of fungi and sulfate reducing bacteria in deep granite fractures. *Nat Commun.* 2017;8(1):55.
178. Suttle CA. Marine viruses – major players in the global ecosystem. *Nat Rev Microbiol.* 2007;5(10):801–812.
179. Suttle CA. Viruses in the sea. *Nature.* 2005;437(7057):356.
180. Anderson RE, Brazelton WJ, Baross J. Is the genetic landscape of the deep subsurface biosphere affected by viruses? *Front Microbiol.* 2011;2:219.

181. Anderson RE, Brazelton WJ, Baross JA. The deep viriosphere: assessing the viral impact on microbial community dynamics in the deep subsurface. *Rev Mineral Geochem.* 2013;75:649–675.
182. Jørgensen BB, Marshall IPG. Slow microbial life in the seabed. *Ann Rev Mar Sci.* 2016;8:311–332.
183. Bird DF, Juniper SK, Ricciardi-Rigault M, Martineu P, Prairie YT, Calvert SE. Subsurface viruses and bacteria in Holocene/Late Pleistocene sediments of Saanich Inlet, BC: ODP holes 1033b and 1034b, Leg 169s. *Mar Geol.* 2001;174(1):227–239.
184. Middelboe M, Glud RN, Filippini M. Viral abundance and activity in the deep subseafloor biosphere. *Aquat Microb Ecol.* 2011;63:1–8.
185. Engelhardt T, Kallmeyer J, Cypionka H, Engelen B. High virus-to-cell ratios indicate ongoing production of viruses in deep subsurface sediments. *ISME J.* 2014;8 (7):1503–1509.
186. Knowles B, Silveira CB, Bailey BA, Barott K, Cantu VA, Cobián-Güemes AG, et al. Lytic to temperate switching of viral communities. *Nature.* 2016;531(7595):466.
187. Labonté JM, Field EK, Lau M, Chivian D, Van Heerden E, Wommack KE, et al. Single cell genomics indicates horizontal gene transfer and viral infections in a deep subsurface Firmicutes population. *Front Microbiol.* 2015;6:349.
188. Engelhardt T, Sahlberg M, Cypionka H, Engelen B. Induction of prophages from deep-subseafloor bacteria. *Environ Microbiol Rep.* 2011;3(4):459–465.
189. Nigro OD, Jungbluth SP, Lin HT, Hsieh CC, Miranda JA, Schvarcz CR, et al. Viruses in the ocean basement. *MBio.* 2017;8(2):e02129-16.
190. Labonte JM, Lever MA, Edward KJ, Orcutt BN. Influence of igneous basement on deep sediment microbial diversity on the eastern Juan de Fuca Ridge flank. *Front Microbiol.* 2017;8:1434.
191. Fredrickson JK, McKinley JP, Bjornstad BN, Long PE, Ringelberg DB, White DC, et al. Pore-size constraints on the activity and survival of subsurface bacteria in a late cretaceous shale-sandstone sequence, northwestern New Mexico. *Geomicrobiol J.* 1997;14(3):183–202.
192. Rebata-Landa V, Santamarina JC. Mechanical limits to microbial activity in deep sediments. *Geochem Geophys Geosyst.* 2006;7(11):Q11006.
193. Cockell CS, Voytek MA, Gronstal AL, Finster K, Kirshtein JD, Howard K, et al. Impact disruption and recovery of the deep subsurface biosphere. *Astrobiology.* 2012;12(3):231–246.
194. Scheibe TD, Hubbard SS, Onstott TC, DeFlaun MF. Lessons learned from bacterial transport research at the South Oyster Site. *Groundwater.* 2011;49:745–763.
195. Bell E, Blake LI, Sherry A, Head IM, Hubert CRJ. Distribution of thermophilic endospores in a temperate estuary indicate that dispersal history structures sediment microbial communities. *Environ Microbiol.* 2018;20(3):1134–1147.
196. Chakraborty A, Ellefson E, Li C, Gittins D, Brooks JM, Bernard BB, et al. Thermophilic endospores associated with migrated thermogenic hydrocarbons in deep Gulf of Mexico marine sediments. *ISME J.* 2018;1:1895–1906.
197. Nakagawa T, Takai K, Suzuki Y, Hirayama H, Konno U, Tsunogai U, et al. Geomicrobiological exploration and characterization of a novel deep-sea hydrothermal system at the TOTO caldera in the Mariana Volcanic Arc. *Environ Microbiol.* 2006;8(1):37–49.
198. Blank JG, Green SJ, Blake D, Valley JW, Kita NT, Treiman A, et al. An alkaline spring system within the Del Puerto Ophiolite (California, USA): a Mars analog site. *Planet Space Sci.* 2009;57(5–6):533–540.

199. Takai K, Moyer CL, Miyazaki M, Nogi Y, Hirayama H, Nealson KH, et al. *Marinobacter alkaliphilus* sp. nov., a novel alkaliphilic bacterium isolated from subseafloor alkaline serpentine mud from Ocean Drilling Program Site 1200 at South Chamorro Seamount, Mariana Forearc. *Extremophiles*. 2005;9(1):17–27.
200. Antunes A, Taborda M, Huber R, Moissl C, Nobre MF, da Costa MS. *Halorhabdus tiamatea* sp. nov., a non-pigmented, extremely halophilic archaeon from a deep-sea, hypersaline anoxic basin of the Red Sea, and emended description of the genus Halorhabdus. *Int J Syst Evol Microbiol*. 2008;58(1):215–220.
201. Hallsworth JE, Yakimov MM, Golyshin PN, Gillion JLM, D'Auria G, de Lima Alves F, et al. Limits of life in $MgCl_2$-containing environments: chaotropicity defines the window. *Environ Microbiol*. 2007;9(3):801–813.
202. Colwell FS, Smith RP. Unifying principles of the deep terrestrial and deep marine biospheres. In *Subseafloor Biosph Mid-Ocean Ridges*. American Geophysical Union, 2004, pp. 355–367.
203. Amend JP, Teske A. Expanding frontiers in deep subsurface microbiology. *Palaeogeogr Palaeoclimatol Palaeoecol*. 2005;219:131–155.
204. Cragg BA, Summit M, Parkes RJ. Bacterial profiles in a sulphide mount (site 1035) and an area of active fluid venting (site 1036) in hot hydrothermal sediments from Middle Valley (Northeast Pacific). In: Zierenberg RA, Fouquet Y, Miller DJ, Normak WR, eds. *Proceedings of the Ocean Drilling Program, Scientific Results Volume 169*. College Station, TX: Ocean Drilling Program, 2000, pp. 1–18.
205. Daniel RM, Cowan DA. Biomolecular stability and life at high temperatures. *Cell Mol Life Sci C*. 2000;57(2):250–264.
206. Baker-Austin C, Dopson M. Life in acid: pH homeostasis in acidophiles. *Trends Microbiol*. 2007;15(4):165–171.
207. Bartlett DH. Pressure effects on *in vivo* microbial processes. *Biochim Biophys Acta*. 2002;1595(1–2):367–381.
208. Hazael R, Meersman F, Ono F, McMillan PF. Pressure as a limiting factor for life. *Life*. 2016;6(3):34.
209. Nesbø CL, Swithers KS, Dahle H, Haverkamp THA, Birkeland N-K, Sokolova T, et al. Evidence for extensive gene flow and *Thermotoga* subpopulations in subsurface and marine environments. *ISME J*. 2015;9(7):1532.
210. Starnawski P, Bataillon T, Ettema TJG, Jochum LM, Schreiber L, Chen X, et al. Microbial community assembly and evolution in subseafloor sediment. *Proc Natl Acad Sci*. 2017;114(11):2940–2945.
211. Anderson RE, Reveillaud J, Reddington E, Delmont TO, Eren AM, McDermott JM, et al. Genomic variation in microbial populations inhabiting the marine subseafloor at deep-sea hydrothermal vents. *Nat Commun*. 2017;8(1):1114.
212. Momper L, Magnabosco C, Hu E, Amend J, Fournier GP. Genomic evidence of chemotrophic metabolisms in deep-dwelling Chloroflexi conferred by ancient horizontal gene transfer events. *American Geophysical Union, Fall Meeting*. 2017;B14D-07.
213. Novichkov PS, Wolf YI, Dubchak I, Koonin EV. Trends in prokaryotic evolution revealed by comparison of closely related bacterial and archaeal genomes. *J Bacteriol*. 2009;191(1):65–73.
214. Lynch M, Conery JS. The origins of genome complexity. *Science*. 2003;302 (5649):1401–1404.
215. Kuo C-H, Moran NA, Ochman H. The consequences of genetic drift for bacterial genome complexity. *Genome Res*. 2009;19(8):1450–1454.

216. Brazelton WJ, Baross JA. Abundant transposases encoded by the metagenome of a hydrothermal chimney biofilm. *ISME J.* 2009;3(12):1420–1424.
217. Paul BG, Bagby SC, Czornyj E, Arambula D, Handa S, Sczyrba A, et al. Targeted diversity generation by intraterrestrial archaea and archaeal viruses. *Nat Commun.* 2015;6:6585.
218. Paul BG, Burstein D, Castelle CJ, Handa S, Arambula D, Czornyj E, et al. Retroelement-guided protein diversification abounds in vast lineages of Bacteria and Archaea. *Nat Microbiol.* 2017;2:17045.
219. Phelps TJ, Fliermans CB, Garland TR, Pfiffner SM, White DC. Methods for recovery of deep terrestrial subsurface sediments for microbiological studies. *J Microbiol Methods.* 1989;9(4):267–279.
220. Sheik CS, Reese BK, Twing KI, Sylvan JB, Grim SL, Schrenk MO, et al. Identification and removal of contaminant sequences from ribosomal gene databases: lessons from the Census of Deep Life. *Front Microbiol.* 2018;9:840.
221. Payler SJ, Biddle JF, Lollar BS, Fox-Powell MG, Edwards T, Ngwenya BT, et al. An ionic limit to life in the deep subsurface. *Front Microbiol.* 2019;10:426.

18

The Genetics, Biochemistry, and Biophysics of Carbon Cycling by Deep Life

KAREN G. LLOYD, CODY S. SHEIK, BERTRAND GARCÍA-MORENO, AND CATHERINE A. ROYER

18.1 Introduction

Much of the microbial life on Earth resides below the surface in the crust (Figure 18.1) (1), either buried in marine sediments (2) and petroleum deposits (3) or entrained in aquifers within oceanic and terrestrial rocks (Figure 18.2) (4–8), fluid inclusions in salt, permafrost, and ice (9–11), as well as hydrothermal and geothermal fluids (12,13). The study of deep subsurface life has defined our understanding of habitability and expanded our knowledge of the mechanisms that enables life to live in these environments (14). While the study of deep life may seem like a philosophical exercise, understanding this enigmatic biosphere has important real-world implications for assessing the safety and feasibility of underground storage of spent nuclear fuel and other toxic compounds, sequestration of atmospheric CO_2, or acquisition of fuels such as tar sands, deep subsurface coal beds, methane hydrates, or fracking (3,5,15).

Organisms inhabiting subsurface environments likely have been isolated from the surface world for hundreds to millions of years (16). Thus, their metabolic lifestyles may differ substantially from those of surface organisms. Even though subsurface environments are diverse (Chapter 16, this volume), subsurface microbes share common biological challenges such as limitations of energy, resources, and space, as well as extremes of pressure, pH, osmolarity, and temperature (Chapter 17, this volume). On the other hand, subsurface environments offer biological advantages, too: environmental stability, protection from UV irradiation, and oxygen. These unique subsurface conditions lead to communities that are often phylogenetically and functionally diverse, with extremely slow population turnover times (14,17,18) and efficient energy metabolisms (14,19). Increasingly, the roles of viruses and eukaryotes, in addition to bacteria and archaea, are being recognized in the deep subsurface biosphere (20–25). Several barriers hamper the study of life in Earth's crust, such as sample acquisition and the difficulty of retrieving sterile, unaltered samples that have not been contaminated by drilling fluid. However, an even bigger hurdle is the difficulty of studying the copious subsurface microbes with no cultured representatives (13,26). Their functional potential must be pieced together from direct assessments of biomolecules or biochemical processes in natural samples. However, even subsurface microbes related to laboratory cultures with "known" functions, may not perform those functions in the natural

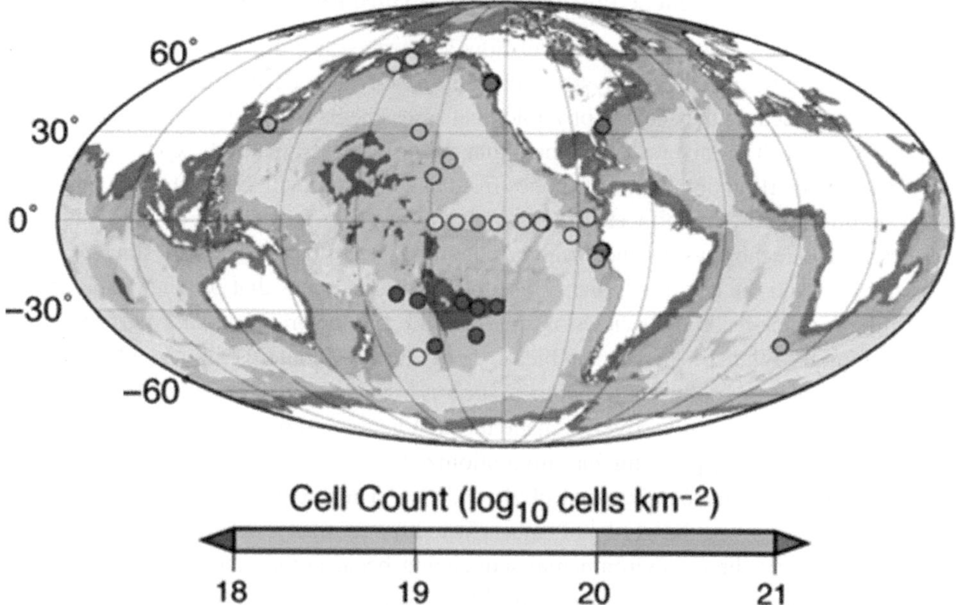

Figure 18.1 Global distribution of marine sediment densities of microbial cells, integrated with depth (adapted from 1).

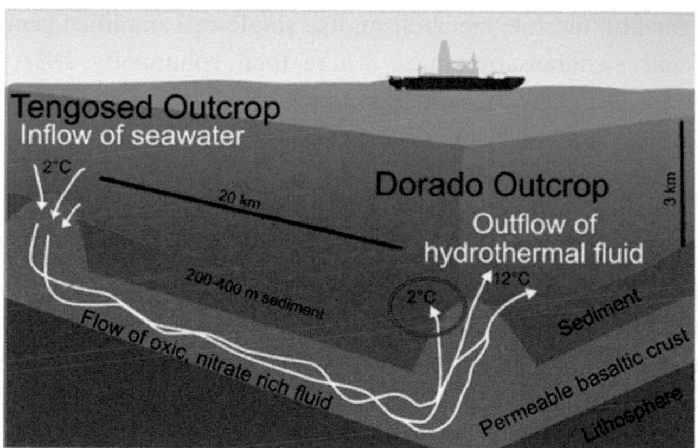

Figure 18.2 Example schematic of a deep subsurface basaltic aquifer. Such environments provide a substantially different habitat for microbial life from sediments or soils (adapted from 33).

environment. Nonetheless, culturing has been extremely important in establishing the existence of a living deep subsurface biosphere (27–32). Researchers must therefore combine direct molecular biological assessments with geochemical and geophysical environmental parameters to describe how carbon is microbially transformed in subsurface environments.

18.2 Genetic Potential of Subsurface Environments

Much research addresses the question of how subsurface life differs from surface life in genetics, biochemistry, and biophysics. Culturing-based methods are still the cornerstone of microbiology because they enable physiological assessments in carefully controlled experiments. However, it is universally accepted that culturing-based methods typically recover a very small proportion of the total microbes from any environment (34–36). Our inability to cultivate microbes is a multifaceted problem that in part has been driven by our inability to perfectly recreate the physical and chemical conditions necessary for growth, slow microbial growth rates, and the rise in popularity of molecular techniques that produce data rapidly. Thus, like the rock record, our views of microbial life and the processes they perform *in situ* are incomplete. While this may seem bleak, there is renewed interest in the cultivation and development of novel culturing techniques, such as high-pressure culture vessels, and this is being spurred, in part, by discoveries made through genome sequencing.

Ribosomal gene sequencing has revolutionized our understanding of the microbial evolution and diversity of microbial phyla (37–39). These culture-independent methods have been a boon for the study of subsurface life, as many of the organisms in the subsurface require direct environmental sequencing because they are slow growing (40), potentially dormant (41,42), or require specific growth conditions or partners (43). Single-gene amplicon sequencing (i.e. 16S rRNA or nitrate reduction genes) remains the most used method for assessing microbial communities, but it lacks the resolution to decipher the metabolic potential of the microbes observed. However, other popular sequencing-based methods can also indicate metabolism, like single-cell amplified genomes (SAGs) or metagenomics and metatranscriptomics, where total community DNA and RNA are sequenced directly from environmental samples (4,44,45). The Census of Deep Life (https://vamps2.mbl.edu/portals/CODL), within the Deep Carbon Observatory, has enabled DNA sequencing from many deep subsurface environments and will provide a great resource as this work continues (46).

The main hurdle for the molecular-based analysis of subsurface environments is the amount of microbial biomass present in the sample, which is often low. Nucleic acid extraction from any environmental sample is a fundamental step for all molecular-based studies (Figure 18.3). Extraction methods are extremely diverse and highly dependent on the samples being processed. Recent work has highlighted that commercial DNA/RNA extraction kit reagents, typically designed for high-biomass environments, contain their own microbiome that can skew downstream amplicon data analyses (46,47). It is also important to remove reagent contamination from SAGs, metagenomes, and metatranscriptomes (48–50). Thus, great care must be taken to screen resulting reads and assembled contigs to identify and remove suspicious sequences, while also allowing that sometimes organisms that were previously thought to represent surface contamination, such as cyanobacteria, may actually be legitimate community members (51).

Post-sequencing analysis of genomes, metagenomes, and metatranscriptomes has made significant advances that have greatly enhanced our knowledge of subsurface life (52–54).

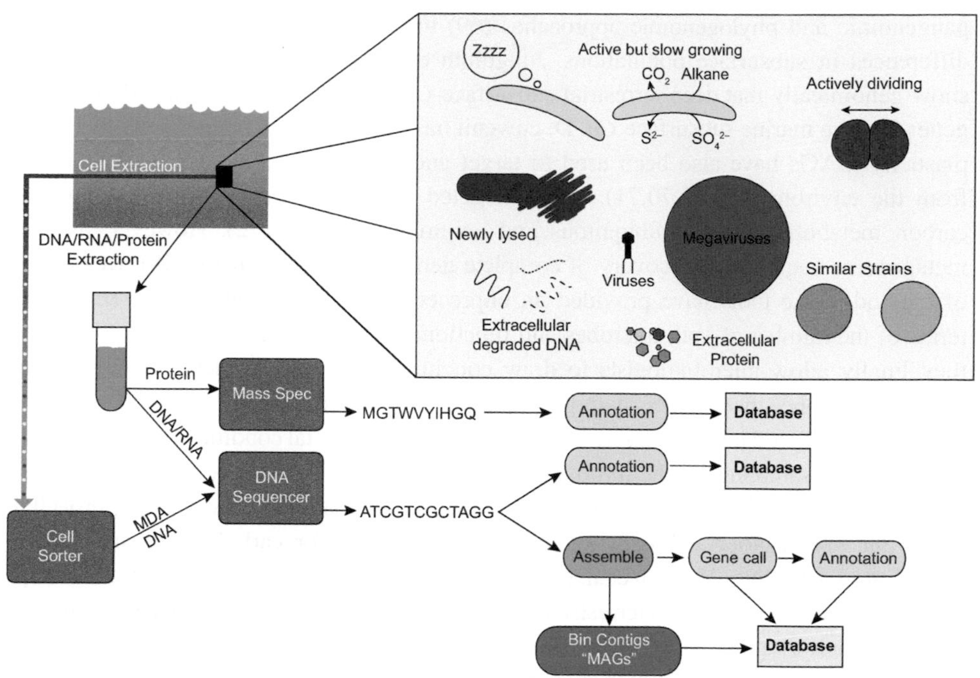

Figure 18.3 Schematic workflow for molecular studies of subsurface ecosystems. MAG = metagenome-assembled genome; MDA = multiple displacement amplification.

The ability to identify and extract microbial genomes from metagenomes (i.e. genome binning, also referred to as metagenome-assembled genomes (MAGs)) (55), along with SAGs, has enabled physiological inferences to be made for uncultured subsurface organisms. Recovery of genomes from metagenomes like *Candidatus* Desulforudis audaxviator (52) or *Leptospirillium* and *Ferroplasma* (56) revealed that microbes living in low-diversity, subsurface environments contain a large amount of genomic diversity, strain heterogeneity, and physiological diversity. *Ca.* D. audaxviator and *Ca.* Desulfopertinax cowenii contain metabolic pathways for heterotrophic growth as well as the potential to fix carbon via the Wood–Ljungdahl (WL) pathway (57). In subsurface environments, the WL pathway has been increasingly identified in MAGs and SAGs. The presence of the WL pathway is intuitive as it requires minimal energy to fix C to biomass (58) and is reversible to oxidize acetate. Interestingly, energetically taxing pathways such as the Calvin–Benson–Bassham cycle are also present (4,59). Metagenome sequencing and subsequent genome binning have been applied to several subsurface environments, such as sediments (marine and freshwater), aquifers, rock, caves, mines, and marine basalts (4,60–68). In these systems, chemolithoautotrophy is always counterbalanced by heterotrophs and fermentative organisms (4,61). As more subsurface environments are sampled and sequenced and their genomes assembled from metagenomes, researchers have the ability to apply pangenomic and phylogenomic approaches (69) to look for functional and evolutionary differences in subsurface populations. Jungbluth et al. (57) applied these techniques to show genomically that deep terrestrial subsurface *Ca.* D. audaxviator lacks carbon uptake genes that the marine subsurface *Ca.* D. cowenii has, indicating the potential for metabolic plasticity. SAGs have also been used to target and recover specific genomes of interest from the environment (22,70,71). These targeted approaches provided insight into the carbon metabolism of two ubiquitous and enigmatic Archaea (72). However, as with metagenome sequencing, recovery of complete genomes can be difficult (48). Regardless of method, these tools have provided an unprecedented glimpse into the diversity – in terms of the number of both microbes and functional genes – of subsurface microbes, and they finally allow microbiologists to draw conclusions as to the functional interactions between microbes that drive system productivity (4,60,64,66).

Metatranscriptomics provides a snapshot of the environmental conditions microbes sense at the time of sampling, as the turnover of RNA in the cell is rapid, being in the order of minutes for many microbes (73). Orsi et al. applied a metatranscriptomic-based approach to show that sub-seafloor sediment microbes express genes for carbohydrate and protein degradation pathways (45). Lau et al. (43) were able to construct metabolic interaction maps of microbial communities in a terrestrial subsurface South African gold mine. These microbial communities appear to use a diversity of carbon fixation and decomposition pathways to cycle carbon. *De novo* assembled metatranscriptomes can be a useful tool for looking at broad patterns of expression in the subsurface (43,74) or the contribution of rare community members to biogeochemical cycles (75). However, this method is highly dependent on the annotation databases and thus could missannotate functions or be unable to give a phylogeny. Even with these caveats, metatranscriptomics coupled to either MAGs or SAGs can link potentially novel microbes to biogeochemical cycles.

Genes and gene transcripts do not always link directly to organismal activity due to post-transcriptional regulation of microbial functions (76). Therefore, proteomics may be a way to probe microbial functions of the entire community more directly. This method has only recently been applied to deep subsurface samples (43). Lau et al. showed that a wide variety of metabolic activities co-occurred at a 1.34-km deep fault zone, implying that the organisms acted syntrophically to support an autotrophic-based ecosystem (43). Proteomics and genomics also suggested metabolic interdependencies among subsurface microbes in an acetate-fed aquifer (77). Another method that pushes closer to measuring microbial activity *in situ* is metabolomics, which uses mass spectrometry to identify all of the small organic molecules present in a sample. This, too, has the potential to point to metabolic pathways that are functional in deep subsurface samples, but it has only recently come into use (78). The greatest challenge facing metabolomics is dealing with the large number of unknown structures in the data set – a problem shared with all complex microbial communities.

The rapid acceleration of data produced from genomic, transcriptomic, and soon proteomic and metabolomic research into deep subsurface environments has opened up a new dimension regarding how we learn about this difficult-to-culture biosphere. Currently, data production far outstrips the ability of a single research lab to analyze each data set to completion. This is already beginning to support a potentially rich research area of meta-analysis studies, where combining data sets from multiple studies allows for novel insights not apparent from a single data set alone (e.g. 46). Through further pushing of the boundaries of what can be achieved by applying biomolecular tools to environmental samples, there is no limit to what researchers can learn about this important biosphere. Such studies not only overcome some of the limitations of whole communities full of uncultured clades, but will also lead researchers to new approaches to achieve culturing of some of these recalcitrant growers.

18.3 Biogeochemistry of Deep Subsurface Life

Microbes are Earth's most prolific chemists. They do this by transforming organic and inorganic elements and compounds using enzymatic protein complexes. Earth's rock record provides a glimpse of some of these processes and their evolution over time, either through isotope signatures or mineral formations, with evidence for the coevolution of minerals and biology (79,80). While the rock record provides a hint of what processes were in operation when they were deposited, interpreting these observations relies on the characterization of extant microbes.

Microbes transform the carbon landscape of Earth's subsurface largely by "eating" or "breathing" carbon-containing compounds (81). Respiring microbes use one molecule as an electron source (reductant) and another molecule as an electron sink (oxidant). Common reductants are organic molecules such as carbohydrates, amino acids, lipids, or small molecules like hydrocarbons, acetate, and formate, as well as inorganic compounds such as reduced sulfur, ammonium, transition metals, or H_2. Carbon-containing oxidants include CO_2 (in the case of methanogenesis), organic molecules like fumarate or oxalate, as well as inorganic compounds like O_2, oxidized sulfur or nitrogen compounds, or oxidized transition metals. An alternative to respiration is fermentation, where microbes split a single molecule of intermediate oxidation state into a more oxidized one and a more reduced one. The reduced product often serves as the reductant for another organism's respiration. Organic matter is the most common fermentative substrate, but microbes can also ferment elemental sulfur through disproportionation. In addition to respiration and fermentation, which are primarily performed to give a microbe energy, microbes can also expend energy to take up carbon compounds. Most notably, primary producers incorporate CO_2 to build biomass. Therefore, every type of microbe transforms carbon, and the presence of life on Earth has had profound effects on its carbon cycle.

To describe transformations of carbon compounds in deep subsurface environments, one must determine what combination of reductants, oxidants, and/or fermentative substrates are utilized, which microbial communities use them, and how quickly they do so (14). In many deep subsurface environments, highly oxidized oxidants such as oxygen and nitrate have already been consumed by shallow organisms, leaving the less powerful (per mole) oxidants such as sulfate and CO_2 to dominate (82,83). However, exceptions occur in environments with conduits connecting to surface-derived fluids such as deep basalts or surface-connected aquifers. In areas of the ocean floor with extremely slow sedimentation rates, O_2 can be detected many meters deep into marine sediments (33,84–86). CO_2 fixation, usually powered by chemolithoautotrophy rather than photosynthesis, is a ubiquitous and important carbon transformation process in deep subsurface environments (87).

Through the application of novel biomolecular techniques that complement the DNA, RNA, and protein-based studies descried above, great strides have been made in the understanding of the biochemistry of microbes performing carbon transformation in Earth's subsurface. These methods include enzyme assays, lipid analysis, stable isotope probing (SIP), metabolomics of small metabolic intermediates, measuring or modeling respiration rates, and heterologous expression of novel enzymes from deep subsurface organisms.

18.3.1 Microbial Metabolism in the Deep Subsurface

Measuring the rate that a microbial function occurs is essential for determining how the function contributes to the carbon cycle. Older methods to study microbial activity in the deep biosphere, such as the incorporation of radioactive thymidine into bulk samples, have largely fallen out of favor since they are insufficiently sensitive to detect microbial activities below a few meters deep into sediments (88). A more sensitive method for estimating *in situ* microbial activity is to measure the turnover rate of trace amounts of radioactive nutrients such as electron donors/acceptors. Such methods have attained widespread use in subsurface environments and have shown that SO_4^{2-} and CO_2 serve as important electron acceptors for deep subsurface biosphere respiration (89). In agreement with this, DNA signatures for microbes capable of these types of respiration have been identified in many subsurface locations (e.g. 90,91). Although it may seem that better oxidants should correspond to faster organic matter degradation, recent work has suggested that the rate of carbon remineralization operates independently of which oxidant is available (92). Here, radiotracer-measured rates of carbon oxidation to CO_2 were constant, whether sulfate or CO_2 was the terminal electron acceptor.

Direct microscopic assessment is often uninformative about either taxonomy or physiology because cells tend to be small (93,94) and their morphology does not indicate their identity or environmental functions (2). In recent years, SIP coupled to fluorescence *in situ* hybridization (FISH) and nanoscale secondary ion mass spectrometry (NanoSIMS) has enabled the direct metabolic assessment of natural populations incubated *ex situ*, but still within their natural microbial community. SIP works on the principle that organisms taking up isotopically labeled substrate will retain it in their biomolecules and can later be identified. After 9 and 405 days of seafloor incubation of a sediment core with ^{13}C-labeled glucose, Takano et al. (95) found that subsurface archaea incorporate fresh glucose in the glycerol headgroups of their membrane lipids, but rely on detrital carbon recycling for the hydrocarbon isoprenoid groups of the lipids. FISH allows the microscopic identification of microbes of a particular taxonomic group by attaching fluorescent probes to short DNA sequences that match the taxonomically informative ribosomal RNA present in cells (96). In places with low activity, such as the deep subsurface, the dim fluorescence signals of FISH often require amplification with catalyzed reporter deposition (CARD) (97). NanoSIMS sputters a cesium ion beam across an area of a few micrometers to detect any isotopic label taken up by the microbe during the SIP incubation. Therefore, when combined with FISH, NanoSIMS SIP gives information about what substrates were consumed by which taxonomic groups of organisms on a nearly single-cell basis.

In 2011, Morono et al. used NanoSIMS SIP with CARD-FISH to show that cells in up to 460,000-year-old sediments, ~200 m deep into the sub-seafloor, were capable of growing on glucose, pyruvate, and amino acids (98). In addition, acetate and bicarbonate were incorporated into the biomass, although they did not promote growth. However, whether the identity of the cells that grew in these experiments reflected the community present at the time of sampling was not assessed, since only a subset of the growing cells hybridized to FISH probes (98). This could be the result of either mismatches to the probe DNA sequences or difficulties of hybridizing cells during CARD-FISH (99,100).

In one of the deepest sub-seafloor drilling operations to date, researchers on Integrated Ocean Drilling Program (IODP) Expedition 337 found relatively high numbers of microbial cells whose taxonomic identities resembled terrestrial communities in 2-km-deep coal beds, suggesting that these cells had persisted through the burial of these terrestrial sediments over tens of millions of years (101). By combining SIP, CARD-FISH, and NanoSIMS, Trembath-Reichert et al. found that these communities incorporated methylated compounds into the biomass, sometimes with concomitant methane production (102). Ijiri et al. combined clumped isotopic measurements and label turnover rates to show that deep fluids stimulate substantial methanogenesis at mud volcanoes (103).

An analogous method to SIP that allows single-cell analysis without requiring an expensive NanoSIMS instrument is called bio-orthogonal noncanonical amino acid tagging (BONCAT) (104). In this method, natural samples are incubated with an amino acid mimic equipped with a functional moiety that can react with a fluorophore after the incubation has ended. In this way, the microbial cells that were making new proteins (and therefore incorporated the amino acid mimic) fluoresce under a microscope and can be counted by eye or physically separated in a flow cytometer for downstream biomolecular study. BONCAT was used in subsurface samples to show stimulation of amino acid incorporation in microbial aggregates incubated with methane (105).

Metabolism can also be controlled by entering an inactive or dormant state. Among cultured bacteria, the Firmicutes and Actinobacteria are capable of forming endospores that can withstand low nutrients and inhospitable conditions for long time periods. These properties make sporulation a potentially advantageous phenomenon for deep subsurface life; however, it is possible that common DNA analyses miss spores since they can be difficult to lyse. Therefore, Lomstein et al. measured dipicolinic acid, a key spore coat protein, in the deep subsurface of the Peru Margin to estimate that spores in this deep subsurface environment may be as numerous as nonsporulated cells and the necromass from other dead cells that may provide food for metabolically active microbes (40).

However, when the decrease in cellular biomass over sediment age is instead used to estimate the microbial necromass production rates, the rate is far outstripped by organic matter buried from ancient pelagic sedimentation (106).

In marine sediments with low organic matter contents and deep terrestrial aquifers, the radiolysis of water may provide microbial food. Natural radioactive decay of Earth's minerals can slowly split water into H_2 and oxidized moieties (107). Hydrogen production from Earth's background radiation may seem inconsequential relative to the power provided by sunlight in surface environments. However, H_2 production rates from water radiolysis have been shown to be sufficient to support life in the deep subsurface (52,107,108). The deep subsurface may have additional sources of energy such as serpentinization, which produces the microbial substrate H_2 and has been shown to support many microbial communities (109,110). Serpentinization also produces a large amount of alkalinity, which inhibits autotrophy by pushing carbonate equilibrium away from CO_2. Therefore, unique types of methanogens have been discovered in serpentinizing systems that have novel mechanisms for ameliorating CO_2 limitation (111,112).

A feature that may be very important to subsurface ecosystems is symbiotic relationships. Historically, microbial communities from deep subsurface aquifers have been studied by examining the cells that are caught on a 1.2-μm filter, but recent work has shown that a wide range of ultrasmall bacteria pass through these filters (4). These cells form previously undiscovered branches on the tree of life called candidate phyla radiations (113). In addition to having extremely small cell sizes ($0.009 \pm 0.002\ \mu m^3$), they have very small genome sizes as well, and they lack many features thought to be important to free-living organisms (4). In addition, their cell structures show that they have cellular appendages that might aid in cell-to-cell communication, similar to those of ultrasmall uncultured archaea called archaeal Richmond Mine acidophilic nanoorganisms (ARMAN) (Figure 18.4) (94). Therefore, these candidate phyla radiations have been proposed to be obligate symbionts of larger microbes (4).

18.3.2 Predicting Functions of Novel Genes

Many of these genes in subsurface organisms cannot be accurately annotated by their homology to characterized gene products, since most of the latter come from microbes in pure culture (114). Given that genomic data are quickly becoming the most accessible type of data to shed light on deep biosphere communities (Section 18.2), the inability to interpret the functions of much of their genomes is a major limitation. For instance, from its genome and phylogeny, *Desulfurivibrio* sp. appear to be anaerobic sulfate reducers. However, when they were obtained in pure culture, they did not reduce sulfate, but instead oxidized it (115). A genome from an uncultured phylum contained a homolog of the methyl coenzyme M reductase gene, from which one might conclude that it performs methanogenesis or anaerobic methane oxidation. However, enrichments of this organism could not perform methane metabolisms, but instead oxidized butyrate anaerobically (116).

One method for overcoming difficulties in accurate annotations of genes in organisms that cannot undergo phenotypic ground-truthing is to amplify target genes of interest

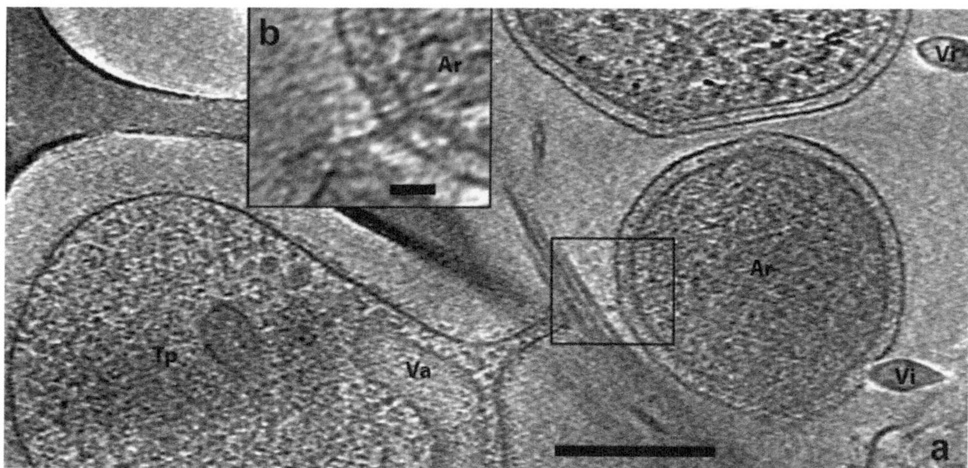

Figure 18.4 Cryo-electron tomographic reconstruction of ARMAN archaea (Ar), Thermoplasmatales (Tp), vacuole (Va), and virus (Vi). (a) shows the larger image, with a 300-nm scale bar, and (b) shows the inset, with a 50-nm scale bar.
Image reproduced from (94).

directly from natural samples or single-cell genomes, ligate them into DNA plasmid vectors, transform them into *Escherichia coli*, express them as proteins, purify them, solve their structures, and characterize their activities. A predicted gene annotated as a peptidase in the uncultured phylum Bathyarchaeota was heterologously expressed in *E. coli* and crystallized in order to determine its structure and catalytic activity (117). Although the gene was annotated as an S15 peptidase with substrate specificity for terminal proline residues, the expressed enzyme, called bathyaminopeptidase, was found to have a much higher affinity for terminal cysteine residues. This distinction illustrates the limitation of inferring enzyme function from homology to known proteins in these uncultured deep subsurface microbes. Homologs of the gene for the II/III form of Rubisco, a well-studied key gene for CO_2 fixation in the Calvin–Benson–Bassham cycle, occur in the candidate phyla radiation among the bacteria in deep terrestrial aquifers. This enzyme had previously only been found in cultured archaea, so its presence in uncultured phyla of deep subsurface bacteria suggested that perhaps it had different catalytic properties in these unknown groups (4,118). However, adding the Rubisco II/III gene from the candidate phyla radiation lineage to a phototroph, *Rhodobacter capsulatus*, which had been engineered to remove its native Rubisco gene and had therefore lost its CO_2 fixation capabilities (119), caused its full CO_2 fixation capabilities to be revived (59).

The limitation of this heterologous gene expression approach is that it has much slower throughput than genetic characterizations. However, it provides hard evidence for function that is lacking when simply inferring enzymatic functions from homology to known proteins. In addition, it presents opportunities for discovering novel functions rather than making assumptions based on previously characterized proteins.

18.3.3 Cellular Bioenergetics

Overcoming energy limitation is a major focus of life in both marine and terrestrial subsurface environments (11,14). Reaction transport modeling of oxygen and nitrate concentration profiles with sediment age showed that deep subsurface communities in the South Pacific Gyre operate at extraordinarily low metabolic rates (85,86). Subsurface microbes are therefore operating in a low energy state, and the majority of these populations may be in a long-term stationary stage in which they require energy only for cell maintenance (120,121). Determining the minimum value for cellular maintenance is difficult, but nanocalorimetry, in which the kinetics and thermodynamics of very small populations of cells can be measured (122), and long incubations with no new substrate additions (123–125) hold promise for studying organisms operating at very low energies. At the very minimum, living cells must replace biomass fast enough to overcome abiotic racemization rates, which can be very slow (126,127). Determining true growth rates for such natural microbial populations outside of carefully controlled laboratory conditions is nearly impossible because the change in the number of cells over time represents the product of growth rate minus death rate. However, total population turnover rates can be determined (14). Braun et al. used racemization rates to estimate total population turnover times of up to tens of years for marine sediment microbes in relatively organic-rich parts of the ocean (17). Slow biomass turnover times of several months to over 100 years have also been measured by measuring uptake of deuterated water (Figure 18.5) (102).

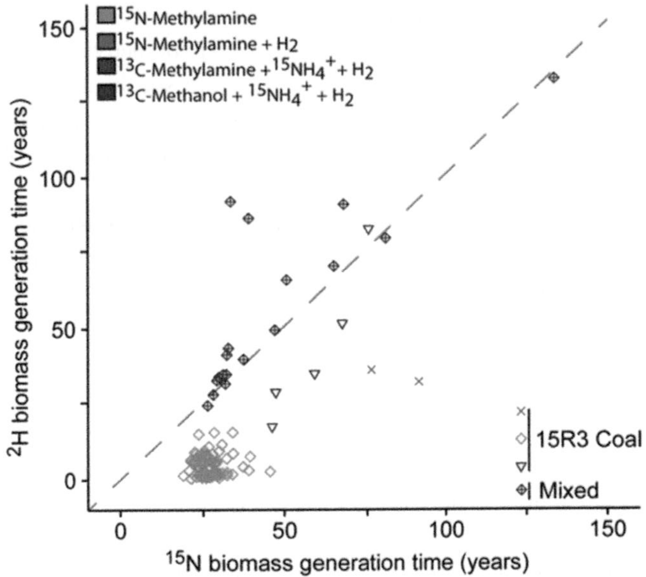

Figure 18.5 Generation times for microbial life in 2-km-deep sub-seafloor coal and shale beds based on uptake rate of either ^{2}H-labeled water or ^{15}N-labeled compounds as described in the graph's key (adapted from 102).

One challenge to such research is the difficulty in making accurate absolute quantifications of specific microbial populations, not just total cell counts, in natural samples that can have very low biomasses or features that disrupt quantification methods. For instance, marine sediments themselves can obscure quantifications with FISH and quantitative polymerase chain reaction (99), and extremely small cells might be missed during filtration of water samples (94). However, a method that accurately quantifies the absolute cellular abundance of particular taxonomic groups in natural deep subsurface samples has yet to be developed.

18.4 Pressure Effects

All lifeforms depend on the chemistry and physics of self-replicating and self-assembling macromolecules, such as proteins, to catalyze the chemical reactions described above. The goal of research in extreme biophysics is to understand how extreme physical conditions affect the fundamental processes responsible for sustaining life. This includes understanding mechanisms of molecular adaptation to extreme conditions and the driving forces of evolution throughout geologic ages.

Much of life on Earth exists under what are considered extreme conditions of pressure: 88% of the volume of the oceans is at an average pressure of 38 MPa, the deepest trenches can exceed 100 MPa, and microbes exist at pressures as much as threefold higher deep in Earth's crust (128). The biomass in high-pressure environments may exceed that on the surface (128). In the oceans, temperature decreases with depth until it reaches an almost constant 3°C with pressure as high as 1000 atmospheres (128). Most of the high-pressure ocean is cold and dark, with minimal sources of carbon and minerals and with a low but constant concentration of oxygen. In contrast, temperatures of the water exiting hydrothermal vents in the deep oceans can reach 400°C, with a temperature gradient so large that the temperature decreases to 3°C only 1 m from the vent. This large temperature gradient on the ocean's floor may be partly responsible for the highly diversified and dynamic deep subsurface microbiome. In the deep subcontinental crust, the total volume in the hydrated fissures in the rock is a large proportion of the biosphere, estimated to be 1016 m^3 and 23–31 Pg of carbon (128,129). If only 1% of that volume is occupied by microorganisms, their biological productivity could be greater than that of Earth's surface (128). It is truly remarkable that life manages to exist in such extreme environments compared to the benign conditions in which humans exist.

18.4.1 Extreme Molecular Biophysics

The structures, interactions, and functions of biological macromolecules are highly sensitive to temperature, pressure, pH, salinity, and the concentration of small osmolytes. Take, for example, proton/sodium-coupled electron transfer (PCET) processes, which are central to energy production. The molecular machinery for H^+-coupled ATP synthesis is essentially the same across most living cells; ATP synthesis succeeds regardless of physical conditions, even though the chemistry of H^+-driven processes in biological macromolecules is highly sensitive to pressure, temperature, and salinity. The adaptive mechanisms for PCET under extreme conditions are unknown. Likewise, the physical properties of both the proteins and nucleic acids that drive the central dogma are highly sensitive to temperature, pressure, pH, and salinity, and yet DNA transcription and translation succeed even in extreme environments. The importance of understanding the molecular mechanisms of adaptation of fundamental biochemical processes to extreme conditions cannot be overstated. Somehow, in these extreme environments, adaptations take place so that the essential processes necessary to sustain life can proceed unimpeded.

Whereas structure–function–thermodynamics relationships in proteins from thermophiles have been studied extensively, studies on the adaptation to high pressure are limited (130). Notable exceptions are pressure-adapted RNA polymerases (131), metabolic enzymes (132–138), the bacterial chromosome organizer, histone-like nucleoid structuring proteins (139), G-protein-coupled receptors in deep-sea fish (140,141), and single-stranded DNA binding proteins (142). In addition to identifiable adaptations at the level of the protein's primary sequence, pressure adaptation at the cellular level has been shown to involve changes in gene expression (143), such as upregulation of specific transport proteins of the outer membrane protein (OMP) family (144,145) and the production of osmolytes (146–149), chaperones (150,151), polyamines (152), and modified lipids (149,150,153–155) (Figure 18.6).

Biophysical aspects of adaptation to extreme environments are poorly understood, although the effects of extreme conditions on mesophilic organisms have been studied extensively. Pressure can modulate the conformation of lipid bilayers, nucleic acids, and proteins (156). In the case of nucleic acids, structures can be either stabilized or destabilized by pressure (157), although in general, RNA and DNA molecules exhibiting tertiary structures, such as hairpins, quadruplexes, and ribozymes, are disrupted by pressure (158,159). Lipid bilayers are compressed and thickened in response to pressure and become much less dynamic. The melting temperature for gel to liquid crystalline transitions increases with pressure, and organisms living at high pressures are known to have a higher proportion of unsaturated lipids in their bilayers to maintain an appropriate degree of fluidity.

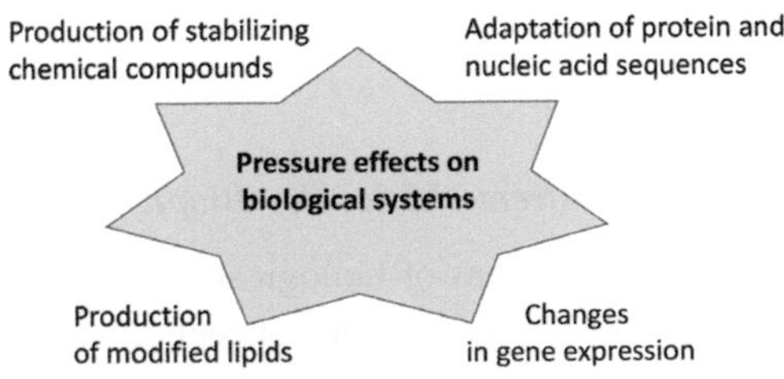

Figure 18.6 Mechanisms of pressure adaptation.

It has long been known that hydrostatic pressure leads to protein unfolding (160) and oligomer dissociation (161,162). Pressure effects on protein structure arise from differences in molar volume between different conformational states. The molar volume of the unfolded states of proteins (U) is generally smaller than that of their folded states (F) (Figure 18.7), as is the molar volume of monomeric subunits compared to oligomeric species. Pressure shifts the folding/association equilibria toward the state that occupies the lowest molar volume: the unfolded or dissociated states. The values of these volume changes are quite small, typically <150 mL/mol for the unfolding of small globular proteins and <300 mL/mol for the dissociation of oligomers.

Although the differences in volume between conformational or oligomeric states of proteins have been well known for some time (165,166), their structural origins remained obscure and were hotly debated until recently. It has been shown repeatedly that substitution of a large hydrophobic amino acid side chain such as leucine, isoleucine, or even valine by a smaller one such as alanine or glycine invariably leads to a larger overall magnitude of the volume change of unfolding and to greater sensitivity to pressure (167–169). A very strong correlation exists between the volume change of unfolding and the packing density (fractional void volume) of a given protein.

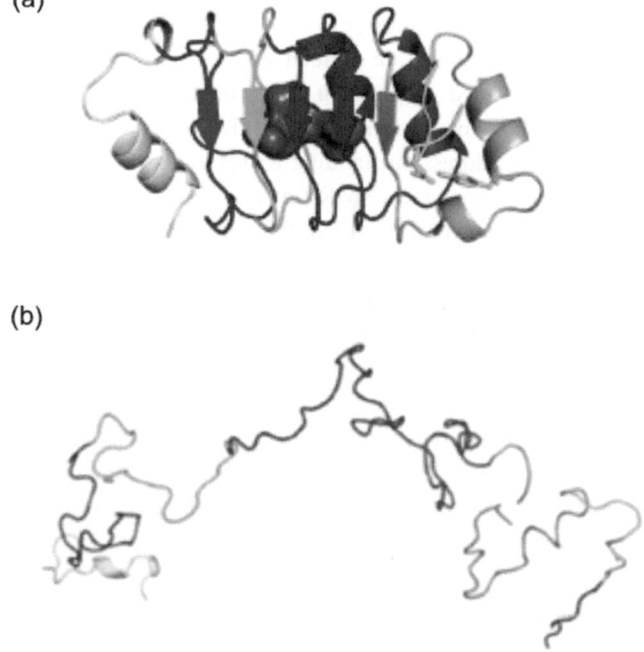

Figure 18.7 Schematic diagrams of a folded (a) and unfolded protein (b), in this case the pp32 leucine-rich repeat protein (adapted from 163). Cavities in (a) were calculated with a 1.4-Å sphere using *HOLLOW* (164).

The destabilization of any given protein by pressure is larger at low temperatures. Chen and Makhatadze proposed that hydration not only does not cause pressure-induced protein unfolding, but actually makes a strong contribution that opposes pressure-induced unfolding (170). However, as temperature increases, the increase in internal void volume becomes smaller than the effects of hydration, and thus the compensation effect diminishes and the overall difference in molar volume between folded and unfolded states diminishes (171), and can even change sign.

Many aspects of the thermodynamic and structural pressure effects on proteins and adaptations to high-pressure environments remain unknown. It was recently shown that there is no pressure-specific "piezolytes" (i.e. there are no specific volumetric effects associated with low-molecular-weight osmolytes). Rather, their effect is simply to stabilize folded proteins against all manner of perturbations, not specifically high hydrostatic pressure (172). It is not well understood whether there are inherent, specific physical attributes of proteins and nucleic acids that allow them to tolerate high-pressure environments. On a larger scale, the ensemble of different organisms in a given environment may modulate biomolecular adaptation through the physical properties of the biomolecules. Large-scale structural adaptive differences may even be identifiable through comparison of whole genomes.

The populations of folding intermediates and low-lying alternative states implicated in the function, degradation, or aggregation of proteins from piezophiles may be different from those of their mesophilic homologs (173,174), Thus, proteins from piezophiles may have evolved to modulate the population of such species.

Thankfully, through sustained progress in instrumentation for fundamental physicochemical studies of the effects of pressure on biological molecules, the biophysics community is well positioned to address outstanding issues in biomolecular adaptation. High-pressure nuclear magnetic resonance (NMR) measurements, once restricted to a few centers with specialized equipment (175–178), have been adapted for low cost and ease of use with high-field NMR spectrometers (179,180). High-pressure small-angle X-ray and neutron scattering devices are available at several facilities in Japan (181), Europe (182), and in the United States (169). Fluorescence, Fourier-transform infrared spectroscopy, and ultraviolet–visible spectroscopy have long been used in high-pressure biophysical research, and electron paramagnetic resonance and circular dichroism spectroscopy have recently been adapted (183,184). Similarly, high-precision modern microcalorimeters are commercially available, as are electronically controlled, temperature-regulated cells for optical spectroscopic equipment. The range of temperature–pressure studies of biological macromolecules that are possible is broad.

18.4.2 Extreme Cellular Biophysics

In the deep-sea bacterium *Photobacterium profundum* (145), pressure affected the expression of many genes that code for proteins implicated in nutrient transport, such as outer membrane proteins (Omps). In the yeast *Saccharomyces cerevisiae* (185), pressure affected the expression of heat-shock and metabolic proteins. Upregulation of heat-shock and cold-shock protein expression in *E. coli* in response to high pressure has also been reported (186). It has also been shown that pressure shock of around 1000 bar leads to the filamentation of *E. coli* (187), and 40 MPa has the same effect on piezophilic *Desulfovibrio indonesiensis* (188), although there are likely multiple mechanisms by which this occurs. Pressure may lead to the dissociation of FtsZ, a tubulin-type GTPase involved in forming the bacterial septum for cell division (189). However, certain strains of *E. coli* bear a pressure-activated Type IV restriction endonuclease, Mrr, which leads to a pressure-dependent SOS response and subsequent filamentation (190,191). The molecular mechanism of the pressure-induced SOS response was shown to be due to pressure-induced dissociation of the inactive Mrr tetramer to an active Mrr dimer that cleaves cryptic sites on the *E. coli* chromosome (Figure 18.8).

Although genetics, genomics, and transcriptomics can provide an overview of the putative mechanisms of adaptation to life under extreme pressures, these studies provide little biophysical insight. The study of extreme conditions on living cells remains a daunting technological challenge and requires the implementation of high-pressure quantitative microscopy approaches. A few high-pressure microscope chambers have been developed, although they have not been widely used. In the 1970s, Salmon and coworkers developed a phase contrast microscope that could withstand about 800 bar pressure

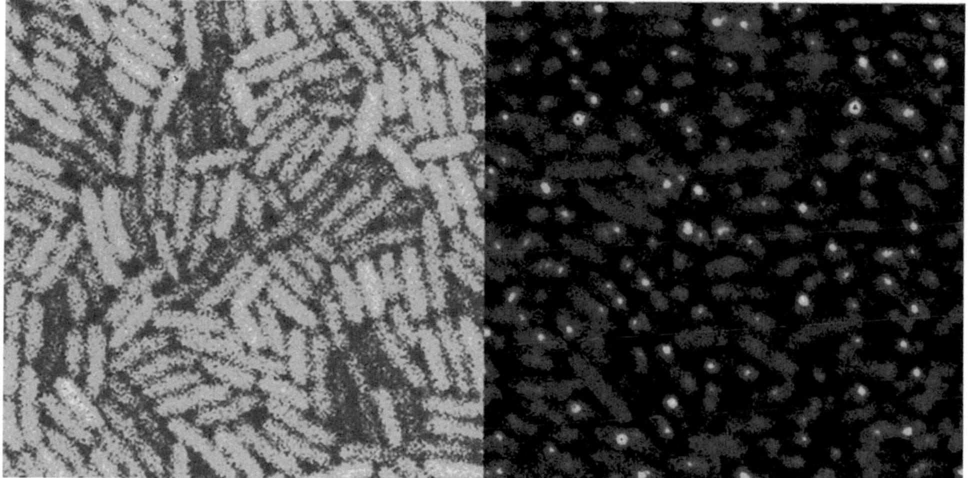

Figure 18.8 Average fluorescence intensity images of GFP-Mrr expressed from an aranbinose-inducible promoter (P_{BAD}) at the natural chromosomal locus in *E. coli* MG1655. (Left) Cells prior to the application of pressure. Full scale is 0–1.48 photon counts per 40 μs. (Right) After 10 minutes at 1 kbar and release of pressure. Full scale is 0.5–6.7 photon counts per 40 μs. Both images are 20 × 20 μm.

Results are similar to those published in (192).

(193,194). Müller and Gratton pioneered the use of capillary tubing for high-pressure microscopy (195) that can withstand pressures up to 6000 bar. However, it has never been applied to live cell imaging. Quantitative analysis of pressure effects on bacterial mobility and flagellar rotation speed were accomplished using a recently designed high-pressure microscopy chamber (196). Unfortunately, owing to the thick windows of this cell, it is not suited for the high-magnification objectives required for advanced quantitative microscopy. This limitation is also the case for two other reported high-pressure microscope chamber designs (197,198). Another hurdle will be the genetic manipulation of extremophile organisms in order to insert fluorescent proteins or other tags into the natural loci of genes of interest.

As in the case of high-pressure molecular biophysics, many unknowns exist regarding the molecular strategies used by cells to adapt to high-pressure or to other extreme conditions. There may be common trends (e.g. dependence on osmolytes) among different types of extremophiles. It is possible that the mechanical properties of high-pressure organisms may be different from those of their mesophilic counterparts or of hyperthermophiles, especially in their membranes and signal transduction pathways. Extreme conditions may affect cell state transitions, growth, differentiation, and development in extremophiles, or even the appearance of multicellularity. Changing geochemical conditions on Earth may have affected cellular phenotypes and the evolution of biological macromolecules.

18.5 Limits to Knowledge and Unknowns

Some of the most fundamental questions about how life on Earth evolved, especially during the first 4 billion years after Earth was formed, relate to how cells and the molecules that keep them alive adapted to changing extreme conditions. Owing to progress in genomics and in molecular and cellular biophysics, the stage is set for a rigorous and systematic examination of some of the most interesting unanswered questions about the evolution of biological systems.

Future work should focus on determining what sorts of physiological properties can be assigned to uncultured subsurface microbes, and perhaps using those insights to develop new techniques to bring them into culture. Linking the experimental work in extreme biophysics with the new frontiers being opened by coupling genomics, transcriptomics, proteomics, and metabolomics with biogeochemical analyses will yield many exiting discoveries about the deep subsurface biosphere in the next few years.

Acknowledgments

Funding provided by the Deep Carbon Observatory, especially within the Deep Life community and Census of Deep Life, has enriched much of the research described herein. Funding for this chapter was provided by an Alfred P. Sloan Research Fellowship

(FG-2015-65399) and NASA Exobiology grant NNX16AL59G to KGL, NSF grants 1745061 and 1660873 to CSS, NSF MCB grant 1514575 to CAR, and NSF MCB grant 1517378 to BG-M.

Questions for the Classroom

1 What novel techniques have enabled the study of uncultured microbes in the deep subsurface?
2 What are the limitations of some of these novel techniques?
3 What are some general features of metabolism in the deep subsurface biosphere?
4 What are some key ways that high pressure impacts biomolecules?
5 What are some of the novel techniques in use for high-pressure biophysics?

References

1. Kallmeyer J, Pockalny R, Adhikari RR, Smith DC, D'Hondt S. Global distribution of microbial abundance and biomass in subseafloor sediment. *Proc Natl Acad Sci*. 2012;109(40):16213–16216.
2. Parkes RJ, Cragg BA, Bale SJ, Getliff JM, Goodman K, Rochelle PA, et al. Deep bacterial biosphere in Pacific Ocean sediments. *Nature*. 1994;371:410–413.
3. Head IM, Jones DM, Röling WFM. Marine microoganisms make a meal of oil. *Nat Rev Microbiol*. 2006;4:173–182.
4. Wrighton KC, Thomas BC, Sharon I, Miller CS, Castelle CJ, VerBerkmoes NC, et al. Fermentation, hydrogen, and sulfur metabolism in multiple uncultivated bacterial phyla. *Science*. 2012;337:1661–1665.
5. Kietäväinen R, Purkamo L. The origin, source, and cycling of methane in deep crystalline rock. *Front Microbiol*. 2015;6:725.
6. Jørgensen SL, Zhao R. Microbial inventory of deeply buried oceanic crust from a young ridge flank. *Front Microbiol*. 2016;7:820.
7. Santelli CM, Orcutt BN, Banning E, Bach W, Moyer CL, Sogin ML, et al. Abundance and diversity of microbial life in ocean crust. *Nature*. 2008;453:5–9.
8. Meyer JL, Jaekel U, Tully BJ, Glazer BT, Wheat CG, Lin H, et al. A distinct and active bacterial community in cold oxygenated fluids circulating beneath the western flank of the Mid-Atlantic ridge. *Sci Rep*. 2016;6:22541.
9. Mackelprang R, Waldrop MP, Deangelis KM, David MM, Chavarria KL, Blazewicz SJ, et al. Metagenomic analysis of a permafrost microbial community reveals a rapid response to thaw. *Nature*. 2011;480(7377):368–371.
10. Borgonie G, Linage-Alverez B, Ojo A, Shivambu S, Kuloyo O, Cason ED, Van Heerden E. Deep subsurface mine stalactites trap endemic fissure fluid Archaea, Bacteria, and Nematoda possibly originating from ancient seas. *Front Microbiol*. 2015;6:833.
11. Parnell J, McMahon S. Physical and chemical controls on habitats for life in the deep subsurface beneath continents and ice. *Philos Trans R Soc A Math Phys Eng Sci*. 2016;374(2059):20140293.

12. Huber JA, Mark Welch DB, Morrison HG, Huse SM, Neal PR, Butterfield DA, et al. Microbial population structures in the deep marine biosphere. *Science*. 2007;318:97–100.
13. Hedlund BP, Murugapiran SK, Alba TW, Levy A, Dodsworth JA, Goertz GB, et al. Uncultivated thermophiles: current status and spotlight on "Aigarchaeota". *Curr Opin Microbiol*. 2015;25:136–145.
14. Hoehler TM, Jørgensen BB. Microbial life under extreme energy limitation. *Nat Rev Microbiol*. 2013;11(2):83–94.
15. Daly RA, Borton MA, Wilkins MJ, Hoyt DW, Kountz DJ, Wolfe RA, et al. Microbial metabolisms in a 2.5-km-deep ecosystem created by hydraulic fracturing in shales. *Nat Microbiol*. 2016;1:16145.
16. Wilhelms A, Larter SR, Head IM, Farrimond P, Di-Primio R, Zwach C. Biodegradation of oil in uplifted basins prevented by deep-burial sterilization. *Nature*. 2001;411:1034–1037.
17. Braun S, Mhatre SS, Jaussi M, Røy H, Kjeldsen KU, Seidenkrantz M, et al. Microbial turnover times in the deep seabed studied by amino acid racemization modelling. *Sci Rep*. 2017;7:5680.
18. Trembath-Reichert E, Morono Y, Ijiri A, Hoshino T, Dawson KS, Inagaki F, et al. Methyl-compound use and slow growth characterize microbial life in 2-km-deep subseafloor coal and shale beds. *Proc Natl Acad Sci*. 2017;114(44):E9206–E9215.
19. Larowe DE, Amend JP. Power limits for microbial life. *Front Microbiol*. 2015;6:718.
20. Takai K, Inagaki F, Nakagawa S, Hirayama H, Nunoura T, Sako Y, et al. Isolation and phylogenetic diversity of members of previously uncultivated Epsilon-Proteobacteria in deep-sea hydrothermal fields. *FEMS Microbiol Lett*. 2003;218:167–174.
21. Edgcomb VP, Beaudoin D, Gast R, Biddle JF, Teske A. Marine subsurface eukaryotes: the fungal majority. *Environ Microbiol*. 2011;13(1):172–183.
22. Labonté JM, Field EK, Lau M, Chivian D, Heerden E Van, Wommack KE, et al. Single cell genomics indicates horizontal gene transfer and viral infections in a deep subsurface Firmicutes population. *Front Microbiol*. 2015;6:349.
23. Anderson RE, Beltrán MT, Hallam SJ, Baross JA. Microbial community structure across fluid gradients in the Juan de Fuca Ridge hydrothermal system. *FEMS Microbiol Ecol*. 2013;83(2):324–339.
24. Sohlberg E, Bomberg M, Miettinen H, Nyyssönen M, Salavirta H, Vikman M, et al. Revealing the unexplored fungal communities in deep groundwater of crystalline bedrock fracture zones in Olkiluoto, Finland. *Front Microbiol*. 2015;6:573.
25. Engelhardt T, Kallmeyer J, Cypionka H, Engelen B. High virus-to-cell ratios indicate ongoing production of viruses in deep subsurface sediments. *ISME J*. 2014;8(7):1503–1509.
26. Lloyd KG, Steen AD, Ladau J, Yin J, Crosby L. Phylogenetically novel uncultured microbial cells dominate Earth microbiomes. *mSystems*. 2018;3:10.1128/mSystems.00055-18.
27. Imachi H, Aoi K, Tasumi E, Saito Y, Yamanaka Y, Saito Y, et al. Cultivation of methanogenic community from subseafloor sediments using a continuous-flow bioreactor. *ISME J*. 2011;5(12):1913–1925.
28. Parkes RJ, Sellek G, Webster G, Martin D, Anders E, Weightman AJ, et al. Culturable prokaryotic diversity of deep, gas hydrate sediments: first use of a continuous high-pressure, anaerobic, enrichment and isolation system for subseafloor sediments (DeepIsoBUG). *Environ Microbiol*. 2009;11:3140–3153.

29. Boone DR, Liu Y, Zhao Z, Balkwill DL, Drake GR, Stevens T, et al. *Bacillus infernus* sp. nov., an Fe(III)- and Mn(IV)-reducing anaerobe from the deep terrestrial subsurface. *Int J Syst Bacteriol.* 1995;45(3):441–448.
30. Russell JA, León-zayas R, Wrighton K, Biddle JF. Deep subsurface life from North Pond: enrichment, isolation, characterization and genomes of heterotrophic bacteria. *Front Microbiol.* 2016;65:678.
31. Bonis BM, Gralnick JA. *Marinobacter subterrani*, a genetically tractable neutrophilic Fe(II)-oxidizing strain isolated from the Soudan Iron Mine. *Front Microbiol.* 2015;6:719.
32. Dodsworth JA, Ong JC, Williams AJ, Dohnalkova AC, Hedlund BP. *Thermocrinis jamiesonii* sp. nov., a thiosulfate-oxidizing, autotropic thermophile isolated from a geothermal spring. *Int J Syst Evol Microbiol.* 2015;65:4769–4775.
33. Zinke LA, Reese BK, Mcmanus J, Wheat CG, Orcutt BN, Amend JP. Sediment microbial communities influenced by cool hydrothermal fluid migration. *Front Microbiol.* 2018;9:1249.
34. Staley JT, Konopka A. Measurement of in situ activities of nonphotosynthetic microorganisms in aquatic and terrestrial habitats. *Annu Rev Microbiol.* 1985;39:321–346.
35. Rappé MS, Giovannoni SJ. The uncultured microbial majority. *Annu Rev Microbiol.* 2003;57:369–394.
36. Hugenholtz P, Goebel BM, Pace NR. Impact of culture-independent studies on the emerging phylogenetic view of bacterial diversity. *J Bacteriol.* 1998;180(18): 4765–4774.
37. Woese CR, Fox GE. Phylogenetic structure of the prokaryotic domain: the primary kingdoms. *Proc Natl Acad Sci.* 1977;74(11):5088–5090.
38. Woese CR. Bacterial evolution. *Microbiol Rev.* 1987;51(2):221–271.
39. Hugenholtz P, Pitulle C, Hershberger KL, Pace NR. Novel division level bacterial diversity in a Yellowstone hot spring. *J Bacteriol.* 1998;180(2):366–376.
40. Lomstein BA, Langerhuus AT, Hondt SD, Jørgensen BB, Spivack AJ. Endospore abundance, microbial growth and necromass turnover in deep sub-seafloor sediment. *Nature.* 2012;484:101–104.
41. Lennon JT, Jones SE. Microbial seed banks: the ecological and evolutionary implications of dormancy. *Nat Rev Microbiol.* 2011;9(2):119–130.
42. Morita RY. Bioavailability of energy and its relationship to growth and starvation survival in nature. *Can J Microbiol.* 1988;34(4):436–441.
43. Lau MCY, Kieft TL, Kuloyo O, Linage-alvarez B, Heerden E Van, Lindsay MR, et al. An oligotrophic deep-subsurface community dependent on syntrophy is dominated by sulfur-driven autotrophic denitrifiers. *Proc Natl Acad Sci.* 2016;113(49):E7927–E7936.
44. Lloyd KG, Schreiber L, Petersen DG, Kjeldsen KU, Lever MA, Steen AD, et al. Predominant archaea in marine sediments detrital proteins. *Nature.* 2013;496:215–218.
45. Orsi WD, Edgcomb VP, Christman GD, Biddle JF. Gene expression in the deep biosphere. *Nature.* 2013;499:205–208.
46. Sheik CS, Reese BK, Twing KI, Sylvan JB, Grim SL, Schrenk MO, et al. Identification and removal of contaminant sequences from ribosomal gene databases: lessons from the Census of Deep Life. *Front Microbiol.* 2018;9:840.
47. Salter SJ, Cox MJ, Turck EM, Calus ST, Cookson WO, Moffatt MF et al. Reagent and laboratory contamination can critically impact sequence-based microbiome analyses. *BMC Biol.* 2014;12:87.

48. Kogawa M, Hosokawa M, Nishikawa Y, Mori K, Takeyama H. Obtaining high-quality draft genomes from uncultured microbes by cleaning and co-assembly of single-cell amplified genomes. *Sci Rep*. 2018;8(1):2059.
49. Olm MR, Butterfield CN, Copeland A, Boles TC, Thomas BC, Banfield JF. The source and evolutionary history of a microbial contaminant identified through soil metagenomic analysis. *MBio*. 2017;8(1):e01969.
50. Heintz-Buschart A, Yusuf D, Kaysen A, Etheridge A, Fritz JV, May P, et al. Isolation of nucleic acids from low biomass samples: detection and removal of sRNA contaminants. *bioRxiv*. 2017;232975.
51. Puente-sánchez AF, Arce-rodríguez A, Oggerin M, Moreno-paz M, Blanco Y, Rodríguez N, et al. Viable cyanobacteria in the deep continental subsurface. *Proc Natl Acad Sci*. 2018;115(42):10702–10707.
52. Chivian D, Brodie EL, Alm EJ, Culley DE, Dehal PS, DeSantis TZ, et al. Environmental genomics reveals a single-species ecosystem deep within Earth. *Science*. 2008;322(5899):275–278.
53. Hug LA, Baker BJ, Anantharaman K, Brown CT, Probst AJ, Castelle CJ, et al. A new view of the tree of life. *Nat Microbiol*. 2016;1:16048.
54. Spang A, Saw JH, Jørgensen SL, Zaremba-niedzwiedzka K, Martijn J, Lind AE, et al. Complex archaea that bridge the gap between prokaryotes and eukaryotes. *Nature*. 2015;521(7551):173–179.
55. Dick G. *Genomic Approaches in Earth and Environmental Sciences*. John Wiley & Sons, 2018.
56. Tyson GW, Chapman J, Hugenholtz P, Allen EE, Ram RJ, Richardson PM, et al. Community structure and metabolism through reconstruction of microbial genomes from the environment. *Nature*. 2004;428:37–43.
57. Jungbluth SP, Glavina del Rio T, Tringe SG, Stepanauskas R, Rappé MS. Genomic comparisons of a bacterial lineage that inhabits both marine and terrestrial deep subsurface systems. *PeerJ*. 2017;5:e3134.
58. Schuchmann K, Müller V. Energetics and application of heterotrophy in acteogenic bacteria. *Appl Environ Microbiol*. 2016;82(14):4056–4069.
59. Wrighton KC, Castelle CJ, Varaljay VA, Satagopan S, Brown CT, Wilkins MJ, et al. RubisCO of a nucleoside pathway known from Archaea is found in diverse uncultivated phyla in bacteria. *ISME J*. 2016;10(11):2702–2714.
60. Baker BJ, Lazar CS, Teske AP, Dick GJ. Genomic resolution of linkages in carbon, nitrogen, and sulfur cycling among widespread estuary sediment bacteria. *Microbiome*. 2015;3(1):14.
61. Crespo-Medina M, Twing KI, Sanchez-Murillo R, Brazelton WJ, McCollom TM, Schrenk MO. Methane dynamics in a tropical serpentinizing environment: the Santa Elena ophiolite, Costa Rica. *Front Microbiol*. 2017;8:916.
62. Brazelton WJ, Nelson B, Schrenk MO. Metagenomic evidence for H_2 oxidation and H_2 production by serpentinite-hosted subsurface microbial communities. *Front Microbiol*. 2012;2:268.
63. Hamilton TL, Jones DS, Schaperdoth I, Macalady JL. Metagenomic insights into S(0) precipitation in a terrestrial subsurface lithoautotrophic ecosystem. *Front Microbiol*. 2015;5:756.
64. Anantharaman K, Brown CT, Hug LA, Sharon I, Castelle CJ, Probst AJ, et al. Thousands of microbial genomes shed light on interconnected biogeochemical processes in an aquifer system. *Nat Commun*. 2016;7:13219.

65. Twing KI, Brazelton WJ, Kubo MDY, Hyer AJ, Cardace D, Hoehler TM, et al. Serpentinization-influenced groundwater harbors extremely low diversity microbial communities adapted to high pH. *Front Microbiol.* 2017;8:308.
66. Probst AJ, Ladd B, Jarett JK, Geller-Mcgrath DE, Sieber CMK, Emerson JB, et al. Differential depth distribution of microbial function and putative symbionts through sediment-hosted aquifers in the deep terrestrial subsurface. *Nat Microbiol.* 2018;3(3):328–336.
67. Evans PN, Parks DH, Chadwick GL, Robbins SJ, Orphan VJ, Golding SD, et al. Methane metabolism in the archaeal phylum Bathyarchaeota revealed by genome-centric metagenomics. *Science.* 2015;350(6259):434–438.
68. Dodsworth JA, Blainey PC, Murugapiran SK, Swingley WD, Ross CA, Tringe SG, et al. Single-cell and metagenomic analyses indicate a fermentative and saccharolytic lifestyle for members of the OP9 lineage. *Nat Commun.* 2013;4:1854.
69. Delmont TO, Eren AM. Linking pangenomes and metagenomes: the *Prochlorococcus* metapangenome. *PeerJ.* 2018;6:e4320.
70. Lloyd KG, Schreiber L, Petersen DG, Kjeldsen KU, Lever MA, Steen AD, et al. Predominant archaea in marine sediments degrade detrital proteins. *Nature.* 2013;496 (7444):215–218.
71. Fullerton H, Moyer CL. Comparative single-cell genomics of Chloroflexi from the Okinawa trough deep-subsurface biosphere. *Appl Environ Microbiol.* 2016;82 (10):3000–3008.
72. Swan BK, Martinez-García M, Preston CM, Sczyrba A, Woyke T, Lamy D, et al. Potential for chemolithoautotrophy among ubiquitous bacteria lineages in the dark ocean. *Science.* 2011;333(6047):1296–1300.
73. Moran MA, Satinsky B, Gifford SM, Luo H, Rivers A, Chan LK et al. Sizing up metatranscriptomics. *ISME J.* 2013;7(2):237–243.
74. Jewell TNM, Karaoz U, Brodie EL, Williams KH, Beller HR. Metatranscriptomic evidence of pervasive and diverse chemolithoautotrophy relevant to C, S, N and Fe cycling in a shallow alluvial aquifer. *ISME J.* 2016;10(9):2106–2117.
75. Baker BJ, Sheik CS, Taylor CA, Jain S, Bhasi A, Cavalcoli JD, et al. Community transcriptomic assembly reveals microbes that contribute to deep-sea carbon and nitrogen cycling. *ISME J.* 2013;7(10):1962–1973.
76. Blazewicz SJ, Barnard RL, Daly RA, Firestone MK. Evaluating rRNA as an indicator of microbial activity in environmental communities: limitations and uses. *ISME J.* 2013;7(11):2061–2068.
77. Wrighton KC, Castelle CJ, Wilkins MJ, Hug LA, Sharon I, Thomas BC, et al. Metabolic interdependencies between phylogenetically novel fermenters and respiratory organisms in an unconfined aquifer. *ISME J.* 2014;8(7):1452–1463.
78. Bird JT, Tague E, Zinke L, Schmidt JM, Steen AD, Reese B, et al. Uncultured microbial phyla suggest mechanisms for multi-thousand-year subsistence in Baltic Sea sediments. *mBio.* 2019;10(2):e02376-18.
79. Giovannelli D, Sievert SM, Hügler M, Markert S, Becher D, Schweder T, et al. Insight into the evolution of microbial metabolism from the deep-branching bacterium, *Thermovibrio ammonificans. Elife.* 2017;6:e18990.
80. Moore EK, Jelen BI, Giovannelli D, Raanan H, Falkowski PG. Metal availability and the expanding network of microbial metabolisms in the Archaean eon. *Nat Geosci.* 2017;10:629–636.
81. Nealson KH. Sediment bacteria: who's there, what are they doing, and what's new? *Annu Rev Earth Planet Sci.* 1997;25:403–434.

82. Froelich PN, Klinkhammer GP, Bender ML, Luedtke NA, Heath GR, Cullen D, et al. Early oxidation of organic matter in pelagic sediments of the eastern equatorial Atlantic: suhoxic diagenesis. *Geochim Cosmochim Acta*. 1979;43(7):1075–1090.
83. D'Hondt S, Rutherford S, Spivack AJ. Metabolic activity of subsurface life in deep-sea sediments. *Science*. 2002;295(5562):2067–2070.
84. Durbin AM, Teske A. Archaea in organic-lean and organic-rich marine subsurface sediments: an environmental gradient reflected in distinct phylogenetic lineages. *Front Microbiol*. 2012;3:168.
85. D'Hondt S, Spivack AJ, Pockalny R, Ferdelman TG, Fischer JP, Kallmeyer J, et al. Subseafloor sedimentary life in the South Pacific Gyre. *Proc Natl Acad Sci*. 2009;106 (28):11651–11656.
86. Røy H, Kallmeyer J, Adhikari RR, Pockalny R, Jørgensen BB, D'Hondt S. Aerobic microbial respiration in 86-million-year-old deep-sea red clay. *Science*. 2012;336 (6083):922–925.
87. Magnabosco C, Ryan K, Lau MCY, Kuloyo O, Lollar BS, Kieft TL, et al. A metagenomic window into carbon metabolism at 3 km depth in Precambrian continental crust. *ISME J*. 2015;10(3):730–741.
88. Cragg B, Harvey S, Fry J, Herbert R, Parkes R. Bacterial biomass and activity in the deep sediment layers of the Japan Sea, Hole 798B. *Proc Ocean Drill Progr*. 1992;128:761–775.
89. Orcutt BN, LaRowe DE, Biddle JF, Colwell FS, Glazer BT, Reese BK, et al. Microbial activity in the marine deep biosphere: progress and prospects. *Front Microbiol*. 2013;4:1–15.
90. Purkamo L, Bomberg M, Kietäväinen R, Salavirta H, Nyyssönen M. Microbial co-occurrence patterns in deep Precambrian bedrock fracture fluids. *Biogeosciences*. 2016;13:3091–3108.
91. Durbin AM, Teske A. Microbial diversity and stratification of South Pacific abyssal marine sediments. *Environ Microbiol*. 2011;13(12):3219–3234.
92. Beulig F, Røy H, Glombitza C, Jørgensen BB. Control on rate and pathway of anaerobic organic carbon degradation in the seabed. *Proc Natl Acad Sci*. 2018;115 (2):367–372.
93. Braun S, Morono Y, Littmann S, Kuypers M, Aslan H, Dong M, et al. Size and carbon content of sub-seafloor microbial cells at Landsort Deep, Baltic Sea. *Front Microbiol*. 2016;7:1375.
94. Baker BJ, Comolli LR, Dick GJ, Hauser LJ, Hyatt D, Dill BD, et al. Enigmatic, ultrasmall, uncultivated Archaea. *Proc Natl Acad Sci*. 2010;107(19):8806–8811.
95. Takano Y, Chikaraishi Y, Ogawa NO, Nomaki H, Morono Y, Inagaki F, et al. Sedimentary membrane lipids recycled by deep-sea benthic archaea. *Nat Geosci*. 2010;3(12):858–861.
96. Amann R, Fuchs BM, Behrens S. The identification of microorganisms by fluorescence *in situ* hybridisation. *Curr Opin Biotechnol*. 2001;12(3):231–236.
97. Pernthaler A, Amann R. Simultaneous fluorescence *in situ* hybridization of mRNA and rRNA in environmental bacteria. *Appl Environ Microbiol*. 2004;70(9):5426–5433.
98. Morono Y, Terada T, Nishizawa M, Ito M, Hillion F, Takahata N, et al. Carbon and nitrogen assimilation in deep subseafloor microbial cells. *Proc Natl Acad Sci*. 2011;108(45):18295–18300.
99. Lloyd KG, May MK, Kevorkian RT, Steen AD. Meta-analysis of quantification methods shows that archaea and bacteria have similar abundances in the subseafloor. *Appl Environ Microbiol*. 2013;79(24):7790–7799.

100. Buongiorno J, Turner S, Webster G, Asai M, Shumaker AK, Roy T, et al. Interlaboratory quantification of Bacteria and Archaea in deeply buried sediments of the Baltic Sea (IODP Expedition 347). *FEMS Microbiol Ecol.* 2017;93(3):fix007.
101. Inagaki F, Kubo Y, Bowles MW, Heuer VB, Ijiri A, Imachi H, et al. Exploring deep microbial life in coal-bearing sediment down to ~2.5 km below the ocean floor. *Science.* 2015;349:420–424.
102. Trembath-Reichert E, Morono Y, Ijiri A, Hoshino T, Dawson KS, Inagaki F, et al. Methyl-compound use and slow growth characterize microbial life in 2-km-deep subseafloor coal and shale beds. *Proc Natl Acad Sci.* 2017;114(44):E9206–E9215.
103. Ijiri A, Inagaki F, Kubo Y, Adhikari RR, Hattori S, Hoshino T, et al. Deep-biosphere methane production stimulated by geofluids in the nankai accretionary complex. *Sci Adv.* 2018;4(6):eaao4631.
104. Hatzenpichler R, Scheller S, Tavormina PL, Babin BM, Tirrell DA, Orphan VJ. In situ visualization of newly synthesized proteins in environmental microbes using amino acid tagging and click chemistry. *Environ Microbiol.* 2014;16:2568–2590.
105. Hatzenpichler R, Connon SA, Goudeau D, Malmstrom RR, Woyke T, Orphan VJ. Visualizing in situ translational activity for identifying and sorting slow-growing archaeal–bacterial consortia. *Proc Natl Acad Sci.* 2016;113(28):E4069–E4078.
106. Bradley JA, Amend JP, Larowe DE. Necromass as a limited source of energy for microorganisms in marine sediments. *J Geophys Res Biogeosci.* 2018;123(2): 577–590.
107. Lin L, Hall J, Lippmann-pipke J, Ward JA, Lollar BS, Deflaun M, et al. Radiolytic H_2 in continental crust: nuclear power for deep subsurface microbial communities. *Geochem Geophys Geosyst.* 2005;6(7):Q07003.
108. Blair CC, D'Hondt S, Spivack AJ, Kingsley RH. Radiolytic hydrogen and microbial respiration in subsurface sediments. *Astrobiology.* 2007;7(6):951–970.
109. Ménez B, Pasini V, Brunelli D. Life in the hydrated suboceanic mantle. *Nat Geosci.* 2012;5(2):133–137.
110. Suzuki S, Wu A, Cheung A, Tenney A, Wanger G, Kuenen JG, et al. Microbial diversity in The Cedars, an ultrabasic, ultrareducing, and low salinity serpentinizing ecosystem. *Proc Natl Acad Sci.* 2013;110(38):15336–15341.
111. Brazelton WJ, Schrenk MO, Kelley DS, Baross JA. Methane- and sulfur-metabolizing microbial communities dominate the Lost City hydrothermal field ecosystem. *Appl Environ Microbiol.* 2006;72(9):6257–6270.
112. Lang SQ, Früh-Green GL, Bernasconi SM, Brazelton WJ, Schrenk MO, McGonigle JM. Deeply-sourced formate fuels sulfate reducers but not methanogens at Lost City hydrothermal field. *Sci Rep.* 2018;8(1):755.
113. Hug LA, Thomas BC, Brown CT, Frischkorn KR, Williams KH, Tringe SG, et al. Aquifer environment selects for microbial species cohorts in sediment and groundwater. *ISME J.* 2015;9(8):1846–1856.
114. Rinke C, Schwientek P, Sczyrba A, Ivanova NN, Anderson IJ, Cheng J-F, et al. Insights into the phylogeny and coding potential of microbial dark matter. *Nature.* 2013;499(7459):431–437.
115. Thorup C, Schramm A, Findlay AJ, Finster KW, Schreiber L. Disguised as a sulfate reducer: growth of the deltaproteobacterium *Desulfurivibrio alkaliphilus* by sulfide oxidation with nitrate. *MBio.* 2017;8(4):e00671-17.
116. Laso-pérez R, Wegener G, Knittel K, Widdel F, Harding KJ, Krukenberg V, et al. Thermophilic archaea activate butane via alkyl-coenzyme M formation. *Nature.* 2016;539(7629):396–401.

117. Michalska K, Steen AD, Chhor G, Endres M, Webber AT, Bird J, et al. New aminopeptidase from "microbial dark matter" archaeon. *FASEB J.* 2015;29:4071–4079.
118. Kantor RS, Wrighton KC, Handley KM, Sharon I, Hug LA, Castelle CJ, et al. Small genomes and sparse metabolisms of sediment-associated bacteria from four candidate phyla. *MBio*. 2013;4(5):e00708–e00713.
119. Smith SA, Tabita FR. Positive and negative selection of mutant forms of prokaryotic (cyanobacterial) ribulose-1,5,bisphosphate carboxylase/oxygenase. *J Mol Biol.* 2003;331:557–569.
120. Lever MA, Rogers KL, Lloyd KG, Overmann J, Schink B, Thauer RK, et al. Life under extreme energy limitation: a synthesis of laboratory- and field-based investigations. *FEMS Microbiol Rev*. 2015;39(5):688–728.
121. Jørgensen BB, Marshall IPG. Slow microbial life in the seabed. *Annu Rev Mar Sci.* 2016;8:311–332.
122. Robador A, Larowe DE, Finkel SE, Amend JP, Nealson KH, Booth V. Changes in microbial energy metabolism measured by nanocalorimetry during growth phase transitions. *Front Microbiol*. 2018;9:109.
123. Timmers PH, Gieteling J, Widjaja-Greefkes HCA, Plugge CM, Stams AJM, Lens PNL, et al. Growth of anaerobic methane-oxidizing archaea and sulfate-reducing bacteria in a high-pressure membrane capsule bioreactor. *Appl Environ Microbiol.* 2015;81(4):1286–1296.
124. Finkel SE. Long-term survival during stationary phase: evolution and the GASP phenotype. *Nat Rev Microbiol*. 2006;4(2):113–120.
125. Kevorkian R, Bird JT, Shumaker A, Lloyd KG. Estimating population turnover rates by relative quantification methods reveals microbial dynamics in marine sediment. *Appl Environ Microbiol*. 2018;84(1):e01443-17.
126. Steen AD, Jorgensen BB, Lomstein BA. Abiotic racemization kinetics of amino acids in marine sediments. *PLoS One*. 2013;8(8):e71648.
127. Onstott TC, Magnabosco C, Aubrey AD, Burton AS, Dworkin JP, Elsila JE, et al. Does aspartic acid racemization constrain the depth limit of the subsurface biosphere? *Geobiology*. 2014;12:1–19.
128. Daniel I, Oger P, Winter R. Origins of life and biochemistry under high-pressure conditions. *Chem Soc Rev*. 2006;35(10):858–875.
129. Magnabosco C, Lin L-H, Dong H, Bomberg M, Ghiorse W, Stan-Lotter H, et al. The biomass and biodiversity of the continental subsurface. *Nat Geosci.* 2018;11:707–717.
130. Cavicchioli R, Thomas T, Curmi PMG. Cold stress response in Archaea. *Extremophiles*. 2000;4(6):321–331.
131. Kawano H, Nakasone K, Matsumoto M, Yoshida Y, Usami R, Kato C, et al. Differential pressure resistance in the activity of RNA polymerase isolated from *Shewanella violacea* and *Escherichia coli*. *Extremophiles*. 2004;8(5):367–375.
132. Kasahara R, Sato T, Tamegai H, Kato C. Piezo-adapted 3-isopropylmalate dehydrogenase of the obligate piezophile *Shewanella benthica* DB21MT-2 isolated from the 11,000-m depth of the Mariana Trench. *Biosci Biotechnol Biochem.* 2009;73(11):2541–2543.
133. Hay S, Evans RM, Levy C, Loveridge EJ, Wang X, Leys D, et al. Are the catalytic properties of enzymes from piezophilic organisms pressure adapted? *ChemBioChem.* 2009;10(14):2348–2353.
134. Nishiguchi Y, Abe F, Okada M. Different pressure resistance of lactate dehydrogenases from hagfish is dependent on habitat depth and caused by tetrameric structure dissociation. *Mar Biotechnol*. 2011;13(2):137–141.

135. Hamajima Y, Nagae T, Watanabe N, Ohmae E, Kato-Yamada Y, Kato C. Pressure adaptation of 3-isopropylmalate dehydrogenase from an extremely piezophilic bacterium is attributed to a single amino acid substitution. *Extremophiles*. 2016;20(2):177–186.
136. Rosenbaum E, Gabel F, Durá MA, Finet S, Cléry-Barraud C, Masson P, et al. Effects of hydrostatic pressure on the quaternary structure and enzymatic activity of a large peptidase complex from *Pyrococcus horikoshii*. *Arch Biochem Biophys*. 2012; 517(2):104–110.
137. Murakami C, Ohmae E, Tate S, Gekko K, Nakasone K, Kato C. Comparative study on dihydrofolate reductases from *Shewanella* species living in deep-sea and ambient atmospheric-pressure environments. *Extremophiles*. 2011;15(2):165–175.
138. De Vos D, Xu Y, Hulpiau P, Vergauwen B, Van Beeumen JJ. Structural investigation of cold activity and regulation of aspartate carbamoyltransferase from the extreme psychrophilic bacterium *Moritella profunda*. *J Mol Biol*. 2007;365(2): 379–395.
139. Jian H, Xu G, Gai Y, Xu J, Xiao X. The histone-like nucleoid structuring protein (H-NS) is a negative regulator of the lateral flagellar system in the deep-sea bacterium *Shewanella piezotolerans* WP3. *Appl Environ Microbiol*. 2016;82(8):2388–2398.
140. Siebenaller JF. Pressure effects on the GTPase activity of brain membrane G proteins of deep-living marine fishes. *Comp Biochem Physiol B Biochem Mol Biol*. 2003;135 (4):697–705.
141. Siebenaller JF, Garrett DJ. The effects of the deep-sea environment on transmembrane signaling. *Comp Biochem Physiol B Biochem Mol Biol*. 2002;131(4):675–694.
142. Chilukuri LN, Bartlett DH, Fortes PAG. Comparison of high pressure-induced dissociation of single-stranded DNA-binding protein (SSB) from high pressure-sensitive and high pressure-adapted marine *Shewanella* species. *Extremophiles*. 2002;6(5):377–383.
143. Boonyaratanakornkit BB, Miao LY, Clark DS. Transcriptional responses of the deep-sea hyperthermophile *Methanocaldococcus jannaschii* under shifting extremes of temperature and pressure. *Extremophiles*. 2007;11(3):495–503.
144. Simonato F, Campanaro S, Lauro FM, Vezzi A, D'Angelo M, Vitulo N, et al. Piezophilic adaptation: a genomic point of view. *J Biotechnol*. 2006;126(1):11–25.
145. Campanaro S, De Pascale F, Telatin A, Schiavon R, Bartlett DH, Valle G. The transcriptional landscape of the deep-sea bacterium *Photobacterium profundum* in both a toxR mutant and its parental strain. *BMC Genomics*. 2012;13:567.
146. Samerotte AL, Drazen JC, Brand GL, Seibel BA, Yancey PH. Correlation of trimethylamine oxide and habitat depth within and among species of teleost fish: an analysis of causation. *Physiol Biochem Zool*. 2007;80(2):197–208.
147. Nagae T, Kawamura T, Chavas LMG, Niwa K, Hasegawa M, Kato C, et al. High-pressure-induced water penetration into 3-isopropylmalate dehydrogenase. *Acta Crystallogr Sect D Biol Crystallogr*. 2012;68(3):300–309.
148. Kaye JZ, Baross J. Synchronous effects of temperature, hydrostatic pressure, and salinity on growth, phospholipid profiles, and protein patterns of four *Halomonas* species isolated from deep-sea hydrothermal-vent and sea surface environments. *Appl Environ Microbiol*. 2004;70(10):6220–6229.
149. Bartlett DH. Pressure effects on *in vivo* microbial processes. *Biochim Biophys Acta Protein Struct Mol Enzymol*. 2002;1595(1):367–381.
150. Campanaro S, Treu L, Valle G. Protein evolution in deep sea bacteria: an analysis of amino acids substitution rates. *BMC Evol Biol*. 2008;8(1):313.

151. Cario A, Jebbar M, Thiel A, Kervarec N, Oger PM. Molecular chaperone accumulation as a function of stress evidences adaptation to high hydrostatic pressure in the piezophilic archaeon *Thermococcus barophilus*. *Sci Rep*. 2016;6(1):29483.
152. Nakashima M, Yamagami R, Tomikawa C, Ochi Y, Moriya T, Asahara H, et al. Long and branched polyamines are required for maintenance of the ribosome, tRNAHis and tRNATyr in *Thermus thermophilus* cells at high temperatures. *Genes Cells*. 2017;22(7):628–645.
153. Michoud G, Jebbar M. High hydrostatic pressure adaptive strategies in an obligate piezophile *Pyrococcus yayanosii*. *Sci Rep*. 2016;6:27289.
154. Eloe EA, Malfatti F, Gutierrez J, Hardy K, Schmidt WE, Pogliano K, et al. Isolation and characterization of a psychropiezophilic alphaproteobacterium. *Appl Environ Microbiol*. 2011;77(22):8145–8153.
155. Jebbar M, Franzetti B, Girard E, Oger P. Microbial diversity and adaptation to high hydrostatic pressure in deep-sea hydrothermal vents prokaryotes. *Extremophiles*. 2015;19(4):721–740.
156. Winter R, Dzwolak W. Exploring the temperature-pressure configurational landscape of biomolecules: from lipid membranes to proteins. *Philos Trans A Math Phys Eng Sci*. 2005;363(1827):537–562; discussion 562–563.
157. Macgregor RB. Effect of hydrostatic, pressure on nucleic acids. *Biopolymers*. 1998;48(4):253–263.
158. Fan HY, Shek YL, Amiri A, Dubins DN, Heerklotz H, Macgregor Jr. RB, et al. Volumetric characterization of sodium-induced G-quadruplex formation. *J Am Chem Soc*. 2011;133(12):4518–4526.
159. Amiri AR, Macgregor Jr. RB. The effect of hydrostatic pressure on the thermal stability of DNA hairpins. *Biophys Chem*. 2011;156(1):88–95.
160. Bridgman. The coagulation of albumin by pressure. *J Biol Chem*. 1914;19:511–512.
161. Weber G. Thermodynamics of the association and the pressure dissociation of oligomeric proteins. *J Phys Chem*. 1993;97(27):7108–7115.
162. Schmid G, Ludemann H-D, Jaenicke R. Dissociationand aggregation of lactic dehydrogenase by high hydrostatic pressure. *Eur J Biochem*. 1979;97:407–413.
163. Fossat MJ, Dao TP, Jenkins K, Dellarole M, Yang Y, McCallum SA, et al. High-resolution mapping of a repeat protein folding free energy landscape. *Biophys J*. 2016;111(11):2368–2376.
164. Ho BK, Gruswitz F. *HOLLOW*: generating accurate representations of channel and interior surfaces in molecular structures. *BMC Struct Biol*. 2008;6:49.
165. Royer CA. Revisiting volume changes in pressure-induced protein unfolding. *Biochim Biophys Acta Protein Struct Mol Enzymol*. 2002;1595(1–2):201–209.
166. Smeller L, Biology R, Smeller L. Pressure–temperature phase diagrams of biomolecules. *Biochim Biophys Acta Protein Struct Mol Enzymol*. 2002;1595 (1–2):11–29.
167. Font J, Benito A, Lange R, Ribo M, Vilanova M. The contribution of the residues from the main hydrophobic core of ribonuclease A to its pressure-folding transition state. *Protein Sci*. 2006;15(5):1000–1009.
168. Roche J, Caro JA, Norberto DR, Barthe P, Roumestand C, Schlessman JL, et al. Cavities determine the pressure unfolding of proteins. *Proc Natl Acad Sci*. 2012;109(18):6945–6950.
169. Ando N, Barstow B, Baase WA, Fields A, Matthews BW, Gruner SM. Structural and thermodynamic characterization of T4 lysozyme mutants and the contribution of internal cavities to pressure denaturation. *Biochemistry*. 2008;47(42):11097–11109.

170. Chen CR, Makhatadze GI. Molecular determinant of the effects of hydrostatic pressure on protein folding stability. *Nat Commun.* 2017;8:14561.
171. Chen CR, Makhatadze GI. Molecular determinants of temperature dependence of protein volume change upon unfolding. *J Phys Chem B.* 2017;121(35):8300–8310.
172. Papini CM, Pandharipande PP, Royer CA, Makhatadze GI. Putting the piezolyte hypothesis under pressure. *Biophys J.* 2017;113(5):974–977.
173. Akasaka K. Probing conformational fluctuation of proteins by pressure perturbation. *Chem Rev.* 2006;106(5):1814–1835.
174. Akasaka K, Kitahara R, Kamatari YO. Exploring the folding energy landscape with pressure. *Arch Biochem Biophys.* 2013;531(1–2):110–115.
175. Jonas J. High-resolution nuclear magnetic resonance studies of proteins. *Biochim Biophys Acta Protein Struct Mol Enzymol.* 2002;1595(1–2):145–159.
176. Peng X, Jonas J, Silva JL. High-pressure NMR study of the dissociation of arc repressor. *Biochemistry.* 1994;33(27):8323–8329.
177. Li H, Yamada H, Akasaka K. Effect of pressure on individual hydrogen bonds in proteins. Basic pancreatic trypsin inhibitor. *Biochemistry.* 1998;37(5): 1167–1173.
178. Kremer W, Arnold MR, Brunner E, Schuler B, Jaenicke R, Kalbitzer HR. High pressure NMR spectroscopy and its application to the cold shock protein TmCsp derived from the hyperthermophilic bacterium *Thermotoga maritima.* In: Winter R, ed. *Advances in High Pressure Bioscience and Biotechnology II, Proceedings.* Springer, 2003, pp. 101–112.
179. Flynn PF, Milton MJ, Babu CR, Wand AJ. A simple and effective NMR cell for studies of encapsulated proteins dissolved in low viscosity solvents. *J Biomol NMR.* 2002;23(4):311–316.
180. Peterson RW, Nucci NV, Wand AJ. Modification of encapsulation pressure of reverse micelles in liquid ethane. *J Magn Reson.* 2011;212(1):229–233.
181. Fujisawa T, Kato M, Inoko Y. Structural characterization of lactate dehydrogenase dissociation under high pressure studied by synchrotron high-pressure small-angle X-ray scattering. *Biochemistry.* 1999;38(20):6411–6418.
182. Krywka C, Sternemann C, Paulus M, Javid N, Winter R, Al-Sawalmih A, et al. The small-angle and wide-angle X-ray scattering set-up at beamline BL9 of DELTA. *J Synchrotron Radiat.* 2007;14(3):244–251.
183. McCoy J, Hubbell WL. High-pressure EPR reveals conformational equilibria and volumetric properties of spin-labeled proteins. *Proc Natl Acad Sci.* 2011;108(4): 1331–1336.
184. Lerch MT, Horwitz J, McCoy J, Hubbell WL. Circular dichroism and site-directed spin labeling reveal structural and dynamical features of high-pressure states of myoglobin. *Proc Natl Acad Sci.* 2013;110(49):E4714–E4722.
185. Iwahashi H, Odani M, Ishidou E, Kitagawa E. Adaptation of *Saccharomyces cerevisiae* to high hydrostatic pressure causing growth inhibition. *FEBS Lett.* 2005;579(13):2847–2852.
186. Welch TJ, Farewell A, Neidhardt FC, Bartlett DH. Stress response of *Escherichia coli* to elevated hydrostatic pressure. *J Bacteriol.* 1993;175(22):7170–7177.
187. Aertsen A, Van Houdt R, Vanoirbeek K, Michiels CW. An SOS response induced by high pressure in *Eschericia coli. J Bacteriol.* 2004;186(18):6133–6141.
188. Fichtel K, Logemann J, Fichtel J, Rulkotter J, Cypionka H, Engelen B. Temperature and pressure adaptation of a sulfate reducer from the deep subsurface. *Front Microbiol.* 2015;6:1078.

189. Ishii A, Sato T, Wachi M, Nagai K, Kato C. Effects of high hydrostatic pressure on bacterial cytoskeleton FtsZ polymers *in vivo* and *in vitro*. *Microbiology*. 2004; 150(Pt 6):1965–1972.
190. Aertsen A, Michiels CW. Mrr instigates the SOS response after high pressure stress in *Escherichia coli*. *Mol Microbiol*. 2005;58(5):1381–1391.
191. Aertsen A, Michiels CW. SulA-dependent hypersensitivity to high pressure and hyperfilamentation after high-pressure treatment of *Escherichia coli* lon mutants. *Res Microbiol*. 2005;156(2):233–237.
192. Bourges AC, Torres Montaguth OE, Ghosh A, Tadesse WM, Declerck N, Aertsen A, et al. High pressure activation of the Mrr restriction endonuclease in *Escherichia coli* involves tetramer dissociation. *Nucleic Acids Res*. 2017;45(9):5323–5332.
193. Salmon ED, Ellis GW. A new miniature chamber pressure for microscopy strain-free optical glass windows facilitate phase-contrast and polarized-light microscopy of living cells. *J Cell Biol*. 1975;65:587–602.
194. Crenshaw HC, Salmon ED. Hydrostatic pressure to 400 atm does not induce changes in the cytosolic concentration of Ca^{2+} in mouse fibroblasts: measurements using fura-2 fluorescence. *Exp Cell Res*. 1996;227(2):277–284.
195. Müller JD, Gratton E. High-pressure fluorescence correlation spectroscopy. *Biophys J*. 2003;85(4):2711–2719.
196. Nishiyama M, Kojima S. Bacterial motility measured by a miniature chamber for high-pressure microscopy. *Int J Mol Sci*. 2012;13(7):9225–9239.
197. Koo J, Czeslik C. High pressure sample cell for total internal reflection fluorescence spectroscopy at pressures up to 2500 bar. *Rev Sci Instrum*. 2012;83(8):085109.
198. Frey B, Hartmann M, Herrmann M, Meyer-Pittroff R, Sommer K, Bluemelhuber G. Microscopy under pressure – an optical chamber system for fluorescence microscopic analysis of living cells under high hydrostatic pressure. *Microsc Res Tech*. 2006;69(2):65–72.

19

Energy Limits for Life in the Subsurface

DOUG LAROWE AND JAN AMEND

19.1 Introduction

Life's demand for energy drives rapid exchanges of carbon between the atmosphere, oceans, and land (1). Photosynthesis and respiration of organic carbon on and near the surface of Earth account for the vast bulk of the transfers of carbon between these reservoirs, processes that dwarf geologic sources and sinks of carbon on short timescales (2; Chapter 16, this volume). In recent years, however, it has become apparent that Earth hosts a vast subsurface biosphere (3–16) that operates much more slowly (17–20) than surface life. It is not clear to what depth this biosphere exists, the rates at which it is active, what reactions are being catalyzed for energy, or what impact it has had on Earth's carbon cycle through time. Extrapolating metabolic rates on Earth's surface to the subsurface is complicated by an essential difference – the types and rates of biological activity on the surface are determined by daily and seasonal cycles driven by the sun, but life in the subsurface seems to be more attuned to geologic processes and timescales. Such slow rates make it difficult to study subsurface life, but the biological demand for energy can be used to better understand the rates at which these organisms are active and how they interact with their geochemical environments. Studying subsurface life can therefore help reveal the energy limits for life and thus its spatial and temporal extent.

In addition to hosting organisms on the low end of the bioenergetic spectrum, the potential size of the subsurface biosphere motivates many researchers studying deep life. Marine sediments alone have been estimated to contain 10^{29}–10^{30} microbial cells (4, 5) and occupy 300 million km^3, over 22% of the volume of Earth's oceans (21). A similar number of microorganisms are thought to inhabit the continental subsurface (9, 22; Chapter 17, this volume). Cells have been found as deep as 3.5 km beneath the surface of land, which, if true of all continental crust, would translate to a potential habitable volume of over 730 million km^3 (taking the total continental crust surface to be 2.1×10^8 km^2; 23). Although there are no estimates of the size of the ocean crustal basement biosphere, if the top 600 m of it is sufficiently hydrologically permeable to host life (24–32), this would correspond to 1.8 billion km^3 (based on the ocean crust covering 68.8% (2.99×10^8 km^2) of Earth's surface; 23). Most recently, it has been estimated that the deep subsurface contains about 13% of the biomass on the planet (10).

Finally, although we have no information about current or past life on extraterrestrial bodies in our solar system, it is most likely that any evidence of it will be in the subsurface (33–35).

Cell counts, microbial cultivation work, and molecular biological efforts are helping to describe the numbers, viability, activity levels, and variety of microorganisms in subsurface environments (Chapters 17 and 18, this volume). However, the relative inaccessibility of these environments, the difficulty of cultivating representative microorganisms, and the long timescales associated with some of their lifestyles are major impediments to obtaining a comprehensive understanding of complex subsurface ecosystems. Hence, modeling approaches that include geochemical data as well as typical microbiological measurements can be useful strategies for quantifying biogeochemical interactions in the subsurface. Because all living things must catalyze redox reactions to obtain energy and the amount of energy available from a chemical reaction depends strictly on the prevailing environmental conditions, energy-based modeling provides a framework for quantifying microbial processes in any setting. In this chapter, we will discuss what is known about the microbial demand for energy in low-energy settings, review recent efforts to quantify the energy available in subsurface habitats, and provide an overview of how these calculations are carried out.

19.2 Microbial States

The physiological state of a microorganism is related to the rate at which it is using energy. In contrast to the four classical physiological states that microbial isolates exhibit in high-energy, high-nutrient, short-timespan laboratory experiments – lag, exponential, stationary, and death phases – microorganisms in nature are often nutrient and energy limited and exist in complex communities that are exposed to varying physiochemical properties such as oscillating temperature, pH, and water activity. Their physiological states do not necessarily correspond to those that have been determined in the laboratory and are in many instances unknown (20, 36). Consequently, the rate at which these organisms are processing carbon and energy in nature is not well constrained. Back of the envelope-style calculations reveal how unlikely widespread microbial growth is in nature: if every microorganism in the subsurface ($\sim 10^{30}$ organisms with ~ 10 fg C cell^{-1}) doubled every day, all of the carbon in Earth's crust (60 million Gt; 37) would become microbial biomass in less than 23 days.

Whatever the physiological status of living microorganisms in natural settings, their use of energy can be partitioned into two broad categories: maintenance and growth. It is difficult to strictly separate maintenance and growth activities because when organisms are growing, they are also carrying out maintenance functions. Similarly, the critical maintenance activity of replacing biomolecules is an aspect of the anabolic processes that lead to growth. Nonetheless, they are commonly viewed as distinct states with particular energetic ramifications.

Maintenance generally refers to the collections of activities that an organism performs to simply stay alive. It can include nutrient uptake, motility, the preservation of charged membranes, excretion of (bio)molecules, and changes in stored nutrient concentrations (see 38 for a review). It is difficult to determine the amounts of energy that each of these and other maintenance functions require, so they are typically lumped together and defined in the negative: all of the activities that a microorganism carries out that are not associated with growth. Traditionally, maintenance energies are determined by *growing* microorganisms at ever-slower rates and extrapolating the amount of energy that they use to a value that would represent zero growth (for a review, see 17). Values reported in the literature range over more than five orders of magnitude, from 0.019 fW (fW ≡ femtowatts, 10^{-15} J s^{-1}) for anoxygenic phototrophy to 4700 fW for aerobic heterotrophy (see 39). This procedure might decipher the energy partitioned into nongrowth activities for a high-energy state – *growth* – but maintenance energies in nature are orders of magnitude lower than these values (17, 18, 39–41).

To distinguish laboratory measurements of maintenance from the survival state of microorganisms in nongrowing and other low-energy environments, the term "basal power requirement" has been introduced (17). The use of the word "power" instead of "energy" is apt (see 39, 41, 42) since so-called maintenance energies are given in units of energy per time. Recent results from modeling studies support the notion that in very-low-energy settings (e.g. deep oligotrophic marine sediments such as those under the South Pacific Gyre), maintenance powers are indeed several orders of magnitude lower than those reported in the literature, from 50 to 3500 zW $cell^{-1}$ (where zW ≡ zeptowatt, 10^{-21} W; 41). Related calculations suggest that the amount of particulate organic matter required to sustain microbes at a basal power level in the same oligotrophic environment is equivalent to about 2% of their biomass carbon per year (36). It should be noted that these latter two studies assume that all of the energy used by these microbes is for maintenance only, not growth. It is not known whether microorganisms in ultra-low-power settings produce daughter cells or merely persist via maintenance (17, 18, 20). It is thought that very slow or no growth is the norm in many environments (43), especially chemically stagnant, low-energy ones such as oligotrophic marine sediments (17, 44–46). However, it should be noted that when conditions improve, metabolic rates can increase dramatically (47). Although values are not well constrained for dynamic ecosystems, maintenance power can vary due to many factors, including temperature (for the same species), substrate identity, redox conditions, and cultivation techniques (18).

Microbial growth can refer to the generation of new cells, the production of biomass accompanying enlarging cells and the synthesis of external structures such as stalks, sheaths, tubes, biofilms, and other polymeric saccharides. Although extracellular biomass can be a significant fraction of the total biomass of a system, most efforts to quantify the amount of energy that it takes to make biomass have focused on the energetics of making cells, and like all chemical reactions, the amount of energy required to synthesize cells depends in large part upon the physiochemical properties of the environment. The first modern effort to explicitly quantify how environmental parameters influence the energetics of biomass synthesis focused on how the redox state of the environment influences the energy required to synthesize the biomolecules that make up cells (48). More recently,

LaRowe and Amend (49) expanded this analysis to quantify the amount of energy required to make biomass as a function of temperature, pressure, redox state, the sources of C, N, and S, and cell mass, while also accounting for the polymerization of monomers into biomacromolecules such as proteins, DNA, RNA, and polysaccharides (see below). These kinds of modeling results depend on both the compositions and the masses of microbial cells. The stoichiometry of biomass varies considerably (see 18, 49 for reviews), with the average nominal oxidation state of carbon (NOSC; see 50) in cell biomass – a tidy numerical representation of the ratio of carbon to hydrogen, nitrogen, oxygen, sulfur, and phosphorous in organic matter – varying from at least from +0.89 to –0.45 (49). Note that the commonly used stand-in for biomass, "CH_2O," fixes the NOSC at 0. As with maintenance power, cell stoichiometry varies with environmental conditions (see 49).

Other attempts to account for the energy required to make biomass rely upon laboratory experiments that are designed to promote growth, conditions that are rarely encountered in the subsurface (see 39 for a review). In addition, numerous energy-based models developed to predict biomass yields have limited applicability because they are restricted to standard and/or reference states (for reviews, see 39, 51). Most of these models fix biomass yield coefficients, which leads to predictions of equal biomass production (for the same synthesis reactions) under potentially very different environmental conditions. Other efforts seeking to determine the amount of energy required to make biomass rely on estimates of the number of moles of ATP that are required to carry out various aspects of biomolecular synthesis. A commonly cited version of this approach (52) reports the ATP requirements for various anabolic and maintenance functions. The reference for these values ultimately cites a paper (53) that effectively assumes the amount of ATP required for a variety of biochemical processes with little to no experimental evidence or references. Even if the amount of ATP required to synthesize biomass were well constrained in natural settings, the amount of energy that is released from the hydrolysis of ATP to ADP and phosphate is a function of the physiochemical characteristics of the environment in which it is happening, and therefore not a constant (54–56).

Microorganisms can also enter into low-energy and dormant states. While some authors use the term "dormancy" to describe low-energy states as well (57), others reserve this term for true endospore-forming microorganisms that are metabolically inactive (58). *Sensu stricto*, dormancy is a reversible, contingent state that is entered into when resources become limiting. Because there is no metabolic activity, critical functions such as DNA and protein repair are not possible. As a result of abiotic hydrolysis, oxidation, and other degradative processes, dormant organisms have a finite lifespan. Although little is known about the energetics or maximum possible length of dormancy (18, 43), there is evidence that revival rates are inversely correlated with the amount of time a microorganism spends in the dormant state (43, 59), an observation that has been included in modeling studies of microbial dormancy (60, 61). One study showed that the number of endospores in frozen samples decreases with sample age since endospores do not have active DNA-repair mechanisms, but that low-energy bacteria in the same samples can survive for up to half a million years (62).

Microorganisms in low-energy states, in contrast to dormant ones, are able to slowly metabolize in order to remain viable for potentially millions of years (19, 40). These low-energy states are thought to be prevalent in natural systems, to exist on a spectrum of levels, and to have multiple entry points (57). In fact, it is possible that $<10\%$ of microbial cells in soil and aquatic environments are active (63). One estimate proclaimed that low-energy cells can exist on three orders of magnitude lower power than typical maintenance power levels (64), though, as pointed out above, maintenance powers determined under relatively high-energy conditions span five orders of magnitude. Perhaps energy usage in the low-energy state is more akin to basal maintenance power as defined by Hoehler and Jørgensen (17): the flux of energy required for a minimum survival state. Some speculate that virtually all microbial cells in deep marine sediments are surviving in ultra-low energy states (20). A recent modeling study represented the cells in an oligotrophic marine sediment section as existing in a series of ever-lower energy states, but none as truly dormant (36).

19.3 Gibbs Energy: Where It Comes from and How to Use It

All organisms catalyze reactions as they acquire energy from the environment and carry out biochemical functions related to maintenance and growth. The amount of energy associated with these chemical transformations depends on the exact identity of the reaction – all of the reactants and products describing the mass- and charge-balanced process – as well as the prevailing temperature, pressure, and composition of the environment in which it is happening. The mathematical approach that is used to quantify the change in energy resulting from a chemical reaction is the Gibbs function, denoted by G. Despite being well known, it is worth noting why this, of several energy functions related to the internal energy of a system, is the one that is used to assess how energy is required by organisms to perform a given task and how much energy results from catalyzing reactions that ultimately fuel life's demands. For textbook-level overviews of thermodynamics, see (65–68).

Because there is no absolute energy scale, we can only discuss changes in the amount of energy that a system contains. The change in the internal energy (U) of a system can be quantified by assessing changes in the temperature (T), pressure (P), volume (V), and entropy (S) of it, quantities that can be modified by the exchange of heat and matter between the system and the surroundings as well as reactions happening within the system. Changes in the internal energy of the system, dU, can be mathematically linked to how these four variables change in different ways (i.e. the various energy functions are effectively different partial derivatives of U with respect to T, P, V, and S). These four variables are known as state variable because they describe the state of the system at any given moment; functions including these variables are known as state functions. For instance, the change in the amount of heat (or enthalpy, H) associated with a system at constant pressure, dH_P, with no chemical reactions occurring, can be defined by $dH_P = dU + PdV$, which is a simple statement of the first law of thermodynamics – the conservation of energy (technically, this and other state functions hold for infinitesimal changes taking place in the time interval t to $t + dt$). Enthalpy is a useful state function that describes the evolution of the heat content of a system, but it cannot predict how the system will evolve. The Gibbs energy function accomplishes this by combining the first law of thermodynamics with the second law, which essentially states that the entropy of an isolated system cannot decrease (see Box 19.1).

Box 19.1 Entropy, affinity, and the reaction progress variable.

Entropy is sometimes wrongly discussed in terms of being equivalent to "disorder." This confusion likely arises from Boltzmann's statistical mechanical interpretation of entropy being related to the number of microstates that a system can occupy – the more microstates, the higher the entropy. The proliferation of the use of the word "entropy" in other fields (e.g. Shannon entropy in information theory, which is more akin to uncertainty than the classical definition of entropy as a quantity of energy) has contributed to the popular conception of entropy as a measure of disorder. From its origins with Carnot and developments by Clausius, de Donder, and Kelvin, entropy is described as a quantity of energy and a state variable; its units are J K^{-1} mol^{-1}. The classical definition of entropy is that it is equal to the change in the heat of a system, Q, at a given temperature: $dS = dQ/T$.

Prigogine and Defay (69) describe two ways in which the entropy of a system can change: (1) entropy can be transported across the boundary of a system and therefore be positive or negative; and (2) entropy can be generated inside of an isolated system due to the occurrence of chemical reactions – these reactions are irreversible and the net change of entropy associated with them is always positive, $dS > 0$. It is this inequality that ultimately specifies the direction in which a system will evolve. So, we can say that the entropy of a system changes due to entropy exchanges with the surroundings (dS_e) and entropy created within the system, dS_i, or $dS = dS_e + dS_i$. Because dS_i is related to the degree to which chemical reactions are occurring, it is in turn related to the extent to which a reaction has occurred. The extent of a reaction is known as ξ, and sometimes is called the reaction progress variable (69–71). Changes in the reaction progress variable for a reaction are related to the stoichiometric coefficients of the ith species in that reaction, v_i, by

$$\frac{dn_i}{v_i} = d\xi, \tag{19.a}$$

where n_i refers to the number of moles of the ith species produced or consumed in the reaction. For example, the extent to which the reaction describing hydrogen-consuming sulfate reduction,

$$SO_4^{2-} + 4H_2 + H^+ \rightarrow HS^- + 4H_2O, \tag{19.b}$$

has progressed, $d\xi_{(19.b)}$, is given by

$$\frac{dn_{SO_4^{2-}}}{-1} = \frac{dn_{H_2}}{-4} = \frac{dn_{H^+}}{-1} = \frac{dn_{HS^-}}{1} = \frac{dn_{H_2O}}{4} = d\xi_{(19.b)}. \tag{19.c}$$

Clearly, when $dn_i = v_i$, $d\xi_{(19.b)} = 1$ and the reaction is said to have turned over.

The reaction progress variable and entropy are connected through what is known as chemical affinity, A, a term that is sometimes used to quantify how far a system is from equilibrium: $dS_i = Ad\xi \geq 0$ (this equation is also known as de Donder's fundamental inequality and can be thought of as another statement of the second law of thermodynamics). Although chemical affinity and the Gibbs energy of a reaction are related simply by $A = -\Delta G_r$, the two quantities have important conceptual differences. A is defined as the change in Gibbs energy resulting from changes in the reaction progress variable while pressure, temperature, and composition (n_k) are held constant (70, 71):

$$A = -\left(\frac{\partial G}{\partial \xi}\right)_{T,P,n_k}. \qquad (19.d)$$

As such, affinity relates irreversible chemical reactions to entropy production, while the Gibbs energy of a reaction is traditionally used in reference to equilibrium states and reversible processes (see 66). Furthermore, chemical affinity can be connected to reaction rate as a measure of the distance that reaction is from equilibrium (69, 70, 72, 73).

Within this conceptualization, thermodynamic rate-limiting terms have been developed in kinetic models that relate the rates of biologically catalyzed reactions to their distance from equilibrium, with reactions closer to equilibrium being slower than those further away (74–78). Finally, affinity and the rate of entropy production have in turn been used to develop more recent developments in the field of nonequilibrium thermodynamics, especially systems far from equilibrium and the establishment of dissipative structures, such as life (see 65).

Changes in the Gibbs energy at constant temperature and pressure are commonly expressed as $dG_{T,P} = dH - TdS$, or in integrated form, $\Delta G = \Delta H - T\Delta S$, but is more clearly linked to changes in the internal energy state of a system by $dG_{P,T} = dU + PdV - TdS$. By incorporating the second law of thermodynamics, the Gibbs energy function quantifies the tendency of a chemical reaction to proceed in a particular direction. That is, for a given chemical reaction, negative values of ΔG indicate that if a reaction occurs, the net result is the formation of products at the expense of reactants; the opposite direction occurs for positive values of ΔG. When $\Delta G = 0$, there is no net reaction and equilibrium has been reached.

Thus far, we have represented thermodynamics functions with simple letters like G and H. However, if we want to use these functions to calculate the energetics of real processes under specific environmental conditions, we must become familiar with the alphabet soup of subscripts and superscripts that modify and specify the meanings of terms such as G and H. As noted above, since we can only know how the energy of a system changes, the upper case Greek letter delta, Δ, is often placed in front of energy functions to signify changes in values of that function (e.g. ΔG and ΔH). These should be thought of in terms of the difference in, for example, the Gibbs energy of the system at one configuration versus another: $\Delta G = G_2 - G_1$. How the system evolved from state 1 to state 2 is irrelevant. When the subscript r is added, ΔG_r, for example, stands for the Gibbs energy of a reaction. These subscripts are straightforward accoutrements decorating thermodynamics functions, but the superscripts

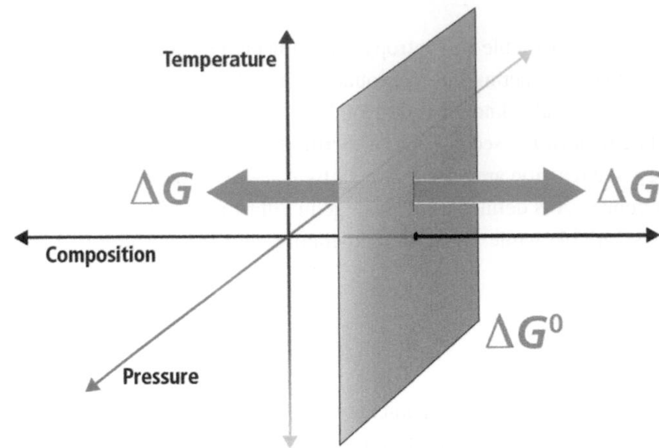

Figure 19.1 Schematic diagram illustrating the difference between standard-state Gibbs energies (ΔG_r^0) and overall Gibbs energies (ΔG_r) in temperature, pressure, and compositional space. For a given standard state, values of ΔG_r^0 refer to a fixed composition at any combination of temperatures and pressures (the orange plane) and that departures from this composition are what distinguish ΔG_r. For gases, pressure is part of the standard-state definition since the state of aggregation of a gas is partially determined by its partial pressure.

are less so, and they are typically more meaningful. When the superscript 0 appears, (i.e. ΔG^0), the symbol is explicitly referring to the change in Gibbs energy at a standard state. Standard states are one of the most frequently misunderstood aspects of thermodynamics.

The standard state used in traditional chemical thermodynamics does not refer to a temperature or pressure (i.e. it does *not* refer to 25°C and 1 bar), but a standard state of aggregation, or composition (see Figure 19.1 for a conceptual overview). The reason for this is that, as discussed above, we cannot know the internal energy, U, of a system, but we can determine changes in it. Both Gibbs energies and enthalpies are partial derivatives of U, so to quantify ΔG^0, ΔH^0, and other such standard state terms, we need a coherent system that relates the thermodynamic functions describing compounds to a set of standards. The system that is used in chemical thermodynamics to define the Gibbs energies and enthalpies of chemical compounds is related to that of the elements. For example, the standard-state Gibbs energy of formation from the elements of gaseous methane, $\Delta G^0_{CH_4(g)}$, is calculated based on the change of Gibbs energy associated with the reaction:

$$C_{graphite} + 2H_{2(g)} = CH_{4\,(g)}, \tag{19.1}$$

or:

$$\Delta G_r^0 = \Delta G^0_{CH_4(g)} - \Delta G^0_{C,graphite} - 2\Delta G^0_{H_2(g)}, \tag{19.2}$$

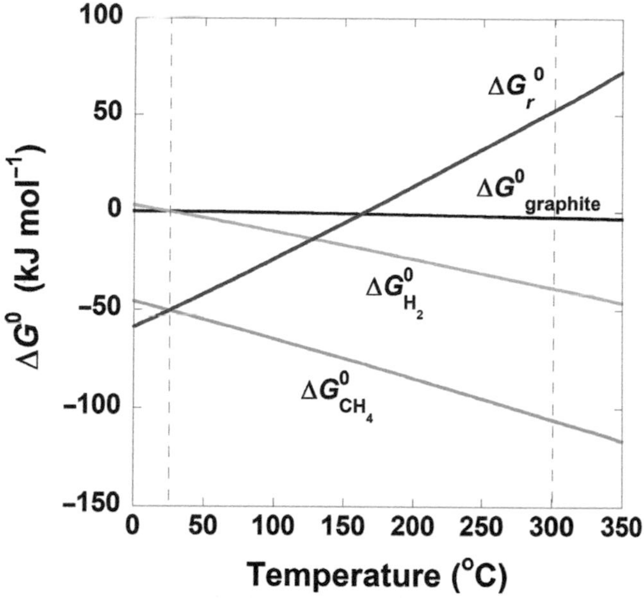

Figure 19.2 Standard-state Gibbs energies (ΔG^0) of H_2 gas, carbon as graphite, CH_4 gas, and the reaction defining the formation of methane from the elements as a function of temperature (see (19.1) and (19.2)). The vertical dashed lines at 25°C and 300°C are marked in reference to the examples discussed in the text.

where $\Delta G^0_{C,graphite}$ and $\Delta G^0_{H_2(g)}$ refer to the standard-state Gibbs energies of formation for elemental carbon as graphite and gaseous dihydrogen, respectively, and ΔG^0_r denotes the standard-state Gibbs energy of (19.1). These compounds are chosen because they are the stable phases of these two elements at the reference conditions of 25°C and 1 bar. The phase and form of the standard-state Gibbs energies and enthalpies of most elements are defined as those that are stable at 25°C and 1 bar, and they are all taken to be zero only at this temperature and pressure – their values are *not* zero at other combinations of temperature and pressure.

Staying with the CH_4 example, since the values of $\Delta G^0_{C,graphite}$ and $\Delta G^0_{H_2(g)}$ are 0 J mol^{-1} at 25°C and 1 bar and the value of ΔG^0_r for (19.1) is –50.4 kJ mol^{-1}, then $\Delta G^0_{CH_4(g)}$ = –50.4 kJ mol^{-1} at 25°C and 1 bar (the value of ΔG^0_r is determined from a series of calorimetric measurements that are beyond the scope of this chapter). At any other combination of temperature and pressure, values of $\Delta G^0_{C,graphite}$ and $\Delta G^0_{H_2(g)}$ are not 0 J mol^{-1} and therefore the value of ΔG^0_r does not equal that of $\Delta G^0_{CH_4(g)}$ (see Figure 19.2). For instance, at 300°C and 86 bars, they are –2.8 and –38.8 kJ mol^{-1}, respectively, and ΔG^0_r for (19.1) is –25.24 J mol^{-1}. Therefore, $\Delta G^0_{CH_4(g)}$ = –105.6 kJ

mol^{-1} at 300°C and 86 bars, more than twice the value at 25°C and 1 bar. Beyond stipulating that the forms of the elements that are used in calculating the Gibbs energies of formation are their stable ones at 25°C and 1 bar, temperature and pressure are not part of the definition of standard-state properties for substances, except for the standard state for gases (see below). Although this is a rather tedious discussion, it is presented to illustrate how temperature and pressure are properly taken into account in thermodynamic calculations, which are particularly important for quantifying the energetics of biogeochemical processes in the subsurface.

In addition to specifying that the definition of the chemical standard state relates thermodynamic properties of chemical species to those of the elements at particular combinations of temperature and pressure, it also specifies the state of aggregation of chemical substances. The commonly used standard state of aggregation differs depending on the phase, and to be accurate, the amount of a substance that is used to define standard states is activity, rather than concentration (discussed below). Similarly, amounts of gases are represented by fugacities instead of partial pressures. The standard state of gases is that of unit fugacity of the pure hypothetical ideal gas at 1 bar and any temperature, that of liquids and solids is unit activity of the pure substance at any temperature or pressure, and that of aqueous species is unit activity in a hypothetical 1 molal solution referenced to infinite dilution at any temperature and pressure. This last one is a bit peculiar owing to its impossibility, but it is necessary since the concentration, and thus the distances between dissolved chemical species, can vary by many orders of magnitude in natural systems on and near Earth's surface. Furthermore, since aqueous species are dissolved in a medium that has its own thermodynamic properties, such as water, the interactions between the solvent and dissolved species must also be accounted for in any standard state of aggregation (see Box 19.2).

Box 19.2 G^0 as a function of temperature and pressure and the equilibrium constant.

Many texts incorrectly state that the definition of standard state for all phases specifies a pressure of 1 bar (e.g. 65). However, a cursory glance at the equation that describes how values of the standard-state Gibbs energy of a system or substance at any temperature and pressure, $G^0_{P,T}$, differ from that at the *reference* temperature of 25°C and pressure of 1 bar, $G^0_{P_r,T_r}$, quickly shows why this is wrong for all phases other than gases:

$$G^0_{P,T} - G^0_{P_r,T_r} = -S^0_{P_r,T_r}(T - T_r) + \int_{T_r}^{T} C^0_{P_r} dT - T \int_{T_r}^{T} C^0_{P_r} d\ln T + \int_{P_r}^{P} V^0 dP, \qquad (19.e)$$

where $S^0_{P_r,T_r}$ refers to the standard molal entropy of the species at the reference pressure and temperature, $C^0_{P_r}$ stands for the isobaric molal heat capacity at the reference pressure, and V^0 designates the standard molal volume. The fourth term on the right-hand side shows how standard-state Gibbs energy changes as a function of pressure when temperature is held

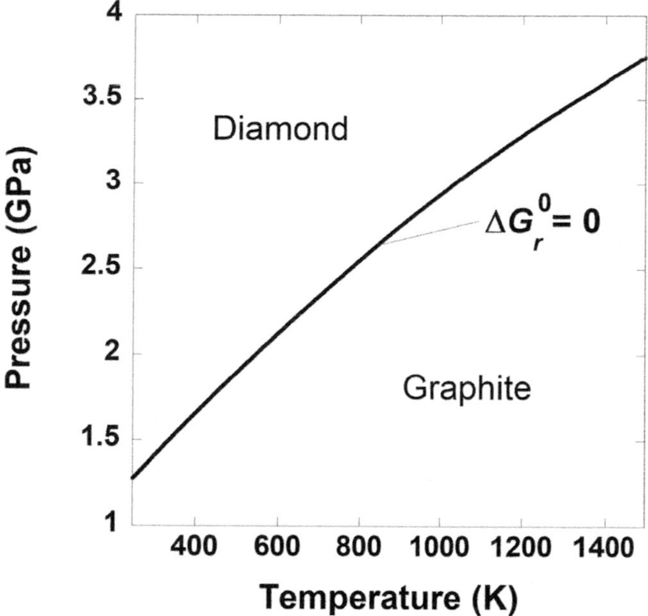

Box Figure 1 A phase diagram for carbon. The curve represents the set of temperatures and pressures where graphite and diamond are in equilibrium, $\Delta G_r^0 = 0$.

constant, $\left(\frac{\partial G^0}{\partial P}\right)_T$. Adding the provision that 1 bar is part of the standard state for solids, liquids, and aqueous species would neglect the volume integral in (19.e) and thus its contribution to the standard-state Gibbs energy. The equilibrium between graphite and diamond further illustrates this point.

The simplified carbon phase diagram presented in Box Figure 1 shows the combinations of temperature and pressure at which graphite and diamond are in equilibrium; that is, $\Delta G_r^0 = 0$ for

$$C_{graphite} \rightleftharpoons C_{diamond}. \qquad (19.f)$$

That is, the *standard-state* Gibbs energies of graphite and diamond are equal on this line. Another way to look at this is to use the more intuitive, and aptly named, equilibrium constant, *K*. The value of *K* for (19.f) is 1 for the temperatures and pressures that define the curve in Box Figure 1. Values of *K* are related to the standard-state Gibbs energy via:

$$K = e^{-\left(\frac{\Delta G_r^0}{RT}\right)}. \qquad (19.g)$$

Again, if a pressure of 1 bar were set for the standard-state values of ΔG_f^0 for graphite and diamond, it would be impossible to account for the impact of pressure on the equilibrium state between graphite and diamond. In fact, there would be no stability field for diamond without elevated pressure, no matter the temperature.

For the interested reader, the first three terms on the right-hand side of (19.e) result from taking into account that both enthalpy and entropy are integrals of C_P^0 and thus represent how these thermodynamic functions vary with temperature at constant pressure:

$$C_P^0 = \left(\frac{\partial H^0}{\partial T}\right)_P = T\left(\frac{\partial S^0}{\partial T}\right)_P. \tag{19.h}$$

If one knows how C_P^0 varies with temperature, then $G_{P,T}^0$ can be calculated for temperatures and pressures other than 25°C and 1 bar. This is rather straightforward for many liquids, gases, and solids; the widely used Maier–Kelley formulation for $C_P^0(T)$ is used: $C_P^0(T) = a + bT + cT^{-2}$. Regression of experimental calorimetry data and/or estimation schemes can be used to determine species-specific values of a, b, and c. However, in the aqueous state, expressions for C_P^0 and V^0 as a function of temperature – and pressure – are more complex due to the interactions of aqueous species with the solvent. The revised Helgeson–Kirkham–Flowers (HKF) equations of state are commonly used to calculate the standard-state thermodynamics properties of aqueous species at temperatures and pressures other than 25°C and 1 bar (see 79–85). The requisite thermodynamic data and equation-of-state parameters for using this model for thousands of aqueous species have been published in the literature (see 50 for a summary of organic compounds, 86 for a summary of minerals, and 87 for inorganic aqueous species, among many others). The *SUPCRT92* software package (80) is commonly used to calculate standard-state thermodynamics properties of species as a function of temperature and pressure using the revised HKF model. In the *R* language and computing environment, *CHNOSZ* has many of the capabilities of *SUPCRT92*, as well as a variety of plotting and other functions (88: chnosz.net).

In many biological applications, the so-called biochemical standard state is used, a term that is typically not rigorously defined (see 56, 89, 90). The Interunion Commission on Biothermodynamics (91) *recommends* that the biochemical standard state should correspond to pH = 7, and the temperature should be 25°C or 37°C, the ionic strength should be set by a 0.1 M KCl solution, concentrations can be used in place of activities, and the concentrations of a group of similar species can be added together and treated as one species (i.e. [ATP] = [ATP^{4-}] + [$HATP^{3-}$] + [H_2ATP^{2-}] + [$MgATP^{2-}$] . . .). Although it is relatively straightforward to convert thermodynamic data reported in the biochemical standard state to the traditional chemical standard state for pH and ionic strength, using it is complicated by the fact that the exact version of the biochemical standard state being used is often not specified (i.e. which of the *recommendations* noted above are being followed). For instance, one of the most widely cited papers on the thermodynamics of chemical reactions in the biochemical literature (92) defines a biological standard state similar to one by the Interunion Commission on Biothermodynamics, but also reports so-called observed Gibbs energies that, in addition to fulfilling the biochemical standard state, also sets the ionic strength to 0.25 and a free Mg^{2+} concentration to 0.001 M. Another problem with the biochemical standard state is that some of the researchers who use it and want to expand it report values of standard Gibbs energies of formation from the elements, ΔG_f^0, for some species to be zero simply because they do not have thermodynamic data for them (see 90). Clearly, this violates the basis of Gibbs energies and enthalpies being based on the Gibbs energy of formation from the elements, which are 0 in their respective reference states at 25°C and 1 bar. This introduces enormous errors that cannot be corrected like pH and ionic strength can be. In addition, many of those who use the biological standard state to compute Gibbs energies of processes rarely use the reaction quotient term, *Q*, which takes into account composition of the solution, the impact of which is seen in the example calculations presented below. Furthermore, it is worth pointing out that pH 7 only defines solution neutrality ($a_{H^+} = a_{OH^-}$) at 25°C and 1 bar in pure water. Neutral pH at 37°C is 6.81, and at 150°C and 5 bars it is 5.82 ($pH = -\log a_{H^+}$). Finally, if the preceding paragraphs have failed to demystify what standard states are, simply remember that they are about purity (composition) and the elements, and that they are only part of calculating the Gibbs energy of a reaction.

The Gibbs energy of a chemical reaction, ΔG_r, is a function of temperature, pressure, and the composition of the system – not just the activities of the reactants and products of the reaction of interest, but also the concentrations of all of the other chemical species in the system:

$$\Delta G_r = \Delta G_r^0 + RT \ln Q_r, \tag{19.3}$$

where *R* represents the gas constant and *T* denotes temperature in Kelvin (see Figure 19.1). The standard-state Gibbs energy of reaction, ΔG_r^0, is simply equal to the difference in the standard-state Gibbs energies of the products and reactants in a reaction, both multiplied by their respective stoichiometric coefficients, v_i:

$$\Delta G_r^0 = \sum_i v_i \Delta G^0_{\text{products}} - \sum_i v_i \Delta G^0_{\text{reactants}}, \tag{19.4}$$

as shown in (19.2) above for reaction (19.1). The reaction quotient, Q_r, a frequently neglected term, is essential to quantifying the Gibbs energy of any reaction that an organism is catalyzing to gain energy. It is responsible for bringing compositional reality into the accurate computation of ΔG_r and is calculated as the product of the activities of the reactants and products, a_i, in a chemical reaction raised to their stoichiometric coefficients, v_i:

$$Q_r = \prod_i a_i^{v_i}. \tag{19.5}$$

Evaluating (19.3)–(19.5) requires that a mass- and charge-balanced reaction has been written to represent a process, and therefore that the identities of all of the product species are known. In some cases, this becomes difficult to ascertain since many common biological elements, such as C, N, and S, have many oxidation states and therefore many different ways to balance the transformation of these elements.

As an example, Q_r for the sulfate reduction shown in Box 19.1, reaction (19.b) is

$$Q_r = \frac{a_{HS^-} a_{H_2O}}{a_{SO_4^{2-}} a_{H_2}^4 a_{H^+}}. \tag{19.6}$$

If only standard-state values of Gibbs energies were used to calculate ΔG_r for (19.b), the resulting calculation would only be valid for activities of all of the species in this reaction being 1 (see Section 19.4 for examples). Since these activities are set by the standard state, it would be equivalent to having 1 molal concentrations of each species in an infinitely dilute solution – an impossible situation. Imagine an environment in which the concentration of all five of these species is 1 molal. At 25°C and 1 bar, this would be a pH of 0 and far beyond the solubility of H_2. Ignoring the Q term when calculating the Gibbs energies of chemical reactions is effectively ignoring physical reality. Using only standard electrical potentials (E^0) to calculate the energetics of reactions is similarly untethered from physical possibilities (see 93).

The way that nonideal conditions (i.e. nonstandard states) are quantified is to use activities in place of concentrations and fugacities instead of partial pressures of gases when evaluating the Q_r term. The concepts of fugacity and activity were developed to account for thermodynamic deviations between ideal and observed behavior (see 94). By close analogy, the ideal gas law only describes the relationship between the amount, temperature, pressure, and volume of a gas ($nRT = PV$) within certain limits of these parameters. When a gas is very concentrated, this simplistic law breaks down and requires additional terms to describe the relationships among the variables in it. Similarly, activity and fugacity account for the nonideal behavior of substances when they are under relatively high concentrations and/or exposed to temperatures and pressures that are far beyond their reference temperatures and pressures. They can be thought of as the effective thermodynamic concentrations of liquids, solids, gases, and aqueous species.

Values of activity are calculated from the concentration of a substance, C, and its activity coefficient, γ: $a = C\gamma$ (a more accurate representation of this relationship is $a = \gamma(C/C^0)$, where C^0 stands for a substance's concentration in the standard state – activity and activity coefficients do not have units). There is a vast and complex literature on how to measure and compute values of γ (e.g. see 66, 83, 85, 95), but suffice to say that they are typically calculated rather than measured for a particular set of conditions. A commonly used model to calculate activities for aqueous species that incorporates elevated temperatures, pressures, and ionic strengths is the extended version of the Debye–Hückel equation (96). Software such as *Geochemist's Workbench* (www.gwb.com) and *PHREEQC* (www.usgs.gov/software/phreeqc) is commonly used to compute activities, as well as to carry out speciation calculations.

Related to the concept of activity is that of speciation. In solution, a given chemical species, especially charged ones, is partitioned among a variety of forms. For instance, in seawater, total sulfate concentration is about 28 mM, and though sulfate exists primarily as SO_4^{2-} at 18.1 mM, the remainder of it is complexed with cations commonly found in seawater. Abundant complexes include $NaSO_4^-$ (7.6 mM), $MgSO_4^0$ (6.2 mM), $CaSO_4^0$

(0.8 mM), and KSO_4^- (0.2 mm). Any calculation of Q_r should take such speciation into account, as well as the activity coefficient of SO_4^{2-}, which in seawater at 25°C and 1 bar is 0.16. Thus, the activity of SO_4^{2-} in seawater at 25°C and 1 bar is 0.0029, nearly an order of magnitude lower than its total concentration of 0.028 M.

Analogous to activity, fugacity (*f*) is the thermodynamic equivalent of pressure, taking into account the differences between the mechanical pressure exerted by a gas (*P*) and its effective pressure: $f = P\chi$, where *f* takes on units of pressure such as bars and the fugacity coefficient, χ, is unitless (see 97).

19.4 Temperature, Pressure, and Composition Affecting *G*

The amount by which variable temperature, pressure, and composition affect the Gibbs energies of chemical reactions can vary tremendously (see 89). Although a number of studies have appeared in the literature demonstrating how different combinations of these variables impact reaction energetics (discussed below), the impact that each of these variables can have alone, and in various combinations, is demonstrated here by using the low-energy metabolism known as hydrogenotrophic acetogenesis:

$$2CO_{2(aq)} + 4H_{2(aq)} \rightarrow CH_3COO^- + H^+ + 2H_2O. \tag{19.7}$$

The curves in Figure 19.3a show the values of $\Delta G^0_{19.7}$ from 0°C to 150°C and the values of ΔG_r for the same reaction under different compositional conditions that characterize high- and low-energy states (the activities of CO_2, H_2, and CH_3COO^- and pH are taken to be 10^{-3}, 10^{-8}, and 10^{-4} and 5 for the low-energy case and $10^{-1.3}$, 10^{-3}, and $10^{-8.5}$ and 9 for the high-energy scenario, respectively). For reference, the diamond shape in Figure 19.3a shows the Gibbs energy of (19.7) under conditions corresponding to the biological standard state, $\Delta G_r^{0'}$, which is more exergonic than even the high-energy scenario, so perhaps it is not very relevant to natural systems. Although the values of ΔG_r^0, the traditional chemical standard-state Gibbs energy, fall between those of the high- and low-energy scenarios, they are much closer to the high-energy state. Notably, ΔG_r for the low-energy case is endergonic throughout the temperature range considered here. Therefore, if one were to use standard-state values of Gibbs energies to quantify how much energy acetogens are gaining by catalyzing (19.7) – at any temperature – when the chemical composition of the system was that specified for the low-energy case, then the direction of the reaction would be wrong: fermentation of acetate (the reverse of (19.7)) would be predicted rather than acetogenesis. Most natural environments could be described by compositions between the high- and low-energy scenarios described here and therefore have values of ΔG_r between the two lines representing them.

The quantitative impact of different pressures on (19.7) is shown in Figure 19.3b. Here, each curve represents ΔG_r of this reaction for activities of CO_2, H_2, and CH_3COO^- and pH equal to $10^{-1.7}$, 10^{-3}, and 10^{-6} and 7, respectively, at saturation pressure (P_{SAT} – just enough pressure to keep water liquid), 250 bars, and 2500 bars. It is clear that values of

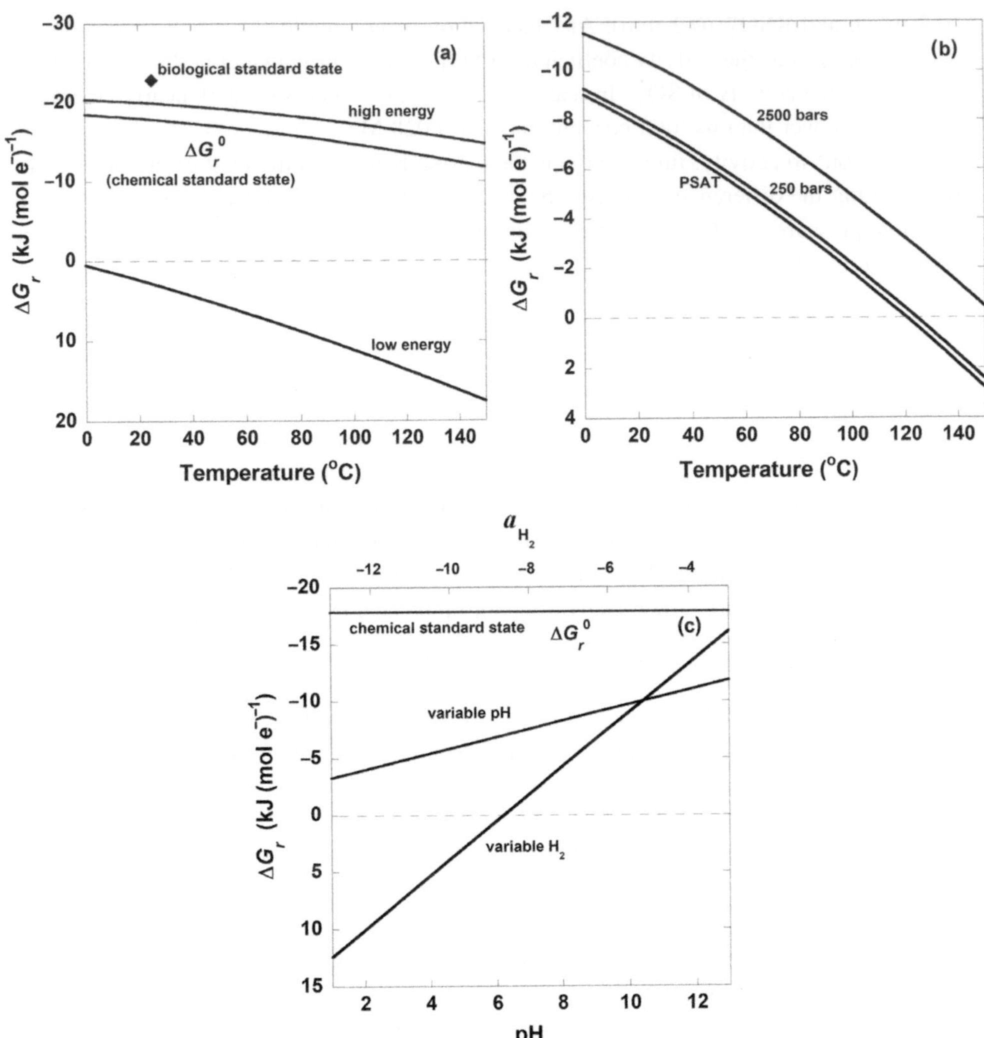

Figure 19.3 Gibbs energies of hydrogenotrophic acetogenesis, $2CO_{2(aq)} + 4H_{2(aq)} \rightarrow CH_3COO^- + H^+ + 2H_2O$, as a function of (a) temperature for the biological and traditional chemical standard states as well as low- and high-energy scenarios, (b) temperature and pressure, and (c) pH and activity of hydrogen, a_{H_2}.

ΔG_r do not differ much between P_{SAT} and 250 bars, and that the difference between P_{SAT} and 2500 bars is just over 2 kJ (mol e$^-$)$^{-1}$. This is a general feature of the quantitative impact of pressure on the Gibbs energies of catabolic reactions. It should be noted that this only takes into account pressure differences on ΔG_r^0 for (19.7) and that pressure can have large effects on the solubility of gases such as CO_2 and H_2, which can lead to higher activities of these compounds at the high pressures often found in the subsurface. These pressure effects on activities would be represented in the Q_r term.

For many microbial processes, the most important variable effecting values of ΔG_r is the Q_r term, or, in other words, the composition of the system. This can clearly be seen in Figure 19.3c, where ΔG_r for (19.7) is shown as a function of pH and H_2 activity, while temperature, pressure, and the activities of the other constituents of (19.7) are held constant (25°C, 1 bar, activities of CO_2 and CH_3COO^- are $10^{-1.7}$ and 10^{-6}; for variable pH, a_{H_2}= 10^{-6} and for variable a_{H_2}, pH = 7). For reference, ΔG_r^0 for this reaction is shown at 25°C and 1 bar and is a flat line at –17 kJ (mol e$^-$)$^{-1}$. Note that both ranges of a_{H_2} and pH result in much less exergonic values of ΔG_r than would be calculated by using standard-state values. What is apparent from the stoichiometry of (19.7) and Figure 19.3c is that for every order of magnitude change in a_{H_2}, there is a much larger change in ΔG_r than for each integer change in pH. In fact, the acetogenesis pathway represented by (19.7) is no longer thermodynamically favored at pH 7 once a_{H_2} falls below about $10^{-8.5}$. As a sidenote, the steep dependence of ΔG_r on pH shown in Figure 19.3c is typically even more exaggerated in reactions describing the oxidation and reduction of iron, with added variability depending on which iron minerals are involved in the reaction (see 98).

19.5 Surveying Gibbs Energies in Natural Systems

Because microorganisms have evolved the ability to catalyze a wide variety of redox reactions to gain energy under a broad set of physiochemical conditions, there are usually multiple catabolic strategies capable of sustaining a given ecosystem, particularly in subsurface environments that light does not reach. As a result, many of the studies that have quantified the Gibbs energies or chemical affinities (see Box 19.1) of plausible energy-sustaining reactions in subsurface settings have examined a large set of potential catabolic reactions: the thermodynamic potentials of hundreds of chemical reactions have been reported for submarine hydrothermal systems (99–109), shallow-sea hydrothermal systems (110–117), terrestrial hydrothermal systems (118–124), marine sediments (125–132), the terrestrial subsurface (133–135), serpentinizing systems (136), specific-element systems such as arsenic (137), the ocean basement (138–141), and even extraterrestrial settings (142–145). The majority of these studies focus on chemolithotrophic catabolic strategies because concentrations of specific organic compounds are not commonly reported (for exceptions, see 112, 146–148). In fact, the pairs of electron donors and acceptors that are considered are typically those whose concentrations have been measured, which tends not only to narrow the list of metabolisms considered, but also to establish a somewhat consistent set of metabolic redox pairs that are evaluated for their catabolic potential.

The Gibbs energies reported in the studies mentioned above range from endergonic ($\Delta G_r > 0$) to nearly –160 kJ (mol e$^-$)$^{-1}$. The amount of energy available from a given reaction varies considerably between study sites and even within them. For example, Shock et al. (120) calculated values of ΔG_r for ~300 reactions using temperature, pressure, and compositional data from dozens of hot springs in Yellowstone National Park, USA. Their results showed that the Gibbs energies for nearly 15% of the reactions that they considered could be exergonic or endergonic, mostly depending on the composition of the individual hot spring fluids. In a similar study, Lu (149) examined the thermodynamic potential of 740 potential catabolic reactions along a two-dimensional transect in sediments next to a shallow-sea hydrothermal vent and found that 559 are exergonic in at least one location.

Although the large number of reactions considered in these and related studies can seem overwhelming, there are a handful of salient points that can be distilled from them beyond the obvious one that many different metabolisms are possible in any given ecosystem. One observation is that the thermodynamic favorability of reactions involving common powerful oxidants such as O_2 and NO_3^- are not necessarily the most exergonic reactions in some environments: the oxidation of Mn^{2+} by O_2, for instance, can easily be endergonic, while the oxidation of CO by NO_2^- (to CO_2 and N_2) can yield more energy than any reaction involving O_2 (149). Composition tends to have a much bigger impact on the energetics of metabolic reactions than either temperature or pressure. The pH of a fluid is responsible for large differences in the values of ΔG_r for the same reaction, especially for those involving iron (120, 127, 135). The wide range of Gibbs energies available from a given electron acceptor shows that the identity of it is not sufficient to predict the order of electron acceptors used by microorganisms (150), a well-known hypothesis in marine science that asserts that organisms in sediments use terminal electron acceptors in the order of their energetic potential (151–153) – an idea that is ultimately based on standard-state Gibbs energies of reaction and restricted to the oxidation of fermentation products. Finally, it is worth pointing out that quantifying the potential energetic landscape in difficult-to-access subsurface environments can be used to predict where novel catabolic reactions could be happening. This is similar to how both the anaerobic oxidation of methane by sulfate and anaerobic ammonium oxidation were predicted based on the thermodynamic calculations demonstrating the Gibbs energies of these reactions (154, 155).

As noted above, Gibbs energy calculations aimed at revealing metabolic potential and involving organic carbon are relatively rare, or are restricted to a small number of compounds such as acetate and other volatile organic acids (e.g. 112, 123, 125, 130, 135, 147, 148). This is certainly an improvement over the common treatment of representing organic carbon as "CH_2O" and thus having a fixed energetic potential that is typically tied to the Gibbs energy of glucose. This is a convenient way of not dealing with the complexity of naturally occurring organic matter – a vast mixture of compounds that vary in their size, charge, oxidation state, structure, and composition. As an extreme example, it is worth noting that Kevlar (polyparaphenylene terephthalamide) and ethanol are both forms of organic matter, yet their properties are wildly divergent.

Although identifying organic compounds in the subsurface is difficult, characterization techniques are improving (156–159), and the thermodynamic properties of thousands of organic compounds have been reported in the literature (see 50 for a review, and, for more recent studies, see 90, 146, 160). Even if organic compounds cannot be identified, if the average NOSC in them can be, then the standard-state Gibbs energy of oxidation can be estimated (50); the range of ΔG_r^0 for the half-oxidation reactions for organic compounds varies by at least 23 kJ (mol e$^-$)$^{-1}$ (50). The reactivity of organic carbon in the subsurface would certainly be better understood if the thermodynamic properties of more organic compounds were known.

19.6 Energy Density

Although a thermodynamic calculation can be useful for determining whether a certain reaction is favored to occur under particular environmental conditions, it does not reveal any information about whether the reaction occurs or at what rates, and therefore how many organisms could be supported by it. Ideally, concentration gradients, rate measurements, and/or modeling tools should be used in conjunction with Gibbs energy calculations (e.g. 106, 134, 161–167) to quantify which reactions are driving the biogeochemistry of a system. However, even in the absence of kinetic information, Gibbs energies of reaction can be scaled to units that reveal information about ecosystems that would be otherwise obscured.

Many of the studies noted in Section 19.5 report the Gibbs energies or affinities of potential catabolic reactions in units of kJ mol^{-1} or kJ (mol e$^-$)$^{-1}$. The latter provides a common basis on which to compare the potential of many redox reactions, though this tends to leave out disproportionation and comproportionation reactions. However, both energy-per-mole units can give a very misleading view of what are the most energy-yielding reactions in an environment. Reactions involving oxygen as the oxidant are especially relevant. For example, ΔG_r for the oxidation of glucose to CO_2 by O_2 is –117 kJ (mol e$^-$)$^{-1}$ (for $\log a_{O_2} = -4$, $\log a_{CO_2} = -1.7$, $\log a_{glucose} = -4$, and $\log a_{H_2O} = 0$ at 25°C and 1 bar). Keeping the activities of everything else constant and lowering the activity of O_2 by ten orders of magnitude ($\log a_{O_2} = -14$) results in $\Delta G_r =$ –103 kJ (mol e$^-$)$^{-1}$, only a 12% change ($\log a_{O_2}$ would have to be lowered to –86.4 to make this reaction endergonic; for reference, this would be equivalent to about 1000 molecules of O_2 in a volume of water equivalent to the volume of the Milky Way). Even if a concentration of O_2 corresponding to $\log a_{O_2} = -14$ could be measured, any environment characterized with this amount of oxygen would be considered anoxic. Yet, a straightforward thermodynamic calculation shows that it is very exergonic, implying that it should be a likely metabolic activity in the environment. If one presents the results of the same calculations in terms of energy densities (e.g. 103, 127), then about 0.01 kJ (kg H_2O)$^{-1}$ would be available in the high-O_2 case and 10^{-12} J (kg H_2O)$^{-1}$ in the latter, if all of the O_2 could be used instantaneously (typically, energy densities are calculated by multiplying the concentration of the limiting reactant in a reaction by its Gibbs energy, taking into account the stoichiometry of the reactants).

On the low end of the energy spectrum, ΔG_r of glucose oxidation by sulfate, for example, is –21 kJ (mol e$^-$)$^{-1}$ (for $\log a_{HS^-} = -5$, pH = 8, $\log a_{CO_2} = -1.7$, $\log a_{glucose} = -4$, $\log a_{H_2O} = 0$, and $\log a_{SO_4^{2-}} = -1.7$ at 25°C and 1 bar), which is about one-fifth of ΔG_r for the analogous reaction with O_2 as the oxidant – when the activity of O_2

is 10^{-14}. The energy density of the sulfate reaction is nearly 0.6 kJ (kg H_2O)$^{-1}$, more than even the high-O_2 activity calculations. Interpreting these results is not straightforward: on a molar or per-electron basis, O_2 yields far more energy than sulfate, no matter what the activity of O_2 is, but the energy density of sulfate plus glucose is greater than that of the analogous high-O_2 scenario. One would not expect sulfate reduction to be a prominent glucose oxidation pathway when oxygen is abundant, but it certainly makes sense that sulfate reduction is a more dominant glucose oxidation pathway when oxygen concentrations are below what is currently measurable. Clearly, some combination of common sense and other data will be useful for interpreting the meaning of thermodynamic calculations of catabolic potential, regardless of what units are used.

The units that have been reported for the Gibbs energy calculations summarized above tend to correlate with the environmental system being examined. For example, values of ΔG_r in deep-sea hydrothermal systems tend to be reported in units of J (kg H_2O)$^{-1}$ because these calculations are based on the *in silico* mixing of different masses of fluids that drastically differ in temperature and composition. With different ratios of hydrothermal fluid to seawater and highly variable concentrations of electron donors and acceptors in hydrothermal fluid, the Gibbs energies available for a given reaction from mixing fluids can vary by many orders of magnitude in these units. Typically, the most exergonic reactions for a particular ratio of hydrothermal fluid to seawater are the ones that are reported in these studies: H_2S, CH_4, H_2, and Fe^{2+} oxidation can provide more than 1000 J/kg H_2O (99), and H_2 oxidation can provide up to 3700 J /kg H_2O when hydrothermal fluids from ultramafic systems mix with seawater (100). However, under unfavorable mixing ratios, the amount of energy available from some potential catabolic reactions, such as Fe^{2+} and H_2S oxidation, can be four to eight orders of magnitude lower than the most optimal conditions (104).

The practice of reporting both molal Gibbs energies of potential catabolic reactions and energy densities for some systems (e.g. shallow-sea hydrothermal, marine sediment, and terrestrial systems) is growing (e.g. 110, 126, 127, 133, 149). These studies all show very different results when Gibbs energies are reported in both molal and density units. For instance, LaRowe and Amend (127) compared the Gibbs energies of 18 reactions in three marine sedimentary environments characterized by different physiochemical conditions, varying mostly in composition. They showed that when values of ΔG_r were normalized by the concentration of the limiting reactant (i.e. Gibbs energies were presented as energy densities), the order of the most energy-rich reactions changed considerably and the energy available from different reactions varied by about six orders of magnitude per cm^3 of sediment. Furthermore, they showed that trends in cell abundance as a function of depth do not follow the most exergonic reaction – per mole of substrate – but by those with the highest energy densities.

A global overview of Gibbs energy densities of chemolithotrophic metabolisms in terrestrial hot springs, shallow-sea hydrothermal systems (<200 m water depth), and deep-sea hydrothermal systems is shown in Figure 19.4. The Gibbs energies of potential catabolic reactions consisting of different combinations of 19 electron acceptors and

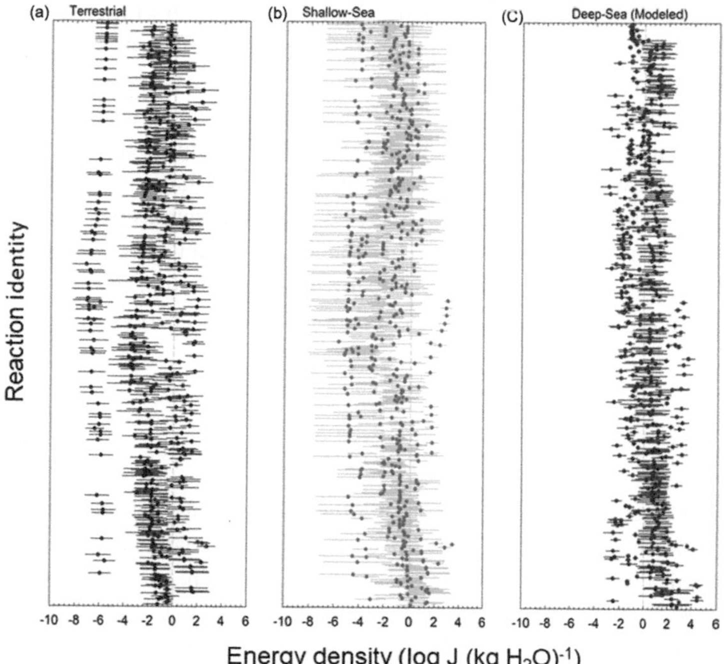

Figure 19.4 Global overview of Gibbs energy densities of chemolithotrophic metabolisms in (a) terrestrial hot springs, (b) shallow-sea hydrothermal systems (<200 m water depth), and (c) deep-sea hydrothermal systems. The Gibbs energies of potential catabolic reactions consisting of different combination of 19 electron acceptors and 14 electron donors were evaluated for 326 data sets describing the geochemistry of 30 distinct systems. The horizontal bars represent the ranges of energy densities for a given reaction and the dots refer to the average energy density of that reaction. Of the 740 reactions considered, 571 are exergonic at one or more sites. The reactions are ordered from the most exergonic to the least based on the Gibbs energies per electron transferred (not shown). Because the compositions of deep-sea hydrothermal systems are often reported as those of calculated end-member hydrothermal fluids, which are typically too hot for life, the results shown in (c) were generated by computing the energy densities of this end-member hydrothermal fluid mixed with enough seawater such that the resulting fluid was 72°C. See (149) for details.

14 electron donors were evaluated for 326 data sets describing the geochemistry of 30 distinct systems. Of the 740 reactions considered, 571 are exergonic at one or more sites. The reactions are ordered from the most exergonic to the least based on the Gibbs energies per electron transferred. Because the compositions of deep-sea hydrothermal systems are often reported as those of calculated end-member hydrothermal fluids, which are typically too hot for life, the results shown in Figure 19.4c were generated by computing the energy densities of this end-member hydrothermal fluid mixed with enough seawater such that the resulting fluid was 72°C, equal to the mean temperature for the terrestrial hot springs and shallow-sea hydrothermal systems represented in the other two

panels in Figure 19.4 (see 149 for details). It can be seen that for many of these reactions, the amount of energy available from a particular combination of electron donors and acceptors varies by many orders of magnitude, reflecting the compositional diversity of global hydrothermal systems. Furthermore, these energy densities span nearly 12 orders of magnitude. The reactions in deep-sea systems tend to have higher energy densities and the reactions in shallow-sea systems tend to show broader ranges. Although there is a sigmoidal pattern for ordering the Gibbs energies of these reactions per electron transferred (see 149), there is no such pattern here. This is because when Gibbs energies are presented in energy density units, directly accounting for the concentration of the limiting electron donor or acceptor, the order of which reaction is most energy yielding can change dramatically.

19.7 Time

Time plays a number of roles in determining the energy limits for life. On a fundamental level, active organisms must be able to catalyze redox reactions faster than they are catalyzed abiotically if they are to gain energy from them. This observation has been more colorfully expressed as "things that burst into flame are not good to eat" (168). Iron oxidation is one such potential metabolism. Although the oxidation of Fe^{2+} with O_2 is a very exergonic reaction under most environmental conditions, the abiotic rate of this process is so fast under certain combinations of pH and temperature that organisms cannot take advantage of the disequilibria. This has been shown to be the case in samples taken from lakes and springs in Switzerland (no biological catalysis for pH $>$7.4) and hot springs in Yellowstone National Park (no biological catalysis for pH > 4.0, at elevated temperatures) (169). Similarly, sulfide oxidation can proceed so quickly in hyperthermophilic settings that isolates capable of catalyzing this reaction, such as *Thermocrinis ruber*, cannot gain energy from it (170).

Microorganisms that are not competing with the abiotic catalysis of potential catabolic reactions still must use energy to combat the slow, abiotic decay of biomolecules such as the depurination of nucleic acids and the racemization of amino acids (18, 20), reactions that accelerate as temperature increases (64, 171). These most basic functions are part of what is known as basal maintenance functions, the absolute minimum flux of energy required to keep a cell viable. Because the rates of biomolecular repair likely approach that at which they are needed, it is difficult to determine the lowest amount of power on which a microorganism can survive. However, marine sediments, with their potential to record environmental data over geologic timescales, can serve as natural laboratories to constrain the basal power limits for life. Geochemical data, cell counts, and modeling efforts have been combined to estimate that microorganisms living in sediments under the South Pacific Gyre are metabolizing organic carbon with oxygen mostly at 50–3350 zW (41). In the same study, the authors hypothesize that an organism could survive on as little at 1 zW.

It is worth clarifying that the lower power limits for life should not be confused with mechanistic arguments that have been made regarding microbial activity and energy minima (see 172). Essentially, it has been proposed that microorganisms require a minimum amount of Gibbs energy from a catabolic reaction (75, 173–175) that is significantly less than $\Delta G_r < 0$. Models that make use of this essentially stipulate that once the Gibbs energy of a particular catabolic reaction dips beneath a given minimum, microorganisms can no longer catalyze the reaction to gain energy (see 78 for an overview). A lower microbial power limit, on the other hand, is simply a flux of energy large enough to keep the cell viable.

19.8 The Cost of Anabolism

As noted above, it is difficult to decipher how microorganisms partition the energy that they get from the environment. However, no matter what state microorganisms are in – growing, maintaining, or some low-level form of dormancy – they must make and/or repair biomolecules to exist for extended periods of time. The identities and quantities of these biomolecules are not well known, nor is the rate at which they must be replaced. What we can determine is the amount of energy associated with the reactions that describe biomolecule synthesis under the temperature, pressure, and compositional conditions prevailing in the deep biosphere. Efforts to do this fall into two groups: (1) those that have sought to determine whether particular organic compounds and biomolecules are thermodynamically favored to be abiotically synthesized under, typically, hydrothermal conditions; and (2) those that quantify the amount of energy required to build the biomolecules that constitute a cell under a variety of environmental conditions (for a review, see 176). Falling into the first group, calculations have shown that the abiotic formation of alkanes, alkenes, alcohols, ketones, aldehydes, carboxylic acids, amino acids, nucleic acid bases, and monosaccharides can be favored to form when particular hydrothermal fluids mix with seawater and kinetic barriers are assumed to prevent carbon transformations to CO_2 or CH_4 (177–181).

The second group of studies focused on the energetics of organic synthesis and quantified the amount of energy that is required to build most of the biomolecules that constitute a cell in the proportions that exist in a model microorganism, *Escherichia coli* (182). McCollom and Amend (48) calculated the Gibbs energy required to make all of the biomonomers that make a cell, starting from inorganic precursor molecules (HCO_3^- (NH_4^+ and NO_3^-), HPO_4^{2-} (H_2S and SO_4^{2-})) under microoxic and anoxic oxidation states. The cost of making biomonomers was 13–15 times higher in the microoxic environment than the anoxic one, depending on the sources of N and S. In an extension of this study, LaRowe and Amend (49) assessed the amount of energy associated with making the same set of biomolecules, but also quantified the role of pressure and used a wider range of redox conditions, temperatures, sources of C, N, and S, and cell sizes. Taken together, environmental variables – and the range of cell sizes – lead to an approximately four orders of magnitude difference between the number of microbial cells that can be made from a Joule of Gibbs energy under the most (5×10^{11} cells J^{-1}) and least (5×10^{7} cells J^{-1}) ideal conditions.

Finally, a few authors have published calculations in which they have quantified the amount of energy that it takes to synthesize microbial biomass due to the disequilibria resulting from mixing seawater with hydrothermal fluids of variable composition. Amend and McCollom (183) focused on the early Earth, while Amend et al. (99) analyzed how hydrothermal fluids whose compositions were influenced by different water–rock interactions impacted the energetics of biomass synthesis. They showed that the most energetically favorable temperatures were between 22°C and 32°C in general, and that hydrothermal fluids that interacted with peridotite and troctolite–basalt hybrid rock systems created the most favorable energetic conditions for producing biomass, *yielding* up to 900 J per g dry cell mass.

19.9 Concluding Remarks

What combination of natural variables defines the limits to life? The discovery of abundant life in the subsurface has expanded this search to: How deep life can exist? How long can it maintain itself under nearly isolated conditions? And what is the minimum power that it can get by on? One way to address these questions is to determine where and when microorganisms no longer get energy as fast as they need it. In the deep subsurface, this threshold will likely be defined by the power required to repair biomolecules. Essential biomolecules such as nucleotides, amino acids, saccharides, and lipids all undergo abiotic decay through a variety of reactions, including racemization, methylation, deamination, isomerization, radiation exposure, and hydrolysis, while biomacromolecules like DNA, RNA, and proteins can become cross-linked and/or unfolded. The rates of all of these processes depend on the environmental context, particularly temperature. Repairing biomolecules requires energy, and the acquisition of energy requires the maintenance of membrane potentials and the ability to transport nutrients and energy substrates across membranes. Although the energetics and rates of amino acid racemization and DNA depurination have received a fair amount of attention (18), it is not known how all of these degradative processes are influenced by the various combinations of physiochemical and temporal extremes that often define subsurface settings. Identifying the biotic fringe in the subsurface will require quantification of how temperature, pressure, and chemical composition impact the rates of biomolecular decay and how much power is available to rectify this decomposition.

Although we do not always have the information needed to evaluate the amount of catabolic power that is used deep below the surface, we often have the requisite information to quantify the Gibbs energy that is available for microbial energy demands as well as the cost of making biomolecules *in situ*. In this chapter, we have reviewed how temperature, pressure, and composition affect the Gibbs energies of chemical reactions associated with staying alive in the deep biosphere and how to accurately quantify it. We have belabored the point that the traditional chemical standard state refers to the composition of a system and not temperature and pressure (except for gases), and that the chemical composition of an environment is typically much more significant for calculating the Gibbs energy of a catabolic or anabolic reaction than temperature and pressure. The examples provided illustrate that there is no cheap shortcut to quantifying how much energy is associated with a given chemical reaction under specified physiochemical conditions and that care must be taken to interpret the results of these calculations: Gibbs energies are potentials that can be used to quantify the direction of natural processes and are therefore useful for constraining what life is possible of doing in the subsurface or in any environment.

Acknowledgments

We appreciate the comments provided by anonymous reviewers. This work was supported by the NSF-sponsored Center for Dark Energy Biosphere Investigations (C-DEBI) under grant OCE0939564; the NASA Astrobiology Institute – Life Underground (NAI-LU) grant NNA13AA92A; the USC Zumberge Fund Individual Grant; the NASA-NSF Origins of Life Ideas Lab program under grant NNN13D466T; and the Alfred P. Sloan Foundation through the Deep Carbon Observatory. This is C-DEBI contribution 453 and NAI-LU contribution 134.

Questions for the Classroom

1 Why does the usage of Gibbs energies and other thermodynamic properties require standard states?
2 What are the limitations of the biochemical standard state?
3 What are the factors influencing the amount of energy an organism gains by catalyzing redox reactions?
4 What is the advantage of using chemical affinities rather than overall Gibbs energies of reactions to quantify the tendency of a chemical reaction to proceed?
5 What is the utility of reporting the results of Gibbs energy calculations in density rather than molar units? What are the shortcomings?
6 How does the redox state of an environment influence the amount of energy that is required to make biomass?
7 Why does one need to know the identities and concentrations of products in a catabolic reaction to quantify the amount of energy an organism gains from it?
8 Use the *CHNOSZ* software package (chnosz.net) to calculate how different fugacities (partial pressures) of CO_2 impact the Gibbs energies of acetoclastic and hydrogenotrophic methanogenesis. Hint – see the vignette on chemical affinity on the *CHNOSZ* website.
9 Calculate how different fugacities of H_2 impact the Gibbs energies of hydrogenotrophic methanogenesis.

References

1. Regnier P, Friedlingstein P, Ciais P, Mackenzie FT, Gruber N, Janssens IA, et al. Anthropogenic perturbation of carbon fluxes from land to ocean. *Nat Geosci.* 2013;6:597–607.
2. Mackenzie FT, Lerman A, Andersson AJ. Past and present of sediment and carbon biogeochemical cycling models. *Biogeosciences.* 2004;1:11–32.
3. Whitman WB, Coleman DC, Wiebe WJ. Prokaryotes: the unseen majority. *Proc Natl Acad Sci.* 1998;95(12):6578–6583.
4. Parkes RJ, Cragg B, Roussel E, Webster G, Weightman A, Sass H. A review of prokaryotic populations and processes in sub-seafloor sediments, including biosphere:geosphere interactions. *Mar Geol.* 2014;352:409–425.
5. Kallmeyer J, Pockalny R, Adhikari RR, Smith DC, D'Hondt S. Global distribution of microbial abundance and biomass in subseafloor sediment. *Proc Natl Acad Sci.* 2012;109:16213–16216.
6. Edwards KJ, Becker K, Colwell F. The deep, dark energy biosphere: intraterrestrial life on Earth. *Annu Rev Earth Planet Sci.* 2012;40:551–568.
7. Edwards KJ, Wheat G, Sylvan JB. Under the sea: microbial life in volcanic oceanic crust. *Nat Rev Microbiol.* 2011;9:703–712.
8. Orcutt BN, Sylvan JB, Knab NJ, Edwards KJ. Microbial ecology of the dark ocean above, at, and below the seafloor. *Microbiol Molec Biol Rev.* 2011;75:361–422.
9. McMahon S, Parnell J. Weighing the deep continental biosphere. *FEMS Microbiol Ecol.* 2014;87:113–120.
10. Bar-on Y, Phillips R, Milo R. The biomass distribution on Earth. *Proc Natl Acad Sci.* 2018;115:6506–6511.
11. Chapelle FH, Zelibor Jr. JL, Jay Grimes D, Knobel LL. Bacteria in deep coastal plain sediments of Maryland: a possible source of CO_2 to groundwater. *Water Resources Res.* 1987;23:1625–1632.
12. Pedersen K, Ekendahl S. Distribution and activity of bacteria in deep granitic groundwaters of Southeastern Sweden. *Microb Ecol.* 1990;20:37–52.
13. Stevens TO, McKinley JP. Lithoautotrophic microbial ecosystems in deep basalt aquifers. *Science* 1995;270:450–455.
14. Parkes RJ, Cragg BA, Bale SJ, Getliff JM, Goodman K, Rochelle PA, et al. Deep bacterial biosphere in Pacific Ocean sediments. *Nature.* 1994;371:410–413.
15. Gold T. The deep, hot biosphere. *Proc Natl Acad Sci.* 1992;89(13):6045–6049.
16. Balkwill DL. Numbers, diversity, and morphological charateristics of aerobic, chemoheterotrophic bacteria in deep subsurface sediments from a site in South Carolina. *Geomicrobiol J.* 1989;7:33–52.
17. Hoehler TM, Jørgensen BB. Micorbial life under extreme energy limitation. *Nat Rev Microbiol.* 2013;11:83–94.
18. Lever MA, Rogers KL, Lloyd KG, Overmann J, Schink B, Thauer RK, et al. Life under extreme energy limitation: a synthesis of laboratory- and field-based investigations. *FEMS Microbiol Rev.* 2015;39:688–728.
19. D'Hondt S, Inagaki F, Zarikian CA, Abrams LJ, Dubois N, Engelhardt T, et al. Presence of oxygen and aerobic communities from sea floor to basement in deep-sea sediments. *Nat Geosci.* 2015;8:299–304.
20. Jørgensen BB, Marshall IPG. Slow microbial life in the seabed. *Annu Rev Marine Sci.* 2016;8:311–332.
21. LaRowe DE, Burwicz EB, Arndt S, Dale AW, Amend JP. The temperature and volume of global marine sediments. *Geology.* 2017;45:275–278.

22. Magnabosco C, Lin L-H, Dong H, Bomberg M, Ghiorse W, Stan-Lotter H, et al. The biomass and biodiversity of the continental subsurface. *Nat Geosci.* 2018;11:707–717.
23. Cogley JG. Continental margins and the extent and number of the continents. *Rev Geophys Space Phys.* 1984;22:101–122.
24. Anderson BW, Zoback M, Hickman S, Newmark R. Permeability versus depth in the upper oceanic crust: In situ measurements in DSDP Hole 504B, eastern equatorial Pacific. *J Geophys Res.* 1985;90:3659–3669.
25. Becker K. Measurements of the permeability of the sheeted dikes in Hole 504B, ODP Leg 111. In: Becker K et al. eds. *Proceedings of the Ocean Drilling Program Science Results.* College Station, TX: Ocean Drilling Program, 1989, pp. 317–325.
26. Becker K, Langseth M, Von Herzen RP, Anderson R. Deep crustal geothermal measurements, Hole 504B, Costa Rica Rift. *J Geophys Res.* 1983;88:3447–3457.
27. Hickman SH, Langseth M, Svitek. *In situ* permeability and pore-pressure measurements near the Mid-Atalntic Ridge, Deep Sea Drilling Project Hole 395A. In: Hyndman RD, Salisbury MH, eds. *Initial Reports of the Deep Sea Drilling Project.* Washington, DC: US Goverment Printing Office, 1984, pp. 699–708.
28. Becker K. Measurements of the permeability of the upper oceanic crust at Hole 395A, ODP Leg 109. In: Detrick R, Honnorez J, Bryan WB, Juteau T, eds. *Proceedings of the Ocean Drilling Program Science Results.* College Station, TX: Ocean Drilling Program, 1990, pp. 213–222.
29. Winslow DM, Fisher AT, Becker K. Characterizing borehole fluid flow and formation permeability in the ocean crust using linked analytic models and Markov Chain Monte Carlo analysis. *Geochem Geophys Geosyst.* 2013;14:3857–3874.
30. Alt JC. Alteration of the upper oceanic crust: mineralogy, chemistry, and processes. In: Davis EE, Elderfield H, eds. *Hydrogeology of the Oceanic Lithosphere.* Cambridge: Cambridge University Press, 2004, pp. 456–488.
31. Teagle DAH, Wilson DS. Leg 206 synthesis: initiation of drilling an intact section of upper oceanic crust formed at a superfast spreading rate at Site 1256 in the eastern equatorial Pacific. In: Wilson DS, Teagle DAH, Acton GD, Vanko DA, eds. *Proceedings of the Ocean Drilling Program, Initial Reports.* College Station, TX, USA: Ocean Drilling Program, 2007, pp. 1–15.
32. Fisher AT, Alt JC, Bach W. Hydrogeologic properties, processes and alteration in the igneous ocean crust. In: Stein R, Blackman D, Inagaki F, Larsen H-C, eds. *Earth and Life Processes Discovered from Subseafloor Environment – A Decade of Science Achieved by the Integrated Ocean Drilling Program (IODP).* Amsterdam/New York: Elsevier, 2014, pp. 507–551.
33. Jakosky B, Shock EL. The biological potential of Mars, the early Earth, and Europa. *J Geophys Res Lett Planets.* 1998;103(E8):19359–19364.
34. Michalski JR, Onstott TC, Mojzsis SJ, Mustard J, Chan QHS, Niles PB, et al. The Martian subsurface as a potential window into the origin of life. *Nat Geosci.* 2018;11:21–26.
35. Vance S, Harnmeijer J, Kimura J, Hussmann H, Demartin B, Brown JM. Hydrothermal systems in small ocean planets. *Astrobiology.* 2007;7:987–1005.
36. Bradley JA, Amend JP, LaRowe DE. Survival of the fewest: microbial dormancy and maintenance in marine sediments through deep time. *Geobiology.* 2019;17:43–59.
37. DePaolo DJ. Sustainable carbon emissions: the geologic perspective. *MRS Energy Sustain.* 2015;2:E9.
38. van Bodegom P. Micorbial maintenance: a critical review of its quantification. *Microb Ecol.* 2007;5:513–523.

39. LaRowe DE, Amend JP. Catabolic rates, population sizes and doubling/replacement times of microorganisms in the natural settings. *Am J Sci.* 2015;315:167–203.
40. Jørgensen BB. Shrinking majority of the deep biosphere. *Proc Natl Acad Sci.* 2012;109:15976–15977.
41. LaRowe DE, Amend JP. Power limits for microbial life. *Front Extr Microbiol.* 2015;6:718.
42. Shock EL, Holland ME. Quantitative habitability. *Astrobiology.* 2007;7:839–851.
43. Morita RY. *Bacteria in Oligotrophic Environments: Starvation–Survival Lifestyle.* New York: Chapman & Hall, 1997.
44. D'Hondt S, Spivack AJ, Pockalny R, Ferdelman TG, Fischer JP, Kallmeyer J, et al. Subseafloor sedimentary life in the South Pacific Gyre. *Proc Natl Acad Sci.* 2009;106:11651–11656.
45. Jørgensen BB, Boetius A. Feast and famine – microbial life in the deep-sea bed. *Nat Rev Microbiol.* 2007;5:770–781.
46. Røy H, Kallmeyer J, Adhikari RR, Pockalny R, Jørgensen BB, D'Hondt S. Aerobic microbial respiration in 86-million-year-old deep-sea red clay. *Science.* 2012;336:922–925.
47. Morono Y, Terada T, Nishizawa M, Ito M, Hillion F, Takahata N, et al. Carbon and nitrogen assimilation in deep subseafloor micorbial cells. *Proc Natl Acad Sci.* 2011;108:18295–18300.
48. McCollom TM, Amend JP. A thermodynamic assessment of energy requirements for biomass synthesis by chemolithoautotrophic micro-organisms in oxic and anoxic environments. *Geobiology.* 2005;3:135–144.
49. LaRowe DE, Amend JP. The energetics of anabolism in natural settings. *ISME J.* 2016;10:1285–1295.
50. LaRowe DE, Van Cappellen P. Degradation of natural organic matter: a thermodynamic analysis. *Geochim Cosmochim Acta.* 2011;75:2030–2042.
51. Heijnen JJ, van Dijken JP. In search of a thermodynamic description of biomass yields for the chemotrophic growth of microorganisms. *Biotech Bioeng.* 1992;39:833–858.
52. Russell JB, Cook GM. Energetics of bacterial growth: balance of anabolic and catabolic reactions. *Microbiol Rev.* 1995;59:48–62.
53. Stouthamer AH. A theoretical study on the amount of ATP required for synthesis of micorbial cell material. *Antonie van Leeuwenhoek.* 1973;39:545–565.
54. LaRowe DE, Helgeson HC. Biomolecules in hydrothermal systems: calculation of the standard molal thermodynamic properties of nucleic-acid bases, nucleosides, and nucleotides at elevated temperatures and pressures. *Geochim Cosmochim Acta.* 2006;70:4680–4724.
55. LaRowe DE, Helgeson HC. The energetics of metabolism in hydrothermal systems: calculation of the standard molal thermodynamic properties of magnesium-complexed adenosine nucleotides and NAD and NADP at elevated temperature and pressures. *Thermochim Acta.* 2006;448:82–106.
56. LaRowe DE, Helgeson HC. Quantifying the energetics of metabolic reactions in diverse biogeochemical systems: electron flow and ATP synthesis. *Geobiology.* 2007;5:153–168.
57. Lennon JT, Jones SE. Micorbial seed banks: the ecological and evolutionary implication of dormancy. *Nat Rev Microbiol.* 2011;9:119–130.
58. Nicholson WL, Munakata N, Horneck G, Melosh HJ, Setlow P. Resistence of *Bacillus* endospores to extreme terrestrial and extraterrestrial environments. *Microbiol Mol Biol Rev.* 2000;64:548–572.

59. Kaprelyants AS, Gottschal JC, Kell DB. Dormancy in non-sporulating bacteria. *FEMS Microbiol Rev.* 1993;104:271–286.
60. Stolpovsky K, Fetzer I, Van Cappellen P, Thullner M. Influence of dormancy on microbial competition under intermittent substrate supply: insights from model simulations. *FEMS Microbiol Ecol.* 2016;92:fiw071.
61. Stolpovsky K, Martinez-Lavanchy P, Heipieper HJ, Van Cappellen P, Thullner M. Incorporating dormancy in dynamic microbial community models. *Ecolog Model.* 2011;222:3092–3102.
62. Johnson SS, Hebsgaard MB, Christensen TR, Mastepanov M, Nielsen R, Munch K, et al. Ancient bacteria show evidence of DNA repair. *Proc Natl Acad Sci.* 2007;104:14401–14405.
63. Locey KJ. Synthesizing traditional biogeography with micorbial ecology: the importance of dormancy. *J Biogeogr.* 2010;37:1835–1841.
64. Price PB, Sowers T. Temperature dependence of metabolic rates for microbial growth, maintenance, and survival. *Proc Natl Acad Sci.* 2004;101:4631–4636.
65. Kondepudi D, Prigogine I. *Modern Thermodynamics: From Heat Engines to Dissipative Structures.* New York: John Wiley & Sons, 1998.
66. Anderson GM, Crerar DA. *Thermodynamics in Geochemistry: The Equilibrium Model.* Oxford: Oxford University Press, 1993.
67. Stumm W, Morgan JJ. *Aquatic Chemistry: Chemical Equilibria and Rates in Natural Waters.* 3rd edn. New York: John Wiley & Sons, Inc., 1996.
68. Garrels RM, Christ CL. *Solutions, Minerals, and Equilibria.* New York: Harper & Row, 1965.
69. Prigogine I, Defay R. *Chemical Thermodynamics.* London: Longmans, Green & Co., 1954.
70. de Donder T, Van Rysselberghe P. *Affinity.* Menlo Park, CA: Stanford University Press, 1936.
71. de Donder T. *Lecons de Thermodynamique et de Chimie-Physique.* Paris: Gauthiers-Villars, 1920.
72. van't Hoff JH. *Études de Dynamique Chimique.* Amsterdam: Frederik Muller & Co., 1884.
73. Aagard P, Helgeson HC. Thermodynamic and kinetic constraints on reaction rates among minerals and aqueous solutions. I. Theoretical considerations. *Amer J Sci.* 1982;282:237–285.
74. Jin Q, Bethke CM. Kinetics of electron transfer through the respiratory chain. *Biophys J.* 2002;83:1797–1808.
75. Jin Q, Bethke CM. A new rate law describing microbial respiration. *Appl Environ Microbiol.* 2003;69:2340–2348.
76. Jin Q, Bethke CM. Predicting the rate of microbial respiration in geochemical environments. *Geochim Cosmochim Acta.* 2005;69:1133–1143.
77. Jin Q, Bethke CM. The thermodynamics and kinetics of microbial metabolism. *Am J Sci.* 2007;307:643–677.
78. LaRowe DE, Dale AW, Amend JP, Van Cappellen P. Thermodynamic limitations on microbially catalyzed reaction rates. *Geochim Cosmochim Acta.* 2012;90:96–109.
79. Tanger JC, Helgeson HC. Calculation of the thermodynamic and transport properties of aqueous species at high pressures and temperatures – revised equations of state for the standard partial molal properties of ions and electrolytes. *Am J Sci.* 1988;288:19–98.
80. Johnson JW, Oelkers EH, Helgeson HC. *SUPCRT92* – a software package for calculating the standard molal thermodynamic properties of minerals, gases, aqueous

species, and reactions from 1 bar to 5000 bar and 0°C to 1000°C. *Comput Geosci.* 1992;18:899–947.

81. Shock EL, Oelkers E, Johnson J, Sverjensky D, Helgeson HC. Calculation of the thermodynamic properties of aqueous species at high pressures and temperatures – effective electrostatic radii, dissociation constants and standard partial molal properties to 1000°C and 5 kbar. *J Chem Soc Faraday Trans.* 1992;88:803–826.
82. Helgeson HC, Kirkham DH. Theoretical prediction of thermodynamic behavior of aqueous electrolytes at high pressures and temperatures: 1. Summary of thermodynamic–electrostatic properties of the solvent. *Am J Sci.* 1974;274:1089–1198.
83. Helgeson HC, Kirkham DH. Theoretical prediction of thermodynamic behavior of aqueous electrolytes at high pressures and temperatures: 2. Debye–Hückel parameters for activity coefficients and relative partial molal properties. *Am J Sci.* 1974;274:1199–1261.
84. Helgeson HC, Kirkham DH. Theoretical prediction of thermodynamic behavior of aqueous electrolytes at high pressures and temperatures: 3. Equation of state for aqueous species at infinite dilution. *Am J Sci.* 1976;276:97–240.
85. Helgeson HC, Kirkham DH, Flowers GC. Theoretical prediction of thermodynamic behavior of aqueous electrolytes at high pressures and temperatures: 4. Calculation of activity coefficients, osmotic coefficients, and apparent molal and standard and relative partial molal properties to 600°C and 5 kb. *Am J Sci.* 1981;281:1249–1516.
86. Helgeson HC, Delany JM, Nesbitt HW, Bird DK. Summary and critique of the thermodynamic properties of rock-forming minerals. *Am J Sci.* 1978;278:1–229.
87. Sverjensky D, Shock EL, Helgeson HC. Prediction of the thermodynamic properties of aqueous metal complexes to 1000°C and 5 kb. *Geochim Cosmochim Acta.* 1997;61:1359–1412.
88. Dick JM. Calculation of the relative metastabilities of proteins using the *CHNOSZ* software package. *Geochem Trans.* 2008;9:10.
89. Amend JP, Shock EL. Energetics of overall metabolic reactions of thermophilic and hyperthermophilic Archaea and Bacteria. *FEMS Microbiol Rev.* 2001;25:175–243.
90. Canovas III PA, Shock EL. Geobiochemistry of metabolism: standard state thermodynamic properties of the citric acid cycle. *Geochim Cosmochim Acta.* 2016;195:293–322.
91. Wadsö I, Gutfreund H, Privalov P, Edsall JT, Jencks WP, Armstrong GT, et al. Recommendations for measurement and presentation of biochemical equilibrium data. *J Biol Chem.* 1976;251:6879–6885.
92. Thauer RK, Jungermann K, Decker K. Energy conservation in chemotrophic anaerobic bacteria. *Bacteriol Rev.* 1977;41:100–180.
93. Amend JP, Teske A. Expanding frontiers in deep subsurface microbiology. *Palaeogeogr Palaeoclimatol Palaeoecol.* 2005;219:131–155.
94. Lewis GN, Randall M. *Thermodynamics and the Free Energy of Chemical Substances*. New York: McGraw Hill, 1923.
95. Pytkowicz RM. *Activity Coefficients in Electrolyte Solutions*. Boca Raton, FL: CRC Press, 1979.
96. Helgeson HC. Thermodynamics of hydrothermal systems at elevated temperatures and pressures. *Am J Sci.* 1969;267:729–804.
97. Appelo CAJ, Parkhurst DL, Post VEA. Equations for calculating hydrogeochemical reactions of minerals and gases such as CO2 at high pressures and temperatures. *Geochim Cosmochim Acta.* 2014;125:49–67.

98. Rowe AR, Yoshimura M, LaRowe DE, Bird LJ, Amend JP, Hashimoto K, et al. *In situ* electrochemical enrichment and isolation of a magnetite-reducing bacterium from a high pH serpentinizing spring. *Environ Microbiol.* 2017;19: 2272–2285.
99. Amend JP, McCollom TM, Hentscher M, Bach W. Catabolic and anabolic energy for chemolithoautotrophs in deep-sea hydrothermal systems hosted in different rock types. *Geochim Cosmochim Acta.* 2011;75:5736–5748.
100. McCollom TM. Geochemical constraints on sources of metabolic energy for chemolithoautotrophy in ultramafic-hosted deep-sea hydrothermal systems. *Astrobiology.* 2007;7:933–950.
101. Shock EL, Holland ME. Geochemical energy sources that support the subseafloor biosphere. The subseafloor biosphere at mid-ocean ridges. In: Wilcock WSD, DeLong EF, Kelley DS, Baross JA, Cary SC, eds. *Geophysical Monograph 144.* Washington, DC: American Geophysical Union, 2004, pp. 153–165.
102. McCollom TM, Shock EL. Geochemical constraints on chemolithoautotrophic metabolism by microorganisms in seafloor hydrothermal systems. *Geochim Cosmochim Acta.* 1997;61:4375–4391.
103. McCollom TM. Geochemical constraints on primary productivity in submarine hydrothermal vent plumes. *Deep-Sea Res Part I Oceanogr Res Pap.* 2000;47:85–101.
104. Houghton JL, Seyfried Jr. WE. An experimental and theoretical approach to determing linkages between geochemical variability and microbial biodiversity in seafloor hydrothermal chimneys. *Geobiology.* 2010;8:457–470.
105. Shock EL, McCollom TM, Schulte MD. Geochemical constraints on chemolithoautotrophic reactions in hydrothermal systems. *Orig Life Evol Biosph.* 1995;25:141–159.
106. LaRowe DE, Dale AW, Aguilera DR, L'Heureux I, Amend JP, Regnier P. Modeling microbial reaction rates in a submarine hydrothermal vent chimney wall. *Geochim Cosmochim Acta.* 2014;124:72–97.
107. Sylvan JB, Wankel SD, LaRowe DE, Charoenpong CN, Huber H, Moyer CL, et al. Evidence for micorbial mediation of subseafloor nitrogen redox processes at Loihi Seamount, Hawaii. *Geochim Cosmochim Acta.* 2017;198:131–150.
108. Reed DC, Breier JA, Jiang H, Anantharaman K, Klausmeier CA, Toner BM, et al. Predicting the response of the deep-ocean microbiome to geochemical perturbation by hydrothermal vents. *ISME J.* 2015;9:1857–1869.
109. Dahle H, Økland I, Thorseth IH, Pedersen RB, Steen IH. Energy landscapes shape micorbial communities in hydrothermal systems on the Arctic Mid-Ocean Ridge. *ISME J.* 2015;9:1593–1606.
110. Price RE, LaRowe DE, Italiano F, Savov I, Pichler T, Amend JP. Subsurface hydrothermal processes and the bioenergetics of chemolithoautotrophy at the shallow-sea vents off Panarea Island (Italy). *Chem Geol.* 2015;407–408:21–45.
111. Amend JP, Rogers KL, Shock EL, Gurrieri S, Inguaggiato S. Energetics of chemolithoautotrophy in the hydrothermal system of Vulcano Island, southern Italy. *Geobiology.* 2003;1:37–58.
112. Rogers KL, Amend JP. Energetics of potential heterotrophic metabolisms in the marine hydrothermal system of Vulcano Island, Italy. *Geochim Cosmochim Acta.* 2006;70:6180–6200.
113. Akerman NH, Price RE, Pichler T, Amend JP. Energy sources for chemolithotrophs in an arsenic- and iron-rich shallow-sea hydrothermal system. *Geobiology.* 2011;9:436–445.

114. Rogers KL, Amend JP, Gurrieri S. Temporal changes in fluid chemistry and energy profiles in the Vulcano island hydrothermal system. *Astrobiology.* 2007;7:905–932.
115. Rogers KL, Amend JP. Archaeal diversity and geochemical energy yields in a geothermal well on Vulcano Island, Italy. *Geobiology.* 2005;3:319–332.
116. Skoog A, Vlahos P, Rogers KL, Amend JP. Concentrations, distributions, and energy yields of dissolved neutral aldoses in a shallow hydrothermal vent system of Vulcano, Italy. *Org Geochem.* 2007;38:1416–1430.
117. Lu G-S, LaRowe DE, Gilhooly III WP, Druschel GK, Fike DA, Amend JP. Chemolithoautotrophic energetics in a shallow-sea hydrothermal system, Milos Island, Greece. *Manuscript in preparation.*
118. Inskeep W, Ackerman GG, Taylor WP, Kozubal M, Korf S, Macur RE. On the energetics of chemolithotrophy in nonequilibrium systems: case studies of geothermal springs in Yellowstone National Park. *Geobiology.* 2005;3:297–317.
119. Inskeep WP, McDermott TR. Geomicrobiology of acid–sulfate–chloride springs in Yellowstone National Park. In: Inskeep WP, McDermott TR, eds. *Geothermal Biology and Geochemistry in Yellowstone National Park.* Bozeman, MT: Montana State University Publications, 2005, pp. 143–162.
120. Shock EL, Holland M, Meyer-Dombard D, Amend JP, Osburn GR, Fischer TP. Quantifying inorganic sources of geochemical energy in hydrothermal ecosystems, Yellowstone National Park, USA. *Geochim Cosmochim Acta.* 2010;74:4005–4043.
121. Spear JR, Walker JJ, McCollom TM, Pace NR. Hydrogen and bioenergetics in the Yellowstone geothermal ecosystem. *Proc Natl Acad Sci.* 2005;102:2555–2560.
122. Vick TJ, Dodsworth JA, Costa KC, Shock EL, Hedlund BP. Microbiology and geochemistry of Little Hot Creek, a hot spring environment in the Long Valley Caldera. *Geobiology.* 2010;8:140–154.
123. Windman T, Zolotova N, Schwandner F, Shock EL. Formate as an energy source for microbial metabolism in chemosynthetic zones of hydrothermal ecosystems. *Astrobiology.* 2007;7:873–890.
124. Costa KC, Navarro JB, Shock EL, Zhang CL, Soukup D, Hedlund BP. Microbiology and geochemistry of great boiling and mud hot springs in the United States Great Basin. *Extremophiles.* 2009;13:447–459.
125. LaRowe DE, Dale AW, Regnier P. A thermodynamic analysis of the anaerobic oxidation of methane in marine sediments. *Geobiology.* 2008;6:436–449.
126. Teske A, Callaghan AV, LaRowe DE. Biosphere frontiers: deep life in the sedimented hydrothermal system of Guaymas Basin. *Front Extr Microbiol.* 2014;5:362.
127. LaRowe DE, Amend JP. Energetic constraints on life in marine deep sediments. In: Kallmeyer J, Wagner K, eds. *Life in Extreme Environments: Microbial Life in the Deep Biosphere.* Berlin: de Gruyter, 2014, pp. 279–302.
128. Wang G, Spivack AJ, D'Hondt S. Gibbs energies of reaction and microbial mutualism in anaerobic deep subseafloor sediments of ODP Site 1226. *Geochim Cosmochim Acta.* 2010;74:3938–3947.
129. Kiel Reese B, Zinke LA, Sobol MS, LaRowe DE, Orcutt BN, Zhang X, et al. Nitrogen cycling of active bacteria within oligotrophic sediment of the Mid-Atlantic Ridge Flank. *Geomicrobiol J.* 2018;35:468–483.
130. Glombitza C, Jaussi M, Røy H. Formate, acetate, and propionate as substrates for sulfate reduction in sub-arctic sediments of Southwest Greenland. *Front Microbiol.* 2015;6:846.
131. Beulig F, Røy H, Glombitza C, Jørgensen BB. Control on rate and pathway of anaerobic organic carbon degradation in the seabed. *Proc Natl Acad Sci.* 2018;155:367–372.

132. Schrum HN, Spivack AJ, Kastner M, D'Hondt S. Sulfate-reducing ammonium oxidation: a thermodynamically feasible metabolic pathway in subseafloor sediment. *Geology*. 2009;37:939–942.
133. Osburn MR, LaRowe DE, Momper L, Amend JP. Chemolithotrophy in the continental deep subsurface: Sanford Underground Research Facility (SURF), USA. *Front Extr Microbiol*. 2014;5:610.
134. Jin Q, Bethke CM. Cellular energy conservation and the rate of microbial sulfate reduction. *Geology*. 2009;37:1027–1030.
135. Kirk MF, Jin Q, Haller BR. Broad-scale evidence that pH influences the balance between microbial iron and sulfate reduction. *Groundwater*. 2015;54:406–413.
136. Canovas III PA, Hoehler TM, Shock EL. Geochemical bioenergetics during low-temperature serpentinization: an example from the Samail ophiolite, Sultanate of Oman. *J Geophys Res Biogeosci*. 2017;122:1821–1847.
137. Amend JP, Saltikov C, Lu G-S, Hernandez J. Micorbial arsenic metabolism and reaction energetics. *Rev Mineral Geochem*. 2014;79:391–433.
138. Edwards KJ, Bach W, McCollom TM. Geomicrobiology in oceanography: microbe–mineral interactions at and below the seafloor. *Trends Microbiol*. 2005;13:449–456.
139. Bach W, Edwards KJ. Iron and sulfide oxidation within the basaltic ocean crust: implications for chemolithoautotrophic microbial biomass production. *Geochim Cosmochim Acta*. 2003;67:3871–3887.
140. Cowen JP. The microbial biosphere of sediment-buried oceanic basement. *Res Microbiol*. 2004;155:497–506.
141. Boettger J, Lin H-T, Cowen JP, Hentscher M, Amend JP. Energy yields from chemolithotrophic metabolisms in igneous basement of the Juan de Fuca ridge flank system. 2013;337–338:11–19.
142. Shock EL. High-temperature life without photosynthesis as a model for Mars. *J Geophys Res Planets*. 1997;102:23687–23694.
143. Zolotov MY, Shock EL. Energy for biologic sulfate reduction in a hydrothermally formed ocean on Europa. *J Geophys Res Planets*. 2003;108:5022.
144. Waite JH, Glein CR, Perryman RS, Teolis BD, Magee BA, Miller G, et al. Cassini finds molecular hydrogen in the Enceladus plume: evidence for hydrothermal processes. *Science*. 2017;356:155–159.
145. Marlow J, LaRowe DE, Ehlman BL, Amend JP, Orphan V. The potential for biologically catalyzed anaerobic methane oxidation on ancient mars. *Astrobiology*. 2014;14:292–307.
146. LaRowe DE, Amend JP. The energetics of fermentation in natural settings. *Geomicrobiology*. 2019;36:492–505.
147. Lever MA. Acetogenesis in the energy-starved deep biosphere – a paradox? *Frontiers in Microbiology*. 2012;2:284.
148. Lever MA, Heuer VB, Morono Y, Masui N, Schmidt F, Alperin MJ, et al. Acetogensis in deep subseafloor sediments of the Juan de Fuca Ridge Flank: a synthesis of geochemical, thermodynamic, and gene-based evidence. *Geomicrobiol J*. 2010;27:183–211.
149. Lu G-S. Geomicrobiology in the Shallow-Sea Hydrothermal System at Milos Island, Greece. PhD thesis. Los Angeles, CA: University of Southern California, 2018.
150. Bethke CM, Sanford RA, Kirk MF, Jin Q, Flynn TM. The thermodynamic ladder in geomicrobiology. *Am J Sci*. 2011;311:183–210.
151. Claypool GE, Kaplan IR. The origin and distribution of methane in marine sediments. In: Kaplan IR, ed. *Natural Gases in Marine Sediments*. New York: Plenum Press, 1974, pp. 99–139.

152. Froelich PN, Klinkhammer GP, Bender ML, Luedtke NA, Heath GR, Cullen D, et al. Early oxidation of organic matter in pelagic sediments of the eastern equatorial Atlantic: suboxic diagenesis. *Geochim Cosmochim Acta.* 1979;43:1075–1090.
153. Stumm W, Morgan JJ. *Aquatic Chemistry: Chemical Equilibria and Rates in Natural Waters*. 3rd edn. New York: John Wiley & Sons, 1996.
154. Reeburgh WS. Methane consumption in Cariaco Trench waters and sediments. *Earth Planet Sci Lett.* 1976;28:337–344.
155. Broda E. Two kinds of lithotrophs missing in nature. *Z Allg Mikrobiol.* 1977;17:491–493.
156. LaRowe DE, Koch BP, Robador A, Witt M, Ksionzek K, Amend JP. Identification of organic compounds in ocean basement fluids. *Org Geochem.* 2017;113:124–127.
157. Hertkorn N, Harir M, Koch BP, Michalke B, Schmitt-Kopplin P. High-field NMR spectroscopy and FTICR mass spectrometry: poweerful discovery tools for the molecular level characterization of marine dissolved organic matter. *Biogeosciences.* 2013;10:1583–1624.
158. Ball GI, Aluwihare LI. CuO-oxidized dissolved organic matter (DOM) investigated with comprehensive two dimensional gas chromatography-time of flight-mass spectrometry (GC × GC-TOF-MS). *Org Geochem.* 2014;75:87–98.
159. Shah Walter SR, Jaekel U, Osterholz H, Fisher AT, Huber JA, Pearson A, et al. Microbial decomposition of marine dissolved organic matter in cool oceanic crust. *Nat Geosci.* 2018;11:334–339.
160. LaRowe DE, Dick JM. Calculation of the standard molal thermodynamic properties of crystalline proteins. *Geochim Cosmochim Acta.* 2012;80:70–91.
161. Dale AW, Regnier P, Van Cappellen P. Bioenergetic controls on anaerobic oxidation of methane (AOM) in coastal marine sediments: a theoretical analysis. *Am J Sci.* 2006;306:246–294.
162. Dale AW, Aguilera DR, Regnier P, Fossing H, Knab NJ, Jørgensen BB. Seasonal dynamics of the depth and rate of anaerobic oxidation of methane in Aarhus Bay (Denmark) sediments. *J Mar Res.* 2008;66:127–155.
163. Dale AW, Regnier P, Knab NJ, Jørgensen BB, Van Cappellen P. Anaerobic oxidation of methane (AOM) in marine sediments from the Skagerrak (Denmark): II. Reaction-transport modeling. *Geochim Cosmochim Acta.* 2008;72:2880–2894.
164. Dale AW, Sommer S, Haeckel M, Wallmann K, Linke P, Wegener G, et al. Pathways and regulation of carbon, sulfur and energy trasnfer in marine sediments overlying methane gas hydrates on the Opouawe Bank (New Zealand). *Geochim Cosmochim Acta.* 2010;74:5763–5784.
165. Dale AW, Van Cappellen P, Aguilera DR, Regnier P. Methane efflux from marine sediments in passive and active margins: estimations from bioenergetic reaction-transport simulations. *Earth Planet Sci Lett.* 2008;265:329–344.
166. Algar CK, Vallino JJ. Predicting microbial nitrate reduction pathways in coastal sediments. *Aquatic Microbial Ecol.* 2014;71:223–238.
167. André L, Pauwels H, Dictor M-C, Parmentier M, Azaroual M. Experiments and numerical modelling of microbially-catalysed denitrification reactions. *Chem Geol.* 2011;287:171–181.
168. Shock EL, Boyd ES. Principles of geobiochemistry. *Elements.* 2015;11:395–401.
169. St. Clair B. *Kinetics, Thermodynamics and Habitability of Microbial Iron Redox Cycling*. Phoenix, AZ: Arizona State Univesity, 2017.
170. Härtig C, Lohmayer R, Kolb S, Horn MA, Inskeep WP, Planer-Friedrich B. Chemolithotrophic growth of the aerobic hyperthermophilic bacterium *Thermocrinus*

ruber OC 14/7/2 on monothioarsenate and arsenite. *FEMS Microbiol Ecol.* 2014;90:747–760.

171. Steen AD, Jørgensen BB, Lomstein BA. Abiotic racemization kinetics of amino acids in marine sediments. *PLoS One.* 2013;8:e71648.
172. Harder J. Species-independent maintenance energy and natural population sizes. *FEMS Microbiol Ecol.* 1997;23:39–44.
173. Schink B. Energetics of synthrophic cooperation in methanogenic degradation. *Microbiol Mol Biol Rev.* 1997;61(2):262–280.
174. Curtis GP. Comparison of approaches for simulating reactive solute transport involving organic degradation reactions by multiple terminal electron acceptors. *Comp Geosci.* 2003;29:319–329.
175. Hoehler TM. Biological energy requirements as quantitative boundary conditions for life in the subsurface. *Geobiology.* 2004;2:205–215.
176. Amend JP, LaRowe DE, McCollom TM, Shock EL. The energetics of organic synthesis inside and outside the cell. *Phil Trans Royal Soc B.* 2013;368:1–15.
177. Amend JP, Shock EL. Energetics of amino acid synthesis in hydrothermal ecosystems. *Science.* 1998;281:1659–1662.
178. Amend JP, Shock EL. Thermodynamics of amino acid synthesis in hydrothermal ecosystems on the early Earth. In: Goodfriend G, ed. *Perspectives in Amino Acid and Protein Geochemistry.* New York: Plenum, 2000, pp. 23–40.
179. Shock EL, Schulte MD. Organic synthesis during fluid mixing in hydrothermal systems. *J Geophys Res Planets.* 1998;103:28513–28527.
180. Shock EL, Canovas III PA. The potential for abiotic organic synthesis and biosynthesis at seafloor hydrothermal systems. *Geofluids.* 2010;10:161–192.
181. LaRowe DE, Regnier P. Thermodynamic potential for the abiotic synthesis of adenine, cytosine, guanine, thymine, uracil, ribose and deoxyribose in hydrothermal systems. *Orig Life Evol Biosph.* 2008;38:383–397.
182. Battley EH. An alternative method of calculating the heat of growth of *Escherichia coli* K-12 on succinic acid. *Biotech Bioeng.* 1991;38:480–492.
183. Amend JP, McCollom TM. Energetics of biomolecule synthesis on early Earth. In: Zaikowski L, Friedrich JM, Seidel SR, eds. *Chemical Evolution II: From the Origins of Life to Modern Society.* Washington, DC: American Chemical Society, 2009, pp. 63–94.

20

Deep Carbon through Deep Time

Data-Driven Insights

ROBERT M. HAZEN, YANA BROMBERG, ROBERT T. DOWNS, AHMED ELEISH, PAUL G. FALKOWSKI, PETER FOX, DONATO GIOVANNELLI, DANIEL R. HUMMER, GRETHE HYSTAD, JOSHUA J. GOLDEN, ANDREW H. KNOLL, CONGRUI LI, CHAO LIU, ELI K. MOORE, SHAUNNA M. MORRISON, A.D. MUSCENTE, ANIRUDH PRABHU, JOLYON RALPH, MICHELLE Y. RUCKER, SIMONE E. RUNYON, LISA A. WARDEN, AND HAO ZHONG

20.1 Introduction: Data and the Deep Carbon Observatory

For most of the history of science, data-driven discovery has been difficult and time-consuming: a lifetime of meticulous data collection and thoughtful synthesis was required to recognize previously hidden, higher-dimensional trends in multivariate data. Recognition of processes such as biological evolution by natural selection (1,2), continental evolution by plate tectonics (3,4), atmospheric and ocean oxygenation by photosynthesis (5,6), and climate change (7,8) required decades of integrated data synthesis preceding the discovery and acceptance of critical Earth phenomena. However, we stand at the precipice of a unique opportunity: to dramatically accelerate scientific discovery by coupling hard-won data resources with advanced analytical and visualization techniques (9,10). Today, Earth and life sciences are generating a multitude of data resources in numerous subdisciplines. Integration and synthesis of these diverse data resources will lead to an abductive, data-driven approach to investigating Earth's mineralogical and geochemical history, as well as the coevolution of the geosphere and biosphere (11–13).

In this chapter, we examine applications of data science in deep carbon research through three "use cases." The first example focuses on geochemical and mineralogical anomalies from a period in Earth history (~1.3 to 0.9 Ga) when the supercontinent Rodinia was being assembled from previously scattered continental blocks. The second case study examines the diversity and distribution of minerals, notably carbon-bearing minerals, through deep time from the contexts of mineral evolution, mineral ecology, and mineral network analysis. The third and most speculative use case considers ways to analyze and visualize data that relate microbial protein expression to growth environments – complex interconnections that may shed light on Earth's coevolving microbial ecosystems and near-surface geochemical environments. In each example, discoveries related to Earth's deep-time evolution have resulted from the analysis and visualization of large data resources fostered by the Deep Carbon Observatory (DCO).

20.2 Use Case #1: Global Signatures of Supercontinent Assembly

Large and growing geochemical and mineralogical data resources facilitate global surveys of trends in crustal evolution through deep time. Over the past 3 billion years, Earth has undergone five periods of supercontinent assembly, during which most continents converged and concentrated into one more or less contiguous landmass. Each of these assembly episodes was followed by intervals of supercontinent stability, rifting, and dispersal (14–16).

In spite of some shared geochemical, mineralogical, and tectonic characteristics, each of these five supercontinent episodes is distinct in detail. The Mesoproterozoic Rodinian supercontinent, in particular, displays several unique mineralogical and geochemical characteristics that point to a unique outcome of collisional events between ~1.3 and 0.9 billion years ago (16–22). Rodinia represents an important transitional period for Earth's carbon cycle in terms of both geochemical and biological evolution. In this section, we examine rapidly growing data resources in mineralogy and geochemistry that shed light on the unique character of this interval of Earth's history.

20.2.1 Mineralogical Evidence

Evidence for five cycles of supercontinent assembly, stability, and dispersal are strikingly preserved in the age distributions of high-temperature minerals (including many igneous, metamorphic, and hydrothermal species), which may be preferentially formed and/or preserved during continental suturing. The most notable mineralogical proxy is detrital zircon grains (23–29). As with other supercontinents (Figure 20.1a), the assembly of Rodinia saw a significant peak in the production and/or detrital preservation of zircon, with a global maximum at ~1.1 to 1.0 Ga (23,29).

Important mineralogical insights into supercontinent cycles are provided by minerals other than zircon (Figure 20.1) (16,30,31), and these mineral species can be explored through deep time thanks to the creation and rapid expansion of the Mineral Evolution Database (MED; rruff.info/evolution), an important contribution of DCO mineralogists. The MED incorporates more than 195,000 mineral/locality/age data, mostly for minerals from well-constrained magmatic, metamorphic, or hydrothermal events (data as of June 10, 2019). Liu et al. (16,22) employed the MED to explore and document age distributions of minerals and found that minerals containing niobium and yttrium (Figure 20.1b and c) exhibit similar trends to those of zircon; these minerals display maxima slightly later than zircon, at ~1.1 to 0.95 Ga. By contrast, minerals of most other elements, including Ni, Co, Au, S, Hg, Li, and C (Figure 20.1d to j), record significant pulses of mineralization during the assembly of Kenorland, Nuna, Gondwana (Pannotia), and Pangaea, but notably indicate decreased mineralization during Rodinian assembly (30–35). From these observations, we conclude that the currently expressed patterns of mineralization associated with the Rodinian assembly are unique relative to those of the other aforementioned supercontinents.

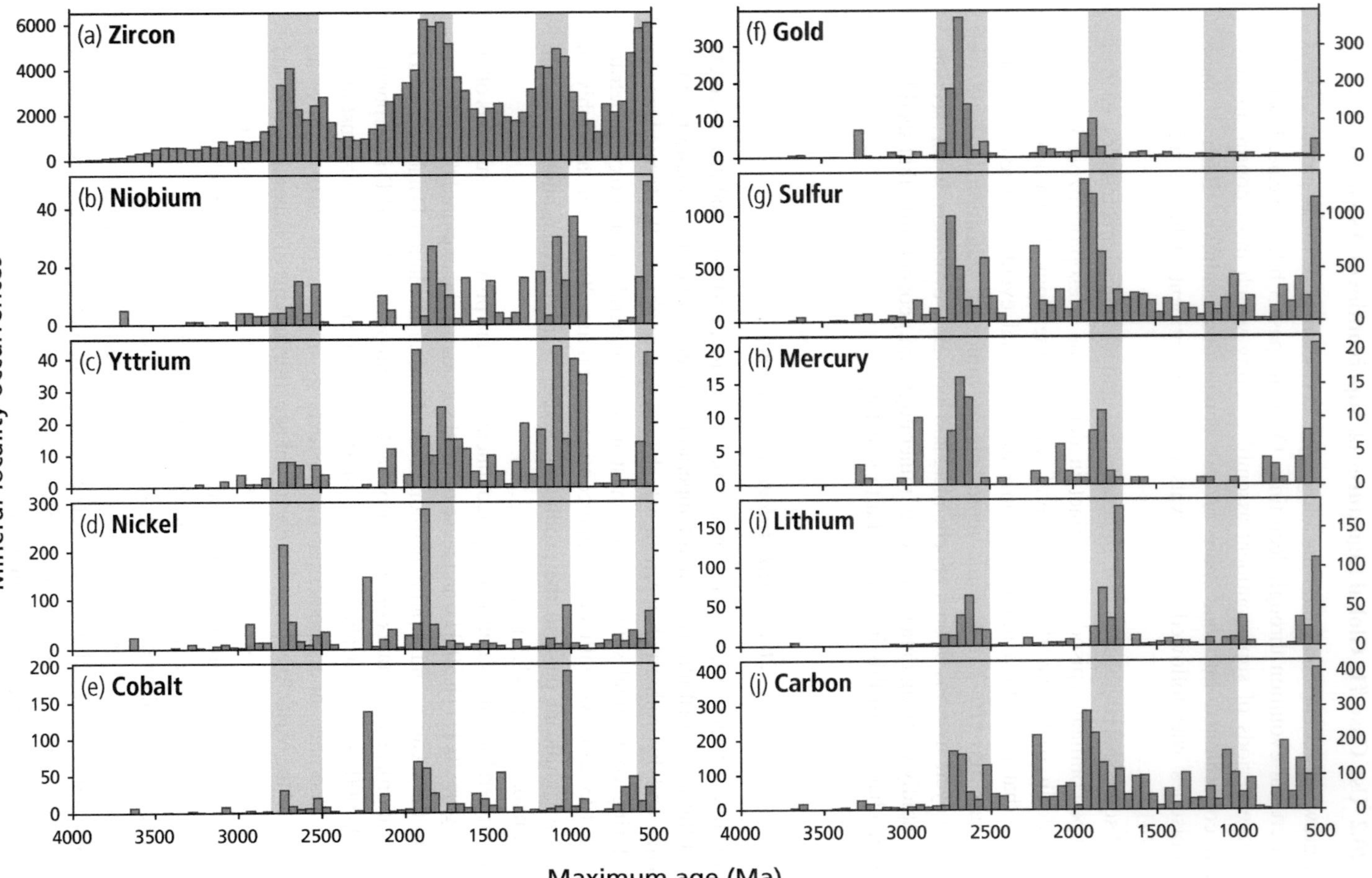

Figure 20.1 Mineral frequency of occurrence through deep time for: (a) detrital zircon (data from 29), (b) minerals with essential niobium, (c) yttrium, (d) nickel, (e) cobalt, (f) gold, (g) sulfur, (h) mercury, (i) lithium, and (j) carbon (data from earthchem.org). Bar graphs display 50-million-year bins. Vertical bars indicate periods of supercontinent assembly.

20.2.2 Trace Element Distributions

Temporal changes in the global averages of trace elements in igneous rocks complement and amplify mineral evolution data. Liu et al. (16,22) compiled trace element data for globally distributed igneous rocks from the EarthChem database (earthchem.org) and the United States Geological Survey (USGS) Mineral Resources Online Spatial database (mrdata.usgs.gov). They compiled age/concentration data for 129,161 samples with reported Zr analyses, 105,045 with Nb analyses, 121,373 with Y analyses, 77,835 with Co analyses, and 82,611 with Ni analyses from igneous rocks, all of which are associated with SiO_2 content (wt.%) and modern geographic coordinates (Figure 20.2).

The period of Rodinian assembly from 1.3 to 0.9 Ga saw significantly greater niobium, yttrium, and zirconium concentrations in igneous rocks than at any other time during the last 3 billion years (Figure 20.2). Furthermore, these trace element maxima apply to both mafic and felsic igneous rocks. By contrast, Liu et al.'s (16,22) survey found that average nickel and cobalt concentrations in igneous rocks display no significant enrichments or depletions during this interval (Figure 20.2).

20.2.3 Why Is Rodinian Assembly Unique?

Rodinia has long been recognized as distinct from other supercontinents. In addition to the mineralogical and geochemical anomalies noted above (Figure 20.1), the time from 1.3 to 0.9 Ga is marked by enhanced anorogenic magmatism, as well as a relative minimum extent of continental margins and collisional belts (21,36–40). Liu et al.'s (16,22) observation of significant maxima in the Nb, Y, and Zr composition of Rodinian igneous rocks (Figure 20.2) amplifies evidence that Rodinian assembly was unique, while pointing to possible reasons for these differences.

The enrichments of Nb, Y, and Zr, coupled with the greater relative abundances of minerals of these three elements, point to a distinctive tectonic setting for Rodinia. Rodinian assembly was dominated by "non-arc" magmatism, in contrast to other intervals of supercontinent assembly when collision-related mineralization and island arc magmatism were of greater relative significance (41–46). In particular, these tectonic conditions at 1.3 to 0.9 Ga led to enhanced production of NYF-type (i.e. Nb-, Y-, and F-enriched) pegmatites, with associated increases in the occurrence and diversity of Nb-, Y-, and Zr-bearing minerals (46–49). This mineralization may have been associated with a warmer mantle and/or a thickened continental crust during Rodinian assembly (50,51) – characteristics that may reduce scavenging of high-field-strength elements by interaction with the depleted mantle during arc magmatism (52,53).

The relative enrichment of Nb, Y, and Zr contrasts with the behavior of many other elements during the period of Rodinian assembly. The minerals of most elements are notably lacking during the 1.3 to 0.9 Ga interval, as manifest in the relatively few ore deposits associated with the time of Rodinian assembly (30,31,36,38). However, the trace element concentrations of Co, Ni, and many other elements in igneous rocks do not show

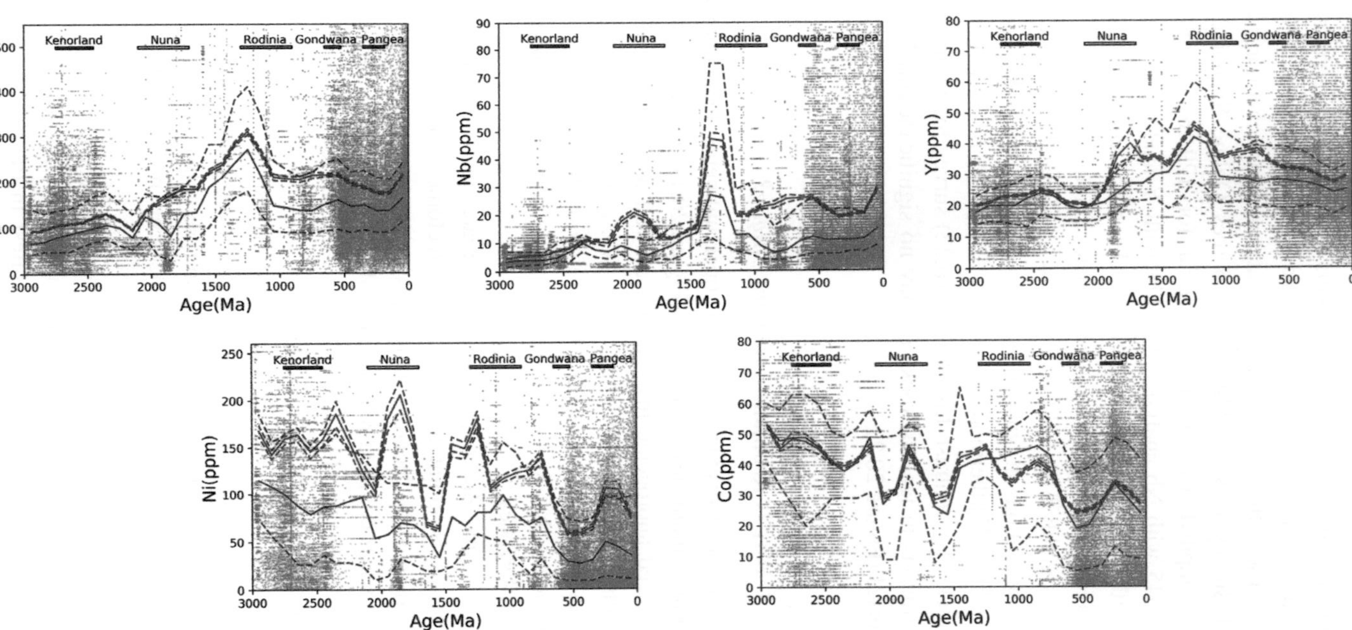

Figure 20.2 Trace element concentrations of Zr, Nb, Y, Ni, and Co in global igneous rocks through the last 3.0 Ga. Maximum values for Zr, Nb, and Y occur during and immediately before Rodinian assembly, in contrast to Ni and Co. Gray-filled circles are data resampled from earthchem.org with bootstrap resampling. Moving averages and medians of samples within ±100 Ma bin sizes are calculated for each 100 Ma. Red solid lines are averages; red dashed lines are 95% confidence intervals of the moving average; blue solid lines are medians; blue dashed lines indicate the lower (25%) and upper (75%) quantiles
(after 16).

corresponding depletions compared to other supercontinent episodes (Figure 20.2) (16). Given this consistency in metal concentrations, reduced Rodinian ore deposition seems unlikely. Rather, the lack of Mesoproterozoic ore deposits may be a consequence of enhanced erosion of near-surface deposits that formed preferentially near active margins. This style of erosion was perhaps more characteristic of Rodinia than other supercontinents for two reasons. First, pre-collisional erosion of Rodinia may have been more aggressive than with other supercontinents, because the accretion of Rodinia is thought to have been both prolonged and "extrovert," with assembly by two-sided subduction (54–56). Such a tectonic context would have caused the loss of most volcanic-hosted massive sulfide deposits, which require rapid accretion of continental margins for preservation (38). Furthermore, the major orogens associated with Rodinian assembly experienced cycles of collisional distension that must have led to enhanced deep erosion. These processes are reflected in the high regional metamorphic grade of many surviving rocks associated with two major Rodinian sutures: the Grenville and Sveconorwegian orogens (20,57–60). Thus, for example, the absence of Rodinian-age gold deposits likely reflects removal of the shallower loci of mineralization, whereas the enhanced production of Grenvillian fluvial sediments led to the abundance of detrital zircon crystals of that age (61–63) Consequently, the observed distribution and diversity of minerals during the period of Rodinian assembly reflects a unique combination of mineralization events and preservational biases.

20.2.4 Implications for the Carbon Cycle

Tectonic events such as supercontinent assembly and dispersal have direct effects on carbon cycling at Earth's surface (64–66; Chapter 11, this volume). How might the distinct aggregation and breakup of Rodinia have influenced the carbon cycle and, related to this, redox conditions and life?

In principle, uplift and erosion associated with supercontinent assembly might have affected both atmospheric pCO_2 and nutrient fluxes into the oceans. Denudation rates of modern active margins (e.g. New Zealand, Taiwan) were reported to be highest on continents/islands – orders of magnitude higher than mountain belts (e.g. Alps, Himalaya) and shields away from the coast (67). The Rodinian supercontinent was proposed to be formed via closure of Pacific-type oceans (62,68), with abundant active but rare passive continental margins (38). On geologic timescales, continental erosion/weathering is the major sink for atmospheric CO_2 (69), and the high erosion/weathering rate of Rodinian active margins could have sequestered CO_2 more rapidly, paving the way for Neoproterozoic global glaciations (36). The fact that global ice ages postdate Rodinian assembly by more than 200 million years indicates that while Rodinian CO_2 drawdown might have contributed to later Proterozoic climate change, other factors must be considered as well.

Enhanced weathering and erosion had the potential to increase P fluxes into the oceans, thus promoting primary production. For example, the later Mesozoic and

Cenozoic uplift of major mountain belts appears to have impacted primary production, driving ecosystem-wide biological changes in the oceans (70). In addition, enhanced formation of rapidly subsiding sedimentary basins during the Rodinian breakup might have increased rates of organic carbon burial, thereby contributing to Neoproterozoic oxygenation (71).

We have several geochemical tools for exploring secular variations in carbon cycling, most notably the carbon isotopic record of carbonate and organic carbon (72). In addition, a variety of proxies permit inferences about changing redox conditions in the oceans and atmosphere (73), and fossils record the course of early evolution (74,75). Interestingly, supercontinental events correlate only weakly with the carbon isotopic, paleo-redox, and fossil records. Rodinian assembly correlates with a moderate increase in the secular variation of carbon isotopes, following a long interval of near-invariant values (76), whereas a much larger amplitude of C-isotopic variations is associated with the Rodinian breakup and its aftermath (77). Proxies for redox conditions show little change in association with either Rodinian assembly or breakup, perhaps because limited P availability (78) muted Earth system responses to these tectonic events. Global changes in oxygen levels and biological complexity occur only near the end of the Proterozoic Era, in association with a state change in P availability linked by some to climate rather than directly to tectonics (79).

Thus, at our present state of knowledge, the momentous tectonic events of Rodinian assembly and dispersal seem to have exerted only a limited influence on the surficial carbon cycle, with dispersal correlating more closely with enhanced organic carbon burial, perhaps minor oxygen enrichment, and protistan diversification (75) than with supercontinent assembly.

20.3 Use Case #2: Carbon Mineral Evolution, Mineral Ecology, and Mineral Network Analysis

Data-driven exploration is built on open-access data resources and the application of advanced analytical and visualization techniques. Databases, such as that of the RRUFF Project (rruff.info), which includes information on all approved mineral species, and that of mindat.org, which documents species found at more than 300,000 localities with greater than 1,000,000 mineral/locality data, provide opportunities to explore mineral data with new analytical tools. The effects of preservational and/or sampling bias in these data are poorly understood and are the subject of further investigation. The DCO has seized this opportunity by facilitating significant advances in the accumulation, analysis, and visualization of mineral data – notably information housed in the MED related to the more than 400 approved carbon-bearing mineral species (80–82). As such, carbon minerals constitute an important test case for new approaches to mineralogy, while providing unique insights into the evolving roles of carbon through deep time (Figure 20.3).

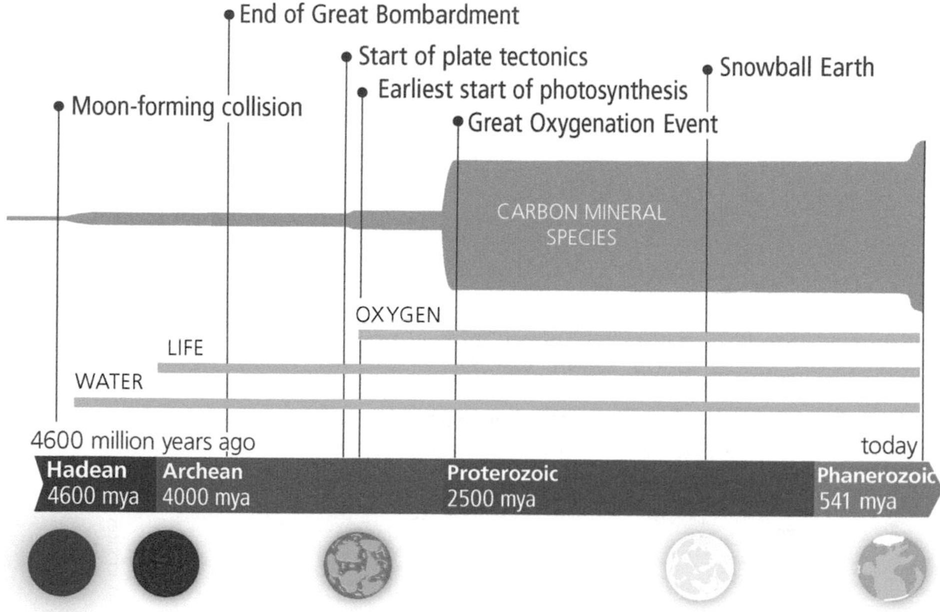

Figure 20.3 Carbon mineral evolution timeline over 4.5 billion years. Carbon played a key role throughout this evolutionary path, with an explosion in carbon mineral diversity in the Proterozoic and Phanerozoic.

20.3.1 Carbon Mineral Evolution

Mineral evolution is the study of the changing diversity and distribution of minerals through deep time – the consequence of varied physical, chemical, and, in the case of Earth, biological processes (11,83–85). Hazen et al. (80) surveyed carbon mineral evolution from a qualitative viewpoint, tracing changes in the nature and extent of carbon-bearing minerals through ten stages of Earth's evolution. From the most primitive Stage 1, characterized by chondrite meteorites, which contain several carbide minerals and allotropes of carbon, to the thriving terrestrial biosphere of Stage 10, with more than 400 approved carbon mineral species, Earth's 4.567-billion-year history saw significant increases in the diversity and complexity of C-bearing phases. The number of crystalline forms of C-bearing compounds has seen a dramatic rise with the creative contributions of chemists in the "Anthropocene Epoch" – an explosion of new mineral-like forms that some observers have dubbed "Stage 11" of Earth's mineral evolution (86,87).

The development of the MED (88), which tabulates 17,455 ages for C-bearing mineral/locality data (data as of May 21, 2018), facilitates a more quantitative examination of carbon mineral evolution. A detailed investigation of these minerals, including their paragenetic modes, associated species, geochemical contexts, tectonic settings, and other parameters, is beyond the scope of this chapter. However, an overview of the temporal

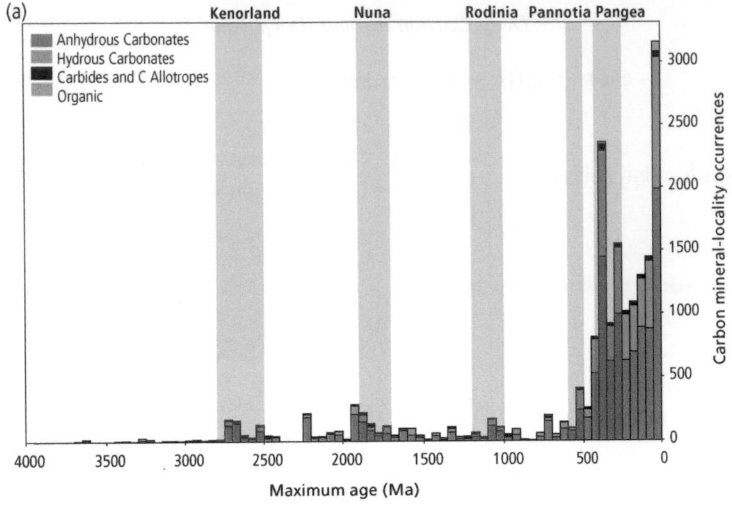

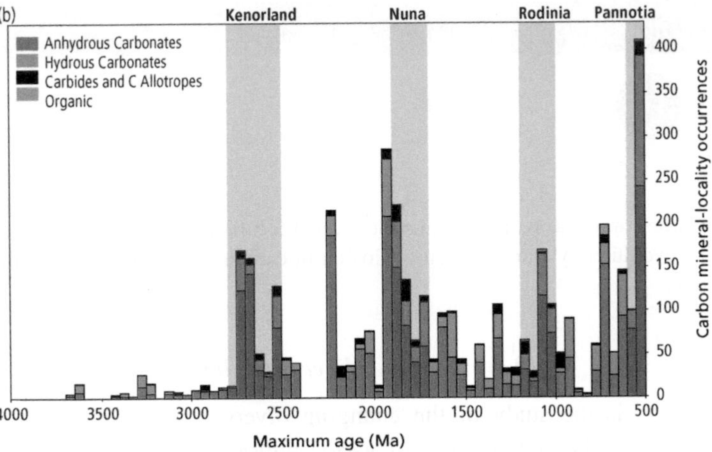

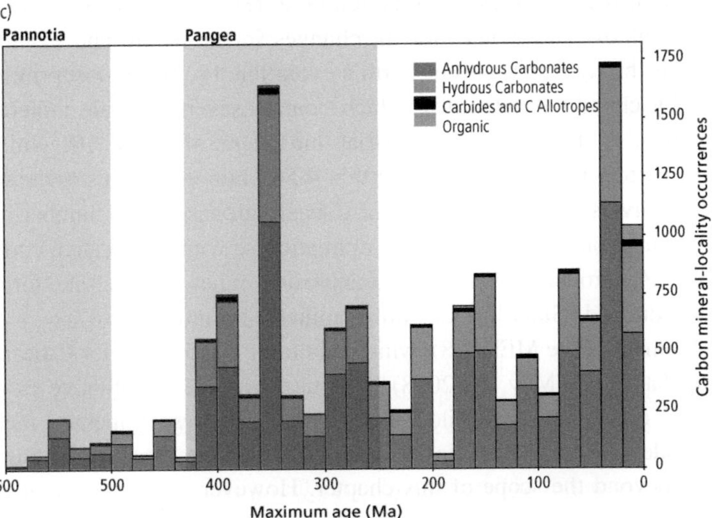

Figure 20.4 Temporal distribution of carbon minerals. (a) The past 4 billion years with 50-million-year bins. (b) Precambrian occurrences (4.0 to 0.5 Ga) with 50-million-year bins. (c) Neoproterozoic

distributions of C-bearing minerals reveals important physical, chemical, and biological processes that influence carbon mineralization. Figure 20.4 illustrates these newly expanded MED carbon mineral data.

The temporal distribution of carbon minerals reveals significant trends. As with most other groups of minerals, C-bearing species display striking episodicity, with pulses of mineralization as well as time intervals with few recorded carbon minerals. For example, significant maxima in preserved carbonate minerals are recorded at 2.75 to 2.70 Ga and at 2.55 to 2.50 Ga, with each interval having more than 150 points of reported carbon mineral/locality/age data. Those two 50-million-year intervals frame the assembly of Kenorland, the earliest well-documented supercontinent. By contrast, the 200-million-year interval from 2.45 to 2.25 Ga, a period of presumed Kenorland stability and generally low mineralization, has fewer than 20 total reported carbon mineral occurrences. As noted in Section 20.2, such a sharp contrast in numbers of mineral occurrences likely reflects a combination of episodic mineralization and preservational biases.

A similar contrast is observed for Nuna, the next widely recognized supercontinent episode in Earth's history. Approximately 800 mineral/locality/age data are recorded for the 250-million-year period of presumed Nuna assembly from 1.95 to 1.70 Ga. By contrast, the 250-million-year interval of Nuna breakup from 1.60 to 1.35 Ga is represented by fewer than 250 reports of C-bearing minerals.

Though less dramatic, the assembly of Rodinia is also reflected in the carbon mineral record. Approximately 400 mineral/locality/age data are recorded for the assembly period from 1.1 to 0.9 Ga, as opposed to fewer than 20 data points from the subsequent 100-million-year interval from 0.9 to 0.8 Ga. As suggested in Section 20.2, the relatively modest mineral inventory from Rodinian assembly likely reflects significant erosional loss of near-surface (i.e. more carbonate-rich) deposits compared to Kenorland and Nuna.

Approximately 80% of reported carbon mineral occurrences in the MED are from the Phanerozoic Eon, which spans the last 540 million years when carbonate biomineralization became an important mode of near-surface carbon mineralization. The greater number of data from the Phanerozoic Eon allows a more detailed examination of carbon mineral evolution during the past 500 million years. Figure 20.4c underscores the nonuniform distribution of documented carbon mineralization during the past 600 million years. Of note is that almost 1700 mineral/locality/age data are recorded from the 20-million-year interval from 360 to 340 Ma, a time of the supercontinent Pangaea's assembly, and thus a plausible time of enhanced mineralization and preservation.

Figure 20.4 (*cont.*) and Phanerozoic occurrences (760 to 0 Ma) with 20-million-year bins. Anhydrous carbonates (orange, lowest segment), hydrous carbonates (blue, next lowest segment), other (i.e. diamond and carbides, black, next lowest segment), and organic minerals (green; topmost segment). Graphs are based on 17,455 mineral/locality/age data tabulated in the MED (rruff.info/ima; as of February 15, 2018). Note that this tabulation is based on mineral specimens collected from specific localities and does not include sedimentary carbonate formations.

An important concurrent event was the expansion of late Paleozoic ice sheets in Gondwana, a scenario linked to enhanced burial of organic matter associated with the evolution of trees and diversification of seed plants, stem group ferns, and lycopods. This interval was also notable for the 359 Ma Devonian–Mississippian boundary, which marks the last pulse of elevated extinctions that occurred through much of the Devonian Period. A notable degree of ecological reorganization also occurred in marine environments, including the complete turnover of rugose corals, a once-abundant order of corals that are now extinct, at the family level. It is not obvious how these paleobiological developments might have led to enhanced mineralization, although it is possible that at least some of the observed paleobiological events might reflect responses to tectonic events and their environmental consequences, as recorded by carbon mineral occurrences.

By contrast, the interval from 200 to 180 Ma is represented by fewer than 15 C-bearing mineral/locality/age data points worldwide. This 20-million-year period occurred at the beginning of Pangaea's breakup and the opening of the modern Atlantic Ocean, a time characterized by tectonic conditions that might be associated with reduced carbon mineralization or deposition and enhanced erosional loss. The beginning of this interval corresponds to the end-Triassic mass extinction associated with massive volcanism, whereas a minor extinction event at 182 Ma is also associated with a large igneous province (89). However, neither of these short intervals of species loss have obvious connections to the mineral record.

Note that the distribution of mineral occurrences during the Precambrian at 50-million-year intervals (Figure 20.4b) is not unlike the peak distributions of the Phanerozoic Eon at 20-million-year intervals (Figure 20.4c). An unresolved question in mineral evolution research is the extent to which the temporal distribution of mineral groups, including C-bearing species, is fractal; in other words, does the same pattern of mineral distribution repeat at finer and finer temporal scales? This question can only be answered by gathering many more mineral/locality/age data with the highest possible time resolution. We are currently limited to the 195,000 mineral/locality/age data compiled in the MED, but there are likely many more data yet to be extracted from the existing literature, as well as many rock and mineral samples that have yet to be analyzed. For instance, rock-forming minerals are particularly underrepresented in the MED simply due to sampling bias.

20.3.2 Carbon Mineral Ecology

Mineral ecology is the study of mineral diversity–distribution relationships of minerals at the global scale –an effort that depends on large and growing data resources on mineral species and their localities on Earth's crust. Hazen et al. (81) applied a large number of rare events (LNRE) formalism (90–93) to model the distribution of 403 approved mineral species of carbon. Using 82,922 mineral species/locality data tabulated in mindat.org (as of January 1, 2015), they demonstrated that all C-bearing minerals as well as several compositional subsets containing C conform to LNRE distributions.

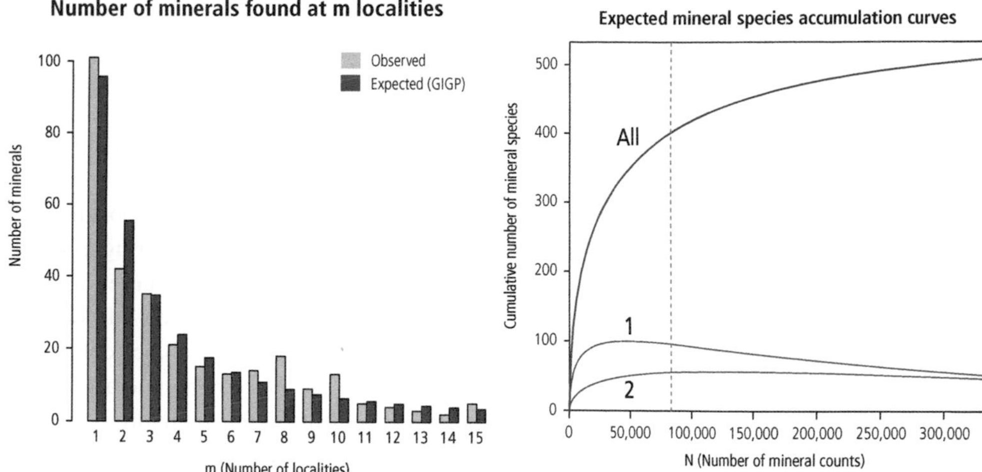

Figure 20.5 (a) Frequency spectrum analysis of 403 C-bearing minerals, with 82,922 individual mineral-locality data (from mindat.org as of January 2015), employing a generalized inverse Gauss–Poisson (GIGP) function to model the number of mineral species for minerals found at between 1 and 15 localities (90). (b) This model facilitates the prediction of the mineral species accumulation curve (upper curve, "All"), plotting the number of expected C mineral species (y-axis) as additional mineral species-locality data (x-axis) are discovered. The vertical dashed line indicates data recorded as of January 2015 in mindat.org. The model also predicts the varying numbers of mineral species known from exactly one locality (curve 1) or from exactly two localities (curve 2). Note that the model predicts that the number of C-bearing mineral species known from only one locality is now decreasing, whereas the number from two localities is now increasing, though it will eventually decrease. We predict that the number of minerals known from two localities will surpass those known from one locality when the number of species-locality data exceeds ~400,000.
Reproduced from Hazen et al. (81) with permission.

The LNRE model is particularly useful because it can be used to determine an "accumulation curve" – a formalism that enables estimations of the probability that the next carbon mineral/locality discovery will represent a new species (Figure 20.5). Figure 20.5a displays the frequency spectrum analysis for 403 C-bearing mineral species based on 82,922 individual mineral-locality data (from mindat.org as of January 2015). We found that 101 minerals – more than 25% of known C-bearing species – have been identified from only one locality worldwide. Another 42 species have been found at exactly two localities. Based on this information, we employed a Generalized Inverse Gauss–Poisson function to model the number of mineral species for minerals found at between 1 and 14 localities (90).

This LNRE model facilitated the prediction of the mineral species accumulation curve (Figure 20.5b). In Figure 20.5b, the upper curve (labeled "All") plots the expected number of approved C mineral species (y-axis) as additional mineral species/locality data (x-axis) are discovered. The vertical dashed line indicates data recorded as of January 2015 in mindat.org. The model also predicts the varying numbers of mineral species known from exactly one locality (curve "1") or from exactly two localities (curve "2"). Note that the model predicts that the number of C-bearing mineral species known from only one locality is now decreasing, whereas the number from two localities is now increasing, though it too will eventually decrease. We predict that the number of minerals known from exactly two localities will surpass those from one locality when the number of species-locality data exceeds ~400,000.

Employing this model, Hazen et al. (81) predicted that at least 548 carbon mineral species occur in Earth's crust today –a result that suggests at least 145 C-bearing minerals exist but have yet to be discovered. Additional hints regarding the nature of these "missing" carbon minerals are gleaned by analyzing compositional subsets of common additional elements in C-bearing minerals, including oxygen, hydrogen, calcium, and sodium. Accordingly, Hazen et al. (81) predicted that 129 missing carbon minerals contain oxygen (primarily carbonates) and 118 species contain hydrogen (mostly hydrous carbonates). In addition, more than 50 of the missing species contain calcium, while more than 60 contain sodium. Additional studies of the distributions of known minerals according to their distinctive sizes, colors, crystal forms, and physical properties (93) suggest that many of the missing carbon minerals may have been overlooked because they are colorless, poorly crystalized, water soluble, and/or occur in minute grains. Similarly, these same factors are likely why nearly 35% of Na minerals have yet to be discovered and, conversely, why fewer than 20% of Cu, Mg, Ni, S, Te, U, and V minerals are still unknown (93). This powerful data-driven approach has allowed the systematic prediction and discovery of large numbers of previously unknown mineral species for the first time.

These newly applied data analytic methods have led to DCO's Carbon Mineral Challenge (mineralchallenge.net), which enlists professional mineralogists and amateur mineral collectors around the world in the search for new species. More than 30 new carbon minerals – roughly 20% of the predicted total missing inventory – have been reported since January of 2016. Two of those species, abellaite ($NaPb_2(CO_3)_2(OH)$) and parasite-(La) ($CaLa_2(CO_3)_3F_2$), were predicted as possible new carbon minerals by Hazen et al. (81). Other new carbon species were not predicted. Of note is the organic mineral tinnunculite ($C_5H_4N_4O_3{\cdot}2H_2O$), which crystallizes when the excrement of the kestrel, *Falco tinnunculus*, bakes in the hot gases of a burning coal fire. Though tinnunculite was not anticipated by our analysis, we did predict that several new organic minerals would be included in the list of new finds.

Mineral ecology and data-driven approaches to predicting and discovering new mineral species (as well as valuable mineral resources identified using similar statistical approaches) are in their infancy. In addition to further studies of carbon mineral ecology on Earth, efforts concentrating outward, focusing on other planetary bodies, will be necessary. Some work has begun, including hypothesizing the mineral diversity of Saturn's moon, Titan (94,95). Maynard-Caseley et al. (94) propose a rich, diverse population of carbon minerals, specifically organic molecular minerals, on Titan's frozen surface. The applications of such data-driven methods as cluster analysis, network analysis, and affinity analysis to mineral systems are poised to revolutionize the way we think about the diversity and distribution of minerals on Earth and other worlds by providing a more complete, multivariate understanding of these systems.

20.3.3 Carbon Mineral Network Analysis

Advances in data-driven discovery rely on application of creative analytical and visualization methods to complex multi-dimensional systems. Mineral network analysis (82) is a particularly powerful approach to understanding complex relationships among mineral species, their localities, paragenetic modes, and varied physical and chemical properties.

Figure 20.6a displays a force-directed network graph in which colored circles (nodes) indicate C-bearing mineral species, while lines between circles (edges) denote coexisting pairs of minerals. The sizes of nodes indicate the relative abundances of the minerals, while colors represent major C-bearing mineral groups. In this force-directed graph, each edge has an optimal length like a spring; edges are stretched or compressed to achieve a "lowest energy" state for the entire network. Similarly, Figure 20.6b shows a bipartite network of 403 C-bearing mineral species from approximately 300 mineralized regions on Earth. These graphs are interactive; each node can be clicked and dragged to more closely examine the number and nature of edges (see dtdi.carnegiescience.edu for interactive renderings).

An important characteristic of network visualizations is that they can be analyzed with numerous metrics, each of which quantifies aspects of the local and global distributions of nodes and links (96–98). For example, the carbon network (Figure 20.6a) has density $D = 0.24$ (i.e. 24% of all possible edges are present) – a value that is intermediate between those of copper minerals ($D = 0.12$) and igneous minerals ($D = 0.64$) (82). The network diameter, which measures the maximum degree of separation between any two network nodes, is $d = 4$, while the network affinity is $a = 0.55$.

One of the surprising findings related to networks of minerals is that they may embed information not coded into the network layout. For example, a slight chemical trend is visible in Figure 20.6a, with nearly all of the anhydrous carbonates not containing transition elements, lanthanides, and/or actinides (orange nodes) plotting on the left side of the network and the majority of the organics and hydrous carbonates containing transition elements, lanthanides, and/or actinides (green and purple nodes, respectively) plotting on the right. In Figure 20.6b, a few trends regarding the diversity and distribution of minerals in space and time are evident. First, the "U-shaped" distribution of black locality nodes, with a few very common carbon minerals "inside" and many more rare carbon minerals "outside," is a visual representation of the LNRE distribution illustrated in Figure 20.5. Second, there is an embedded timeline, with the oldest minerals in the center of the locality "U" radiating outward as the mineral species' age of first occurrence becomes younger.

Mineral network analysis, a direct outgrowth of interactions among diverse members of the DCO community, is in its infancy. We anticipate that open-access data resources, as well as freely available analytical and visualization software, will lead to a transformation in the ways that we study complex mineral systems on Earth and other worlds.

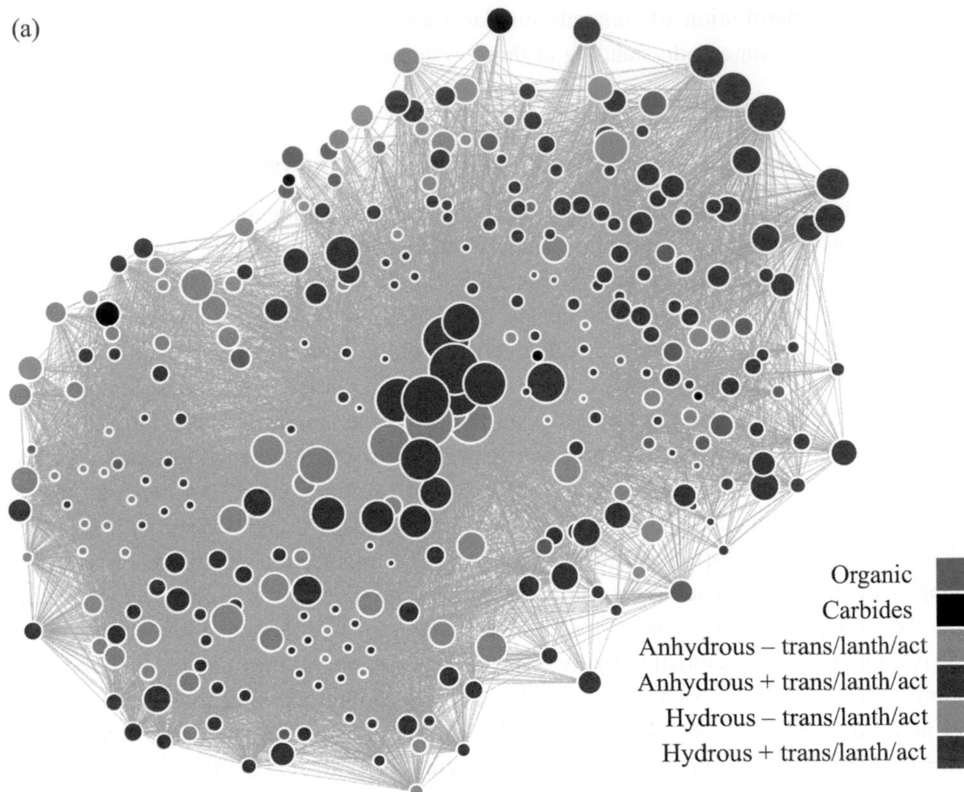

Figure 20.6 (a) Force-directed, unipartite network graphs of 403 C-bearing mineral species. Nodes represent C-bearing mineral species, while lines between nodes denote coexisting pairs of minerals. Node diameters indicate the relative abundances of the minerals, while colors represent compositional groups (dark blue = hydrous carbonates with transition elements, lanthanides, and/or actinides; light blue = hydrous carbonates without transition elements, lanthanides, and/or actinides; red = anhydrous carbonates with transition elements, lanthanides, and/or actinides; orange = anhydrous carbonates without transition elements, lanthanides, and/or actinides; black = carbon allotropes and carbides; green = organic minerals). (b) Force-directed, bipartite network of 403 C-bearing mineral species and their localities on Earth (see also http://dtdi.carnegiescience.edu/node/4557 for an interactive version). Colored nodes represent carbon mineral species, with node size corresponding to the frequency of occurrence and color corresponding to the age of earliest known occurrence of each mineral species. Black nodes represent regional localities, with diameter corresponding to the relative numbers of distinct C-bearing mineral species found at each locality. The network rendering reveals important information regarding the diversity and distribution of carbon minerals through space and time. In particular, the "U-shaped" distribution of black locality nodes, with a few very common carbon minerals "inside" and many more rare carbon minerals "outside," is an alternative visual representation of the LNRE distribution illustrated in Figure 20.5.

(b)

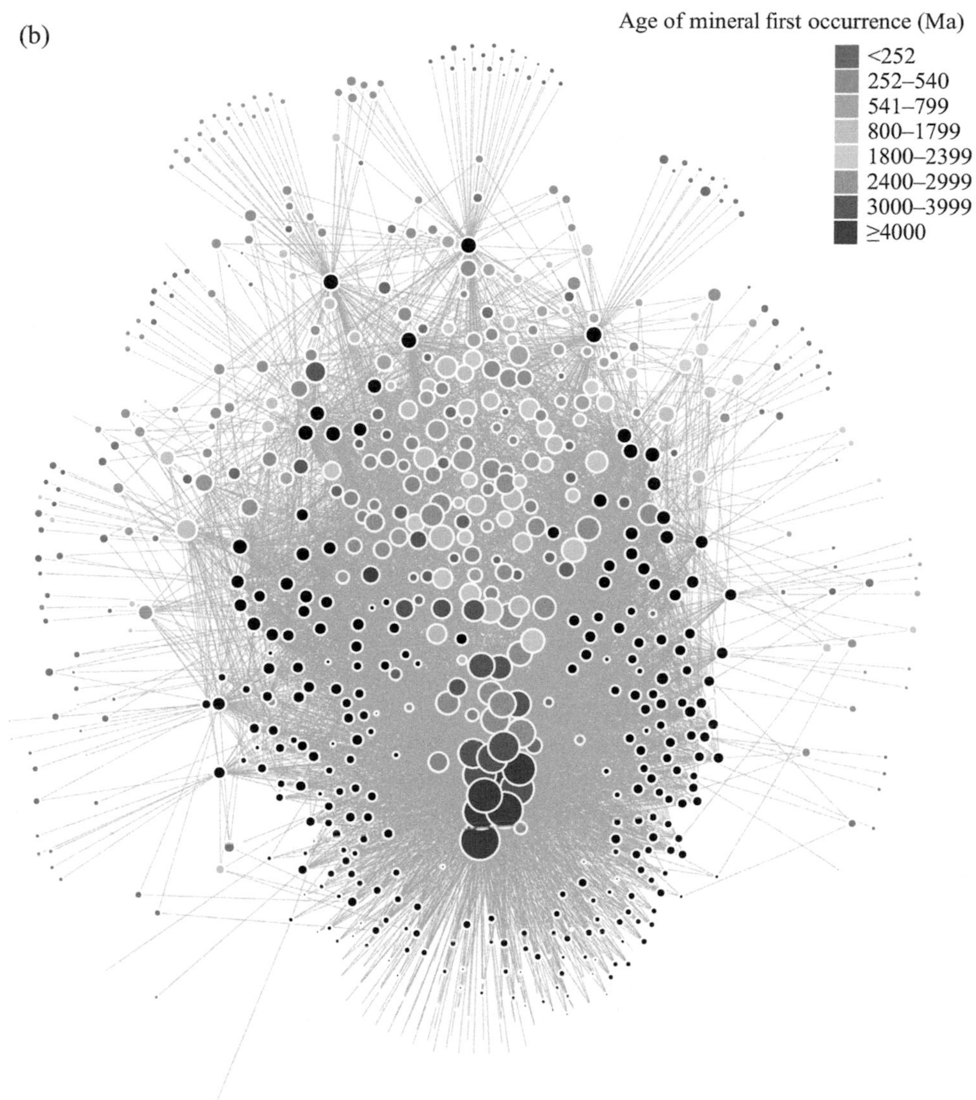

Figure 20.6 (*cont.*)

20.4 Use Case #3: Enzyme Evolution and the Environmental Control of Protein Expression

Microbes in Earth's crust have played key roles in the carbon cycle throughout space and time (99; Chapters 17 and 18, this volume). In order to better understand "whole-Earth carbon," we must examine the relationships among: (1) the physical and chemical characteristics of varied microbial environments (Chapters 16 and 19, this volume); (2) the metabolic strategies adopted by microbial consortia in these environments (Chapter 17, this volume); and (3) the consequent variation of microbial gene molecular function and expression (Chapter 18, this volume). The exploration of the complex interconnections among the physical, chemical, and biological aspects of microbial ecosystems represents an as yet unrealized opportunity for understanding the coevolving geosphere and biosphere.

A fundamental stumbling block in documenting the role of microbes in Earth's carbon cycle through deep time is the lack of relevant data on the nature and expression of proteins in ancient microbial ecosystems. In spite of the occasional preservation of Precambrian microfossils, scant biomolecular traces survive in ancient rock formations (100–102). Therefore, an understanding of the biochemical evolution of microbes might seem beyond our reach.

A promising strategy to understand aspects of the coevolution of geochemical and biochemical systems is based on the analysis of the large and growing data resources describing microbial ecosystems. Extant microbial communities span a wide range of physical and chemical environmental conditions (e.g. high and low pH, temperature extremes, high salinity and pressure, low consumable resource availability, and low water activity), some of which likely mimic a range of ancient conditions extending back to the dawn of life (103). While extant microorganisms living in these ecosystems are modern organisms that coevolved with our planet and adapted to its changing conditions, they still harbor ancestral metabolic traits. Consequently, today's microorganisms contain both inherited traits as well as recently acquired ones.

Considering that ancient protein structures and functions are at least to some extent conserved in modern organisms, then modern analogs of presumed ancient environments may resemble life's earliest enzymatic systems. For instance, extant strict anaerobes that inhabit anoxic, geothermal environments must have inherited the metabolic machinery necessary to conserve energy using redox couples abundant in geothermally influenced environments (e.g. hydrogen and sulfur) and to fix carbon dioxide of magmatic origin (103). These same organisms also must have acquired the ability to cope with reactive oxygen species in order to adapt as atmospheric oxygen levels on Earth increased over the last 700 million years. However, being unable to accurately differentiate new adaptations of older functions from truly new innovations complicates the process of reconstructing the emergence and evolution of metabolisms. The integration of large data sets obtained from the study of extant microorganisms and their protein structures, coupled with detailed environmental, geochemical, and mineralogical information, may allow us to better understand the emergence and evolution of microbial metabolism. In particular, it may provide new insights into how the geosphere and biosphere have coevolved, ultimately resulting in the complex network of metabolic reactions we see today (104,105).

Here, we propose strategies for applying methods of data analysis and visualization in order to answer questions about microbial ecology, protein evolution, and their relationship to carbon mineralization through deep time.

20.4.1 Network Analysis of Protein Structures: Geo–Bio Interactions on Evolutionary Scales

Methods of network analysis are well suited to the exploration of the evolution of and relationship between protein structure and function (106–108). The combination of geochemically identifiable timescales with biologically determined timelines permits glimpses into the history of life on Earth. For example, Bromberg and colleagues have employed similarity networks to analyze relationships among the structures of nearly 4700 oxidoreductases from varied microbial and multicellular organisms. Since electron transfer reactions are necessary to fulfill the energy requirements of all life-forms, the ability to carry out redox reactions must have been among the first functions acquired by early life. Understanding the evolution of biological redox machinery can thus shine light on the history of life and on its interactions with Earth's environment.

Ideally, the evolution of redox abilities could be traced through the analysis of the relevant enzyme sequences. However, the origins of biological redox, which likely correspond to the origins of life, as well as the dramatic environmental changes that have since taken place (e.g. the Great Oxidation Event and the "fold explosion" of protein structures), are ancient. This fact makes the exploration of the mutations in sequence space that led to the current biological "state of the art" nearly impossible (109,110). Protein three-dimensional structures, on the other hand, retain evolutionary evidence for significantly longer stretches of time. Note that the process of the divergent evolution of folded structures implies that existing folds emerged from prior ones. However, functionally similar folds may also arise independently via convergent evolution. Using network analysis, augmented by metrics of sequence similarity in structural alignments, it is possible to trace distant relations between redox proteins to estimate whether they have common ancestors or whether they developed independently.

Bromberg et al. have created a method, *sahle* (structure-annotated homology, ligand-extended), for evaluating the reliability of *structural* similarity of transition metal binding sites in proteins, defined as spheres of 15-Å radius from the active metal-containing site (111). A *sahle* score, ranged 0–100, gives weight to an edge between two spheres/nodes in the resulting network (Figure 20.7). The color of the nodes indicates the primary metal at the active site of a given sphere in a protein. Interestingly, network connectivity illustrates that the biological use of metals may be traceable through evolutionary time; in other words, the earliest proteins preferentially incorporated Fe, with later proteins using Mn and then Cu –the same sequence seen in the network graph – although metal information was not explicitly encoded in the network topology. This network reinforces previous findings from geochemistry (112) and biochemistry (104,105) that suggest that Fe proteins are ancient, whereas Cu-bearing proteins evolved later, possibly related to the presence and bioavailability of Fe and Cu in Earth's oceans through deep time (113).

An important finding of these and other network applications is that graphs of evolving systems (i.e. fossil taxa or mineral species) inevitably embed a time axis (Figure 20.8). This discovery points to possible data-driven approaches to gaining insights into the evolution

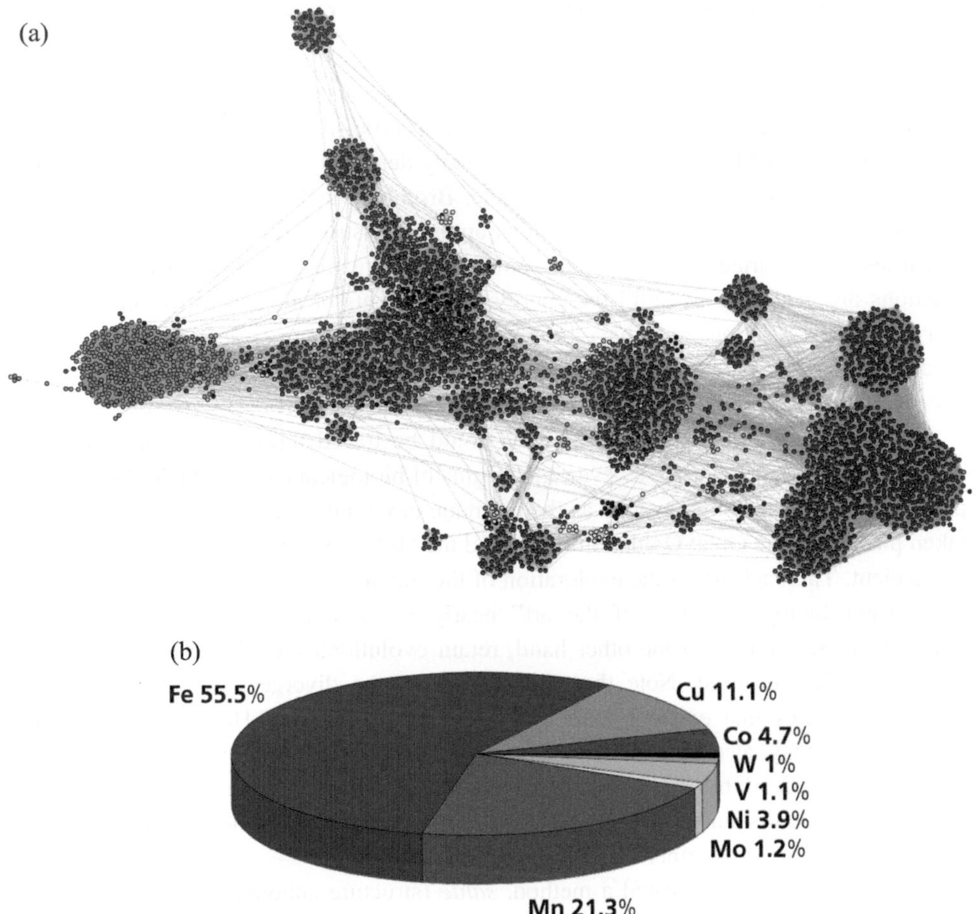

Figure 20.7 (a) Similarity network diagram of relationships among protein structures. The 4686 circular nodes represent oxidoreductases for which the three-dimensional structure is known. The linking and therefore distribution of nodes relates to similarities in protein fold structure in a 15-Å radius from the active metal-containing site. Nodes are colored according to the principal metal cation at the active site. Network connectivity illustrates that functional similarity of spheres may be traceable through evolutionary time (i.e. although metal information was not used in the building of this network, the time-related sequence Fe to Cu is embedded). This network indicates that Fe proteins are ancient, whereas Cu proteins evolved later. The pie chart (b) shows the relative abundance of metals in the graph.

of specific protein groups. For example, clustering of spheres in the network provides a means for reducing experimental bias in favor of generating a more naturally representative set of nodes and edges, which can be further used to build evolutionary trees of redox reactions on global timescales. These approaches can also inform synthetic biology, directing possible experimental mutagenesis efforts for designing and evaluating evolutionary intermediates that no longer exist in nature.

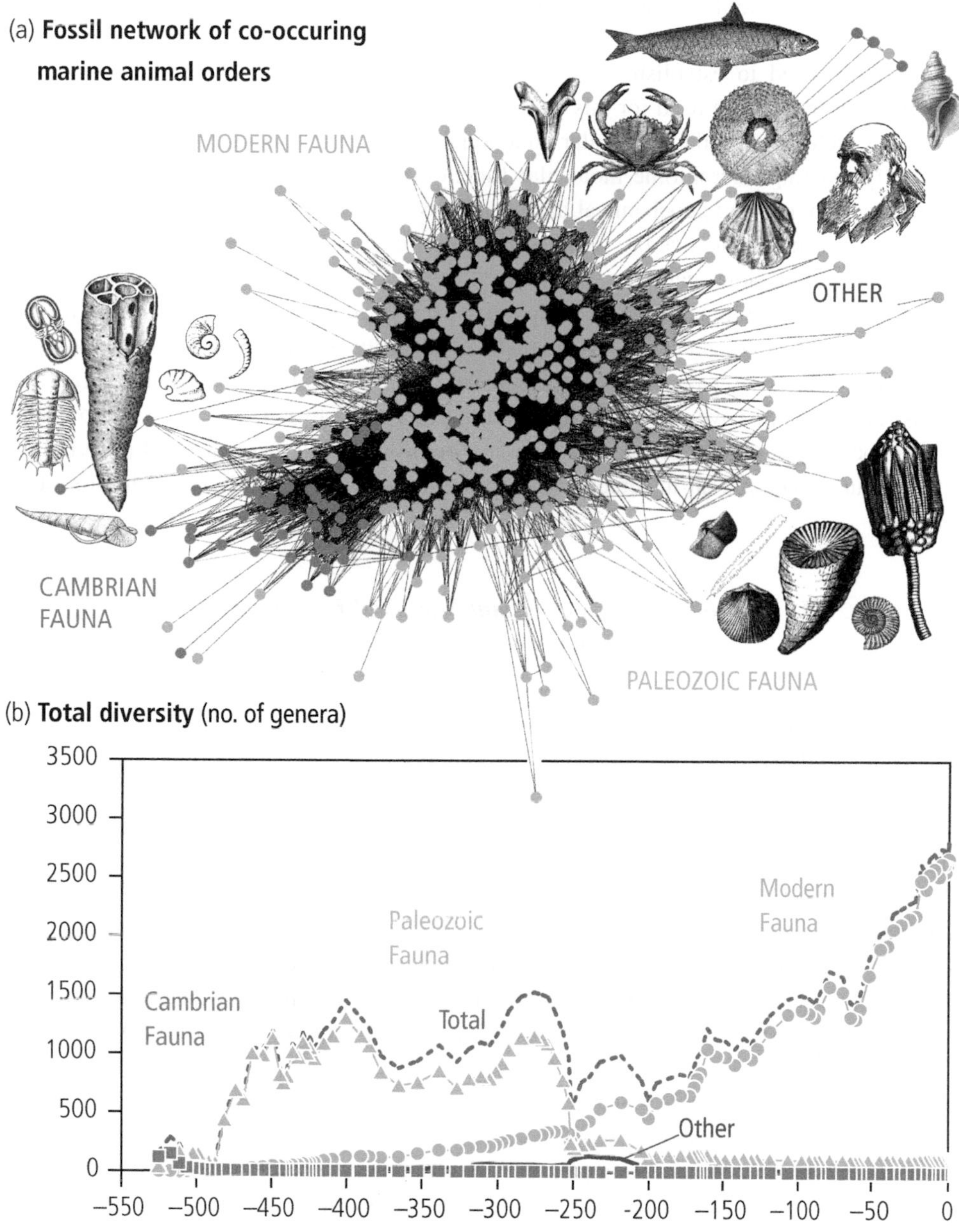

Figure 20.8 Networks that illustrate structural or coexistence relationships among individual members of an evolving system (i.e. mineral species, fossil taxa, or protein structures) inevitably embed a time axis, even though no age information is used in the generation of the graphs. (a) Phanerozoic fossil animals: nodes represent family-level taxa, while lines indicate coexisting fauna. The network was partitioned using the Louvain (multilayer) algorithm for community detection (138), resulting in the discovery of five modules, or "evolutionary paleocommunities." An embedded time axis is visible from the Cambrian to modern fauna and each partition represents a major extinction event. (b) Plot of diversity (total number of genera) versus time for each of the modules in (a).

The inherent flexibility of network approaches allows for the incorporation of additional data, thus strengthening any inferences made. For example, as there are no protein fossils that can be used to establish dates of redox protein existence, one reliable piece of information that can be used for this purpose is transition metal availability, which would drive the selection of the molecular functionality necessary for life. By matching the currently existing microbiome molecular function (114) and metal cofactor annotations with mineralogical and geochemical data, it is possible to reveal the relationships between the presence and abundance of specific enzymatic functionalities and metal availabilities. Functional annotations can thus be mapped to metal availability and, further, to the corresponding evolutionary age. Additionally, using machine learning techniques to recognize patterns in molecular function to metal availability relationships, it is likely possible to pinpoint any discrepancies between expectations and existing annotations, suggesting areas for more extensive research. As a result, protein structure networks, in combination with geochemical evidence, could provide a glimpse into the emergence and evolution of life on our planet and an understanding of the principles that could govern life on other planets.

20.4.2 Network Analysis of Extant Microbial Ecosystems: Geo–Bio Interactions on Ecological Scales

Investigations of the relationships between individual microbial taxa, microbiomes, and environmental conditions are complicated by the large number of contributing physical, chemical, and biological parameters, culminating in a complexity that is not easily representable by two-dimensional graphical methods. It has been suggested that new analytical techniques will be necessary to explore the large data sets produced by high-throughput DNA sequencing to discover new connections between microbiomes and the environment (115). Quantitative gene content analysis of terrestrial and marine microbial communities has already revealed habitat-specific fingerprints that reflect known characteristics of the sampled environments (116). Metagenomic and amplicon sequencing of diverse environments and microbial communities are now paving the way toward outlining the global ecosystem network and the development of ecosystem-wide dynamic models (117,118).

Network analysis and machine learning can be used to investigate microbial communities from all types of ecosystems and are useful approaches for examining and determining patterns in large, complex data sets, and they provide predictive power in the absence of mechanistic models (115,119–121). Since microbes are notoriously difficult to culture, the primary source of information on their diversity and evolution comes from the environmental distribution of microbiome data (122,123). Metagenomics – the study of genetic material obtained directly from environmental samples – has opened the door to the incredible diversity of microbial communities in the biosphere. The large-scale analysis of metagenomes, in concert with a wide range of environmental characteristics and geological diversity, will allow for the identification of unknown geo–bio interactions in the near future. This opportunity may lay the foundations for better understanding the geosphere and biosphere and their coevolution on this planet. As of the time of writing (January 2018), there were 6983 metagenomes available on the Department of Energy Joint Genome Institute public database (https://img.jgi.doe.gov), covering a variety of environments. Identifying relationships among physical and chemical parameters, such as temperature, pH, salinity, geochemistry, and the diversity in microbial communities, can reveal microbial responses to changing environmental conditions, and such information is critical to understanding microbial adaptations to different environments and their functions within those environments. Many studies have already shown the strong links between environmental conditions and microbial populations, a number of which did so with network analytical approaches (115,118,122,124–131). We suggest that the application of advanced analytical approaches to the microbial metagenomes and their corresponding environments, coupled with geochemical, geological, and mineralogical information, could transform the way we understand the role of microbial diversity in ecosystems.

Sharing and relating data sets between different disciplines, however, remains a great challenge. One way to deal with this challenge is through ensuring online availability of data. Currently, large amounts of sequenced data that represent a substantial portion of the total environmental diversity of Earth reside in online databases (e.g. MG-RAST, NCBI, JGI IMG, CAMERA). However, the quality of the associated metadata is generally low, with essential information like pH, temperature, salinity, redox state, and organic load often missing (132). Moreover, the links among sequence data, metadata, and any geochemical, geological, or other environmental data collected during the study are difficult or impossible to establish. Numerous attempts are being made by the scientific community to standardize the quality and type of metadata collected along with each sequenced sample in order to increase interoperative power. For example, efforts from the Genomic Standard Consortium (gensc.org) such as the Minimum Information about a Metagenomic Sequence (133), initiatives like the Earth Microbiome Project (earthmicrobiome.org), and the release of metadata-curated metagenomes (134) are pointing the metagenomics community in the right direction. Pioneering data sets of interdisciplinary, colocated data have been collected by the International Continental Drilling Programs (icdp-online.org), the International Ocean Discovery Program (iodp.org), and the DCO Integrated Field Site Initiatives (deepcarbon.net). These sampling programs will provide unprecedented environmental, geological, and geochemical metadata to analyze along with the associated metagenomes. Expansion of these efforts is crucial for advancing this important work in the future toward understanding geo–bio interactions on a global scale.

Our ability to generate predictive models of the relationships between -omic data and environmental data is further hindered by the varying data structures specific to the different fields of study (135,136). The poor resolution of our current understanding of the relationship between functional diversity and redundancy, biodiversity, ecosystem roles, and niche partitioning also presents challenges. A possible way to overcome this problem is by using predictive models that are not linked to specific hypothesis but take advantage of big data approaches that allow data-driven discoveries. Tools such as network analysis and machine learning can identify hidden patterns in large-scale data and provide predictive power in the absence of mechanistic models (115,119–121). Similar techniques have been used in metagenomic modeling to predict microbial assemblages and their metabolic properties (e.g. 113–115,137), and they can be applied to the investigation of the interaction between the geosphere and biosphere.

Recently, we have attempted a preliminary exploration of large-scale patterns in the relationships among oxidoreductase metalloproteins and the mineral diversity present at the same location (Figure 20.9). Based on publicly available metagenomic data from

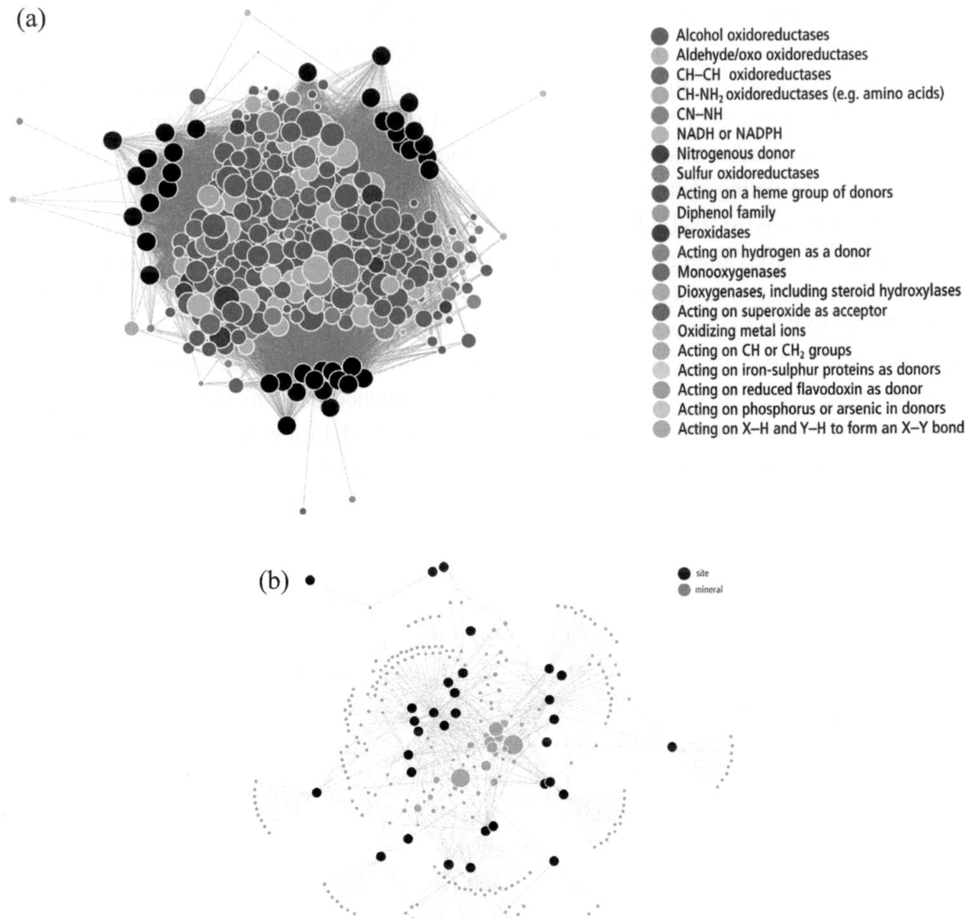

Figure 20.9 Bipartite networks of our preliminary analysis of geo–bio interactions based on 40 random metagenomes downloaded from MG-RAST and the mineral composition of the same site obtained from the Mindat database. (a) Bipartite network of the metalloprotein oxidoreductases (enzyme commission EC1 class) and the sites where they were found (in black). Enzyme nodes sized according to their counts and colored by their subclass. (b) Bipartite network of the mineral diversity at the same sites. Mineral nodes in gray, sized according to their mineral diversity; site nodes in black.

40 randomly selected microbial ecosystems (including samples from shallow-water and deep-sea hydrothermal vents, hot springs, permafrost, mines, soils, arctic soils, marine sediments, and salt marshes), our analysis reveals distinct patterns in the association between specific metalloprotein functions and the mineral settings where those functions are commonly abundant. In particular, geochemistry and redox conditions govern oxidoreductase gene diversity distribution in the observed environments. The microbial communities of certain locations had few or no distinctively expressed oxidoreductase proteins within the network, thus exhibiting overlap with other communities with similar environmental conditions. However, microbial communities from most locations expressed unique oxidoreductases that were not present in the communities of the other environments. This information is crucial to understanding niche partitioning among environmental taxa and may reveal key details regarding how environmental conditions and metal availability shape microbial community function.

We expected a great deal of overlap in gene expression between the microbial populations of many environments as we observed in our initial analysis. These functions will shed light on the expected and unexpected core functions of diverse communities. Additionally, numerous genes that are exclusively expressed in particular environments or under distinctive physical/chemical conditions will reveal geo–bio interactions that evolved in systems that are ancient Earth analogs to the modern day. We conclude that expanding data resources on microbial communities and ecosystems and better integration with geochemical, mineralogical, and geological databases will provide opportunities for documenting the effects of environmental parameters on gene distribution and functional diversity.

20.5 Conclusions: The Future of Data-Driven Discovery

Among the DCO's enduring legacies, and a tremendous opportunity for future advances, is the continued development and exploitation of data resources in the geosciences and biosciences. Our experiences over the decadal adventure of the DCO have convinced us that further advances in data-driven discovery will rest on three coequal pillars. The first ongoing demand is the creation and enhancement of comprehensive data resources, including those in geochemistry, petrology, mineralogy, paleobiology, paleotectonics, microbiology, proteomics, and other deep time aspects of carbon's global cycles in space and time.

Hand in hand with database enhancement, we require the development and adaptation of established and new methods for data analysis and visualization. Ongoing advances include new techniques to exploit geochemical data, novel LNRE formulations designed for specific applications to mineralogical and paleobiological systems, modified approaches to visualizing networks of varied geological and biological systems, and applications of affinity analysis to Earth systems.

Thirdly, data-driven discovery will advance through continued creative application of data resources and analytical methods targeted to answer complex problems related to

Earth's evolution through space and time. Our ambitions for the coming years include: estimating the erosional bias of the ancient rock record from differential mineral preservation through deep time; investigating the completeness of the fossil record with LNRE methods applied to the Paleobiology Database (paleobiodb.org); creating interactive networks of all known mineral species, fossil genera, and microbes and their environmental contexts; and applying affinity analysis to the discovery of new mineral and ore deposits.

The DCO has fostered the beginning of the era of data-driven discovery in carbon mineral science and has promoted the collection and assembly of a wide range of data resources. The DCO has employed existing analytical and visualization methods while developing new approaches and has raised and refined a suite of fundamental questions about Earth's carbon from crust to core – its forms, movements, quantities, and origins. Looking forward to the next decade of exploration, we predict that data-driven discovery will play an ever-greater role in our emerging understanding of carbon in Earth.

Acknowledgments

This chapter is a contribution of the Deep Carbon Observatory. This work was supported by the W.M. Keck Foundation's Deep-Time Data Infrastructure project (dtdi.carnegiescience.edu), with additional support by the Alfred P. Sloan Foundation, the Templeton Foundation, a private foundation, the Carnegie Institution for Science, NASA NNX11AP82A – Mars Science Laboratory Investigations, and NSF grant MCB 15-17567. Any opinions, findings, and conclusions or recommendations expressed in this material are those of the authors and do not necessarily reflect the views of the National Aeronautics and Space Administration.

Questions for the Classroom

1 What are the "three pillars" of data-driven discovery and why are all three important?
2 What are some of the visualization methods that can enhance discovery and how many different parameters can be displayed simultaneously with each of these methods?
3 Why are time axes embedded in network graphs of evolving systems, even though no age information is used in the generation of these graphs?
4 What was "Rodinia" and what is the evidence for its unique signature in Earth's history?
5 What are some of the preservational biases likely affecting the rock record and how do these biases scale with time?
6 How many carbon mineral localities are in the MED today and how many of those localities are dated? Which locality has the most carbon mineral species?
7 What are the biases in sampling the carbon minerals listed in the text and what are additional biases not covered in the chapter?

8 What is an LNRE distribution and why is it a useful model for mineral distributions?
9 To what other systems could you apply an LNRE model and associated accumulation curve?
10 What factors might be important in describing a microbial ecosystem, such as a community of microbes living beneath the ocean floor?
11 What is a metagenome and how is it sequenced? Why is shotgun metagenomics used instead of pure cultures?

References

1. Darwin C. *On the Origin of Species by Means of Natural Selection, or the Preservation of Favoured Races in the Struggle for Life*. London: John Murray, 1859.
2. Beddall BG. Wallace, Darwin, and the theory of natural selection. *J Hist Biol*. 1968;1 (2):261–323.
3. Muir Wood R. *The Dark Side of the Earth*. Illus. by M Woodhouse London: Allen & Unwin, 1985.
4. Hazen RM. *The Story of Earth*. New York: Viking, 2012.
5. Holland HD. *The Chemical Evolution of the Atmosphere and Oceans*. Princeton, NJ: Princeton University Press; 1984.
6. Canfield DE. *Oxygen: A Four Billion Year History*. Princeton, NJ: Princeton University Press, 2014.
7. Intergovernmental Panel on Climate Change. *Fifth Assessment Report (4 volumes)*. New York: Cambridge University Press, 2013.
8. Weart SR. *The Discovery of Global Warming*. Cambridge, MA: Harvard University Press, 2008.
9. Fayyad U, Piatetsky-Shapiro G, Smyth P. From data mining to knowledge discovery in databases. *AI Mag*. 1996;17(3):37.
10. Bolukbasi B, Berente N, Cutcher-Gershenfeld J, Dechurch L, Flint C, Haberman M, et al. Open data: crediting a culture of cooperation. *Science*. 2013;342(6162):1041–1042.
11. Hazen RM, Bekker A, Bish DL, Bleeker W, Downs RT, Farquhar J, et al. Needs and opportunities in mineral evolution research. *Am Mineral*. 2011;96(7):953–963.
12. Keller CB, Schoene B. Statistical geochemistry reveals disruption in secular lithospheric evolution about 2.5 Gyr ago. *Nature*. 2012;485(7399):490.
13. EarthCube. *A Community Roadmap for EarthCube Data: Discovery, Access, and Mining*. Washington, DC: National Science Foundation, 2012.
14. Spencer CJ, Hawkesworth C, Cawood PA, Dhuime B. Not all supercontinents are created equal: Gondwana–Rodinia case study. *Geology*. 2013;41(7):795–798.
15. Nance RD, Murphy JB, Santosh M. The supercontinent cycle: a retrospective essay. *Gondwana Res*. 2014;25:4–29.
16. Liu C, Knoll AH, Hazen RM. Geochemical and mineralogical evidence that Rodinian assembly was unique. *Nat Commun*. 2017;8(1):1950.
17. Bogdanova SV, Pisarevsky SA, Li ZX. Assembly and breakup of Rodinia (some results of IGCP project 440). *Stratigr Geol Correl*. 200917(3):259–274.
18. Huston DL, Pehrsson S, Eglington BM, Zaw K. The geology and metallogeny of volcanic-hosted massive sulfide deposits: variations through geologic time and with tectonic setting. *Econ Geol*. 2010;105(3):571–591.
19. Persson S, Eglington B, Huston D, Evans D. Why was Rodinia underendowed? Comparing the effects of paleogeography versus lithosphere thickness on secular ore

deposit preservation. *Mineralogical Magazine*, 2003. Available from: https://goldschmidtabstracts.info/abstracts/abstractView?id=2013006335.
20. Cawood PA, Hawkesworth CJ. Earth's middle age. *Geology*. 2014;42(6):503–506.
21. Bierlein FP, Groves DI, Cawood PA. Metallogeny of accretionary orogens – the connection between lithospheric processes and metal endowment. *Ore Geol Rev*. 2009;36(4):282–292.
22. Liu C, Runyon SE, Knoll AH, Hazen RT. The same and not the same: ore geology, mineralogy and geochemistry of Rodinia assembly. *Earth Sci Rev*., 2019;196:102860, DOI: 10.1016/j.earscirev.2019.05.004.
23. Kemp AIS, Hawkesworth CJ, Paterson BA, Kinny PD, Kemp T. Episodic growth of the Gondwana supercontinent from hafnium and oxygen isotopes in zircon. *Nature*. 2006;439(7076):580–583.
24. Campbell IH, Allen CM. Formation of supercontinents linked to increases in atmospheric oxygen. *Nat Geosci*. 2008;1(8):554–558.
25. Rino S, Kon Y, Sato W, Maruyama S, Santosh M, Zhao D. The Grenvillian and Pan-African orogens: world's largest orogenies through geologic time, and their implications on the origin of superplume. *Gondwana Res*. 2008;14(1–2):51–72.
26. Condie KC, Belousova E, Griffin WL, Sircombe KN. Granitoid events in space and time: constraints from igneous and detrital zircon age spectra. *Gondwana Res*. 2009;15:228–242.
27. Condie KC, Aster RC. Episodic zircon age spectra of orogenic granitoids: the supercontinent connection and continental growth. *Precambrian Res*. 2010;180 (3–4):227–236.
28. Hawkesworth CJ, Dhuime B, Pietranik AB, Cawood PA, Kemp AIS, Storey CD. The generation and evolution of the continental crust. *J Geol Soc London*. 2010;167 (2):229–248.
29. Voice PJ, Kowalewski M, Eriksson KA. Quantifying the timing and rate of crustal evolution: global compilation of radiometrically dated detrital zircon grains. *J Geol*. 2011;119(2):109–126.
30. Hazen RM, Golden J, Downs RT, Hystad G, Grew ES, Azzolini D, et al. Mercury (Hg) mineral evolution: a mineralogical record of supercontinent assembly, changing ocean geochemistry, and the emerging terrestrial biosphere. *Am Mineral*. 2012;97 (7):1013–1042.
31. Hazen RM, Liu X-M, Downs RT, Golden J, Pires AJ, Grew ES, et al. Mineral evolution: episodic metallogenesis, the supercontinent cycle, and the coevolving geosphere and biosphere. *Soc Econ Geol Spec Publ*. 2014;18(18):1–15.
32. Golden JJ, McMillan M, Downs RT, Hystad G, Goldstein I, Stein HJ, et al. Rhenium variations in molybdenite (MoS_2): evidence for progressive subsurface oxidation. *Earth Planet Sci Lett*. 2013;366:1–5.
33. Grew ES, Hazen RM. Beryllium mineral evolution. *Am Mineral*. 2014;99 (5–6):999–1021.
34. Grew ES, Krivovichev SV, Hazen RM, Hystad G. Evolution of structural complexity in boron minerals. *Can Mineral*. 2016;54(1):125–143.
35. Hummer D, Golden J, Hystad G, Downs R, Eleish A, Liu C, et al. The oxidation of Earth's crust: evidence from the evolution of manganese minerals. *Nat Geosci*., in revision.
36. Hoffman PF, Kaufman AJ, Halverson GP, Schrag DP. A neoproterozoic Snowball Earth. *Science*. 1998;281(5381):1342–1346.
37. Goldfarb RJ, Bradley D, Leach DL. Secular variation in economic geology. *Econ Geol*. 2010;105(3):459–465.

38. Bradley DC. Secular trends in the geologic record and the supercontinent cycle. *Earth Sci Rev*. 2011;108(1–2):16–33.
39. Van Kranendonk MJ, Kirkland CL. Orogenic climax of Earth: the 1.2–1.1 Ga Grenvillian superevent. *Geology*. 2013;41(7):735–738.
40. Dickinson WR. Impact of differential zircon fertility of granitoid basement rocks in North America on age populations of detrital zircons and implications for granite petrogenesis. *Earth Planet Sci Lett*. 2008;275(1–2):80–92.
41. Nicholson SW, Cannon WF, Schulz KJ. Metallogeny of the midcontinent rift system of North America. *Precambrian Res*. 1992;58(1–4):355–386.
42. Upton BGJ, Emeleus CH, Heaman LM, Goodenough KM, Finch AA. Magmatism of the mid-Proterozoic Gardar Province, South Greenland: chronology, petrogenesis and geological setting. *Lithos*. 2003;68(1–2):43–65.
43. Iriondo A, Premo WR, Martínez-Torres LM, Budahn JR, Atkinson WW, Siems DF, et al. Isotopic, geochemical, and temporal characterization of Proterozoic basement rocks in the Quitovac region, northwestern Sonora, Mexico: implications for the reconstruction of the southwestern margin of Laurentia. *Bull Geol Soc Am*. 2004;116 (1–2):154–170.
44. Greentree MR, Li ZX, Li XH, Wu H. Late Mesoproterozoic to earliest Neoproterozoic basin record of the Sibao orogenesis in western South China and relationship to the assembly of Rodinia. *Precambrian Res*. 2006;151(1–2):79–100.
45. McLelland J, Selleck B, Bickford M. Review of the Proterozoic evolution of the Grenville Province, its Adirondack outlier, and the Mesoproterozoic inliers of the Appalachians. In: Tollo R, Bartholomew M, Hibbard J, Karabinos P, eds. *From Rodinia to Pangea: The Lithotectonic Record of the Appalachian Region: Geological Society of America Memoir*. Boulder, CO: Geological Society of America, 2010, pp. 21–49.
46. Prol-Ledesma RM, Melgarejo JC, Martin RF. The El Muerto "NYF" granitic pegmatite, Oaxaca, Mexico, and its striking enrichment in allanite-(CE) and monazite-(Ce). *Can Mineral*. 2012;50(4):1055–1076.
47. Baadsgaard H, Chaplin C, Griffin WL. Geochronology of the Gloserheia pegmatite, Froland, southern Norway. *Nor Geol Tidsskr*. 1984;64(2):111–119.
48. Foord EE, Černý P, Jackson LL, Sherman DM, Eby RK. Mineralogical and geochemical evolution of micas from miarolitic pegmatites of the anorogenic pikes peak batholith, Colorado. *Mineral Petrol*. 1995;55(1–3):1–26.
49. McCauley A, Bradley DC. The global age distribution of granitic pegmatites. *Can Mineral*. 2014;52(2):183–190.
50. Dhuime B, Wuestefeld A, Hawkesworth CJ. Emergence of modern continental crust about 3 billion years ago. *Nat Geosci*. 2015;8(7):552–555.
51. Dijkstra AH, Dale CW, Oberthür T, Nowell GM, Graham Pearson D. Osmium isotope compositions of detrital Os-rich alloys from the Rhine River provide evidence for a global late Mesoproterozoic mantle depletion event. *Earth Planet Sci Lett*. 2016;452:115–122.
52. Kelemen PB, Johnson KTM, Kinzler RJ, Irving AJ. High-field-strength element depletions in arc basalts due to mantle–magma interaction. *Nature*. 1990;345 (6275):521–524.
53. Woodhead J, Eggins S, Gamble J. High field strength and transition element systematics in island arc and back-arc basin basalts: evidence for multi-phase melt extraction and a depleted mantle wedge. *Earth Planet Sci Lett*. 1993;114 (4):491–504.

54. Cawood PA, Strachan RA, Pisarevsky SA, Gladkochub DP, Murphy JB. Linking collisional and accretionary orogens during Rodinia assembly and breakup: implications for models of supercontinent cycles. *Earth Planet Sci Lett*. 2016;449:118–126.
55. Evans DAD, Li ZX, Murphy JB. Four-dimensional context of Earth's supercontinents. *Geol Soc London Spec Publ*. 2016;424(1):1–14.
56. Cawood PA, Pisarevsky SA. Laurentia–Baltica–Amazonia relations during Rodinia assembly. *Precambrian Res*. 2017;292:386–397.
57. Hoffman PF, Grotzinger JP. Orographic precipitation, erosional unloading, and tectonic style. *Geology*. 1993;21:195–198.
58. Bingen B, Stein HJ, Bogaerts M, Bolle O, Mansfeld J. Molybdenite Re–Os dating constrains gravitational collapse of the Sveconorwegian orogen, SW Scandinavia. *Lithos*. 2006;87(3–4):328–346.
59. Rivers T. Assembly and preservation of lower, mid, and upper orogenic crust in the Grenville Province – implications for the evolution of large hot long-duration orogens. *Precambrian Res*. 2008;167(3–4):237–259.
60. Möller C, Andersson J, Dyck B, Antal Lundin I. Exhumation of an eclogite terrane as a hot migmatitic nappe, Sveconorwegian orogen. *Lithos*. 2015;226:147–168.
61. Rainbird R, Cawood P, Gehrels G. The great Grenvillian sedimentation episode: record of supercontinent Rodinia's assembly. In: Cathy B, Antonio A, eds. *Tectonics of Sedimentary Basins: Recent Advances*. Chichester: John Wiley & Sons, Ltd., 2012, pp. 583–601.
62. Cawood PA, Hawkesworth CJ, Dhuime B. The continental record and the generation of continental crust. *Bull Geol Soc Am*. 2013;125(1–2):14–32.
63. Spencer CJ, Prave AR, Cawood PA, Roberts NMW. Detrital zircon geochronology of the Grenville/Llano foreland and basal Sauk Sequence in west Texas, USA. *Bull Geol Soc Am*. 2014;126(7–8):1117–1128.
64. Brune S, Williams SE, Müller RD. Potential links between continental rifting, CO_2 degassing and climate change through time. *Nat Geosci*. 2017;10(12):941–946.
65. Lee CTA, Caves J, Jiang H, Cao W, Lenardic A, McKenzie NR, et al. Deep mantle roots and continental emergence: implications for whole-Earth elemental cycling, long-term climate, and the Cambrian explosion. *Int Geol Rev*. 2018;60(4):431–448.
66. Aulbach S, Creaser RA, Stachel T, Heaman LM, Chinn IL, Kong J. Diamond ages from Victor (Superior Craton): intra-mantle cycling of volatiles (C, N, S) during supercontinent reorganisation. *Earth Planet Sci Lett*. 2018;490:77–87.
67. von Blanckenburg F. The control mechanisms of erosion and weathering at basin scale from cosmogenic nuclides in river sediment. *Earth Planet Sci Lett [Internet]*. 2006;237(3–4):462–479.
68. Silver PG, Behn MD. Intermittent plate tectonics? *Science*. 2008;319(5859):85–88.
69. Berner RA. *The Phanerozoic Carbon Cycle: CO_2 and O_2*. Oxford: Oxford University Press, 2004.
70. Knoll AH, Follows MJ. A bottom-up perspective on ecosystem change in Mesozoic oceans. *Proc R Soc B Biol Sci*. 2016;283(1841):20161755.
71. Knoll AH. Biological and biogeochemical preludes to the Ediacaran radiation. In: Lipps JH, Signor PW, eds. *Origin and Early Evolution of the Metazoa*. Boston, MA: Springer, 1992, pp. 53–84.
72. Hayes JM, Waldbauer JR. The carbon cycle and associated redox processes through time. *Philos Trans R Soc B Biol Sci*. 2006;361:931–950.
73. Lyons TW, Reinhard CT, Planavsky NJ. The rise of oxygen in Earth's early ocean and atmosphere. *Nature*. 2014;506:307–315.

74. Knoll AH. Paleobiological perspectives on early eukaryotic evolution. *Cold Spring Harb Perspect Biol*. 2014;6(1):a016121.
75. Knoll AH. Paleobiological perspectives on early microbial evolution. *Cold Spring Harb Perspect Biol*. 2015;7(7):a018093.
76. Kah LC, Sherman AG, Narbonne GM, Knoll AH, Kaufman AJ. $\delta^{13}C$ stratigraphy of the Proterozoic Bylot Supergroup, Baffin Island, Canada: implications for regional lithostratigraphic correlations. *Can J Earth Sci*. 1999;36(3):313–332.
77. Halverson GP, Hoffman PF, Schrag DP, Maloof AC, Rice AHN. Toward a Neoproterozoic composite carbon-isotope record. *Bull Geol Soc Am*. 2005;117:1181–1207.
78. Reinhard CT, Planavsky NJ, Gill BC, Ozaki K, Robbins LJ, Lyons TW, et al. Evolution of the global phosphorus cycle. *Nature*. 2017;541(7637):386–389.
79. Laakso TA, Schrag DP. Regulation of atmospheric oxygen during the Proterozoic. *Earth Planet Sci Lett*. 2014;388:81–91.
80. Hazen RM, Downs RT, Kah L, Sverjensky D. Carbon mineral evolution. *Rev Mineral Geochem*. 2013;75(1):79–107.
81. Hazen RM, Hummer DR, Hystad G, Downs RT, Golden JJ. Carbon mineral ecology: predicting the undiscovered minerals of carbon. *Am Mineral*. 2016;101(4):889–906.
82. Morrison SM, Liu C, Eleish A, Prabhu A, Li C, Ralph J, et al. Network analyses of mineralogical systems. *Am Mineral*. 2017;102(1):1588–1596.
83. Hazen RM, Papineau D, Bleeker W, Downs RT, Ferry JM, Mccoy TJ, et al. Mineral evolution. *Am Mineral*. 2008;93(11–12):1693–1720.
84. Hazen RM, Ferry JM. Mineral evolution: mineralogy in the fourth dimension. *Elements*. 2010;6(1):9–12.
85. Zhabin AG. Is there evolution of mineral speciation on Earth? *Dokl Earth Sci Sect*. 1981;247:142–144.
86. Zalasiewicz J, Kryza R, Williams M. The mineral signature of the Anthropocene in its deep-time context. *Geol Soc London Spec Publ*. 2014;395(1):109–117.
87. Hazen RM, Grew ES, Origlieri MJ, Downs RT. On the mineralogy of the "Anthropocene Epoch." *Am Mineral*. 2017;102(3):595–611.
88. Golden JJ, Pires AJ, Hazen RM, Downs RT, Ralph J, Meyer M. Building the mineral evolution database: implications for future big data analysis. *Geol Soc Am Abstr Prog*. 2016;48:7.
89. Ernst RE. *Large Igneous Provinces*. Cambridge: Cambridge University Press, 2014.
90. Hystad G, Downs RT, Hazen RM. Mineral species frequency distribution conforms to a large number of rare events model: prediction of Earth's missing minerals. *Math Geosci*. 2015;47(6):647–661.
91. Hystad G, Downs RT, Grew ES, Hazen RM. Statistical analysis of mineral diversity and distribution: Earth's mineralogy is unique. *Earth Planet Sci Lett*. 2015;426:154–157.
92. Hazen RM, Grew ES, Downs RT, Golden J, Hystad G. Mineral ecology: chance and necessity in the mineral diversity of terrestrial planets. *Can Mineral*. 2015;53(2):295–324.
93. Hazen RM, Hystad G, Downs RT, Golden JJ, Pires AJ, Grew ES. Earth's "missing" minerals. *Am Mineral*. 2015;100(10):2344–2347.
94. Maynard-Casely HE, Cable ML, Malaska MJ, Vu TH, Choukroun M, Hodyss R. Prospects for mineralogy on Titan. *Am Mineral*. 2018;103(3):343–349.
95. Hazen RM. Titan mineralogy: a window on organic mineral evolution. *Am Mineral*. 2018;103(3):341–342.
96. Abraham A, Hassanien A-E, Snasel V. *Computational Social Network Analysis: Trends, Tools and Research Advances (Computer Communications and Networks)*. New York: Springer, 2010.

97. Scott J, Carrington PJ, eds. *The SAGE Handbook of Social Network Analysis*. Thousand Oaks, CA: SAGE Publications Ltd., 2011.
98. Newman M. *Networks: An Introduction*. Oxford: Oxford University Press, 2010.
99. Colwell FS, D'Hondt S. Nature and extent of the deep biosphere. *Rev Mineral Geochem*. 2013;75(1):547–574.
100. Tashiro T, Ishida A, Hori M, Igisu M, Koike M, Méjean P, et al. Early trace of life from 3.95 Ga sedimentary rocks in Labrador, Canada. *Nature*. 2017;549 (7673):516–518.
101. Dodd MS, Papineau D, Grenne T, Slack JF, Rittner M, Pirajno F, et al. Evidence for early life in Earth's oldest hydrothermal vent precipitates. *Nature*. 2017;543 (7643):60–64.
102. Stüeken EE, Buick R, Anderson RE, Baross JA, Planavsky NJ, Lyons TW. Environmental niches and metabolic diversity in Neoarchean lakes. *Geobiology*. 2017;15 (6):767–783.
103. Giovannelli D, Sievert SM, Hügler M, Markert S, Becher D, Schweder T, et al. Insight into the evolution of microbial metabolism from the deep-branching bacterium, *Thermovibrio ammonificans*. *Elife*. 2017;6:e18990.
104. Jelen BI, Giovannelli D, Falkowski PG. The role of microbial electron transfer in the coevolution of the biosphere and geosphere. *Annu Rev Microbiol*. 2016;70(1):45–62.
105. Moore EK, Jelen BI, Giovannelli D, Raanan H, Falkowski PG. Metal availability and the expanding network of microbial metabolisms in the Archaean eon. *Nat Geosci*. 2017;10:629–636.
106. Greene LH. Protein structure networks. *Brief Funct Genomics*. 2012;11(6):469–478.
107. Zhu C, Delmont TO, Vogel TM, Bromberg Y. Functional basis of microorganism classification. *PLoS Comput Biol*. 2015;11(8):e1004472.
108. Zhu C, Mahlich Y, Miller M, Bromberg Y. Fusion DB: assessing microbial diversity and environmental preferences via functional similarity networks. *Nucleic Acids Res*. 2018;46(D1):D535–D541.
109. Harel A, Falkowski P, Bromberg Y. TrAnsFuSE refines the search for protein function: oxidoreductases. *Integr Biol*. 2012;4(7):765–777.
110. Harel A, Bromberg Y, Falkowski PG, Bhattacharya D. Evolutionary history of redox metal-binding domains across the tree of life. *Proc Natl Acad Sci*. 2014;111 (19):7042–7047.
111. Senn S, Nanda V, Falkowski P, Bromberg Y. Function-based assessment of structural similarity measurements using metal co-factor orientation. *Proteins*. 2014;82 (4):648–656.
112. Anbar AD, Knoll AH. Proterozoic ocean chemistry and evolution: a bioinorganic bridge? *Science*. 2002;297:1137–1142.
113. Dupont S, Lundve B, Thorndyke M. Seawater carbonate chemistry and biological processes during experiments with a sea star *Crassaster papposus*. Supplement to: Dupont S et al. (2010): Near future ocean acidification increases growth rate of the lecithotrophic larvae and juveniles of the sea star *Crossaster papposus*. *Journal of Experimental Zoology Part B – Molecular and Developmental Evolution*. *Pangaea*. 2010;314B:382–389.
114. Zhu C, Miller M, Marpaka S, Vaysberg P, Rühlemann MC, Wu G, et al. Functional sequencing read annotation for high precision microbiome analysis. *Nucleic Acids Res*. 2018;46(4):e23.
115. Barberán A, Bates ST, Casamayor EO, Fierer N. Using network analysis to explore co-occurrence patterns in soil microbial communities. *ISME J*. 2012;6(2):343–351.

116. Tringe SG, Von Mering C, Kobayashi A, Salamov AA, Chen K, Chang HW, et al. Comparative metagenomics of microbial communities. *Science*. 2005;308 (5721):554–557.
117. Faust K, Raes J. Microbial interactions: from networks to models. *Nat Rev Microbiol*. 2012;10:538–550.
118. Delgado-Baquerizo M, Oliverio AM, Brewer TE, Benavent-González A, Eldridge DJ, Bardgett RD, et al. A global atlas of the dominant bacteria found in soil. *Science*. 2018;359(6373):320–325.
119. Weiss SM, Kapouleas I. An empirical comparison of pattern recognition, neural nets and machine learning classification methods. Shavlik JW, Dietterich TG, eds. *Readings in Machine Learning*. San Mateo, CA: Morgan Kaufmann Publishers, 1990, pp. 177–183.
120. Hand DJ. Data mining: statistics and more? *Am Stat*. 1998;52(2):112–118.
121. Proulx SR, Promislow DEL, Phillips PC. Network thinking in ecology and evolution. *Trends Ecol Evol*. 2005;20:345–353.
122. Lozupone CA, Knight R. Global patterns in bacterial diversity. *Proc Natl Acad Sci*. 2007;104(27):11436–11440.
123. Solden L, Lloyd K, Wrighton K. The bright side of microbial dark matter: lessons learned from the uncultivated majority. *Curr Opin Microbiol*. 2016;31:217–226.
124. Giovannelli D, D'Errico G, Fiorentino F, Fattorini D, Regoli F, Angeletti L, et al. Diversity and distribution of prokaryotes within a shallow-water pockmark field. *Front Microbiol*. 2016;7:941.
125. Fierer N, Jackson RB. The diversity and biogeography of soil bacterial communities. *Proc Natl Acad Sci*. 2006;103(3):626–631.
126. Lauber CL, Hamady M, Knight R, Fierer N. Pyrosequencing-based assessment of soil pH as a predictor of soil bacterial community structure at the continental scale. *Appl Environ Microbiol*. 2009;75(15):5111–5120.
127. Auguet JC, Barberan A, Casamayor EO. Global ecological patterns in uncultured Archaea. *ISME J*. 2010;4(2):182–190.
128. Barberán A, Casamayor EO. Global phylogenetic community structure and β-diversity patterns in surface bacterioplankton metacommunities. *Aquat Microb Ecol*. 2010;59(1):1–10.
129. Caporaso JG, Lauber CL, Walters WA, Berg-Lyons D, Lozupone CA, Turnbaugh PJ, et al. Global patterns of 16S rRNA diversity at a depth of millions of sequences per sample. *Proc Natl Acad Sci*. 2011;108(Suppl. 1):4516–4522.
130. Hoppe B, Kahl T, Karasch P, Wubet T, Bauhus J, Buscot F, et al. Network analysis reveals ecological links between N-fixing bacteria and wood-decaying fungi. *PLoS One*. 2014;9(2):e88141.
131. Ruff SE, Biddle JF, Teske AP, Knittel K, Boetius A, Ramette A. Global dispersion and local diversification of the methane seep microbiome. *Proc Natl Acad Sci*. 2015;112(13):4015–4020.
132. Gilbert J. Metagenomics, metadata, and meta-analysis. In: Karen EN, ed. *Encyclopedia of Metagenomics*. Boston, MA: Springer US, 2015, pp. 439–442.
133. Field D, Garrity G, Gray T, Morrison N, Selengut J, Sterk P, et al. The minimum information about a genome sequence (MIGS) specification. *Nat Biotechnol*. 2008;26:541–547.
134. Pasolli E, Schiffer L, Manghi P, Renson A, Obenchain V, Truong DT, et al. Accessible, curated metagenomic data through ExperimentHub. *Nat Methods*. 2017;14:1023–1024.

135. Reed DC, Algar CK, Huber JA, Dick GJ. Gene-centric approach to integrating environmental genomics and biogeochemical models. *Proc Natl Acad Sci.* 2014;111(5):1879–1884.
136. Schimel J. Microbial ecology: linking omics to biogeochemistry. *Nat Microbiol.* 2016;1:15028.
137. Blondel VD, Guillaume J-L, Lambiotte R, Lefebvre E. Fast unfolding of communities in large networks. *J Stat Mech.* 2008;P10008.

Index